Understanding and controlling the hydrothermal growth process
and the electronic properties of rutile TiO$_2$ nanorods

Understanding and controlling the hydrothermal growth process and the electronic properties of rutile TiO_2 nanorods

Dissertation submitted for the degree
of Doctor of Natural Sciences (Dr. rer. nat.)

by

Julian Kalb

at the

Universität Konstanz

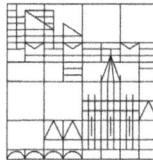

Faculty of Mathematics and Natural Sciences
Department of Physics

Date of the oral examination: February 26th, 2018
First referee: Professor Dr. Lukas Schmidt-Mende
Second referee: Professor Dr. Paul Leiderer

Bibliografische Information der Deutschen Nationalbibliothek
Die Deutsche Nationalbibliothek verzeichnet diese Publikation in der
Deutschen Nationalbibliografie; detaillierte bibliografische Daten
sind im Internet über http://dnb.d-nb.de abrufbar.
1. Aufl. - Göttingen: Cuvillier, 2019
Zugl.: Konstanz , Univ., Diss., 2019

© CUVILLIER VERLAG, Göttingen 2019
  Nonnenstieg 8, 37075 Göttingen
  Telefon: 0551-54724-0
  Telefax: 0551-54724-21
  www.cuvillier.de

  ISBN 978-3-7369-7126-4
  eISBN 978-3-7369-6126-5

Contents

Abstract

Nanomaterials provide different or additional features compared to bulk material. Their properties are often controlled by their size and shape. For several decades, more and more nanomaterials have found their way into various applications. Nanostructures are employed due to their specific properties or enhanced surface area, but also provide routes towards decreasing the size of devices. An outstanding material for nanostructures is titania (TiO_2). It is a non-toxic, abundantly available semiconductor capable of resisting intensive chemical and thermal stress and its optoelectronic features are tunable over a broad range through relatively inexpensive methods. Hence, TiO_2 is useful for a variety of applications such as surface functionalization, photocatalysis, photovoltaics, photodetectors, gas sensing, data storage, and energy storage. There are many methods to generate TiO_2 nanostructures. This thesis deals with controlling the fabrication and properties of hydrothermally grown rutile TiO_2 nanorods.

First, a detailed introduction reviews known properties, fabrication methods, and manipulation techniques of these nanostructures.

Second, the hydrothermal growth process on selected substrates such as fluorine tin oxide (FTO), anatase and rutile TiO_2 films is described. Especially, anatase substrates exhibit a deviation from simple epitaxial growth processes. It is strongly assumed that anatase nanoparticles become chemically unstable after reaching a critical size and are converted into rutile growth seeds. In addition, the annealing temperature affects the density of grown nanorods significantly. Films annealed at lower temperatures consist of smaller grains and provide a higher density of nanorods. The films are fabricated using a sol-gel method, spray pyrolysis, sputter deposition, and electron-beam evaporation. Features are investigated by SEM, AFM, XRD, and TEM.

Third, besides nucleation and the early stage of nanorods, the growth itself is investigated in more details. Especially, the fine structure of these rods riddles scientists and their origin is not completely understood. An effect of the fibrous fine structure on optical and electronic properties has to be considered. In this work, a model is introduced describing the origin of the fine structure by a thermodynamic disequilibrium between the crystallization and energy minimization process. Several features of nanorods grown under different conditions are predicted by the presented model such as the reduction of the fine structure at elevated growth temperatures.

Fourth, the main challenge of this thesis was the finding of methods for position-controlled fabrication of nanorods. Here, new, fast, and inexpensive techniques are presented and compared with well-established techniques such as optical or electron-beam lithography.
A 2D lattice without employing lithography is achieved by placing a monolayer of polystyrene spheres (PSML) on a blank substrate. The PSML acts as a mask during the seed layer deposition and the nanorods grow on highly ordered seed islands.
A rather simple method is to sputter a thin SiO_2 layer on a rutile TiO_2 film that was deposited on a substrate with some trenches. Inside the trenches, the film becomes much thinner than on the flat surface. While thick SiO_2 films prevent the growth of nanorods on the untreated surface, the SiO_2 shell inside the trenches is too thin. Consequently, nanorods grow inside the trenches only. The trenches have been created by focused ion beam (FIB) milling.
Another newly implemented technique is a scanning probe method. In order to do so, anatase TiO_2 particles are generated as a primary stage for rutile seeds by scratching with a silicon tip across an anatase film. Since the particles do not move across the film, the growth of nanorods is triggered on scratched positions only.
Furthermore, a laser is employed to melt mixed phases containing titanium, silicon, and oxygen which are situated at the interface between anatase TiO_2 films and silicon substrates. These mixed phases are created by diffusion of atoms across the TiO_2/silicon interface during annealing. As the melt solidifies, rutile TiO_2 is formed promoting the local growth of rutile TiO_2 nanorods. The melting is performed using a frequency-doubled Nd:YAG laser at $532\,nm$. A continuous wave laser is applied to write individual structures on a wafer, while a pulsed laser is applied to create a large-area pattern by two-beam interference.

Fifth, optical characterization offers insights in light reflection and scattering of highly anisotropic TiO_2 nanostructures. These results give rise to the assumption that TiO_2 nanorods act as tiny dielectric antennas. In this context, the transition from single particle to effective medium behavior is discussed based on measurements performed with an optical microscope run in bright and dark field mode.

Sixth, current over voltage characteristics demonstrate the variety of adjustable electronic features achieved by different growth conditions or annealing in different atmospheres. It is known that the variation of the oxygen vacancy density changes the number of mobile electrons and even the band structure. This results in either highly insulating or highly conductive nanorods as shown in this thesis. Switching between ohmic and rectifying contacts at the nanorod/electrode interface is described as well. Both, charge trapping and ion migration, change the band structure, the spatial, and energy distribution of charges and trap states in TiO_2 nanorods during intensive voltage or current stress. These effects are characterized by time-dependent transient currents. All experiments were performed in a controlled and dry atmosphere.

Based on physical models and descriptions, this work gives a detailed instruction for the position-controlled fabrication of rutile TiO_2 nanorods and how to manipulate their optical and electronic properties.

Zusammenfassung

Nanomaterialien bieten unterschiedliche und zusätzliche Eigenschaften gegenüber weniger fein gegliederten Materialien. Ihre Eigenschaften werden oft durch ihre Größe und Form festgelegt. Seit vielen Jahren werden immer mehr Nanomaterialien in verschiedenen Anwendungen verwendet. Einerseits werden Nanostrukturen aufgrund ihrer spezifischen Eigenschaften oder relativ großen Oberfläche eingesetzt. Andererseits bieten Nanostrukturen die Möglichkeit Geräte und Bauteile immer kleiner zu bauen.
Ein hervorragendes Material für Nanostrukturen ist Titandioxid (TiO$_2$). Es ist ein ungiftiges, reichlich in Erzlagerstätten vorhandenes Halbleitermaterial, welches eine hohe chemische und thermische Beständigkeit aufweist. Seine optoelektronischen Eigenschaften sind über ein breites Spektrum mit relativ kostengünstigen Methoden abstimmbar. Somit eignet sich TiO$_2$ für eine Vielzahl von Anwendungen wie Oberflächenfunktionalisierung, Photokatalyse, Photovoltaik, Photodetektoren, Gassensoren, Daten- und Energiespeicherung. Es gibt viele Methoden, um Nanostrukturen aus TiO$_2$ zu erzeugen. Diese Arbeit beschäftigt sich mit der kontrollierten Herstellung und den Eigenschaften von hydrothermal gewachsenen Rutil-TiO$_2$-Nanostäbchen.

An erster Stelle wird eine detaillierte Einführung zu bereits bekannten Eigenschaften, Herstellungs- und Optimierungstechniken dieser Nanostrukturen gegeben.

Zweitens wird das Verfahren des hydrothermalen Wachstums auf ausgewählten Substraten, wie z.b. Fluor-Zinnoxid- (FTO) oder Dünnschichten aus Anatas- und Rutil-TiO$_2$, beschrieben. Insbesondere Anatas-Substrate zeigen eine Abweichung von einfachen epitaktischen Wachstumsprozessen. Es wird angenommen, dass Nanopartikel aus Anatas nach Erreichen einer kritischen Größe chemisch instabil und in Wachstumskeime aus Rutil umgewandelt werden. Zusätzlich beeinflusst die Temperatur beim Tempern der Schicht aus Anatas die Dichte der gewachsenen Nanostäbchen erheblich. Filme, die bei niedrigeren Temperaturen getempert werden, bestehen aus kleineren Körnern und liefern eine höhere Dichte von Nanostäben. Die Filme können mittels Sol-Gel-Verfahren, Sprühpyrolyse oder Sputterabscheidung hergestellt werden. Die Eigenschaften werden durch REM, AFM, XRD und TEM untersucht.

Drittens wird neben der Keimbildung und dem frühen Wachstumsstadium der Nanostäbe das Wachstum selbst näher untersucht. Vor allem die Feinstruktur dieser Nanostäbe wirft Wissenschaftlern Rätsel auf und ihre Entstehung ist bei Weitem nicht vollständig verstanden. Ein absehbarer Einfluss der Faserstruktur auf die optischen und elektronischen Eigenschaften ist zu berücksichtigen. In dieser Arbeit wird ein Modell eingeführt, das den Ursprung der Feinstruktur durch ein thermisches Ungleichgewicht zwischen den Prozessen der Kristallisation und der Energieminimierung beschreibt. Mehrere Merkmale von Nanostäben, die unter unterschiedlichen Bedingungen gezüchtet wurden, werden durch das dargestellte Modell vorhergesagt, wie beispielsweise das Verschwinden der Feinstruktur bei erhöhten Wachstumstemperaturen.

An vierter Stelle bestand die Hauptaufgabe darin, Methoden zur lokalen Herstellung von Nanostäben zu finden. Hierbei werden neue, schnelle und kostengünstige Techniken vorgestellt und mit bewährten Verfahren wie optischer oder Elektronenstrahl-Lithographie verglichen.
2D-Gitter lassen sich ohne Verwendung einer Lithographie herstellen, indem eine Monolage aus Polystyrolkugeln (PSML) auf ein Substrat aufgebracht wird. Die PSML wirkt als Maske während die Saatschicht aufgetragen wird. Somit wachsen die Nanostäbe auf hochgradig geordneten Saatinseln.
Ein weiteres, einfaches Verfahren zur Herstellung individueller Strukturen besteht darin, eine dünne Schicht aus SiO$_2$ auf eine Stelle auf einem Substrat aufgetragen wurde. Innerhalb der Gräben werden die Filme viel dünner als auf der unbehandelten Oberfläche. Während dicke Schichten aus SiO$_2$ das Wachstum von Nanostäben auf der Oberfläche verhindern, ist die Hülle aus SiO$_2$ innerhalb der Gräben zu dünn dafür. Folglich wachsen Nanostäbe nur innerhalb der Gräben, welche durch Fräsen mit einem fokussierten Ionenstrahl (FIB) erzeugt wurden.
Als weitere neue Technik wird ein Rastersondenverfahren demonstriert. So werden Partikel aus Anatas als Vorstufe für Keime aus Rutil durch Kratzen mit einer Siliziumspitze über eine Dünnschicht aus Anatas

3

erzeugt. Da sich die Partikel nicht frei bewegen können, wird das Wachstum von Nanostäben nur an den verkratzten Positionen ausgelöst. Außerdem wird ein Laser verwendet, um Mischphasen, die Titan, Silizium und Sauerstoff enthalten, zu schmelzen. Diese Mischphasen befinden sich an der Grenzfläche zwischen der Schicht aus Anatas und dem Siliziumsubstrat. Sie werden durch die Diffusion von Atomen über die TiO_2/Silizium-Grenzfläche erzeugt, während die Schichten aus Anatas getempert werden. Während des Verfestigungsprozesses der Schmelze wird Rutil gebildet, wodurch das lokale Wachstum der Nanostäbe aus Rutil gefördert wird. Das Schmelzen wird unter Verwendung eines frequenzverdoppelten Nd:YAG-Lasers bei einer Wellenlänge von 532 nm durchgeführt. Es wird sowohl ein cw-Laser zur Herstellung individueller Strukturen als auch ein gepulster Laser zur Herstellung großflächiger Muster durch Zweistrahlinterferenz verwendet.

Die optische Charakterisierung bietet Einblicke in die Lichtreflexion und Streuung der stark anisotropen Nanostrukturen. Diese Ergebnisse führen zu der Annahme, dass die Nanostäbe als winzige dielektrische Antennen wirken. Und so kommen sie als Nano-Streuer, Nano-Resonatoren, Nano-Polarisatoren, Nano-Gitter und Wellenleiter in Frage. In diesem Zusammenhang wird das Übergangsverhalten von Einzelpartikeln zu einem effektiven Medium diskutiert. Die Eigenschaften werden durch ein optisches Mikroskop im Hell- und Dunkelfeldmodus untersucht.

Zuletzt zeigen Strom-Spannungs-Kennlinien die Vielfalt an erreichbaren elektronischen Eigenschaften, die durch Tempern in verschiedenen Gasen oder durch die Herstellung mit unterschiedlichen Wachstumsbedingungen erreicht werden. Es ist bekannt, dass die Variation der Sauerstofffehlstellendichte die Menge der mobilen Elektronen und sogar die Bandstruktur ändert. Dies führt entweder zu stark isolierenden oder gut leitfähigen Nanostäben, wie in dieser Arbeit gezeigt wird. Ein Schalten zwischen ohmschen und Gleichrichtungs-Kontakten an der Nanostäbchen/Elektroden-Grenzfläche wird ebenfalls beschrieben. Sowohl das Einbringen von Ladungsträgern als auch die Ionenwanderung verändern die Bandstruktur sowie die räumliche und energetische Verteilung von Fallenzuständen in Nanostäben aus TiO_2 während hoher Spannungs- oder Strombelastungen. Diese Effekte werden durch zeitabhängige Messungen untersucht. Alle Messungen werden in einer wohldefinierten und trockenen Atmosphäre durchgeführt.

Auf der Grundlage von physikalischen Modellen und Beschreibungen liefert diese Arbeit eine detaillierte Anleitung für die kontrollierte Herstellung von Nanostäben aus Rutil und die Anpassung deren optischer und elektronischer Eigenschaften.

1 Introduction

As a nontoxic, chemically stable, and frequently occurring material with various valuable electronic and optoelectronic properties TiO_2 has great potential for innumerable applications. The broad range of properties and applications can even be extended by using nanostructured TiO_2 [1]. In addition, nanostructured TiO_2 systems offer a high surface area, single crystal structures, and shape related scattering facilities. In spite of intensive research on this material in the past, many properties which are essential for various applications are not yet understood or even controllable. In this thesis, fundamental questions of the growth mechanism, the position-controlled fabrication, and the optical and electronic properties of rutile TiO_2 nanorods$^\diamond$ are discussed. For a better understanding, terms marked with a diamond symbol ($\diamond$) are explained briefly in the glossary. Framed paragraphs emphasize important information and introduction graphics in the beginning of "results and discussions" chapters help the reader to keep the track. A lot of additional information and graphics are given in the appendix. The detailed chapters summarizing the state of research are supposed to help other young researchers to find a faster access to this topic.

TiO_2 was brought into the industry for the first time in 1923. In those days, it was used as a white pigment only, but soon its applications became more and more versatile. Today TiO_2 is indispensable in many branches of industry as shown in the first chapter. In addition, researchers opened various doors to new potential applications in recent decades [2]. A quite promising trend is to use nanostructured TiO_2 such as rutile nanorod array (NRA). Due to their appropriate electronic and optoelectronic properties, their simple fabrication methods, and high chemical stability [3–5], these arrays are beneficial for photocatalysis devices [3, 6–14] (using both anatase [15–19] and rutile [20–23] TiO_2), photocatalytic disinfection [24–28], intimately coupled photobiocatalysis (ICPB) [29], hydrophobic/superhydrophilic$^\diamond$ materials [30–36], gas sensors [37–46], gasochromic displays [47–49], electrochromic displays [50], biosensing [51, 52], transparent electrodes [50], photodetectors [53, 54], field emission devices [55, 56], nonlinear optical devices [57], hybrid solar cells$^\diamond$ [58–62], organic light-emitting diodes (OLEDs) [63–66], fuel cells [67], data storage devices [68], flexible data storage devices [69], field-effect transistors [64, 70, 71], flexible capacitors [72], supercapacitors$^\diamond$ [73], and lithium batteries [74–84]. Since TiO_2 is a non-toxic material there are applications in medical engineering as well. It was shown by Bjursten et al. [85] that TiO_2 nanostructures are a promising candidate for implant coatings to obtain a stronger adhesion between osteoblasts$^\diamond$ and the implant surface compared with conventional microstructured coatings. In many current applications, NRAs are fabricated by simple, fast, and inexpensive hydrothermal methods resulting in homogeneously distributed NRAs. Applications using defined distributions of rutile TiO_2 nanorods are rare, largely because of the absence of suitable techniques for the fabrication of locally confined arrangements. Confined geometries for NRAs are of great interest for locally increased gas-sensing, photocatalysis, hydrophobic/superhydrophilic characteristics, light scattering, or surface roughening as it is useful for lab-on-a-chip$^\diamond$ for instance [86–88].

The findings shown in this thesis are not only important for the integration of rutile TiO_2 nanorods into devices, but a significant part of the achieved knowledge could be transferred to other rutile nanostructures consisting of different metal oxides. In the following a brief overview of various rutile metal oxides and their applications is given to demonstrate the great significance of this research for the technical development in many fields. VO_2 shows a metal-insulator transition at 65 °C [89]. CrO_2 nanorods show temperature-dependent magnetotransport properties [90]. GeO_2 is used for infrared (IR) transparent glasses [91] and as a catalyst in the production of polyethylene terephthalate [92] and also available as nanorods [93]. MoO_2 has a low electrical resistivity of $88\,\mu\Omega \cdot cm$ [94, 95], high chemical stability [96] and thus, it is used in a wide range of applications such as anode material in lithium batteries [97–102], field emission devices [96, 103], nanorod based photodetector devices [104] and catalysis processes such as the dehydrogenation of alcohols [105], the reformation of hydrocarbons [106] and biodiesel° [107]. Furthermore, several fabrication techniques for nanorods are reported [94, 96, 96]. RuO_2 is used for catalyzing the Deacon process° for chlorine production [108], as an active material in supercapacitors due to its high charge transfer capability [109–112], and as a resistive temperature sensor for low-temperature experiments [113]. Different recipes for RuO_2 nanorod fabrication are reported [111, 114, 115]. SnO_2 nanorods are used for resistivity based gas sensors [116–118], transparent electrodes for dye-sensitized solar cells (DSSCs)° [119], and lithium ion batteries [120] as well as fully transparent transistors [121]. TeO_2 is used for fiber optics and waveguide applications due to its high refractive index [122] and as an acousto-optic material [123–125]. OsO_2 single crystals show metallic resistivity of about $15\,\mu\Omega \cdot cm$ [126] and IrO_2 is used for microelectrodes for electrophysiology [127, 128].

In this thesis, the research focuses on rutile TiO_2 nanorods as a potential model system for all other rutile metal oxide nanostructures. In the beginning, a general overview on TiO_2 is given starting from the economic meaning, natural mineral deposits, extraction from natural minerals, morphologies, chemical and mechanical stability, and electronic and optical properties to the point of confinement effects on low dimensional systems (Chapter 2,3).

In the experimental chapter (Chapter 4), fabrication and characterization techniques are described. This includes the hydrothermal growth method of rutile TiO_2 nanorods as well as the processing of seed layers promoting the growth of such nanorods. For the methods applied in this thesis, the processing is described in detail. For the sake of completeness, further typical fabrication techniques are described briefly as well. Characterization techniques include methods to determine the surface topography (atomic force microscopy (AFM), scanning electron microscopy (SEM)), the atomic structure (X-ray diffraction (XRD), transmission electron microscopy (TEM), energy dispersive X-ray spectroscopy (EDX)), the optical properties (ellipsometry, ultraviolet-visible-infrared (UV-Vis-IR) absorption spectroscopy, optical microscopy), and the electronic properties (current over voltage (I-V), current over time (I-t), impedance spectroscopy (IS)).

To promote the growth of rutile TiO_2 nanorods on substrates without a rutile surface an appropriate rutile seed coating is needed [61, 129–131]. Typically, these seed layers are made of polycrystalline TiO_2 films that can be fabricated employing sputter deposition [132–134], electron-beam evaporation [132, 135, 136], atomic layer deposition (ALD) [137–141], sol-gel methods [142, 143], spray pyrolysis [144–147], or $TiCl_4$ treatments [148–150]. Besides TiO_2, fluorine tin oxide (FTO) appears as a polycrystalline rutile film as well. Usually, FTO films are made by sputter deposition [151] or spray pyrolysis [152]. Due to its high conductivity and transparency, FTO is often used as an electrode for photovoltaic applications [61, 153]. There are a lot of reports about the application of rutile TiO_2 nanorods, but only a few articles about the details of their growth process [61, 129, 153–167]. Usually, the as-grown nanorods consist of numberless smaller crystallites forming a mesocrystal rather than a single crystal [168]. So far, it is assumed that the grain boundaries separating these crystallites originate from point defects implemented during the crystal growth.

In this context, the new knowledge delivered by this work is a detailed overview of the structural properties of differently fabricated seeds (Chapter 5). In addition, a detailed relationship between the shape of the seed, the growth time, and the shape of the grown nanorod is described (Chapter 6). Besides rutile seeds, particular crystal formations on anatase surfaces are introduced as an alternative growth facilitator of rutile TiO_2 nanorods. A model is introduced explaining the creation of small crystallites inside the as-grown nanorods. It is demonstrated that the consequences of this model allow to control the density and shape of these crystallites and hence expand existing models with respect to important problems (Chapter 7).

A basic question in many electronic devices is to locate individual semiconductor structures precisely on a substrate. The position-controlled fabrication of semiconductor nanorods was applied successfully for different semiconductor materials in the past. The most common technique is to use lithography for selective seed layer deposition. A typical example is the patterning of ZnO NRAs, which share many properties and applications with TiO_2 NRAs. Common techniques to achieve submicron structures are optical immersion, deep UV, X-ray, nanoimprint, or electron-beam lithography [169–172]. These techniques can be used to create very localized seeds for the hydrothermal growth process. A simple option is to use optical lithography and accept a rather large minimal structure for a low purchase price. However, this technique is absolutely suitable for mass production [173, 174]. The smallest structures that are achievable with electron-beam lithography are about 2 nm wide [175–179], but this technique is too expensive and time consuming for industrial mass production. The resolution of extreme ultraviolet (EUV) lithography is 10 nm so far [180]. Although this technique is more suitable for commercial applications, it is also extremely expensive. Mi-Hee and Hyoyoung introduced nanoimprint lithography (NIL) to fabricate stripes and ordered islands consisting of ZnO nanorods [181]. Park et al. developed this technique further and demonstrated a pattern of parallel arranged stripes covered with ZnO nanorods by combining ultraviolet-assisted nanoimprint lithography (UV-NIL) and hydrothermal growth [182]. Another quite advanced technique is to generate local wetting gradients before the growth of ZnO nanorods [183]. Techniques of position-controlled growth of TiO_2 NRAs are found at rare intervals and dealing mostly with large area patterns [184–186]. However, it is possible to transfer many techniques listed for ZnO to TiO_2 nanorods.
In this context, the new knowledge delivered by this work includes mainly two further techniques – a scanning probe and a thermal lithography – to achieve submicron structures in a relatively fast and inexpensive way (Chapter 8-9). The new techniques are compared with optical and electron-beam lithography and a few derived techniques are discussed.

Related to their high refractive index, TiO_2 nanorods are efficient light scatterers. Because of this feature, they are applicable as nano-antennas [187, 188], nano-cavities [189–192], and nano-polarizers [193–196]. But there are hardly reports describing the scattering of small anisotropic accumulations of TiO_2 nanorods.
In this context, the new knowledge delivered by this work is the documentation of the transition from single nanorod scattering through collective nanorod scattering into non-scattering effective media consisting of dense NRAs (Chapter 10). To investigate collective nanorod scattering, highly anisotropic NRAs were grown on structures made by the above described position-controlled seed deposition techniques.

The ability to switch the electronic properties of TiO_2 in a controlled way turns this typically n-type semiconductor into an interesting building block for all kinds of electronic devices. The dominant intrinsic electron donors in TiO_2 are oxygen vacancies [197–203]. Their density corresponds to the number of mobile electrons and influences the position of the Fermi level. As a consequence, the limiting conduction mechanism is also fixed by the distribution of oxygen vacancies. The number and location of oxygen vacancies can be influenced by electrical fields

[204, 205], plasma [206], chemical [207–209], radiative [30, 210, 211], or thermal [212–214] treatments. Especially, the effect of the electrical field is used for data storage based on resistive switching [215–224]. Besides their electron donating nature, crystal defects can act as electron traps and charge the TiO_2 locally [225]. Of course, this feature affects the limiting charge transport mechanisms as well, but such charging is highly valuable for nanocrystal (NC) memory devices, where the electrical field of the trapped charges control the current flow through the channel of a field-effect transistor (FET) [226]. This effect was already applied in experiments employing silicon nitride as a charge trapping material and it is essential for decreasing the size of FETs [227, 228].

In this context, the new knowledge delivered by this work is a detailed description how the conduction mechanisms are influenced by certain chemical and thermal treatments of as-grown rutile TiO_2 nanorods (Chapter 11). In particular, the role and control of the density and distribution of intrinsic electron donors and traps are discussed.

In its entirety, this thesis can be split into two parts. On the one hand, there is the "hardware" part which covers new insights of the physical and chemical understanding of the hydrothermal growth of rutile TiO_2 nanorods, as well as the position controlled growth, which is essential for the fabrication of devices needing a spatially resolved functionality on a micro- or nanoscale. On the other hand, there is the "software" part, which deals with the physical understanding of optical and electronic properties of rutile TiO_2 nanorods and thus offers pathways to design optoelectronic properties for user-defined applications.

Introduction	C. 1	
State of Research	C. 2 C. 3	structural, optical, electronic properties hydrothermal growth of rutile TiO_2 nanorods
Experimental Methods	C. 4	fabrication and characterization
Results and Discussions: **Seeds and** **Growth Mechanisms**	C. 5 C. 6 C. 7	seed layers hydrothermal growth process on seed layers fine structure of rutile TiO_2 nanorods
Results and Discussions: **Nanorods Grown in Controllable** **Confined Geometries**	C. 8 C. 9	established position-controlled techniques, advanced scanning probe lithography, LASER-induced hydrothermal growth
Results and Discussions: **Optical and Electronic** **Properties**	C. 10 C. 11	from single nanorod to many body scattering conduction mechanisms in defect-rich nanorods
Conclusion and Outlook	C. 12	

Table 1.1: *Thematic overview of this thesis (C. = Chapter).*

2 Basics of Titanium Dioxide

2.1 Exploitation and Economic Meaning of TiO_2

A very good and detailed review about the mining and economic meaning of TiO_2 is given by Gázquez et al. [229]. In this paragraph, only the most important fact from this review are summarized and complemented with further information.

Since titanium is one of the most abundant elements in the Earth's crust (0.63 wt%) TiO_2 is found in thousands of natural mineral deposits worldwide and hence, it is available in huge amounts [230, 231]. TiO_2 was separated firstly from the mineral rutile by the German chemist M.H. Klaproth around 1800 [229]. It is used frequently from the beginning of the 20th century as white pigment in numberless products. The consumption is still rising and the German consultant Ceresana eK in Konstanz estimates the yearly consumption at 7.5×10^6 t in 2019 (February, 2013).

TiO_2 is extracted from titanium-containing ores such as ilmenite ($FeTiO_3$), leucoxene ($Fe_2O_3 \cdot nTiO_2$), which is an oxidation product of ilmenite and composed of finely crystalline rutile [232], rutile, anatase, and brookite. Although the three last named minerals have all the formula TiO_2, they differ in their crystal structure [229]. Further minerals with less economic meaning are titanium-rich (2-20%) magnetite, so-called titaniferous magnetite or titanomagnetite, pseudobrookite (Fe_2TiO_5), perovskite ($CaTiO_3$), geikielite ((Mg, Fe)TiO_3), pyrophanite ($MnTiO_3$), and titanate, so-called sphene [233]. Details about the mineralogical properties are shown in Table 2.1. It should be noted that the term "rutile" has a slightly different meaning in mineralogy and crystallography. In mineralogy, rutile is a crystal having mainly a rutile morphology and consisting of TiO_2 with a relatively high content of iron and other impurities. In crystallography, rutile describes the ideal rutile lattice – independent from the elements it is made of. The extraction of TiO_2 from minerals results in impurities such as chromium, manganese, vanadium, magnesium, aluminum, calcium, silicon and others [234] even in high purity TiO_2 sputter targets. As it will be discussed later these impurities play an important role for the electronic properties of TiO_2.

Most of the titanium is produced from ilmenite and rutile in shoreline placer deposits in Australia, South Africa, USA, India, and Sri Lanka. Other sources are magmatic ilmenite deposits in Canada, Norway, Finland and the USA [238]. In 2009 the total production of ilmenite was 5.3×10^6 t (from South Africa 1.05×10^6 t, Australia 1.02×10^6 t, Canada 0.65×10^6 t, China 0.50×10^6 t, India 0.45×10^6 t, Vietnam 0.412×10^6 t, Norway 0.30×10^6 t, Ukraine 0.30×10^6 t), while the production of rutile was 0.55×10^6 t only (Australia 0.266×10^6 t, South Africa 0.127×10^6 t, Sierra Leone 0.06×10^6 t) [239]. It is estimated that today's reserves of ilmenite and rutile are about 650×10^6 t and 42×10^6 t, respectively [229]. For the next years an increasing demand for TiO_2 with a global growth rate of 3% is expected [240, 241]. It is expected that the demand will increase significantly above average in the Asia region, especially in China and India [229].

In a first processing step titanium ores have to be purified for the final fabrication methods of high purity TiO_2. The purification of ilmenite into titania "slag" (about 70-90 wt% TiO_2) is performed by smelting in electrical arc furnaces. The product can be already used to fabricate TiO_2 by the sulfate process [242]. But for applying the chloride process the slag contains too

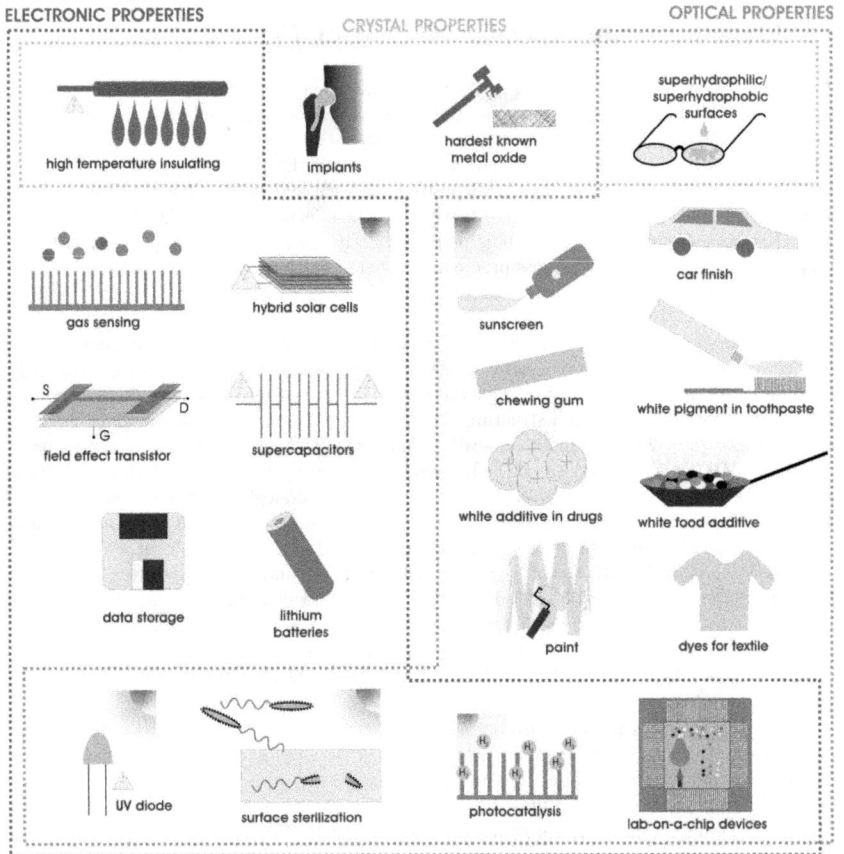

Figure 2.1: *Applications of TiO₂ separated by the properties of TiO₂ that are relevant for each application.*

Name (Formula)	% TiO_2	Color	Hardness	Density	Crystal Form
Ilmenite ($FeTiO_3$)	52.6	black	5-6	4.5-5.0	hexagonal
Perovskite ($CaTiO_3$)	58	black, brown, reddish-brown or yellow	5.5	4.26-4.48	monoclinic (pseudocubic)
Rutile, Anatase, Brookite (TiO_2)	95	reddish-brown, red, yellowish or black	6.0-6.5 (rutile) 5.5-6.0 (anatase and brookite)	4.23-5.5 (rutile) 3.82-3.97 (anatase) 4.08-4.18 (brookite)	tetragonal (rutile and anatase) orthorhombic (brookite)
Titanate (sphene) ($CaTiSiO_5$)	35-40	brown, green grey, yellow or black	5,0-5.5		monoclinic

Table 2.1: *Properties of typical titanium ores used for titanium and TiO_2 production [235]. It is reported by Moore that synthetically fabricated rutile single crystals have Mohs values of 7.0–7.5 Mohs [236]. The record holder among the hardest polycrystalline oxide materials is cotunnite-structured titanium oxide. Its hardness (38 GPa) is close to sintered diamond (50 GPa). For comparison, the hardness of silicon carbide is 29 GPa only [237].*

much calcium and magnesium and has to be purified once more by thermal treatment [243–245] and pressure leaching in hydrochloric acid (HCl) resulting in the upgraded slag (UGS) (about 95 wt% TiO_2). Besides the chloride process is used to produce titanium metal as well. Synthetic rutile (90-96 wt%) is made from ilmenite by reducing the iron oxide and leaching with mineral acids. The most common processes for the fabrication of synthetic rutile are the Becher and the Benilite process [246].

High-purity TiO_2 is fabricated from the purified ores by the sulfate or chloride process [242]. The base product of the sulfate process is ilmenite (40-60 wt% TiO_2), titanium slag (72-85 wt% TiO_2) or controlled blends. The raw material is digested with concentrated sulphuric acid (98%) triggering the dissolution into titanyl sulfate ($TiOSO_4$) and iron sulfate ($FeSO_4$) ($FeTiO_3$ + $2H_2SO_4 \rightarrow TiOSO_4 + FeSO_4 + H_2O$). As a next step, the solution is hydrolyzed to titanium dioxide hydrate ($TiOSO_4 + H_2O \rightarrow TiO_2n\cdot H_2O + H_2SO_4$) and separated from the acid by filtration. By calcination, the residual water is removed and the crystal shape and size of TiO_2 particles is defined ($TiO_2n\cdot H_2O \rightarrow TiO_2 + nH_2O$). For the fabrication of 1 kg TiO_2 anatase pigments, about 2 kg ilmenite are needed within the sulfate process. Compared to the sulfate process the chloride process is environmentally safer, causes less waste, needs less energy and results in a higher quality product [247]. Due to the high iron content of ilmenite, huge amounts of iron sulfate have to be disposed within the sulfate process. The base product of the chloride process is mainly rutile and synthetic rutile (90-95 wt% TiO_2). It is exposed to gaseous chlorine at 900–1000 °C in a fluidized bed reactor using coke as a reducing agent [248] ($2TiO_2 + 3C + 4Cl_2 \rightarrow 2TiCl_4 + 2CO + CO_2$). Besides titanium tetrachloride ($TiCl_4$) the gas contains oxides of carbon and further metal chlorides, but silica and zirconium are supposed to be left unchlorinated in the furnace [247, 248]. The glass is cooled down in recycled liquid $TiCl_4$ until other metal chlorides condense and settle down. Purities of up to 99.999 wt% $TiCl_4$ could be achieved with this treatment. The purified $TiCl_4$ reacts at 1500 °C with oxygen resulting in TiO_2 and chlorine, which is reused in the process ($TiCl_4 + O_2 \rightarrow TiO_2 + 2Cl_2$). Finally, residual chloride is extracted by aqueous hydrolysis. If needed, there are further chemical surface treatments, milling and drying steps. Nowadays also the waste products of both processes are used partly for industrial applications [229].

Besides intensively investigated TiO_2 based nanomaterials for future applications listed in the introduction, TiO_2 has already an important meaning in today's economy and currently rising technologies. The relatively high refractive index of TiO_2 results in effective scattering and hence, it is suitable for thin film coatings. Automotive industry requires the most critical performances, since color pigments have to show high photodurability, corrosion protection, and chip resistance. All these requirements are delivered by TiO_2 outstandingly. For the same characteristics, it is applied in opacity plastic materials to limit translucence and increase photo-durability. Involving chemical reactions with organic compounds of plastics, TiO_2 is used for yellowing [229, 247]. For printing inks, very fine TiO_2 particles are used. In particular, such TiO_2 particles are suitable for white hiding and high brightness printing. Rheology°, abrasiveness°, gloss°, and redispersibility° are influenced by the type of TiO_2 particles. Further applications for TiO_2 in inks are wood molding, marking pens, decorative sheets, ink correction fluids, and concealed writing (scratch-off lottery tickets). TiO_2 is an additive for high quality or thin papers to improve brightness and opacity [247]. Rutile pigments are favored in coated paperboard due to their higher optical scattering efficiency. It is also used in pharmaceuticals and cosmetics such as lotions, creams, shampoos, soaps, stretch marks and cellulite treatments, hair treatments, balms, lotions, creams, sunscreens, hair dyes, toothpaste, lipsticks, anti-wrinkle treatment, soaps, and much more in order to provide pleasant colors for final customer. For textile applications such as polyamide, polyethylene terephthalate (PET), polyester fibers, and acrylic anatase TiO_2 pigments are used as a dye component because of the high whiteness, strong hiding power, good dispersibility, outstanding heat and UV light resistance [247]. Due to its low toxicity TiO_2 is a common material for food coloring (food additive E171). But usually not more than 1 wt% is added to foods such as cheese, baked, toppings, icings, confectionery, skim milk, codfish, and food supplements [229]. If TiO_2 is exposed to UV light it acts as a photocatalyst and is used for water splitting and the production of hydrogen, which is supposed to be an important energy carrier for future electrical power-trains in vehicles. In addition, the ability to split high energetic molecular bonds under UV light exposure makes TiO_2 an efficient medium for disinfection and hence, it becomes useful for medical devices, food preparation tools, air conditioning filters and sanitary ware surfaces. Furthermore, UV light makes TiO_2 coated surfaces superhydrophilic and thus, it works as anti-fogging coating [249]. The multitude of applications of TiO_2 nanomaterials was listed in the introduction already and is part of this thesis as well.

By far the most amount of TiO_2 is used in paints and coatings, papers and paperboards, and plastics by consumers in the mature sectors in the developed world. Thus, the consumption of TiO_2 follows mainly general economic trends. The UK consultant Artikol estimates the global growth for TiO_2 to be 2.7% per year until 2019. Besides China, also India is expected to have a great impact on the growing consumption [249]. Besides the demand, the market price depends on the manufacturer, grade, particle size, and morphology. Thus, the price margin is quite large and it is ranging from 90 ct/kg (industry grade for white pigments, no specific quality characteristics, Suzhou Baisede Chemical Co., Ltd., China) to 19,560 €/kg (99.99% trace metals basis, brookite TiO_2 nanopowder, <100 nm particle size, Sigma-Aldrich, Switzerland) (May 22nd, 2016).

2.2 Crystal Properties of Rutile and Anatase TiO_2

In this paragraph, basic crystal properties of rutile and anatase TiO_2 are described and up to date, known correlations between the crystal structure and its electronic and optical properties are discussed. Pure and modified TiO_2 structures show a broad range of electronic properties in particular. Thus, the following overview broaches characteristics, which are either very fundamental or crucial for the understanding of the results in this work. For some issues, further reading is quoted.

2.2.1 Morphologies and Crystal Structure

Crystalline TiO_2 material appears in three different morphologies in nature, which are rutile, anatase and brookite [250]. In addition, there are three metastable, synthetically fabricated phases: $TiO_2(B)$, which is monoclinic and gained from hydrolyzing of $K_2Ti_4O_9$ followed by calcination [251], $TiO_2(H)$, which is tetragonal with a hollandite-like form and gained from the oxidation of potassium titanate bronze ($K_{0.25}TiO_2$) [252], and $TiO_2(R)$, which is orthorhombic with a ramsdellite-like form and gained from the oxidation of lithium titanate bronze ($Li_{0.5}TiO_2$) [253]. Furthermore, there are five high-pressure forms: $TiO_2(II)$, which is orthorhombic with a α-PbO_2-like form [254], seven coordinated titanium, which is monoclinic with a baddeleyite-like form [255], TiO_2-OI, which is orthorhombic [256], TiO_2-OII, which is orthorhombic as well with a cotunnite($PbCl_2$)-like form [237] and a cubic form [257]. Some of these modifications are used in research as well. In this thesis, only the rutile and anatase morphology are investigated. Rutile TiO_2 is assigned to the space group $P4_2/mnm$ (No. 136) and its X-ray lattice parameters are $a = 4.5937\,\text{Å}$ and $c = 2.9587\,\text{Å}$ [258]. Anatase TiO_2 belongs to the space group $I4_1/amd$ (No. 141) and its X-ray lattice parameters are $a = 3.7845\,\text{Å}$ and $c = 9.5143\,\text{Å}$ [259]. As a consequence of the significantly anisotropic crystal structure many mechanical, electronic, and optical properties depend on the crystal orientation.

Rutile and anatase structures are classified in detail by Burdett et al. [260]: For metal oxide structures having a MeO_2 stoichiometry, a preferred arrangement for occupying half of the octahedral holes in a hexagonal close-packed lattice consisting of oxide ions is given by the rutile structure, which is a tetragonal variant of the orthorhombic $CaCl_2$ structure. Within the $CaCl_2$ structure there are chains of edge-sharing octahedra, which have common vertices. Each oxide ion is linked with three metal atoms forming a distorted trigonal-pyramidal complex. While this structure has a C_s point symmetry, the pyramids are arranged in a plane with C_{2v} point symmetry. The unique angle for the O–Ti–O complex is about 99°. The metal octahedra are compressed with two long and four short Ti–O bonds [260].

The anatase structure is built up similarly to the rutile structure but based on a cubic closed-packed lattice of oxide ions. The octahedra share four edges with each other [261–265]. The sharing of such edges result in short O–O bonds [260].

The rutile crystal has ionic character consisting of Ti^{4+} cations and O^{2-} anions with ionic radii of $r_{Ti^{4+}} = 0.60\,\text{Å}$ and $r_{O^{2-}} = 1.46\,\text{Å}$ [197, 266]. The interatomic distance in ionic crystals is given by the sum of the cation radius r_c, the anion radius r_a, and a correction Δ_N depending on the coordination number N. Since the cation Ti^{4+} is surrounded by six O^{2-} anions N is equal to six and Δ_6 is found to be zero [197]. Based on this the interatomic distance is supposed to be 2.06 Å, but the experimentally observed values are 1.944 and 1.988 Å [197]. The deviation results from covalent forces. It was found by Baur that the electron density between the cations and anions is larger than zero at any point [267, 268]. This is a strong hint for a covalent contribution to the bonding. Also, the temperature-dependent paramagnetism results from the intermixing of ionic and covalent bonding character, since both the tetravalent titanium and the divalent oxygen ion have noble-gas configuration and thus, the temperature-dependent paramagnetic susceptibility

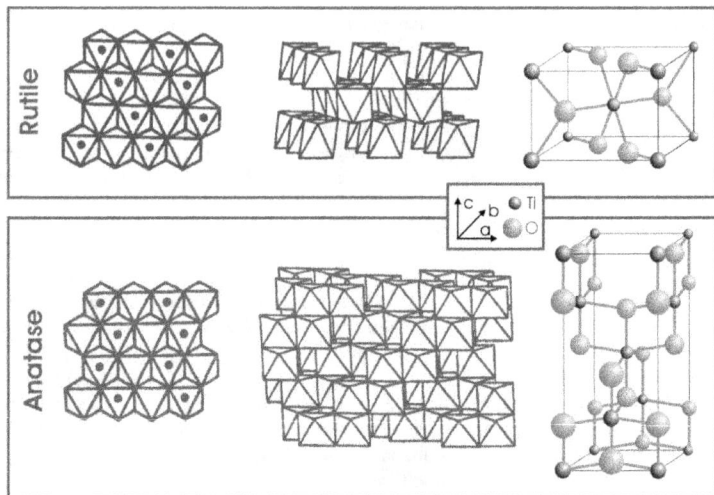

Figure 2.2: *Crystal structure of rutile and anatase TiO₂. Reprinted with permission from Burdett et al. [260]. Copyright ©1987 American Chemical Society.*

is supposed to vanish for pure ionic or covalent bonding [197]. A consequence of the covalent contribution to the Ti–O bonding is the low solubility of rutile in water and other polar solvents [197]. The temperature dependence of the bonding forces between titanium and oxygen atoms and the resulting vibration dynamics have been discussed intensively [197, 260, 269]. Also, anatase TiO_2 shows ionic character, but it is less influenced by covalent contributions [270]. The effect of different electronic arrangements and band structure in both morphologies will be discussed in more detail in Paragraph 2.6 about the electronic properties.

2.2.2 Crystal Defects

Crystal defects such as oxygen vacancies, titanium interstitials or impurities are not uncommon in TiO_2 and play a major role in thermal stability, conductivity, and light absorption.

Typical thermodynamically reversible point defects are oxygen and titanium vacancies, oxygen and titanium interstitials, and electron (Ti^{3+}) and hole (O^-) vacancies [271, 272]. The latter two defects are responsible for the charge transport in TiO_2 [273]. Yoon et al. report even an oxygen-defect-induced magnetism in anatase titania films [274]. In general, crystal defects are supposed to influence the photoreactivity of a sample strongly [273]. Very typical and intensively discussed defects are oxygen vacancies [275], since they are experimentally unavoidable and result in n-type TiO_{2-x} structures [11, 199, 271, 272, 276–291]. They are generated either during the fabrication process of TiO_2 material, annealing at sufficiently high temperatures in vacuum or reduction of TiO_2 with hydrogen at elevated temperatures [212, 292]. Furthermore, oxygen vacancies are generated and injected by high electrical fields as it is discussed in conjunction with the resistive switching° effect [205]. Annealed in oxygen rich atmosphere TiO_2 becomes oxidized and p-type by forming titanium vacancies [288–291].

If defects cluster and the clusters affect larger regions they are called extended defects. Within these extended defects, the point defects could be arranged orderless or form Magnéli phases° with a defined order and stoichiometry° Ti_nO_{2n-1} [293, 294]. A typical Magnéli phase of titanium oxide is Ti_2O_3. The diffusion of the listed point defects differs strongly. Trivalent titanium interstitials move much more easily through the crystal than titanium vacancies for instance [287, 288]. As a consequence, defect gradients between the surface and bulk rise.

2.2.3 Surface Science of TiO_2

Diebold wrote a very detailed and comprehensive report about surface properties of rutile and anatase TiO_2 in 2003 [295]. The surface of TiO_2 crystals differs from the bulk due to reconstruction [213]. One of the most famous techniques for investigating the surface structure is scanning tunneling microscopy (STM) [296, 297]. Especially the $TiO_2(110)$ facet is a prototypical model system of metal oxides in general and was investigated extensively [213].

Regarding the rutile bulk structure, it is expected that titanium atoms are five-fold-coordinated with one dangling bond on the (1×1) surface. Oxygen atoms are two-fold-coordinated forming bridges with their third bond. These surface oxygen atoms desorb at elevated temperatures easily. The observed relaxation of the bridging oxygen atoms of $0.27 \, \text{Å}$ is much stronger than expected from first-principles and total energy calculations [298]. M. Harrison et al. suggested that the strong relaxation originates from anharmonic surface phonons excited at room temperature [299].

While step edges seem to play a minor role in surface chemistry, point defects such as vacancies among bridging oxygen atoms appear in high concentrations and might be hydroxylated from surrounding water [300–303]. Oxygen vacancies are generated simply by annealing at high temperatures in vacuum and they diffuse into the bulk, where they create bulk oxygen vacancies and titanium interstitials [213]. Even at room temperature, a significant exchange process of oxygen vacancies between surface and bulk takes place [304]. High densities of titanium interstitial tend to aggregate resulting in extended defects [305–307]. A typical surface stoichiometry for such a reduced TiO_2 surface is Ti_2O_3 [308] and re-oxidizing results in a huge variety of surface structures [300–303, 305–311]. At a given temperature, annealing time, gas pressure and reduction state highly mobile trivalent titanium interstitial react with oxygen forming different surface formations such as (1×2) "rosette"-like structures [297, 312–315]. In general, different structures show different adsorption and catalytic reactivity [316–319]. It is reported that water is able to stick on TiO_2 surfaces up to $327\,°C$ [320, 321]. At elevated temperature water dissociated at oxygen vacancies desorbs or decomposes resulting in a healed vacancy and a hydrogen molecule [301, 322–324].

TiO_2 surfaces with low defect densities are hydrophobic and oleophilic° [213]. Exposing the surface to UV light with photon energies above the bandgap energy, Ti^{4+} ions are photoreduced resulting in Ti^{3+} [211]. This reaction is preferred at oxygen bridging sites [30, 210]. In air, either water or oxygen could bind to the defect and the equilibrium of both effects is reflected by the wetting angle [325]. Since hydroxyl adsorption is preferred from the defect sites, the wetting angle decreases to $0°$ [30, 210]. It is remarkable that the surface is still oleophilic and thus amphiphilic° under UV light exposure. By extensive water adsorption electronic properties, surface structure, and stability are changing. The effect is withdrawn in sufficient oxygen-rich atmosphere [30, 211].

There is also an exchange between surface and bulk defects such as the adsorption of sulfur, which replaces an oxygen site on the surface and diffuse into the bulk [213, 295, 326, 327]. As discussed in Paragraph 2.1, even highly pure TiO_2 contains some natural impurities. These impurities diffuse from bulk to surface during annealing and desorb or form planar crystallites such as calcium [328].

An important issue for the electronic measurements in this work is the structure of metal/TiO_2 interlayers. Thereby, the temperature-dependent oxide formation of the involved metal is a crucial factor for the interface quality [329–331]. During deposition or post-deposition annealing an oxidation/reduction process is going on, which scales with the reactivity of the metal. This reaction results in a rearrangement of the crystal structure and composition in the interface region. Thus, different material properties such as additional electronic states in the bandgap are expected. Diebold suggested a virtual line across the periodic table connecting cobalt and rhenium. On the left side there are reactive metals and on the right side, metals showing no oxidation/reduction reaction with TiO_2 [213]. The wetting for reactive metals is improved compared to inert metals. The mentioned non-stoichiometric interface region appear even for noble metals such as platinum and gold [332–336]. In the case of gold, the new interfacial material constellation provides the oxidation of carbon monoxide (CO), although neither gold nor TiO_2 is able to trigger this reaction [337–340].

2.2.4 Crystal Dynamics and Thermal Properties

The most fundamental characteristic of the crystal dynamics is the dispersion relation of phonons. In combination with the phonon density of states (phDOS), it is the starting point for calculating many crystal properties such as the dielectric function, Young's modulus°, elastic constants, Debye temperature°, heat capacity, Curie temperature°, speed of sound, and thermal conductivity. There are several techniques for determining the phonon dispersion relation [341]. The most complemented results are achieved by inelastic scattering of thermal neutrons, inelastic X-ray scattering and high-resolution electron energy loss spectroscopy (HREELS). Due to the large mean free path in condensed matter inelastic neutron scattering is useful for the bulk dispersion relation, while the latter methods are more surface sensitive. Phonons around the center of the Brillouin zone having a rather small momentum are determined by Raman spectroscopy°, infrared (IR) spectroscopy and Brillouin scattering°. The surface sensitivity of such techniques could be increased with slight modifications such as the surface enhanced Raman spectroscopy. Although the optical characterization techniques are limited to a certain part of the phonon dispersion, they are used much more frequently compared to inelastic neutron scattering since they are much less laborious [342]. As shown in this paragraph, comprehensive information can be extracted from Raman- and IR spectroscopy as well. This paragraph summarizes the most important results gained from the phonon dispersion relation and the density of phonon states for anatase and rutile TiO_2.

Dispersion Relation of Bulk TiO_2

The phonon dispersion of rutile and anatase was investigated extensively in theory and experiments. There are six atoms in the unit cell of rutile and anatase. Thus, 15 optical and three acoustic modes are expected for both morphologies. In case of rutile, the optical modes around the Γ point are

$$\Gamma_{op,(rutile)} = A_{1g} + A_{2g} + A_{2u} + B_{1g} + 2B_{1u} + B_{2g} + E_g + 3E_u, \tag{2.1}$$

which follows from group analysis of the space group $P4_2/mnm$ [343, 344]. In this representation Raman active modes are labeled with g, infrared IR active modes with u and E are degenerated modes. longitudinal optical phonons (LOs)-transversal optical phonon (TO) splitting for the IR active modes at the Γ point is supposed to originate from the long-range dipole–dipole interactions in the ionic crystal [343]. The macroscopic electrical fields of LO modes cause a splitting of the polar A_{2u} and E_u modes into LO and TO modes [343]. A_{2g} and B_{1u} phonons are silent modes (Raman and IR inactive) [343, 344].

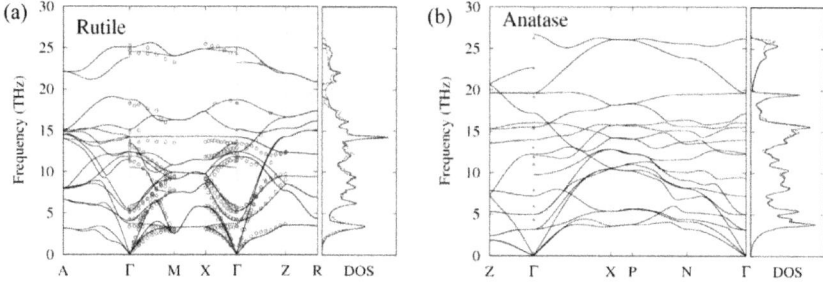

Figure 2.3: *Phonon dispersion relation and phonon density of states of rutile and anatase TiO_2. Solid lines were calculated by Mei et al. [343], blue open cycles are data from inelastic neutron scattering from Traylor et al. [341], and the solid red triangles are Raman and IR spectroscopy data from Porto et al. [345] and Eagles [346] [12], respectively. There are no dipole–dipole interactions considered in the calculated DOSph [343]. Reprinted with permission from Mei et al. [343]. Copyright ©2011 American Chemical Society.*

In case of anatase, the phonons at the Γ point resulting from group analysis of the space group $I4_1/amd$ are

$$\Gamma_{op,(anatase)} = A_{1g} + A_{2u} + 2B_{1g} + B_{2u} + 3E_g + 2E_u. \quad (2.2)$$

Here B_{2u} is a silent phonon [343, 344]. Theoretical modeling can be hardly proved for anatase TiO_2, since there is no complete inelastic neutron scattering data for the whole Brillouin zone up to date [343].

The dipole–dipole interactions influence the density of phonon states (DOSph) only for high frequencies [343]. Thus, those interactions have to be considered for free energy calculations and phase stability studies at high temperatures only [343]. It is worth noting that the phonon dispersion relation and DOSph of high-pressure polymorphs of TiO_2 show no imaginary phonon frequency [343]. This feature suggests that they are dynamically stable under ambient conditions and thus, they are a promising candidate for ultrahard materials [343].

A detailed measurement of the phonon dispersion relation of bulk rutile TiO_2 using inelastic scattering of thermal neutrons is given by Traylor et al.[341]. Theoretical models describing the dispersion relation are usually based on density functional theory (DFT) [343]. A very detailed and comprehensive overview of bulk rutile and anatase TiO_2 lattice dynamics comparing experimental data from neutron scattering, Raman- and IR spectroscopy with DFT calculations is given by Mei et al. [343].

Dielectric Functions and Curie Temperature

Of course, the static dielectric constant could be achieved from the capacity of a well defined TiO_2 film. But this method is limited and inaccurate if the investigated sample is porous or nanostructured. Here, the ratio between TiO_2 and TiO_2-free volume has to be measured, the electrode has to be placed on the TiO_2 complex without changing or penetrating it. For spatially resolved measurements the electrical field is highly inhomogeneous and influenced by complex edge effects. Furthermore, due to intrinsic defects or impurities in TiO_2 there are free charge carriers such as electrons, holes, and ions, which could accumulate and reduce the homogeneity of the electrical field. Especially the ion migration, which plays an important role in the resistive switching effect of thin TiO_2 films, changes the crystal structure and therefore the dielectric constant during the measurement.

Thus, extracting the dielectric constants from the phonon dispersion is supposed to be much more accurate. The ratio between the static and high-frequency dielectric constants ϵ_0 and ϵ_∞ for rutile TiO_2 is given by

$$\frac{\epsilon_{0,c}}{\epsilon_{\infty,c}} = \frac{[\tau_l^2(\vec{q} \to 0)]_{\Gamma_1^-}}{[\tau_t^2(\vec{q} \to 0)]_{\Gamma_1^-}}, \tag{2.3}$$

where $\tau_{l/t}(\vec{q} \to 0)$ is the longitudinal (l)/ transverse (t) optical phonon mode frequency at the zero wave vector. This expression is derived from a generalized Lyddane-Sach-Teller relation for cubic crystals having more than two atoms per unit cell [341, 347]. The index Γ_1^- indicates that only one pair of longitudinal and transverse optical (LO and TO) polar modes exist having displacements parallel to the c-axis. The dielectric constants in rutile TiO_2 at room temperature are $\epsilon_{0,a} = 86$ and $\epsilon_{\infty,a} = 6.843$ along the a-axis and $\epsilon_{0,c} = 170$ and $\epsilon_{\infty,c} = 8.427$ along the c-axis using $\epsilon_\infty = n^2$ and taking the refractive index n from Cronemeyer [348, 349]. The high value of the static dielectric constant is much larger than the dielectric constant in the visible light regime, which is typical for ionic crystals [197].

The temperature dependence of the dielectric constants is given by Parker between 1.6 and 1060 K [349]. The squared frequency of the TO mode $\tau_t^2(T)$ is proportional to the difference between the temperature T and the Curie temperature T_C [350]. Using this relation and the experimental data from neutron scattering, Traylor et al. estimated T_C to be -540 K, which means that rutile TiO_2 is not expected to be ferroelectric at any temperature [341]. The possibility of a ferroelectric transition at low temperatures is proposed by several groups due to the strong softening of the A_{2u} mode around the Γ point, which is reflected by the remarkable temperature coefficient of the static dielectric constant at 300 K [343, 349, 351, 352], but Parker excluded such a transition between 1.6 and 1060 K [349]. However, to combine the outstanding electronic properties of TiO_2 with ferroelectrical properties core-shell structures are employed [353].

A very detailed study determining the dielectric function in rutile TiO_2 from IR reflectivity measurements is given by Schöche et al. ($\epsilon_{0,a} = 86.65$, $\epsilon_{0,c} = 157.34$) [354]. The extraction of the temperature-dependent static dielectric constant from Raman spectroscopy is shown in detail by Samara and Peercy ($\epsilon_{0,c} = 173$, $\epsilon_{\infty,c} = 8.42$ at room temperature) for rutile TiO_2 and by Gonzalez et al. ($\epsilon_{0,a} = 45.1$, $\epsilon_{0,c} = 22.7$, $\epsilon_{\infty,a} = 5.82$, $\epsilon_{\infty,c} = 5.41$) for anatase TiO_2 [355–358].

Debye Temperature in Bulk TiO_2

The temperature-dependent Debye temperature Θ_D of bulk rutile TiO_2 is measured by heat capacity measurements and presented by Traylor et al. [341]. At room temperature they found $\Theta_{D,rt} = 740$ K and it is decreasing with decreasing temperature to $\Theta_{D,30K} = 430$ K at 30 K. Towards 0 K Θ_D is increasing to 660 K again.

A model which predicts the experimental data very well is demonstrated by Sikora [359]. They use an ab initio method to calculate the structure, lattice dynamics and thermodynamic properties of rutile TiO_2 and compare their model with the experimental data from Traylor et al. [341]. Their calculated value for the Debye temperature in bulk rutile TiO_2 using the classical limit $T \to \infty$ is $\Theta_{D,\infty} \approx 815$ K, and the Debye frequency is $\omega_D \approx 17$ THz [359]. Howard et al. calculated the Debye temperature for 15 K and 295 K [269]. They estimated the Debye temperatures to be $\Theta_{D,15K} \approx 582(10)$ K and $\Theta_{D,295K} \approx 600(10)$ K for rutile TiO_2 and $\Theta_{D,15K} \approx 522(10)$ K and $\Theta_{D,295K} \approx 520(10)$ K for anatase TiO_2. Interestingly, the room temperature value for rutile is $\Theta_{D,295K,calo} \approx 778(5)\,$K from specific heat measurements [360]. This discrepancy was also discovered for CaF$_2$ [361] and commented by Willis and Pryor [362].

Heat Capacity of Bulk TiO$_2$

The heat capacity is the sum of the energies of all available phonon states at a given temperature and thus, it is calculated from the Debye temperature. The heat capacity of rutile TiO$_2$ at 298.15 K is 54.50 J/(mol · K) [363, 364]. Compared to other standard metal oxides it is slightly higher than for α-quartz (43.5 J/(mol · K) [365]) and lower than for sapphire (79.0 J/(mol · K) [366]). Since the heat capacity is the derivative of the entropy with respect to temperature, the thermodynamic functions can be calculated from the phonon dispersion relation as well. A detailed and temperature-dependent overview on the thermodynamic functions for rutile TiO$_2$ is given by Grant and de Ligny et al. [197, 364]. The heat capacity calculated from the phonon dispersion is supposed to be similar for anatase and rutile TiO$_2$ [367, 368].

Above room temperature, the temperature-dependent heat capacity is given by the Shomate equation [363]. Below room temperature, the heat capacity has to be modeled by different equations. For $T < 15$ K lattice vibrations are modeled by a polynomial containing odd powers in temperature [369]. A linear dependency is usually said to originate from electronic contributions. However, for insulators, it results from oxygen vacancies and defects [283, 370]. If there are interacting energy levels one has to consider the appearance of Schottky anomalies [370]. This effect seems to appear when water is adsorbed on the surface and is related to tunneling of electrons between hydroxyl groups, which are situated between neighboring titanium atoms on contorted TiO$_2$ surfaces [371]. For $10\,$K $< T < 75\,$K, the heat capacity is described with a polynomial function containing odd and even powers. For $T > 35$ K, the heat capacity is a sum of Debye and Einstein functions including information on the lattice specific heat [372]. For nanoparticles, the heat capacity differs slightly from the bulk value [372].

Thermal Conductivity of Bulk TiO$_2$

The thermal conductivity in bulk material depends mainly on the mean free path of the phonons and the grain size. In a perfect crystal, the mean free path and thus, the thermal conductivity are increasing with decreasing temperature. In a polycrystalline material, the mean free path follows the same trend as in a single-crystalline material. But the mean free path is limited by the grain size due to the high scattering efficiency of such extended crystal defects. For sufficiently short times the grain size is supposed to be temperature independent in polycrystalline TiO$_2$ films and thus, the mean free path remains constant for a further decrease of temperature. As a consequence, the thermal conductivity follows the trend of the heat capacity and decreases for decreasing temperature. This is a typical feature of glasses for instance. Since the grain size depends strongly on the fabrication technique, the thermal conductivity is controllable by the processing parameters [373].

For bulk TiO$_2$, it is roughly 11 W/(m · K) at room temperature [374]. It is remarkable that the thermal conductivity of rutile bulk TiO$_2$ (6.531 W/(m · K) at 100 °C [295] and 5.0–3.4 W/(m · K) for 200–800 °C [197]) is about one-third of the thermal conductivity of polycrystalline metallic titanium (20.7 W/(m · K) at 100 °C [375]). On the one hand, it is still far below the thermal conductivity of well heat conducting materials such as crystalline Al$_2$O$_3$ (29 W/(m · K) at 100 °C [376]), graphite (125 W/(m · K) at 100 °C [376]), BeO (209 W/(m · K) at 100 °C [376]), annealed high purity silver (426 W/(m · K) at 100 °C [375]), or diamond ($\approx$ 1500 W/(m · K) at 100 °C [377]). On the other hand, it is still slightly better heat conducting as crystalline quartz (2 W/(m · K) at 100 °C [378]). But as discussed above the relative thermal conductivity compared with the listed materials changes with temperature. The thermal conductivity of bulk is still increasing below 100 °C and thus, the measured value of anatase ($8.5\,W/(m \cdot K)$ [379]) at room temperature is in the same order as rutile TiO_2.

With decreasing film thickness, the thermal conductivity decreases due to increased phonon scattering at grain boundaries and point defects [380]. Consequently, it increases with increasing grain size and shows the lowest values for amorphous films.

Compared to polyamide insulators such as Kapton[®◇] ($\approx 2\,W/(m \cdot K)$ at 100 °C [381]) the thermal conductivity of TiO_2 is larger. Due to this fact and in combination with the high melting temperature and breakdown voltage, TiO_2 is a promising candidate to enhance the quench behavior for superconductor insulation [382].

Plasmon-Phonon Coupling in Non-stoichiometric TiO_2

As discussed later, the density of free electrons is significantly increased for non-stoichiometric TiO_2. Gervais and Baumard performed IR reflectivity measurements on rutile TiO_2 and introduced a model considering plasmon-phonon coupling to describe their data [383]. In Raman spectra, Betsch et al. observed the broadening of Raman lines caused by the randomizing effects of oxygen vacancies in a reduced TiO_2 crystal [384].

Crystal Dynamics and Thermal Properties at TiO_2 Surfaces

Nearly all crystals show the effect of reconstruction on their surfaces. This results in surface phonons with a different phonon dispersion relation between surface and bulk phonon modes. Hence, all thermodynamic properties are supposed to be different in the surface regime. Akhadov et al. investigated the surface lattice dynamics using helium atom surface scattering on rutile $TiO_2(110)$ planes between 100 and 1000 K [385]. They found no temperature dependence of the reconstruction. From their results, they estimated the Debye temperature of the rutile TiO_2 surface to be $\Theta_D = 310\,K$.

Crystal Dynamics and Thermal Properties in TiO_2 Nanostructures and Polycrystalline Films

Today, nanostructured TiO_2 is investigated usually by Raman or IR spectroscopy instead of time-consuming and expensive neutron scattering. For sufficiently small anatase and rutile particles with a size smaller than roughly 20 lattice parameters ($\approx 1 \times 10^{-8}\,m$) phonons are reflected very frequently at grain boundaries and thus, they remain often in a single particle. This effect is called phonon confinement [386]. Besides this effect, the Raman peak position and intensity is affected by the defect density, chemical environment, stress, anharmonicity, nano-size distribution, and temperature of the sample [387, 388, 388–392].

It has to be considered that the sample is heated during Raman spectroscopy measurements by photon-phonon coupling. Thus, Raman peaks are broadened and shifted by the measurements itself. Since the phonon dispersion relation depends on the heating, it is less precise for nanocrystallites than for larger TiO_2 structures [388]. Especially for complex nanostructured layers with high densities of grain boundaries and sometimes even polymorph° structures, the Raman signal differs significantly from bulk TiO_2. Arora et al. suggested that lattice dynamics of a layer consisting of roughly 10 nm sized rutile nanoparticles depends on the particle size [393].

The best fitting model describing Raman scattering data from anatase nanoparticles with sizes between 5 nm and 20 nm is achieved if anharmonic coupling and phonon confinement are considered [386, 393–395]. The main coupling process is three-phonon processes between an optical phonon and two low-energy phonons. Compared to larger particles or bulk TiO_2 the anharmonic phonon-coupling is slowed down by the surface-stress-induced decrease of the thermal expansion but promoted by the phonon confinement effect in small nanocrystallites. Another consequence of the reduced thermal expansion of small nanoparticles is that the contribution to the phonon decay from the confinement effect is temperature independent and the temperature dependence of the phonon lifetime is given by anharmonic-phonon-coupling only [386, 396]. Of course, all considerations of nanoparticles are applicable to polycrystalline TiO$_2$ films as well [397].

2.3 Phase Transitions in TiO$_2$/Substrate Systems

The melting point of highly pure rutile TiO$_2$ is 1870 °C and the boiling point at an oxygen pressure of 1001.325 kPa is 2927 °C. There is no melting point for amorphous or anatase TiO$_2$ given, since amorphous TiO$_2$ crystallizes at several hundred degree Celsius and anatase TiO$_2$ converts into rutile TiO$_2$ between 600 °C and 1000 °C [197, 398, 399]. There is no direct transformation from rutile to anatase known, but it could be performed by melting and recrystallization applying a fast temperature drop to the melt as described in Chapter 9.

Phase transitions in TiO$_2$ are triggered by several parameters such as annealing temperature, crystal structure, chemical environmental conditions, light exposure and impurities. There is some literature explaining the effects that influence the phase transitions. For instance, Nam et al. observed an increased transition temperature from amorphous to anatase sol-gel TiO$_2$ films due to sodium ions diffusing from the glass substrate into the TiO$_2$ film [400]. In this work, the knowledge of controlling phase transitions in TiO$_2$ is brought into application extensively.

2.3.1 Anatase-to-Rutile Phase Transition in Bulk TiO$_2$

On the one hand, the anatase-to-rutile phase transition (ART) is an interesting system for investigating processes that trigger the transition between two phases in a solid crystal. On the other hand, this phase transition is important for applications. Due to its electronic and optical properties anatase is often preferred. Thus, the phase transition has to be suppressed if the fabrication or application temperature exceeds the transition temperature of pure bulk anatase TiO$_2$. Such high-temperature applications could be gas sensors and porous gas separation membranes [401–403]. But if rutile is preferred in an application, it might be advantageous due to substrate specifications and economic reasons to lower the phase transition temperature. Both, the controlled decrease and increase of the transition temperature is discussed in this paragraph. The crystallization of amorphous TiO$_2$ will be discussed on the sidelines of this paragraph. It is a common opinion that the reverse phase transition from rutile to anatase is excluded due to energetic reasons in the solid phase [262, 404–408]. But as described in detail in Chapter 9, it is possible to convert rutile into anatase TiO$_2$ via a solid-liquid-solid phase transition.

Processing Techniques

There are two routes to induce phase transitions in TiO$_2$. The chemical route is via peptization° at low pH values. Here, it is used that anatase dissolves in low pH value solutions easily and the TiO$_2$ is able to recrystallize as rutile [409, 410]. This route is suitable for low-temperature phase transitions. The physical route is a thermally or pressure-induced transition. Pressure is usually applied using a diamond-anvil cell [411]. For thermally induced processes several techniques are used in case of TiO$_2$. The most commonly applied technique is calcination in an oven or hotplate. Fortunately, most of the reported transition processes appear below 1200 °C, which is achieved by special commercially available ovens by default. This method is not suitable for temperature sensitive substrates and processes that need large heating and cooling rates. Using a halogen lamp driven oven limits the heating rate at roughly 100 °C/s. But even if faster rates were possible the effective heating and cooling rate is limited by the heat capacity and thermal conductivity of the whole sample. There are alternative techniques where the input energy is used to generate molecular vibrations or hot electrons that transfer their energy to phonons. Since the photon-phonon and electron-phonon coupling appears at much shorter time scales heating rates of up to 1×10^{16} °C/s could be achieved [412]. Typical methods are microwave irradiation [413, 414], laser irradiation in the visible light regime [415] or swift heavy ion exposure [412]. All these methods provide position-controlled sample manipulation as well.

Characterization Techniques

The most common technique to investigate the ART is **X-ray diffraction (XRD)** [416–424] using the Spurr and Myers method [425]. Here the ratio of both phases W_{rut}/W_{ana} is determined by the ratio of the anatase (101) peak at $2\Theta = 25.176°$ and rutile (110) peak at $2\Theta = 27.355°$ $I_{n,rut}/I_{n,ana}$ [399]. The relation between both ratios is given empirically by $(W_{rut}/W_{ana}) = 1.22 \cdot (I_{n,rut}/I_{n,ana}) - 0.028$ [399]. Although XRD is the most common method it has to be considered that its accuracy is influenced by morphology or sample preparation-dependent preferred crystal orientations, encapsulation of anatase in rutile or rutile in anatase [426, 427], inhomogeneously dissolved dopants resulting in peak broadening, phase-dependent crystallinity, phase-dependent grain size such as small anatase grains and large rutile grains [428, 429], phase-dependent shapes such as acicular rutile grains [430], and surface nucleation of rutile resulting in an inhomogeneous distribution of the two phases in thin films [399].

Another frequently used method is **Laser Raman spectroscopy** [269, 423, 426, 431–437]. Compared to XRD this method is space-resolved with a resolution of less than 1 μm, nondestructive, available for mapping, fast, independent from preferred crystal orientations, more sensitive especially to nanoscale phases and the sample fabrication is easier as well [399].

A related method to Raman spectroscopy with similar advantages is **infrared (IR) spectroscopy**. Compared to Raman spectroscopy, this method offers to probe complementary phonon modes.

Ohno et al. used the **solubility of anatase** in hydrofluoric acid (HF) to dissolve anatase TiO_2 [417, 438, 439] in an anatase/rutile mixture. Since this method provides a health risk, loss of material, and slight solubility of rutile, it is more suitable for rutile purification than for determining the ratio of anatase and rutile in a sample [399]. Furthermore, it cannot be applied as an *in situ* method.

Impedance spectroscopy is a more exotic method to determine the ratio of anatase and rutile TiO_2 by their different resistivity [440]. Grain connectivity and impurities are responsible for a rather low accuracy and thus limit the operation of this method [399]. But this method could be used in conjunction with XRD and differential thermal analysis (DTA) to investigate the effect of doping such as silica [441].

Phase Formation During the Fabrication of TiO_2 Samples

For most of the synthesis methods, anatase is the preferred resulting TiO_2 phase [442]. This might be due of the less-constrained molecular constellation of anatase compared to rutile [443]. For small crystallites, the lower Gibbs free energy of bulk TiO_2 is more than compensated by the significantly lower surface energy of anatase resulting in the preferred generation of anatase nucleation sites [406, 444]. In spite of these considerations, there are procedures known to generate rutile TiO_2 close to room temperature conditions [445–448]. Hanaor and Sorrell give a detailed overview of common synthesis methods of TiO_2 and their resulting phases [399].

If rutile is required it might be more economic to avoid anatase as an intermediate step in the fabrication process. The direct fabrication of rutile TiO_2 is often realized by solvothermal

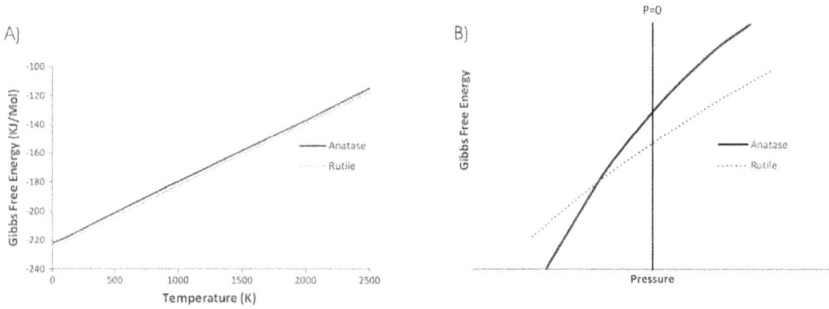

Figure 2.4: A) Gibbs free energy of anatase and rutile versus temperature; B) Schematic plot of Gibbs free energy versus pressure (assumed to be at room temperature). Reprinted from Journal of Materials Science, Review of the anatase-to-rutile phase transformation, 46, 2010, 855–874, Dorian A. H. Hanaor and Charles C. Sorrell, ©Springer Science+Business Media, LLC 2010, with permission of Springer [399].

methods containing titanium precursors such as titanium alkoxides [420, 446, 449–455]. In low pH environments a hydrolysis converts alkoxides into $[Ti(OH)_x(OH_2)_{6-x}]^{(4-x)+}$ hydroxide octahedra [447, 447, 448, 450–456]. The same result comes from dissolving amorphous TiO_2 particles [453]. For sufficiently slow reaction rates the octahedra are aligned linearly in order to minimize the total free energy. This alignment results in grains, which are elongated in the [001] direction [399, 453, 454]. A more detailed description of such a formation process is given in Paragraph 3.2.

Kinetics of the Anatase-to-Rutile Phase Transition in Pure Anatase TiO_2

Rutile is the thermodynamically stable morphology of TiO_2, which is reflected in a relatively low Gibb free energy. Thus, the phase transition from anatase to rutile is driven by a negative change of enthalpy at sufficiently high-temperatures and pressures [407, 408, 457, 458]. In air, the ART appears at around 600 °C in pure bulk anatase [444, 447, 459]. In perfect anatase bulk TiO_2 the transition temperature decreases with increasing pressure [399]. There is a triple point at around 500 °C and 1 GPa, where anatase, rutile, and high-pressure phases of TiO_2 exist in equilibrium [399, 404].

The phase transition is reconstructive, which means that bonds have to be cracked in order to form new kind of chemical bonds [404, 460–462]. Thus, the phase transition is a time-dependent nucleation and growth process [402, 460]. If the transition is not influenced by any of the listed influencing factors, rutile {100} planes grow on and parallel to {112} anatase planes due to their similar structure [402, 461, 463]. Thus, anatase {112} twin planes are supposed to act as nucleation sites in pure anatase TiO_2 [402, 463]. The rearrangement of atoms results in an increase of the density and a volume reduction of roughly 8% [460, 464, 465].

There are significant discrepancies among the reported thermodynamic properties of the ART including the change of enthalpy ΔH_{298}, entropy ΔS_{298}, and Gibbs free energy ΔG_{298} at room temperature. For example Navrotsky and Kleppa reported $\Delta H_{298} = -5.19\,kJ/mol$, $\Delta S_{298} = 0.42\,kJ/mol$, and $\Delta G_{298} = -5.32\,kJ/mol$ [407], Stull and Prophet reported $\Delta H_{298} = -11.7\,kJ/mol$, $\Delta S_{298} = 0.42\,kJ/mol$, and $\Delta G_{298} = -11.84\,kJ/mol$ [466], and Smith et al. reported $\Delta H_{298} = -1.70\,kJ/mol$, $\Delta S_{298} = 0.556\,kJ/mol$, and $\Delta G_{298} = -1.87\,kJ/mol$ [399, 405].

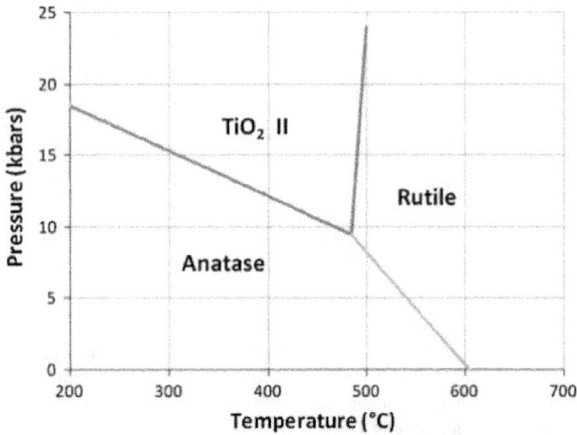

Figure 2.5: *Reaction boundaries of phase transitions in TiO_2. Reprinted from Journal of Materials Science, Review of the anatase-to-rutile phase transformation, 46, 2010, 855–874, Dorian A. H. Hanaor and Charles C. Sorrell, ©Springer Science+Business Media, LLC 2010, with permission of Springer [399].*

Effect of Crystal Defects

In modified anatase TiO_2 the transition temperature is adjustable between 400 and 1200 °C [427, 459, 460, 467–470]. Influencing factors are heating rate, soaking time, particle size, particle shape, surface area, volume, density of grain boundaries, brookite inclusions, impurities, atmosphere, composition of sample container [399], and substrate material and roughness. Furthermore, the reported values depend on the measurement technique as well [399]. The variety of influencing factors emphasizes the role of fabrication processing parameters. In sol-gel methods a strong dependence of the precursor solutes and solvents [445, 456, 471], the annealing temperature [407, 469, 472, 473] and pH value [440, 445, 446, 463, 474, 475] on the influencing factors such as grain size, surface area, morphology, and phase distribution is reported [399]. Of course in other fabrication techniques, the resulting TiO_2 properties depend on the processing parameters as well. In case of sputter deposition and evaporation, critical parameters are the deposition rate, purity and kind of initial material, processing temperature and atmosphere, for instance.

The energy cost of a rearrangement in TiO_2 is lowered significantly by introducing oxygen vacancies [460, 476–481]. Thus, a low oxygen partial pressure is supposed to lower the ART temperature [476, 482, 483]. But it has to be noted that annealing in vacuum conditions might show the opposite behavior due to less convective heat transfer compared to gas atmosphere [399, 460]. Annealing in reducing atmosphere such as hydrogen is supposed to promote the ART due to the efficient generation of plenty oxygen vacancies.

Besides point defects, the density of grain boundaries plays an important role. As discussed above, anatase {112} twin planes are supposed to act as nucleation sites for rutile. Thus, the more grain boundaries exist in an anatase sample the faster the ART takes place since the transformation can "attack" from more sites at the same time.

Effect of Impurities and Doping

Impurities play a major role for the ART temperature and either promote or inhibit the transition [484]. In substitutional solid solutions, dopant ions increase or lower the density of oxygen vacancies. In interstitial solid solutions, dopant ions stabilize or destabilize the anatase crystal by lattice constraint, which is defined by the size and valence of the dopant ions [399]. There are three main routes for doping: dry mixing (mixing anatase with solid dopant-containing powder such as a metal oxide powder), wet impregnation (adding anatase powder to a dopant containing solution such as dissolved salts or metal-alkoxides [476, 485]), and molecular-level mixing (mixing soluble titanium and dopant precursors in a solution such as a sol-gel) [399]. The method influences the efficiency and speed of doping and therefore the effect on the ART as well. In this paragraph, effects promoting and inhibiting the ART are discussed only briefly, since this topic is quite complex and not all aspects are relevant for this work. A very detailed review about influencing factors of the ART is given by Hanaor and Sorrell [399].

Cationic dopants are supposed to substitute Ti^{4+} ions [464, 469, 480]. For ions with a lower valence than four, the substitution has to result in a formation of oxygen vacancies or titanium interstitials with a lower valence in order to keep charge neutrality [442, 486]. For ions with a higher valence, the number of oxygen vacancies is increasing or titanium interstitials with the same or lower valence are formed [399]. For substitutional solid solubility, small cations with a valence lower than four are expected to promote the ART and large cations with a higher valence should inhibit the transition.

Besides substitutional titanium, interstitials are a common defect structure especially in reduced TiO_2 as well [487]. In case of interstitial solid solubility, injected cations cause constraints for the lattice contraction during the phase transition [488]. Such interstitials usually stabilize the anatase lattice and inhibit the ART [489–491].

Anionic dopants are supposed to occupy oxygen vacancies. Promoting or inhibiting the ART depends on their size and charge effects. Ions with ionic radii within 15% of the host ion are supposed to be soluble [492]. This limitation is fulfilled by N^{3-} ions ($\approx 6\%$), F^{1-} ions ($\approx 6\%$) and trivially by O^{2-} ions. In contrast, Cl^{1-} ions ($\approx 33\%$) are not expected to enter the anatase lattice [399]. But the charge effect has to be considered as well. If two N^{3-} ions are injected into the lattice forming Ti^{5+} ions, three O^{2-} ions have to be removed to keep charge neutrality. Thus, nitrogen doping is increasing the number of oxygen vacancies and promotes the ART [399, 462, 478]. Applying the same argumentation fluorine doping is an ART inhibitor [399, 493, 494]. Although chlorine is unlikely to enter the lattice it is reported that chlorine doping inhibits the ART [493, 494], but atmospheric chlorine promotes the ART [482].

It has to be considered that the valence of dopants can change and reduction-oxidation reactions take place [481, 495]. For example, in contrast to Al^{3+} dopants, Fe^{3+} can change at elevated temperatures and in oxygen rich atmosphere to Fe^{2+} by generating oxygen vacancies in the anatase crystal. As a consequence, Al^{3+} inhibits and Fe^{3+} promotes the ART [399]. It was shown with Mössbauer° that the inclusion of Fe^{2+} ions that are already in the two valence state by using $FeCl_2$ inhibit the ART [413, 496–498]. A similar behavior is observed for Mn^{4+} dopants that are reduced to Mn^{2+} above 400 °C in oxygen rich atmosphere creating oxygen vacancies and thus promote the ART [399, 495].

Carbon doping plays a special role among the dopants. Carbon impurities on samples are not unusual in unsterile environments, but carbon is likely to be oxidized at temperatures below the ART temperature [399]. But if carbon remains on or inside anatase it acts as a strong reducing agent increasing the level of oxygen vacancies and thus promotes the ART [399]. In addition, the carbon could form titanium carbide which prevents the formation of rutile as well [399].

In general oxide dopants such as Al_2O_3, SiO_2, and ZrO_2 inhibit the ART [419, 459, 468, 490, 495, 499]. This could be due to interstitial solid solubility as discussed above or by forming oxide doping layers at grain boundaries preventing efficient ion diffusion [419, 442, 491]. For example, Francisco and Mastelaro suggested that T–O sites at grain boundaries are blocked by cerium as a result of CeO_2 doping [433]. This blocking prevents efficient material diffusion and reduces the level of rutile nucleation sites. This effect becomes very important for elements that are absolutely not soluble in TiO_2 solid solutions at all. A famous example is phosphorous which gets bound to TiO_2 surface by chemisorption forming bidentate ligands [464, 500]. There is no injection of phosphorous into bulk anatase TiO_2 known, but the ligands lower the mobility of ions at the surface drastically and thus inhibits the ART.

As described in the introduction, there are a lot of rutile metal oxides. If such a rutile metal oxide is present it could promote the ART by offering nucleation sites. A metal oxide offering suitable nucleation sites for an efficient epitaxial growth, in particular, is SnO_2 [452, 501]. This is because of the low mismatch of the lattice parameters between rutile SnO_2 and rutile TiO_2. As mentioned in the beginning, the theory of influencing factors on the ART is quite complex. An overview of the effect of different dopants is shown in Figure 2.6. Dopants that are expected to inhibit the ART based on the considerations described above and by Hanaor and Sorrell are labeled with a red background color. Dopants that showed the inhibition of the ART in experiments are labeled with a purple background color. Dopants that are expected to promote the ART are labeled with a green background color and dopants that promoted the ART in experiments already are labeled with a blue background color. Dopants that show under certain condition an inhibition and under different conditions a promotion of the ART are labeled with violet background color. For each element, the references for the experimental proof are given.

Effect of Atmosphere

The most obvious influence of a gaseous atmosphere is the heat flux between the sample and the oven. Due to the specific heat, the heat flux for a constant flow rate and pressure is trivially minimal for annealing in vacuum, nearly equal for air, oxygen, and nitrogen, and by far maximized for hydrogen. This results in different thermal conditions and consequently the ART could be shifted to higher oven temperatures in vacuum atmosphere [399, 460].

Apart from that, the influence of the atmosphere is related to the effect of dopants. Annealing in vacuum, inert or reducing atmosphere is supposed to increase the number of oxygen vacancies while annealing in high partial oxygen pressure reduces the level of oxygen vacancies [460, 476, 496, 498, 542]. In hydrogen atmosphere, the reduction of TiO_2 results in a high level of oxygen vacancies and thus this kind of atmosphere promotes the ART [460, 543]. In vacuum, the creation of Ti^{3+} interstitials is favored over the formation of oxygen vacancies and these interstitials stabilize the anatase phase and inhibit the ART [460, 543]. Ahonen et al. decreased the ART temperature of nanocrystalline anatase TiO_2 from 580 °C to 500 °C by annealing in nitrogen [504]. Gamboa and Pasquevich reported that the ART is 300 times faster in an argon-chlorine atmosphere compared to air at 950 °C. They suggested that oxygen vacancies are created by the adsorption of chlorine and a surface reaction [482].

2.3.2 Anatase-to-Rutile Phase Transition in Nanocrystalline TiO_2

There were a lot of investigation in the ART of nanosized anatase TiO_2 particles in the last 20 years. Although mainly general surface and bulk influencing factors are discussed in the previous paragraph, there were already some hints that the effect of influencing factors is different for anatase TiO_2 nanomaterial. Above all, the ratio between bulk and surface atoms is increased drastically for nanocrystals [544]. As a consequence, surface energies and interface forming energies compete with the Gibbs free energy of bulk anatase [544]. As a result, the ART is more or less favored for nanoparticles compared to macroscopic anatase particles

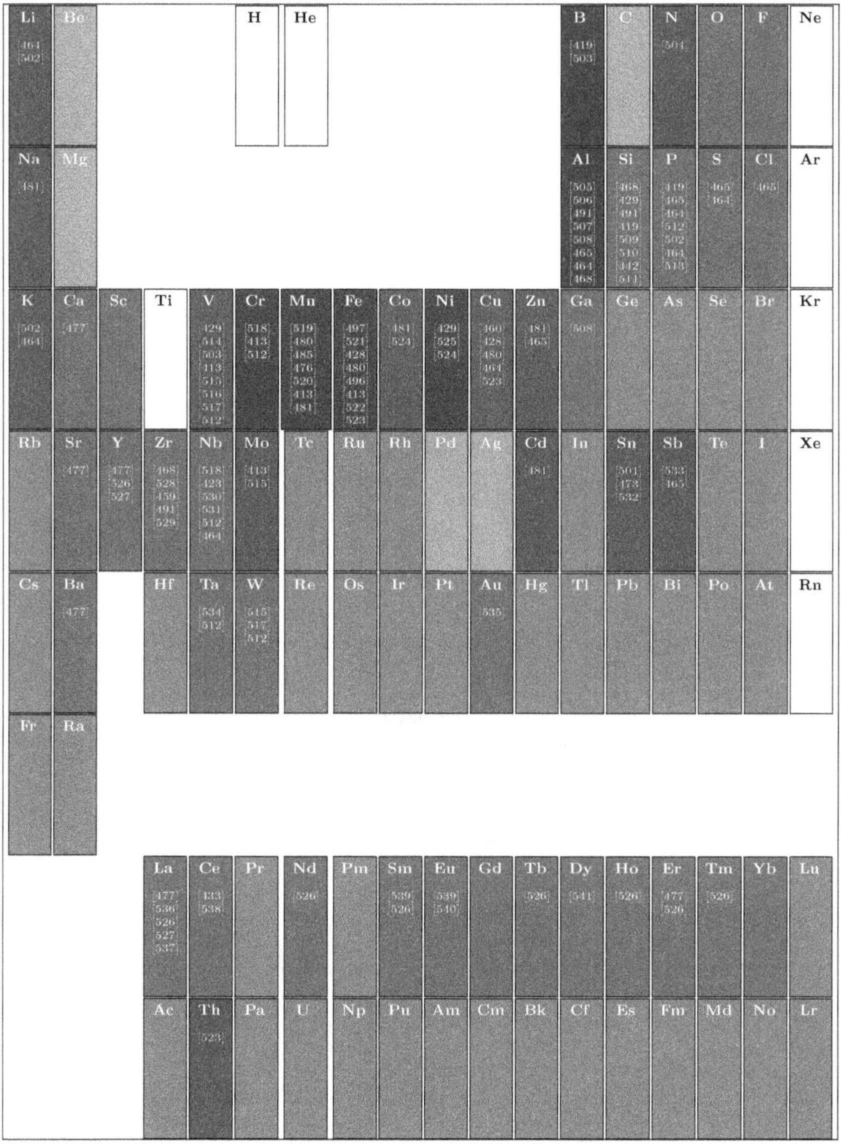

| Li | Be | | | H | He | | | | | | | | B | C | N | O | F | Ne |
| [464] [502] | | | | | | | | | | | | | [419] [503] | | [504] | | | |

| Na | Mg | | | | | | | | | | | | Al | Si | P | S | Cl | Ar |
| [481] | | | | | | | | | | | | | [505] [506] [491] [507] [508] [465] [464] [468] | [468] [429] [491] [419] [509] [510] [464] [611] | [419] [465] [491] [512] [502] [464] [513] | [465] [464] | [465] | |

| K | Ca | Sc | Ti | V | Cr | Mn | Fe | Co | Ni | Cu | Zn | Ga | Ge | As | Se | Br | Kr |
| [502] [464] | [477] | | | [429] [514] [503] [413] [515] [516] [517] [512] | [518] [413] [512] | [519] [480] [485] [476] [520] [413] [481] | [497] [521] [428] [480] [496] [413] [522] [523] | [481] [524] | [429] [525] [428] [524] | [360] [428] [480] [464] [523] | [481] [465] | [508] | | | | | |

| Rb | Sr | Y | Zr | Nb | Mo | Tc | Ru | Rh | Pd | Ag | Cd | In | Sn | Sb | Te | I | Xe |
| [477] | [477] [526] [527] | [468] [528] [459] [491] [529] | [518] [423] [530] [531] [512] [464] | [413] [515] | | | | | [481] | | [501] [473] [532] | [533] [465] | | | | | |

| Cs | Ba | | Hf | Ta | W | Re | Os | Ir | Pt | Au | Hg | Tl | Pb | Bi | Po | At | Rn |
| [477] | | | [534] [512] | [515] [517] [512] | | | | | [535] | | | | | | | | |

| Fr | Ra | | | | | | | | | | | | | | | | |

| | | La | Ce | Pr | Nd | Pm | Sm | Eu | Gd | Tb | Dy | Ho | Er | Tm | Yb | Lu |
| | | [477] [536] [526] [527] [537] | [413] [538] | | [526] | | [539] [526] | [539] [540] | | [526] | [541] | [526] | [477] [526] | [526] | |

| | | Ac | Th | Pa | U | Np | Pu | Am | Cm | Bk | Cf | Es | Fm | Md | No | Lr |
| | | | [523] | | | | | | | | | | | | | |

Figure 2.6: *The effect of dopants on the ART:* experimentally confirmed inhibitors, predicted inhibitors, experimentally confirmed promoters, predicted promoters, dependent on conditions either promoter or inhibitor. *This overview was established by using the review from Hanaor and Sorrell [399] and own extensive literature research.*

29

when applying different conditions. For example, the enthalpy of ART at approximately 700 °C increases from -3.26 kJ/mol for macroscopic (100–200 µm [408]) particles to -6.57 kJ/mol for nanoparticles (50–250 nm [407]). Using suitable ART promoters and inhibitors the ART temperature is adjustable between 350 and 1200 °C for anatase nanoparticles [532, 545–548]. In this paragraph, the most important influencing factors for the ART in anatase nanoparticles are discussed briefly.

Critical-Nuclei-Size-Effect in Pure Anatase TiO$_2$ Nanoparticles

Surface energy: The surface energy of anatase and rutile is 0.4 J/m^2 and 2.2 J/m^2, respectively [549]. The Gibbs free energy of bulk anatase is significantly higher than for rutile and thus, anatase is transformed to rutile for sufficiently high-temperatures and pressures. For the surface energies, the situation is reversed and for small crystallites, the anatase phase is preferred and stabilized by its low surface energy. But due to the high activation energy, the backward transition from rutile to anatase for such nanoparticles is still very unlikely and was not observed so far. The competition between surface and bulk energy results in the existence of a critical particle diameter [422, 534, 550–556]. This effect was called "critical-nuclei-size-effect" by Kumar [422]. The expression critical-growing anatase particles are supposed to convert into rutile particles much more easily at sufficiently high-temperatures. For smaller particles, the superior surface energy works as an inhibitor of the phase transition [557, 558]. The ART is driven by a negative change of the total Gibb free energy, which is given by

$$\Delta G_{ART} = \Delta G_{V,R}(T) - \Delta G_{V,A}(T) + (A_R \gamma_R - A_A \gamma_A) + (\Delta P_R V_R - \Delta P_A V_A). \quad (2.4)$$

Here $\Delta G_{V,R/A}(T)$ are the changes for the Gibbs free energy of bulk, $A_{R/A}$ are the surface areas, $\gamma_{R/A}$ are surface tensions, and $\Delta P_{R/A} V_{R/A}$ are the excess pressures of rutile and anatase, respectively [559]. For polycrystalline films or agglomerated nanoparticles, an additional term considering the interface energy (e.g. anatase/anatase, anatase/rutile, anatase/brookite, anatase/matrix, anatase/substrate) is necessary. For nanoparticles, the contribution from surface energy and excess pressure becomes significant since their surface area is larger compared to macroscopic particles with the same volume, and the pressure $\Delta P = 2\gamma/r$ is increasing for decreasing particle diameter r [550]. As a consequence, a nonequilibrium transition is favored for high anatase surface energy and surface stress energy. This condition is fulfilled for small nanoparticles in particular. But for authoritative forecasts, the components of Equation 2.4 have to be treated carefully. The Gibbs free energy for bulk anatase is significantly lower than for anatase resulting in a negative contribution from the bulk energies. This fact is the reason why bulk rutile TiO$_2$ is the thermodynamically stable phase of TiO$_2$. As mentioned in the beginning of this paragraph the surface energy of rutile is much larger than for anatase. Thus, there is a positive contribution from the surface energies and in general also for interface energies in case of monodisperse anatase and rutile particles. But for blends with small rutile and large anatase particles with an equal total volume of both phases, the surface energy contribution is lower and could even become negative. With the same argument, a system with large rutile and small anatase particles is relatively stable. At first sight, the contribution from the excess pressure is expected to be positive for equally sized anatase and rutile particles as well. But the $1/r$-dependence suggests that it becomes small or even negative for small anatase and large rutile crystallites. However, since the density of rutile is larger compared to anatase the volume shrinks during the ART by approximately 10%, which results in a negative contribution in the excess pressure term and thus promotes the ART [550]. All these considerations emphasize that there is a strong size and size distribution dependence of the ART. It should be noted that the ratio between anatase and rutile crystallite sizes is changing during the ART making the kinetics of the transition time-dependent.

Size dependence: Ranade et al. calculated the critical diameter to be 14 nm [406, 549]. One has to consider that this value depends on all influencing factor that affects the surface energy of the particles. Barnard and Zapol demonstrated that hydrogenated and dehydrogenated particles show different surface energies [560]. Thus, the phase stability is controlled by hydrogenation as well [561]. Hu et al. introduced an intermediate range between 11 and 35 nm where the transition results not directly in rutile, but in a first step from anatase to brookite and in a second step from brookite to rutile [420]. Although anatase phase is favored for smaller nanoparticles the nucleation rate becomes slower with increasing particle size. This is because polycrystalline material with large grain sizes has less surface and interface area compared to particles with smaller diameters. And due to the fact that nucleation is favored on surfaces and interfaces, there are more nucleation sites for anatase with small grains [559].

Pressure dependence: There are similar considerations for the high-pressure phase of TiO_2 [562]. Also here anatase particles with less than 15 nm diameter are stabilized while large particles are likely to be converted into the α-PbO_2 structure [563]. For macroscopic anatase particles, the ART pressure is reported to be 2.6–4.5 GPa [411, 564–569]. But for anatase nanoparticles the anatase-to-α-PbO_2 transition (APT) pressure is increasing with decreasing particle size by more than one order of magnitude [570, 571]. Wang et al. reported for 30 nm sized anatase particles an APT pressure of 16.4 GPa [562]. In comparison, the phase transition pressure needed for the transformation of rutile into the baddeleyite (ZrO_2) structure is only 8.7 GPa for 30 nm sized rutile particles [562]. By releasing the pressure the ZrO_2 structure is transformed into the α-PbO_2 structure [562]. For macroscopic rutile particles, the pressure for the ZrO_2-α-PbO_2 transition is higher ($\approx$ 13 GPa) [411, 570]. For anatase nanoparticles with a size of just 9 nm, the APT pressure is even 24 GPa [562].

Particle shape: Since the ratio between surface energy and bulk energy is changing with increasing anatase particle diameter, the shape of nanoparticles differs from the macroscopic Wulff construction significantly [561]. Vice versa, the shape of existing anatase nanoparticles defines their surface energy and thus, the ART becomes size-dependent as well. In general, small particles above the critical size with an asterisk-like structure perform the phase transition more likely than large, elongated particles [409].

Nucleation Sites

In pure anatase TiO_2 polycrystalline films or nanoparticles the nucleation appears on surfaces and interfaces [469]. Around 700 °C the most favored nucleation is surface nucleation, followed by interface and bulk nucleation [559]. If there are tiny rutile nanocrystals from the beginning, they act as nucleation sites and promote the ART. Ocaña et al. fabricated such rutile embryos by using organic impurities at the outer part of anatase particles and they were able to decrease the ART temperature for 0.1–6 μm particles down to 660 °C [432]. Fonzo et al. were able to achieve an ART by rutile impurities at 400 °C [572]. Instead of rutile embryos, brookite is used to lower the ART temperature as well [420, 553]. Ovenstone and Yanagisawa fabricated such brookite inclusions by peptization at 100–300 °C [573]. Under certain conditions, the film thickness defines the crystal structure of the resulting film as well. Hwang et al. reported that they achieved amorphous, anatase, and rutile TiO_2 films on rutile nanorods by varying the film thickness [574]. At elevated temperatures amorphous and anatase films could be easily transformed into rutile at lower temperatures due to the presence of the rutile substrate.

The ART appears in a well-defined temperature range. In the lower temperature regime, nucleation appears at interfaces and surfaces, but with increasing temperature, the bulk nucleation becomes more and more dominant [575]. Hence, the nucleation process spreads out in the whole sample with increasing temperature and thus, the ART becomes faster. An additional acceleration results from the increasing diffusion rate with increasing temperatures.

Effect of Doping on Phase Stability in Anatase Nanoparticles

The effect of doping is similar to bulk materials, but surface and interface-dependent effects are boosted by the larger surface-to-volume ratio. An example for an extremely low ART temperature of 350 °C is given by Cao et al. for 10 nm small Sn^{4+} ion doped anatase particles [532]. The ART temperature for the same particles without doping is 550 °C [532]. The references for doped nanoparticles are listed in Figure 2.6 as well. Investigated dopants promoting the ART in anatase TiO_2 nanoparticles are molybdenum [515] and vanadium [516]. Dopants that are known for inhibiting the ART are aluminum [508], gallium [508], niobium [530], tantalum [534], and tungsten [515].

Effect of Silicon and Aluminum Doping on Phase Stability in Anatase Nanoparticles

In this work, the silicon or SiO_x/TiO_x interface is used frequently and relatively high-temperatures are applied. Thus, the diffusion of silicon into TiO_2 is likely and the effect of silicon doping on the crystal stability and phase transition have to be considered.
As discussed above Si^{4+} ions are likely to substitute Ti^{4+} ions in the crystal and stabilize the anatase phase [511, 576]. Since the volume of the unit cell is decreasing by this substitution it is favored over the formation of dopant interstitials [419, 442, 510]. SiO_2 particles are often used as a silicon source for doping anatase TiO_2 [442]. The ART temperature is increasing for increasing silicon doping. Especially for low concentrations, this effect rises quickly as shown in Figure 2.7. Silicon is known to stabilize TiO_2 and for doping concentrations of 5% the crystallization of amorphous TiO_2 into anatase is increased from 300 °C to 500 °C and the ART temperature is increased from 600 °C to 800 °C [429, 577]. In the extreme case of 70% silicon concentration, the crystallization of amorphous TiO_2 into anatase is increased to 800 °C and the ART temperature is increased to 1200 °C [429]. The mentioned temperatures are the lowest temperatures where the transitions are observed [429, 577]. One has to consider that two phases coexist at slightly higher temperatures and the transition is completed for the whole sample at much higher temperatures. So anatase was even found at 1600 °C for high silicon doping concentrations [429]! The system of Si^{4+} ions and TiO_2 forms a solid solution above 1000 °C [578, 579]. It is supposed that surface defects are generated on anatase/rutile interfaces in such solutions preventing particle diffusion into rutile crystallites. As a consequence, the threshold barrier for the phase transition is increased. And the Si_{4+} ions are supposed to prevent the formation of large crystallites in a polycrystalline film for instance [578]. Using first-principles calculation Shi et al. showed that the bandgap of anatase TiO_2 is narrowed by silicon doping [580]. Amongst others, bandgap designing is interesting for photocatalysis applications. Su et al. demonstrated an enhanced photocatalytic degradation of phenol on silicon-doped TiO_2 films [581].
Diffusion from a substrate results in a gradient of the dopant density perpendicular to the interface. Thus, the ART temperature becomes position-dependent in a sufficiently thick TiO_2 film on silicon. This feature results in a gradient of the ART temperature and thus, films could be fabricated with a high anatase content close to the interface and rutile morphology in a sufficiently large distance to the interface. Besides the ART temperature Okada et al. observed an increasing temperature for the crystallization of amorphous TiO_2 from 415 °C to 609 °C [442]. An interesting effect appears for $ZrSiO_4$ doping. While silicon or zirconium doping results in a homogeneous growth of rutile grains, $ZrSiO_4$ doping results in abnormal grain growth (AGG) bearing some extraordinary large grains in the middle of relatively small grains [582]. A common explanation of this phenomenon is the existence of significant anisotropy in the surface energies resulting in fast crystal growth in certain crystallographic directions [582].

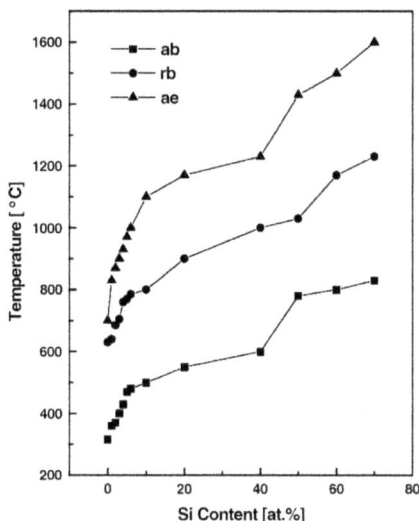

Figure 2.7: *Dependence of the ART temperature on silicon doping level. "ab" labels the emerge of anatase phase; "rb" labels the emerge of the rutile phase; and "ae" labels the accomplishment of the ART. Reprinted from Materials Science and Engineering: C, 19, Yu-Hong Zhang and Armin Reller, Phase transformation and grain growth of doped nanosized titania, 323–326, Copyright ©2002, with permission from Elsevier. [429]*

The ART temperature depends on the crystallite size in case of silicon doped anatase TiO_2 nanoparticles as well [510]. Besides stabilizing the bulk anatase the Si^{4+} ions are expelled at higher temperatures and form SiO_2 core-shell layers suppressing material transport nucleation sites at the surface [442, 510]. Since the latter effect increases with increasing surface and interface area silicon doping is a strong and efficient ART inhibitor for nanoparticles and polycrystalline films in particular. The same effect is supposed to inhibit the transformation of γ-Al_2O_3 into α-Al_2O_3 in alumina [583]. In addition, silicon doping influences thermodynamic functions in anatase nanoparticles significantly. As a result, the critical size for stable anatase particles is expected to be 70–100 nm [509], which is much larger than the critical size for undoped nanoparticles. For 14–16 nm sized anatase particles Hirano et al. observed an ART temperature of 1300 °C. The same particles without silicon doping performed the ART already at 850 °C [509].

Another material that was used as a substrate in this work is sapphire (Al_2O_3). Battiston et al. reported that the ART temperature increased to 1100 °C on alumina substrates [584]. This indicates that there is a diffusion of aluminum ions into TiO_2 since Al^{3+} is expected to be an inhibitor of the ART [468]. Hence, the influence of alumina substrates is quite similar to silicon or SiO_2 substrates.

It could be concluded that changes on structural, chemical, electronic and optical properties have to be considered for highly annealed TiO_2 films on silicon and aluminum containing substrate always.

33

Effect of Substrate

As discussed so far, diffusion has a great impact on the morphology during annealing processes. Diffusion appears mainly between the substrate and the TiO_2 layer. For instance, it was reported that anatase is dominant on SiO_2 after annealing at 800 °C. But employing the same fabrication and annealing conditions for TiO_2 on a platinum substrate, the resulting morphology is rutile [585]. Besides the diffusion of impurities from the substrate, additional substrate properties influence the ART behavior. The heat capacity and thermal conductivity limit heating and cooling rates. This becomes important for high rates as they appear in laser heating. Samples with similar crystallinity such as SnO_2 evoke epitaxial growth or at least additional nucleation sites for rutile TiO_2. The wetting angle is a measure of the adhesion force between the substrate and a TiO_2 film. Strong adhesion in combination with significantly different thermal expansion coefficients cause strains in TiO_2. Strains change the thermodynamic functions and thus the ART temperature. Sufficiently high strains result in crystal defects increasing the level of grain boundaries and thus nucleation sites. The film roughness results in strain and crystal defects as well. An important example is the crystallization of TiO_2 deposited by chemical vapor deposition (CVD) on SiO_2 and platinum resulting in anatase and rutile films, respectively [585]. In conclusion, features of the ART are controllable by the kind of substrate and surface treatment of the substrate.

Effect of Fabrication Technique

Most influencing factors of the ART characteristics such as grain size and doping concentration are set by the fabrication technique. An important parameter in sputter deposition, evaporation, and CVD for defining the crystallite size, defect density and crystallization of anatase TiO_2 films is the processing temperature [136, 584, 586–592].
The addition of impurities inhibits the material transport between anatase and rutile crystallites and thus increases the ART temperature. Suitable additives are metal oxides such as alumina [575] or organic material. It was shown by Nolan et al. that the phase transition temperature from anatase to rutile depends on the content of formic acid and water in anatase [593]. Djaoued et al. investigated the heat-induced ART in thin TiO_2 films prepared by dip-coating as a function of the complexing agent. Diethanolamine (DEA) is promoting the phase transition at 750 °C but acetylacetone (AcAc) in combination with acetic acid stabilize the anatase phase up to 1000 °C [398].
Since the nucleation is limited by mass transfer it is retarded by inserting clear areas [594]. This could be realized by loosely packed nanoparticles for instance. The opposite effect is achieved by improving the contacts between single anatase crystallites [595]. Pillai et al. perform such an improvement by using urea for generating a better sol-gel network with fewer voids [595, 596]. Zhang et al. achieved the same effect by increasing the aging time of the sol-gel from one to eight weeks. By this simple treatment, they decreased the crystallization temperature of anatase from 345 to 315 °C and the ART temperature from 630 to 500 °C [544]. The choice of precursor defines the resulting crystallite size and thus the ART temperature as well [597]. Byun et al. presented a metal-organic chemical vapor deposition (MOCVD) technique for the fabrication of anatase films with a relatively high transition rate due to a high level of nucleation sites [598]. It is possible to implant nucleation sites within a pretreatment. Hu et al. created brookite inclusions by applying suitable pH values during the fabrication process of hydrous TiO_2 powders [420]. As discussed above brookite acts as a catalyst for the ART and thus lowers the ART temperature. Crystal defects offering nucleation sites could be induced by sonication [558].

Low-Temperature Methods for the Anatase-to-Rutile Transition in Nanoparticles

There are many applications such as organic substrates where high-temperatures are not usable. Thus, also low-temperature methods for the ART were developed. A simple method is peptization. Here anatase TiO$_2$ is dissolved in solution and usually, an intermediate product is formed that recrystallizes as rutile. Of course, the process depends strongly on the peptization agent and further processing parameters [409]. Jung et al. succeeded to perform the ART via peptization at 30 °C only [410]! Zhu et al. applied wet-chemical reactions to transform anatase in rutile via the intermediate steps hydrogen-titanate and proton titanate [599]. The energy barriers for this reactions are so low, that it is possible to perform the transition below 200 °C [599]. As already discussed, rutile embryos decrease the ART temperature already a lot. Beyond that, Gole et al. suggested the creation of spinel-like structures by high spin Co^{II} and Ni^{II} doping and reported an ART at room temperature [524]. A rude method to generate as much as possible crystal defects is ball milling. Also with this method, the ART appears already at room temperature [600].

Swift Heavy Ion-Induced Anatase-to-Rutile Transition

Instead of using an oven or hotplate for annealing thermal energy can be transferred to anatase by the absorption of photons by electronic excitations. An electron-phonon coupling results in sample heating as discussed in the next paragraph. But the energy carriers must not be photons necessarily. Rath et al. exposed an anatase film to swift heavy argon ions with a kinetic energy of 200 MeV [412]. These ions lost their energy mainly to the electrons of the anatase material. Phonons are excited by the high energetic electron system via electron-phonon coupling that appears on a timescale of 1×10^{-14}–1×10^{-12} s resulting in local temperatures of more than 1000 °C [412]. The temperature is obviously high enough to melt the TiO$_2$ locally. Since the heat is restricted to the middle of the ion beam the cooling rate is quite fast after the exposure and the TiO$_2$ recrystallizes again [412]. Cooling rates of approximately 1×10^{13} K/s result in rutile TiO$_2$ while fast cooling rates lead to amorphous TiO$_2$ [601].

Light-Induced Anatase-to-Rutile Phase Transition

In this paragraph, only general aspects of light-induced ART are described. A detailed discussion of this technique related to the investigated samples is found in Chapter 9.

The most obvious approach to induce crystallization and phase transitions in TiO$_2$ by light exposure is to use photon energies above the bandgap. This is supposed to result in a high absorption of UV light and thus, the film is heated up. Mitrev et al. investigated the effect of 193 nm and 248 nm irradiation on pure TiO$_2$ films with different thicknesses and intensities. For lower UV intensities the amorphous TiO$_2$ crystallized into anatase [415]. For higher intensities, they observed an ART. For thinner films, the change of the crystal structure appears at lower threshold intensities [415].

Besides thickness and light intensity the phase transition is controllable via the oxygen pressure as well. For high oxygen pressures during intra-bandgap excitation, the adsorption of oxygen molecules on anatase surfaces is enhanced and they stabilize the anatase phase by bridging two neighboring titanium sites [602]. The adsorption process is supposed to originate from recombination of electrons trapped in the monovalent anionic oxygen molecules and photoexcited holes in the TiO$_2$ valence band [6, 602, 603]. Ricci et al. demonstrated that the ART happens quite fast under illumination with 488.0 nm [604, 605]. Thus, the phase transition speed is controllable via the oxygen pressure.

Besides intrinsic defects, impurities cause additional electronic states in the bandgap and hence, absorption of visible light. Additional light absorption excites phonons and could heat up the

TiO_2 until its melting temperature is achieved. Beyond that, a sufficiently high impurity density influences the stability of the crystal structure and could reduce the melting temperature. Zhang and Reller reported that iron doping increases the phase transition temperature from anatase to rutile in a nanocrystalline mesoporous TiO_2 film from 500 °C to 600 °C even at low doping concentrations [606]. For densities above 3% iron, they observed an additional phase transition from anatase to pseudobrookite $Fe_2Ti_2O_5$ [606].

An obvious approach to increase the absorption of TiO_2 in the visible light regime is to cover the TiO_2 with a suitable dye. Parussulo et al. were able to convert anatase into rutile in a thick mesoporous anatase layer covering the surface with the [Ru(4,4'-dicarboxy-2,2'-bipyridine)$_2$(NCS)$_2$] (N3) dye and exposing the sample to 532 nm for one second at 25–41 mW/cm² [607]. The absorption is increased by dopants and defects offering electronic states within the bandgap. Placing carbon on the anatase surface generates oxygen vacancies by forming CO or CO_2 under irradiation [608]. And thus, the phase transition is increased in this case as well [609]. The phase transition is expected at a power density of 40 W/cm² at 488 nm [604]. In case of carbon or iron doping, the phase transition works with 532 nm green laser light and relatively low power densities compared with undoped anatase samples as well [609–611].

Effects of the Anatase/Rutile Interface on the Anatase-to-Rutile Phase Transition

Anatase/rutile composites are an interesting material for some applications. Due to the band alignment between rutile and anatase a fast electron transport from rutile to anatase [417, 612] is expected. Furthermore, rutile TiO_2 has a slightly smaller bandgap and is able to absorb more visible light resulting in catalytic "hot spots" at the rutile/anatase interface [612]. By comparing bulk sensitive powder X-ray diffraction (PXRD) and surface sensitive Raman spectroscopy Hurum et al. found small rutile inclusions in anatase grains [612, 613]. Applying molecular dynamics (MD) simulations Deskins et al. showed that the anatase/rutile interface is disordered to some extent, but the disorder affects a narrow interface region only [614]. Furthermore, they identified stable low-index surfaces between anatase and rutile such as $(100)_r/(100)_a$ or $(110)_r/(101)_a$. Phase transitions in TiO_2 have been investigated intensively using Raman spectroscopy [344, 345, 613, 615–617]. The linewidth of Raman modes increases with increasing nanocrystal size [554, 616, 618–620], density of crystal defects [389], and temperature [613]. It is also possible to determine the percentage of exposed crystal surfaces from the E_g and A_{1g} vibrational modes [621]. According to their Raman measurements Zhang et al. suggested that anatase resists higher temperatures at the surface compared to the bulk [622]. Thus, the ART begins in the bulk. As a consequence, the transition temperature decreases with decreasing particle size or grain boundary density [608, 623]. This is in contrast to the models and experimental data discussed in previous paragraphs. This discrepancy might originate from neglecting thermodynamic functions in the model deviated from the Raman measurements. In addition, the ART characteristics depend on the chemical surroundings and impurities [399, 551, 557, 605, 624, 625]. The more common opinion is that the anatase/rutile TiO_2 interface lower the threshold energy for the ART. Tsai et al. reported that anatase nanoparticles coalesce at {112} surfaces under pulsed laser ablation conditions resulting in twinned bicrystals [626]. In case of thermally induced ART, the nucleation of rutile crystallites starts exactly at these twin boundaries [608, 627, 628]. Thus, the density of nucleation sites increases with decreasing particle size and increasing oxygen vacancy density [629, 630]. In case of a thin La_2O_3 buffer layer covering anatase particles, the phase transition was slowed down significantly or even canceled [622, 631].

Light-Induced Anatase-to-Rutile Phase Transitions at the Anatase/Rutile Interface

Samples are heated by laser irradiation during Raman measurements [388, 613, 616, 617, 632–634]. The temperature is extracted from the position and width of the phonon peaks. Xia et al. fabricated a rutile/anatase nanocomposite with 65% 17 nm anatase and 35% 20 nm rutile crystallites by applying a sol-gel method and annealing at 550 °C afterward [613]. Their amazing result was that the anatase parts achieved a temperature of 777–907 °C while the rutile parts were heated up to 252–328 °C only. Because of this temperature gradient, heat dissipation becomes directional and is perpendicular to the anatase/rutile interface. Thus, the hot interface with its neighboring rutile crystallite is supposed to be a preferred site for the phase transition, which in good agreement with experimental observations [402, 463, 469, 635]. Energetically preferred interfaces between anatase and rutile are rutile (101) and anatase (001) surfaces, which show a relatively close bonding arrangement [213, 613, 636–639]. Using density functional theory (DFT) Xia et al. demonstrated that there are only minor relaxation effects at the assumed interface structure [613]. But they found significant changes in the electronic band structure. The bandgap of the interface was calculated to be 2.1 eV only [613]. Experimental data from UV-Vis reflectance spectroscopy suggested the bandgap to be as small as 1.2 eV [640]. Thus, these interfaces absorb light with wavelengths up to 785 nm and the energy is converted partly in the excitation of phonons heating up the interface [613]. The Young's modulus of [001] axis is significantly smaller than in any other direction of rutile and anatase TiO$_2$ [641, 642]. As a consequence, the amplitude of phonons moving in parallel to the anatase [001] axis is large compared to the amplitude of phonons moving along the rutile [101] axis. Hence, anatase parts heat up much faster than rutile crystallites [613]. Based on transmission electron microscopy (TEM) studies Penn and Banfield suggested rutile-like objects on anatase interfaces [643]. For temperature-induced phase transition these rutile-like elements act as nucleation sites for rutile crystals [635]. By in situ TEM experiments and kinetic data from X-ray diffraction (XRD) Lee and Zuo and Zhang and Banfield observed an ART at the interface between anatase particles [463],[469].

2.4 Extended Definition of Optical, Electronic, and Optoelectronic Properties of TiO$_2$ Semiconductors

In this work, an extended definition is proposed for optical, electronic, and optoelectronic properties in semiconductors. This is done in order to express the whole range of physical properties of TiO$_2$ nanostructures in a much better way. Generally, effects that describe the interaction between photons and the sample without the excitation of electrons are related to optical properties. Effects that describe the behavior of mobile electrons in the absence of photon-derived excitations are related to electronic properties. Finally, effects that involve the quantum-mechanical excitation of electrons by photons are related to optoelectronic properties. Hence, properties of the electronic system probed with frequencies up to the terahertz regime are electronic properties. Properties of the electronic system probed with excitation frequencies between IR and UV light are optoelectronic properties. In this context, the term "electronic system" includes all electrons except strongly bound core electrons. An overview of the grouping of electronic and optoelectronic properties according to their shape- and frequency-dependence is given in Figure 2.8. Interactions between TiO$_2$ and photons with energies higher than 10 keV (X-rays, γ-photons) are not discussed in this thesis.

2.4.1 Optical Properties

Optical (dielectric) properties described with Lorentz oscillators° are discussed in Section 2.5. The description of these properties is mainly classical and based on Maxwell's equations. On the one hand, this covers the reflection and transmission on interfaces as well as interference caused by interflection and absorption phenomena such as photon-phonon interaction. The latterly mentioned effect increases the sample temperature. Consequently, the temperature rise depends on the light intensity, the applied wavelength, the internal quantum efficiency for phonon excitation, the angle of incidence, the heat capacity as well as the thermal conductivity of the sample, and the coupling to a thermal reservoir. On the other hand, dielectric scattering is an important phenomenon, especially for micro- or nanostructured TiO_2 due to the large difference between the refractive index of air and rutile TiO_2. In particular, a dielectric antenna effect appears on anisotropic dielectric nanostructures. Finally, the surface roughness and dense conglomeration of nanostructures such as nanorod arrays form the transition between scattering and interface reflections. In this regime, nanostructured materials behave often as "smooth" interfaces and effective medium theories have to be applied.

2.4.2 Electronic Properties

Electronic properties of TiO_2 might be the most comprehensive properties and they are of great interest for a large variety of electronic applications. Also in this work, a wide range of electronic properties is investigated. Hence, Section 2.6 to 2.8 review the electronic properties of TiO_2 in detail. Since TiO_2 is usually an n-type semiconductor, mainly the properties of electrons are discussed. These properties are based on interactions between mobile electrons and any kind of applied external electrical, magnetic, or electromagnetic field with frequencies which are significantly lower than that of NIR radiation. This covers effects appearing for constant voltage stress, in $I - V$ characteristics, during impedance spectroscopy, for the absorption of terahertz radiation, and in temperature-dependent measurements, as long as all properties are investigated in the absence of UV-Vis-NIR radiation. Pure magnetic properties are not part of this work and are discussed elsewhere [197, 644, 645]. Typically, magnetic properties of TiO_2 nanostructures are enhanced by doping with selected transition metals [646–649].

In Section 2.6, size effects are discussed in the same way as presented for the structural properties in Section 2.2. First, bulk properties such as the electronic band structure, charge carrier density, and charge carrier mobility give an overview of relevant features. Second, reducing the size and dimension changes bulk features. Now, additional effects such as quantum confinement and surface properties have to be considered. Third, nanoparticles that are interconnected in grainy films or mesoporous layers share grain boundaries with each other. Similar to surfaces, these boundaries have different electronic properties compared to bulk TiO_2. Fourth, highly anisotropic nanoparticles such as nanorods show direction-dependent properties.

In Section 2.7, the effect of doping and defects on the electronic properties is described. As a consequence, the fabrication technique and post-deposition treatments influence the electronic properties, too.

Furthermore, in Section 2.7, charge transport mechanisms through the TiO_2 interface and bulk are discussed. Interfaces cause a band bending inside the TiO_2 and an energy barrier is formed that depends strongly on the electrode material and other influencing factors. Typically, the conduction mechanism depends not only on the structural properties but also on the applied voltage and temperature.

As listed in Figure 2.8, information about conduction mechanisms are gained by frequency-dependent properties which are probed by applying electrical or electromagnetic fields with several frequencies ω_e. For $\omega_e = 0$, time-dependent changes such as defect generation at a

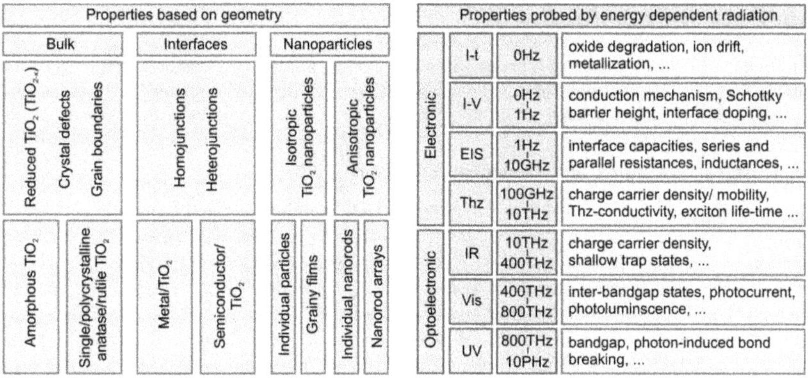

Properties based on geometry		
Bulk	**Interfaces**	**Nanoparticles**
Reduced TiO$_2$ (TiO$_{2-x}$) / Crystal defects / Grain boundaries	Homojunctions / Heterojunctions	Isotropic TiO$_2$ nanoparticles / Anisotropic TiO$_2$ nanoparticles
Amorphous TiO$_2$ / Single/polycrystalline anatase/rutile TiO$_2$	Metal/TiO$_2$ / Semiconductor/TiO$_2$	Individual particles / Grainy films / Individual nanorods / Nanorod arrays

Properties probed by energy dependent radiation		
I-t	0Hz	oxide degradation, ion drift, metallization, ...
I-V	0Hz / 1Hz	conduction mechanism, Schottky barrier height, interface doping, ...
EIS	1Hz / 10GHz	interface capacities, series and parallel resistances, inductances, ...
Thz	100GHz / 10THz	charge carrier density/ mobility, Thz-conductivity, exciton life-time ...
IR	10THz / 400THz	charge carrier density, shallow trap states, ...
Vis	400THz / 800THz	inter-bandgap states, photocurrent, photoluminscence, ...
UV	800THz / 10PHz	bandgap, photon-induced bond breaking, ...

(Electronic / Optoelectronic)

Figure 2.8: *Most common structure- and frequency-dependent optical, electronic, and optoelectronic properties. On the left side (greenish background), size- and shape-dependent influencing factors are listed. On the right side, properties (blueish background) probed by different frequencies (reddish background) are listed. Free electrons have a limited mobility and they are too slow to follow frequencies above the plasma frequency. Here, the plasma frequency is used to separate electronic and optoelectronic properties considering that dielectric properties are important in the regime of electronic properties as well.*

fixed applied potential are determined. A slow sweep of the applied potential results in $I - V$ characteristics giving insight into the energy barrier and the conduction band landscape close to the interface. Frequencies of up to a few gigahertz are applied during impedance spectroscopy. From these measurements, interface capacities, inductances, and shunt resistances are determined. Terahertz spectroscopy probe the electron mobility for short distances. In Section 2.8, time-dependent charge transport effects are discussed. These effects are mainly based on a change of the crystal structure or a locally charging of the oxide.

2.4.3 Optoelectronic Properties

Optoelectronic effects are based on the quantum-mechanical excitation of an electron into a higher energy state. For instance, IR radiation and elevated temperatures are able to excite electrons from shallow traps into the conduction band and thus increase the conductivity. Intra-bandgap states caused by defects or doping are exited usually by visible light and result in tunable absorption and photoconductivity. UV light is able to excite electrons from the valence band to the conduction band changing the density of mobile electrons and as a consequence, the conductivity and photocatalytic activity significantly. Plasmonic excitations appear for high electron densities and cause often a strong coupling between the incident light and the local electronic properties.

Hence, typical experiments to determine optical properties are UV-Vis-NIR absorption spectroscopy, transient absorption spectroscopy, photo- and electroluminescence spectroscopy, and photocurrent measurements.

Optoelectronic properties are not investigated in this work. However, the results obtained from investigating optical and electronic effects are an important base to explore a variety of phenomena in confined TiO$_2$ nanostructures.

2.5 Optical Properties of TiO_2

For semiconductors, optical properties are often closely related to electronic properties. The optical and electronic properties of TiO_2 nanostructures depend on many features and the overall behavior is quite complex.

2.5.1 Refractive Index and Dielectric Function

The wavelength-dependent dielectric function of TiO_2 is quite complex, especially if there are crystal defects or impurities. However, for this work, it is not necessary to know the dielectric function over the full frequency range. The most important values are the dielectric constant for static electrical fields ϵ_{stat} and for visible light ϵ_{opt}. Besides the wavelength, the dielectric constant depends on the temperature and crystallographic direction as shown in Figure 2.9 C [349]. Along the c-axis the average bond length is shorter and stronger. Hence, the dielectric function has larger values along the c-axis than along the a- and b-axis. Due to rotational symmetry the values for the a- and b-axis are equal and thus, the b-axis is not mentioned in the further discussion anymore. An overview of the dielectric constants is given in Table 2.2.

temperature	1.6 K	300 K	1000 K	reference
$\epsilon_{stat, a\text{-axis}}$, rutile	111	86	58	[349]
$\epsilon_{stat, a\text{-axis}}$, rutile		89		[650–652]
$\epsilon_{stat, c\text{-axis}}$, rutile	257	170	97	[349]
$\epsilon_{stat, c\text{-axis}}$, rutile		173		[650–652]
$\epsilon_{stat, a\text{-axis}}$, anatase		31		[650–652]
$\epsilon_{stat, c\text{-axis}}$, anatase		48		[650–652]
$n_{vis, a\text{-axis}}$, rutile		2.60		[653]
$n_{vis, c\text{-axis}}$, rutile		2.89		[653]
$n_{vis, a\text{-axis}}$, anatase		2.55		[653]
$n_{vis, c\text{-axis}}$, anatase		2.48		[653]

Table 2.2: *Dielectric constants of rutile TiO_2 along the a- and c-axis for selected temperatures from Parker [349].*

Rutile TiO_2 crystals offer a high polarizability. Consequently, high local internal electrical field amplitudes appear. In (anti)ferroelectric materials the internal field remains partly if the external field is switched off. Parker did not observe an (anti)ferroelectric transition between 1.6 and 1060 K but the ionic polarizability is close to the value of a ferroelectric polarization catastrophe at all temperatures [349]. Such a transition might appear at high oxygen vacancy or doping densities. The temperature coefficient of the ionic polarizability is slightly negative at room temperature and saturates below 100 K.

Cronemeyer specified the refractive index for visible light as $n_a = 2.616$ and $n_c = 2.903$ [348]. In the case of pure rutile TiO_2, absorption can be neglected and the dielectric constant is approximated by squaring the refractive index resulting in $\epsilon_{opt,a} \approx 6.84$ and $\epsilon_{opt,c} \approx 8.43$. Besides these constant values, there are some models taking the dispersion in the visible light regime into account. The wavelength-dependent refractive indices of rutile TiO_2 single crystals $n_a^2 = 5.913 + 2.441 \cdot 10^7/(\lambda^2 - 0.803 \cdot 10^7)$ and $n_c^2 = 7.197 + 3.322 \cdot 10^7/(\lambda^2 - 0.843 \cdot 10^7)$ are given by DeVore and are in good agreement with the results from Bärwald for wavelengths between 425 and 1500 nm [654, 655]. Here, the unit of the wavelength λ is Ångström. The refractive index is plotted in Figure 2.9 A. Due to electronic excitations there is a significant absorption for UV light as shown in Figure 2.9 B.

For thin films and nanorods, the dielectric function and hence, the refractive index differ from the values measured in bulk material. Dielectric constants between 4 and 86 are reported [658, 659]. As it is discussed later, nanorod arrays (NRAs) contain a lot of air and are treated as an effective

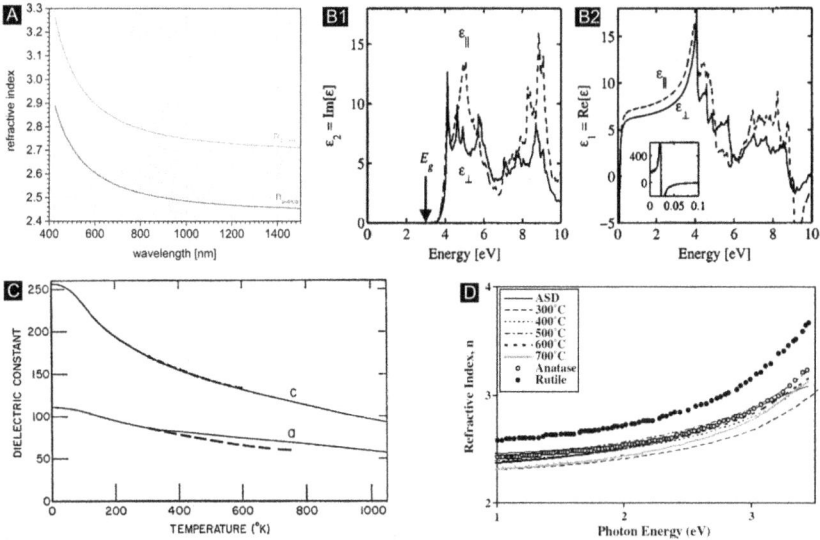

Figure 2.9: *Dielectric function and refractive index of rutile TiO$_2$: A) Refractive index of single-crystalline rutile TiO$_2$ as a function of the wavelength and crystallographic direction based on the equation given by DeVore for visible light and NIR radiation [654]. B) Imaginary (B1) and real (B2) part of the dielectric function as a function of the photon energy and crystallographic direction in the visible and UV light regime; reprinted from [656], with permission of AIP Publishing. C) Calculated static dielectric constant as a function of the temperature and crystallographic direction. The dashed lines are experimental data; reprinted Figure 1 with permission from [349], copyright (2017) by the American Physical Society. D) Measured refractive index of thin TiO$_2$ films. It should be noted that films annealed at more than 400 °C contain a significant amount of rutile fragments; reprinted from [657], Copyright (2017), with permission from Elsevier.*

medium. But besides this obvious effect, the refractive index depends on impurities, crystal defects, and grain boundaries. Hence, the fabrication technique and post-deposition treatment is essential as demonstrated in Figure 2.9 D [657, 660–664]. In thin films, birefringence is usually suppressed by the disordered orientation of the grains. The refractive index and the absorption coefficient of thin TiO$_2$ films are usually measured by the envelope method applying transmission spectroscopy [665–667] or by wavelength-dependent ellipsometry as used in this work (see Chapter 4).

In NRAs, there is a gradient of the crystal orientation between the bottom layer and the surface. In the upper part of the arrays, the nanorods grow vertically and thus, the c-axis is perpendicular to the substrate plane. But inside the bottom layer, the rods are oriented horizontally and thus, the c-axis is parallel to the substrate surface.

2.5.2 Light Scattering on Spherical TiO_2 Nanoparticles

Dielectric scattering plays an important role in TiO_2 systems. This is demonstrated by comparing the refractive index n, tinting strength° t_{in}, and specific gravity s_{gr} of industrial white pigments such as China clay ($n = 1.57$, $t_{in} = 100$, $s_{gr} = 2.6$), white lead ($n = 2.00$, $t_{in} = 100$, $s_{gr} = 6.7$), zinc oxide ($n = 2.02$, $t_{in} = 200$, $s_{gr} = 5.6$), lithopone 30% ($n = 1.84$, $t_{in} = 300$, $s_{gr} = 4.3$), antimony oxide ($n = 2.30$, $t_{in} = 400$, $s_{gr} = 5.7$), zinc sulfide ($n = 2.37$, $t_{in} = 900$, $s_{gr} = 4.0$), anatase TiO_2 ($n = 2.55$, $t_{in} = 1350$, $s_{gr} = 3.7 - 3.85$), and rutile TiO_2 ($n = 2.70$, $t_{in} = 1850$, $s_{gr} = 3.7 - 4.2$) [229]. In respect to their specific gravity, TiO_2 pigments are among the best industrial available scatterers. In photovoltaic or photocatalysis devices, visible light scattering and large surface area NRAs are essential for high efficiencies [668–670].
Tyndall scattering denotes the scattering of light at dielectric particles in general [671, 672]. Based on Tyndall's rather qualitative model and Maxwell's equations, the Mie theory describes the scattering of electromagnetic waves at spherical absorbing and non-absorbing dielectric particles [673]. The cross-sections for scattering and total absorption are given by [674]

$$\sigma_{M,scat} = \frac{\lambda^2}{2\pi} \sum_{n=0}^{\infty} (2n + 1) Re\{a_n + b_n\} \tag{2.5}$$

and

$$\sigma_{M,abs} = \frac{\lambda^2}{2\pi} \sum_{n=0}^{\infty} (2n + 1) \left(|a_n|^2 + |b_n|^2 \right), \tag{2.6}$$

respectively. Here, the parameter a_n and b_n consist of Ricatti-Bessel functions, which depend on the refractive index and the diameter of the particles. For large particles, Mie scattering° can be described by geometric optics whereas for small spherical particles, Rayleigh scattering° is employed [675].
Rayleigh scattering is applicable if $\beta \ll 1$ and $|\hat{n}|\beta \ll 1$, with $\beta = \pi d/\lambda$. d is the diameter of the scatterer, λ is the wavelength of the incident light, and $\hat{n}$ is the complex refractive index of the scatterer. The cross-section of Rayleigh particles for scattering and absorption is given by [668, 676]

$$\sigma_{R,scat} = \frac{2\pi^5}{3} \left| \frac{\hat{n}^2 - 1}{\hat{n}^2 + 2} \right|^2 \frac{d^6}{\lambda^4} \tag{2.7}$$

and

$$\sigma_{R,abs} = \pi^2 Im \left\{ \frac{\hat{n}^2 - 1}{\hat{n}^2 + 2} \right\} \frac{d^3}{\lambda}, \tag{2.8}$$

respectively. Due to polarization effects, the scattered light intensity depends on the scattering angle:

$$I_{scat}(d, \lambda, \phi) = I_0 \frac{\sigma_{R,scat}}{|\mathbf{r}|^2} (1 + \cos^2(\phi)). \tag{2.9}$$

The shown equation includes the assumption that the refractive index of the environment is 1. Rayleigh scattering is engraved by its λ^{-4} dependency promoting the scattering of blue light in the visible regime. Mie scattering is much less wavelength-dependent. But also the drop of scattering efficiency with the sixth power of the diameter for decreasing particles sizes is remarkable for Rayleigh scattering. In addition, on Rayleigh scattering particles, either nondirectional scattering or backscattering appears while large particles show forward scattering.

The transition from Mie to Rayleigh scattering depends on the difference of the refractive index of the particle and its environment. For materials with a high refractive index of $n \lesssim 3$ such as TiO$_2$ or silicon, Mie scattering appears for radii much smaller than the wavelength λ [677, 678]. Fu et al. observed Mie scattering of light for wavelengths between 500 and 660 nm ($n_{\alpha\text{-Si}} \approx 4.0 - 4.5$ [680]) on spherical silicon particles with a diameter of 75 nm only [679]. The most effective light scattering for TiO$_2$ particles is reported for diameters which are equal to half of the wavelength in air [229, 681]. This corresponds to diameters between 200 and 350 nm for visible light scattering.

Popov et al. applied Mie theory for approximately spherical and rutile TiO$_2$ particles with a diameter between 35 and 200 nm and wavelengths between 305 and 500 nm [682]. In general, Rayleigh scattering is applicable for TiO$_2$ particles in air with a diameter of less than 10% of the applied wavelength [670]. Typical diameters for good rutile TiO$_2$ Rayleigh particles are in the order of 20 nm [683]. For wavelengths between 200 and 500 nm, it is observed that the scattering efficiency of rutile TiO$_2$ nanoparticles drops significantly from a 21 to a 10 nm particle diameter according to the d^6-dependence [674]. But one should keep in mind that for such small nanoparticles the dielectric function differs from the bulk properties due to the influence of surface effects [676]. Interestingly, for dense films made of mainly anatase TiO$_2$ particles with a diameter of 23.7 nm, the features of Rayleigh scattering are conserved [684].

2.5.3 Dielectric Light Scattering on Highly Anisotropic Nanoparticles

For theoretical models, spheres have the simplest shape amongst all dielectric scattering particles. But even for spheres, the scattering cross-section determined from the Mie theory is quite complex as mentioned above. Describing the scattering on anisotropic particles requires powerful simulation tools and some approximations. To calculate the scattering behavior of electromagnetic scattering by anisotropic particles, the T-matrix method is often used as a convenient tool [685–688]. Non-spherical particles exhibit edges, summits, tips, holes, voids, surface roughness, and other topological features that influence dielectric scattering significantly. Anisotropic shapes result in directional scattering properties.

Anisotropic particles have been investigated by many researchers and several theoretical approaches were developed. Singham and Nousiainen et al. described the relationship between the scattered light and the shape of the scatterer in a more general way [689, 690]. Often, the idea is to replace the real particles by an effective particle with a simpler shape that is easier to handle in calculations. A typical feature of these effective particles is the equivalent spherical radius $r_{eq} = (3abc/4\pi)^{-3}$ where a, b, and c are the semiaxes of an ellipsoid with the same volume as the scattering particle. This approximates oblate scatterers very well [691–693]. It is also applied for molecules [694], aggregates [695], and grains [696] with dimensions of even less than 10% of the incident wavelength. More instrumental in describing the scattering of nanorods are reports utilizing an equivalent spherical radius for rectangular scatterers [697, 698]. The scattering from finite, three-dimensional bodies with one large dimension related to the others (such as nanorods) is calculated using the generalized Rayleigh-Gans (GRG) approximation [699–701]. In particular, GRG is applicable if the large dimension is in the order of the wavelength and the other dimensions are small compared to the wavelength [702]. Here, nanorods are approximated to be spheroidal nanoparticles with different polarizability along the three principal axes which are given by [676, 703]

$$\alpha_i = V \frac{\epsilon_{TiO2,\lambda} - 1}{1 + L_i(\epsilon_{TiO2,\lambda} - 1)} \quad (i \in [1, 2, 3]), \quad (2.10)$$

with the geometric factors

$$L_i = 0.5 \int_0^\infty \prod_{j=1}^3 a_j (a_j^2 + q)^{\delta_{ij}+0.5} dq. \tag{2.11}$$

Here, i is the principal axis, V the volume of the ellipsoid, $a_1 \leq a_2 \leq a_3$ the semiaxes, and δ_{ij} the Kronecker delta. Related to the factor V in Equation 2.10, α is approximately proportional to the length of each semiaxis for small dimensions. The highest polarizability appears along the long semiaxis. For larger dimensions, terms containing a^x in Equation 2.11 become more and more important. Liu et al. discussed different solutions of Equation 2.11 for gold particles with $a_1 = a_2 = 14\,\mathrm{nm}$ and $a_3 = 56\,\mathrm{nm}$. Another detailed investigation on the scattering properties of a cylindrical nanoparticle was performed by Grießhammer and Rohrbach in order to simulate the tracking behavior of such a particle in optical tweezers [704].

Approximating a dielectric scatterer as a spheroidal particle is sufficient for many applications, but accuracy is lost for increasing deviations from the modeled shape. For highly anisotropic nanoparticles, the orientation in relation to the incident light becomes very important. Enhanced light scattering on highly anisotropic nanostructures is well known and denoted as the dielectric "antenna" effect [705]. In contrast to metallic antennas bearing plasmonic effects, electrical *and* magnetic resonances might occur at the same time in dielectric antennas. In semiconductors with high electron densities, the antenna effect can arise from dielectric and metallic properties. It is well described in III-V semiconducting nanorods such as GaN, GaAs, and InP [706–709]. The intensity of this effect is often tuned by a core-shell structure as it was applied several times for silicon nanorods [710, 711].

There are also a few reports about the antenna effect in metal oxides. For example, Choi et al. investigated the position- and orientation-controlled interaction between individual indium tin oxide (ITO) nanorods and polarized light [207]. Terakawa et al. simulated the electrical field intensity distribution in the surroundings of ZnO nanorods lying on a ZnO film [712]. In addition, they applied Mie theory to calculate the near- and far-field scattering efficiencies as a function of the rod diameter for light with a wavelength of 532 nm. There is no far-field scattering for diameters less than roughly 100 nm predicted. Compared to ZnO ($n_{\mathrm{ZnO,532nm}} = 2.02$) rutile TiO$_2$ nanorods have a much higher refractive index ($n_{\mathrm{TiO2,539nm}} = 2.95$) and hence, the scattering efficiency is supposed to be higher for rutile TiO$_2$ nanorods. Also the position of electrical and magnetic resonances might be different for rutile TiO$_2$ nanorods. Likodimos et al. emphasized the role of the orientation of a TiO$_2$ nanotube towards an incident electromagnetic wave for Raman scattering$^\circ$ [713]. The scattering efficiency for subwavelength dielectric cylinders depends on the angle of incident light due to polarization-dependent light scattering [714]. For instance, Raman scattering is relatively efficient if the electrical field of the incoming light is parallel to the long axis of GaN nanowires [714], WS$_2$ nanotubes [715], or CuO nanorods [716].

Light as a transversal wave is polarized in parallel to the surface for normal incidence. Most nanorods grown homogeneously on a flat substrate are oriented vertically. It is assumed that in case of normal incidence, an electromagnetic wave is scattered similarly to a spheroidal particle with an equivalent radius r_{eq}. This radius increases with increasing geometrical cross-section of the nanorod from the point of view of the incident light [717]. For vertically aligned nanorods, the quadratic, geometrical cross-section corresponds to the diameter of the nanorod ($\approx 20\,\mathrm{nm}$). Hence, the equivalent radius is small and scattering does not depend on the polarization as demonstrated in Figure 2.10. For horizontally aligned nanorods, the geometrical cross-section corresponds to the diameter/length ($\approx 400\,\mathrm{nm}/\approx 20\,\mathrm{nm}$) for light which is polarized perpendicularly/in parallel to the long axis of the nanorod. In this case, the scattering depends strongly on the polarization. For the applied dimensions of the nanorods, it might happen that wavelength-dependent Rayleigh scattering appears for the smallest equivalent radii and the rather wavelength independent Mie scattering describes the scattering for larger equivalent radii within the same nanorod and for the same applied wavelength.

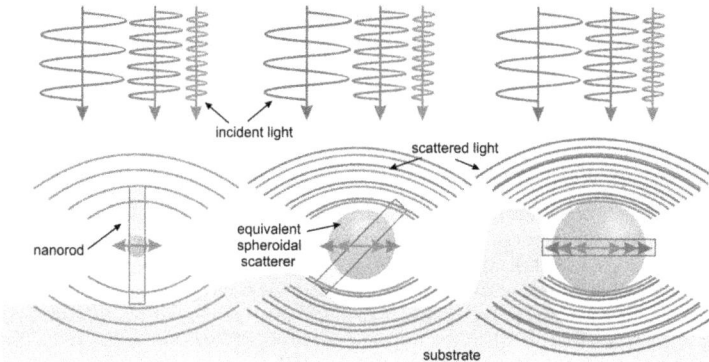

Figure 2.10: *Schematic drawing of white light scattering at dielectric nanorods. The effective cross-section (represented by a gray sphere with an equivalent radius) depends on the angle between the rod and the incident light. The cross-section increases with an increasing angle and reaches its maximum at 90°. The angle-dependent effective cross-sections correspond roughly to spherical dielectric scatterers with different radii.*

The orientation-dependent light scattering of the investigated rutile TiO$_2$ nanorods determines not only the behavior of individual nanorods but also the scattering behavior of NRAs in confined geometries. Nanorods in a homogeneous NRA are aligned mainly perpendicular on the substrate. However, rods grown at the edge of NRAs as well as on thin linear or point-shaped seed structures are omnidirectional. Hence, light scattering is supposed to depend on the position and the shape of the NRA.

It should be noted that most of the investigated rutile TiO$_2$ nanostructures are grown on seed layers with grains causing a rough surface. The grains have diameters between 5 nm and 5 μm and hence, as individual particles, they would scatter light like the above-mentioned white pigments. But the grains are packed so densely and without interspaces inside the films that their individual scattering behavior is suppressed. The surface roughness might also contribute to light scattering [718]. For efficient light scattering, a minimal surface roughness is needed, which is not achieved on the investigated samples within this work as discussed in Chapter 9 and 10.

2.5.4 An Ordered Array of Nanoparticles as a Basic Module of Metamaterials

The advantage of dielectric scattering particles compared to plasmonic light coupling is the low energy loss by absorption. As a consequence, the material is less heated up and resists higher energies of the incident light. In addition, dielectric scatterers can have electrical and magnetic resonances at the same time which is an important requirement for metamaterials [677, 719–723]. Such Mie resonances are tunable by using individual nanorods or small nanorod assemblies as building blocks for superlattices [677, 724]. Evlyukhin et al. investigated the features of arrays made of silicon nanoparticles [677].

Furthermore, these superlattices are advantageous for increasing the sample surface and enhancing the scattering of longer incident wavelengths if scattering on individual buildings blocks is mainly dominated by Rayleigh scattering. Already TiO_2 spheres without the hedgehog structure (TiO_2 sphere covered with TiO_2 nanorods) enhance light scattering in dye-sensitized solar cells significantly [725]. An efficient approach to create highly ordered monolayers of spheroidal nanoparticles is to use polystyrene spheres as a template for further functional layers [726]. A brief insight into this method is given in Chapter 8.

2.5.5 Disordered Assemblies of Nanoparticles as an Effective Medium

For structures equal or smaller than the wavelength of the incident light, geometric optics fails to describe the interaction between light and matter. In this work, air is applied as a host medium for the nanoparticles. Sufficiently long and thick nanorods as they grow on FTO are diffuse scatterers and NRAs appear milky. This kind of scattering is described by the Mie theory. For rods shorter than 1 µm, the milky appearance, which does not depend on the wavelength and incident angle, is replaced by thin film interference effects. This behavior has to be described with an effective medium model. Typically, three simple models based on dipole interaction are discussed: the Lorentz-Lorenz (LL), Maxwell-Garnett (MG), and Bruggeman effective-medium approximation (EMA) model [727, 728]. All models assign a single parameter to a heterogeneous dielectric mixture and hence, they are valuable to describe a nanostructure. The best fitting model depends on the system which is at hand. For TiO_2 NRAs, the EMA seems to respect the properties of the real system best [729] (the reference investigates voids in α-silicon) and is used for similar systems such as ZnO NRAs as well [730]. Bruggeman described the EMA for disordered objects such as spheres, circular cylinders, and lamellae in a surrounding medium. The NRAs investigated in this work can be approached by parallel cylinders, although the quadratic cross-section of the TiO_2 rods results in discrepancies [731]. The cylinders must not be monodisperse and the air in the interspace is treated as significantly small air cylinders in order to fill the whole volume which is not occupied by TiO_2. The dielectric constant of the effective medium is given by

$$\epsilon_{eff}^2 - (\delta_{TiO2} - \delta_{air})(\epsilon_{TiO2} - \epsilon_{air})\epsilon_{eff} - \epsilon_{TiO2}\epsilon_{air} = 0, \qquad (2.12)$$

where $\delta_{TiO2/air} \in [0; 1]$ and $\epsilon_{TiO2/air}$ is the volume fraction and optical dielectric constant of TiO_2 and air, respectively [731]. When neglecting absorption and any dependence on the wavelength in the visible light regime, the refractive index is given by $n = \sqrt{\epsilon_{opt}}$.

2.6 Electronic Properties of Bulk and Nanostructured TiO_2

In this section, bulk properties such as the electronic band structure and conductivities are discussed and how these properties are influenced by crystal defects, dopants, and nanostructuring.

2.6.1 Band Structure in Bulk TiO_2

The eigenfunctions of the Schrödinger equation for a weak periodic potential are the product of plane waves $\Psi'_k(r) = \exp(i\mathbf{k}r)$ with wave vector $\mathbf{k}$ and a periodic function $u_k(\mathbf{r}) = u_k(\mathbf{r} + \mathbf{R})$ with the lattice vector $\mathbf{R}$. The resulting waves

$$\Psi_k(\mathbf{r}) = u_k(\mathbf{r}) \exp(i\mathbf{k}r) \tag{2.13}$$

are known as **Bloch waves**° propagating with a speed of $v_{n,k} = \partial E_n(k)/\hbar \partial k$. Constructive and destructive interference of wave and lattice functions result in allowed and forbidden states. For a strong background potential, a band structure is predicted which includes energy regimes that are forbidden for electrons. For simplification, the term "bandgap" is used for the specific bandgap above the valence band maximum (VBM). The VBM is the highest occupied energy level and the conduction band minimum (CBM) is the lowest unoccupied energy level for electrons at 0 K.

There are several calculations for the band structure and density of states in rutile, anatase, and brookite TiO_2 performed on the basis of density functional theory. The calculated values differ from experimental results, but they are self-consistent and reflect the experimentally observed difference of the bandgap widths for rutile and anatase [732–734]. These calculations indicate that rutile has an indirect and anatase a direct bandgap.

According to experiments and calculations performed by Pascual et al. and Daude et al., rutile TiO_2 has a direct forbidden bandgap of 3.03 eV being nearly degenerated with an indirect allowed gap of 3.05 eV [735, 736]. Dependent on the measuring technique or the theoretical model, also larger values for the indirect (3.1 eV) and direct (3.6 eV) bandgap are reported [737, 738]. The bandgap of anatase TiO_2 has been found to be between 3.15 and 3.98 eV [739]. The obtained value depends on the fabrication and characterization technique [739]. So far, it is not clear if anatase has a direct or indirect bandgap [740–742]. The bandgap of brookite TiO_2 (3.4 eV) is even larger than that of rutile or anatase TiO_2 [743]. However, the largest bandgaps are found in amorphous TiO_2 [744, 745]. The electron affinity, which is the difference between the CBM and the vacuum level, is about 3.9 eV in rutile TiO_2 [746].

The bandgap is slightly decreasing with increasing temperature because of phonon effects and thermal expansion as demonstrated by Tang et al. [747, 748]. Towards low temperatures, the bandgap is increasing quadratically while the increase is linear at room temperature.

Rogers et al. drafted the contributions of atomic orbitals to the band structure of rutile TiO_2. While the VBM is formed by a Ti–O π-bond, the CBM consists of Ti–Ti σ-bonds [94]. In general, the valence band in rutile and anatase TiO_2 consists of bonding and non-bonding occupied $2p$ states of oxygen [134, 749–751], which are hybridized with the $3d$ orbitals of titanium [752]. A parabolic conduction band results from the anti-bonding states formed by the s orbital of titanium and the p orbital of the oxygen. Due to the parabolic band formation, a wide bandgap occurs between the VBM and the CBM [50]. p states of titanium ions contribute to higher empty electronic states in the conduction band [753].

Related to the anisotropic crystal structure, the bandgap depends on the crystallographic direction which is indirect along the c-direction and direct along the a-direction [754]. The values are given in Table 2.3. The electronic band structure is shown in Figure 2.11.

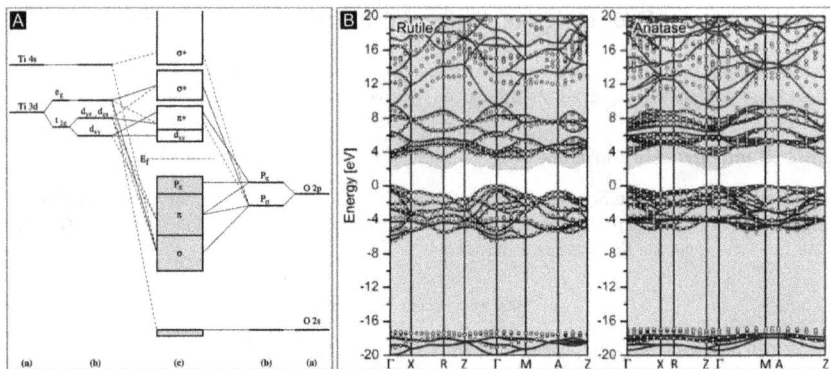

Figure 2.11: *A) Comparing the electronic states of TiO_2 and its components: a) Selected states of oxygen and titanium atoms, b) split levels in the crystal-field, and c) states resulting from interatomic interactions in TiO_2. The dashed and thin-solid lines display small and large contributions, respectively [755]. Reprinted with permission from [755], Copyright 2000 by the American Physical Society. B) The electronic band structure of rutile and anatase TiO_2 calculated by different models: DFT-HSE06 (solid lines), PBE-G0W0 (dots), and DFT-PBE (grey shaded areas above the CBM and below the VBM). Details are found in [756]. Reproduced with permission from [756], ©IOP Publishing. All rights reserved.*

The band structure of defect-free TiO_2 is supplemented or even changed by the presence of crystal defects, surface states, or impurities. As discussed in previous sections, perfect rutile TiO_2 crystals do not exist. Dependent on the fabrication technique and post-treatment, there is always a certain amount of crystal defects. And some fabrication techniques provide even significant amounts of impurities [757]. Any deviation from the defect-free crystal structure influences the electronic band structure and might result in charge carrier traps, donors, or acceptors. For instance, calculations predict that the band structure of TiO_2 differs from reduced crystalline oxides such as Ti_2O_3 [758] and TiO [733, 759, 760]. Table 2.3 shows a small collection of a few band structure properties for different TiO_2 samples and characterization techniques. Data of materials used often in combination with TiO_2 such as transparent conductive oxides (TCOs) are listed as well. In this section, static properties on the band structure are discussed only. Due to its complexity effects on the charge carrier mobility are discussed separately in the following section.

2.6.2 Chemical Potential and Electron Density in Intrinsic Semiconductors

In this work, all experiments were performed at room temperature. Hence, the Fermi level has to be considered rather than the Fermi energy. The Fermi level equals the electrochemical potential μ and is temperature-dependent. In the absence of acceptor and donor sites, the chemical potential is given by

$$\mu = \frac{E_c + E_v}{2} + \frac{k_b T}{2} \ln\left(\frac{N_v}{N_c}\right) = E_v + \frac{E_g}{2} + \frac{3}{4} k_b T \ln\left(\frac{m_{dh}}{m_{de} M_c^{2/3}}\right) \quad (2.14)$$

where E_c is the energy level of the CBM, E_v is the energy level of the VBM, $E_g = E_c - E_v$ is the bandgap energy, k_b is the Boltzmann constant, T is the absolute temperature, N_c is the effective density of states in the conduction band, N_v is the effective density of states in the valence band, $m_{de} = (m_{e,1}^* m_{e,2}^* m_{e,3}^*)^{1/3}$ is the density-of-state effective mass for electrons, m_{dh} is the density-of-state effective mass for holes, $m_{e,i}^*$ is the effective electron mass along a certain crystallographic direction, and M_c is the number of equivalent minima in the conduction band [794, 795]. To keep the notation simple, m_e^* is used instead of m_{de} for the effective mass in the following. If properties along a certain crystallographic direction are considered, m_e^* has to be replaced by $m_{e,i}^*$. The density of thermally activated charge carriers in the case of $E_F \ll E_c$ is given by [795]

$$\eta = N_c \exp\left(-\frac{\mu - E_c}{k_b T}\right) \qquad (2.15)$$

where N_c is given by [795, 796]

$$N_c = 2\left(\frac{m_e^* k_b T}{2\pi \hbar^2}\right)^{3/2} M_c. \qquad (2.16)$$

Typically, N_c is determined by Mott-Schottky plots that are mentioned later. At room temperature, only $10^{-26} \cdot N_c$ electrons are expected in the bandgap of a semiconductor with a bandgap of 3.0 eV. Hence, perfectly crystalline TiO_2 shows a high resistivity of approximately 10^{15} Ωcm [797].

2.6.3 Chemical Potential and Electron Density in n-type TiO_2

In TiO_2, there is often a huge number of electron donor levels caused by oxygen vacancies. The presence of **donors** affects the chemical potential μ and the electron density η_e. The chemical potential in an n-type semiconductor is expressed by

$$\mu_n = E_c + k_b T \ln\left(\frac{\eta}{N_c}\right) \qquad (2.17)$$

where η equals the effective density of ionized states [798]. For instance, considering N_c equal 10^{21} cm^{-3} and η equal 10^{17} cm^{-3} the difference between μ and E_c is 0.23 eV [799]. The chemical potential is listed for different TiO_2 structures in Table 2.3.
The density of mobile electrons in an n-type semiconductor for $N_d \gg \frac{1}{2} N_c \exp\left(-\frac{E_d}{2k_b T}\right) \gg N_a$ is given by

$$\eta_e \approx \sqrt{\frac{N_d N_c}{2}} \exp\left(-\frac{E_d}{2k_b T}\right) \qquad (2.18)$$

where N_d is the electron donor density, N_a is the electron acceptor density, and E_d is the difference between the CBM and the donor level [795]. E_d results from a hydrogen-atom like potential with the ionization energy of

$$E_d = \left(\frac{\epsilon}{\epsilon_s}\right)^2 \left(\frac{m_{ce}}{m_e}\right) E_H \qquad (2.19)$$

where

$$m_{ce} = 3\left(\frac{1}{m_1^*} + \frac{1}{m_2^*} + \frac{1}{m_3^*}\right)^{-1} \qquad (2.20)$$

is the conductivity effective mass and

$$E_H = \frac{m_e q^4}{32\pi^2 \epsilon^2 \hbar^2} = 13.6\,\text{eV}. \tag{2.21}$$

is the ionization energy of a hydrogen atom [794].

2.6.4 Band-Derived Effective Mass of Electrons

In the semi-classical description, electrons are treated as particles described by wave packets. As a consequence, an effective mass is assigned to the electrons in the lattice that differs from their rest mass. The effective mass

$$\left(\frac{1}{m_e^*}\right)_{i,j} = \frac{1}{\hbar^2} \frac{\partial^2 E_n(\mathbf{k})}{\partial k_i \partial k_j} \tag{2.22}$$

depends on the flection of the band structure [795]. The effective mass m_e^* calculated from the band structure is called conduction band mass. In rutile and anatase TiO_2, m_e^* is usually between 0.7 and 10 m_e [800–803]. For particular defect densities, assumptions, and theoretical models, the effective conduction band mass can differ from this range [803].

2.6.5 Conductivity and Mobility of Electrons in TiO$_2$

TiO_2 appears mainly as an n-type semiconductor for two reasons. On the one hand, the extended spherical shaped s-orbitals of the titanium ions support the carrier transport rather in the conduction band than in the valence band. As a consequence, the electron mobility is larger than the hole mobility [753, 804]. On the other hand, the high electronegativity of oxygen ions that form the valence band with their occupied $2p$-orbitals result in a strongly localized valence band edge [50, 753]. Thus, holes have a higher effective mass and lower mobility compared to electrons. Holes that are localized strongly around oxygen ions form deep acceptor levels. Thus, the activation energy for hole hopping is relatively high resulting in a low p-type conductivity in TiO_2 [753, 805, 806]. A third reason for the high n-type conductivity in TiO_2 is the typically high density of oxygen vacancies that donate electrons to the conduction band at room temperature as described later [200]. Hence, the n-type behavior in amorphous TiO_2, which can be seen as highly defective, is more distinct compared to crystallized TiO_2 [744, 745]. As a consequence, the following explanations treat only the properties of electrons.

Electron transport is driven by external fields, internal fields caused by charge density inhomogeneities, and dissipative effects due to scattering [807–813]. A general description is given by the Boltzmann equation. A detailed description of charge scattering and a discussion of the Boltzmann equation is found elsewhere [814–819].

In the absence of an electrical field, electrons move via a temperature-driven diffusion. The Einstein relationship links the electron diffusion constant D_e with the electron mobility for $\eta_e \ll N_c$ as follows [795]:

$$D_e = \left(\frac{k_b T}{q}\right) \mu_e^* \tag{2.23}$$

Here, μ_e^* is the electron mobility and q is the elementary charge. Related to the anisotropic crystal structure of TiO_2, the conductivity and mobility depend on the crystallographic direction. In certain temperature regimes, the mobility along the a-axis is more than twice as large as along the c-axis [820].

In the presence of an electrical field, the conductivity according to the Drude model is given by

$$\sigma = \eta_e q \mu_e^*. \tag{2.24}$$

In case of undoped semiconductors, it is given by

$$\sigma = \sigma_0 \exp\left(-\frac{\mu - E_c}{k_b T}\right), \tag{2.25}$$

where σ_0 depends on the density of mobile electrons η_e in the conduction band and is discussed below in the case of polaron conductivity [197, 795]. The simplest way to measure the conductivity is the four-point-probe method [821, 822]. The mobility is given by

$$\mu_e^* = \frac{v_D}{\mathcal{E}} = \frac{q \tau_s}{m_e^*}, \tag{2.26}$$

where v_D is the drift velocity, $\mathcal{E}$ is the amplitude of the applied electrical field, $\tau_s = l/v_F$ is the mean momentum scattering time, l is the mean free path, and v_F is the Fermi velocity of the electrons. Hence, the electron mobility depends on the band structure [795].
Typically, the mobility is affected by different electron-lattice interactions such as scattering on phonons, impurities, defects, grain boundaries as well as inter- or intravalley scattering [823–833]. The total scattering time τ_s is composed of the different scattering effects like

$$\frac{1}{\tau_s} = \frac{1}{\tau_{ph}} + \frac{1}{\tau_{def}} \tag{2.27}$$

where τ_{ph} and τ_{def} results from scattering on phonons and crystal defects, respectively. For temperatures above the Debye temperature Θ_D, electrons scatter with optical phonons and the electron conductivity σ is proportional to the temperature T [834, 835]. For temperatures significantly below Θ_D, the phonon scattering-dependent conductivity is proportional to T^3. Acoustic and optical phonon scattering rates are different and each rate depends on the crystal orientation in TiO_2 [836]. For instance, the scattering rate for optical phonon scattering along the c-axis of excited electrons in rutile TiO_2 is $2 \times 10^{12} \, s^{-1}$ at $50 \, K$ and $1 \times 10^{14} \, s^{-1}$ at $300 \, K$. The scattering rates along the a-axis are in the same order but slightly lower [836]. In polaronic semiconductors such as TiO_2, the acoustic phonon scattering-derived mobility μ_{ac}^* is proportional to $(m_e^*)^{-5/2} T^{-3/2}$ [837] and the ionized impurity scattering-derived mobility μ_i^* is proportional to $(m_e^*)^{-1/2} N_i T^{3/2}$ [838], where N_i is the density of ionized impurities. Hence, in the presence of non-ionized defects and impurities, the total momentum scattering rate does not vanish for $T \to 0 \, K$ since there is a finite and in general temperature-independent defect scattering rate [834, 835]. Considering all mentioned scattering processes the total mobility μ_e^* is proportional to $(m_e^*)^{-3/2} T^{1/2}$ [839]. Deviations from this relation indicate the presence of further scattering mechanisms.

Using Equation 2.25 and the assumption that the temperature dependence of the mobility and $T^{3/2}$ is small compared to the exponential term, the conductivity for non-interacting donor sites is given by [840–843]

$$\sigma = \sigma_0 \exp\left(-\frac{E_d}{k_b T}\right). \tag{2.28}$$

For sufficiently high donor densities, donor wave functions overlap and form donor bands that are comparable to hydrogen molecule-like donor sites. The effective mass in such donor bands is much higher than in the conduction band [197].
An empirical and temperature-dependent relation between the conductivity of highly pure sintered rutile TiO_2 and temperatures between $380 \, K$ and $490 \, K$ is given by Gorelik [844]:

$$\ln \sigma = 2.5 - 11700/T. \tag{2.29}$$

Here, the activation energy is $1.02\,\text{eV}$ only and thus much smaller than the bandgap energy. Between room temperature and $1400\,°\text{C}$, the conductivity increases five orders of magnitude [841].

Polaron Conductivity

In contrast to typical covalently-bonded crystals, in ionic and highly polar crystals such as TiO_2 the displacement of ions in presence of an electron or hole cannot be neglected. Even at low temperatures, when electrons are barely scattered off phonons, the effect of Coulomb interaction between mobile charges and lattice ions has to be taken into account. Since the dynamics of the ion displacements are determined rather by the velocity of mobile charges than by a frequency, the lattice formation is described by virtual phonons. The quasiparticle melding a mobile charge and its surrounding virtual phonons is called a polaron° [845, 846]. In this work, only polarons based on longitudinal-optical (LO) phonons are discussed. For the sake of completeness, it is mentioned that there are also other kinds of polarons, such as spin polarons (interaction between electrons and atomic magnetic moments) and piezopolarons (interaction between electrons and acoustic phonons) [845, 847].

One distinguishes between weak electron-phonon coupling resulting in large polarons and strong coupling resulting in small polarons [848]. The diameter of **large (or Fröhlich) polarons** exceeds the lattice constant. These polarons are described by the unitless Fröhlich constant°

$$\alpha_F = \frac{q^2}{\hbar c} \sqrt{\frac{m_e^* c^2}{2\hbar\omega_{\text{LO}}}} \left(\frac{1}{\epsilon_o} - \frac{1}{\epsilon_s} \right) \tag{2.30}$$

where c is the speed of light, m_e^* is the effective electron mass determined by the band structure, ω_{LO} is the LO-phonon angular frequency, ϵ_o is the optical dielectric constant, and ϵ_s is the static dielectric constant [134, 656, 849, 850]. The phonon energy $\hbar\omega_{\text{LO}}$ is approximately $45\,\text{meV}$ and was determined by thermal neutron scattering [341]. The Fröhlich coupling constant equals roughly twice the amount of virtual phonons encircling a single electron [851]. Hence, polaronic effects become important for values larger than one. Typical values for α_F are between 0.02 (III-V semiconductors) and 4.5 ($SrTiO_2$) [845]. Due to the significant interactions between the lattice and the electron, the effective mass of the polaron is larger than the mass of the underlying electron. A rough estimation for the relation between the effective polaron mass m_e^{**} and electron mass m_e^* was given by Richard Feynman:

$$m_e^{**} \approx m_e^*(1 + \frac{\alpha_F}{6}) \tag{2.31}$$

for $\alpha_F \gg 1$ and

$$m_e^{**} \approx m_e^*(0.02\alpha_F^4) \tag{2.32}$$

for $\alpha_F \ll 1$ [845]. For large polarons, the mobility is limited by phonon-scattering and hence, it is expressed by

$$\mu_l^{**} = F_{\alpha_F} \exp\left(\frac{\Theta_D}{T} \right) \tag{2.33}$$

where F_{α_F} is a theoretically predicted function [845, 851]. Since scattering at grain boundaries dominates often the temperature-dependent mobility described in Equation 2.33, it is observed in high-purity crystals only [852]. There are several electronic states within the Coulomb potential

of a polaron. The excitation frequencies of electrons in large polarons are in the terahertz regime [845].
Polarons fitting in one unit cell are called **small (or Holstein) polarons**. Due to their strong confinement, internal excitation energies are higher compared to large polarons [850]. To describe the properties of small polarons the motion of each neighboring atom has to be considered and this requires usually *ab initio* methods. For $T > 1/2\Theta_D$, small polarons are driven by thermally activated hopping [853]. Often, the mobility follows an Arrhenius-like behavior expressed by

$$\mu_{s,drift}^{**} = \frac{\sigma_0}{T^\beta} \exp\left(-\frac{W}{k_b T}\right) \tag{2.34}$$

for the drift mobility and

$$\mu_{s,Hall}^{**} = \frac{\sigma_0}{T^\beta} \exp\left(-\frac{W}{3k_b T}\right) \tag{2.35}$$

for the Hall mobility [850, 854]. The exponent β is equal 1 for adiabatic and equal $3/2$ for anti-adiabatic hopping. The activation energy W is $W_H + 1/2 W_D$ with the hopping energy W_H and the disorder energy W_D for adiabatic and W equals W_D for anti-adiabatic hopping. The pre-exponential factor σ_0 is expressed by

$$\sigma_{0,a} = \frac{2\pi\hbar\eta_e q^2 a_p^2}{k_b} \tag{2.36}$$

for adiabatic and

$$\sigma_{0,aa} = \frac{\eta_e q^2 a_p^2}{k_b} \frac{G^2}{2\hbar\omega_o} \left[\frac{2\pi}{W_p k_b}\right]^{1/2} \tag{2.37}$$

for anti-adiabatic hopping [854]. Here, ω_o is the optical phonon frequency, a_p is the hopping distance of the polaron, G is a transfer integral, and $W_p \approx 2W_H$ is the binding energy of the polaron. A detailed description of this term is performed by Yildiz et al. [854]. For $T < 1/2\Theta_D$, the motion of small polarons is driven by the band mechanism. Similar to large polarons, the electron is tunneling through the crystal with the effective mass

$$m_e^{**} \approx m_e^* \exp\left(D_c \frac{2W_H}{\hbar\omega_o}\right) \tag{2.38}$$

where D_c is a constant [853]. There is no abrupt transition from band to hopping conduction. Adiabatic hopping occurs when the electron is always relaxed in its potential well. If the potential well with the electron is ionized and the electron relaxes as soon as the potential well is formed in the new electron position the hopping is called anti-adiabatic or non-adiabatic. The behavior of real polarons is a mixture of adiabatic and anti-adiabatic hopping.
In some solids, the attraction force between the lattice distortion of two polarons becomes so high that polaron pairs, so-called bipolarons, are formed [845]. A pair consisting of an electron and a hole polaron is called polaron exciton [850]. The temperature-dependent mobility displayed in
Figure 2.12 A reflects the nature of polaron conductivity in rutile TiO$_2$. Austin and Mott observed a minimum of the drift mobility around room temperature [853]. The decrease/increase of the mobility with increasing temperature for low/high temperatures is described by tunneling/hopping of small polarons [754, 855–860]. The relatively high electron mobility in anatase TiO$_2$ indicates the presence of thermally excited electrons in the conduction band which is described by an Arrhenius-type activated conduction [134, 754, 861].
Small polarons are typically found in TiO$_2$ and are detected by measurements of the electron transport, thermopower, photoemission, and infrared absorption [862, 863]. However, there are

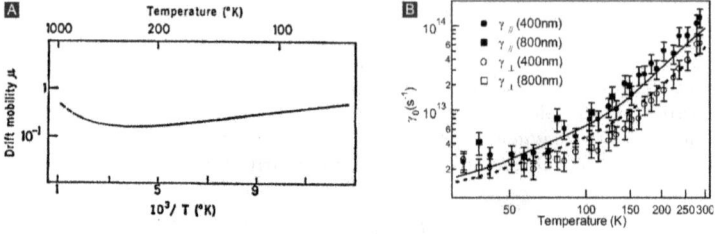

Figure 2.12: *A) Temperature-dependent electron mobility in TiO_2 [853]. Reproduced with permission from [853], ©IOP Publishing. All rights reserved. B) Temperature-dependent momentum scattering rate determined by THz time-domain spectroscopy in TiO_2 [836]. Reproduced with permission from [836], ©IOP Publishing. All rights reserved.*

ongoing discussions about the nature of polarons in rutile TiO_2 under certain conditions. Besides arguments supporting the presence of small polarons [202, 857, 864], there are indications for large polarons as well [865, 866].

In intrinsic rutile TiO_2, small polarons with a hopping energy of $W_H = 0.13\,eV$ are observed [857, 858]. The radius of small polarons in rutile is $r_{pol,s} \approx 5\,\text{Å}$ [848]. In contrast, the large polaron radius in anatase is reported to be roughly 20 Å [867].

Applying an excitation energy of $\hbar\omega_{abs} = 0.8\,eV$ in rutile TiO_2, the electron leaves its polaron potential well via Franck-Condon excitation [868, 869]. Assuming $\hbar\omega_{abs} \approx W_p$, the polaron energy

$$W_p = -\frac{q^2}{\epsilon_\infty r_{pol,s}} \qquad (2.39)$$

is roughly 0.4 eV, where ϵ_∞ is the high frequency dielectric constant and $r_{pol,s}$ is the effective radius of the potential well [853, 857, 858]. The polaron effective mass m_e^{**} is larger than the effective electron mass m_e^* in the "rigid-crystal" deduced from the band structure and m_e^* is larger than the rest mass of a free electron m_e [853]. In experiments, polaron masses between $8m_e$ and $190m_e$ are reported for rutile TiO_2 [202, 853, 865, 870–872].

Besides lattice distortion caused by the electron itself, also donor centers act as polaronic potential wells [873]. Such donor sites could be Ti^{3+} ions [874]. In an anisotropic crystal such as TiO_2, the hopping energy W_H depends on the crystal orientation. Based on density functional theory, Deskins and Dupuis listed values for the hopping energy in rutile ($W_H([001]) = 0.288\,eV$, $W_H([111]) = 0.307\,eV$) and anatase ($W_H([100]) = 0.304\,eV$, $W_H([201]) = 0.297\,eV$) TiO_2 as well as for the polaronic mobility in rutile ($\mu_{s,drift}^{**}([001]) = 5.24 \times 10^{-2}\,cm^2/Vs$, $\mu_{s,drift}^{**}([111]) = 7.42 \times 10^{-7}\,cm^2/Vs$) and anatase ($\mu_{s,drift}^{**}([100]) = 1.06 \times 10^{-5}\,cm^2/Vs$, $\mu_{s,drift}^{**}([201]) = 4.96 \times 10^{-5}\,cm^2/Vs$) TiO_2 [874]. Due to the importance of atomic bonds and their distances, the polaron conductivity is pressure-dependent as well [843].

In thin polycrystalline films, grain boundaries, different crystal orientations, and crystal defects effect the conductivity additionally and more simple models such as that expressed in Equation 2.34 need to be modified. One example is the Schnakenberg model expressing the polaron conductivity [875]

$$\sigma = \frac{\sigma_0}{T} \left[\sinh\left(\frac{\hbar\omega_o}{k_bT}\right) \right]^{1/2} \exp\left[\left(-\frac{4W_H}{\hbar\omega_o}\right) \tanh\left(\frac{\hbar\omega_o}{4k_bT}\right) \right] \exp\left(-\frac{W_D}{k_bT}\right). \qquad (2.40)$$

The complex situation in thin rutile films is discussed in detail by Yildiz et al. and Yagi et al. [202, 854]. Small Polaron hopping is the dominating transport mechanism in doped TiO$_2$ below the critical doping density for band conduction. In particular, there are reports about small polaron hopping in niobium- or tantalum-doped rutile TiO$_2$ [876, 877]. Yildiz et al. reported about an anti-adiabatic hopping above room temperature with a polaron hopping energy of $W_H = 0.3\,$eV in thin niobium-doped TiO$_2$ films [878].

Field-Dependent Conductivity

Cronemeyer reported that the conductivity becomes field-dependent if the field strength exceeds $10\,$V/cm in single-crystalline intrinsic rutile TiO$_2$ and even $0.1\,$V/cm in non-stoichiometric samples [841]. In defective samples, there are a lot of traps with different trap levels. With increasing electrical field, more and more electrons are emitted from these traps and hence, the electron density in the CBM becomes field-dependent. In bulk TiO$_2$, this effect is described by Poole-Frenkel emission. In addition, charged defects and impurities can be mobilized at sufficiently high fields and change the band bending and local Fermi level.

Measurement-Dependent Conductivity

To measure the conductivity of electrons in a semiconductor, mobile electrons are usually accelerated by an external field and their mobility is determined form the electronic inertia. If electrons are accelerated by an alternating external electrical field, they move forward and backward. In inhomogeneous materials such as polycrystalline films, the mobility becomes strongly frequency-dependent. Any kind of crystal defect causes electron scattering and in extended defects such as grain boundaries, the band structure differs from the one in bulk. These effects influence the mean free path and the effective mass of electrons. Since the grain boundary or defect plane is in thermodynamic contact with the surrounding single-crystalline TiO$_2$, the local change of the Fermi level is causing a depression or elevation of the conduction band for n- or p-type defect planes, respectively. As a consequence, electrons are reflected or trapped in grain boundaries and the mobility of the whole polycrystalline TiO$_2$ system is lowered. As long as the electron is moving between grain boundaries, the mobility is relatively high. But as soon as the electron has to cross several grain boundaries or defect planes its mobility decreases. Four-point-probe, field-effect transistor (FET) devices, time-of-flight (ToF) measurements, impedance spectroscopy (IS) and Hall measurements probe the effective mobility for rather large distances while terahertz and IR-spectroscopy probe the mobility for relatively short distances and hence, different values for electron mobilities are expected [879].

Hall conductivity: The type of semiconductor, charge carrier density, and charge carrier mobility are determined from Hall measurements. The Hall mobility is expressed by

$$\mu_H = -\frac{\mathcal{E}_y}{B_z \mathcal{E}_x} = -R_H \sigma = -\frac{V_H}{I_x B_z} \frac{A}{D} \sigma \qquad (2.41)$$

where $\mathcal{E}_{x,y}$ is the applied field in x- and y-direction, B_z is the magnetic field in z-direction, R_H is the Hall coefficient, σ is the Hall conductivity, V_H is the measured voltage, I_x is the applied current, A is the electrode area where the Hall voltage is measured, and D is the cross-section of the conductor in the direction of the applied current [197]. The average effective mass determined with Hall measurements is about $m_e^{**} \approx 25 m_e$ in single-crystalline rutile [880] and $m_e^{**} \approx m_e$ in anatase TiO$_2$ [134]. The term "averaged" means that the Hall mobility is averaging over several grains with several crystallographic orientations.

The Hall mobility depends strongly on the nature of the TiO_2 sample such as crystallinity, defects, grain boundaries, and morphology. Values between $0.01\,cm^2/Vs$ and $10\,cm^2/Vs$ were reported [197, 202, 834, 841, 881]. In slightly reduced (non-stoichiometric) rutile TiO_2, Cronemeyer found a Hall mobility of $10\,cm^2/Vs$ and an electron density of $10^{20}\,cm^{-3}$ at room temperature [841]. This is only two orders of magnitude smaller than typical densities of free electrons in metals.

Hall measurements on reduced rutile TiO_2 from Breckenridge and Hosler show a temperature-dependent scattering effect [834]. For low temperatures around $200\,K$, scattering on impurities is dominant [838]. At higher temperatures, scattering on optical phonons becomes more and more important [882, 883]. Grant found out that the resulting total Hall mobility follows a $T^{-5/2}$ trend for temperatures above $200\,K$ [197].

Compared to the mobility in rutile TiO_2, a qualitatively similar behavior is obtained in anatase TiO_2. Forro et al. investigated highly mobile electrons from shallow donor states with a density of $10^{18}\,cm^{-3}$ and an activation energy of $4.2 \times 10^{-3}\,eV$. At $50\,K$, the resistivity is less than $0.01\,\Omega cm$ and the electron Hall mobility is roughly $600\,cm^{-2}/Vs$ in this system. This emphasizes the high conductivity of anatase compared to rutile TiO_2. By increasing the temperature to $300\,K$ the Hall mobility drops by two orders of magnitude [871, 884, 885].

In niobium-doped rutile TiO_2, Frederikse confirmed the dominance of hopping conduction above $300\,K$ [880]. They measured Hall mobilities of 0.2 and $1.0\,cm^2/Vs$ along the a- and c-axis at $300\,K$, respectively. But at $20\,K$, it is roughly $10^3\,cm^2/Vs$. The Hall mobility becomes proportional to $\exp(0.10/k_bT)$ for temperatures above $150\,K$ [880]. Dependent on the sample fabrication they obtained effective masses m_e^{**} between 12 and $32\,m_e$.

A detailed discussion of the electronic properties of rutile TiO_2 determined by Hall measurements is given by Yagi et al., for instance [202].

FET conductivity: One of the simplest techniques to determine the semiconductor type and mobility of a nanodevice is the transistor device. Since the charge carrier density depends on the applied gate voltage U_g, the charge carrier mobility μ_{FET}^* is directly linked with U_g. Hence, μ_{FET}^* is determined by measuring the gate and source-drain voltage-dependent source-drain current I_{sd}:

$$\mu_{FET}^* = \frac{I_{sd}}{C_{ox}} \frac{L}{Z} \left[U_g U_{sd} - \frac{1}{2}U_{sd}^2 \right]^{-1} . \tag{2.42}$$

C_{ox} is the oxide capacity, L is the channel length, Z is the channel width, and U_{sd} is the source-drain voltage [886]. The method is limited by the velocity saturation at high fields [795], the influence of the contact resistance [887], short-channel effects for large depletion layer widths [795], and other non-linear effects [879, 888].

ToF conductivity: This is a direct method to determine the drift velocity and the mobility of charge carriers. A constant but small voltage U is applied across the oxide with the thickness d_{ox}. By a short but high voltage pulse charge carriers are injected on one side into the oxide. A time-resolved measurement of the current on the other side of the oxide is employed to measure the time of flight Δt and hence the drift velocity and the mobility

$$\mu_{ToF}^* = \frac{d_{ox}^2}{U \cdot \Delta t} \tag{2.43}$$

of the charge carriers. Instead of the short voltage pulse, a light pulse can be used to excite electrons from the Fermi level of the electrode into the conduction band of the oxide, too [886].

Terahertz conductivity: In general, alternating electrical fields in the terahertz regime cause the fastest possible oscillations of a mobile electron in semiconductors. For higher frequencies,

the electrons are not able to follow the applied field anymore. Of course, the plasma frequency of a specific material depends on its electron density and mobility. As a consequence, the amplitude of such an oscillation is quite small and is typically in the order of a few nanometers. This means that most of the electrons in polycrystalline TiO_2 never cross interfaces or grain boundaries, which is in contrast to mobilities determined from Hall measurements or observed in $I - V$ characteristics. The mobilities in low-frequency measurements are influenced by surface scattering, trapping, and interparticle charge transfer [889]. Terahertz time-domain mobility measurements of polycrystalline TiO_2 films or nanocrystals have the ability to determine the mobility inside grains and other building blocks [889].
A detailed **review on terahertz spectroscopy** in general, and on nanostructures in particular, is given by Lloyd-Hughes and Jeon [890].

In rutile TiO_2, electron-phonon scattering rates of up to $10^{14}\,\mathrm{s}^{-1}$ are observed [848]. The change of the permittivity related to momentum scattering rate γ^* of electrons predicted by the Drude model is given by [891]

$$\epsilon(\omega) = \epsilon_\infty - \frac{\omega_p^2}{(\omega^2 + i\omega\gamma^*)} \tag{2.44}$$

with the plasma frequency

$$\omega_p^2 = \frac{4\pi q^2 \eta_e}{\epsilon_0 m_e^{**}}. \tag{2.45}$$

The complex conductivity

$$\sigma_{el}(\omega) = i\omega\epsilon_{el}(\omega) \tag{2.46}$$

depends on the dielectric function of the electron gas ϵ_{el} [891]. The momentum scattering rate

$$\gamma^* = \gamma_{ac}^* + \gamma_{opt}^* = \beta T^{3/2} + \gamma_{opt}^*(\Upsilon, T) \tag{2.47}$$

depends on the temperature and hence, it is composed of scattering on acoustic (γ_{ac}^*) and optical (γ_{opt}^*) phonons [837]. Here, β is the strength of the acoustic-phonon-electron interaction. The scattering on LO-phonons dominates at high temperatures and depends on Υ which is the strength of the optical-phonon-electron interaction. Υ is determined from temperature-dependent measurements. In TiO_2, $\Upsilon \ll 6$ and $\Upsilon \gg 6$ correspond to large and small polarons, respectively [202, 857, 892]. The polaron nature effects the determined electron mobilities in terahertz measurements as well. Since this nature is supposed to depend on the morphology, defect and impurity density, and the crystallographic direction of TiO_2, a formalism is needed which does not depend on the electron-phonon coupling strength. This nonperturbative treatment is provided by Feynman's polaron theory [893-896]. Based on this theory, Hendry et al. found intermediate-sized polarons with mobilities of roughly $1\,\mathrm{cm}^2/\mathrm{Vs}$ in bulk single-crystalline rutile TiO_2 [836]. They obtained anisotropic electron-LO-phonon coupling constants of $\Upsilon_\parallel = 4.0\pm0.5$ and $\Upsilon_\perp = 6.0 \pm 0.5$ parallel and perpendicular to the c-axis, respectively. Hence, the higher polaron scattering rates appear along the crystallographic direction providing a higher mobility [202, 881]. This is explained by the anisotropic polaron effective mass m_e^{**} that contributes to the terahertz mobility

$$\mu_{tz} = \frac{q}{m_e^{**}\gamma^*} \tag{2.48}$$

as well [891]. Since the dielectric constants and band structures along the c-axis and a,b-axis differ from each other significantly, the electron mobility becomes direction-dependent. The

terahertz mobility parallel to the c-axis is $\mu_{tz\parallel} = 0.6\,\mathrm{cm^2/Vs}$ and perpendicular to the c-axis $\mu_{tz\perp} = 0.2\,\mathrm{cm^2/Vs}$ [848].
This anisotropy is reflected by the effective masses of $m_\parallel^{**}$ and $m_\perp^{**}$ as well [202]. *Ab initio* calculations result in band masses of $m_\parallel^* = 0.6m_e$ and $m_\perp^* = 1.2m_e$ [897]. Applying the Feynman model, the anisotropy becomes even more distinct yielding polaron effective masses of $m_\parallel^{**} = 2m_e$ and $m_\perp^{**} = 15m_e$ for $\Upsilon_\parallel = 4.0\pm0.5$ and $\Upsilon_\perp = 6.0\pm0.5$, respectively [836]. In single crystals with low defect densities, the determined room-temperature terahertz mobilities of $\mu_\parallel = 8\,\mathrm{cm^2/Vs}$ and $\mu_\perp = 1.4\,\mathrm{cm^2/Vs}$ are in the same order as the corresponding anisotropic Hall mobilities [202, 836]. Based on the calculations suggested by Schultz, Hendry et al. obtained a polaron radius in the order of the lattice spacing and thus, the polarons have an intermediate size and cannot be assigned clearly to small or large polarons [836, 895].

IR conductivity: In previous sections, the interaction of infrared radiation and crystalline metal oxides was introduced to determine structural lattice properties. In case of investigating the electronic properties, infrared absorption spectroscopy is applied in a very similar way to terahertz spectroscopy. It is often used to determine the number of free charge carriers in semiconductors with high mobile electron densities [898].

material	Fermi level	VBM	CBM	bandgap	reference
TiO_2, film, amorphous				3.5 eV	[70]
TiO_2, film, amorphous		−8.1 eV	−4.3 eV	3.8 eV	[761]
TiO_2, bulk, rutile		−7.83 eV	−4.80 eV	3.03 eV	[762]
TiO_2, bulk, rutile		−8.84 eV	−5.82 eV	3.02 eV	[763]
TiO_2, bulk, rutile, $c_\parallel$-axis				3.051 eV	[735, 764]
TiO_2, bulk, rutile, $c_\parallel$-axis				3.05 eV	[765]
TiO_2, bulk, rutile, $c_\perp$-axis				3.035 eV	[735, 764]
TiO_2, bulk, rutile, $c_\perp$-axis				3.04 eV	[765]
TiO_2, nanocrystals, rutile(100)	−4.9 eV			3.06 eV	[766]
TiO_2, bulk, anatase		−8.30 eV	−5.10 eV	3.20 eV	[762]
TiO_2, bulk, anatase		−7.41 eV	−4.21 eV	3.20 eV	[767, 768]
TiO_2, bulk, anatase		−8.98 eV	−5.75 eV	3.23 eV	[763]
TiO_2, bulk, anatase, $c_\parallel$-axis				3.46 eV	[134]
TiO_2, bulk, anatase, $c_\perp$-axis				3.42 eV	[134]
TiO_2, bulk, anatase, 4 K				3.30 eV	[747]
TiO_2, anatase (001)	−5.1 eV			3.13 eV	[766]
TiO_2, nanoparticles, anatase				3.23 eV	[738]
TiO_2, nanoparticles, amorphous				3.4 eV	[738]
TiO_2, mesoporous anatase	−5.15 eV	−7.65 eV	−4.45 eV	3.20 eV	[769, 770]
TiO_2, nano-film	−5.6 eV	−7.64 eV	−4.4 eV	3.24 eV	[771]
TiO_2, nanocrystalline (nc) anatase	−5.20 eV	−8.50 eV	−5.20 eV	3.30 eV	[772]
TiO_2, nc-anatase	−5.15 eV	−7.65 eV	−4.45 eV	3.2 eV	[770]
TiO_2, nc-anatase, pristine	−5.15 eV	−8.45 eV	−5.15 eV	3.30 eV	[773]
TiO_2, nc-anatase, after UV-exposure	−4.10 eV	−7.50 eV	−4.20 eV	3.30 eV	[773]
TiO_2, nc-anatase, after annealing	−5.08 eV	−8.51 eV	−5.21 eV	3.30 eV	[773]
ITO, pristine	−4.25 eV	−7.77 eV	−4.20 eV	3.60 eV	[773]
ITO, after UV-exposure	−3.92 eV	−7.17 eV	−3.47 eV	3.60 eV	[773]
ITO, after annealing	−4.77 eV	−8.02 eV	−4.29 eV	3.60 eV	[773]
ITO	−4.7 eV	−7.4 eV	−3.9 eV	3.5 eV	[774]
ITO	−4.20 eV	−7.4 eV	−3.8 eV	3.60 eV	[772]
ITO	−4.5 eV				[775]
WO_3				2.8 eV	[776]
FTO	−4.8(1) eV				[769, 777, 778]
					[779, 780]
FTO	−4.85 eV				[770]
FTO	−5.0 eV				[781]
FTO	−4.3(1) eV			4.25(25) eV	[782]
Pt/Ir tip	−5.2 eV				[783]
Pt/Ir tip	−5.25 eV				[784]
Pt/Ir tip	−5.4 eV				[785, 786]
Pt/Ir tip	−5.5 eV				[787]
Pt/Ir tip	−5.7 eV				[788]
Au	−5.1 eV				[789–791]
Pt	−5.56 eV				[789]
Ti	−4.33 eV				[789]

Table 2.3: *Semiconductor properties of employed and selected related materials. VBM is the valence band maximum and CBM is the conduction band minimum. The electronic properties depend on the fabrication technique [792, 793].*

59

2.6.6 Electronic States and Conductivity in Doped TiO_2

The conductivity of a semiconducting metal oxide is linked with the density of free electrons in the conduction band. Related to the large bandgap, charge carriers cannot be excited thermally in pure and perfect TiO_2. To manipulate the electronic properties, crystal defects and dopants are employed to create donor and acceptor levels within the bandgap. Crystal defects are point defects such as vacancies and interstitial atoms, or line defects such as displacements and edges on the surface, or plane defects such as stacking faults, grain boundaries, and surfaces [899, 900]. There are extensive theoretical investigations of the deep and shallow characteristics of defect states in rutile and anatase TiO_2 [901]. In metals, wave functions of $3d$ electrons are overlapping resulting in an outstanding electronic conductivity. In metal oxides, the cations are separated by oxide ions and the overlap of $3d$ electron wave functions is less distinct compared to metals, which results in a bandgap as demonstrated in Figure 2.13. Thus, the conductivity is expected to be smaller compared to metals [902–905]. This means that small changes of the stoichiometry change the conductivity significantly. Indeed, Table 2.5 shows this effect clearly for reduced titanium dioxide. Hence, the simplest way of doping is the generation of crystal defects that act as donors or acceptors. Usually, as-prepared TiO_2 nanostructures contain a lot of defects. Point defects are often discussed as electron donors or enhance ion movement [900]. Typical characterization methods for trap levels are deep transient spectroscopy (DLTS) and thermally stimulated currents (TSC) [795, 879].

Doping via Point Defects

The generation probability of defects depends on their forming enthalpy. In anatase TiO_2, it is 4.55 eV per oxygen vacancy and 9.11 to 9.24 eV per titanium interstitial [201, 906, 907]. Hence, titanium interstitials are formed at higher temperatures (above 800 °C) preferentially compared to oxygen vacancies (below 600 °C) [2, 797]. At temperatures above 1100 °C, the formation of titanium interstitials is preferred even at high oxygen pressures [271]. The formation of intrinsic defects depends on the partial oxygen pressure $P(O_2)$ as well. This is related to the equilibrium constants $K_1 = [V_o^{2+}]\eta_e^2 P(O_2)^{1/2}$ [271], $K_2 = [T_i^{3+}]\eta_e^3 P(O_2)$ [271], $K_3 = [V_{ti}^{4-}]\eta_h^4 P(O_2)^{-1}$ [50], and $K_i = np$ [908], where V_o are oxygen vacancies, T_i titanium interstitials, V_{ti} titanium vacancies, and $\eta_{e/h}$ is the electron and hole concentration [909].

Since the investigated nanostructures have extended dimensions, the equilibrium defect concentration given by the environmental conditions is found firstly on the surface. The diffusion of defects into the bulk or the other way round (in case of crystal healing) is time-consuming and temperature-dependent [50]. Amongst other crystal defects, two defects are standing out related to their frequency and importance: oxygen vacancies and titanium interstitials. As demonstrated in Table 2.4, defects seem to cause distinct levels in the bandgap. However, in many experiments an exponential distribution of trap states is found in undoped but defective TiO_2 [739, 910–914]. This originates from the fact that point defects are often in the neighborhood of extended defects such as surfaces, grain boundaries, stacking faults, or agglomerations of point defects which affect their level energies.

Oxygen vacancies: An intensively discussed defect is the oxygen vacancy which is an electron donor and hence, TiO_2 appears mostly as an n-type semiconductor [203, 769, 915, 916]. In perfectly stoichiometric TiO_2, no mobile electrons exist in the conduction band at any temperature below the melting point due to the large bandgap. However, oxygen vacancies are electron donors with donor states one- or two-hundred milli-electron volts below the CBM [197]. Electrons rising from oxygen vacancy donor sites might be linked to the oxygen vacancies forming helium-like donor sites or they convert Ti^{4+} into Ti^{3+} anions [197]. Thus, reducing TiO_{2-x} to some extent makes it applicable as a transparent oxide [917]. Heat-treatments in hydrogen

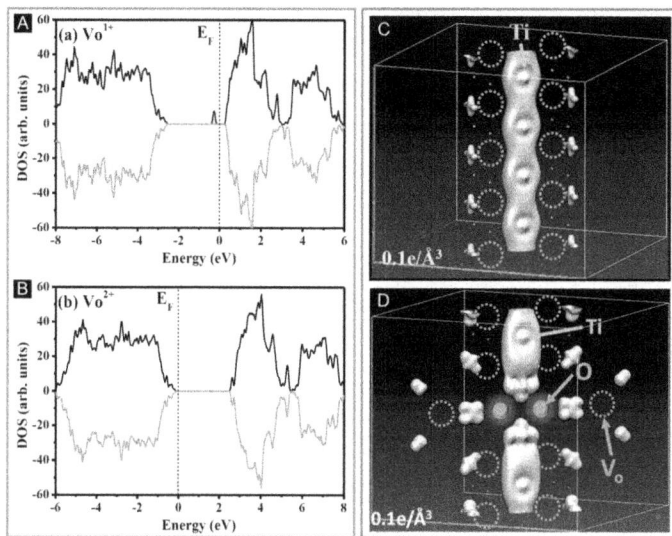

Figure 2.13: *Calculated total density of states for single (A) and double (B) ionized oxygen vacancies in rutile TiO₂ calculated by $LDA+U^d+U^p$ using $U^d = 8$ eV and $U^p = 6$ eV. Calculated charge density distribution of stringed oxygen vacancies (green cycles) pairs along the [001] direction (C) and the same structure with a missing pair (D) displayed by the iso-surface. Details found in [204]. Figures adapted from [204].*

plasma [206] or in a rapid annealing processing (RTP) oven [918, 919] as well as electrochemical [920, 921] and chemical reduction treatments [208, 292, 922, 923] are employed to reduce TiO₂ precisely. Below 1100 °C, rutile is not reduced significantly in an evacuated RTP oven [871]. If hydrogen is added, the temperature for an appreciable reduction decreases [206].

Ohmori et al. investigated the increase of the conductivity in relation to the oxygen vacancy density in a sprayed rutile shell structure quantitatively. Their maximum oxygen vacancy density was 6.5×10^{21} cm⁻³ resulting in a resistivity of 1.25×10^{-2} Ωcm [200]. On the one hand, oxygen vacancies provide up to two charges to the conduction band [199, 201]. On the other hand, oxygen vacancies lower the electron mobility because of impurity scattering [924]. However, dielectric materials with a high dielectric constant provide a high electron exchange frequency between oxygen vacancy sites [834]. Slightly reduced and often as-prepared rutile TiO₂ contains oxygen deficiency densities of more than 1.3×10^{19} cm⁻³ already [202]. Interestingly, the conduction performs a phase transition at 130 K in reduced rutile TiO₂ films. This effect is supposed to originate from an ordering at Ti^{3+} and Ti^{4+} ions [223, 925]. Below 40 K, the conductivity of slightly reduced TiO₂ is lower than for stoichiometric TiO₂, since there are almost no excited donors anymore and impurity scattering becomes dominant [202].

Titanium interstitials: Interstitial titanium contributes with three to four electrons to the conduction band [199, 201]. The enthalpy for oxidation is between 2.2 and 3.3 eV and depends on the crystallinity [927, 928]. The enthalpy for reduction is between 9.3 and 10.6 eV and hence, the annihilation of a titanium interstitial is much easier than its creation [928–932].

61

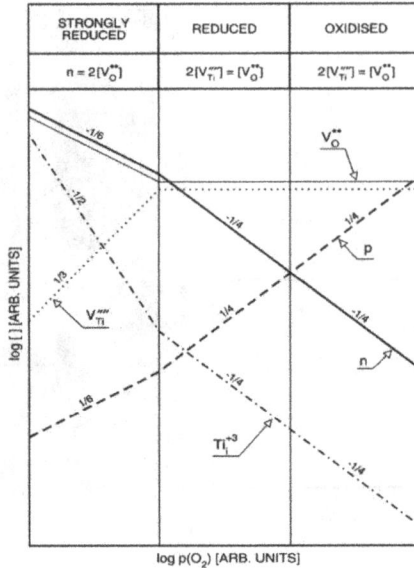

Figure 2.14: *Simplified temperature-dependent ion and electronic defect diagram in dependence of the oxygen partial pressure $P(O_2)$ and based on charge neutralities in undoped TiO_2. Reprinted with permission from [926]. Copyright (2006) American Chemical Society.*

As a consequence, the generation of titanium interstitials is dominant at very high annealing temperatures, whereas at lower annealing temperatures oxygen vacancies prevail [209]. Often, titanium interstitials are formed by Ti^{3+} ions causing states within the bandgap [933].

Electronic Properties Resulting from Specific Dopants

There is a wide range of dopants available for TiO_2 [953]. Employing specific dopants, the conductivity of TiO_2 can be increased or reduced and even the kind of majority charge carriers can be turned from electrons into holes. In this work, the electronic properties were mainly controlled by defect doping and hence, impurity doping is not discussed in detail. However, it has to be considered that atoms from the substrate and the gaseous environment might diffuse into the TiO_2 structure during high-temperature treatments. This kind of diffusion was already discussed in Section 2.3. Related to applied substrates and annealing atmospheres, a rough overview of silicon, gold, platinum, nitrogen, and hydrogen doping is given here. For the sake of completeness, the most important dopants used for highly conductive TiO_2 such as niobium and tantalum are mentioned [954, 955]. An overview of a wide range of elementary dopants influencing the conductivity of TiO_2 is given in Figure 2.15. Based on charge neutrality, the approximated equilibrium is given by

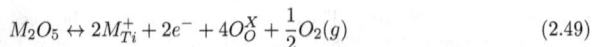

$$M_2O_5 \leftrightarrow 2M_{Ti}^+ + 2e^- + 4O_O^X + \frac{1}{2}O_2(g) \tag{2.49}$$

for pentavalent metal ions resulting in n-type doping [50, 909] and

sample	defect	energy level	reference
rutile TiO_2, bulk	O-vacancy, shallow	E_c- 0 to 200 meV	[797, 834, 934, 935]
rutile TiO_2, bulk	O-vacancy, deep	E_c- 600 to 750 meV	[797, 834, 934, 935]
rutile TiO_2, bulk	O-vacancy, deep	E_c- 750 to 1180 meV	[936]
rutile TiO_2, bulk	Ti-interstitial	E_c- 7 to 80 meV	[880, 928]
rutile TiO_2, bulk	Ti_2O_5	E_F- 600 meV	[937]
rutile TiO_2, (110) surface	Ti^{3+}, interstitial	E_F- 850 meV	[938]
rutile TiO_2, (110) surface	Ti^{3+}, interstitial	E_F- 700 meV	[939–941]
rutile TiO_2, (110) surface	Ti^{3+}-pairs	E_c- 700 meV	[942]
rutile TiO_2, (110) surface	Ti^{3+}-pairs	E_F+ 600 meV	[937]
rutile TiO_2, (110) surface	O-vacancy, deep	E_c- 300 meV	[943]
rutile TiO_2, (110) surface	hydroxyl group	E_c- 1600 meV	[944]
rutile TiO_2, (110) surface	hydroxyl group	E_c- 1200 meV	[944]
TiO_2, polycrystalline film	before annealing	E_c- 510(40) meV	[945]
TiO_2, polycrystalline film	after annealing	E_c- 620(10) meV	[945]
rutile TiO_2, polycrystalline	O-vacancy, shallow	E_c- 190 meV	[209]
rutile TiO_2, polycrystalline	Ti-interstitial	E_c- 500 meV	[209]
rutile/anatase TiO_2, polycrystalline		E_c- 300 and 360 meV	[946]
rutile TiO_2, nanorods	O-vacancy, deep	E_c- 730 meV	[918, 947]
rutile TiO_2, nanorods	O-vacancy, deep	E_c- 1180 meV	[918, 947]
rutile TiO_2, nanorods	hydroxyl group	E_v- 700 meV	[918]
rutile TiO_2, nanorods	hydroxyl group	E_v- 2600 meV	[918]
rutile TiO_2, nanotubes	O-vacancy, deep	E_c- 750 to 1180 meV	[203]
anatase TiO_2, bulk	O-vacancy, shallow	E_c- 300 meV	[933]
anatase TiO_2, bulk	Ti^{3+}-OH	E_c- 400 meV	[933]
anatase TiO_2, bulk	O-vacancy, deep	E_c- 700 meV	[933]
anatase TiO_2, bulk	Ti^{3+}, interstitial	E_c- 800 meV	[933]
anatase TiO_2, polycrystalline	Ti^{3+}, interstitial	E_c- 500 meV	[948]
anatase TiO_2, polycrystalline	O-vacancy, deep	E_c- 2000 meV	[948]
anatase TiO_2, (101) surface	Ti^{3+}, interstitial	E_F- 1000 meV	[939–941]
anatase TiO_2, powder	O-vacancy, deep	E_c- 750 to 1180 meV	[478, 949–951]
anatase TiO_2, N-doped nanobelts	Ti^{3+}, interstitial	E_c- 1280 meV	[952]
anatase TiO_2, N-doped nanobelts	Ti^{3+}, interstitial	E_c- 1200 meV	[952]

Table 2.4: *Electronic states of defects in different TiO_2 samples.*

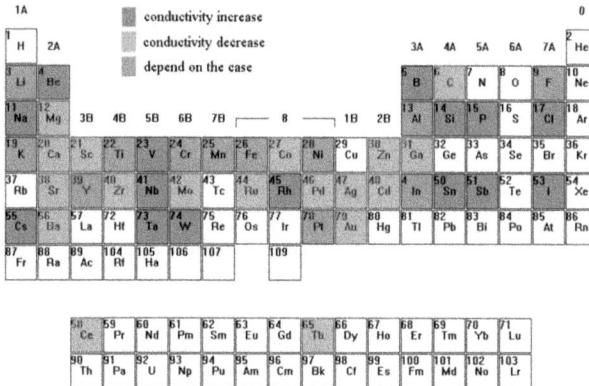

Figure 2.15: *The periodic table showing which elementary impurity increases or decreases the conductivity in TiO_2. Figure adapted from [900].*

oxide formula	TiO_2	$TiO_{1.9995}$	$TiO_{1.995}$	$TiO_{1.75}$	Ti_2O_3	TiO
resistivity	$<10^{10}\,\Omega cm$	$10\,\Omega cm$	$1.25\,\Omega cm$	$0.01\,\Omega cm$	$1.06\,\Omega cm$	metallic

Table 2.5: *Resistivity of the different $Ti_xO_{2\text{-}y}$ oxides adapted from [205].*

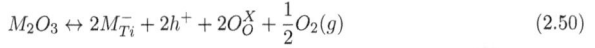

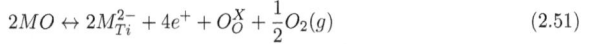

$$M_2O_3 \leftrightarrow 2M_{Ti}^- + 2h^+ + 2O_O^X + \frac{1}{2}O_2(g) \tag{2.50}$$

$$2MO \leftrightarrow 2M_{Ti}^{2-} + 4e^+ + O_O^X + \frac{1}{2}O_2(g) \tag{2.51}$$

for tri- and divalent cation doping resulting in p-type doping [956–961]. A detailed overview is given by Grant [197].

Nitrogen: Some of the investigated rutile TiO_2 nanorods were exposed to nitrogen at elevated temperatures and hence, nitrogen atoms might have been diffused into the nanocrystals resulting in changed electronic properties. Nitrogen doping results in N^{3-} species forming nitrogen $2p$ states a few hundred milli-electron volts above the VBM in rutile and anatase TiO_2 [270, 462, 962], e.g. 700 [948] and 750 meV [963]. Hence, nitrogen dopants cause acceptor levels and the TiO_2 becomes more p-type. However, the electron transfer between TiO_2 and FTO is improved by nitrogen doping [964]. Additional states are related to Ti^{3+} which is created during the thermally activated doping [462]. In addition, the implementation of nitrogen into TiO_2 promotes the creation of oxygen vacancies and titanium $3d$ states within the bandgap. The increased number of oxygen vacancies contributes significantly to the 1×2 reconstruction of the rutile (110) surface, which affects electronic surface states [462]. However, nitrogen which is induced at high temperatures is able to substitute oxygen ions close to vacancies creating vacancy-nitrogen complexes with no intra-bandgap states [965]. This results in a decreased amount of trap states, which improves the trapping/detrapping-related electron diffusion in TiO_2. For high nitrogen densities and in certain process conditions, even highly conductive Ti–N complexes appear [283].

Hydrogen: Hydrogen is used to reduce TiO_2 and hence, the resulting states after thermal hydrogen treatment correspond usually to the properties described for defect doping in TiO_2. But hydrogen links also to TiO_2 and forms TiH or $TiO_{2\text{-x}}H_x$ above 400 °C [966]. As a consequence, Ti^{3+} is created providing the same intra-bandgap states as described for doping with oxygen vacancies. The intra-bandgap state introduced by the TiH constellation is slightly lower than the Ti^{3+}-derived state. In general, hydrogen implantation results in deep trap states between 0.92 and 1.37 eV below the CBM and electron densities of more than $10^{20}\,cm^{-1}$ are observed [966]. However, also shallow traps with a depth of only 20 meV acting as donors are observed in hydrogen-doped TiO_2 [871]. By intense hydrogen doping, the bandgap could be reduced drastically and absorption takes place in the visible light or even in the NIR regime [947, 966]. If the reduction does not result in the release of water molecules as it happens likely on the surface, excess protons and electrons are left in the TiO_2. The H^+ ions create Ti–OH species causing additional states and contributing to hydrogen release during current stress as it will be discussed later in Section 2.8 [933]. Hydrogen which is implemented during the fabrication process is removed by oxygen annealing at 800 °C [871].

Carbon: Often, a carbon-containing contamination exists on the sample surface. In particular, the absence of oxidation during oxygen-free post-annealing processes prohibits the reaction of carbon-containing adsorbates into volatile organic compounds. As a consequence, carbon diffuses into TiO_2 and intra-bandgap states such as carbon $2p$ are created [967]. Wu et al. observed an intra-bandgap state 0.46 eV above the VBM of anatase TiO_2 [968]. Applying *ab initio* density functional theory Wang and Lewis obtained a bandgap of 2.35 and 2.76 eV for a dopant content of 5% carbon in rutile and anatase TiO_2, respectively [969]. For anatase TiO_2, the experimental results confirm this calculated bandgap energy [970]. Similar to hydrogen doping, an absorption of NIR is achieved by intensive carbon implantation [967]. Reducing TiO_2 carbothermally results in titanium carbide (TiC) [971]. Compared to rutile TiO_2 (10^{22} Ωcm [936]), the resistivity of TiC (56×10^{-6} Ωcm [972]) is significantly lower.

Fluorine: Fluorine doping might appear during post-annealing of nanorod arrays grown on FTO. The effect of fluorine doping is similar to nitrogen or carbon doping: An F⁻ anion substitutes an O^{-2} anion reducing a Ti^{+4} ion into a Ti^{+3} ion [933]. This generates additional states in the bandgap and additional absorption in the visible light regime appears [948, 973]. Fang et al. was able to reduce the bandgap in rutile TiO_2 nanorods from 3.05 to 2.58 eV by a controlled increase of the fluorine concentration achieving electron densities of up to 5.92×10^{19} cm^{-1} [974].

Tin: Similar to fluorine, the insertion of tin results in the substitution of Ti^{4+} by Sn^{4+} ions, especially in oxygen-rich rutile and anatase TiO_2 [975], and has to be considered in case of post-annealing of TiO_2 nanostructures on FTO. Sn^{4+}-derived states are located 0.4 eV below the CBM [976]. Density functional theory calculations predict a drop of the conduction band by the above-mentioned substitution [975]. Xu et al. and Sun et al. reported a significant increase of the charge carrier density in tin-doped rutile TiO_2 nanorod arrays [977] from 5.5×10^{17} to 1.25×10^{19} cm^{-3} [978]. By extensive doping, Duan et al. lowered the conduction band edge by 0.45 eV and achieved a donor density of 2.18×10^{19} cm^{-3} [979]. But they described tin dopant sites also as charge traps.

Silicon: There are a lot of reports describing the effect of silicon-doped TiO_2 on the photo-catalytic activity qualitatively. Similar to fluorine, silicon atoms diffuse from substrates such as glass or pure silicon into TiO_2 at elevated temperatures. Especially the substitution of titanium atoms in oxygen-rich TiO_2 occurs very likely [980]. Hence, interstitial Si^{4+} and Si–O–Ti bonds are found frequently in silicon-doped TiO_2 [419, 981]. This substitution decreases the bandgap and Yang et al. reported a narrowing of 0.25 and 0.2 eV in anatase and rutile TiO_2, respectively [980]. Vice versa, titanium and oxygen atoms diffuse into the silicon substrate and cause additional excitable states which are discussed in more details in Chapter 9.

Niobium: A typical dopant for highly conductive TiO_2 is niobium [954, 955]. Typically, these layers are used as transparent conductive oxides (TCOs) [982]. In addition, niobium-doped rutile films are used for photocatalytic applications [983]. But also niobium-doped nanorods employed in dye-sensitized solar cells are of great interest [984].

Niobium dopants substituting a titanium ion have the oxidation state 5+ and are able to release an electron into the conduction band with an activation energy of only 4.4 meV [871, 933]. Band calculations predict a strong hybridization of niobium-4d orbitals with titanium-3d orbitals resulting in a d-nature conduction band. Here, no states within the bandgap are formed. Thus, the niobium-doped TiO_2 is still transparent in the visible regime, but it has a high electron density at the same time [985]. Both properties are required for TCO applications and were confirmed by Hirose et al. [50, 986]. However, above a critical doping density, crystal defects result in states within the bandgap compensating the advantages needed for TCO applications [987]. Furubayashi et al. achieved an electron density of more than 10^{21} cm^{-3} in combination with $\epsilon_\infty = 5.9$ and an effective mass of roughly one m_e [954]. They were able to reduce the resistivity of rutile TiO_2 applying niobium doping by more than three orders of magnitude [841, 954, 988].

In anatase TiO_2, the effect is similar but less intense. Mulmi et al. determined a resistivity of $5 \times 10^{-1}\,\Omega\,cm$, Forro et al. reported a resistivity of $1 \times 10^{-1}\,\Omega\,cm$ and an electron mobility of about $600\,cm^2/Vs$ at $50\,K$ in single-crystalline anatase TiO_2 [884, 989]. For anatase TiO_2 films deposited on different substrates such as $SrTiO_3$ (100), GaN(0001), and $LaAlO_3$ (LAO), resistivities between 3 and $8.1 \times 10^{-4}\,\Omega\,cm$ and Hall mobilities of up to $22\,cm^2/Vs$ were observed [954, 990, 991]. The mobility in electrospun anatase TiO_2 nanowires was raised to $0.16\,cm^2/Vs$ by niobium doping which is only two orders of magnitude below the mobility in macroscopic niobium-doped anatase single crystals [992].

An interesting feature is that the activation energy determined from the Hall coefficient and the conductivity of niobium-doped rutile TiO_2 is larger compared to simply reduced TiO_2 at temperatures below $60\,K$. At higher temperatures, there are no discrepancies between doped and reduced TiO_2 samples [880]. This behavior is explained with the support of electron spin resonance (ESR) measurements to determine the temperature-dependent dominant donor sites. At low temperatures, Nb^{4+} ions act as donors in the niobium-doped TiO_2 and Ti^{3+} ions are donors in the irradiated or reduced TiO_2. In reduced TiO_2, different Ti^{3+} donors with different activation energies appear. There are Ti^{3+} ions on regular sites, interstitial Ti^{3+} ions, and Ti^{3+} ions on regular lattice sites but influenced by neighboring lattice defects. In niobium-doped TiO_2, most of the Ti^{3+} ions seem to be deep traps on regular lattice sites without the influence of nearby defects at $4\,K$. Hence, electrons in niobium-doped TiO_2 are released in the conduction band at lower temperatures compared to reduced TiO_2. Subsequently, they become trapped by these deep "regular" Ti^{3+} ions and need to be released with a relatively high activation energy. In reduced TiO_2, electrons are excited from different kind of Ti^{3+} ions and some of these donors have lower activation energies [880].

p-type doping: p-type doping of TiO_2 is motivated by the fabrication of FETs [64], OLEDs [63–66, 993, 994], photodetectors [53], and "invisible electronics" [995]. From the considerations above, it is clear that n-type doping is more common than p-type doping in TiO_2. In the same way as oxygen vacancies result in n-type behavior, non-stoichiometric excess oxygen causes p-type conduction [996]. As a consequence, the conductivity is increasing with increasing oxygen partial pressure during post-annealing of TiO_2 [909]. Usually, p-type doping in metal oxides is done by adding a metal cation with a closed d^{10} orbital having a similar energy than the $O2p^6$ orbital in order to form a degenerated hybridized $O2p$ level [806]. Although copper and silver satisfy these requirements, structural properties have to be considered to avoid a bandgap narrowing by the interaction between d^{10} cations [753, 805]. A detailed description of p-type doping of metal oxides is given by Banerjee and Chattopadhyay [805]. Sufficient p-type doping of TiO_2 is achieved with Al^{3+}, Cr^{3+}, Fe^{3+}, Ni^{2+}, or Co^{+2} [63, 997–1003].

Other important dopants: Similar to niobium, tantalum is used to achieve high electronic conductivities of TiO_2 films and rutile nanorods [954, 955, 1004]. In highly transparent anatase TiO_2 films, resistivities of less than $1 \times 10^{-3}\,\Omega\,cm$ were achieved [955, 1005]. Employing iron dopants a donor density of $4.55 \times 10^{22}\,cm^3$ was realized [1006]. Besides tantalum and iron, tungsten was applied in rutile TiO_2 nanorods in order to increase the conductivity as well [1007, 1008]. Vanadium-doped TiO_2 is used to enhance the photocatalytic activity of TiO_2 [983]. And there are other metal dopants used for various purposes in TiO_2 such as chromium, manganese, cobalt, molybdenum, rhodium, indium [1009], cadmium [1009], copper, yttrium, zirconium [467, 1010–1015], zinc [1016, 1017], antimony [1018], europium [1019], silver [1020, 1021], neodymium [1022], ytterbium [1023], platinum [1024], palladium [1024], lead [1025], and some lanthanides [1026] such as cerium [1012], samarium [1027], and erbium [1028]. They act either as an electron donor or narrow the bandgap. Besides the above mentioned non-metallic dopants, sulfur, bromine, and chloride are used to enhance the absorption of visible light of TiO_2 [467, 1029–1035]. Similar to charged hydrogen ions, lithium ions are known to reduce TiO_2 as well by creating Ti^{3+} species [841, 936, 1036–1038].

Multi-element doping: The semiconductor type is influenced by employing different dopants. In rutile TiO_2 films, terbium in combination with palladium results in p-type and europium and palladium in n-type doping [1039]. Using the right dopant concentrations, both films become highly transparent and conductive. $Cu_{2-x}O$ doping narrows the bandgap and the TiO_2 becomes colored [1040]. A large variety of intra-bandgap states are realized by combining nitrogen and transition metal doping [1041].

Effect of ultra-high doping and defect concentrations

Mott Transition of Impurity-Band Conduction: The bandgap in strongly reduced TiO_2 is much smaller and visible light absorption is enhanced. Thus, the oxide becomes dark and is often called black TiO_2 [1042]. Chen et al. found a bandgap of only 1.54 eV in black TiO_2. At a certain donor density, the wave functions of the donor state begin to overlap sufficiently and the conductivity is increased significantly. This metal-nonmetal transition of the impurity band conduction is known as Mott transition° [765, 861, 1043, 1044]. The critical donor concentration is given by $N_d^{1/3} a_B \approx 0.25$ where $a_B \propto \epsilon_s \propto 1/m_e^*$ is the effective Bohr radius° of a hydrogen-like donor state [134, 1045, 1046]. Assuming an average dielectric constant of 31 [650, 651], a_B is 15 Å, the effective mass is $m_e^* \approx m_e$, and N_d is calculated to be 5×10^{18} cm^{-3} in anatase TiO_2 [134]. The band conduction is related to the $3d$ electrons and the band is situated directly below the CBM. This bears a strongly temperature-dependent conductivity [954, 986]. For rutile TiO_2, no Mott transition is predicted. Assuming an average dielectric constant of 100 [197], a_B is only 2.6 Å which equals the ionic distances [134]. The effective mass is $m_e^* \approx 25m_e$ and its high value results from the flat band structure in rutile and polaronic effects [880]. Due to the small effective Bohr radius, there is no overlap of the donor electron wave functions in rutile and thus, the Mott transition fails to appear [8].

Change of crystal structure: For very high defect or dopant densities, the crystal structure deviates from the pure crystal structure forming new kinds of morphologies. This results in a change of the electronic band structure and thus, the effective mass is changing as well [1047]. For instance, a transition from insulator to metal is observed for the reduction of TiO_2 to Ti_2O_3 and TiO is even superconducting [853]. Additionally, the formation of the Ti_2O_5 Magnéli phase provides an insulator-metal transition [937, 942]. Naldoni et al. observed a completely disordered shell around black (highly reduced) anatase nanoparticles [1048].

Formation of traps: Exceeding a certain niobium density, it is observed that the electron density decreases [987]. This is due to interstitial oxygen ions interacting with niobium atoms which results in intra-bandgap states that act as electron traps [987].

Burstein-Moss effect: At a certain electron donor concentration, the Fermi level shifts above the CBM. As a consequence, the conduction band is filled with electrons level by level. Hence, electrons that are excited from the VBM need higher energies to achieve free states in the conduction band as the bandgap energy E_g. This results in a blue shift of the absorption spectra [1049, 1050]. Furthermore, the density of mobile electrons depends on the DOS above the CBM.

Plasmon effect: The plasma frequency depends on the electron density in the conduction band. Hence, for sufficiently high electron densities, the plasma frequency is shifted into the NIR regime and the reflectivity and absorption for this kind of radiation is increased significantly. This effect limits the application of highly doped TiO_2 in transparent electrical devices [50].

2.6.7 Electronic States and Conductivity on Surfaces and Grain Boundaries in TiO_2

The electronic properties in TiO_2 are connected intimately with the inner structure. This fact was reviewed comprehensively by Chen et al. [1051]. Surfaces and grain boundaries represent extended two-dimensional defects. Their properties differ from the bulk and these deviations are important to understand the properties of nanoparticles that are discussed in the next section. On surfaces, the electronic states are affected by crystal relaxation, a specific surface point defect density, and adsorbates. Trivially, the importance of surface properties is increasing with increasing surface-to-volume ratio of the TiO_2 structure. In addition, they play a crucial role in the charge transport between electrodes and TiO_2 (as discussed in Section 2.7) as well as for the charge transfer between single-crystalline TiO_2 grains. Treating amorphous TiO_2 structures as the most "grainy" form of TiO_2, the different electronic properties of amorphous and single-crystalline TiO_2 become comprehensible. As a consequence of all the different deviations from a highly crystalline stoichiometric bulk, real TiO_2 does not show one or a finite number of discrete trap levels only, but rather an exponential distribution of trap energies below the CBM or above the VBM [1052–1058]. The exact origin of this kind of distribution is not clear, so far [1053, 1059]. However, the exponential trap distribution can be superposed by discrete surface states [913]. An experimentally access to the interface state density is proposed by Terman [1060]. Adsorbates on surfaces usually act as electron traps or donors and hence, they also change the local charge density and influence space-charge regions.

Properties of TiO_2 Surfaces

The surface science of TiO_2 is a complex story and has been discussed extensively in the past [295, 749]. Here, only the most important features for TiO_2 surfaces are described.

The atomic arrangement on surfaces is often affected by relaxation of surface atoms or reconstruction [1061]. This results in states within the bulk bandgap [900, 1061–1064]. In particular, the effect on the intra-bandgap states caused by defects and impurities differ from the bulk. Similar to the bulk, oxygen vacancies are the most important crystal defects on TiO_2 surface at low oxygen pressure causing states within the bandgap of bulk TiO_2 [940, 941]. If the reduced rutile (110) TiO_2 surface reacts with molecular oxygen, interstitial titanium is created likely causing a state $0.85\,eV$ below the Fermi level [938, 939]. This effect was also observed in vanadium-, chromium-, manganese-, and iron-doped TiO_2 where surface states are centered $2.0\,eV$ below the CBM and bulk states range from the VBM to roughly $1.9\,eV$ below the CBM [935].

The localization of defects on the surface cause inhomogeneous electronic properties in TiO_2 [1065], especially in nanostructures with a large surface-to-volume ratio. In slightly reduced rutile TiO_2 nanorods, a trap-limited conduction at the surface and a trap-free conduction in the core is suggested [1066]. In black TiO_2 nanoparticles, the surface becomes metallic and most of the current is focused there [1067]. The same effect appears in the subinterface region in crystalline reduced TiO_2 covered with amorphous TiO_2 and results from electronic interface reconstruction [1067].

Effect of Adsorbates on the Electronic Properties of TiO_2

In real TiO_2 systems and in particular, under atmospheric measurement conditions, the surface is occupied by adsorbates such as elementary gas atoms, water, or organic molecules. Adsorbates on grain boundaries and surfaces play an important role for the detection of gases such as hydrogen. The adsorption of atoms or molecules from the gas phase can add further donor or acceptor states resulting in a change of the electronic conductivity. A detailed review on adsorbates on TiO_2 is given by Linsebigler et al. [1068]. Here, only the most important adsorbates for this thesis are introduced.

Oxygen: One of the most important adsorbates is oxygen, which contributes to the annihilation of oxygen vacancies. The adsorbed oxygen is reduced to the superoxide anion O^{2-} and hence, it was observed that the electronic conductivity depends on the partial pressure of O_2 [734, 869, 1069]. Rutile adsorbs oxygen better than anatase [1070]. On rutile (110) TiO_2, it is adsorbed between 105 and 400 K [1071]. The highest adsorption rate appears at 348 K [1072]. In particular, the enthalpy for reduction of nanoporous TiO_2 is significantly lower compared to bulk single-crystalline rutile TiO_2. Hence, in contrast to bulk rutile TiO_2, nanoporous TiO_2 shows an oxygen pressure dependence at temperatures below 200 °C [1069]. In coarse-grained TiO_2, which is permeable to oxygen, the depletion of oxygen vacancies is so significant that a transition from n-type to p-type behavior takes place at a high partial oxygen pressure [928]. According to their nature as electron donors, the reduction of oxygen vacancies results in a very low density of mobile electrons. Hence, the conductivity is decreasing with increasing oxygen pressure [840]. At high oxygen pressures, this could lead to the effect that ion migration contributes much more to the charge transport than the drift of electrons [928]. The ion migration is supported by the presence of the huge amount of grain boundaries in nanoporous TiO_2 films [1073]. However, adsorbed oxygen which is not used for vacancy healing act as an electron trap, because the conduction band of TiO_2 (-4.3 eV) is much higher than that of molecular oxygen (-5.7 eV) and hence, electrons move likely from TiO_2 to the adsorbed O_2 molecules [38, 792, 1074–1082]. Thus, a change of the oxygen concentration in the environment changes the conductivity of the oxide [1083]. As a consequence of the decreased number of free electrons, the depletion layer is increased [1077]. The effect of oxygen adsorption is enhanced if niobium, chromium, or tin are added which is used for gas sensing [1084].

Water: Water is present in all measurements in atmospheric conditions and provides an additional transport path for charge carriers [769]. Electrons can tunnel into the donor states of the water molecules and move via hopping to the anode [1085–1087]. Hence, water infiltrating polycrystalline TiO_2 along grain boundaries reduce the high electronic resistance of these boundaries simply by adding an additional tunneling path [769]. Alivov et al. investigated the effect of water vapor pressure on the electronic properties. Within the investigated pressure range, they observed an increase of the electron density by two orders of magnitude with increasing pressure, but a decrease of the electron mobility by a factor of two [1077]. The additional current promoted by adsorbed water is called the Grotthuss effect and has to be avoided if only the intrinsic properties of TiO_2 should define the charge transport [1088]. The pressure in the applied vacuum measurement chamber is decreased to roughly 1×10^{-6} mbar. At this pressure and at room temperature, the physisorbed water is removed.

Hydroxyl groups: H_2O reacts with oxygen vacancies and is chemisorbed partly by forming Ti^{+3}-OH hydroxyl groups [38, 734, 1076, 1089, 1090]. The chemisorbed hydroxyl groups remain due to the strong binding forces and long surface relaxation times [769, 1091]. It is reported that samples need to be kept at least for one hour at 450 K to get rid of the hydroxyl groups [769]. On rutile (110) TiO_2, hydroxyl groups remain below 400 K [734].
It was shown for rutile (110) TiO_2 surfaces that these hydroxyl groups trap electrons with trap levels between 1.2 and 1.6 eV below the CBM. In contrast to shallow trap levels introduced by oxygen vacancies, these trap levels do not contribute to the charge transport [769, 944, 1092, 1093]. However, in case of a dense layer of hydroxyl groups, electrons are able to tunnel between the electron states, which might increase the conductivity [1085–1087].
Nevertheless, it has to be considered that the number of electrons donated by oxygen vacancies decreases with increasing hydroxyl groups. Since oxygen vacancy levels are much closer to the

Fermi level in TiO_2, the annihilation of these vacancies decreases the conductivity of TiO_2 much more than it is increased by tunneling between hydroxyl molecules [769, 770, 944, 1092, 1093]. As mentioned above, Ti^{+3} centers are formed during hydroxylation and these defects support additional states with fast trapping and detrapping kinetics supporting the electron transport strongly [487, 1093–1097]. Elementary hydrogen, which is created during the reaction with water and TiO_2, binds to bulk lattice sites forming oxygen vacancies that provide the previously discussed shallow donors [792, 1040, 1076].

Carbon: Carbon on TiO_2 surfaces occupies substitutional and interstitial lattice site and creates additional localized states. As a consequence, an effective bandgap of only 1.7 eV is observed for TiO_2 nanotube arrays with surface carbon doping [1098].

Often, not only single atoms or simple molecules are applied as adsorbates, but larger complexes. In particular for photocatalysis applications, adsorbate particles made of gold [1099, 1100], graphene [1100], NiO [1101], WO_3 [1101] or transition metal dichalcogenides [1102] are employed. The drift mobilities in rutile TiO_2 nanorods, which is said to result from deep surface traps [1103, 1104], are improved by a surface passivation using octadecylphosphonic acid (ODPA). For rutile TiO_2 nanorods, Mohammadpour et al. observed an increase of the electron mobility from $1 \times 10^{-6}\,cm^2/Vs$ to $1 \times 10^{-3}\,cm^2/Vs$ and for the hole mobility from $8.2 \times 10^{-5}\,cm^2/Vs$ to $7.1 \times 10^{-3}\,cm^2/Vs$ [1105]. To increase the absorption and charge transport in hybrid solar cells special dye molecules are placed on the surface of TiO_2 [1106].

Properties and Charge Transfer Across Grain Boundaries

Grain boundaries are similar to surface bilayers and appear in polycrystalline films as well as in many TiO_2 nanostructures. Due to enhanced boundary scattering [1051], the electron mobility in polycrystalline and porous TiO_2 is significantly lower than in single-crystalline TiO_2. Hence, the electron Hall mobility at room temperature in polycrystalline thin rutile and anatase TiO_2 films is only $0.1\,cm^2/Vs$ and 0.1 to $4\,cm^2/Vs$, respectively (compared to up to $10\,cm^2/Vs$ and $550\,cm^2/Vs$ in bulk crystals, respectively) [765, 834, 884].

The effect of a grain boundary is not limited to the geometric position of the boundary only. Similar to a surface [1107, 1108], interface [1109, 1110], or dislocations [1111, 1112], grain boundaries are surrounded by a space-charge region in ionic crystals. It is assumed that negative titanium vacancies are dominant on grain boundaries in TiO_2 [1113]. Thus, mobile positive charges such as holes, oxygen vacancies, or titanium interstitials are attracted and electrons are repelled from grain boundaries. As a consequence, a space-charge region in the close neighborhood of grain boundaries is created. The p-type character and low electron mobility enhance the reflection of electrons on grain boundaries and reduce the electron conductivity perpendicular to the boundaries [209, 1113].

The decreased electron mobility in polycrystalline films is well known. In TiO_2 with an electron density of $10^{19}\,cm^3$, an electron mobility of $20\,cm^2/Vs$ and $4\,cm^2/Vs$ was reported for an anatase single crystal and a polycrystalline film with a grain diameter of 30 to 40 nm, respectively [134, 884]. The drift electron mobility of photo-excited electrons in porous TiO_2 determined with time of flight measurements is 10^{-4} to $10^{-7}\,cm^2/Vs$ [739]. Thereby, the mobility in weak fields below $55\,kV/cm$ is 10^{-6} to $10^{-7}\,cm^2/Vs$ and as already mentioned for bulk TiO_2, the mobility depends strongly on the fabrication technique [1114]. For such structures, time of flight measurements give more reasonable values than terahertz spectroscopy, since terahertz spectroscopy probes only a small volume around each electron. Thus, many electrons might not cross any grain boundary and the drift mobility is overestimated.

Even in highly niobium-doped TiO_2 films, scattering on grain boundaries exceeds scattering at impurities [1115]. Besides scattering, grain boundaries provide shallow trap states causing additional scattering and trapping [1116]. Related to the band structure, it is reasonable that the effect of grain boundaries on electrons depends on the crystal facets, which are forming the grain boundary. In contrast to the mobility in single-crystalline rutile TiO$_2$, the mobility in bulk porous rutile TiO$_2$ is 10^{-3} cm^2/Vs only [739]. Later, Hendry et al. improved the fitting by applying the Maxwell-Garnett effective medium theory for porous TiO$_2$ and confirmed the trends observed by Dittrich [1117–1121]. In addition, Hendry et al. were investigating the time-resolved mobility in porous films [1117]. They injected charge carriers via bandgap excitation into the conduction band. Within the first 10 ps, the electron mobility was found to be 10^{-2} cm^2/Vs. In steady-state conditions, grain boundary crossings and charge trapping influence the electrons and the mobility drops down to 7×10^{-6} cm^2/Vs [739, 1122–1124]. Furthermore, electrical field screening reduced the mobility in macroscopic devices in steady-state conditions [1117]. This experimental result emphasizes the importance of grain boundary scattering on charge transport in rutile TiO$_2$ [836, 1117].

In general, it is assumed that electrons tunnel through grain boundaries in porous or polycrystalline TiO$_2$ and hence, deviations from the Drude behavior appear [1125, 1126]. As a consequence, the complex optical conductivity of nanostructured TiO$_2$ is approximated with the Drude-Smith model [1125, 1127, 1128]. Here, the reciprocal total conductivity

$$\frac{1}{\sigma^*(\omega)} = \frac{b}{\sigma_f^*(\omega)} + \frac{1-b}{\sigma_t^*(\omega)} \qquad (2.52)$$

is given by the sum of the reciprocal conductivities of free ($\sigma_f^*(\omega)$) and tunneling ($\sigma_t^*(\omega)$) carriers. The volume fraction of grains compared to bulk TiO$_2$ is b. In particular, the conductivities are given by

$$\sigma_f^*(\omega) = \frac{q^2 \eta_e \tau_c / m_e^{**}}{1 - i\omega\tau_c} \qquad (2.53)$$

where η_e is the number of free charge carriers and τ_c is the collision time [1129–1131] as well as

$$\sigma_t^*(\omega) = \frac{\eta_t(qr_t)^2}{2k_bT\tau_t} \frac{i\omega\tau_t}{\ln(1 + i\omega\tau_t)}, \qquad (2.54)$$

where $\eta_t = N_F k_b T$ is the number of tunneling electrons, N_F is the DOS at the Fermi level, r_t is the tunneling distance, and τ_t is the tunneling time [1126, 1132, 1133]. Compared to other metal oxides, the difference between bulk and non-bulk mobilities such as in thin metal oxide layers used in solar cell devices is less distinct in TiO$_2$. For instance, Tiwana et al. found that TiO$_2$ has the highest device (0.017 cm^2/Vs, nanoporous films) and lowest bulk (1 cm^2/Vs, single-crystalline) mobility compared to ZnO and SnO$_2$ while SnO$_2$ has the lowest device (0.003 cm^2/Vs) and highest bulk (250 cm^2/Vs) mobility [889].

2.6.8 Electronic States and Conductivity in TiO$_2$ Nanoparticles

In this section, electronic properties of isotropic and anisotropic TiO$_2$ nanoparticles are discussed. In particular, quantum effects appear and influence the electronic properties of one-dimensional structures such as nanorods or zero-dimensional structures such as quantum dots. In nanoparticles, which are too large to show quantum confinement, localized states affect the electronic properties of nanostructured semiconductors [1134–1136]. Hence, size-effects depend on the density, energy levels, and distribution of these surface states [1052].

Electronic Properties of Nearly Isotropic TiO$_2$ Nanoparticles

As discussed above, dopants are often employed to increase the electronic conductivity and enhance the absorption in the visible regime. However, there are also ways to cause a blue shift of the absorption. One approach is the Burstein-Moss effect, which needs very high donor densities [50]. Another way is to decrease the size of particles until the band structure is stretched on the energy scale by quantum confinement [1046].

Quantum confinement: This effect appears typically for particles having a size in the order of the bulk excitonic Bohr radius [1137–1139]. Only rutile particles with a size in the order of 2.5 nm or smaller are expected to show quantum confinement effects [1140, 1141]. The bandgap broadening in nanoparticles via quantum confinement is based on the Brus effect. It can be calculated by

$$E_{g,rod} = E_{g,bulk} - \frac{1.786q^2}{4\pi\epsilon\epsilon_s}\frac{1}{r} + \frac{h^2}{m_0}\left(\frac{1}{m_e^*} + \frac{1}{m_h^*}\right)\frac{1}{r^2} - 0.248E_{RY} \qquad (2.55)$$

where $E_{g,bulk}$ is the bulk bandgap of TiO$_2$, r the radius of the nanoparticle, m_0 the rest mass of the electron, $m_{e/h}^*$ the effective mass of the electron/hole, and $E_{RY} = q^4/2\epsilon\hbar^2\pi^2 \cdot (1/m_e^* + 1/m_h^*)$ the Rydberg energy [1137, 1142–1146]. The second term displays the Coulomb energy [1137, 1142], the third term the localization energy [1068], and the fourth term the correlation effect [1138]. The expectations were satisfied by experimental results [1, 1141, 1147, 1148]. This effect is also observable in nanocomposite films [1149–1151].

Band bending: Usually, band bending appears on semiconductor surfaces. The radius-dependent potential for spherical particles can be written as

$$\phi(r) = \frac{1}{6L_D^2}\left[r^2 - 3r_w^2 + 2\frac{r_w^3}{r}\right], \qquad (2.56)$$

where r_w is the radius of the inner spherical region which is not affected by band bending, $L_D = \sqrt{\frac{\epsilon\epsilon_s k_b T}{N_d q^2}}$ is the extrinsic Debye length, and a is the particle radius ($0 < r_w < r < a$) [1152]. As soon as r_w becomes zero, the bandgap of the whole particle rises.

Surface states: In addition, the interplay between bulk and surface traps needs to be considered. For decreasing particle size, defects cause delocalization of molecular orbitals in the CBM generating shallow and deep traps. As a consequence, the bandgap energy is supposed to be lowered [1141]. Lin et al. showed that decreasing the diameter of anatase TiO$_2$ nanoparticles from 29 to 17 nm reduces the bandgap from 3.239 to 3.173 eV due to localized surface states [1141]. However, a further decrease of the diameter from 17 to 3.8 nm results in an increase of the bandgap from 3.173 to 3.289 eV, which might be related to quantum confinement although the size is supposed to be too large for such effects [1140, 1141].

Surface scattering: At room temperature, polaron scattering is dominated by bulk properties in single-crystal rutile TiO$_2$. In contrast, below 77 K surface scattering becomes important [1153].

Electronic Properties of Highly Anisotropic TiO$_2$ Nanorods

The effects discussed for isotropic nanoparticles appear in nanorods as well. However, they depend on the orientation. Effects resulting from quantum confinement and surface states are much more distinct along the short axis. In contrast, effects affecting the diffusion and drift of charge carriers such as the mean free path for momentum scattering or Anderson localization in doped or defective nanorods become more distinct with increasing length [71].

Quantum confinement: Peng and Li calculated the effect of quantum confinement on the bandgap in rutile TiO$_2$ nanorods [1140]. They predicted that quantum confinement appears for rods thinner than 50 nm and becomes significant for diameters less than 25 nm, which is larger than the dimensions predicted for quantum confinement in isotropic nanoparticles. To perform a quantitative assumption of the bandgap broadening they introduced fitting parameters α and β and suggested that the relation between the bandgap of thin nanorods $E_{g,rod}$, the bulk bandgap $E_{g,bulk}$, and the diameter of the rods d can be expressed as

$$E_{g,rod} = E_{g,bulk} + \frac{\beta}{d^\alpha}. \tag{2.57}$$

The last term, which describes the diameter-dependent gain of bandgap, is inversely proportional to the crystallographic orientation-dependent reduced effective electron mass [1140].

Band bending: The surface provides a different band structure compared to the bulk resulting in a space-charge region. In nanorods, the surfaces of opposing facets are very close and hence, the space-charge region can penetrate the whole rod [1066]. As a consequence, the bulk properties disappear completely in thin rods even without the influence of quantum confinement. The width of the space-charge region is calculated similarly to the width in macroscopic TiO$_2$ [1154, 1155]. The conduction band bending along the short axis without tip effects is given by

$$\phi(r) = \phi_s - \frac{1}{L_D^2}\left[\frac{1}{4}(a^2 - r^2) - \frac{r_w^2}{2}\ln\left(\frac{r}{a}\right)\right], \tag{2.58}$$

where the parameters are defined in the same way as in Equation 2.56 [1152]. A detailed model including tip effects was introduced by Bisquert et al. [1152].

Surface states: Often, the band bending is affected by surface states, adsorbate-induced surface depletion, and defect density gradients between the shell and core [1156]. This band bending facilitates or hinders the charge transport through the surface and causes an inhomogeneous current density inside the nanorods.

In rutile TiO$_2$ nanorods, the charge transport is mainly reduced by deep surface traps and hence, the electron transport follows a trap-assisted diffusion model with the diffusion constant

$$D = C_1\eta_t^{-1/\alpha+1/3} + \eta_e^{1/\alpha-1}, \tag{2.59}$$

where C_1 is a constant pre-factor and α ($0 < \alpha < 1$) is related to the shape of the distribution of the intra-bandgap states [1104, 1157]. Shallow traps resulting from oxygen vacancies are decreasing during oxygen annealing at sufficiently high temperatures [311, 1158]. However, the observed deep traps cannot be removed by a simple oxygen annealing completely [1104]. It is suggested that the deep traps result from surface reconstruction induced by Ti$_i^{3+}$ defects, which diffuse from the bulk to the surface at annealing temperatures larger than 400 K [311, 1159–1162]. Hence, oxygen annealing increases the density of deep surface traps and does not passivate the TiO$_2$ nanorod completely [1104]. Sheng et al. showed a method to passivate these surface traps by a wet-chemical treatment using H$_2$O$_2$ as reactant [1104, 1163]. Resulting H$^+$ ions are removed by an aqueous NH$_3$ solution [1104]. α values of 0.37 and 0.26 were determined for rutile TiO$_2$ nanorods with and without passivation treatment [1104]. Hence, the trap energy distribution is affected by the passivation treatment.

Grain boundaries: Another property of nanorod and nanotube arrays is their high crystallinity and low density of grain boundaries parallel to the long axis. It was already mentioned that grain boundaries in the electron pathway increase the resistance for electrons. As a consequence, the trap-free electron diffusion coefficient is much larger compared to nanoparticle films [1164].

Rods vs tubes: The effective electron drift mobility in rutile TiO_2 nanorods measured by time-of-flight (TOF) and space-charge-limited current (SCLC) experiments is $1.95 \times 10^{-5} \, cm^2/Vs$ for a free electron density of $1 \times 10^{15} \, cm^{-3}$ and a trap density of $3.5 \times 10^{16} \, cm^{-3}$ [1103]. The resistivity in single-crystalline rutile nanorods ($\rho = 1.4 \times 10^2 \, \Omega cm$ [1165]) is significantly lower than in polycrystalline nanotubes ($\rho = 1 \times 10^8 \, \Omega cm$ [1166]). The large difference cannot be explained by the smaller cross-section of the tubes, but it is likely a result of the high grain boundary density. This is emphasized by the observation of donor levels at 8 and 28 meV below the CBM, which provide more mobile electrons to the conduction band as the donor levels at 58 meV observed in single-crystalline rutile nanorods [1165, 1167]. Hence, the conductivity is increased by a larger number of electrons in the conduction band, but it is decreased significantly by a high scattering rate on grain boundaries at the same time [1167].

Electronic Properties in TiO_2 Nanoparticle Compounds

For nanoparticle compound layers, many properties discussed for grain boundaries can be adopted. In porous rutile TiO_2, the drift mobility is decreased by the barrier between the particles significantly ($<10^{-3} \, cm^2/Vs$) compared to a TiO_2 single crystal ($\approx 0.5 \, cm^2/Vs$) [848]. In general, the electron mobility is inversely proportional to the surface area in nanostructured metal oxide semiconductors (n-MOS) [1168]. This emphasizes the importance of surface effects on the electronic properties in nanostructured TiO_2. Since some of the various surface states are thermally excitable, the conductivity is much more temperature-dependent compared to bulk TiO_2 with the same defect density [911, 1062]. Due to the multiple trapping events on nanoparticles forming a percolation pathway between the electrodes, the electron mobility is much less compared to single-crystalline TiO_2 [1062]. For nanoparticles dimensions below roughly 25 nm, space-charge regions are much less distinct and almost depleted. Hence, their effect on electrons drifting in an external field is much less compared to bulk TiO_2 [1169].

The low transmission rates at the interfaces between nanoparticles cause an increased reflection of electrons. Because of that, self-trapping and long-lived polaron states in sufficiently small TiO_2 particles appear [1170–1173]. This gives rise to a photoluminescence signal at 500 nm (2.25 eV) caused by trapped excitons [1174, 1175].

2.7 Static Transient Current Effects in TiO_2

The interface between an electrode and TiO_2 is affected by surface, defect, and impurity (dirt, water)-induced electronic states. But additionally, connecting a metal with a semiconductor causes an exchange of charge carriers between the two materials resulting in a band bending close to the interface. Thus, the effects taking part in the formation of the electronic landscape at the interface depend on the choice of electrode material and fabrication method of the TiO_2. Besides that, a lot of other effects influence the transient current through a metal/TiO_2 interface.

In Figure 2.16, an overview of selected charge transport phenomena in a metal/TiO_2/metal device is drafted. For space reasons, the labeling is shifted from the caption to the text. The following effects are shown in Figure 2.16: *(1) Schottky emission, (2) cathode Fowler-Nordheim (FN) tunneling, (3) Trap-assisted tunneling (TAT), (4) cathode interface trap states, (5) Poole-Frenkel (PF) emission, (6) electron trapping results in negative space charge, (7) high energetic electron transfer to another conduction band resulting in a negative differential resistance (NDR), (8) thermally activated conduction, (9) band conduction mechanism, (10) high energetic electrons excite electron-hole pairs at the anode resulting in anode hole injection, (11) anode interface trap states, (12) anode Fowler-Nordheim (FN) tunneling, (13) hole trapping results in positive space charge, (14) electron-hole recombination; **Insets: top left:** metallization by a large number of electron donors close to the interface; **top right:** most important energy levels (vacuum level E_{vac}, Fermi energy E_F, conduction band minimum CBM_{TiO2}, valence band maximum VBM_{TiO2}) and energy differences (work function Φ_W, Schottky barrier φ_{sb}, built-in potential φ_{bi}, lowering of φ_{bi} by image charge potential $\Delta\varphi_{bi}$, deep/shallow traps states $\varphi_{deep}/\varphi_{shallow}$ below CBM_{TiO2}, bandgap E_g, electron affinity χ, chemical potential μ, ionization potential Ξ); **down left:** Trapping appears not instantly everywhere in the oxide but the main trapping zone drifts from the cathode to the anode after applying a potential. This results in a time-dependent transient current; **down right:** for a significantly low work function of the electrode, the contact becomes ohmic; **Bottom:** This equivalent circuit displays the electronic components of the investigated system.*

2.7.1 Band Bending at the Metal-Semiconductor Interface

Dependent on the electrode material, there is either an ohmic or a rectifying contact between the electrode and the TiO_2. Typically, the Schottky contact is used to model the charge transport across a rectifying metal-semiconductor interface [1176]. The presented equations refer to homogeneous doping distributions.

Space-Charge Region and Schottky Barrier

If a semiconductor with the chemical potential μ is connected to a metal with the work function Φ_W charges are exchanged between the two materials until the two energy levels are equalized in thermodynamic equilibrium. For $\Phi_W > \mu$, electrons move from the semiconductor to the metal, for instance. As a consequence, ionized donors are left in the semiconductor and a space-charge region builds up. The width of the space-charge region is given by

$$W_{dep} = \sqrt{\frac{2\epsilon\epsilon_s}{qN_d}\left(\varphi_{bi} \pm U - \frac{2k_bT}{q}\right)}, \tag{2.60}$$

where ϵ is the vacuum permittivity, ϵ_s is the relative permittivity of the semiconductor, $\varphi_{bi} = 1/q(\Phi_W - \mu)$ is the built-in potential, q is the elementary charge, N_d is the donor concentration, U is the applied potential, k_b is the Boltzmann constant, and T is the absolute temperature

75

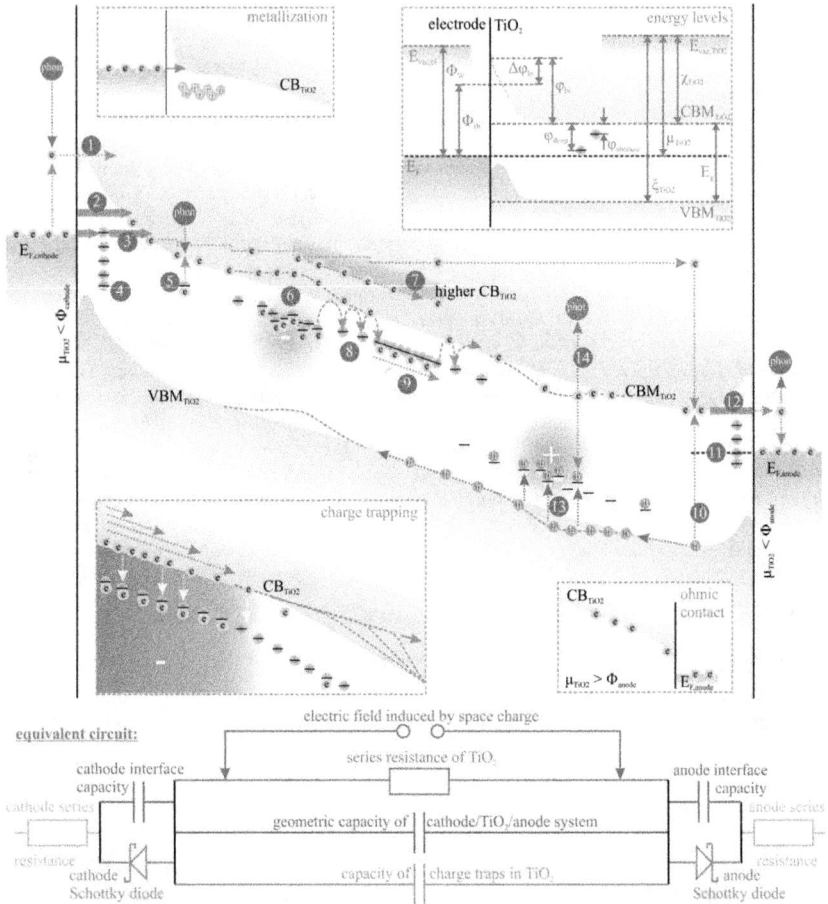

Figure 2.16: *Static mechanisms that affect the charge transport in a metal/TiO_2/metal sandwich. Details are found in the text.*

[795, 1152, 1177]. The width of the space-charge region can reach a few microns for donor densities below 15×10^{15} cm^{-3}. In a metal, the space-charge region is in the order of a few Ångström due to Thomas-Fermi screening [1178]. It is reported that in defective TiO$_{2-x}$ nanorods with electron densities of 3.31×10^{20} cm^{-1} and higher, the band bending appears on such a short distance that the CBM on the surface and in the core differ significantly [1066, 1156, 1179, 1180]. Hence, doping plays an important role for the height of the Schottky barrier and consequently the limiting charge transport mechanism in TiO$_2$. In niobium-doped TiO$_2$, the space-charge region is tightened due to the higher electron density in the bulk TiO$_2$. Thus, the Schottky barrier becomes thinner and the transient current is dominated by thermionic field emission with increasing niobium concentration [984, 1181, 1182]. This is observed for electrons moving from the nanorod into the FTO, in particular [984].

The internal electrical field plays an important role for ion diffusion and detrapping. Within the space-charge region, the amplitude of the internal electrical field is not constant. In first proximity, the strongest field is found directly at the interface and is given by [795]

$$E_{max,x=0} = \frac{2}{W_{dep}} \left(\varphi_{bi} \pm U - \frac{2k_b T}{q} \right). \tag{2.61}$$

From the space charge

$$Q_{sc} = q N_d W_{dep}, \tag{2.62}$$

the depletion-layer capacitance

$$C \equiv \frac{|\partial Q_{sc}|}{\partial U} = \frac{\epsilon \epsilon_s}{W_{dep}} \tag{2.63}$$

is obtained. Hence, $C - U$ measurements (Mott-Schottky plots) are often used to determine the charge carrier density [795, 918]

$$N_c = \frac{2}{q \epsilon \epsilon_s} \left[- \left(\frac{d(1/C^2)}{dU} \right) \right]. \tag{2.64}$$

For electrons, there is an energy barrier between the metal and the semiconductor which is called Schottky barrier and its potential height is given in case of abrupt junctions by

$$\varphi_{sb} = \frac{1}{q} (\Phi_W - \chi), \tag{2.65}$$

where χ is the electron affinity of the semiconductor [795]. If the energy barrier $q\varphi_{sb}$ is larger than $k_b T$, the contact is rectifying and it is called a Schottky contact. Applying a positive potential to the metal lowers the built-in potential and at a certain bias U_{open}, a significant number of electrons is able to move from the semiconductor to the metal. This bias is called the forward voltage of the Schottky diode. Applying a negative potential to the metal lowers and narrows the Schottky barrier. At a certain applied potential and barrier thickness, tunnel currents increase significantly. This bias is called the reverse voltage of the Schottky diode. If a forward biased Schottky contact is ohmic or rectifying depends on the relation between Φ_W and μ. An overview for rectifying and ohmic contacts in relation to the Fermi levels is shown in Table 2.6.

relation of work functions	majority carriers	contact characteristics
$\Phi_W > \mu$	electrons	rectifying
$\Phi_W > \mu$	holes	ohmic
$\Phi_W < \mu$	electrons	ohmic
$\Phi_W < \mu$	holes	rectifying

Table 2.6: *Rectifying and ohmic contact characteristics for different metal-semiconductor combinations biased in forward direction.*

Schottky Lowering by Image Charge Potential

An electron inside the semiconductor moving close to the metal-semiconductor interface causes a positive image charge in the metal and hence, an image potential is built up. This potential superposes the potential in the space-charge region and lowers the Schottky barrier [1183, 1184]. The decrease of the Schottky barrier by the image charge potential is usually small in relation to the height of the Schottky barrier. But since the image charge effect is extremal at the interface it has a relatively large effect in semiconductors with high donor densities where the space-charge region is limited to a small sub-interface region. The strength of the image charge depends on the electron density in the semiconductor and hence, it increases with increasing current across the metal-insulator interface. As a consequence, the effective potential close to the interface decreases with increasing applied potential as illustrated in Figure 2.16.

The image charge model assumes that there are no mobile charge carriers in the closest neighborhood of the electrons and that the polarization of the semiconductor caused by moving electron happens on a much faster timescale than the transit time through the affected semiconductor volume. The so-called Schottky lowering of the energy barrier in homogeneously doped semiconductors is given by

$$\Delta\varphi = \sqrt{\frac{q\mathcal{E}}{4\pi\epsilon\epsilon_i}} = \sqrt[4]{\frac{q^2 N_d(qU + qV_d - k_bT)}{8\pi^2\epsilon^3\epsilon_i^2\epsilon_s}}, \qquad (2.66)$$

where

$$\mathcal{E} = \sqrt{\frac{2qN_d}{\epsilon\epsilon_s}\left(\varphi_{bi} \pm U - \frac{k_bT}{q}\right)} \qquad (2.67)$$

is the amplitude of the electrical field at the interface, ϵ_i is the image force dielectric constant, U is the externally applied reverse bias across the space-charge region, and V_d is the diffusion potential. Considering Schottky lowering, the maximum of the potential electron energy is shifted from the interface inside the semiconductor by [1185]

$$\Delta x_\varphi = \frac{\Delta\varphi}{2\mathcal{E}} = \sqrt{\frac{q}{16\pi\epsilon\epsilon_o\mathcal{E}}}. \qquad (2.68)$$

It should be noted that the optical dielectric constant ϵ_o is often taken for the image force dielectric constant ϵ_i to respect the fast injection time compared to the dielectric relaxation time [1186, 1187].

Effect of Surface States on the Potential Landscape

In real systems, the barrier height depends on more complex parameters and the experimentally determined dependence on the difference between the chemical potential of TiO_2 and the Fermi level of the electrode is usually weaker [1188, 1189]. So far, intra-bandgap states resulting from surface states are not considered, yet. And self-consistent-charge density functional tight binding calculations predict that these surface states depend on the crystal orientation [1190]. Unsaturated dangling bonds are a typical reason for surface states which are within the bandgap as it is observed for silicon [1191]. These states might act as trap states and increase the charge carrier density at the surface and hence, the Fermi level is shifted resulting in a different band bending. Interface states with similar consequences arise if wave functions of electron states from the metal and semiconductor are overlapping [1192, 1193].

Creation of new Compounds: Mott Barriers and Metallization

New materials could be created during chemical reactions between the metal and the semiconductor resulting in thin insulating layers such as oxides. Typically, charge carriers need to tunnel through these so-called Mott barriers and the current is reduced significantly. If ignoble metals are deposited on TiO$_2$, an oxidation or reduction takes place. Therefore, noble metals such as gold, palladium, or platinum are preferred as electrodes [213, 329–331]. But these chemical reactions can also be used to design specific contact properties such as metallization.

Adsorbates and Surface Dipoles

Adsorbed molecules such as water and hydrocarbon cause dipoles at the interface [773]. Surface dipoles are also induced by compounds consisting of elements with different electronegativity. Dipoles cause potential steps at the interface which change the Schottky barrier [1178]. Gutmann et al. observed an increase of the CBM by 0.15 eV after polarizing hydrocarbon contamination on nanoporous TiO$_2$ via UV exposure [773]. The oriented dipoles are generated most likely by UV-induced photochemical hydroxylation of the TiO$_2$ surface [1194, 1195]. Thereby, the positively charged hydrogen ion is directing outwards and increases the electron affinity [773].

Influence of Surface-Bulk Diffusion Processes

Heating could change bands and foreign atoms diffuse into the bulk TiO$_2$. By diffusion, the interface region is broadened and an impurity gradient is established. At elevated temperatures or for a sufficiently high applied voltage, ignoble metals such as aluminum or yttrium are oxidized easily by oxygen atoms from the TiO$_2$ resulting in an inhomogeneous interface region. Noble metals can diffuse from the surface into the bulk and change the dopant concentration in the TiO$_2$.

But impurities can also diffuse from the bulk TiO$_2$ to its surface and accumulate at the interface. Dulub et al. [757] reported that impurities such as magnesium, potassium, and calcium diffuse from the bulk TiO$_2$ to the surface during annealing. Some of these impurities desorb, but others remain and form two-dimensional crystals on the TiO$_2$ surface. Therefore, the surface of even highly pure bulk TiO$_2$ single crystals can be covered extensively with foreign elements. For organic semiconductors, the situation is even more complex and described elsewhere [1196].

Band Bending on Selected Interfaces between Conductors and Semiconductors

Electronic properties of several selected semiconductors are collected in Table 2.3. In Table 2.7, selected reports investigating TiO$_2$ layer systems are listed. It has to be considered that the barrier between the electrode and TiO$_2$ depends on the doping-dependent Fermi level of the TiO$_2$ and hence, it depends on the fabrication technique.

Consequently, the contact between titanium and TiO$_2$ is either ohmic [1238] or rectifying [1215]. Due to the high heat of formation of alumina, aluminum reacts with oxygen more easily compared to titanium [1239]. As a consequence, a highly conductive reduced TiO$_2$ layer is created at the aluminum/TiO$_2$ interface [1227]. Such layers are not created for gold, palladium, and platinum electrodes.

interface system	environmental influences	analysis	reference
ITO/WO$_3$/TiO$_2$-colloids	dark/UV	CAFM	[776]
ITO/TiO$_2$-thin anatase film		Mott-Schottky analysis	[658]
ITO/TiO$_2$-nanocrystalline anatase		UPS, LIXPS	[772]
ITO/TiO$_2$-nanofilm	light	photo-KFM	[771]
ITO-nanowire/TiO$_2$-core-shell	dark/light	$I - V$	[1197]
ITO/TiO$_2$-nanocrystalline anatase	dark/UV	UPS, LIXPS	[773]
ITO/TiO$_2$-nanoparticles, anatase/Au		$I - V$	[1040]
ITO/TiO$_2$-nanotubes/Si	dark/UV	$I - V$	[1198]
ITO/TiO$_2$/ITO		$I - V$	[1199]
ITO/TiO$_2$/Ti		$I - V$	[1198]
ITO/TiO$_2$/Pt		$I - V$, $I - T$	[1200]
FTO/TiO$_2$-mesoporous layer		Theory	[777]
FTO/TiO$_2$-nanorod, Nb-doped	dark/light	$I - V$, $C - V$, EIS	[984]
FTO/TiO$_2$-nanocrystalline		$I - V$	[1201]
FTO/TiO$_2$-nanotube, anatase/Au		$I - V$	[1202]
FTO/TiO$_{2-x}$-layer/TiO$_2$-NRA, rutile/Al		$I - V$	[1203]
Au/TiO$_2$-rutile/Au		$I - V$, $C - V$	[1204]
Au/TiO$_2$/Au		$I - V$	[946]
Au/TiO$_2$/Ti/Au	dark/UV	$I - V$, $R - T$	[1083]
Pt/TiO$_2$/Pt		$I - V$, $R - T$	[1205]
Pt/TiO$_2$/Pt		$I - V$	[222, 1206–1208]
Pt/TiO$_2$/Pt		$I - V$ (CAFM)	[1209]
Pt/TiO$_2$-single rutile nanorod/Pt	dark/light	$I - V$, $I - T$	[1167]
Pt/TiO$_2$-rutile/Pt		$I - V$	[1210]
Pt/TiO$_{2-x}$TiO$_2$/Pt		$I - V$	[1211]
Pt/TiO$_x$/Pt		$I - V$	[1208, 1212]
Pt/43 nmTiO$_2$/Pt		$I - V$	[224]
Pt/40 nmALD-TiO$_2$/Pt		$I - V$	[223]
Pt/TiO$_2$, nanoporous		$I - V$	[8]
Pt/TiO$_2$-nanotube/Pt		$I - V$	[1083]
Pt/TiO$_2$/Au		$I - V$	[1213]
Pt/TiO$_2$/Ti		$I - V$	[1214, 1215]
Pt/TiO$_2$/Ti/Pt		$I - V$	[1216]
Pt/8 nmTiO$_2$/Ti		$I - V$ (CAFM)	[1217]
Pt-Ir/TiO$_2$-nanotube, anatase/Ti		$I - V$	[1218]
Pt/MeO$_x$/TiO$_2$		$I - V$	[1068]
Pt/TiO$_2$/Ru/SiO$_2$/Si		$I - V$	[220]
Ir/TiO$_2$/Pt		$I - V$, $C - V$	[1219]
Ti/TiO$_2$-nanotube/Ti		$I - V$	[1077]
Ti/TiO$_2$/Pt		$I - V$	[1217]
Pt/TiO$_2$/Ti/Pt/SiO$_2$/Si	Ar$^+$-ion irradiation	$I - V$	[1220]
Ti/TiO$_2$/Si	dark/light	$I - V$	[1198]
TiN/single-crystal rutile TiO$_2$/Pt		$I - V$	[1221]
Al/TiO$_2$/Al		$I - V$, $C - V$	[945]
Al/TiO$_2$-amorphous/Al		$I - V$	[744]
Al/TiO$_2$/Pt		$I - V$, $C - V$	[585]
Al/TiO$_2$/Ti		$I - V$, $C - V$	[945]
Al/TiO$_2$-nanowire networks, anatase/Ti		$I - V$	[1222]
Al/Ti/TiO$_2$/Si/Al		$I - V$, $C - V$	[1223]
Al/TiO$_2$/Si(p-type)/Al		IS	[1224]
Al/TiO$_2$/Si		$I - V$, $C - V$	[283, 1225, 1226] [70, 1227, 1228]
Al/TiO$_2$/SiO$_2$/Si		$I - V$, $C - V$, DLTS	[585, 1229]
Al/TiO$_2$/Ru/SiO$_2$/Si		$I - V$	[220]
Cu/TiO$_2$-nanotube/Ti		$I - V$	[208]
Ag/TiO$_2$/Si		$I - V$	[1230]
Ag/TiO$_2$/Mo-doped In$_2$O$_3$		$I - V$	[1231]
Si/SiO$_x$/TiO$_2$/Au		$I - V$	[1232]
Si(n-type)/TiO$_2$/Pd		$I - V$	[1233]
Si,n,p-type/TiO$_2$/electrolyte	dark/light	$I - V$	[1234]
Si/TiO$_2$/Pt		$I - V$, $C - V$	[1219, 1227, 1235]
ZnO/TiO$_2$, core-shell	dark/light	$I - V$ of solar cell	[1236]
ZnO/TiO$_2$, core-shell	dark/light	EIS, $I - V$ of solar cell	[1237]

Table 2.7: *Electronic properties of applied and related interfaces determined by $I - V$, $C - V$, DLTS (deep level transient spectroscopy), impedance spectroscopy (IS), conductive atomic force microscope (CAFM), ultraviolet photoemission spectroscopy (UPS), and low-intensity x-ray spectroscopy (LIXPS).*

interface	height	reference
FTO/TiO_2, nanoporous film	0.1 eV	[777]
Pt/TiO_2	1.7 eV	[1091, 1240, 1241]
Pt/TiO_2	1.2–1.3 eV	[1214]
Pt/TiO_2, 8 nm thin film	0.73 eV	[1215]
Ti/TiO_2, 8 nm thin film	0.13 eV	[1215]
Ti/TiO_2, nanotube	0.043 eV	[1077]
Au/TiO_2, rutile single crystal	0.87–0.94 eV	[1204]
Ag/TiO_2-nanorod	0.9 eV	[1230]
metal/TiO_2, metal independent	0.56 eV	[1077]

Table 2.8: *Schottky barrier heights of applied and selected interfaces. Different reported heights for the same electrode material might originate from different defect densities and distributions.*

Rühle and Cahen investigated the FTO/TiO_2 barrier intensively. They found a barrier height of 100 meV and that 80% of the height drop within the first nanometer behind the interface inside the TiO_2 [777]. This indicates that the internal field is extremely high compared to deeper parts of the depletion layer. In their case, the total depletion depth was calculated to be approximately 20 nm for nanoporous TiO_2 films consisting of TiO_2 particles with 20 nm diameter.

Band Structure at the Anatase/Rutile Interface

The band structure at the anatase-rutile interface differs from the bulk structure significantly. On the one hand, surface states reduce the bandgap at the anatase/rutile interface down to 1.2 eV [613]. On the other hand, there are many trap states at the interface. As a consequence, electrons are excited by visible light from these states into the conduction band of rutile or anatase. The CBM of anatase is about 0.2 eV higher compared to rutile TiO_2 [417]. Hence, the interface becomes rectifying at sufficiently low temperatures and photoexcited charges are separated at this interface, which is used in photovoltaic and photocatalytic devices to reduce recombination [612, 1242–1244]. Unfortunately, the calculated positions of the valence and conduction band depend on the applied model and the measurement technique. Consequently, it is also reported that the CBM of rutile appears to be above the CBM of anatase TiO_2 [762].

2.7.2 Electrode-Limited Conduction Mechanisms

The transient current density in a metal is always dominated by at least one limiting conduction mechanism, which affects the shape of the $I - V$ characteristics. This limiting mechanism acts as a bottleneck for the electron flow through a device and might be related to interface or bulk effects.

In the same TiO_2 structure, different conduction mechanisms appear for different voltage and temperature ranges. For instance, a bias-dependent transition from tunneling to Poole-Frenkel emission is observed between low and high electrical fields in TiO_2 films [945, 1223, 1228]. Furthermore, Kim et al. observed a temperature-dependent transition from tunneling to thermionic emission at roughly 100 °C in thin TiO_2 films [1235]. Chong et al. found a hopping conduction at low bias, trap-assisted tunneling at medium bias, and Fowler-Nordheim tunneling at high bias in a thermally grown thin TiO_2 film [1227].

In this section, interface-derived limiting conduction mechanisms such as thermionic emission, Fowler-Nordheim tunneling, thermionic field emission, or trap-assisted tunneling are presented. Thermionic emission appears rather for low doping concentrations, thick space-charge regions, and elevated temperatures while tunneling is dominant for extensive n-type doping concentrations, thin space-charge regions, and low temperatures [1189, 1191, 1245].

Independent from the conduction mechanism, the specific contact resistance

$$R_c \equiv \left(\frac{\partial J}{\partial U} \right)_{U=0}^{-1} \tag{2.69}$$

in dependence of the applied voltage U and the measured transient current density J is a fundamental property of a semiconductor-metal interface. R_c depends on the conduction mechanism and hence, it is influenced by the temperature, barrier height, effective Richardson constant, effective mass, and charge carrier concentration [1246].

Schottky Emission

In metals and for finite temperatures, there are electrons with energies above the Fermi level. In particular, electron energies could be in the order of the work function or Schottky barrier height and hence, electrons become able to leave the metal. The emission of electrons from a metal surface into the vacuum is described by the Richardson equation. Exchanging the work function and the rest mass of the electrons in the Richardson equation by the Schottky barrier height and effective electrons mass in the semiconductor, respectively, one gets a first approximation of the thermionic emission-driven current from a metal into a semiconductor. In this section, some additional effects such as the image charge potential are considered to obtain a more precise expression of the Schottky emission. Assuming that diffusion processes as well as phonon scattering do not play a role, $q\varphi_{sb}$ is much larger than $k_b T$, the system is in thermal equilibrium, and the presence of a current flow does not affect this equilibrium, the current density J_{se} from thermionic emission is given by

$$J_{se} = \left[R^* T^2 \exp\left(-\frac{q(\varphi_{sb} - \Delta\varphi_{sb})}{k_b T} \right) \right] \cdot \left[\exp\left(\frac{qU}{n_{id} \cdot k_b T} \right) - 1 \right] \tag{2.70}$$

where U is the applied voltage,

$$n_{id} \equiv \frac{q}{k_b T} \frac{\partial U}{\partial(\ln J_{se})} \tag{2.71}$$

is the ideality factor, and

$$R^* = \frac{q m_e^* k_b^2}{4\pi^2 \hbar^3} \approx 120 \frac{m_e^*}{m_e} \tag{2.72}$$

is the Richardson constant for thermionic emission. A low value of the Richardson constant is related to a significant charge carrier diffusion [1247]. Equation 2.70 considers the total current density including the current from the metal into the semiconductor and the current in the opposite direction. The current density from the metal into the semiconductor is similar to Equation 2.70 but there is no "−1" in the right parenthesized term. The current from the semiconductor into the metal is given by the left parenthesized term only [795, 1248]. Based on Equation 2.70, the barrier height

$$\varphi_{sb} = \frac{k_b T}{q} \ln\left[J_{se} \cdot \frac{2\pi^2 \hbar^3}{q m_e^* (k_b T)^2} - \left(\exp\left(\frac{qU}{k_b T} \right) - 1 \right)^{-1} \right] + \sqrt{\frac{q\mathcal{E}}{4\pi\epsilon\epsilon_o}} \tag{2.73}$$

is determined by the $I-V$ characteristics [795, 884, 1248]. An exemplary study on electrode/TiO$_2$ interfaces and the determination of the temperature-dependent ideality factor based on Schottky emission measurements were presented by Pakma et al. [1249].

Extended Schottky emission model: In Equation 2.73, an ideality factor of 1 is assumed. A more realistic model for J_{se} is given by

$$J_{se} = J_0 \exp\left(qU\left[\frac{1}{k_bT} - \frac{1}{E_0}\right]\right),$$ (2.74)

where J_0 is the left term in Equation 2.70 and E_0 is given by

$$E_0 = E_{00} \coth\left(\frac{E_{00}}{k_bT}\right)$$ (2.75)

with

$$E_{00} = \frac{\hbar q}{2}\sqrt{\frac{\eta_e}{m_e^* \epsilon \epsilon_s}},$$ (2.76)

where η_e is the electron density [1040, 1077, 1218, 1250, 1251].

Effect of trapping and scattering: In the presence of trapping and scattering, the mean free path of the electrons might become shorter than the oxide thickness and Equation 2.70 has to be modified [1252–1254]. In this case, the factor R^*T^2 in Equation 2.70 is replaced by

$$\delta T^{3/2} \mathcal{E} \mu_e^* \left(\frac{m_e^*}{m_e}\right)^{3/2}$$ (2.77)

where δ is a constant [1248]. An exemplary study on the mean free path of electrons in ZrO$_2$ was performed by Chiu et al. [1252].
In case of a dominant diffusion, the factor R^*T^2 in Equation 2.70 is replaced by [795]

$$\frac{q^2 D_n N_c}{k_bT}\sqrt{\frac{2N_d q(V_{bi} - U)}{\epsilon \epsilon_o}}.$$ (2.78)

The combination of thermionic emission and diffusion theory results in a different Richardson constant and is described elsewhere [795].

Schottky emission with scattering and quantum reflections: Considering scattering on optical phonons and quantum mechanical reflections, the Richardson constant has to be replaced by the effective Richardson constant

$$R^{**} = \frac{f_p f_q R^*}{1 + f_p f_q v_r / v_d},$$ (2.79)

where $f_p = \exp\left(\frac{\Delta x_\phi}{l^*}\right)$ is the probability of an electron emission regarding reflections by optical phonons [1255, 1256], f_q considers quantum mechanical reflections and tunneling [1257, 1258], $v_r = \frac{R^*T^2}{qN_c}$ is the thermal velocity, $v_d = \mu_e^* \mathcal{E}$ is the effective drift velocity, Δx_ϕ is given by Equation 2.68, l^* is the carrier mean free path, and N_c is the effective density of states in the conduction band [795].

Fowler-Nordheim Tunneling

Especially at low temperatures in the absence of thermally driven processes and for very low trap densities close to the interface, tunneling is the dominant charge injection mechanism. Tunneling through the approximately triangular energy barrier is described by Fowler-Nordheim tunneling (or field emission) current, given by

$$J_{fn} = \frac{q^3 \mathcal{E}^2}{16\pi^2 \hbar \varphi_{sb}} \cdot \exp\left(-\frac{\sqrt{2qm_t}\varphi_{sb}^{3/2}}{3\hbar\mathcal{E}}\right).$$ (2.80)

Here, m_t is the tunneling effective mass of the electron. The dominance of Fowler-Nordheim tunneling is proved by a linear slope in the Fowler-Nordheim plot (I/U^2 versus $1/U$) [1227]. By determining m_e^* from thermionic emission at high temperatures and m_t from Fowler-Nordheim tunneling at low temperatures it was found that m_e^* is approximately equal to m_t [1248]. For instance, this correlation was demonstrated by Chiu in HfO$_2$ [1259].

Thermionic Field Emission

Thermionic field emission is a combination between thermionic emission and Fowler-Nordheim tunneling. It appears when electrons in the metal electrode with an energy larger than the Fermi energy but lower than the Schottky barrier maximum hit the metal/TiO$_2$ interface. Hence, the electron needs to tunnel through the relevant barrier but the barrier height and thickness is less than for Fowler-Nordheim tunneling. This conduction mechanism appears likely in a temperature and doping regime between Fowler-Nordheim tunneling and Schottky emission. The current density for thermionic field emission is given by [1248, 1260]

$$J_{tf} = \frac{q^2 \sqrt{m_e^* k_b T} \mathcal{E}}{8\hbar^2 \pi^{5/2}} \cdot \exp\left(-\frac{q\varphi_{sb}}{k_b T}\right) \cdot \exp\left(\frac{\hbar^2 q^2 \mathcal{E}^2}{24 m_e^* (k_b T)^3}\right). \tag{2.81}$$

2.7.3 Bulk-Limited Conduction Mechanisms

The conduction in the bulk is limited either by ohmic conduction, Poole-Frenkel emission, hopping conduction, space-charge-limited currents, hole injection, recombination in the space-charge region, or diffusion of charge carriers in the depletion region [1178, 1248, 1261, 1262]. Conduction mechanisms which are not based on the electron transport such as ionic conduction are not discussed here and detailed descriptions are found elsewhere [1248].

Ohmic Conduction

Ohmic conduction is based on electron drift and affected by the number of electrons and the electron mobility only. Hence, the current density depends on temperature, phonon-scattering, and the effective electron mass. Trapping, impurity scattering, space-charge-limited effects, recombination, or diffusion do not play an important role in case of an ohmic charge transport [1263]. It is often found for low applied voltages or in relation to the low resistance state within resistive switching behaviors [215–218]. Based on Equations 2.15 and 2.24, the ohmic current density is expressed by

$$J_\Omega \approx \sigma\mathcal{E} = \eta_e q \mu_e^* \mathcal{E} = N_c \frac{q^2 \mathcal{E}}{m_e^* \gamma^*} \exp\left(-\frac{E_c - \mu}{k_b T}\right) \tag{2.82}$$

for low donor densities and

$$J_\Omega \approx \sqrt{\frac{N_d N_c}{2}} \frac{e^2 \mathcal{E}}{m_e^* \gamma^*} \exp\left(-\frac{E_c - E_d}{2 k_b T}\right) \tag{2.83}$$

for donor densities N_d much higher than the effective density of states N_c in the conduction band [795, 1248]. Here, $E_d = E_c - q\varphi_d$ is the donor level of a donor with the depth of $q\varphi_d$. In TiO$_2$, the latter equation is much more reasonable due to its large bandgap. In double-logarithmic plots, the linear slope for ohmic conduction is close to 1 [1248]. Since $\ln \sigma$ is proportional to $1/T$ for ohmic conduction, the donor depth E_d is determined from the slope in the Arrhenius plot. According to Equation 2.15, N_c is proportional to $(m_e^* \cdot T)^{3/2}$ and hence, the effective electron mass m_e^* and mobility μ_e^* is determined by Equation 2.83 as well [1191, 1248, 1264]. It is noted that for a complete description of the charge current, the electron diffusion current $J_D = eD\frac{\partial \eta_e}{\partial x}$ needs to be added [914]. However, in case of a small Schottky barrier, which allows ohmic conduction at room temperature, the electron density is assumed to be constant and hence, the diffusion current can be neglected.

Space-Charge-Limited Conduction

When electrons are emitted from the cathode into the semiconductor and the electrons are not withdrawn to the anode immediately, an electron cloud is formed close to the cathode interface. An electrical field is established by this electron cloud which repels electrons passing the cathode/semiconductor interface. As a consequence, the electron cloud acts as a flow regulator and the current density does not depend on the mobile electron density η_e anymore. This is called a space-charge-limited conduction (SCLC). In general, the density of electrons injected from the interface must be significantly larger than the density of thermally activated donor-assisted electrons to observe SCLC [1265]. Similar to the Richardson equation in metal/vacuum and the Schottky emission in metal/semiconductor systems, the space-charge-limited effect is described by the Child's law in metal/vacuum and the Mott-Gurney equation which is given by

$$J_{mg} = \frac{9}{8}\epsilon\epsilon_s\mu_e^*\frac{\mathcal{E}^2}{d_{ox}} \tag{2.84}$$

in metal/semiconductor systems. Here, d_{ox} is the thickness of the semiconductor. So far, the focus was on metal/semiconductor systems but often, a metal/semiconductor/metal system is employed. The existence of two interfaces affects the SCLC considerably and it becomes proportional to the applied bias. In this case, the transient current density is given by

$$J_{msm} = \frac{2\epsilon\epsilon_s v_s}{W^2}U = qv_sN_d\frac{U}{\varphi_{fb}} \tag{2.85}$$

where v_s is the scattering-limited velocity of electrons, $W = W_1 + W_2$ is the sum of the widths of the two space-charge regions, and φ_{fb} is the flat-band potential [795, 1266, 1267].

Trap-Filled-Limited Conduction

Trap-filled-limited conduction (TFLC) is the space-charge-limited conduction in a semiconductor with a high trap density such as pristine TiO₂ nanorods [910, 1066, 1268] and doped TiO₂ [8, 1269]. In the presence of trapping and detrapping, the transient current depends additionally on the ratio of trap (N_t) and conduction band (N_c) states, the depth of the trap states E_d, and the temperature. As a consequence, the current is described either by a trap-free or a trap-limited mode, which depends on the parameters and properties of the measurement and sample, respectively [914, 1066, 1270–1274]. Especially for resistive switching devices with thin TiO₂ films, TFLC becomes a dominant conduction mechanism [915, 916]. In general, trapping lowers the SCLC. The TFLC is expressed by

$$J_{tf} = \frac{9}{8}\epsilon\epsilon_s\mu_e^*\frac{\mathcal{E}^2}{d_{ox}}\cdot\Gamma = \frac{9}{8}\epsilon\epsilon_s\mu_e^*\frac{\mathcal{E}^2}{d_{ox}}\cdot\frac{N_c}{gN_t}\exp\left(\frac{E_d}{k_bT}\right) \tag{2.86}$$

where g is the conduction band state degeneracy [1263, 1275–1278].

Important material properties that define the applied voltage range in which SCLC is observed are the carrier transit time τ_c and the dielectric relaxation time τ_d. τ_c decreases with increasing voltage. As long as τ_c is smaller than τ_d, the amount of thermally activated electrons is larger than the amount of injected electrons. Hence, the injected charges are neutralized by charge redistribution in the semiconductor and an ohmic conduction is dominant [1279]. Above a specific threshold voltage U_{tr}, τ_c becomes smaller than τ_d and a space charge builds up. For voltages slightly larger than U_{tr}, the specific amount of injected charges are trapped for a while and their transition time is larger compared to trap-free semiconductors. But since the SCLC does not depend on the charge carrier density, a reduced transition time results in a lower transient current which is a typical feature of TFLC. At the voltage U_{tf}, the amount of injected charge carriers is so high, that the Fermi level is raised to the conduction band and all traps are filled. Hence, for higher applied voltages, trapping disappears and the $I - V$ characteristic follows Equation 2.84 [1248]. The mentioned dimensions are given by

$$\tau_c = \frac{d_{ox}^2}{\mu_e^* \Gamma U_{tr}}, \qquad \tau_d = \frac{\epsilon\epsilon_s}{q\eta\mu_e^*\Gamma}, \qquad U_{tr} = \frac{9}{8} \cdot \frac{q\eta_0 d_{ox}^2}{\epsilon\epsilon_s\Gamma}, \qquad U_{tf} = \frac{qN_t d_{ox}^2}{2\epsilon\epsilon_s} \qquad (2.87)$$

where η_0 is the electron density in the conduction band at thermal equilibrium and accordingly the density of thermally excited electrons [1248]. A very clear and exemplary study of ohmic conduction, TFLC, and SCLC on La$_2$O$_3$ is presented by Chiu et al. [1265].

There might be not always a significant current jump between TFLC and SCLC in real samples. On the one hand, U_{tf} might be so high that a significant material degradation would happen before the transition. On the other hand, the trap density might be so low that TFLC does not appear. But one can distinguish TFLC and SCLC also by the slope in the double-logarithmic $I - V$ plot. Without the influence of traps and with monoenergetic traps, the slope is roughly 2 [1248].

Exponential trap distribution: An exponential trap distribution results in a power-law with a slope larger than 2 [1280–1282]. In this case, the power-law for a trap-affected SCLC is given by

$$J_{tf,exp} \approx q\mu_e^* N_c \left(\frac{\epsilon\epsilon_0}{qN_t k_b T_t}\right)^l \frac{U^{(l+1)}}{d_{ox}^{(2l+1)}} \qquad (2.88)$$

where N_t is the trap concentration, $l = T_t/T$ is the power coefficient, and $k_b T_t$ is the characteristic energy of the exponential trap distribution, which is given by

$$\tilde{\eta}_t(E) = N_t \exp\left(\frac{E - E_c}{k_b T_t}\right). \qquad (2.89)$$

Here, E is the energy and E_c the edge of the conduction band [1214, 1279]. An example for an exponential trap distribution is nanoporous TiO$_2$. Here, an ohmic conduction appears for low electrical fields and a power-law dependence with $l \approx 2.3 - 2.9$ dominates at high field amplitudes [8, 946, 1091].

Frenkel effect: In case of high electrical fields, the effective barrier for detrapping is significantly decreased by the Frenkel effect. Taking this into account, the transient current density is expressed by

$$J_{fr} = J_{tf} \cdot \exp\left(\frac{0.891}{k_b T}\sqrt{\frac{q^3\mathcal{E}}{\pi\epsilon\epsilon_s}}\right) \qquad (2.90)$$

where J_{tf} is given by Equation 2.86 [371].

Poole-Frenkel Emission

Poole-Frenkel emission describes the thermionic emission of charge carriers from trap states slightly below the conduction band. It can be seen as the equivalent of the Schottky emission in the bulk. Hence, it is observed at rather high temperatures and strong electrical fields. The transient current for Poole-Frenkel emission is given by

$$J_{se} = q\mu_e^* N_c \mathcal{E} \exp\left(-\frac{q(\varphi_t - \Delta\varphi_t)}{k_b T}\right) \tag{2.91}$$

where $q\varphi_t$ is the depth of the trap and $\Delta\varphi_t$ is the same expression as $\Delta\varphi_{sb}$ [1227, 1248]. Similar to the Schottky emission, the local dielectric constant ϵ_d around the traps is determined from the barrier lowering $\Delta\varphi_t$ and the Poole-Frenkel plot ($\ln(I/U)$ versus $\sqrt{U}$). Only a dynamic dielectric constant ϵ_d between the optical dielectric constant ϵ_o and the static dielectric constant ϵ_s is self-consistent and indicates that the conduction is mainly related to Poole-Frenkel emission [1283].

Exemplary studies on metal oxides based on Poole-Frenkel emission measurements are presented by Chiu et al. for Pr_2O_3 and ZnO_2 [1252, 1284]. Chong et al. determined $\epsilon_o = 7.56$, $\epsilon_d = 7.89$, and $\epsilon_s = 12.5$ for an 18 nm thin TiO_2 film between platinum and p-type silicon [1227]. The small difference between the optical dielectric constant ϵ_o and ϵ_d obtained from the Poole-Frenkel plot is caused by the fast electron transit time from the trap site across the barrier maximum. This transit time is shorter than the dielectric relaxation time. In addition, the depth of the traps φ_t and the electronic drift mobility μ_e^* can be determined from the Poole-Frenkel plot.

In Equation 2.91, the electrons are affected by the donor concentration N_d only. But also the concentration of trap states N_t influences the $I - V$ characteristics. As long as N_t is smaller than N_d, the Poole-Frenkel emission is called normal. If N_t is in the order of N_d, the Poole-Frenkel emission is called anomalous and the slope in the Poole-Frenkel plot is halved and hence, it becomes equal to the slope in the Schottky plot [1285].

Hopping Conduction

Hopping conduction describes the tunneling between trap states and hence, it is the equivalent for Fowler-Nordheim tunneling in the bulk. It is expected at rather low temperatures and the transient current for hopping conduction is expressed by

$$J_{hc} = qa\eta_e \nu \exp\left(\frac{qa\mathcal{E} - E_d}{k_b T}\right), \tag{2.92}$$

where a is the mean hopping distance and ν is the thermal vibration frequency of the trapped electrons [1248, 1263, 1284]. It should be noted that different assumptions result in different expressions for the hopping conduction [1277, 1286]. Exemplary studies on metal oxides are presented by Chiu et al. on Pr_2O_3 and MgO [1284, 1287].

Variable range hopping: With increasing temperature, the average potential electron energy becomes larger and hence, the tunnel barrier between two neighboring traps shrinks. Expressed in other words, the same tunnel probability at two different temperatures is observed for a larger trap distance. As a consequence, the hopping distance R_h becomes larger with increasing temperature, which is described by

$$R_h(T) = R_0 \exp\left(\left[\frac{T_F}{T}\right]^{(D+1)^{-1}}\right), \tag{2.93}$$

where D is the dimension of the structure ($D = 2$ for nanotubes and nanorods), R_0 is a material parameter, and T_F is a temperature scale based on the DOS at the Fermi level [1205, 1288, 1289]. Hence, this conduction mechanism is also called variable range hopping (VRH). Is was observed in amorphous TiO_2 nanotubes, for instance [1083]. However, for larger trap distances, direct hopping is neglected and the electron current is dominated by trapping and detrapping events only [1268].

Grain Boundary Conduction

In polycrystalline films, the electron conduction is affected by grain boundaries. Due to the n-type characteristics of defective TiO_2, the CBM in grain boundaries is most likely lower than the CBM of bulk TiO_2. As a consequence, the current is focused on the grain boundaries, if the grain boundaries are parallel to the applied electrical field or form percolation paths. For high electrical fields, the Schottky barriers of the electrode/grain or boundary/grain interface might be lowered sufficiently to allow electrons to enter the grains. The grain and grain boundary currents are superposed and the total $I - V$ characteristic might differ from the pure grain boundary currents at low bias. This change is reasonable since the electron mobility and shape of the Schottky barrier are supposed to differ between grain boundaries and single-crystalline grains.

If electrons need to cross grain boundaries, the boundaries behave similarly to traps. The grain boundary trap level depth is given by

$$\varphi_{gb} = \frac{qN_{gb}^2}{2\epsilon\epsilon_s N_d}, \tag{2.94}$$

where N_{gb} is the grain boundary trap density and N_d the dopant concentration [1248, 1263, 1290]. Since the average distance of grain boundaries in polycrystalline TiO_2 is usually much larger than the tunneling distance, the resulting grain boundary-limited current is best described with a Poole-Frenkel emission-like behavior. Hence, the $I - V$ characteristic is most likely given by

$$J_{gb} = q\mu_e^* N_c \mathcal{E} \exp\left(-\frac{q(\varphi_{gb} - \Delta\varphi_{gb})}{n_{id}k_b T}\right). \tag{2.95}$$

The ideality factor n_{id} considers deviations from pure thermal emissions [1291]. Similar to Schottky and Poole-Frenkel emission, $\Delta\varphi_{gb}$ can be approximated by the Schottky effect given by Equation 2.66 or determined by a comparison with the experimentally determinable surface barrier φ_{sb} [1291–1294], which results in

$$\Delta\varphi_{gb} = \Delta\varphi_{sb}\frac{1}{4n_{id}}. \tag{2.96}$$

Trap-Assisted Tunneling

For sufficiently thin oxides, direct tunneling from the cathode to the anode metal appears. Typically, direct tunneling is described with the Tsu-Esaki formula [1295] and requires film thicknesses of less than $10\,nm$ [1296]. Details about the $I - V$ characteristics of direct tunneling are found elsewhere [1248]. In this work, the oxide thickness is too large for direct tunneling. But another tunneling effect, that appears also for thicker films, plays an important role for the conduction mechanism in metal oxides.

In the presence of traps, electrons can tunnel via tunnel states into or through the TiO_2 resulting in a significant transient current even at low applied voltages [1297, 1298]. If traps are generated by voltage or current stress, the increased trap-assisted tunneling (TAT) current is called stress-induced leakage current (SILC) [1299, 1300]. TAT is rather a generic term describing various trap-assisted tunneling phenomena at interfaces and in the bulk. For instance, there are single and multiple trap models [1301, 1302].

For single trap models, it is assumed that the electron tunnels from the cathode into a trap state in the dielectric material which is the capture process and then, it tunnels from this trap state into the anode which is the emission process [1303, 1304, 1304, 1305]. The TAT probability depends on the position of the trap states, their energetic level, and their degree of occupation [1303]. Hence, some trap states are accessible only for hot metal electrons and electrons might tunnel into excited trap states and relax afterward. Such inelastic processes have to be modeled using phonon-assisted tunneling theories [1304, 1305].

For increasing trap densities and material thickness, models taking two or more traps into account become more suitable [1301, 1306–1308]. An example for TAT including many traps is the hopping conduction described above [1309].

For low electrical fields, the slope of the conduction band is so low that the TAT is affected by a trapezoidal-like potential barrier. But for large fields, electrons are emitted to the conduction band and the shape of the potential barrier becomes triangular. Hence, Houng et al. suggested a generalized trap-assisted tunneling (GTAT) model to consider the transition between the two kinds of potential barrier shapes [1297].

In more realistic assumptions, the transient current is affected by several conduction mechanisms at the same time such as thermionic emission, direct tunneling, and TAT [1310]. In addition, trapping in defect-rich dielectric materials changes the space charge and thus, the threshold voltages and other parameters are changed as well [1311, 1312].

For the majority of TAT models, there is no simple relation between the applied voltage and the transient current given. Typically, numerical fits are employed to describe the $I - V$ characteristics for specific samples [1296]. Besides that, kinetic Monte Carlo charge transport simulations can be used to model the TAT in metal oxides [1313].

For some applications, there are advanced structures consisting of two metal oxide layers or more, which show a complex TAT behavior [1296]. Such bilayer systems could appear on TiO$_{2\text{-x}}$/TiO$_{2\text{-y}}$ core-shell structures as well.

One of the simplest TAT models assumes a fixed distance between the cathode and the trap d_t and a fixed trap depth E_t. By neglecting the Schottky effect a trapezoidal-like energy barrier is employed. Furthermore, the thermionic emission rate from interface traps into the conduction band by Poole-Frenkel emission should be much lower than the TAT [221]. Any other interface-limited conduction mechanism is neglected. In this case, an analytical solution for the transient current density exists and is given by

$$J_{tat} = N_t \cdot q \cdot \nu \tag{2.97}$$

where N_t is the density of the nearest traps and ν is the transition rate. In an electrode/TiO$_2$/electrode system, where the current is limited always on the same interface, electrode electrons have to be captured by traps for reverse bias and emitted from traps to the electrode for forward bias. ν is given by

$$\nu_r = \nu_0 \cdot f \dot{T}^* \tag{2.98}$$

for reverse and

$$\nu_f = \nu_0 \cdot (1 - f) \dot{T}^* \tag{2.99}$$

forward bias where

$$f = \cfrac{1}{1 + \exp\left(\cfrac{q\varphi_{sb} - E_t + \mathcal{E}d_t}{k_b T}\right)} \tag{2.100}$$

is the Fermi-Dirac distribution of electrons in the cathode and $q\varphi_{sb}$ is the height of the Schottky barrier without the image charge effect and $(1 - f)$ is the occupation of empty states in the anode [1262]. The transmission probability T^* through a trapezoidal-like barrier according to the Wigner-Kramers-Brillouin (WKB) approximation is given by [1314]

$$T^* = \exp\left(-\frac{4}{3\hbar q\mathcal{E}}\sqrt{2m_e^*}\left[E_t^{3/2} - (E_t - \mathcal{E}d_t)^{3/2}\right]\right). \tag{2.101}$$

Another simple TAT model is to assume that tunneling from trap states into the conduction band through a triangular energy barrier is the limiting conduction process. This implies that the interface charge injection is high, which is usually realized by a low Schottky barrier and deep trap states in order to keep Poole-Frenkel emission negligible. In this case, the formula is derived simply from Fowler-Nordheim tunneling by exchanging the Schottky barrier height φ_{sb} by the trap depth φ_{deep} [1227]:

$$J_{tat} = \frac{q^3\mathcal{E}^2}{16\pi^2\hbar\varphi_{deep}} \cdot \exp\left(-\frac{\sqrt{2qm_t}\varphi_{deep}^{3/2}}{3\hbar\mathcal{E}}\right). \tag{2.102}$$

TAT tunneling was observed in TiO_2 thin films besides hopping conduction [1227]. Particularly in TiO_2 bilayer systems, inelastic TAT has to be considered as shown by Bera and Maiti [1315].

Ionic Conduction

In doped or defective semiconductors or insulators, charged ions contribute to the total current additionally. For sufficiently high electrical fields, charged ions are released from their lattice sites and move to the next potential minimum and so on [1263]. Hence, they need to pass a potential barrier φ_{io}. This conduction is similar to Poole-Frenkel emission for electrons and it is given by

$$J_{io} = J_0\exp\left(-\left[\frac{q\varphi_{io}}{k_bT} - \frac{q\mathcal{E}d}{2k_bT}\right]\right), \tag{2.103}$$

where J_0 is a constant and d is the average distance between two neighboring jumping sites [1248].
In undoped but defective TiO_2, positively and negatively charged ions contribute to the charge transport as well. In particular, negatively charged oxygen ions move towards the anode and hence, positively charged oxygen vacancies move towards the cathode. However, in highly defective, usually pristine TiO_2, the current resulting from drifting oxygen ions and vacancies is insignificant compared to the effect of the changed ion and vacancy distribution on the electronic landscape. In contrast, in highly crystalline and stoichiometric TiO_2, there are almost no charged defects that are able to drift. As a consequence, the contribution of ionic conduction in the investigated TiO_2 structures is neglected.

Negative Differential Resistance

The band structures of most semiconductors consist of several local CBMs. Typically, the global CBM is found in the Γ-point, but sometimes, there is another local CBM which differs only by a small amount of energy and momentum from the global CBM. At a sufficiently high electrical field, electrons are injected more and more into the higher conduction band with increasing external field amplitude. In a few semiconductors, the electron mobility $\mu_{e,2}^*$ in the higher band is lower than the mobility $\mu_{e,1}^*$ in the lower band. As a consequence, the conductivity is decreasing with increasing field as long as the majority of electrons drift inside the higher conduction band [1316, 1317]. For very high electrical fields, the current increases with increasing field again as it is known for the usual conduction behavior. The part of the $I - V$ curve with a negative slope is called the negative differential resistance (NDR) and the model including the two CBMs is called two-valley model. The main principle of the two-valley model is shown in Figure 2.17. This effect is used in Gun diodes to amplify terahertz radiation, for instance [1318].

Indeed, rutile TiO$_2$ has a relatively flat conduction band (Figure 2.11) and there are several local CBM which are just slightly higher than the global CBM [820]. This enables the possibility of a NDR behavior. Yagi et al. and Breckenridge and Hosler reported about a second conduction band just 30–50 meV above the global CBM which becomes populated above 100 K [202, 834]. Similar results were obtained by Becker and Hosler, who found a conduction via a band 50 meV higher than the global CBM above 40 K. However, the upper conduction band provides a smaller effective electron mass, a higher mobility, and a larger anisotropy which does hardly provide NDR. There are no reports in the literature about an NDR behavior in TiO$_2$ based on the two-valley model so far. Nevertheless, NDR appears often in TiO$_2$ systems resulting from mechanisms that provide a time-dependent transient current. Hence, NDR in TiO$_2$ systems is reported usually in the combination with resistive switching [1319–1323]. These mechanisms are discussed in Section 2.8.

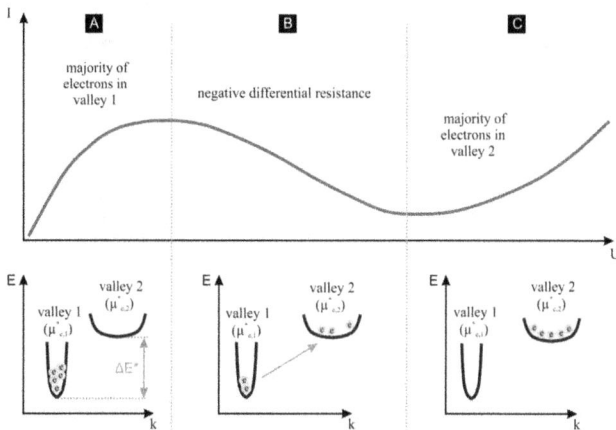

Figure 2.17: *The two-valley model for NDR: At a certain bias, electrons begin to occupy a slightly higher conduction band, which provides a different effective mass and hence, a different electron mobility. The transition in a higher conduction band results in a negative differential resistance.*

2.7.4 Effects of Structure Size on the Transient Current

Quantum confinement: The size influences the transient current in different ways. One of the most important effects is quantum confinement. As described in Section 2.6.8, the bandgap broadening based on quantum confinement becomes significant for particle sizes below 2.5 nm. Usually, the enlarged bandgap results in an increased Schottky barrier as well. As a consequence, a film consisting of tiny TiO$_2$ particles could show a completely different $I - V$ characteristic compared to dense TiO$_2$ films and a mesocrystalline rutile nanorod could provide a different limiting conduction mechanism compared to a highly crystalline nanorod with no separated segments.

Surface effects: Another effect has to be considered for particles which are significantly larger than the critical size for quantum confinement. Surface effects become important as soon as the

	SE	FN	TFE	OHM/SCLC/TFLC	TAT	PE	HC	GB
as-deposited films						[283]	[1223]	
amorphous films	[283]			[1324] [1214] [222] [1325] [1213]	[1315] [1227]			
mixed films	[1235]	[945] [1223] [1235] [946]		[945] [1223]			[945] [1223]	
rutile single crystal	[1204]			[1326]				
rutile films	[1228]							[1228]
anatase films		[1228]		[1223] [739]		[1228] [283]	[1223]	
anatase nanorods				[79, 1220]				
rutile nanorods	[984]			[1203]				
rutile nanorods, Nb-doped			[984]					

Table 2.9: *Conduction mechanisms observed in different TiO$_2$ structures: SE (Schottky emission), FN (Fowler-Nordheim tunneling), TFE (thermionic field emission), OHM (ohmic conduction), SCLC (Space-charge-limited current), TFLC (trap-filling-limited current), PE (Poole-Frenkel emission), HC (hopping conduction), GB (grain boundary conduction).*

particle size falls below the depletion length [1327–1329]. In those particles, there is no position where bulk properties such as the bandgap are found. In case of an n-type semiconductor with a depletion layer on the surface, mobile electrons are pulled towards the core by the internal field. If there is an accumulation at the surface, mobile electrons are preferentially found in the shell of the particle.

Often, the electronic structure in small particles cannot be described with analytical models and numerical simulations need to be applied [1329]. For instance, the square-root relation between the width of the depletion layer W_{dep} and the applied voltage U turns into a linear or sub-linear relation for ZnO nanorod Schottky barriers [1327]. In addition, the $1/\sqrt{N_d}$ dependence of W_{dep} turns into an exponential behavior [1329].

2.7.5 Effects of the Fabrication Technique on the Transient Current

Each fabrication technique provides specific crystallographic, mechanical, optical, electronic, and optoelectronic properties which are related to technique-dependent morphologies, impurities, and defects. These properties are directly related to doping and trap concentrations and might even affect the band structure. This affects likely the conduction mechanism in a TiO$_2$ sample, as explained in the previous section. Here, a few concrete experimental observations should be given.

For instance, thermally grown TiO$_2$ shows much smaller transient currents compared to plasma-oxidized [1330], plasma-enhanced chemical vapor deposition (PECVD)-grown [1331], metal-organic chemical vapor deposition (MOCVD)-grown [659], and electron-beam deposited TiO$_2$ [1223]. Aduda et al. found fabrication technique-dependent effective drift mobilities between 10^{-7} and 10^{-6} meV for sputter deposition, doctor blading of sol-gel dispersions, and electrostatic spray assisted vapor deposition (ESAVD) [1114].

2.7.6 Effects of Post-Treatment on the Transient Current in TiO$_2$

Post-treatment methods such as annealing or chemical surface reactions, which change, enhance, or diminish the properties of TiO$_2$, affect the conduction mechanisms.

Chemical Reduction

In particular on samples which cannot resist high temperatures, a chemical reduction is used rather than annealing in a reducing atmosphere. For instance, the field emission from TiO$_2$ nanotubes were enhanced via a reduction in NaBH$_4$ solution [208]. The disadvantage of chemical reduction is the high probability of residues from the chemical reaction on the surface of the nanoparticles.

Annealing in Oxygen Atmosphere

It is known that annealing as-deposited TiO$_2$ in oxygen atmospheres at more than 400 °C affects the electronic properties [283, 585, 1219, 1223, 1225, 1229, 1332].

One of the most important effects is the increase of the **conductivity** [283, 1219]. This is in contrast to annealing in nitrogen atmosphere, where the conductivity is not changing significantly [1219]. Yang et al. found an exponentially decreasing conductivity below 600 °C for annealing in oxygen [1228]. Mikhelashvili and Eisenstein observed a reduction of the transient current by four orders of magnitude after annealing the 25 nm thick film, fabricated by electron-beam assisted evaporation, at 700 °C for 60 min in oxygen atmosphere [659, 945, 1223]. This effect originates most likely from the annihilation of oxygen vacancies. The transient current is dominated by hopping conduction and they obtained a change of the impurity conductivity and average hopping distance from 10^{-11} (Ωcm)$^{-1}$ and 4 to 5 nm before annealing to 10^{-16} (Ωcm)$^{-1}$ and 6 nm after annealing, respectively.

However, above 700 °C annealing temperature, the conductivity increases significantly [585, 1226]. This is most likely caused by the well-shaped and extended grain boundaries resulting from the large grains which are generated at such high annealing temperatures [1228, 1333, 1334].

The **Schottky barrier** height is also affected by the proposed grain boundary conduction. It drops from roughly 0.53 eV in films annealed in air between 600 and 700 °C to 0.1 eV in films annealed at 800 °C [1228]. Due to high interface defect densities, the barrier height in as-deposited TiO$_2$ is 0.34 eV only [1228, 1332]. An even larger increase of the barrier height between as-deposited (0.5 eV) and annealed (0.94 to 1.0 eV) films was found at the silicon/TiO$_2$ film interface [1223, 1235].

The **interface impurity state density** drops from roughly 10^{12} cm^{-2}eV^{-1} to 10^{11} cm^{-2}eV^{-1} during annealing in oxygen atmosphere [283, 1225, 1229]. This impurity density is determined by applying the Terman method [1335] and the disorder-induced gap-state (DIGS) continuum model [1229, 1336]. This procedure is based on a hysteresis phenomena in the $C - V$ characteristic, which is explained by a spatial and energy distribution of interface defect states [1337–1339]. According to the spatial distribution of the traps, also the conductivity becomes position-dependent which is drafted by Dueñas et al. [1229]. Besides interfaces, grain boundaries between anatase and rutile grains are supposed to offer intra-bandgap states, which were observed in films fabricated by sputter deposition and ALD [1229].

During annealing, not only the total number and spatial distribution of impurity states is changing, but also the **energy levels of the trap states**. Yang et al. observed a decrease of the trap level from 0.51(4) eV to 0.62(1) eV during annealing in an oxygen atmosphere. Additionally, even deeper trap levels at 0.78, 0.84, and 0.9 eV below the conduction band were observed for annealed films [1219, 1225, 1340].

All the mentioned effects configure the conduction mechanism. As a consequence of the high trap density in as-deposited TiO$_2$ films, the conduction is dominated by Poole-Frenkel emission

[283]. In annealed anatase films, Fowler-Nordheim tunneling and Poole-Frenkel emission and in annealed rutile films, Schottky emission and grain boundary conduction are dominant [1228]. A brief overview on the sample-dependent conduction mechanisms is given in Table 2.9.

Zhang et al. found that annealing in air and oxygen-rich atmosphere have similar effects on the transient current [208]. There are reports that annealing in oxygen is not sufficient enough to remove all surface oxygen vacancies. It is suggested that surface passivation is more efficient with NO annealing or wet-chemical post-treatments [370, 1104, 1163].

Annealing in Vacuum and Nitrogen Atmosphere

The importance of the huge amount of defects and impurities in as-deposited TiO_2 films is emphasized by observations in single-crystalline rutile TiO_2. Even after reducing such single crystals in vacuum at 800 °C, the conduction mechanism is still Schottky emission across a barrier height between 0.87 to 0.94 eV [1204]. This result is in contrast to the low barrier height of 0.1 eV in grainy TiO_2 films annealed at 800 °C in an oxygen atmosphere. As discussed above, the only reasonable explanation for this observation is grain boundary conductivity in grainy films. The donor density and resistivity in annealed rutile single crystals is $7 \times 10^{20} \, cm^{-1}$ and 20–30 Ωcm, respectively [1204].

Annealing in Hydrogen Atmosphere

Annealing in hydrogen reduces TiO_2 [1066]. Low annealing temperatures provide the creation of defects on the surface since the diffusion constant of defects might be sufficiently small. Kim et al. found that the hydrogen concentration in TiO_2 correlates with the interface state density for annealing temperatures around 450 °C in hydrogen atmosphere [1235]. Hydrogen which is close to the TiO_2 surface is removed by annealing at 750 °C in oxygen atmosphere [1332].

2.8 Time-Dependent Transient Current Effects in TiO$_2$

Experimental conditions and a sufficiently high applied bias or current cause electronic or even structural changes in a metal oxide resulting in a temporary, reversible, or permanent change of the electronic properties. For applications, these effects could be either advantageous or catastrophic. Scientific investigations demonstrated the complexity of effects contributing to time-dependent transient currents during voltage or current stress. Even if no specific investigations on the current over time behavior are performed, a discussion of these effects is absolutely necessary, because whenever an instability of the metal oxide is present, it results in a voltage sweep-dependent hysteresis of the $I - V$ characteristics. Hence, the charge injection and conduction processes described in the last section are not sufficient anymore to describe the electronic properties of a metal oxide system. One of the most intensively investigated oxides is SiO$_2$ due to its outstanding meaning as a gate oxide in MOSFETs. So, there are many review articles and reports about time-dependent transient currents in this material demonstrating that the physical processes in metal oxides exposed to any kind of electrical stress are quite complex [1341, 1342]. There are models describing the time-dependent transient current in dependence of the applied bias, injected charges, oxide thickness, irradiation, defect density, temperature, and alternating fields, for instance [70, 1341]. Many effects described in SiO$_2$ can be transferred to TiO$_2$. However, it is often not clear which effect is dominant.

The property of showing a time-dependent resistivity is called **resistive switching**. One distinguishes between **nonpolar** and **bipolar** switching. Nonpolar switching means that the resistance is switched independently from the sign of the applied electrical field. In a bipolar switching device, the resistance is increased for a defined sign of the applied field and decreased for the opposite field direction [216, 223]. There are miscellaneous microscopic reasons for resistive switching such as ion migration [1343, 1344], charge trapping [1207, 1345, 1346], or Mott transitions [216]. A detailed overview on resistive switching is found elsewhere [1347].

In this section, some of the most important origins for time-dependent transient currents are described. An overview on some effects causing time-dependent transient currents and conduction mechanisms in TiO$_2$ nanorods is presented in Figure 2.18.

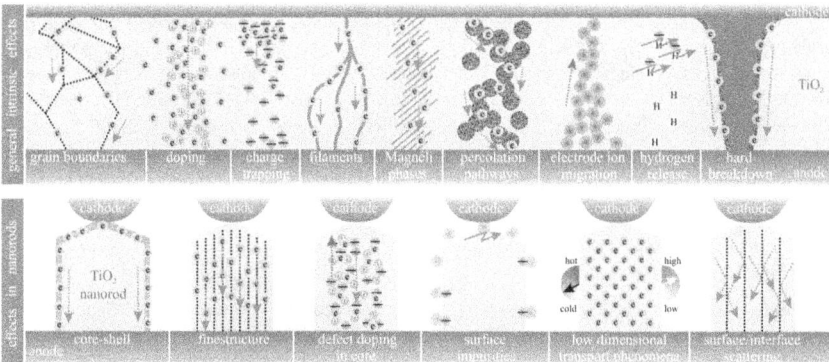

Figure 2.18: *Mechanisms causing a time-dependent transient current in TiO$_2$ films and nanorods. The defects are discussed in Sections 2.6, 2.7, and 2.8.*

2.8.1 Capacitive Effects

Geometric Capacitance

TiO_2 nanorod arrays (NRA) sandwiched between two electrodes form a geometric capacitor. Hence, one expects that this geometric capacitor is charged or discharges during any changes of the applied bias. The geometric capacity C_g is given by

$$C_g = \frac{\epsilon}{d_{ox}} \left(\epsilon_s A_{ox} + A_g \right) \tag{2.104}$$

where ϵ is the dielectric constant of vacuum, ϵ_s is the static dielectric constant of the applied TiO_2 structure, $A = A_{ox} + A_g$ is the electrode size, A_{ox} is the fractional electrode area covered with nanorods, A_g is the fractional electrode area corresponding to the interspace cross-section, and d_{ox} is the thickness of the NRA. One has to consider that differently fabricated TiO_2 structures have different static dielectric constants. The time constant τ_{ch} of a charging process is given by

$$\tau_{ch} = R_C C_g = \frac{U}{J_0} C_g \tag{2.105}$$

where R_C is the effective internal resistance of the whole sample for a certain applied voltage U and J_0 is the measured current at the beginning of the charging process. R_C depends on the applied voltage and the temperature.

Polarization Effect

In a capacitor filled with a perfectly insulating dielectric medium, the polarization of the atoms or molecules appears instantly when an external electrical field between the capacitor plates is applied. This assumption is reasonable since oscillation frequencies of the valence electrons around the atomic core can be up to several hundred terahertz and hence, these oscillations dominate the frequency-dependent behavior of the refractive index for visible light. As a consequence, a speed limit of the atomic polarization is excluded for electronic applications. If the polarization is mainly based on much slower effects such as lattice reconstructions or charge migration, a limiting polarization speed has to be considered, namely for the charging *and* discharging process [70]. Consequently, there is no exponential time dependence between the charging and discharging current. In stoichiometric TiO_2, the polarization is caused mainly by Ti–O bonds which generate dipoles in an applied electrical field [70, 1348]. The discharge current decays according to the Curie-von-Schweidler law [1349]. It is given by

$$J_p(t) = u \cdot P \cdot t^{-z}, \tag{2.106}$$

where u and z are constants. P is the total polarization or surface charge density.

Interface Capacitance

Bonds in the TiO_2 break if they are stressed by an electrical field or irradiation too much. This results in the creation of defects and charged defects are able to drift along the electrical field. Several effects of this ion migration will be discussed later. For the interface capacity, only the effect of charge accumulation at electrode/TiO_2 interfaces or extended crystal defects is relevant. The charge accumulation and its image charge cause dipoles affecting the capacity of the device. Accumulation processes appear on a rather long timescale. Often, there are several accumulation processes including different defects and accumulation regions and hence, a distribution of the relaxation time τ_i needs to be considered in mathematical models [1350].

Interface Capacitance in Nanostructured Devices

In nanostructured devices where the counter electrode is placed inside the interspace of the TiO_2 nanostructure, the electrode/TiO_2 interface and hence the interface capacitance is much larger compared to flat devices. This is the case, when the TiO_2 structure is used in an electrolyte or when the top electrode is deposited by certain PVD or CVP techniques. In particular, employing electrolytes generate a Helmholtz layer which increases the interface capacity a lot [1351]. The formation of such a layer provides another time-dependent behavior. In this thesis, the nanostructure-related capacitance is avoided by contacting the nanorods using a tip with a large tip radius as a top electrode. Thus, the capacity is given by Equation 2.104.

Displacement current

A displacement current occurs when the stored charge Q of the capacitor is changing. This is described by

$$I_{dis} = \frac{dQ}{dt} = \frac{d(CU)}{dt} = C\frac{dU}{dt} + U\frac{dC}{dt} = C\frac{dU}{dt} + U\frac{dC}{dU}\frac{dU}{dt} \qquad (2.107)$$

Trivially, a displacement current appears when the applied voltage is changed. Besides that, it has to be considered that the capacity C depends on the applied voltage U, too [879]. Such a behavior could be linked with nonlinear polarization mechanisms in the dielectric material and the drift of charged defects and impurities.

Additionally, trapping and detrapping of electrons affect the relationship between the stored charge and applied voltage. The emission rate of trapped electrons for a given trap level depends on the applied electrical field via the Schottky effect which was introduced in Equation 2.66. Hence, at a higher electrical field electrons from deeper trap levels are emitted. These emitted charges are extracted from the dielectric material and do not contribute to the total charge of the capacitor anymore. The current of emitted electrons is not proportional to the electrical field. Furthermore, neither exponential nor discrete trap levels promote a linear relation between the field and the detrapping current.

Another important effect that has to be considered is the short Debye length, which is only $1\,nm$ in TiO_2 due to its high polarizability. As a consequence, charges trapped in the bulk are shielded completely by neighboring atoms and hence, they do not influence the density of surface charges on the metal electrodes. Charges trapped close to the interface are not shielded completely by the surrounding TiO_2 and their electrical field penetrates the electrode surface. In the case of trapped electrons, negative surface charges in the neighboring electrode are decreased with an increasing number of trapped charges. If the trap density at the interface is changed by ion migration in a constant electrical field, the charge on the two electrodes is changing and a time-dependent displacement current appears [70].

2.8.2 Inelastic Electron-Phonon Scattering and Joule Heating

Inelastic electron-phonon scattering is the energy exchange mechanism between the electron system and the crystal lattice [1352]. In a closed system, there is no effective energy exchange in the thermal equilibrium. However, if energy is pumped into the electron system from an external power source, additional phonons are excited by electron-phonon scattering and the material is heated. External power sources could be electromagnetic radiation which is absorbed by electronic excitations or a bias-induced electrical current which results in Joule heating.

Thin metal electrodes with a low heat capacity and a bad thermal connection to the surrounding heat reservoir are heated up quickly and their electronic resistivity is increasing. At room temperature, the injected heat reduces the resistance of TiO_2 and increases the resistance of the metal electrodes. If the total resistance of the electrode is small, the letter effect is negligible. A decreasing resistance in TiO_2 is caused by almost all temperature-dependent conduction mechanisms such as Schottky emission, thermionic field emission, and Poole-Frenkel emission. Hence, the thickness and composition of the metal electrodes, the morphology, the composition and pressure of the surrounding gas as well as the geometry of the sandwiched TiO_2 in case of sub-micrometer electrodes are important parameters affecting the influence of heating. The heating rate of the lattice depends on the absorbed power, the heat capacity, the thermal conductivity, the thermal contact to the external heat reservoir, and the environmental temperature. The thermalization of the electrons happens within some nanoseconds and cannot be resolved in the analyzed $I - V$ characteristics.

2.8.3 Trapping and Detrapping of Electrons

The potential of a neutral bulk trap is given by

$$\varphi(r) = -\frac{\alpha q^2}{2(4\pi\epsilon\epsilon_s)^2}\frac{1}{r^4}, \tag{2.108}$$

where α is the polarizability and r is the distance from the trap center [1275]. The trap cross-section

$$\sigma_t \propto \left(\frac{\alpha}{(\epsilon\epsilon_s)^2}\right)^2 \tag{2.109}$$

is related to the high dielectric constant of TiO_2 relatively small [1232, 1353]. Trapping and detrapping cause time-dependent transient currents [1213, 1324, 1354, 1355]. In case of a constant trap density, the trap filling is saturated if the trapping rate is larger than the escape rate. Typically, this results in a time-dependence of the transient current. On the one hand, a high trapping rate reduces the transient current until all traps are filled and a high escape rate increases the transient current until all traps are emptied. Often this effect lasts not long or is hardly detectable since the number of electrons which are trapped or emitted is small compared to the number of injected electrons. On the other hand, trapped electrons charge the TiO_2 and subsequently change the electronic landscape inside the structure [1211, 1356]. In this case, a single electron can influence the behavior of many injected electrons. Hence, this effect is supposed to be more distinct compared to trapping losses.

The formation of traps results in an increase of trapped electrons and thus, the creation of an inhomogeneously distributed space charge is brought forward. The inhomogeneity is increased if the potential drops not uniformly across the TiO_2 structure due to boundary effects, preexisting defects, and charge accumulation. The consequence is a more complex time-dependent transient current which is affected by buildup charges and an oxide degradation.

Trapping in the Bulk

Bulk trapping is considered to be one of the most important driving factors of oxide breakdown, especially in SiO_2 which is extensively used as a gate oxide [1357–1361]. Trapped electrons reduce the mobility of free electrons via Coulomb scattering [370]. The time-dependent current in case of charge trapping in thin metal oxides at high electrical fields is given by [1362]

$$\Delta J(t) = A\left[1 - \exp\left(-\frac{t}{\tau}\right)\right] + \beta(U)t^\delta. \tag{2.110}$$

The first term describes the buildup charges and contains the time constant $\tau = \frac{q}{\sigma_c J^*}$ [1363], the trap cross-section σ_c, and the mean current J^* injected into the TiO_2 structure during electrical stress. The second term describes the stress-induced leakage current (SILC) which is characterized by the trap generation rate β and an exponent δ in the order of 0.3 to 0.5 [70, 1364, 1365]. The trap generation is linked with τ by

$$\ln\left(1 - \frac{N_t(t)}{N_t(0)}\right) = -\frac{1}{\tau}, \tag{2.111}$$

where the trap density N_t has the following frame conditions: $N_t(t) = 0$ for $t = 0$ and $N_t(t) = N_t(0)$ for $t \to \infty$ [70]. Houssa et al. determined a trap cross-section of $\sigma_c = 2.3 \times 10^{-19}$ cm^2 for thin TiO$_2$ films made by atomic layer deposition (ALD) [1232]. In this case, the traps were created by a hydrogen release process which is described later. However, the cross-section found in thin TiO$_2$ films made by plasma enhanced chemical vapor deposition is one order of magnitude larger (2.7×10^{-18} cm^2) emphasizing the meaning of the fabrication process once more [70]. Comparing the characteristic energy of the exponential tails determined from photoresponse measurements (E_t) and $k_b T_t$ obtained from SCLC measurements offer information about shallow traps [1214]. Dittrich et al. found a strong increase of E_t for intensive charge injection from platinum into TiO$_2$ and a temperature dependence of $k_b T_t$. Both effects are related to the formation of defects since the chemical properties of metal oxides are linked strongly to the presence of electrons [1366]. For instance, the trapping of an electron at a Ti^{4+} lattice site resulting in a Ti^{3+} site shows a completely different chemical environment compared to the empty trap [1214].

Trapping at the Interface

Typically, traps are not distributed homogeneously inside TiO$_2$ nanocrystals. For instance, the surface structure of TiO$_2$ nanorods differs between the bottom and the tip [168]. Hence, it is assumed that also the trap density is not distributed homogeneously inside a real TiO$_2$ structure. Because of surface and interface crystal defects, the trap state density is higher close to the surface [8, 1162, 1228, 1367, 1368]. Trap-assisted tunneling (TAT) results in positively and negatively charged traps close to the anode and cathode, respectively [70, 1369, 1370]. In nanoparticle networks, the electron transport is described by percolation where the diffusion coefficient increases exponentially with an increasing electrical field [1052]. This increase of the diffusion coefficient is related to trap filling [1052]. Efficient simulations of percolation in nanostructured metal oxides with exponentially distributed trap levels include the Miller-Abrahams hopping rates in random walk numerical simulations (RWNS) [1371, 1372]. Assuming that the density of deep traps is constant, the deep traps are filled almost completely if a sufficient number of electrons is injected or excited in the metal oxide. As a consequence, the electron transport is restricted to shallow traps which outnumber the deep traps due to the exponential trap distribution. Hence, the diffusion coefficient increases with the electron density. That is why it depends on the trap density and distribution, the electrical field, and the illumination density [1052, 1373–1376]. Including additional interface or surface states, the electron transport is usually described by the multiple-trapping (MT) model [1377]. Some defects such as oxygen vacancies [915, 916] are electron donors which are excited at room temperature and the mobile electrons diffuse randomly if no bias is applied [369]. Thus, a certain number of traps is filled all the time. And if the trap state density is particularly high close to the interface, an accumulation layer is created. For high charged trap densities, the effective potential close to the interface is lifted and consequently, the Schottky barrier is increased. Thus, while the new space-charge region builds up the transient current is time-dependent.

Anode Hole Injection

For large electrical field amplitudes in the order of $1\,MV/cm$, electrons gain a lot of kinetic energy within their mean free path [1378–1380]. If their kinetic energy is sufficiently high, inelastic scattering results in the formation of electron-hole pairs [1381]. The holes move anti-parallel to the electrons inside the TiO_2 and give rise to an additional stress-induced leakage current (SILC) [1342, 1382–1385]. It is assumed that these holes break bonds and contribute intensively to the electrical breakdown process [1261, 1381, 1386].

Furthermore, these holes can be trapped in the TiO_2 film in the same way as electrons and thus, the hole trapping causes a time-dependent transient current behavior [1385, 1387]. The holes are trapped and lower the field close to the injection interface [1388]. As a consequence, the Schottky barrier lowering caused by the initially applied field becomes less and the transient current decreases [1389, 1390]. Schwerin et al. and Khosru et al. found a hole injection-related drop of more than 50% of the transient current through SiO_2 within one minute [1384, 1385]. For SiO_2, it is predicted that the applied voltage stress must be much larger than the oxide bandgap energy divided by the elementary charge q [1387]. In case of TiO_2, this is more than $3\,eV$ which is not the case in the experiments performed in this work. The effect is observed in SiO_2 for thicknesses above $4\,nm$. Below this limit, direct valence band tunneling becomes dominant [1378].

The time-to-breakdown t_b in the anode hole injection model is given by

$$t_b = \frac{Q_b}{I_t} \tag{2.112}$$

where

$$Q_b = \frac{Q_p}{a_p \Lambda_p} = \frac{Q_p}{a_p} \exp\left(\frac{B^*}{\mathcal{E}} \phi_p^{3/2}\right) \tag{2.113}$$

is the charge-to-breakdown, Q_p is the critical hole fluence at the breakdown which is $0.1\,C/cm^2$ in SiO_2, a_p is the hole generation probability, Λ_p is the probability that a hole which is generated at the anode interface tunnels back through the barrier into the oxide [1356, 1391]. Typically, a_p is determined by curve fitting and all other parameters are calculated [1348].

For high constant voltage stress, the trap generation is described by a dispersive transport (DT) model [70]. The positively charged species, which are generated at the cathode/TiO_2 interface, drift into the TiO_2 by random hopping which is known as dispersive transport. Related to their positive charge these species act at least as Coulomb traps for electrons and their generation provides a time-dependent transient current [1392]. This model is described in detail for TiO_2 by Bera and Maiti [70].

An empirical relation between the trap generation probability Λ_t and the applied voltage U is given by

$$\Lambda_t = \frac{1}{Q_i^p} \frac{\Delta J}{J_0} = a \cdot \exp\left(\frac{U}{b}\right) \tag{2.114}$$

where Q_i is the number of injected charges, p is the exponent of the power law, and a and b are fitting parameters [70].

2.8.4 Drift Velocity Effects

Injected electrons drift through the TiO$_2$ from the cathode to the anode with a limited drift mobility which is usually related to electron trapping [1354]. In single-crystalline bulk TiO$_2$, this limit is given by $1\,cm^2/Vs$ [848, 871]. The drift mobility depends strongly on the morphology and fabrication technique. For instance in porous TiO$_2$, the drift mobility is only 1×10^{-7} to $1 \times 10^{-4}\,cm^2/Vs$ [739, 1114, 1214]. As a consequence, the transit time of electrons from the cathode to the anode becomes relatively large in certain TiO$_2$ structures. Hence, as soon as electrons are injected from the cathode by a sufficiently high bias, an electron cloud is moving across the TiO$_2$ and traps are filled gradually from the cathode side to the anode side [914]. As a consequence, the space-charge region and internal field amplitudes are changing until the transient current permeates the whole TiO$_2$ structure. The time constant of this effect is given by the transit time which is in the order of microseconds for drift mobilities of $1\,cm^2/Vs$ in a $1\,\mu m$ long TiO$_2$ nanostructure, but it is in the order of seconds for drift mobilities of $1 \times 10^{-4}\,cm^2/Vs$! The gradual change of the space-charge region causes a time dependence of the transient current because it affects the effective height of the potential barrier at the electrode/TiO$_2$ interface. In addition, the internal electrical field, which drives the electron current, is gradually changed as well.

And there are further consequences accompanying the change of the internal fields. First, the drift velocity is an average value. Due to the velocity distribution, the transient current at the TiO$_2$/anode interface is increasing over a certain time period. Second, besides the trapping rate also the detrapping rate is influenced by the change of the internal field. Third, the moving electron front is changing the capacity of the electrode/TiO$_2$/electrode system. This results in a displacement current that appears right after electrons are injected into the TiO$_2$ and superposes the transient current until the steady-state condition is accomplished.

Thus, a precise analytical formula describing the time dependence is difficult to find and numerical simulations might be the only approach. However, some parameters such as the saturation time are determined quite easily from the measurements.

An important feature of the drift velocity model is that the behavior of the first measurement is expected to differ from further measurements. Because as soon as the traps are filled throughout the whole TiO$_2$, a certain number of traps remain filled after the bias is switched off again. Hence, these localized charges and their effect on the space-charge region will be present when the bias is applied for the next time. After the bias is switched off, mobile electrons are only pulled by internal fields and the TiO$_2$ space-charge region relaxes partly and spare electrons are ejected at the electrode/TiO$_2$ interfaces.

The slow diffusion rate of electrons in defective metal oxides results in long relaxation times which are typically between a few seconds and up to one hour [1373, 1393, 1394].

2.8.5 Oxide Degradation

So far, the crystal structure was seen as immutable but it is not. A variety of time-dependent transient current studies were performed with thin gate oxides in order to understand leakage currents and oxide degradation during stress [205, 1395]. All discussed oxide degradation phenomena resulting in a time dependence of the transient current are supposed to be statistical effects [1261]. Thus, the time-dependent behavior of different similar samples or on different positions of the same sample can differ from each other in terms of time constants and current magnitudes.

Stress-Induced Leakage Current (SILC) and Time to Breakdown

In the beginning, a phenomenological description of oxide degradation is given. In the following sections, microscopic effects for the oxide degradation are discussed.

There are different ways to apply stress to an oxide: by voltage, current, high energetic irradiation, or heat [1261, 1396–1398]. Different degradation mechanisms show different voltage dependencies. An increase in temperature results in a faster degradation rate, which shows Arrhenius° as well as non-Arrhenius dependencies [1342, 1399, 1400]. For electrical stress, a general description of silicon oxide degradation is described by Ghetti including stress-induced trap generation, anode hole injection, charge carrier trapping, change of the flat band structure, and shift of the threshold voltage [1261, 1396]. As a consequence of these effects, the transient current is increasing over time. This additional stress-induced leakage current (SILC) is supposed to be mainly a result of an increased trap-assisted tunneling (TAT) [1401, 1402]. Often the SILC is composed of an electron and hole current [1382]. This part of the oxide degradation is called wear-out period. Above a critical defect density, the current jumps to significantly higher values and becomes noisy [1403–1408]. This is called the quasi-breakdown [1409], SILC-B-mode, [1410] or soft breakdown [1411]. These events are correlated with the trapped charges in the oxide [1408]. Afterward, the $I - V$ characteristic changes from an exponential behavior to a power law [1412]. In a defective sample, the charge transport is described as hopping between nearby traps and thus related to a percolation phenomenon as discussed in Section 2.7.2 for TAT [1411, 1413, 1414]. The noisy behavior indicates that soft breakdowns are reversible effects. It was observed in very thin oxide layers so far. Finally, the film is damaged irreversibly by a hard breakdown and the $I - V$ characteristic turns into an ohmic behavior [1415]. The current over time at constant voltage and current over voltage graphs are shown in the report of Ghetti [1261]. Intrinsic defects support the oxide degradation. In contrast, metal oxides with an initially low defect density show the highest reliability [1341].

The $\mathcal{E}$ and $1/\mathcal{E}$ model: There are two field-related time-dependent dielectric breakdown (TDDB) models: the $\mathcal{E}$ and $1/\mathcal{E}$ model [1396]. The $\mathcal{E}$ model describes a thermodynamical process which is driven by voltage stress. The median time to breakdown t_b is given by

$$t_{b,50} = \beta \exp\left(-\kappa \mathcal{E}_{ox}\right) \exp\left(\frac{E_b}{k_b T}\right) \qquad (2.115)$$

where β is a constant, κ is the field acceleration factor, $\mathcal{E}_{ox}$ is the electrical oxide field, and E_b the thermal activation energy of the breakdown process [1397, 1416]. The index "50" means that a breakdown is observed in 50% of the investigated samples after $t_{b,50}$. For a polycrystalline anatase TiO_2 film, the experimental value of E_b is 0.5–0.7 eV [1217, 1397]. κ depends on the dielectric constant, the applied external electrical field, the temperature, and it is decreasing with decreasing oxide thickness [1417–1419]. McPherson et al. discussed the relation

$$\kappa = -\left[\frac{\partial \ln(t_b)}{\partial \mathcal{E}}\right]_T = \frac{P_0(2 + \epsilon_s)}{3k_b T} \qquad (2.116)$$

between the field acceleration parameter κ and the dielectric constant in detail [1418]. Here, P_0 is the molecular dipole-moment. The electrical breakdown appears at a certain field amplitude $\mathcal{E}_b$ which has a typical relation to the static dielectric constant ϵ_s expressed by

$$\mathcal{E}_b \propto \frac{1}{\sqrt{\epsilon_s}}. \qquad (2.117)$$

This approximation fits for almost three decades of the dielectric constant within the same dielectric material [1420, 1421]. The $1/\mathcal{E}$ model is related to current stress and hence, it includes a term describing the Fowler-Nordheim tunneling and t_b is expressed as

$$t_{b,50} = \tau_0 \exp\left(\frac{G}{\mathcal{E}_{ox}}\right) \exp\left(\frac{E_b}{k_b T}\right) \tag{2.118}$$

where τ_0 and G are constants [1397, 1416].
If bond breaking is involved into the observed time dependence of the current, the effect should not relax by 100%. The reason for that is the forming of alternative bonds, that would have to be broken down in order to end up with the defect-free crystal again [1381, 1397]. Since this relaxation process needs energy a complete defect healing after removing the bias is very unlikely.

Ion Migration

Ions released from their lattice sites drift towards one of the electrodes [216, 1395, 1422, 1423]. The drift velocity and hence, the change rate of the transient current depends on the applied bias, intrinsic fields, extended crystal defects, and temperature [216, 1221, 1395, 1422, 1423]. Compared to titanium ions, oxygen ions or oxygen vacancies are more mobile. In a sufficiently high electrical field, weakly bound electrons leave their donor states and positively charged ions are created. Besides the generated ions, there are constantly charged species such as negatively charged oxygen ions. A homogeneous ion migration would change the charge carrier density and the electronic properties in the whole oxide equally [1423–1426]. However, for rather low electrical fields, only weakly bound ions are able to leave their lattice sites and hop to the next site. Weakly bound ions are mostly found in amorphous materials, on grain boundaries, and on other extended crystal defects [744, 1221]. The number of drifting ions is increased at higher electrical fields [1423]. The ion drift changes the local lattice composition and ion density. Hence, the number of drifting ions becomes time-dependent. The time-dependent ion current is often described by

$$J_{io}(t) = J_0 \alpha_{io} \exp\left(\left[\frac{t}{\tau}\right]^{\beta_{io}}\right), \tag{2.119}$$

where α_{io} and β_{io} are constants and τ is the relaxation time. The stretched exponential behavior results from the variety of ion release and transport processes in the material [1427–1436]. Each transport process has its own activation energy for the ion transport and relaxation time. The Kohlrausch stretching exponent β_{io} is between 0 and 1 and displays the correlation between the threshold energy for ion release E_{th} and average activation energy for the ion transport E_a ($E_a = \beta_{io} E_{th}$) [1430–1432]. The relaxation time τ is proportional to $\epsilon \epsilon_s / \sigma_{io}$, where σ_{io} is the ion conductivity in TiO_2 [1430].
As discussed in Section 2.7.3, the total ion current is most likely small and superposed by much stronger effects [1212]. A changing ion distribution modifies the electronic landscape in particular at interfaces. For instance, an increased number of electron donors lowers the Schottky barrier which promotes charge injection. As a consequence, the transient current becomes time-dependent. Ion drift is responsible for all oxide degradation related processes which are described in the following sections.

Reduction of the Oxide and Generation of Donors, Acceptor, and Traps

The defect and trap density is often not constant. There are a few mechanisms generating crystal defects during voltage or current stress. These defects affect the donor, acceptor, and trap density as well as the band structure as discussed in Section 2.6.1. Especially the formation of traps is well known in oxide layers [1397, 1437, 1438]. The Ti–O couples are highly polarizable and cause intensive stress to bonds. At field amplitudes above $8.6\,MV/cm$, the force is strong enough to break Ti–O bonds and the negative oxygen ion is released from the lattice site [70, 1418]. At such high fields, bond breaking is induced by phonon scattering or hole absorption [1418]. The remaining electrons in the generated oxygen vacancy form a hole trap or contribute to the conductivity [199, 201, 1046]. On the one hand, the increased number of donor states provide more electrons in the conduction band. This enhances the conductivity and the transition to SCLC is shifted to a higher voltage. On the other hand, the increasing defect concentration raises the defect scattering rate which lowers the conductivity. However, the overall conductivity becomes time-dependent. Furthermore, the Fermi level is elevated by the increased number of mobile electrons and hence, the Schottky barrier becomes smaller and thinner supporting all interface-limited conduction mechanisms. The in-situ generation of acceptor levels in TiO_2 is rather unlikely and was not reported so far.

Metallization

As a consequence of ion generation and migration, oxygen vacancies accumulate at the TiO_2 surface. Hence, the sub-surface region becomes highly n-type doped and the Fermi-level is elevated. As soon as the energy gap between the Fermi-level and the conduction band becomes smaller than $k_b T$, the Schottky barrier is negligible and the contact becomes ohmic. This effect is called metallization.

Even if the vacancies do not accumulate on the surface but in a close sub-surface layer, the metallization pulls the conduction band down towards the pinned Fermi level and the remaining Schottky barrier is getting thinner and thinner [219, 1439, 1440]. Below a certain thickness, the tunnel resistance becomes negligible and the contact becomes ohmic as well.

The density of oxygen vacancies is limited by Coulomb repulsion or a constant recombination rate with oxygen atoms from the environment [1217]. However, the time-dependent current must not saturate at any time. Because in the presence of an ohmic contact the main resistance is given by the series resistance of the bulk. Since it is a common opinion that electrical breakdown is just a matter of time and injected charges for any sample, it is reasonable that the generation of oxygen vacancies does not end after the metallization is completed.

Oxidation of the Electrode Material

When oxygen ions reach the electrode/TiO_2 interface, the electrode material can be oxidized [744]. This reaction is likely for ignoble metals such as aluminum, copper, iron, titanium, or silver. Hence, noble metals such as gold, palladium, and platinum are preferred as an electrode material. And there might be an exchange of oxygen between TiO_{2-x} and conductive oxides such as ITO or FTO as well. According to the reaction kinetics, the TiO_{2-x} and electrode material are oxidized or reduced.

An oxidation of an ignoble metal electrode increases the interface resistance and the whole device might become insulating. An oxidation of the TiO_{2-x} by a highly doped metal oxide reduces the conductivity since the density of oxygen vacancies is reduced. Similarly, the conductivity of TiO_{2-x} is increased if it is reduced by the electrode material.

Since the mentioned reactions appear mainly at the electrode/TiO_{2-x} interface, the transient current reacts very sensitively to this effect. In general, it is unlikely that this effect is reversible, in particular after the formation of a thick oxide layer at an ignoble metal electrode.

Generation of Filaments and Magnéli Phases

Oxygen vacancies are generated along crystal defects preferentially and drift along these interruptions towards the cathode interface [1216, 1439]. After a certain time, oxygen vacancy-rich

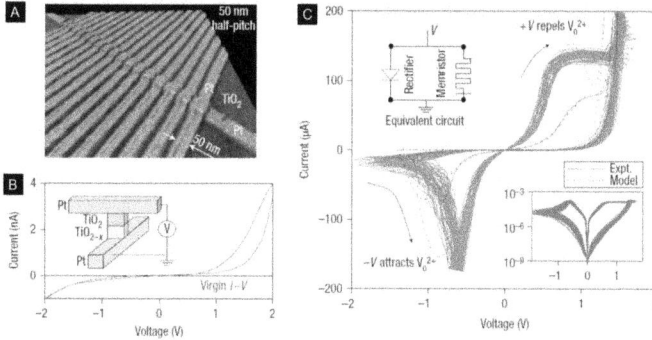

Figure 2.19: *A) AFM image of a resistive switching nano-crosspoint device made of a $Pt/TiO_{2-x}/Pt$ structure. B) Hysteresis observed for the initial I-V measurement on a single crosspoint. C) Experimental (solid, 50 loops, red: standard switching, blue: virgin state) and calculated (dotted) I-V characteristics of a single crosspoint (V_O: oxygen vacancy). Top inset: The equivalent circuit of a resistive switching crosspoint device. Bottom inset: semilogarithmical plot to emphasize the high switching ratio between ON and OFF state. Details, especially concerning the applied model, are found in [219]. Reprinted by permission from Macmillan Publishers Ltd: Nature Nanotechnology ([219]), Copyright (2008).*

and highly conductive paths are formed in the oxide [70, 1215, 1438]. As soon as these paths form a continuous connection between the electrodes as shown in Figure 2.13 C, the electrical resistance drops significantly and a low resistance state (LRS) is created. Typically, these conductive extended defects are called filaments. For sufficiently high vacancy densities along these paths, the charge transport mechanism changes from hopping to band conduction [204, 1441–1443]. Hence, the expected conduction mechanisms are ohmic for low applied electrical fields and SCLC for high current densities. In the absence of continuous filaments, Poole-Frenkel emission is likely [1444].

The filaments are formed between a few nanoseconds [1405] and several seconds [224]. Intensive current stress results in a fast degradation because Joule heating destroys the aligned oxygen vacancy structure and a high resistance state (HRS) is formed [220, 222, 224, 1445, 1446]. A typical resistive switching device with its $I - V$ characteristic is shown in Figure 2.19. The most striking feature of the $I - V$ characteristic is the hysteresis behavior caused by the switching.

The filaments are stable at room temperature for many days or even years [220, 1199, 1342]. The radius of filaments is discussed in literature and proposed to be between 5 and 100 nm [223, 1447–1449]. Typically, filament-based resistive switching is reported for thin TiO_2 films. However, the formation of filaments is also predicted in anisotropic nanostructures such as TiO_2 nanotubes [1202].

If the extended defects become larger, they lose their 1D character. Intensive TEM analysis showed that Wadsley defects and Magnéli phases are established in these reduced TiO_2 volumes [223, 1221].

Molecular dynamics simulations indicate that the models of a homogeneous ion migration and the formation of highly localized filaments are two extreme cases of one stochastic process [1450].

Percolation Model and Electronic Breakdowns

Similar to filaments, percolation pathways open highly conductive channels in TiO_2 films [70, 1347, 1348, 1405, 1406]. Percolation appears above a critical defect density [1451]. These percolation pathways are larger than filaments and the crystal structure is much less defined compared to Magnéli phases. Because of their rather large size, these events result in much higher current density changes as filaments would do [1437, 1438]. Usually, the effect of percolation pathways is described by a simple model of spherical defect accumulations that oppose no high resistance to electrons in an electrical field [70, 1347]. Inside these spherical defect volumes, a stress-induced trap generation takes place [1361, 1452–1454]. It is assumed that the conduction mechanism in percolation pathways is based on slow trapping and detrapping events [1407, 1411, 1455]. Since trapping and detrapping rates are not constant the transient current is fluctuating [70]. Often, the formation of such traps is linked with the generation of positive charges in the oxide [1406]. This phenomenon is explained in more detail in the anode hole injection (Section 2.7.3) and hydrogen release model (Section 2.8.7). For high defect generation rates, the probability of a continuous row of defects between the electrodes ascends [1406]. The formation speed of such a continuous path depends on the oxide material and the oxide thickness. In case of 9.3 nm SiO_2, it is about 2×10^6 cm/s which is almost equal to the sound velocity in silicon [1405]. The defect volume is not fixed in the percolation model and experimental values of 10^{-13} to 10^{-12} cm^2 are reported [1405]. So, either many small or a few huge defects establish a connection between the electrodes [1261]. As soon as there is a connection between the electrodes, a current is flowing through the percolation pathway which heats up the film locally. This effect results in healing or further growth of the percolation pathway [1406]. As a consequence, fluctuations appear in the measured current. Typically, the formation of such a pathway ends up in soft and hard electronic breakdowns [1411]. The breakdown is defined as the time when complex fluctuations of the transient current appear [1406].

2.8.6 Localized Resistive Switching Behavior

There are many reports claiming that the microscopic process for resistive switching is localized in a small defined volume of the oxide [68, 1203, 1213, 1222, 1324, 1456–1458]. Within these switching TiO_2 fractions, there are various resistive switching mechanisms suggested. The resistance can be switched by trapping and detrapping in a defective region of the TiO_2 structure while the remaining defect-free region does not contribute to the switching behavior [1213, 1324, 1456]. A commonly used explanation for the trapping-related resistive switching is a switching of the conduction mechanisms between a high resistive TFLC and a low resistive SCLC. Hence, the resistivity is small for filled traps. For very large trap densities, a reduction of the conductivity caused by Coulomb scattering by filled traps is reasonable as well. However, for the latter explanation, the resistance is high for filled traps. A different approach is to explain the resistive switching with the creation and extinction of filaments in a nanometer-thick sub-interface layer [68, 222, 1221, 1457, 1458]. The advantage of such a thin layer is the fast switching process even in the case of ion migration. A similar approach for a fast resistive switching device is the creation of silver filaments in an Ag/TiO_2 system. As soon as the filaments are formed, the resistive switching occurs only in a tiny volume close to the electrode resulting in switching times of a few hundred nanoseconds [1325].

2.8.7 Effect of Adsorbates

Condensed Physisorbed Water

If the water is removed from the environment of the sample, the $I - V$ characteristics change not immediately but it needs a settling time of more than one minute [769]. Water or other molecules on the surface of TiO$_2$ promote an electrolysis process [1459]. During such an electrochemical process free charges are generated or vanished which affects the transient current through the TiO$_2$ nanostructure. The remains of the electrolysis might be electron donors or acceptor and hence, the electronic properties of the surface changing continuously.

Hydroxyl-Groups and Hydrogen Release Model

In general, hydrogen gets into TiO$_2$ films during the fabrication, by adsorption of gas atoms from the environment, and by electrochemical reactions on the surface. Wet techniques such as spray pyrolysis, sol-gel spin coating, atomic layer deposition, or hydrothermal growth provide water as a source of hydrogen. Although post-annealing reduces the amount of water, there is a significant chance that hydrogen from water molecules remains in the TiO$_2$. But even sputtered TiO$_2$ might contain some hydrogen since any substrate exposed to the atmosphere is covered with a monolayer of hydroxyl groups from water molecules [1202, 1460–1462]. Without pre-annealing of the substrate in the vacuum chamber of the sputter deposition device, these hydrogen atoms remain on the surface and are implemented into the substrate/TiO$_2$ interface. Besides pre-annealing, the water content is reduced by starting the sputter process with an insulating layer such as SiO$_2$ or Al$_2$O$_3$ on the substrate. Continuing with the deposition of the electrode/TiO$_2$/electrode structure without exposing the sample to the atmosphere in between keeps the active TiO$_2$ nearly hydrogen free.

Injected electrons with sufficiently high energies generate H$^+$ ions via impact ionization. These ions are accelerated by the applied electrical field towards the cathode and are able to break bonds and create traps with the voltage-dependent generation rate

$$\alpha(U) = \alpha_0 \exp\left(\frac{qU - E_{th}}{E_{th}}\right) \exp\left(\frac{(ql_hU)/(2d_{ox}) - E_a}{k_bT}\right), \tag{2.120}$$

where α_0 is a constant, E_{th} is the threshold energy for the release of H$^+$ ions, l_h is the hopping distance, d_{ox} is the oxide thickness, and E_a is the activation energy for the H$^+$ transport [1463]. Hence, the left term describes the release and the left term the transport of H$^+$ ions. H$^+$ ions create usually neutral trapping centers such as TiOH sites [1464]. Thus, the trap state density inside the TiO$_2$ film and on the cathode/TiO$_2$ interface is increasing. A similar effect is reported for SiO$_2$ [1465, 1466]. Electron and proton trapping results in a time-dependent transient current density behavior [1261].

In contrast to the anode hole injection model, hydrogen release is observed already for 1.2 eV in ultra-thin SiO$_2$ [1467] which is far below the bandgap energy of SiO$_2$. Thus, this effect might appear in the investigated TiO$_2$ structures, too.

Houssa et al. performed time-dependent current measurements through thin TiO$_2$ films made by ALD and embedded in an Au/TiO$_2$/SiO$_x$/Si structure [1232]. They suggested that the water molecules in the as-deposited ALD layer are decomposed by hot electrons. For a hole excitation, the electrical field is not high enough. Hence, only mobile H$^+$ protons are generated which break bridging oxygen bonds and form TiOH neutral centers. These neutral centers are electron traps with a rather small trap cross-section [1232].

Electrolysis Model

In an atmospheric environment, there are water molecules, OH-groups, and organic molecules on the surface. Hence, electrochemical reactions resulting in CO_2, water, or other molecule degradation could take place at "open" interfaces such as the tip contact used on NRAs in this work [1468–1470]. The reactions happen violently at high bias causing changes in the resistivity of the interface. After heavy electrocatalytic reactions, incoming molecules are arranged differently at the interface and the resistivity is changed frequently. Several driving forces for incoming molecules are supposed. The thermodynamical driving force is based on the density gradient of molecules between the surrounding atmosphere and the contact area, where the density of educts is lowered and the density of products is increased by electrochemical reactions. Here, molecules move from high density to low-density regions. The second driving force is due to wetting. In ambient atmosphere, a humidity, temperature, and wetting angle-dependent water film covers almost all surfaces. If this water film is reduced by water splitting, surface tension pulls water from the environmental area into the contact. In addition, there are charged molecules in the atmosphere, which are attracted by Coulomb interaction. This driving force does not depend on the polarization of the applied electrical field.

2.9 Optoelectronic Properties of TiO_2

As mentioned in Section 2.5, it is distinguished between optical, electronic and optoelectronic properties in this work. In contrast to optical and electronic characteristics, optoelectronic effects were not investigated here. However, optoelectronic features are often directly linked with the optical and electronic properties. Hence, the most important optoelectronic mechanisms are addressed in this section. An overview of various mechanisms affecting the electron system in TiO_2 in general and in TiO_2 nanorods is presented in Figure 2.20.

2.9.1 Excitation of Optical Phonons

Discrete absorption lines related to electron-phonon coupling appear in the infrared (IR) light regime due to the excitation of vibrational modes of chemical bonds inside and on the surface of TiO_2 [1471]. This is used to distinguish bonds of pure TiO_2 and bonds involving impurities and defects. Thus, IR and Raman spectroscopy are powerful tools to investigate the molecular structure of bulk defects as well as surfaces and interfaces between TiO_2 and neighboring materials. The excitation of optical bulk phonons is used to determine the morphology of TiO_2.

2.9.2 Excitation of Bound Charge Carriers

The absorption coefficient $\alpha \propto K \left(\frac{\Delta E}{\hbar \omega} \right)^y$ depends on the energy $\hbar \omega$ of the incident light and the available excitation energies ΔE in the medium [1472]. The pre-factor K scales with the effective electron mass and ΔE refers to the bandgap energy E_g or trap level depth E_t. y equals 0.5 and 2 for direct and indirect semiconductors, respectively [1472]. Direct and indirect transitions in TiO_2 have been discussed intensively in literature [736, 897]. Calculations indicate that rutile has an indirect and anatase a direct bandgap [732–734].

The photoluminescence (PL) spectrum is derived directly from the electronic band structure. There are several direct and indirect transitions between states at the Γ and X point [736]. Pure TiO_2 is transparent in the visible regime but shows a high absorption of UV radiation due to the bandgap energy of roughly 3 eV. Thus, TiO_2 nanostructures with a large surface and high crystallinity such as nanotubes are valuable for photodetector devices [1083]. However, absorption and PL for visible light is caused by intra-bandgap states originating from defects and impurities. Hence, TiO_2 with appropriate defect and doping densities can be used as a photodetector in the visible light regime as well. It has to be considered that UV light is powerful enough to break bonds in metal oxides and consequently, generate further crystal defects [867], in particular in anatase TiO_2 with its weaker bonds [229]. The polaron binding energy results in another excitation energy which is around 0.8 eV [859].

Defect- and Dopant-Induced Absorption, Color Centers, and Quantum Size Effect

In TiO_2, impurity absorption overlaps bandgap excitation below 410 nm [834]. In particular, anatase TiO_2 has significant defect-related PL [1473]. The continuous distribution of trap states below the conduction band results in an exponentially decreasing absorption profile of the absorption spectrum, which is known as Urbach tail [748, 1474–1476]. This is why anatase is the preferred morphology for photocatalysis applications [229]. These defects are mainly related to grain boundaries and surface defects [1477].

Post-deposition annealing lowers the defect density and consequently, the absorption of defective initial materials such as sol-gel TiO_2 films [985] or hydrothermally grown rutile TiO_2 nanorods [1478] depends on the annealing temperature. Even in single crystalline anatase TiO_2, oxygen annealing affects the PL which is a strong hint for the importance of surface defects in

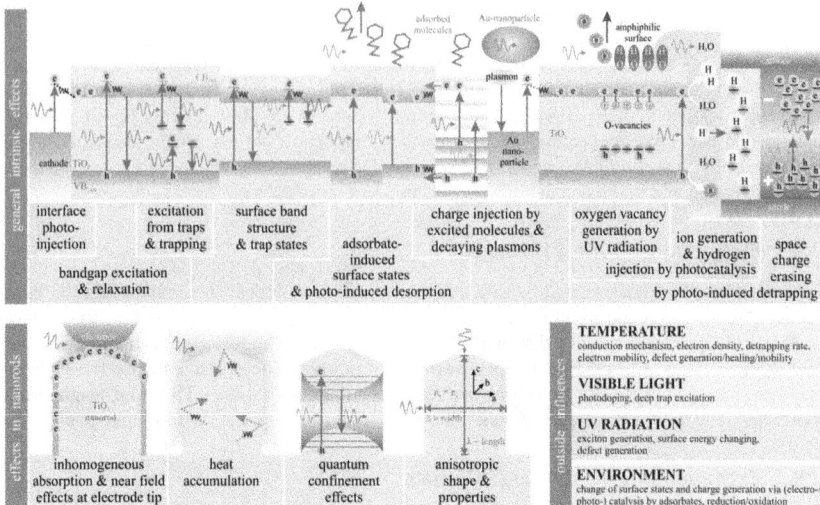

Figure 2.20: *Draft of mechanisms affecting the transient current during UV-Vis-NIR exposure in TiO₂ films and nanorods. Electronic properties which are affected by temperature, visible light, UV-light, and environmental conditions are listed in the lower right part of the graphic.*

anatase TiO$_2$ [1479]. There are several intra-bandgap states resulting from oxygen vacancies [736]. However, there are additional PL signals in the visible regime which do not result from crystal defects. For instance, self-trapped excitons cause a PL response at 2.4 eV [134].

For metallic titanium oxide, a blue shift of the absorption related to the Burstein-Moss effect appears [1049, 1050].

Localized trap accumulations show an increased recombination of charge carriers in TiO$_2$ [1477]. The radiation from the recombination process is in the visible light regime and hence, these defect sites are called color centers. Such color centers were found in as-grown [1098, 1480, 1481] as well as in reduced anatase TiO$_2$ nanotubes [919, 922] and nanoparticles [1482] in combination with a reduced bandgap energy. Mercado et al. give a deep insight into the charge generation, diffusion, trapping, and charge carrier recombination in single TiO$_2$ nanotubes [1483].

It is clear that all discussed effects for defective TiO$_2$ are transferable to doped TiO$_2$. The choice of dopant and its density is used to adjust the required absorption spectra [1025, 1484]. As mentioned in Section 2.8, the bandgap increases for small TiO$_2$ nanoparticles due to quantum size effects, which result in a blueshift of the absorption spectra [687].

Photogenerated Charge Carriers and Their Relaxation

The injection and excitation of electrons increase the number of mobile electrons in TiO$_2$ [1153]. Besides other effects discussed in Section 2.8, trapping and detrapping is not in equilibrium at the beginning and a time-dependent transient current appears. A feasible graphical explanation for this behavior is given in Figure 2.21. The time period until equilibrium depends on the injection or excitation rate [1485, 1486].

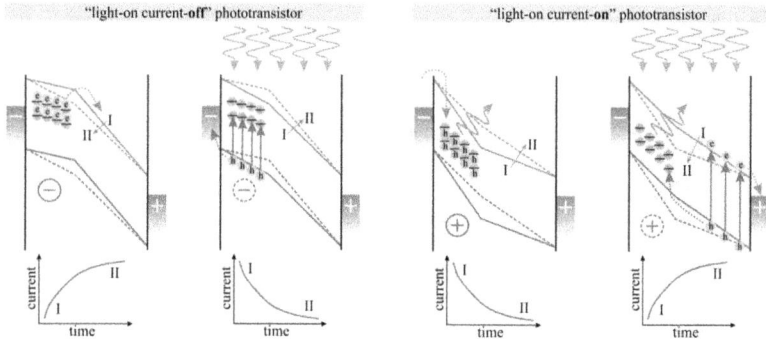

Figure 2.21: *Different time-dependent processes appear for illuminated TiO_2 structures. A reasonable explanation is charge trapping and detrapping. Under certain conditions, illumination results in an increase or decrease of the transient current ("off" and "on" transistor logic). However, this is not the only possible scenario describing the response of the transient current on light exposure.*

Often, intensive UV exposure generates more mobile electrons than injection from an electrode and hence, it causes a slightly faster current change. However, these processes are relatively slow since long-range diffusion is necessary to guide the system into an equilibrium state. As long as UV-induced excitation or charge injection is going on, charges are trapped and detrapped continuously. The trapping process of a single electron in a metal oxide appears on a much shorter timescale as the system needs to achieve a constant electronic landscape and charge distribution. This is demonstrated in Figure 2.22.

Short Time Processes

Short time processes last between a few femtoseconds and less than a microsecond and include typically exciton generation and splitting as well as charge carrier recombination [1201, 1487]. Diffusion processes are only relevant on these timescales for small diffusion distances and high diffusion coefficients D_c. Recombination events limit the diffusion length $L_d = \sqrt{D_c \tau_r}$ of free charge carriers as a function of the recombination time τ_r (charge carrier lifetime) [1052, 1488]. The lifetime of charge carriers in TiO_2 depends on various factors such as the morphology, defect and impurity density, and the shape. The quantitative determination of the lifetime is found elsewhere [1489–1491]. The trapping of electrons is directly linked with the recombination. Diffusing electrons are trapped in shallow traps faster (100–200 ps) than in deep traps (500 ps) [1487].

Long-Time Processes

Long-time processes appear on a timescale between milliseconds and hours. They are related typically to slow diffusion processes and macroscopic diffusion lengths [223, 1209, 1217, 1397, 1492]. A typical example of the appearance of short and long relaxation processes are TiO_2 nanoparticle films, where the fast process is linked with recombination in particles and the long process is addressed to hopping of charge carriers between the nanoparticles [1397]. The excitation of charge carriers in TiO_2 cause the same time-dependent effects that are reported in Section 2.8 for electrons which are injected from an electrode. Hence, it takes some time until

111

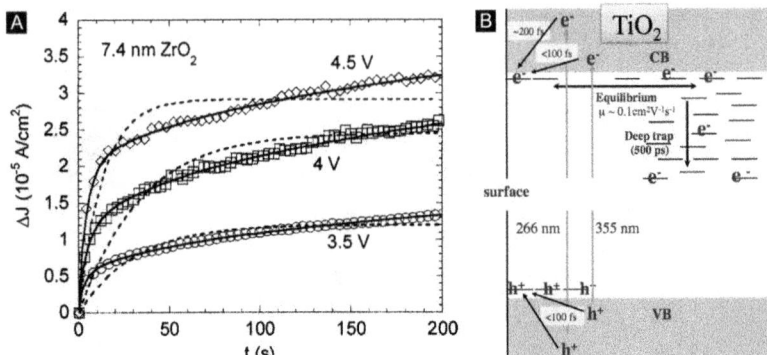

Figure 2.22: *A) Experimental (symbols) and fits (solid and dotted) of the time-dependent transient current evolution in a 7.4 nm thin ZrO$_2$ film in case of a constant bias. Details about the models are found in [1232]. Reprinted from [1232], with the permission of AIP Publishing. B) Trapping times of excited electrons and holes in TiO$_2$ probed by terahertz time-domain spectroscopy. Details are found in [1487]. Reprinted with permission from [1487]. Copyright (2009) American Chemical Society.*

an equilibrium between trapping and detrapping and a steady charge distribution is established. This results in a typical time-response of the transient current after the UV light source is turned on and a discharge current is observed after it is turned off again. This is observed for several TiO$_2$ nanostructures [948, 1198, 1493].

2.9.3 Electron Injection via Adsorbates

Besides doping, adsorbates are used to achieve high electron densities via visible light excitation [1068]. This strategy is often employed for photocatalysis and photovoltaics. There are mainly two approaches using adsorbates.

The first approach includes organic dye molecules which increase the light absorption and transfer electrons into the conduction band of TiO$_2$. The second approach is the use of quantum nanodots (QD) [1494]. The bandgap in semiconducting QDs and hence the absorption is adjustable by varying their size and shape.

Another effect is based on plasmonic excitations in metallic quantum dots. Decaying plasmons generate hot electrons, which are able to cross the energy barrier between the metal and the TiO$_2$ [1495, 1496]. Consequently, different shapes and materials were investigated in the past such as gold [1497–1502], palladium [1503, 1504], silver [1505], CdS [1506], and carbon [1507–1511].

2.9.4 Light-Controlled Density of Mobile Electrons

Besides the charge carrier injection by photoexcited adsorbates, additional surface atoms or molecules can reduce the number of electrons, too. A famous electron scavenger is oxygen [8, 739, 928, 1078, 1090]. Hence, the conductivity of TiO$_2$ nanostructures depends on the oxygen-pressure. Light has enough energy to remove the adsorbed oxygen molecules from the TiO$_2$ surface. Consequently, the density of adsorbed oxygen is reduced by light exposure. After

turning the light on and off, the system needs a while until the equilibrium electron density is reached, which is seen in a time-dependent transient current [1080–1082, 1512].
An additional effect that promotes the desorption of oxygen is the accumulation of holes at the surface [1083]. In nanostructures, these holes result often from the splitting of photogenerated excitons in the space-charge region close to the surface. The space-charge pulls electrons to the core and holes to the surface [1083]. Adsorbates are able to annihilate or create surface states and consequently, a large number of adsorbates might even change the color of a semiconductor. This effect is also known for TiO$_2$ and is valuable for gasochromic devices [48–50]. In particular, TiO$_2$ doped with palladium or tantalum is a promising candidate for such devices [47].

2.9.5 Excitation of Mobile Electrons

Oscillations of mobile electrons are excited by incident electromagnetic waves as well. The plasma frequency $\omega_p = \sqrt{\frac{\eta_e q^2}{\epsilon_{opt} m_e^*}}$ is a measure of the electron density η_e [1513]. For highly doped TiO$_2$ with a huge density of mobile electrons, the plasma frequency is in the terahertz regime [865]. Hence, the density of free charge carriers, their short-range mobility, photoconductivity, and the relaxation rate of photoexcited charge carriers is determined from terahertz absorption characteristics.

2.9.6 Photoconductivity and Photocurrent

In optoelectronic devices, the interaction between electrons and holes has to be considered. The decay of excitons into free charge carriers is promoted by an internal electrical field which is provided by a space-charge layer close to the interface. Consequently, the internal quantum efficiency is enhanced if a larger volume of the TiO$_2$ is affected by the space-charge region. For instance, a mesoporous film made of single TiO$_2$ nanoparticles [1514] has a lot of surface and interface areas that provide space-charges and hence, the photoconductivity in such a film is larger than in bulk TiO$_2$ [8].
For high electrical fields, the TiO$_2$ becomes strongly polarized by several effects such as a stretching of the Ti–O bonds, ion drift, or local trapping and donor ionization as discussed in Section 2.8. This is supposed to affect the recombination of charge carriers and reduces the increase of the photocurrent at high applied voltages as it is observed for TiO$_2$ nanotubes [1198]. For very intensive UV exposure, the charge carrier density becomes so high that the interaction between charge carriers becomes significant [836]. An additional effect is observed in SnO$_2$ nanowires where the built-in field separates the charges and holes move towards the surface and support the desorption of adsorbed oxygen [1288]. Consequently, the transient current is increasing [1083]. The long processes discussed above affect the response of the photocurrent. When the light source is switched on, it takes a while until a constant current is obtained and when the light is turned off a discharge current appears.

2.9.7 Internal Photoemission

A metal/TiO$_2$ interface provides another absorption mechanism, namely the excitation of electrons from the metal into the conduction band of the semiconductor. This effect is also reported for TiO$_2$/metal contacts and generates a photocurrent under visible light exposure even in the absence of intra-bandgap states [1214].

3 Manufacturing of TiO$_2$ Nanostructures

This section illustrates the state of the research of anatase and rutile TiO$_2$ thin film deposition and rutile TiO$_2$ nanorod fabrication techniques applied in this work. The results of this work upgrade the knowledge and control in certain parts of the listed techniques. Thus, some issues will be picked up in the "results and discussion" chapters again.

3.1 Thin TiO$_2$ Films

There are plenty methods for depositing thin TiO$_2$ films [1515]. New methods arising, other methods have been replaced by improved new methods. Most of the thin film fabrication methods have been developed in the last decades of the 20th century in conjugation with characterization techniques reaching resolutions down to the nanometer regime. The following overview shows the most important features of the techniques used or developed further in this work. Often, the application makes the choice of the technique. It should be mentioned that this list does not contain all available techniques including their modifications.

3.1.1 Sputter Deposition

One of the most common methods is **sputter deposition** [1516–1521]. The basic idea is to accelerate ionized noble gas atoms and guide the ion beam on a target consisting of the material that should be deposited. The knocked out target atoms move with a high kinetic energy through the process chamber and are settling down on the substrate. Sputter targets are available with standard purities of up to 99.99% (product sheet of TiO$_2$ sputter target, Testbourne Ltd., June 2016). Typical impurities in TiO$_2$ sputter targets are iron (16ppm), silicon (10ppm), copper (8ppm), cobalt (8ppm), tin (8ppm), manganese (7ppm), nickel (7ppm), vanadium (7ppm), and tungsten (6ppm) (exemplary certificate of analysis, Testbourne Ltd., December 18th, 2015). Sputtered atoms have a relatively high kinetic energy resulting in enhanced adhesion to the substrate. Sputter deposition at room temperature usually results in amorphous TiO$_2$, but polycrystalline films can be achieved by calcination. Alternatively, titanium is deposited by sputter deposition using target purities up to 99.9995% (product sheet of titanium sputter target, Testbourne Ltd., June 2016) in combination with an addition annealing in oxygen-rich atmosphere. Typical impurities in titanium sputter targets are carbon (8ppm), oxygen (8ppm), iron (6ppm), hydrogen (6ppm), nitrogen (5ppm), nickel (4ppm), aluminum (3ppm), magnesium (2ppm), molybdenum (2ppm), and tungsten (2ppm) (exemplary certificate of analysis, Testbourne Ltd., December 18th, 2015). Compared to the sputter deposition of TiO$_2$, the deposition of titanium offers significantly lower impurity densities. Doping gradients are fabricated by depositing from several targets at the same time applying time-dependent deposition rates.

3.1.2 Laser Ablation

Another common deposition method for TiO$_2$ is **laser ablation** [797, 1522–1524]. It is related to sputter deposition, but the setup is less expensive and the target material is consumed more efficiently. This technique uses the high energies gained with laser beams to remove atoms from the target material. The film composition and morphology is controlled by the process

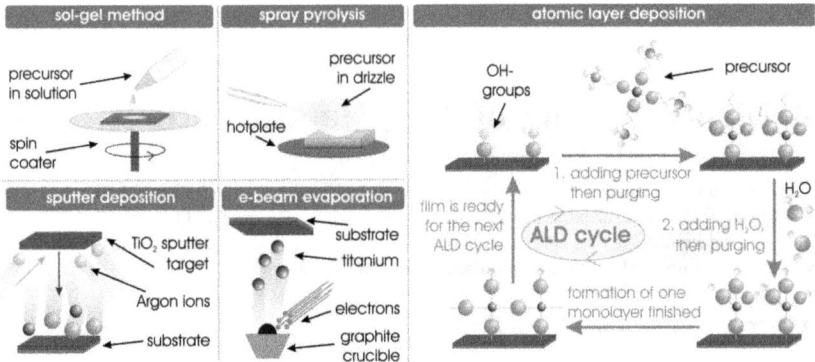

Figure 3.1: *Selected methods for the fabrication of TiO$_2$ films.*

parameters [483, 1525]. Besides TiO$_2$, titanium is used as a target material in combination with oxygen-rich atmosphere or post-deposition annealing as well [1526]. By immersing the TiO$_2$ target into a liquid during laser ablation nanoparticles with diameters down to 3 nm can be achieved [1527–1529].

3.1.3 Electron-Beam Evaporation

Another well controllable method resulting in TiO$_2$ films is **electron-beam evaporation** [1530, 1531]. In this case, the target material is heated up with an intense electron-beam until the material evaporates. Either titanium is deposited in oxygen-rich atmosphere resulting in TiO$_2$ directly (reactive electron-beam evaporation) or the deposition process is performed in vacuum with a following annealing step in oxygen-rich atmosphere similar to the sputter deposition of titanium [135, 589, 1532, 1533]. Since the temperature of the target material is above 3200 °C [1534] the kinetic energy of titanium atoms is quite high and thus, the hopping rate h on the substrate is large as well. But with increasing electron-beam power, the deposition rate ξ is increasing rapidly and thus, the ratio ξ/h is increasing. It is known that the density of islands formed on a substrate is proportional to $(\xi/h)^{1/3}$ [1535]. Therefore, the film roughness is very sensitive to the deposition rate. The stoichiometry and morphology of TiO$_x$ deposited by evaporation in oxygen-rich atmosphere depend strongly on the applied oxygen pressure [1536]. A technical advantage of the evaporation in vacuum is the perpendicular impact of the atoms on the substrate due to the low scattering at gas atoms. This feature is essential for high resolution optical and electron-beam lithography.

3.1.4 Spray Pyrolysis

There are several inexpensive, chemical deposition methods. A very simple method is **spray pyrolysis**. A titanium-containing organic precursor is sprayed with an aerosol can on a heated substrate in ambient or defined atmosphere [1537]. Dependent on the applied recipe, TiO$_2$ films or particles are gained with this method [1538]. In this work, titanium isopropoxide (Ti(OPr^i)$_4$) is used as a precursor for thin film fabrication. The substrate is kept at elevated temperatures and Ti(OPr^i)$_4$ decomposes by reacting with surrounding oxygen resulting in a polycrystalline anatase TiO$_2$ film [1539].

3.1.5 Sol-gel Method

Sol-gel methods use titanium alkoxides such as titanium n-butoxide, titanium isobutoxide, titanium isopropoxide, and titanium 2-ethylhexoxide as a precursor [1540]. The sol-gel, which contains the precursor, is deposited by spin coating on the substrate. Topography, morphology and structure are controlled by the kind of solvent, acid addition, molar ratios of alcohol/alkoxide and water/alkoxide, reaction temperature and post-deposition annealing temperature [1540]. Also, certain sol-gel methods promote porous structures and nanoparticle layers [1541]. The highly reactive alkoxide precursors are stabilized by alkanolamines such as monoethanolamine (MEA), diethanolamine (DEA), or ethylenediamine (ED) [1542]. In this work, ethanol is used as alcohol, titanium(IV) butoxide (Ti(OBu)$_4$) as the alkoxide precursor and DEA as stabilizer. Ethanol offers the advantage of causing films with high density [1540]. During the final calcination step in air or oxygen-rich atmosphere the gel decomposes into TiO$_2$. This kind of film fabrication results in very smooth films, but the thickness range is limited by the spin coating process.

3.1.6 Atomic Layer Deposition

Another deposition method resulting in a homogeneous TiO$_2$ layer on porous structures is **atomic layer deposition (ALD)** [1543–1547]. Hence, it is useful for many applications. Because of this importance, some basic results from the hydrothermal growth on ALD TiO$_2$ films will be mentioned. But since no physical results were gained from the growth on these films, results are not discussed in detail.

Here, the substrate is placed in a process chamber and exposed to the precursor containing gases alternately. ALD is a growth method resulting in rutile films [1548–1550]. The substrate is exposed periodically to a gaseous titanium precursor and gaseous water or hydrogen peroxide (H$_2$O$_2$) [1551]. Typical precursors for TiO$_2$ film fabrication are TiCl$_4$ and titanium alkoxides [1543]. In this work, TTiP is used as a precursor. The precursor reacts with OH-groups on the substrate resulting in a substrate–O–TiCl layer. In the next step, this layer reacts with water resulting in a substrate–O–Ti–OH layer [1552]. This process is repeated as long as the final film thickness is achieved. Thus, the film thickness is controlled precisely by the number of exposure steps and, as a consequence, ALD is the preferred method for the deposition of ultra-thin TiO$_2$ films. Schuisky et al. showed that rutile and anatase TiO$_2$ is achieved on MgO and α-Al$_2$O$_3$ by applying 275 °C and 375 °C, respectively [1550]. Dependent on the kind of precursor and processing temperature different temperature windows occur for amorphous, anatase, and rutile TiO$_2$ films [137]. It is noteworthy that the deposition at these low temperatures results in much better crystallinity compared to higher temperatures [1543].

3.1.7 TiCl$_4$ Treatment

A low cost method for very thin films is the **TiCl$_4$ treatment**. It is usually applied to fabricate core-shells with high electron mobilities [1368]. Hereby, the substrate is emerged in a solution containing TiCl$_4$ at slightly elevated temperatures. Under these conditions, TiCl$_4$ is hydrolyzed easily with water to TiO^{2+} ions by releasing hydrochloric acid. The TiO^{2+} ions react with water to TiO$_2$ and hydrogen ions [1553]. Because the solution penetrates even tiny pores, this method works for porous substrates as well [150, 1368]. For improving the crystallinity and reducing remaining water, the films have to be annealed. In contrast to ALD this technique results in less smooth films and the thickness cannot be controlled precisely, but the setup is much cheaper than ALD devices.

3.2 Hydrothermal Growth of Rutile TiO$_2$ Nanorods

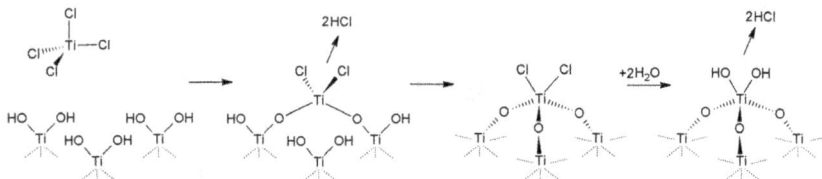

Figure 3.2: *The simplified reaction of TiCl$_4$ to TiO$_2$ assumed from Wisnet et al. [168].*

Due to their outstanding functionality in many applications, various fabrication techniques of TiO$_2$ nanostructures have been developed in the past [162]. These nanostructures include 3D structures such as nanocubes [1554] and inverse opals [1555], 2D structures such as nanoflakes [1556] and honeycombs [1557], 1D structures such as nanorods (or nanowires) [153, 1558, 1559], nanotubes [1560–1562], and nanofibers [1563–1565] as well as 0D structures such as nanoparticles [1566] and nanodots [1567, 1568]. There are quite a lot of fabrication methods for those structures such as template-assisted methods [1569], chemical vapor deposition (CVD) [1570, 1571], electrochemical anodic oxidation methods [1572], electrospinning [1573], solvothermal and hydrothermal methods [153].

In the following, the focus will be on hydrothermally grown rutile TiO$_2$ nanorods as they are used in this work. The hydrothermal method is inexpensive, fast and simple [1574]. Besides rutile nanorods, various different structures consisting of anatase [1575, 1576] and rutile [443] TiO$_2$ are feasible with hydrothermal methods. The hydrothermal growth of rutile TiO$_2$ nanorods is described partly in the literature [61, 1577], but it is by far not understood completely yet. A detailed overview of the influence of growth parameters such as precursor concentration, acidity, growth temperature, and growth time on the length and array density is given by Iraj et al. [1574]. The growth process is subdivided into two competing parts. One part is certainly the chemical reaction converting the precursor into TiO$_2$. The other part is the minimization of the surface energy of the whole crystal. The interplay between these two parts is investigated and discussed extensively in Chapter 7. Two growth mechanisms will be reported for the growth of straight rods in the following: particle attachment and ion-by-ion growth [159, 615, 1578, 1579].

3.2.1 The Conversion of the Precursor into TiO$_2$

The growth solution used in this work consists of water, hydrochloric acid (HCL) and Ti(OBu)$_4$ as the organic titanium precursor. A detailed description of the growth is given by Wisnet et al. [168]. The author suggests that at an appropriate temperature, the hydrocarbon chains of Ti(OBu)$_4$ are replaced by chloride atoms resulting in titanium tetrachloride (TiCl$_4$) according to the following reaction:

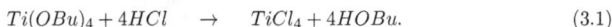

$$Ti(OBu)_4 + 4HCl \quad \rightarrow \quad TiCl_4 + 4HOBu. \tag{3.1}$$

In the following hydrolysis

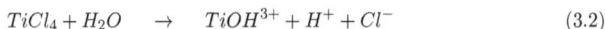

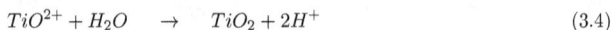

$$TiCl_4 + H_2O \quad \rightarrow \quad TiOH^{3+} + H^+ + Cl^- \tag{3.2}$$

$$TiOH^{3+} \quad \rightarrow \quad TiO^{2+} + H^+ \tag{3.3}$$

$$TiO^{2+} + H_2O \quad \rightarrow \quad TiO_2 + 2H^+ \tag{3.4}$$

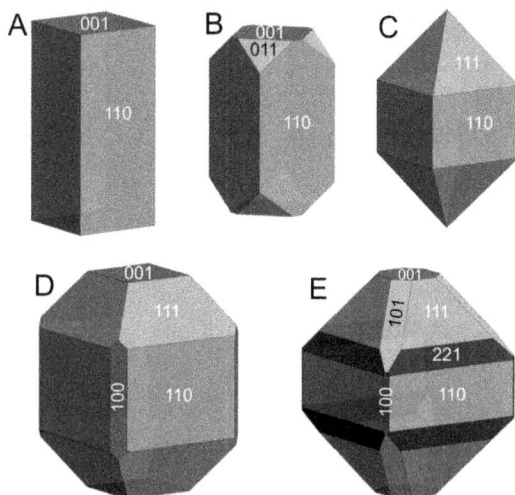

Figure 3.3: *A few possible Wulff shapes for the resulting rutile TiO_2 nanorods created with Wulff-maker [1581] and based on Goldschmidt [1582].*

$TiCl_4$ reacts with two $(OH)^-$ groups resulting in an attached titanium atom with two dangling Cl^- ions. These ions react with water generating two HCl molecules and two attached $(OH)^-$ groups [1567, 1580].

Related to the high oxophilicity° of titanium, it should be noted that the intermediate $TiCl_4$ molecules might not be formed in significant amounts. Hence, it is also reasonable that the titanium atoms forms directly Ti – O bonds instead of Ti – Cl bonds when it is released from the butoxide molecule.

Hydrochloric acid decreases the hydrolysis rate and results in a relative slow growth process avoiding the creation of high crystal defect densities [449, 450]. This reaction, illustrated in Figure 3.2, lasts until an equilibrium between growth and dissolution has set. Due to aggregates and dirt particles nanorods are able to arise in solution and on seed free substrates as well. The nucleation stage is determined by the growth conditions. In this work, the nucleation stage appears in a short time slot only. Afterwards, it is energetically unfavorable to create new nucleation sites and the crystallization takes place at already existing nanorod surfaces only. As a consequence, the resulting nanorods are nearly monodisperse.

Both, anatase and rutile TiO_2 consists of TiO_6 octahedra, but in anatase, the octahedra share their faces and in rutile, they share their edges [450, 1574]. The growth rate of a crystal facet depends on the amount of exposed and hence, available corners and edges of the polyhedral. In case of rutile, the slowest growth is expected on {110} facets, followed by {100} and {101} facets. The fastest rate appears on {001} facets and thus, the long axis of the resulting nanorods is parallel to the [001] direction [451, 1574].

3.2.2 The Rutile TiO₂ Crystal Shape at Thermodynamic Equilibrium Condition

Another important aspect besides the growth mechanism at an atomic scale is the resulting crystal shape. Goldschmidt is giving a detailed overview on possible rutile crystal shapes [1582]. The actual resulting shape depends on the surface energy of all formed facets. Detailed TEM studies [159, 168, 1583–1585] were performed at the resulting rutile TiO₂ nanocrystals when using the above-described recipe for the hydrothermal growth solution. The sidewalls of the nanorods consist of {110} surfaces. The facet facing towards the preferred growth direction is the {001} surface. Furthermore, the {111} and {101} surface were observed in the close neighborhood of {001} facets. DFT calculations have similar results [298, 561, 1586–1591] and tell us about the surface energy of these facets. Since the absolute calculated value depends on the applied model [1589, 1590], different values for the same facet are reported as shown in Table 3.1.

	PBE+U [1590]	PBE [1590]	LDA [298]	LDA [1587]	PBE [1587]	PBE0 [1587]	PW91 [1589]	US10 [1589]	PAW4 [1589]	PAW10 [1589]
{110}	0.83	0.58	0.89	0.89	0.42	0.55	0.50	0.54	0.50	0.48
{100}	0.92	0.67	1.12	1.20	0.69	0.83	0.69	0.76	0.69	0.67
{101}	1.18	0.99	1.40				1.03	1.08	1.03	1.01
{001}	1.54	1.31	1.65	1.88	1.39	1.59	1.25	1.32	1.25	1.21

Table 3.1: *Stoichiometric surface energies in J/m² for four rutile TiO₂ facets calculated using different density functional theories.*

For all models, there is a clear tendency, which explains the resulting shape of the rutile TiO₂ nanocrystal: The {110} facets have the lowest surface energy, thus, their surface area is supposed to be maximized during the hydrothermal growth process. The {001} surface has the highest surface energy and therefore there is no energetic motivation to enlarge these facets. To satisfy both conditions the growth in the [001] direction should be much faster than in the ⟨110⟩ direction. Since both facets are perpendicular to each other, this prevents the {001} facet to become significantly larger, but the {110} facets have a high growth rate at the same time. This behavior results in a nanorod with {110} facets as sidewalls and {001} facets at the two tips. The surface energies and thus, the exact shape of the resulting crystals is controlled by the pH-value [156, 643, 1592]. Possible Wulff constructions related to different environmental conditions are shown in Figure 3.3. The resulting shape, which is formed under defined experimental condition, depends strongly on growth conditions such as pH-value and reaction temperature [61, 129, 162]. A low pH value slows down the hydrolysis of the precursor [153, 159].

3.2.3 Formation of Branches

Another considerable effect is observed for the solution based nanorods and plays also a major role in the growth mechanism of complex nanorod structures on seed layers. Dependent on the composition of the growth solution, branching is energetically preferred continuously [1593] or at the beginning only as it is the case in this work [1593]. A branching process of a developed nanorod would firstly form a new {001} facet with the above mentioned fixed size, before the {110} sidewalls begin to increase due to the growth in the (001) direction. The newly formed {001} facets would cover a part of the former {110} facet of the primary nanorod. Thus, a branching would begin with an increase of the high energetic {001} facet and a decrease of the low energetic {110} facet. Hence, this effect is characterized by an activation energy barrier for branching events. If this barrier becomes too high for increasing nanorod sizes, branching events decrease dramatically. But a different composition of the growth solution and growth parameters makes also continuous branching possible [159]. Branching is supposed to originate from {101} [154, 159, 165, 166, 1578, 1594–1602], {111} [1603], {200} [1594], {110} [1594, 1604], and {301} [1578, 1601, 1602] twins. The existence of {101} facets indicates that a crystal shape similar to Figure 3.3 E is present at least at the beginning of the growth process. According to Li et al., anatase particles play an important role in the growth and branching process [159]. Fine anatase TiO₂ particles with {100} and {103} facets are observed in the growth solution. The anatase particles attach with their {103} facet to {101} facets of the rutile nanocrystals. There are two different docking orientations of anatase nanocrystals likely resulting either in a straight nanorod growth or in branching. Hence, it is reasonable that the branching depends on the shape of the nuclei, too [167]. The primary nanorod and the branch include an angle of about 8° [165] and 65° for {101} twins [154, 159, 165, 166, 1578, 1594, 1595] and 55° for {301} [1578] twins. In conclusion, there are two growth mechanisms reported for the growth of straight rods so far: particle attachment and ion-by-ion growth [159, 615, 1578, 1579]. Branching is a common feature during crystallization and was observed for anatase nanorods as well [1605].

3.2.4 Substructure of Full-grown TiO₂ Nanorods

Within this study, new insights concerning possible nucleation sites, growth mechanisms and inner structures of rutile TiO₂ nanorods are found. For instance, a typical feature of full grown TiO₂ nanorods is the substructure consisting of tiny fingers as shown in Figure 3.5 [159, 162, 168, 978, 1606]. This feature is seen clearly on the tip of the nanorod [767, 1607]. Here, the nanorod splits off in a multitude of small fingers that have a rectangular – or in some cases even a quadratic – cross-section with a side length of a few nanometers and are arranged as a quadratic lattice giving the nanorod its quadratic cross-section. Their long axis is parallel to the long axis of the nanorod. Hence, they are like tiny rutile TiO₂ nanorods. This means that each full-grown nanorod is not a real single crystal, but consists of a mesocrystalline upper part [168]. Wisnet et al. reported that the crystal facets of the nanofingers are similar to those shown in Figure 3.3 D [168], while the shape of the mother rod is more related to the crystal shown in Figure 3.3 A. Thus, there are two kinds of rutile crystal shapes constructed at the same time. However, it should be noted that the shape of the bottom side of the nanorod is fixed by the topography of the substrate. Wisnet et al. suggested that defects caused by chloride inclusions into the crystal are responsible for the creation of the fine structure and that the fibrous substructure influences device properties such as the efficiency of dye-sensitized solar cells (DSSCs)° [1608]. The size of the substructure depends on the composition of the growth solution and hydrothermal processing parameters [1609]. Similar substructures appear in many other nanorods as described for rutile VO₂ [1610], monoclinic ZrO₂ [1611], CuO [1612], graphene-MoO₃ [1613], In₂O₃ [1614], Sb₂S₃ [1615], ZnTe [1616], Bi₂Te₃ [1617], Bi₂Te₃/Bi₂S₂ [1618], α-Fe₂O₃ [1619–1621], as well as for single wall carbon nanotubes (SWCNTs) [575] and for fluorapatite-gelatine [1622]. Two systems that are very similar to the substructure of rutile TiO₂ nanorods are C₆₀ nanorods [1623] and monoclinic WO₃ nanorods [1624, 1625]. Most of the listed nanostructures were grown hydrothermally and a meaningful selection is shown in Figure 3.6.

In many cases, the formation of these fibrous substructures is not understood completely and by far not controllable. Beyond that, there might be different origins in different materials. In this work, a model describing the formation of the observed substructure in rutile TiO₂ nanorods based on a non-equilibrium state between two competing processes is introduced. The two processes are crystallization and surface energy minimization. This model will be used to explain experimental results and to identify new routes to control the properties of the substructure.

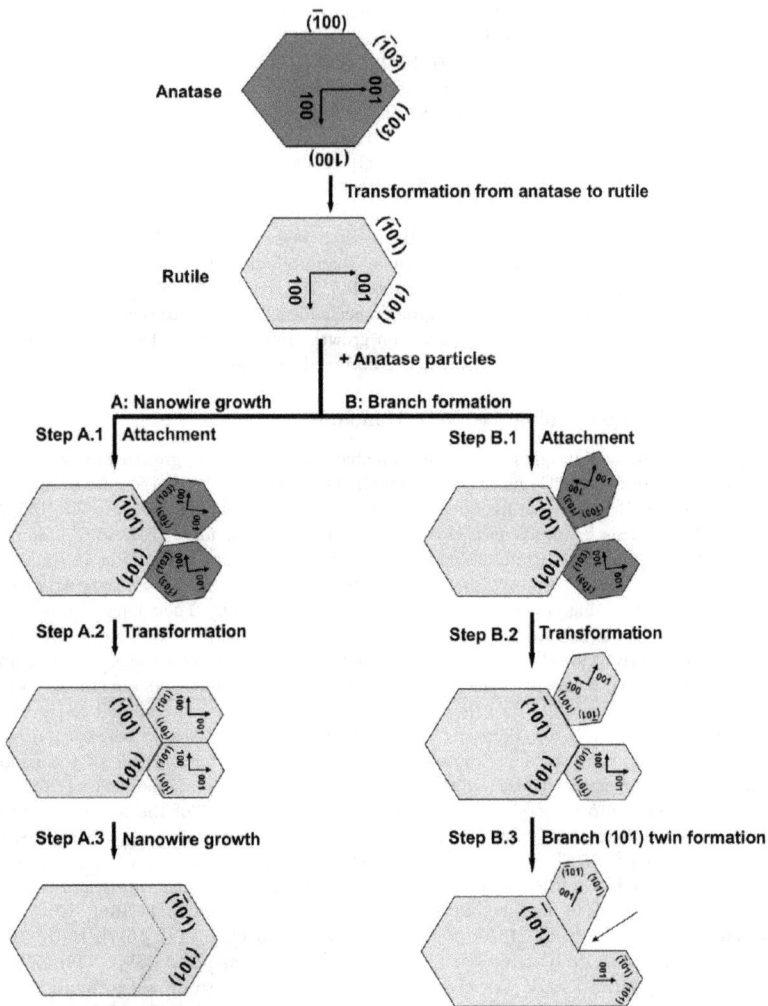

Figure 3.4: *The phase transition from anatase and rutile TiO₂ nanocrystallites and their preferred attachment orientation during the suggested growth of single and branched rutile nanorods. Reprinted with permission from Li et al. [159]. Copyright ©2013 American Chemical Society.*

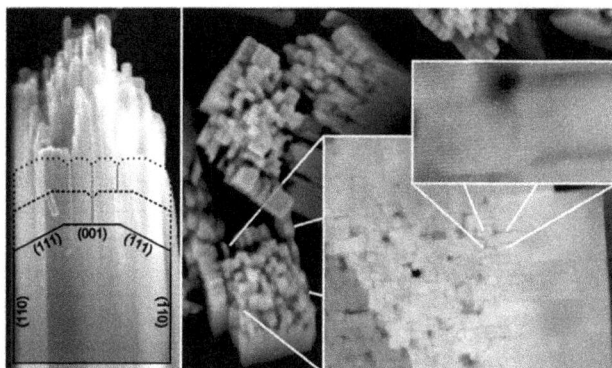

Figure 3.5: *Nanofingers in hydrothermally grown rutile TiO$_2$ nanorods. Reprinted with permission from [168]. Copyright (2014) American Chemical Society.*

3.2.5 Substrates for the Hydrothermal Growth of Rutile TiO$_2$ Nanorods

Different reports indicate that rutile TiO$_2$ nanorods can be grown on different substrates such as silicon [129, 158, 1627], SiO$_2$ [129, 1627], indium tin oxide (ITO) [129, 158], fluorine tin oxide (FTO) [61, 129, 1627], glass [129, 158], stainless steel [158], carbon cloth [158], and sapphire [1627, 1628]. For non-rutile substrates, usually an additional seed layer is needed [1627]. On FTO the growth works without any additional seed layer since the mismatch of the rutile lattice parameters between FTO and TiO$_2$ is 2% only [158, 1574, 1629]. Although Kim et al. have managed to prepare a TiO$_2$ seed layer on a ruthenium electrode by depositing a TiO$_2$ by ALD and annealing it at 500 °C, reports about TiO$_2$ NRAs on metal electrodes are very rare [1630]. To the best of my knowledge, detailed investigations of the relation between seed layer properties and the growth of rutile TiO$_2$ nanorods have not been done so far.

3.2.6 Toxicity of the Investigated TiO$_2$ Nanostructures

Since TiO$_2$ is chemically inert, no negative interaction is expected and documented with human bodies so far [1631].
But if TiO$_2$ is exposed to UV light close to the skin chemical reactions are triggered by high energetic electrons in the conduction band of TiO$_2$. Thus, there is photocatalysis resulting in cellular toxicity [1632].
An obvious threat for the functionality of the lung originates from inhaling huge amounts of TiO$_2$ dust. Thus, it is assumed that bulk TiO$_2$, TiO$_2$ films and most – especially spherical – TiO$_2$ particles are absolutely not toxic for humans. An exception are particles with a certain geometry. If particles are longer than 5 µm and have a diameter of about 1 µm they enter the lung and destroy huge amounts of macrophages resulting in the cicatrization° of the lung surface and thus, the risk of asbestosis° is given [1633]. Furthermore, lung inflammation, pulmonary and cardiac edema as well as systemic inflammation can appear [1634]. The nanorods used in this work are more than one order of magnitude smaller. An interaction of this kind of TiO$_2$

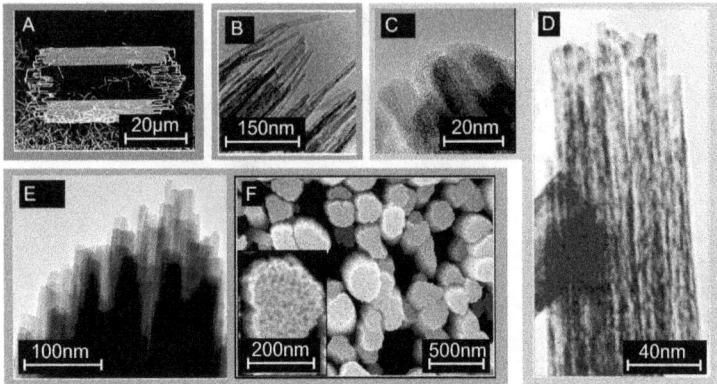

Figure 3.6: *Selected nanorods with similar substructure compared to the hydrothermally grown rutile TiO$_2$ nanorods: A) C$_{60}$ nanorods grown in a solvent (Reprinted with permission from Zhang et al. [1623] Royal Society of Chemistry, copyright ©Royal Society of Chemistry 2009), B) hydrothermally grown CuO nanorods (Reprinted with permission from Yang et al. [1612], Elsevier, Copyright ©2008 Elsevier B.V. All rights reserved.), C) tetragonal α-Fe$_2$O$_3$ (Reprinted with permission from Vayssieres et al. [1620], John Wiley and Sons, Copyright ©2005 WILEY-VCH Verlag GmbH & Co. KGaA, Weinheim), D) α-Fe$_2$O$_3$ nanorods similar to (C) (Reprinted with permission from Vayssieres et al. [1619], American Chemical Society, Copyright ©2001, American Chemical Society), E) WO$_3$ nanorods prepared by hot-filament thermal deposition (Reprinted with permission from Chi et al. [1626], IOP Publishing, copyright ©2006 IOP Publishing LTD), F) hydrothermally grown WO$_3$ nanorods (Reprinted with permission from Shen et al. [1625], Elsevier, Copyright ©2009 Elsevier B.V. All rights reserved.).*

particles with the human body is not known. Furthermore, the investigated nanorods are fixed on substrates and do not exist as single particles. Also, any suggested application needs strongly bound TiO$_2$ nanorods only.

4 Experimental Methods

In this chapter, basic fabrication and characterization techniques employed in this work are described. Fabrication techniques that have been developed in this thesis are explained in detail in the following chapters containing results and discussions.

4.1 Fabrication Techniques

This section includes the choice of substrates, TiO_2 film and seed layer fabrication, the hydrothermal growth of rutile TiO_2 nanorods, and fabrication of confined geometries using optical and electron-beam lithography.

4.1.1 Substrates

The TiO_2 seed layers were deposited on a polished, boron-doped, p-type (100) silicon wafer. The low sheet resistance of 5–$25\,\Omega/cm^2$ helped to achieve high-quality scanning electron microscope (SEM) images without depositing any additional conductive material. Ellipsometry measurements showed that the silicon is covered with a native $2\,nm$ thin SiO_2 layer. For optical measurements, both-sided polished, $1 \pm 0.1\,mm$ thin fused silica (from Präzision Glas & Optik GmbH, Iserlohn, Germany) was used. Commercial fluorine tin oxide (FTO) (TCO10-10, Solaronix SA, Aubonne, Switzerland) on aluminoborosilicate glass with a sheet resistance of $10\,\Omega/cm^2$ is used as an alternative growth promoting substrate. Furthermore, the hydrothermal growth process was performed on commercial rutile TiO_2 single crystals (Latech Scientific Supply Pte. Ltd., Singapore) with {100}, {001}, {110} and {111} facets fabricated by a float zone (FZ) crystal growth method. According to manufacturers' data, the surface roughness is less than $5\,nm$ and the purity is above 99.99%. For high-resolution transmission electron microscopy (HRTEM) images, Si_3N_4 grids covered by a $30\,nm$ thin carbon carrier film (DuraSiN DTF-03523 from Science Service GmbH, Germany) were employed. Seed layer deposition, annealing, and hydrothermal growth was performed on these grids directly, since they are able to resist hot hydrochloric acid and temperatures up to $1000\,°C$. This method was used to avoid time-consuming lamella cutting and there is no transfer process of grown nanorods from seeds to transmission electron microscope (TEM) grids needed that could damage the nanocrystals. Except for the TEM grids, all substrates were sonicated for $10\,min$ in acetone and $10\,min$ in isopropanol afterward. Finally, they were rinsed with ethanol and dried with pressurized nitrogen.

4.1.2 Rutile and Anatase TiO_2 Films/ Seed Layers

The films were fabricated using various methods. They were labeled according to their final morphology, deposition technique and annealing temperature using a "XXX-YYY-ZZZ" label code. "XXX" encodes the major morphology of the film and is either AMO (amorphous), RUT (rutile), or ANA (anatase). "YYY" encodes the film deposition technique and is either SPU (sputter deposition), EBE (electron-beam evaporation), SG (sol-gel), PYR (spray pyrolysis), or ALD (atomic layer deposition). "ZZZ" encodes the temperature of the post-deposition annealing process in degree Celsius and is typically 000 (no annealing), 450 (450 °C), 600 (600 °C) or

850 (850 °C). Basically, the mentioned crystal structure is achieved not until post deposition annealing.

Sol-gel TiO$_2$ Films

Sol-gel TiO$_2$ films were made from a mixture of 154 mg diethanolamine (DEA) and 2 mL ethanol. After stirring for some minutes, 510 μL titanium(IV) butoxide was dropped slowly into the mixture. After 3 h stirring, 260 μL ethanol and 26 μL distilled water were added and the solution was stirred for another while. Finally, it was kept in a dark place at room temperature for at least 12 h. To reduce the water film on the samples, the substrates were placed a few minutes on a hotplate at 120 °C before spin coating. To perform spin coating at room temperature, the substrates were cooled down on a metal plate for about one minute. Then the sample was fixed in a spin coater and 50 μL sol-gel was dropped on the substrates during an acceleration of 500 rpm/s to a final speed of 2000 rpm applied for 60 s. Finally, the film was kept at 120 °C on a hotplate for more than 24 h resulting in an amorphous film.

Sprayed TiO$_2$ Films

The solution for the spray pyrolysis consists of titanium diisopropoxide bis(acetylacetonate) and ethanol in a ratio of 1:10 and was stirred for some minutes before the first use. The substrate was heated to 450 °C on a hotplate and the solution was sprayed manually on the sample, while the temperature was kept at 450 °C for another 15 min. After spraying, the sample was kept at 120 °C for 24 h. The film thickness depends strongly on the number of spray pulses, the distance between the spray head and the sample, and the spray head geometry.

Sputtered Anatase TiO$_2$ Films

Sputter deposition was performed with a *Gamma 1000C sputter system* (from Surrey NanoSystems LTD, Newhaven, U.K.). Sputtered anatase TiO$_2$ films were deposited by direct current (DC) sputter deposition at room temperature using a TiO$_2$ sputter target with 99.99% purity (from Testbourne Ltd, Basingstoke, England). Argon was employed as process gas (20 sccm Ar) at a pressure of 6.67×10^{-3} mbar. Applying a sputter power of 100 W (370–380 V and 0.26–0.28 A) a sputter rate of 0.25 Å/s was achieved.

Sputtered Rutile TiO$_2$ Films

Here, the same device as used for sputtered anatase TiO$_2$ films was employed. Sputtered rutile TiO$_2$ films were deposited by DC sputter deposition at room temperature and a titanium sputter target with 99.995% purity (from Testbourne Ltd, Basingstoke, England). Process gas was argon (20 sccm Ar) and the pressure was 6.67×10^{-3} mbar. Using a sputter power of 300 W (380–390 V and 0.75–0.80 A) a sputter rate of 1.67 Å/s was achieved. The oxide formation takes place in the post deposition annealing process.

Evaporated rutile TiO$_2$ Films

A custom-built deposition chamber equipped with an electron-beam evaporator (from Sentech Instruments GmbH, Berlin, Germany) was employed for electron-beam assisted film deposition. The deposition of titanium was performed at a pressure of around 1.5×10^{-6} mbar and a deposition rate of 0.8–0.9 Å/s. The purity of the titanium pellets is denoted with 99.995% (from Kurt J. Lesker Ltd, East Sussex, England). The oxide formation takes place in the post deposition annealing process.

Atomic Layer Deposition (ALD) of TiO_2 Films

The films made by atomic layer deposition (ALD) were fabricated using the TTiP-water process in a self-made ALD chamber as shown in Figure 3.1 [1635]. As precursors, titanium tetraisopropoxide (99.8% purity, from Sigma-Aldrich, Switzerland) heated up to 45-50 °C and deionized water at room temperature were employed. The process temperature was 180 °C and the deposition rate 0.6 Å/cycle. The cycle includes a 2 s TTiP pulse, 10 s exposure time, several purging steps using dry nitrogen, and finally a 1 s water pulse, 10 s exposure and several purging steps with nitrogen again. A detailed description of the chemical process during ALD is given by Rahtu and Ritala [1636].

Spatial Atmospheric Atomic Layer Deposition (SALD) of TiO_2 Films

A detailed overview of the differences between ALD and spatial atmospheric atomic layer deposition (SALD) (or AALD) is given by Poodt et al., and Muñoz-Rojas and MacManus-Driscoll [1637–1639]. For SALD, titanium tetraisopropoxide (TTIP) ($\geq$ 97% purity, from Sigma-Aldrich, Switzerland) and deionized water were used as precursors. A nitrogen flow through the heated and separated liquid precursors was employed to transfer vaporized precursors through a heated pipe into the SALD head. The bubbling rates depend heavily on the geometry of the setup. In the employed setup, the bubbling rate was set to 150 Nml/min for TTIP and 100 Nml/min for water. The gases were released via nozzles about 250 μm above the substrate in a defined order. The head consists of alternating TTIP and water nozzles. Between the precursor supplying nozzles, there are additional nozzles for nitrogen purging to prevent a reaction of the two precursors before they reach the sample. The sample, which was placed on a hotplate (here 300 °C), was moved forth and back below the head using a scan speed of about 50 mm/s and consequently, each position on the sample ran through the same procedure as used for conventional ALD. The time needed to change the moving direction was reduced to a minimum. The sample was moved 1500 times below the SALD head. A similar setup is described in detail by Muñoz-Rojas et al. [1639]. There was no further post-annealing for the investigated samples.

$TiCl_4$ Treatment Processed TiO_2 Films

The $TiCl_4$ bath was performed in 210 mL of a 0.1 molar $TiCl_4$ (99.99%, from Sigma-Aldrich, Switzerland) solution using deionized water as solvent. The mixing process was performed after the $TiCl_4$ and the deionized water were cooled down in an ice bath over night. During the $TiCl_4$ treatment the solution was placed in a water bath at 70 °C for 3 h.

WO_3 and Reduced WO_3 Films

Resistive evaporation was performed with a *TVA 01* (from Theva Dünnschichttechnik GmbH, Ismaning, Germany). 40 nm thick WO_3 films were deposited at a pressure of around 5×10^{-6} mbar and a deposition rate of 0.3 Å/s. As target material sintered yellow-green WO_3 pieces with a diameter of 3–12 mm (T9-5006-M, 99.99%, from Testbourne Ltd, UK) were taken. The films were fabricated by reducing WO_3 in forming gas (95%N_2, 5%H_2) at 700 °C for 10 min. The reduction is supposed to result in pseudo-rutile WO_2 partly.

Deposition of Further Thin Films

Metal electrodes: For metallic electrodes different materials were taken. **Titanium** as an electrode material was deposited either by sputter deposition or electron-beam evaporation using the same parameters as described above. **Platinum** was deposited using electron-beam evaporation. The deposition of platinum was performed at a pressure of around 2×10^{-6} mbar and a deposition rate of 0.5 Å/s. The purity of the employed platinum pellets was denoted with 99.99% (from Testbourne Ltd, UK). A crucible made of graphite was taken. **Gold** was deposited using a resistive evaporator with tungsten boats at a pressure of around 2×10^{-5} mbar and a deposition rate of 0.5 Å/s. As target material granulate containing pieces with a diameter of 1–3 mm (V30927-4, 99.99%, from Testbourne Ltd, UK) were used. **Yttrium** was deposited by DC sputter deposition at room temperature and a yttrium sputter target with 99.99% purity (from Testbourne Ltd, Basingstoke, England) in an argon atmosphere (20 sccm Ar) with a pressure of around 6.67×10^{-3} mbar was achieved. Applying a sputter power of 50 W (250 V and 0.20 A) a sputter rate of roughly 0.71 Å/s.

SiO₂: Thin films were deposited by radio frequency (RF) sputter deposition at room temperature and a SiO_2 sputter target with 99.995% purity (from AJA International, Inc., Massachusetts, U.S.A.) in argon and oxygen atmosphere (17 sccm argon, 1 sccm oxygen) with a pressure of around 2.67×10^{-3} mbar. Applying a sputter power of 60 W a sputter rate of roughly 0.065 Å/s was achieved.

Silicon: Thin films were deposited by RF sputter deposition at room temperature and a silicon sputter target with 99.999% purity (from AJA International, Inc., Massachusetts, U.S.A.) in argon atmosphere (17 sccm) with a pressure of around 2.67×10^{-3} mbar. Applying a sputter power of 100 W a sputter rate of roughly 0.35 Å/s was achieved.

Post Deposition Annealing

After film deposition, each film was annealed in a rapid thermal processing (RTP) oven in 500 sccm oxygen flow for 2 h using a temperature slew rate of $\pm$ 1 °C/s. The annealing temperature was set to 450 °C for ANA-SPU-450 and 850 °C for ANA-SPU-850, RUT-SPU-850 and RUT-EBE-850, for instance. Sol-gel and sprayed films annealed at 450 °C are supposed to become anatase [1640], which is also observed for the other films. In contrast, the morphology of films annealed at 850 °C depends on the deposition technique. The oxygen rich atmosphere kept the formation rate of oxygen vacancies low and turned the titanium film into TiO_2 [869, 915, 1641, 1642].

Frequently Used Samples

notation	film deposition	thickness	annealing	morphology	seed for NRAs
ANA-SPU-450	sputter, TiO_2	40 nm	at 450 °C for 2 h in oxygen	anatase	yes
ANA-SPU-850	sputter, TiO_2	40 nm	at 850 °C for 2 h in oxygen	anatase	no
RUT-SPU-850	sputter, Ti	40 nm	at 850 °C for 2 h in oxygen	rutile	yes

Table 4.1: *Overview of most frequently used samples.*

Fabrication of Gold Nanoparticles

As a primary material for gold nanoparticles, a 2 nm thin film was deposited by RF sputter deposition at room temperature using a gold sputter target with 99.99% purity (from AJA International, Inc., Massachusetts, U.S.A.) in argon atmosphere (17 sccm) and a pressure of around 13.3×10^{-3} mbar. Applying a sputter power of 30 W a deposition rate of roughly 1.11 Å/s was achieved. To gain gold nanoparticles the dewetting effect of gold on TiO_2 at elevated temperatures was used. Hereby, the sample was annealed at 300 °C for 3 min in oxygen-rich atmosphere to prevent the creation of additional oxygen vacancies. The resulting gold particles had diameters between roughly 1 nm and 10 nm. To achieve a better coverage on nanostructured surfaces, gold was deposited via sputter deposition due to the broader impact angle distribution compared to evaporation techniques.

4.1.3 Hydrothermal Growth of Rutile TiO_2 Nanorods

Rutile TiO_2 nanorods were fabricated using a hydrothermal method in an acid aqueous solution. 20 mL concentrated (37%) hydrochloric acid (HCl) (VWR Chemicals BDH Prolabo) diluted in distilled water were mixed with 350 µL titanium(IV) butoxide ($C_{16}H_{36}O_4Ti$, reagent grade 97%, Sigma-Aldrich, Switzerland) and stirred for some minutes at room temperature (10 min before and after adding the precursor to the solution, respectively). Unless otherwise noted, 14.5% HCl solution was taken as a standard concentration (12 mL H_2O and 8 mL 37% HCl). The substrates were placed vertically into the growth solution in a Teflon liner and autoclaved in an oven at a standard temperature of 180 °C unless otherwise noted. To investigate the seedless growth in solution, a blank silicon wafer was placed horizontally on the bottom of the Teflon liner. To exclude an influence of the silicon surface on the growth, it was performed once without any substrate and after the hydrothermal growth, a drop of the growth solution was transferred on a silicon substrate by pipetting. The resulting nanorods looked the same in both cases. To achieve defined stop of the growth process, the autoclaves were submerged in cool water. Thereby, the temperature dropped within a few minutes down to room temperature. Finally, the samples were rinsed with distilled water and dried using pressurized nitrogen.

The experiments showed that drop-like areas with nearly no nanorods appeared on samples, which were covered with exsiccated drops of soap. It might be that the hydrochloride acid reacts with the soap into carbon acids and is no longer driving the nucleation process. Therefore, the use of soap was deleted from the cleaning process.

Temperature Development in the Growth Solution

It has to be brought up that the temperature inside the Teflon liner is increasing slowly due to the low heat conductivity of Teflon and it does not reach the temperature inside the oven within the first hours as shown in Figure 4.1. Thus, temperature and pressure were continuously changing within the applied growth periods of 40–240 min.

The temperature inside the Teflon liner was measured using a type K thermocouple taking room temperature as a reference. At the beginning, the temperature inside the autoclave without a Teflon liner was measured with the same thermocouple as used in further experiments. This measurement showed that there is no mismatch between the internal thermometer of the oven and the type K thermocouple employed inside the Teflon liner. The feedthrough was made by drilling a small hole into the lids of the autoclave and the Teflon liner and sealing the gap between the cable and the Teflon liner with two-component adhesive. The temperature measurement was performed with pure distilled water to avoid corrosion and the thermocouple was immersed into the water completely. Altogether, the temperature development was reported for several hours and three different oven temperatures. Afterward, a run with the acid growth solution was performed and it was verified that there is no difference between water and the growth solution. After a few hours, the thermocouple was dissolved by the acid.

4.1.4 Lithography

In this work, two kinds of lithography techniques were applied: optical and electron-beam lithography. The film deposition takes place after the development of the photoresist and the

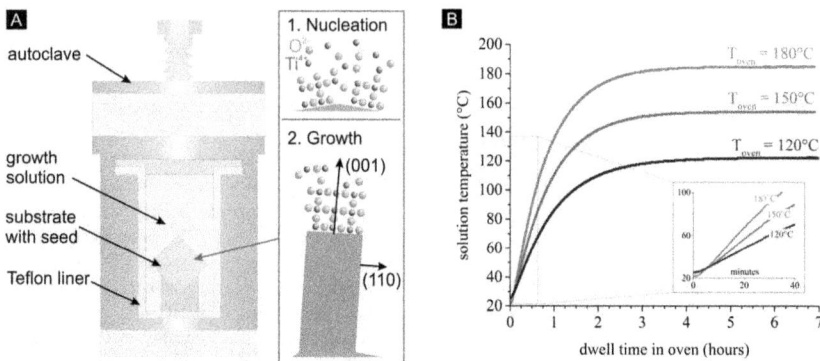

Figure 4.1: *A) Setup for hydrothermal growth and B) temperature devolution of distilled water inside the autoclave for temperatures of 120, 150 and 180 °C inside the oven. The measurements were performed using a type K thermocouple.*

annealing is performed after the lift-off. Deviations and extensions of the standard lithography methods are explained in detail in the appropriate chapters.

Optical Lithography

Optical contact lithography was performed by using an approximately 1.2 µm thick layer of S1813 photoresist spin coated at 5000 rpm for 30 s. Softbaking was performed on a hotplate at 120 °C for 60 s. Samples were exposed to a mercury lamp using a *Karl Süss MJB 3 mask aligner* for 54 s. The mask offered a pattern consisting of parallel stripes with 2 µm width and 2 µm pitch. Samples were developed using MF-319 developer for 30 s. The development was stopped in a water bath and the samples were dried with pressurized nitrogen. After titanium deposition, the lift-off process was done by placing the samples for 12 h in acetone and applying sonication for 10 min. Finally, the samples were cleaned for 10 min in isopropanol and dried with pressurized nitrogen.

Electron-Beam Lithography

For electron-beam lithography, poly(methyl methacrylate) (PMMA) 950 A2 was used as a resist. After keeping the substrates at 120 °C on a hotplate for some minutes, the resist was deposited while the substrates were rotating at 400 rpm for 4.5 s. During the second step, the rotation speed was increased by 1000 rpm/s to 3000 rpm and kept for 90 s. The samples were placed at 120 °C on a hotplate for some minutes again and put in an oven for 30 min at 170 °C for prebaking. This process results in a 40 nm thick resist layer. To prevent charging, a 10 nm thick aluminum layer was deposited using resistive evaporation at an initial pressure of 1×10^{-5} mbar and a deposition rate of 0.3 Å/s. Electron-beam exposure was performed with the field emission scanning electron microscope (FESEM) *Zeiss CrossBeam 1540XB*. Electrons with an energy of 10 keV, an aperture of 10 µm, an electron-beam current of 25 pA, an exposure dose of 200 µAs/cm², an area step size of 1.6 nm and an area dwell time of 200 ns were employed. The aluminum was removed within 12 s in 0.5 molar NaOH. Then, the samples were rinsed with water, developed in methylisobutylketon (MIBK) for 25 s and cleaned in isopropanol for some minutes. After titanium deposition, the lift-off was performed by placing the samples for 12 h in acetone and in an ultrasonic bath for 1 min. Afterward, the samples were cleaned in isopropanol 10 min and dried with pressurized nitrogen.

4.1.5 Fabrication of Polystyrene Sphere Monolayers (PSML)

Fabrication of Polystyrene Spheres

Conventional emulsion polymerization of styrene was employed to fabricate polystyrene spheres. Deionized water (1029 g, 57.12 mol, 23.8 Eq) is mixed with acrylic acid used as co-monomer (5.19 g, 0.072 mol, 0.03 Eq, not stabilized; 99% purity; 180-200 ppm MEMQ; Sigma-Aldrich Co. LLC), sodium dodecyl sulfate (SDS) used as emulsifier (0.9808 g, 3.40×10^{-3} mol, 1.42×10^{-3} Eq, 3 mol% relative to styrene; ≥99% purity; biochemistry grade; Carl Roth GmbH + Co. KG), potassium persulfate as initiator (0.9839 g, 3.64×10^{-3} mol, 1.52×10^{-3} Eq; ≥99% purity; Sigma-Aldrich Co. LLC), and styrene (249.96 g, 2.40 mol, 1.00 Eq, stabilized; ≥99% purity; Merck Chemicals GmbH). The reaction lasted 6 h and took place at 80 °C in nitrogen atmosphere and under continuous stirring (320 rpm). The resulting spheres had a diameter of 160 nm and were purified by exhaustive ultrafiltration against water.

To gain larger spherical particles, seed polymerization was employed. A purification by serum replacement was followed by heating the polystyrene latex made by the emulsion polymerization described above in nitrogen atmosphere. Before heating, deionized water (180 g, 9.97 mol, 19.36 Eq), acrylic acid (0.56 g, 8.0 mmol, 0.015 Eq), and SDS (0.0527 g, 1.83×10^{-4} mol, 3.55×10^{-4} Eq) were added. The initiator solution (0.4920 g, 1.82×10^{-3} mol, 1.52×10^{-3} Eq) and styrene (53.6 g, 515 mmol, 1.00 Eq) were dropped to the seed particles (47.51 g of 13.42 wt%) over a period of 2 h simultaneously in order to run the second-stage polymerization. Afterward, the reaction was continued for another 4 h resulting in spheres with a diameter of 393 nm.

In an additional step, particles generated by the first seed polymerization were enlarged by a second polymerization step. Within this second step, the polystyrene latex samples were kept at 60 °C in nitrogen atmosphere and stirred continuously (200 rpm). Then, the deionized water (20.00 g, 1.110 mol, 10.66 Eq) with the initiator (0.1294 g, 0.479 mmol), and styrene (10.84 g, 0.104 mol, 1.00 Eq) were dropped over a period of 16 h to the seed particles (120 g of 4.1 wt% solution) simultaneously in order to trigger the second-stage polymerization. After the whole amount of styrene was added, the solution was mixed with acrylic acid (0.2150 g, 0.003 mol, 34.9 Eq) and the reaction was continued for another hour at 60 °C, 4 h at 70 °C and finally, 1 h at 80 °C resulting in spheres with a diameter of 532 nm.

The diameter of all polystyrene spheres was measured by diffusive light scattering (DLS). The polymerization process was performed by Maxim Schlegel (AG Wittemann, Department of Chemistry, University of Konstanz).

Fabrication of a Monolayer

The water subphase consisted of 0.1 mmol/l SDS. To facilitate spreading, the colloidal dispersion (5 wt%) was diluted with 50 wt% ethanol. The dispersion was placed on the water via a tilted (approximately 45°) glass slide. A polished silicon substrate was immersed into the water phase and pulled out at a small angle in order to absorb the monolayer. The samples dry at a tilted angle of about 45°. The formation of polystyrene sphere monolayer (PSML) is discussed in detail by Vogel et al. [1643]. The fabrication of monolayers was performed by Dr. Simone Plüisch (AG Wittemann, Department of Chemistry, University of Konstanz).

4.1.6 Focused Ion Beam (FIB) Milling

The focused ion beam (FIB) milling was performed in a *Zeiss CrossBeam 1540XB* FESEM. The gallium ion current was set to 50 pA at a chamber pressure of about 3×10^{-6} mbar and the energy of the ions was 30 keV. The milling power was fixed at a value that corresponds to a milling depth of 0.5 μm in pure silicon. Trenches were made by two steps in order to avoid asymmetric sidewalls due to sedimentary depositions of sputtered material. The milling direction was always perpendicular to the long side of the trench. In the beginning, the milling direction was from right to left and afterward from left to right. The milled area in both steps overlapped by roughly 50% in the middle of the trench.

4.1.7 Scanning Probe Lithography

Within this method, an AFM tip was used to scratch across a substrate. The AFM device and the AFM-tips used for this technique were the same as employed for topography characterization described in Section 4.2. For scanning probe lithography, the device was run in contact mode with different forces that were usually stronger than used for surface characterization. Also, the number of writing events in one position were varied from 1 to 128.

4.1.8 Laser Lithography

For laser lithography, different setups for continuous wave (cw) and pulsed laser exposure were used and shown in Figure 4.2.

For cw laser exposure, a frequency doubled Nd:YAG laser (from Laser Quantum GmbH, Konstanz, Germany) at a wavelength of 532 nm and an output power of 1 W was used. The combination of a rotary half-wave plate and a Glan-Taylor polarizer◇ was employed to adjust the beam power. To focus the laser beam Plan Apo Mitutoyo objectives with magnification factor of 20X, 50X, and 100X used. To keep the focal point on the substrate surface the distance between the objective and the sample was adjustable. The sample was mounted on an individually controllable, linear xy positioning stage (PLS-85, from PI miCos GmbH, Freiburg, Germany) in order to fabricate individual structures. The minimal step size of the stage was 50 nm and applied displacement speeds were 0.01–14 mm/s. Live control was achieved with a charge-coupled device (CCD) camera integrated into the light path. The beam power was measured with a shiftable power meter.

The pulsed laser setup was used for two beam interference employed to induce a temperature pattern on the TiO_2 surface [1644–1647]. This pattern is defined by the respective incident angles, polarization, and intensities of the two interfering beams. Hence, the intensity depends on the position and is given by $I_n(x) = 4I_{n,0} \cos^2 (kx \cdot \sin (\Theta))$ [1648]. Of course, the accessible temperatures depend on reflectivity, absorption, and heat conductivity of the sample as well as the pulse length. The pattern shown in Figure 4.2C is characterized by the nearly Gaussian intensity distribution of the laser beam and the $\cos^2\Theta$-distribution of the interference pattern [1648]. Θ is the incident angle shown in Figure 4.2 B and $d = \lambda/(2\sin\Theta)$ the period of the pattern for a given wavelength λ.

In this work, the second harmonic of an injection-seeded Nd:YAG laser pulse at 532 nm was employed for the two beam interference. The pulse was 12 ns long and the period of the pattern was $d = 2.5$ μm. Similar to the cw laser setup, the laser power was adjusted by the combination of a half-wave plate and a Glan-Taylor polarizer.

It should be noted that running and optimizing the two presented laser systems was not part of this work and was done by previous Master and Ph.D. students. This thesis focuses on the physical interpretation of the effects induced by laser light exposure on specific samples as described in detail in Chapter 9.

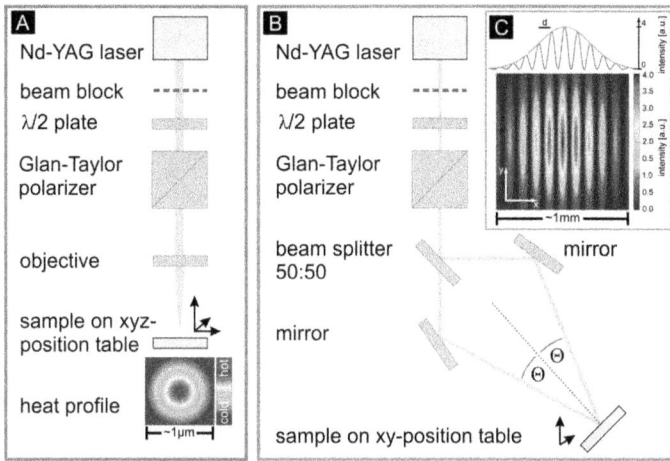

Figure 4.2: *Setups used for laser lithography: A) cw laser used for individual structures. The expected intensity profile is drafted (not measured). B) Pulsed laser employed for large area patterning. To keep clarity, the representation of both setups is greatly simplified. C) The measured position-dependent intensity of the interference pattern with a relatively large period d. The period employed in this work was much smaller. Figure (C) is reprinted with the permission from [1644] (Figure 1b). The scale bar at the bottom was added later. Copyright ©2015 IOP Publishing Ltd, Creative Commons Attribution 3.0 license.*

4.2 Characterization Techniques

Here, the most important characterization techniques and their typical measurement parameters are listed. Characterization techniques that have been developed during this thesis will not be explained here but in the following chapters.

4.2.1 Standard Characterization Techniques

Atomic force microscopy (AFM) measurements were performed with a *Bruker (former Vecco) Innova* (from Bruker, Massachusetts, U.S.A.) device. As AFM-tip an OTESPA ($1\,\Omega$ silicon, Cantilever: 3.6–$5.6\,\mu m$ thick, 140–$180\,\mu m$ long, 48–$52\,\mu m$ wide, resonance frequency: 281–$346\,kHz$, spring constant: 12–$103\,N/m$, no front side coating, $50(10)\,nm$ thick aluminum coating on the backside; Tip: height: 7–$15\,\mu m$, front angle: $0(1)°$ back and side angle: $35(1)°$, from Bruker, Massachusetts, U.S.A.) with a tip radius of $7\,nm$ was employed. The lateral resolution is given by the tip radius. The height resolution is roughly $1\,Å$. Typical scan parameters were $1\,\mu m \times 1\,\mu m$ scan size, $0.2\,Hz$ scan rate, and 512 dots per line scan resolution.

X-ray diffraction (XRD) measurements were done with a *Bruker AXS D8 Advance diffractometer* in the group of *Prof. Sebastian Polarz* (department of chemistry) at the *University of Konstanz* using CuKα radiation and Bragg-Brentano geometry.

The data were compared with known diffraction patterns from literature. XRD-characterization was performed by Melanie Gerigk (AG Polarz, Department of Chemistry, University of Konstanz).

Variable angle spectroscopic ellipsometry (VASE) was performed with a *V-VASE Ellipsometer* (from Woolam, Nebraska, U.S.A.). The light source was a xenon lamp (from Hamamatsu, Japan) in combination with a monochromator. Wavelengths between 250 nm and 1100 nm were applied. ANA-SPU-850 and RUT-SPU-850 samples were deposited on a polished silicon (100) wafer.

The following procedure was employed: Firstly, the layer was modeled as a Cauchy-layer in the transparent spectral range resulting in a film thickness of 42.64(2) nm and 61.21(3) nm for ANA-SPU-850 and RUT-SPU-850, respectively. Secondly, the optical constants were fitted for the whole investigated spectral range (point-by-point) in order to gain initial values. Thirdly, an oscillator model was established using a Tauc-Lorenzt dispersion formula. Here, it was assumed that the absorption of energies below the bandgap energy is negligible. A Kramers-Kronig consistent model followed from this procedure.

The mean squared error (MSE) for anatase and rutile TiO_2 is approximately 20 and 5, respectively. Hence, it appears reasonable that the previous assumption is not justifiable, in particular for the spectral regime slightly below the bandgap energy.

Ellipsometry characterization was performed by Johannes Rinder (AG Hahn, Department of Physics, University of Konstanz).

Scanning electron microscope (SEM) images were made employing a *Zeiss CrossBeam 1540XB* FESEM in the *Nanostructure Laboratory* of the *University of Konstanz* employing an electron energy of 5 keV. The working distance was set to 4–6 mm. Since the samples were sufficiently conductive, no additional conductive layer was necessary. The highest resolution for samples used in this work was achieved with rutile TiO_2 films on p-type, boron-doped silicon substrate. Here, reasonable images with a magnification of up to 200 kx were achieved and structures as small as 5 nm were observed clearly.

Transmission electron microscope (TEM) imaging for microstructural characterization was performed employing two devices.

Polycrystalline films were fabricated on Si_3N_4 grids having central windows covered with amorphous, 30 nm thick, levitating carbon layers. A Jeol JEM-2200FS field emission gun instrument operated at 200 kV was employed for TEM characterization. The films were investigated using bright field (BF) imaging. The crystallinity of the films was investigated by applying selected area diffraction (SAD) and zero-loss filtered (10 eV slit) high-resolution transmission electron microscope (HRTEM).

In order to characterize single nanorods, a multitude of nanorods was scraped off from samples with dense NRAs and different post-treatments. Then, the resulting powder was dispersed on a copper grid covered with a holey carbon film. SAD and conventional BF TEM was performed with a *Philipps CM 20* operated at 200 kV.

The entire TEM characterization was performed by Alena Folger (Max-Planck-Institut für Eisenforschung GmbH in Düsseldorf).

Energy-dispersive X-ray spectroscopy (EDX) was applied for elemental analysis taking an *INCA Energy Dispersive X-ray Spectroscopy Detector* (Oxford Instruments, Buckinghamshire, UK) in the *Nanostructure Laboratory* of the *University of Konstanz*. The device is attached to the *Zeiss CrossBeam 1540XB*. For titanium detection, an acceleration energy of 13 keV and a working distance of 10 mm was employed. Depending on the question either line or area scans were performed. Due to the pear-shaped scattering volume, this technique is neither surface sensitive nor highly resolving compared to SEM imaging. Hence, a strong signal from the substrate such as silicon appears always. Here, this analysis technique is not used for quantitative investigations but more for comparative studies.

4.2.2 Electronic Properties

To gain some information about the electronic properties the transient current was investigated by current over voltage, current over voltage and time, current over time, and impedance spectroscopy measurements.

Current over Voltage and Time Characteristics

Measurements using a conductive atomic force microscope (CAFM) under ambient conditions were not successful due to water condensed on the metal tip [45, 1649]. Hence, a self-made setup was established in a cube filled with nitrogen. The basic idea is shown in Figure 4.3. A platinum/iridium (80/20) scanning tunneling microscope (STM)-tip was approached to the sample manually and kept on a certain position by its own weight. A similar approach was reported by Patil et al. [1650]. The bottom electrode was contacted with silver past and thin copper wires. After the sample was placed in the chamber, it was evacuated and purged with nitrogen several times in order to get rid of water.

The tip and bottom electrode were connected via a BNC-cable with a *SourceMeter Keithley 2401* (from Tektronix GmbH, Germering, Germany). Maximum voltages between $-20\,\mathrm{V}$ and $20\,\mathrm{V}$ could be applied to this device. The Keithley was run by remote control using a PC and a self-made MATLAB program. For fast data transfer rates, a GPIB connector was used. The MATLAB program managed the data measured by the Keithley as well. Basically, three different measurement modes were applied: Firstly, a simple current over voltage characteristics was measured with a defined slew rate and small voltage step sizes (in the order of $1 \times 10^{-2}\,\mathrm{V}$). The minimum and maximum value for the applied biases was set by de- and increasing the voltage step by step until the absolute value of the current achieved an order of $1 \times 10^{-6}\,\mathrm{A}$. Due to asymmetric sample compositions, the absolute values for the extremal applied biases differ between negative and positive applied voltages usually. The current over voltage characteristics was measured five times without respite in each tip position. Secondly, the current over voltage and time characteristics was measured with two repetitions. Here, the voltage step size was in the order of $1 \times 10^{-1}\,\mathrm{V}$. For each bias value, the applied potential was kept for a defined time (typically $2\,\mathrm{min}$). The extremal values for the applied voltages remain unchanged during this measurement. Thirdly, the extremal potentials were applied for a quite long time (typically $20\,\mathrm{min}$ or longer). Then, potentials slightly above and below the opening voltage gained from the first measurement were applied as long as the extremal potentials before. For most samples, a diode-like current over voltage characteristic was measured with clear opening voltages. Finally, the simple I-V characteristics as described above were measured with five repetitions again in order to see the influence of long-time voltage and current stress. To gain a reasonable number of reliable data, the described procedure was employed to two or three tip positions on each sample. All measurements were performed in the dark at room temperature. From time to time the platinum/iridium tip was cleaned with acetone and a nitrogen gun.

Impedance Spectroscopy (IS)

Impedance spectroscopy (IS) is similar to the current over voltage and time measurements. But instead of a constant bias, an alternating potential was applied for each voltage step. Hence, an alternating current was induced inside the sample probing capacities, inductances,

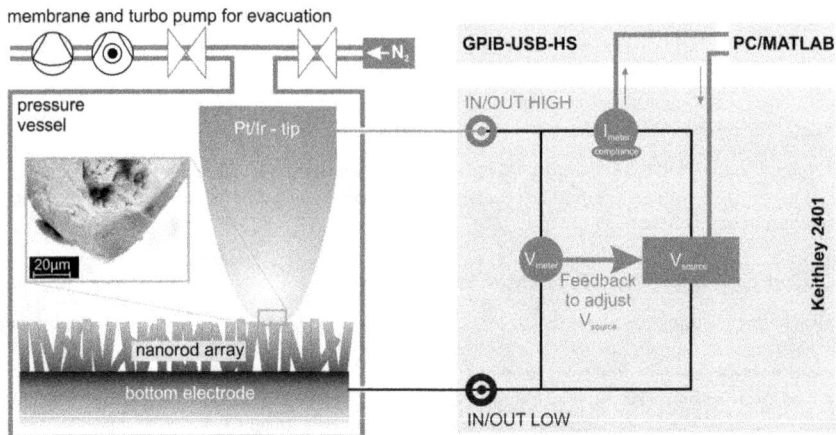

Figure 4.3: *Setup for electronic measurements. The inset shows a SEM image of the platinum/iridium tip on a NRA.*

shunt resistances, and series resistances. IS was performed employing an *Ecochemie Autolab Potentiostat/Galvanostat* (from Metrohm Autolab B.V., Utrecht, The Netherlands) with a dark sample chamber under ambient conditions. Frequencies between 10 Hz and 1 MHz were applied and the software *ZView* was employed for data modeling. A detailed description of IS with the described setup is found in the Master thesis of Eugen Zimmermann and Michael Puls [1651, 1652]. A comprehensive IS study on TiO_2 films is presented by Capan and Ray, for instance [1224].

4.2.3 Optical Properties

Basically, two optical properties have been investigated within this work. Firstly, the dielectric scattering of rutile TiO_2 nanostructures was mainly examined by an optical microscope. Secondly, electronic bandgap and trap state excitation was examined by UV-Vis-IR absorption measurements.

Optical Microscopy

For optical imaging an *Axio Imager 2 M2* (from Carl Zeiss Microscopy GmbH, Jena, Germany) with an *Axiocam 506 color* camera was employed. The simplified light path is shown in Figure 4.4. The microscope is equipped with a set of light emitting diodes (LEDs) called the *Colibri* system including 10 different LEDs between UV and dark red light. The total spectrum is shown in Figure 4.4 as well. The light was polarized before it reaches the sample optionally. As objective an *EC Epiplanar – NEOFLUAR* (0.9 HD, ∞/0, from Carl Zeiss Microscopy GmbH, Jena, Germany) with an amplifying factor of 100 and optimized for dark field (DF)-mode was used for imaging. All samples were measured in reflection mode. Dependent on the scientific question BF- or DF-mode was applied. Within the DF-mode, light with low incident angles is approaching the sample in a way that the directly reflected is not captured by the objective lens. Hence, only light that is reflected by steep surfaces such as sidewalls of 3D objects or light that

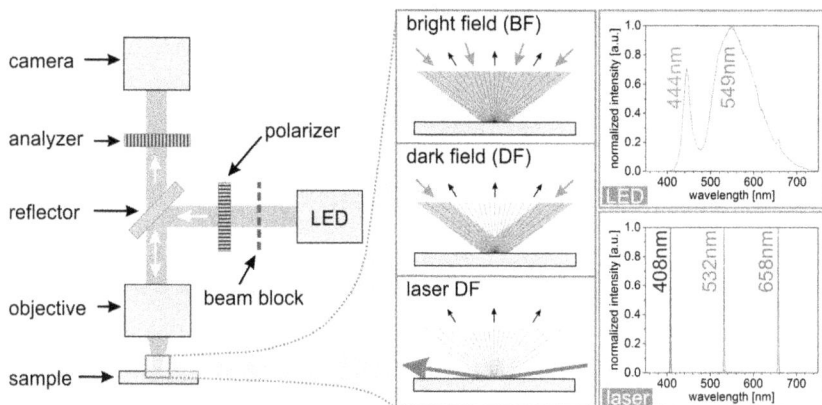

Figure 4.4: *Light path inside the optical microscope for BF, DF, and laser DF-mode as well as the spectra of the microscope LEDs and applied lasers.*

is scattered by surface roughness and dielectric or metallic scatterers. For the laser DF-mode, light from the LEDs was blocked completely and three different cw lasers (laser pointers) were applied with roughly 100 mW beam power and wavelengths of 405 nm, 532 nm, and 650 nm. The light was detected with a CCD camera as shown in Figure 4.4. The imaging software *ZEN* was employed for data management. The brightness for fixed LED powers was adjusted by an auto exposure function. For investigating polarization-dependent effects on the same sample, this functions was switched off. Images of faint samples were improved by a digital noise filter.

In many images taken by the microscope the polarization of the light is important. The light is linearly polarized before entering the objective. Since light is a transversal wave, additional polarization directions are accruing in the focused light beam as demonstrated in Figure 4.5. In particular, there are polarizations components, which are parallel to the initial light beam and hence, perpendicular to the substrate plane. This feature enhances the interaction between light and 3D structures on the sample.

UV-Vis-IR Absorption Measurements

Total absorption measurements were performed in an integrating sphere (150 nm diameter) using a *CARY 5000 UV-VIS-NIR* (from Agilent, California, U.S.A.). The samples were mounted on a *Center Mount Sample Holder*. The illuminated area was roughly 0.5 cm^2. For the measurements, a sample and a reference beam (double-beam mode) were applied and a default zero/baseline correction was conducted. The silicon substrates do not transmit any light and hence, light which is not reflected or scattered must have been absorbed by the sample. Thus, the total absorption $A_{abs,tot}$ is given in units of absorbance $\tilde{A}(\lambda)$.

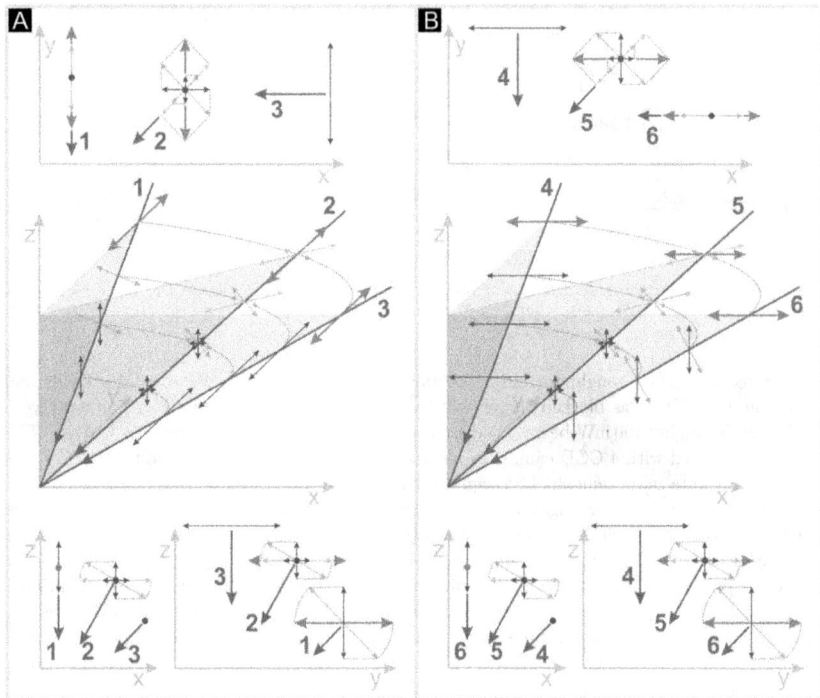

Figure 4.5: *Polarization in a focused light beam: The drawing represents a quarter of a linearly polarized light beam focused by a lens. All components of the electrical field vector for three selected directions are shown. The polarization of the initial light beam is tilted from (A) to (B) by 90°. In the focused beam, components which are perpendicular to the polarization of the initial non-focused light beam appear as well as components which are even in parallel polarized in relation to the initial beam. Hence, out-of-plane vector components appear on the sample surface that interact with 3D structures on the sample more efficiently.*

5 Thin TiO₂ Films and Seed Layer for the Hydrothermal Method

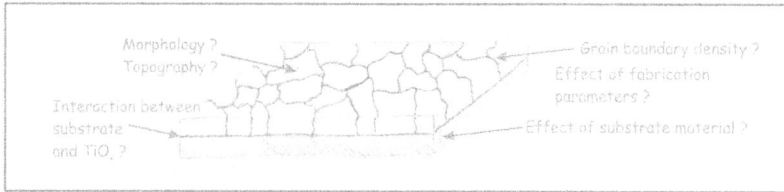

Figure 5.1: *Scientific objectives investigated in this chapter. In which way does the morphology, topography, grain boundary density, substrate adhesion, and diffusion of atoms between the substrate and deposited layer depend on the substrate material and fabrication parameters?*

One of the most fundamental issues for the growth of TiO₂ nanorods is to find a suitable seed layer. Amorphous and anatase TiO₂ usually dissolve in hydrochloric acid. Rutile TiO₂ films are much more stable in acid environments. In this work, rutile TiO₂ *and* anatase TiO₂ seed layers were developed that could resist the hot hydrochloric acid during the hydrothermal growth process. In this chapter, specific properties of these seed layers are discussed. Seed layers are fabricated by sol-gel, spray pyrolysis, sputter deposition, and electron-beam evaporation and annealed at different temperatures in oxygen rich atmosphere. Topography and cross-sections were investigated using AFM and SEM, morphology was investigated by TEM and XRD. The outcome of the hydrothermal growth process differs for different seed layers. Especially, the density of nanorods is controllable by the suitable choice of a seed layer, which will be discussed in detail in the following chapters.

AFM, SEM, and XRD studies are performed on films that were deposited on silicon substrates with a native oxide film on top. The native SiO₂ layer prohibits the formation of titanium silicide, but additional SiOₓ could be formed during annealing[585], which might strengthen the adhesion between the forming polycrystalline TiO₂ film and the silicon substrate.

The challenge for TEM investigations was to find a substrate that resists high annealing temperatures of up to 850 °C and hot hydrochloric acid, and which is transparent for high energetic electron-beams. These specifications are fulfilled by TEM grids consisting of a 30 nm thick carbon film that bears on silicon nitride (Si₃N₄) at the edges. For meaningful data analysis, it is important to ensure that there is no significant difference between silicon and levitating carbon films as substrate. SEM images in Figure 5.2 show that the grain structure does not differ for films fabricated on silicon, Si₃N₄ covered with a 30 nm amorphous carbon film and a levitating carbon film considerably. Besides silicon and carbon layers, TiO₂ films can be fabricated on many other substrates such as organic layers and self-assembled monolayers as well [1653, 1654].

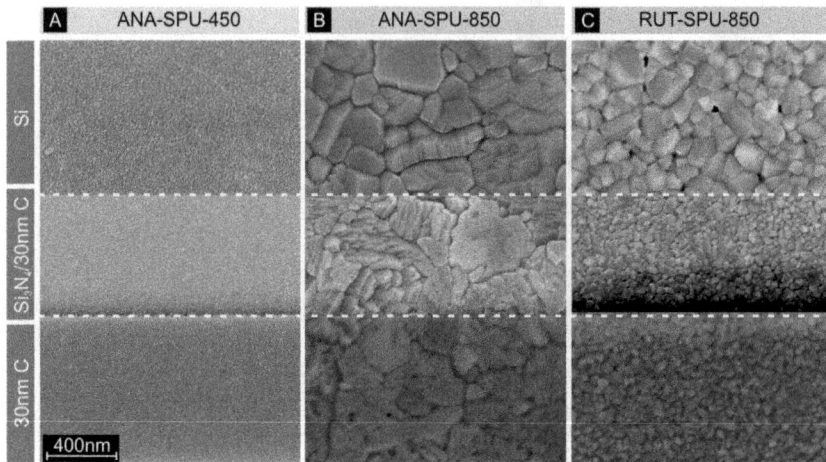

Figure 5.2: *ANA-SPU-450 (A), ANA-SPU-850 (B), and RUT-SPU-850 (C) deposited on silicon with a native oxide layer, Si₃N₄ covered with* 30 nm *carbon and levitating* 30 nm *carbon film.*

5.1 Thin Anatase TiO₂ Seed Layers

Anatase TiO₂ films evolved into the most important seed layer in this work. Amongst others, growth characteristics will be controlled by fabrication techniques and processing parameters. Applying an annealing temperature of 450 °C to sputtered amorphous TiO₂ films, the resulting film consists of fine anatase crystallites. Annealing at 850 °C is used to achieve relatively large grains. Films, fabricated by spray pyrolysis at 450 °C are already anatase without any further annealing. Also SALD films deposited at 300 °C are anatase without any further annealing. Their grains are even larger than grains in films deposited with sputter deposition and annealed at 850 °C. The grain size will play an important role for the seed layer performance including the density of nucleation sites and solubility in hydrochloric acid.

5.1.1 Anatase Polycrystalline TiO₂ Films Made by Sputter Deposition (ANA-SPU)

Sputter deposition is one of the most common deposition techniques for the fabrication of thin films and thus, it was used extensively in this work. Since sputter deposition was performed at room temperature, the films are supposed to be amorphous which is in good agreement with the XRD measurements in Figure 5.4 A. Annealing sputtered TiO₂ at 450 °C turns the film into a polycrystalline anatase film with small grains as shown in Figure 5.3 and 5.4. The grains have a diameter of approximately 4 nm. Thus, the anatase grains are stabilized by their surface energy as discussed in Paragraph 2.3.1, and they are expected to resist hot and acid environments longer than macroscopic anatase particles. However, any dissolving process begins at surfaces or grain boundaries. Hence, the higher stability of small grains might be compensated partly by the huge amount of grain boundaries in fine grained films.

It is assumed that silicon diffusion from the substrate is low at 450 °C and the only impurities originate from the TiO₂ sputter target and water molecules that were attached to the silicon substrate before film deposition. According to Paragraph 2.3.1, the formation of anatase TiO₂

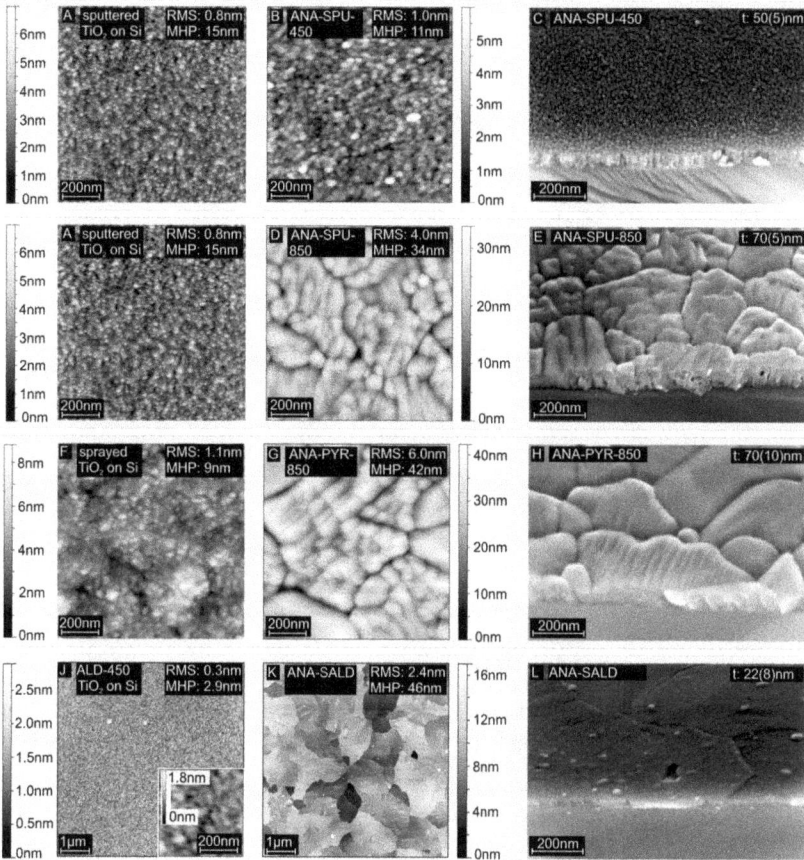

Figure 5.3: *AFM images (A,B,D,F,G,J,K), SEM cross-section images (C,E,H,L) of anatase seed films on silicon substrates: sputtered TiO$_2$ films before annealing (A), and after annealing at 450 °C (B,C) and 850 °C (D,E), respectively. Image (A) appears twice since the initial film before annealing is the same for the seed layers shown in image (B) and (D). Films made by different atomic layer deposition procedures are compared in image (J) and (K). Both procedures (ALD in (J) and SALD in (K)) are described in Paragraph 4.1.2. The SALD film (K) fabricated at 300 °C consists of much larger grains than the annealed but amorphous ALD film (J). This emphasizes the importance of crystallization during the ALD deposition process. Hence, the amorphous films made by sputter deposition and ALD look very similar after annealing at 450 °C (see image (B) and inset in (J)) while all polycrystalline anatase films show similar grain shapes independent from the fabrication technique. Image (L) shows the SEM cross-section of the SALD layer displayed in image (K). The root-mean-square roughness (RMS) and maximum height of the profile (MHP) are calculated from AFM-measurements and the film thickness (t) is determined from SEM cross-section images.*

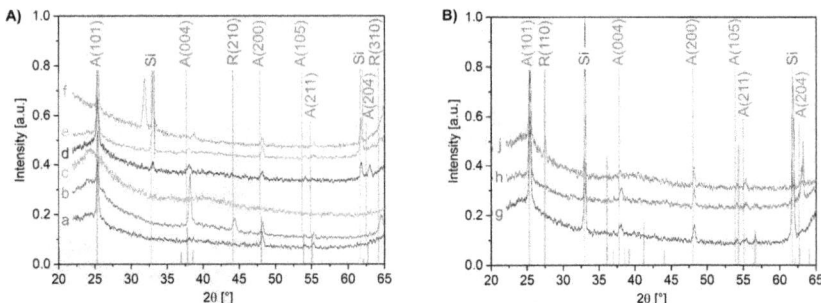

Figure 5.4: *XRD pattern of anatase seed layers. A) Sprayed TiO_2 before (a) and after (b) annealing at 850 °C, sputtered TiO_2 before (c) and after annealing at 450 °C (d) and 850 °C (e), SALD film (shown in image 5.3 K) (f). The peak at 44.5° for sprayed films arises from the rutile (210) reflection. For SALD films, a peak at 32° occurs that belongs most likely to a crystalline hydrogen titanate morphology. B) ANA-SPU-850 (g), ANA-SPU-1050 with a temperature slew rate of 5 °C/s (h) and 90 °C/s (j). For peak determination, the XRD pattern published by Howard et al. [269] is used. Reflexes from the silicon substrate are marked as well.*

is expected at 450 °C for pure and amorphous TiO_2. Hence, the mentioned possible impurities do not play a significant role for the crystallization temperature of the amorphous films, but an effect on the grain size cannot be excluded. For lower annealing temperatures the films are less crystalline and they dissolve in hot hydrochloric acid more easily.

Besides chemical stability, a seed layer should provide crystal facets that promote the growth of rutile nanorods. The TEM images in Figure 5.5 show that the polycrystalline anatase films is exposing many different crystal facets. From Wang and Li, it is known that nanorods appear also on anatase crystal facets such as the {101} and {301} facet [1578]. In the presence of the multitude of clearly observable crystal planes, it is conceivable that there are some growth promoting facets in a few places. But even in the absence of such growth promoting planes, the fine grain structure offers a lot of free-standing vertices and edges that could transform into rutile nanoparticles sitting on the anatase grains as described by Li et al. [159]. Such rutile crystallites act as nucleation seeds for hydrothermal growth as well.

Applying an annealing temperature of 850 °C results in a polycrystalline anatase film showing grain diameters between 0.1 and 1 µm. The surface of the grains is smooth and steps on crystal facets are pronounced. According to the TEM images in Figure 5.6 the large crystallites show voids in their inner part. These voids might result from the volume contraction during annealing. Furthermore, the formation of additional surfaces reduces the total Gibbs free energy of the anatase TiO_2 film. This would explain why these voids appear mainly in the center of grains and not on the low energetic surfaces and interfaces. A notable feature is seen in the diffraction pattern: The anatase films, annealed at 850 °C, lose their variety of facets that are observed for the films annealed at 450 °C. (102) is the most dominant facet in nearly all grains. This homogeneity of the crystals is suggested to result in a much lower density of nucleation sites.

The anatase morphology is taken during the temperature rise at the beginning of the annealing. The transition from anatase to rutile TiO_2 needs for pure and dry anatase TiO_2 approximately 900 °C [593]. For sputtered TiO_2 films it is reasonable to assume that they have a relatively

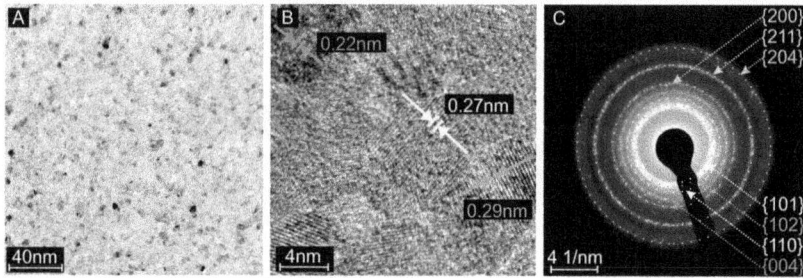

Figure 5.5: *TEM measurements on ANA-SPU-450 films. A) The fine grain structure of this film is supposed to be a strong promoter of the hydrothermal growth of rutile TiO_2 nanorods. B) The polycrystalline anatase film offers a lot of different closely packed crystal planes. The d-spacing of 0.29 nm, 0.27 nm, and 0.22 nm belong to the (102), (110), and (104) facets, respectively. C) The diffraction pattern shows the high variety of crystal planes in that sample. The well-shaped cycles result from the fine grain structure of the films.*

low water content. However, the XRD measurements in Figure 5.4 show that even at annealing temperatures of up to 1050 °C the film remains completely anatase, although there is no surface energy-induced stabilization effect for such large crystallites anymore. By applying a nearly 100 times higher heating rate of 90 °C/s the crystallization of amorphous TiO_2 is not completed before reaching 850 °C and thus, small amounts of remaining amorphous TiO_2 could directly convert into rutile TiO_2.

The fact that these films remain anatase at temperatures far above the reported ART temperature for pure anatase films indicate that the TiO_2 is contaminated by stabilizing impurities. An obvious impurity is the silicon of the subjacent substrate, which is according to Paragraph 2.3.1 known to inhibit the ART. From Figure 2.7, a silicon concentration of roughly 50% is assumed as a lower limit considering silicon doping as the only ART inhibiting factor. Such high doping concentrations appear to be only reasonable within a narrow volume close to the substrate/TiO_2 interface. Krabbes et al. reported that Ti-Si bonds are created at such high temperatures [37] and might stabilize the anatase phase by suppressing ion movement close to the interface in addition. However, the native SiO_2 layer might suppress this effect slightly [37]. Instead, additional SiO_x could be formed during the annealing process improving the adhesion of the polycrystalline films and the silicon substrate [585].

A similar explanation fits to sapphire substrates where aluminum could diffuse into the TiO_2 prohibiting the phase transition as well. In the case of the levitating carbon film substrate, this explanation fails since carbon is an ART promoter and there are no other phase stabilizing impurities present – except impurities in the sputter target. But most of the impurities in TiO_2 sputter targets are ART inhibitors. It should be noted that silicon diffusion on a silicon substrate is not excluded by this argument and will play an important role for laser-induced ART applied in Chapter 9. Since only a few crystal orientations are observed in highly annealed anatase TiO_2 films it might be also possible that the crystallites form surfaces with extremely low surface energies shifting the phase transition to higher temperatures according to Equation 2.4.

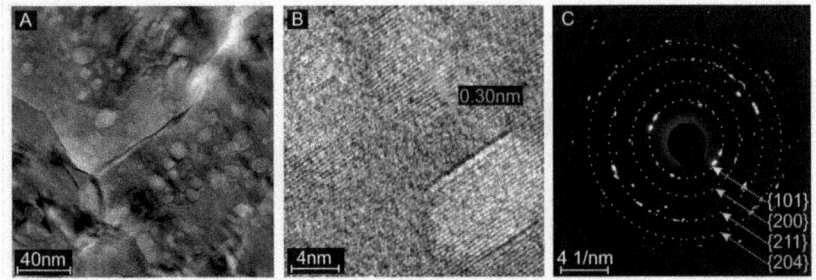

Figure 5.6: *TEM measurements on ANA-SPU-850 films. A) The grains are relatively large and have plenty of small voids in their center. Due to the low level of interfaces, the density of nucleation sites is expected to be low as well. B) The crystals are very homogeneous and only very few dominant crystal planes exist. The d-spacing of 0.30 nm belongs to the (102) facet. The structure is interrupted by some voids that have clear straight-lined borders. C) The variety of facets is reduced compared to ANA-SPU-450.*

5.1.2 Anatase Polycrystalline TiO₂ Films Made by Spray Pyrolysis (ANA-PYR)

A quite inexpensive deposition method for anatase TiO₂ films is spray pyrolysis. As the process takes place at 450 °C the resulting film is already anatase. However, grains are not yet developed clearly. After annealing, the sprayed film annealed at 850 °C looks quite similar to the sputtered TiO₂ film annealed at the same temperature. The grains are even a little bit larger and smoother. In contrast to sputtered anatase films, sprayed films show a single rutile reflex corresponding to the (210) facet. This indicates that the sprayed films are less stable than the sputtered films. This might be due to a higher water content which is brought by the spraying process into the film as well as precursor remains. The organometallic precursor contains a lot of carbon, which is known to destabilize the anatase phase. Three times thicker films (approximately 210 nm) have the same appearance. Hence, the effect of the substrate is negligible or at least homogeneous in even much thicker films. On the one hand, this observation makes an influence of the crystal orientation of the silicon substrate on the phase stability unlikely. On the other hand, diffusion of silicon ions into TiO₂ films might still be an important factor for thicker films and sufficiently long annealing times [585].

5.1.3 Anatase Polycrystalline TiO₂ Films Made by SALD (ANA-SALD)

ALD has several advantages and hence, it was added to the presented seed fabrication techniques as well. The method is inexpensive, dopants are easily added, and the thickness is controlled precisely. The low process temperature enables the use of organic substrates such as plastic foils. One of the most important advantages is that this method is able to create homogeneous films on porous and high aspect-ratio nanostructures [1546]. In contrast to the relatively slow deposition process achieved with ALD, SALD offers fast deposition rates and turns this technique into an interesting method for commercial applications.

The SALD process was performed at 300 °C on a polished silicon wafer. Obviously, the applied temperature is high enough to promote the epitaxial growth of a crystalline TiO₂ film. The morphology depends on the process parameters such as the process temperature and the substrate [137, 1548–1550]. As expected, the SALD films used for this thesis are anatase since they were deposited at 300 °C [137]. The importance of crystallization during the deposition process is demonstrated by comparing the SALD film with an ALD film that was fabricated with different process parameters including a lower temperature. The ALD film did not crystallize during the deposition and afterward, it was not possible to convert the film into observable anatase grains, although the film was annealed at 450 °C. The reason for the different morphologies of both presented films is found in the diverse process conditions. The ALD process did not provide the conditions for the formation of crystalline TiO_2 at all. The SALD process promoted epitaxial growth resulting in extended crystal structures although the applied temperature was below the theoretical transition temperature of pure and completely amorphous bulk TiO_2 into the anatase crystal structure. The resulting grains are extremely large and their surface is very smooth compared with grains gained by other techniques as demonstrated in Figure 5.3.

Anatase seed layers are created with different techniques on various substrates. The grain size is controlled by the post-annealing temperature. It is suggested that the phase transition from anatase to rutile is suppressed at high post-annealing temperatures (850 °C) by impurities.

5.2 Thin Rutile TiO₂ Seed Layers

Trivially rutile substrates provide the growth of rutile nanorods and are chemically stable. The minimal processing temperature for creating pure rutile TiO_2 films depends strongly on the fabrication technique and precursor impurities as reported in detail in Chapter 2.3.1. In this work, the processing temperature for rutile TiO_2 seed layers was 850 °C, but it should be noted that this was not the minimal process temperature for gaining pure rutile films with the applied techniques.

5.2.1 Rutile Polycrystalline TiO₂ Films Made by Sputter Deposition (RUT-SPU)

The fabrication of rutile TiO_2 films by sputter deposition has to be performed without TiO_2 as raw material. As discussed above, sputter deposition of TiO_2 would result in anatase films. Hence, the creation of the anatase phase has to be avoided. The route that was developed in this work is to sputter pure titanium and oxidize the film afterward. Due to the low oxidation rate the anatase morphology cannot be formed since the material is still far away from stoichiometric TiO_2 during the temperature rise. Composites with less oxygen could be amorphous or form different crystal structures such as cubic TiO or orthorhombic Ti_2O_3 [1655, 1656]. These morphologies remain until the temperature is too high for the formation of anatase TiO_2. When the intermediate morphologies became unstable due to the rising temperature and oxygen content, rutile will be formed. The model, that the ART depends on the oxygen partial pressure employed during laser-ablated TiO_2 thin film deposition is supported by the observations of Czanderna et al. [1657]. Because of the different morphology, it is expected that the grains have different shapes as well. Rutile grains offer a pyramid-like surface with clear facets for all investigated rutile TiO_2 films the grain size is between 20 and 200 nm as shown in Figure 5.7. The diffraction pattern in Figure 5.8 C shows that there are several crystal orientations present in the polycrystalline rutile film. However, according to the high-resolution image in Figure 5.8 B the [001] direction (0.29 nm d-spacing) is dominant. Since all facets except the (110) facet are expected to promote the growth of rutile nanorods the investigated films work as an efficient seed layer.

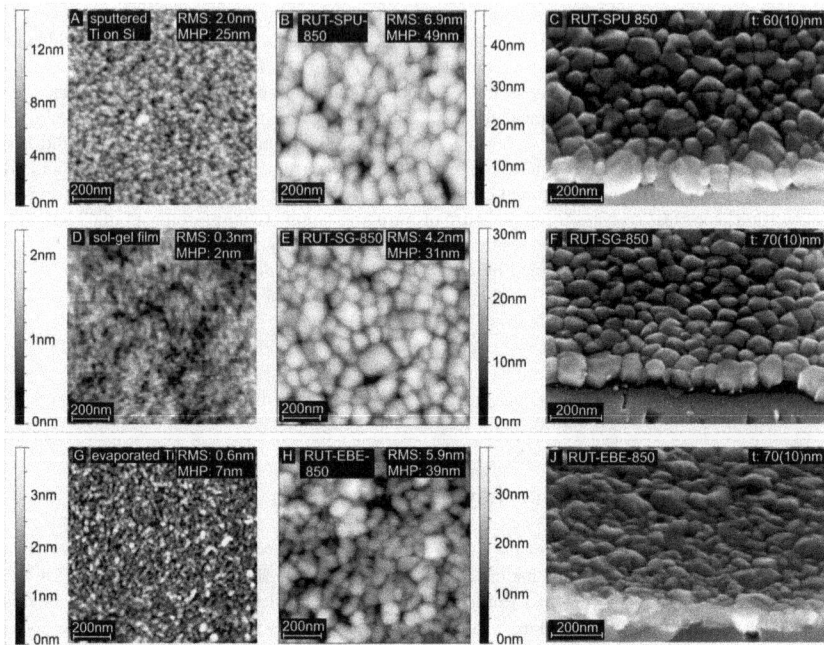

Figure 5.7: *AFM images (A,B,D,E,G,H), SEM cross-section images (C,F,J) of rutile seed films on silicon substrates: sputtered (A,B,C), evaporated (G,H,J), and sol-gel (D,E,F) films before (A,D,G) and after (B,C,E,F,H,J) annealing at 850 °C. The (RMS) and (MHP) are calculated from AFM-measurements and the film thickness t is determined from SEM cross-section images.*

5.2.2 Rutile Polycrystalline TiO$_2$ Films Made by Electron-Beam Evaporation (RUT-EBE)

The principle of polycrystalline rutile film formation starting with an evaporated titanium layer is the same as for sputtered titanium layers. In both cases, the source material is pure titanium that is available in much higher purities than TiO$_2$. Thus, the resulting rutile films are supposed to contain fewer impurities compared to anatase films. In particular, there is a low level of impurities that cannot be refilled from the environment such as silicon or carbon. As a consequence, there might be less stabilizing impurities in the film that prevent the survival of anatase crystallites. This effect sums up with the creation of unstable crystal structures for titanium films with relatively low oxygen content as discussed for the sputtered titanium film in Section 5.2.1 already. If evaporated films are turned into anatase TiO$_2$ at low annealing temperatures it needs more than 1100 °C to convert them into rutile TiO$_2$ [1658].

Compared to sputter deposition the adhesion of evaporated films is lower. But due to the narrow angle distribution of atoms hitting the sample, evaporated films are more suitable for photoresist

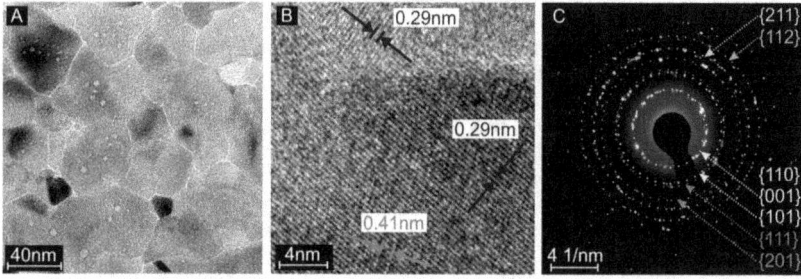

Figure 5.8: *TEM measurements on RUT-SPU-850 films. A) Clearly visible grains with interfaces perpendicular to the surface. These grains contain some voids, but they are much smaller than the voids observed in ANA-SPU-850 films. According to their crystal orientation, some grains appear darker than others. B) The zoom-in confirms the high crystallinity of the grains. The d-spacing of 0.29 and 0.41 nm belong to the (001) and (100) facet, respectively. C) Polycrystalline rutile TiO_2 film contains many facets that promote the growth of rutile TiO_2 nanorods.*

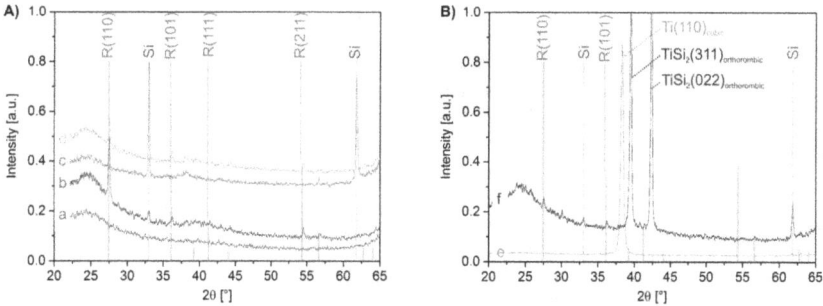

Figure 5.9: *XRD pattern of rutile TiO_2 seed layers. A) As-deposited sol-gel film (a) and RUT-SG-850 (b); as-deposited sputtered titanium film (c) and RUT-SPU-850 (d). B) as-deposited evaporated titanium film (e) and RUT-EBE-850 (f). For peak determination the XRD pattern published by Abrahams and Bernstein [258] is used for rutile TiO_2. For rutile Calvert and Häglund et al. were used as the primary reference. The pattern was calculated from ICSD using POWD-12++. The report of Eppelsheimer and Penman is the reference for the unit cell of titanium. Cotter et al. was used as reference for $TiSi_2$.*

147

lithography than sputtered films. Both, sputter deposition and electron-beam evaporation can be used to generate very thin films.

If the as-deposited film is thinner than 5 nm the available material is not enough to form a dense polycrystalline TiO_2 film during oxidation. Instead, the material forms individual grains or small grain accumulations and the silicon substrate remains uncovered partially. Thus, decreasing the film thickness is a simple technique to lower the density of nucleation sites. Since even nanoparticles act as nucleation sites there is no technical limit for scaling down the grain size. However, it should be considered that the total interface area between substrate and TiO_2 film is decreasing with decreasing grain size as well, and thus, the adhesion of the resulting nanorods is significantly lower than for nanorods grown on relatively thick and dense seed layers. Furthermore, the anatase phase might be preferred for grain diameters below 14 nm due to the low surface energy of anatase compare to rutile TiO_2 as discussed in Section 2.3.1.

5.2.3 Rutile Polycrystalline TiO₂ Films Made by a Sol-gel Method (RUT-SG)

Obviously, as-deposited sol-gel films do not consist of stoichiometric TiO_2 yet. A lot of organic materials and water remain in the film. Hence, there are even two clear factors that prevent the formation of anatase TiO_2 at an annealing temperature of 850 °C. The first issue is the same as for sputtered or evaporated titanium films: The film does not consist of TiO_2 as raw material and hence, anatase TiO_2 cannot be formed quickly. If anatase crystallites are formed during the temperature rise at the beginning of the annealing process, they form a compact polycrystalline layer. And compact layers of small anatase crystallites are supposed to convert easily into rutile since they have a high level of interface areas and offer sufficient material transport as discussed in Section 2.3.1 [595, 596]. The second issue is the high water content that lowers the ART temperature significantly [593].

A sol-gel technique is introduced as a fabrication technique for rutile seed layers as well, since it is a very simple and inexpensive method which is frequently used as a chemical film deposition technique. However, this method features some limitations. The film thickness is not adjustable arbitrarily and especially film thicknesses of less than 20 nm are hard to achieve. Very thick sol-gel films tend to form cracks during annealing due to the enormous volume loss and thermomechanical stress. The adhesion of sol-gel films is quite low and they are not useful for photoresist lithography. But the low effort of this method legitimates its use considerably.

Rutile seed layers are fabricated with various techniques. The initial layer before annealing does not consist of TiO_2 but contains at least titanium. While the temperature is increased at the beginning of the post-annealing process, intermediate structures consisting of highly reduced TiO_2 are formed which prevent the creation of anatase. After the final temperature (850 °C) is achieved, the intermediate structures decay and the film turns into rutile TiO_2.

6 Hydrothermal Growth of Rutile TiO$_2$ Nanorods on Rutile and Anatase Films

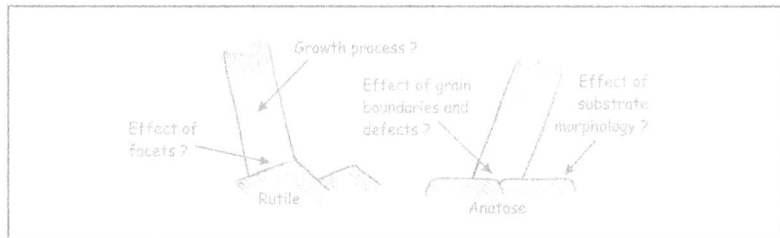

Figure 6.1: *Scientific objectives investigated in this chapter. How is growth going on? In which way is the growth affected by the morphology, facets, and grain boundary density of the seed layer?*

As already discussed in Chapter 3.2, the hydrothermal growth process of rutile TiO$_2$ nanorods is based on a complex chemical and thermodynamically driven reaction. There are reasonable models for many aspects of the hydrothermal growth, but there are still some features that are not fully understood or where several competing models exist. In this chapter, the growth process on rutile and anatase substrates is described from nucleation to full-grown nanorods. Doing so, several questions are answered: Why is the visual nature of a NRA on rutile films as it is? What is the structure of a NRA? How do rutile nanorods grow on anatase TiO$_2$ films? One outstanding feature of the full-grown nanorods is their substructure [1663]. Thus, it is addressed in an extra chapter (Chapter 7). The content of this chapter is also dealt with in the articles [1664–1666].

In order to test the stability of different seed layers, the hydrothermal treatment was performed without precursor. The results are shown in Figure 6.2. All rutile and anatase samples annealed at 850 °C resist the hydrothermal conditions without any significant change, but most films annealed at 450 °C dissolve or peel off. Sputtered anatase films annealed at 450 °C (ANA-SPU-450) reshape during the hydrothermal treatment. The only applied anatase seed layer that remains unchanged is sprayed anatase (ANA-PYR-450). To work as a seed layer, the desired annealing temperature has to be applied for at least 3 min. Films without annealing or annealing at significantly lower temperatures peel off or decay during the hydrothermal growth process. This indicates the forming of a strong bond between the TiO$_2$ and the silicon during annealing. Exposures to the hydrothermal growth solution longer than 12 h showed that the polycrystalline film is dissolved partly – in particular at the grain boundaries. This effect originates from the lower chemical stability of anatase in acids compared to rutile TiO$_2$. The smaller the grain size the more grain boundaries are present and the fast the film dissolves.

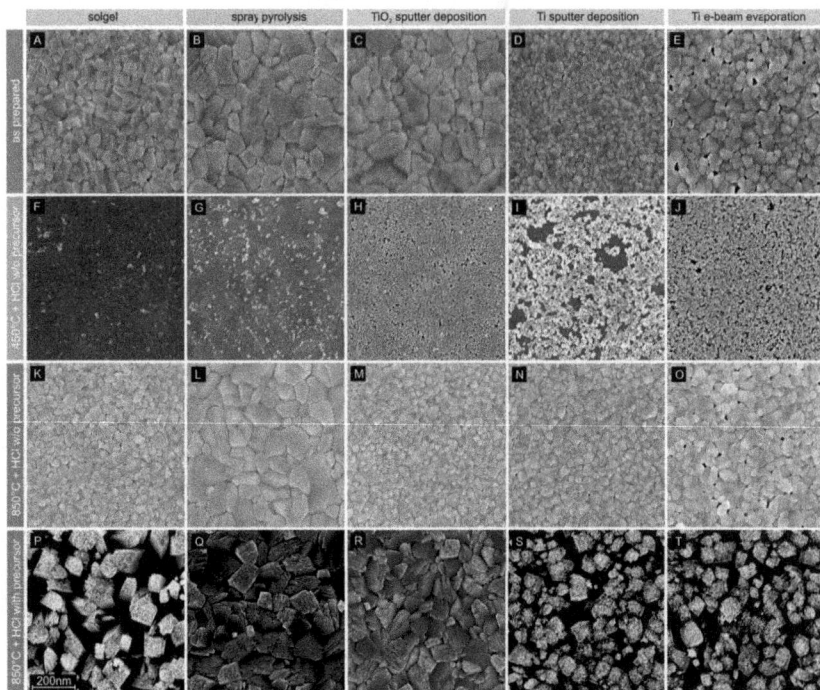

Figure 6.2: *SEM images of substrates fabricated by different techniques. A-E) TiO$_2$ films after annealing at 850 °C for two hours, but before hydrothermal treatment; F-J/K-O) TiO$_2$ films after annealing at 450 °C/850 °C for two hours and after hydrothermal treatment in 14.8% hydrochloric acid solution at 180 °C for three hours without any precursor; P-T) Rutile TiO$_2$ films after hydrothermal growth at 180 °C for three hours with 350 µL titanium(IV) butoxide as a precursor.*

The chemical stability of the employed seed layers is achieved by high post-annealing temperatures. Thus, only highly crystalline TiO$_2$ resists the hot hydrochloric acid.

6.1 Hydrothermal Growth in Solution

A consequence of the time-dependent process temperature in the first hours as demonstrated in Section 4.1.3 is that the growth process starts at different points in time for different oven temperatures. This results in different growth durations for the same storage period in the oven. During the first hour, growth conditions are changing continuously and thus, the result is not related to a structure created at a certain defined temperature. The quickly changing temperature implicates a quick increase of the growth rate at the beginning of the growth process. As a consequence, there are only a few minutes between the initial nucleation and the appearance of a few hundred nanometer long TiO_2 rods. This fact makes the investigation of growth steps challenging and one has to hold growth process times with an accuracy of a few seconds! The "growth times" stated below are always the retention times of the autoclave inside the oven, except it is specified differently. It should be noted that the temperature development depends on the geometry of the autoclave and heating source. That is, why the temperature development differs certainly between heating with an oven or a hotplate. For time-resolved growth experiments as described in this report, a precise measurement of the temperature development is absolutely necessary for characterizing the growth steps.

To investigate nanorods grown in solution a polished silicon substrate is placed horizontally on the bottom of the Teflon liner. Rods grown in solution settle out on the substrate. The first rods appear after approximately 60 min. Rutile TiO_2 nanorods grow even in the absence of any seed layer and are found in solution or on seed free substrates for instance. A possible explanation is that the initial nucleation takes place at floating dirt particles or precursor agglomerates in the growth solution [155]. It is reasonable that such particles appear next to full-grown nanorods or are even embedded by them. In the early growth stage, nanorods grown in solution are often embedded in some kind of organic material that might be the residue of a dirt particle or precursor agglomeration. As shown in Figure 6.3, these remains degrade quickly when exposed to the electron-beam of the SEM and are not seen in later growth stages anymore. Due to the fast decay in an electron-beam, the determining of the content of the remains applying EDX and TEM are not possible. However, in numerous investigations using SEM no unambiguous candidate for the nucleation triggering particles was found so far and thus, the concrete origin of the initial nucleation in seed layer free solutions remains unknown. Beyond that, it seems that they do not play any major role in the following growth process. Hence, it can be concluded that seed particles are either much smaller than the smallest investigated rod diameter (10 nm) and thus, they are embedded by the nanorod completely, or the particles move away from the nucleation site or dissolve in the hot hydrochloric acid.

Figure 6.3: *Remains of supposed nucleation sites in solution marked by yellow arrows. Such remains are observed within the first 90 min only.*

Compared to full-grown rutile TiO_2 rods, dissolving rods show much less electron reflection. But also highly electron reflecting rods with the same size as low electron reflecting rods were observed. Therefore, it is assumed that the reflectivity is not a size-dependent property, but depends on the composition of the structure. Especially in the low-temperature regime after the first hour in the oven, hybrid nanostructures such as hydrogen pentatitanate nanorods might be formed although mainly reported for alkaline environments [157, 158, 163]. Titanates are converted into rutile TiO_2 during further exposure to the hydrothermal growth solution [161, 164]. Organic material can be decomposed in focused electron-beams easily and electron-beam-induced deposition (EBID) is even based on this effect [1667]. Unstable small nanorods are observed in solution only, which indicates that nucleation and growth on seed layers result immediately in rutile TiO_2 without any long living intermediate stages. After 80 min the majority of nanorods is stable and show a high electron reflectivity similar to full-grown rutile TiO_2 nanorods. It is reasonable that nanorods have turned completely into rutile single crystals at this point in time. This phase transition is probably associated with the quickly increasing temperature in the Teflon liner. Now, each rod acts as a seed crystal for the surrounding precursor molecules and the transformation from $TiCl_4$ into rutile TiO_2 is so efficient that the formation of intermediate stages have become energetically unlikely and no further rods are created. Furthermore, the precursor concentration needed for nucleation is larger than the concentration needed for the growth on preexisting nanocrystals. Hence, when the precursor concentration drops below a critical value nucleation stops but growth continues. This is the reason why nanorods are mostly monodisperse after a defined growth time.

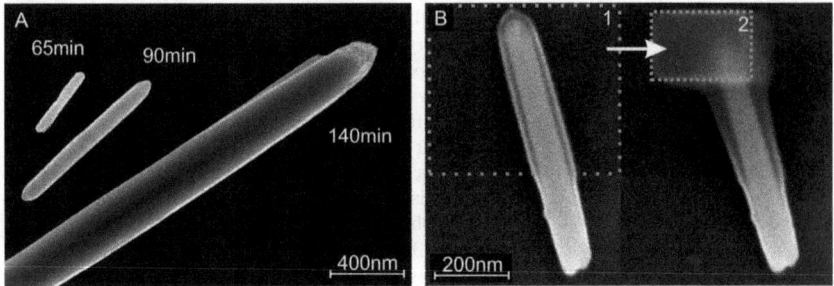

Figure 6.4: *A) SEM image of TiO₂ nanorods grown in solution at different growth stages. B) Effect of electron-beam with different exposure intensities on the early growth state of solution based nanorods: large exposure area with a low areal energy density (1) and small exposure area with a high areal energy density (2).*

Nanorods dropping out from solution will be present on all seed layer substrates as well and it has to be distinguished between seed layer and solution based nanorods. A trivial fact is that seed layer based rods can grow in one direction only, since the bottom tail is connected with the seed and cannot be supported with precursor molecules. Hence, solution based rods are supposed to be longer than seed layer based structures, which is seen clearly in the experimental results. The adhesion between the silicon surface and the nanorod is not very strong since there is no chemical bond. As a consequence, nanorods are removed from the silicon by rinsing with ethanol only. This is advantageous for transferring nanorods on TEM grids. For all seed layer based nanorods, there are only rough methods for removing them from the substrate such as scraping or sonication. Thus, the risk of damaging is quite high and solution based nanorods are preferred for TEM investigations, for instance. Nevertheless, the growth has to be stopped at the right time in order to keep the thickness small enough for TEM characterization.

Besides extended seed layers, nanorods grow on small seeds in solution. Before they obtain a critical size, they decay during intensive electron-beam exposure. This effect might be related to an incomplete transformation of the precursor into crystalline TiO₂ in the early growth stage. This instability is not observed for nanorods grown on seed films.

6.2 Hydrothermal Growth on Rutile Seed Layers

Although the diffraction pattern of sputtered rutile seed layers shown in Figure 5.9 reveals the crystal planes being present in the film, it is not clear which facets are exposed on the surface of the film. Of course, there are expensive experimental techniques to determine the surface structure of such films. STM is one of the most accurate techniques to determine the surface structure with a high spatial resolution [295]. Here, a more simple method is presented that estimates the crystal orientation of surface facets. Firstly, the hydrothermal growth is applied on macroscopic rutile TiO$_2$ single crystals with well-defined crystal orientations. Secondly, the growth angles observed on macroscopic rutile crystals are compared with the angle being included by a nanorod and the corresponding seed facet on the polycrystalline film. Same angles indicate same crystal orientation.

Finally, the growth on different polycrystalline rutile seed layers made by different fabrication techniques and made of different materials are compared.

6.2.1 Hydrothermal Growth on Macroscopic Rutile TiO$_2$ Single Crystals

The hydrothermal growth process on rutile TiO$_2$ single crystal wafers is essential for the understanding the hydrothermal growth on all thin film seed layers discussed in the following. Due to their appearance in nanorods macroscopic single crystals with the following crystal facets are used: (001), {111}, {110}, and {100}. The result of hydrothermal growth on macroscopic rutile crystals is predictable from the crystal structure of rutile TiO$_2$ nanorods. According to Wisnet et al. [168] the tip consists of (001) and {111} facets, the side walls of {110} facets and the edges point towards the ⟨100⟩ direction. [001] is the growth direction. Hence, for polycrystalline rutile TiO$_2$ films, it is expected that growth is promoted on (001) facets and any facet which can be projected on (001) facets such as {111} and {011} facets. But facets being perpendicular to the (001) facet such as the {110} and {100} facets should not promote the growth. This assumption is checked easily by using macroscopic rutile single crystals with different facets as seeds for the hydrothermal growth.

The prospects fit perfectly with the experimental results as shown in Figure 6.5. For the (001) facet, a dense, crystalline film is grown on the crystal. A closer look on this film reveals the same nanofingers as a substructure that is observed in nanorods. In the case of the (001) surface, the fingers are perpendicular to the substrate surface. This is expected since the growth is parallel to the [001] direction. The diameter of the nanofingers is between 10 and 15 nm. Cracks in the film exist from stress caused by a thinner finger diameter at the top compared to the bottom. The decreasing finger width from the bottom to the top of the film might be caused by the constantly changing growth conditions such as the increasing temperature and decreasing precursor concentration. This is in good agreement with the observation that the nanorod width decreases with decreasing initial precursor concentration as reported by Li et al. [159]. From the average distance and width of the cracks, it is assumed that the tips are 4-6% thinner than their base. The cracks form not straight lines but more random courses comparable to cracks in a desiccated mud. Usually, they are formed at grain boundaries of the substructure. Besides these cracks, there are a few perfectly linear ruptures that are supposed to originate from fine, atomic fractures in the wafer as shown in the insets of Figure 6.5 A and C.

Since stepless {111} facets include an angle of 45° with the (001) facet, nanofingers of the substructure are tilted on the {111} crystal by 45° as well. As demonstrated in Figure 6.5 cracks caused by finger narrowing are formed likely at grain boundaries between the nanofingers and here, the cracks are in line with the growth direction projected on the substrate plane. Because of the formation of crystal steps on the investigated polycrystalline rutile TiO$_2$ films, the inclination must not be 45° necessarily.

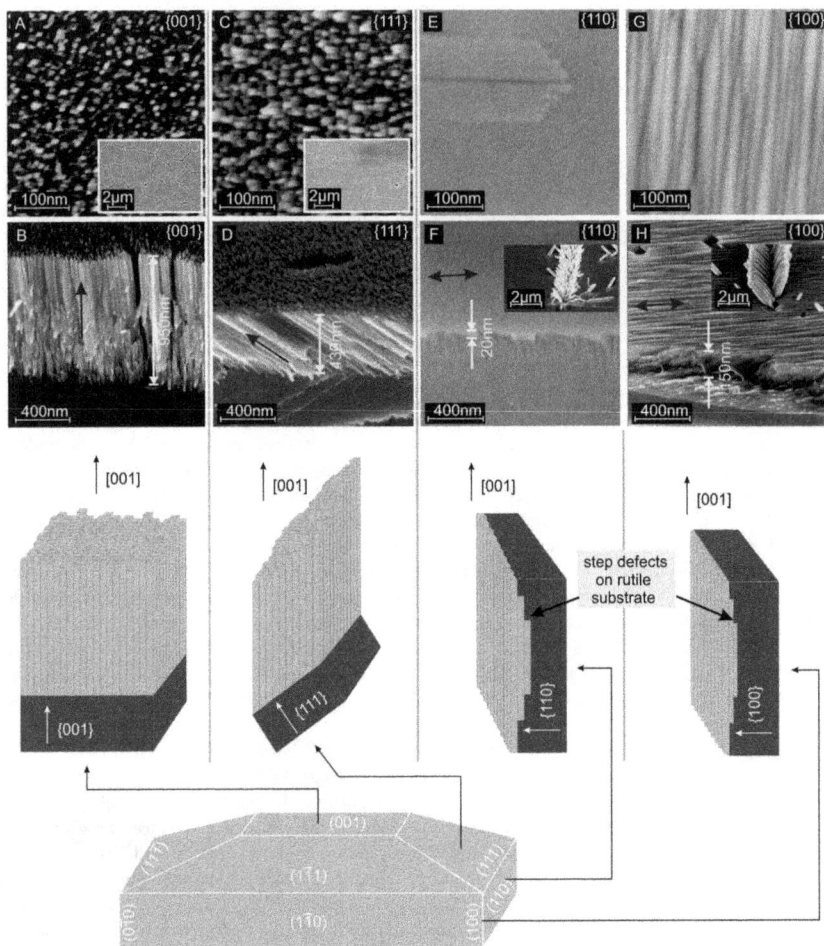

Figure 6.5: *Top view (A,C,E,G) and cross-section (B,D,F,H) SEM images of hydrothermally grown structures on rutile single crystals with (001), {111}, {110} and {100} facets. The schematic drawing makes up a relationship between these structures and the observed facets on a rutile TiO$_2$ nanorod: Growth on single crystal facets appearing also on the tip of a nanorod results in the same fine structure as it is seen in nanorods. Growth on facets perpendicular to the (001) facet is expected to be much less pronounced, which is in good agreement with the cross-section images. Growth on the {110} facet, which corresponds to the flat side walls of a nanorod results in a flat structure as well. In contrast, growth on the {100} facet, which corresponds to the edges of a nanorod results in dense parallel gables. The growth direction is marked with a blue arrow.*

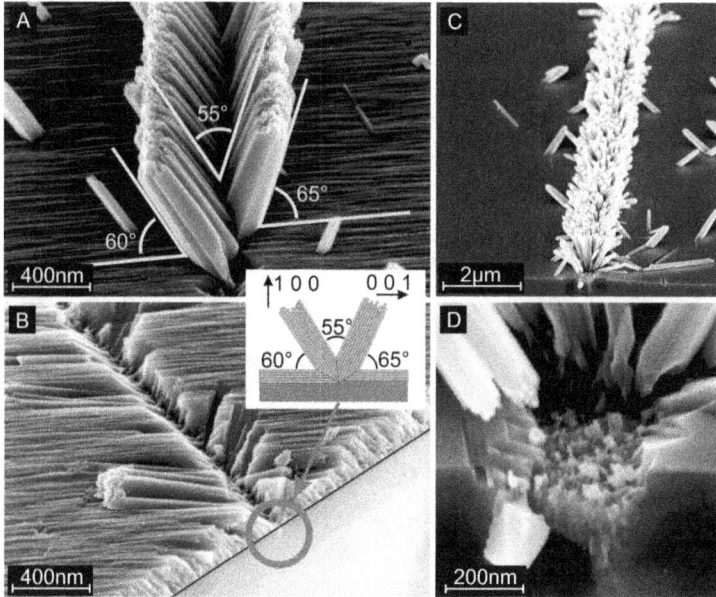

Figure 6.6: *SEM images of growth on one-dimensional crystal defects on rutile single crystals. Inset: Schematic drawing of the hydrothermal growth on {100} single crystals including a double branching event resulting in extended nanorod fences. A) SEM cross-section image of an extended branching site on a {100} single rutile crystal. B) SEM cross-section image of an extended branching row, where the branches were removed mechanically. The blue line marks the interface between the single-crystalline rutile TiO$_2$ {001} substrate and the TiO$_2$ layer, which was grown during the hydrothermal growth process. The red cycle marks a rupture inside the substrate that might have triggered the branching above. The visible defect inside the substrate might have become large due to mechanical stress that was applied to break the sample for the cross-section image. C) SEM cross-section image of an extended defect on a {110} single-crystalline rutile TiO$_2$ substrate, which is significantly broader than those shown in (A). This defect might come from scratches generated during sample handling and storing. D) Zoom of (C) showing that this structure is not caused by a fine one-dimensional crystal defect.*

{110} facets form the side walls of nanorods. Thus, growth on {110} facets appear in parallel to the surface and no bottom-up growth along the [001] direction is possible. However, a thin film is formed since there is still a slow growth perpendicular to the preferred growth direction. As demonstrated in Figure 6.5 nanofingers are parallel to the crystal surface.

{100} facets would be found at the vertical edges of nanorods, but these facets are not strongly pronounced. Again, the slow growth perpendicular to the preferred growth direction allows the horizontal growth of nanofingers parallel to the crystal surface in the [001] direction similar to {110} crystals. The resulting surface is not flat, but corrugated since the growth structure recreates the edges of the nanorods on the substrate as shown in Figure 6.5.

In conclusion, the experimental results shown in Figure 6.5 are in good agreement with the presented prediction. In general, any facet being the origin of a nanorod sticking out of the seed layer is either a (001) facet, if the rod grows perpendicularly on the facet or a projection of the (001) facet, if the rod is inclined on the facet.

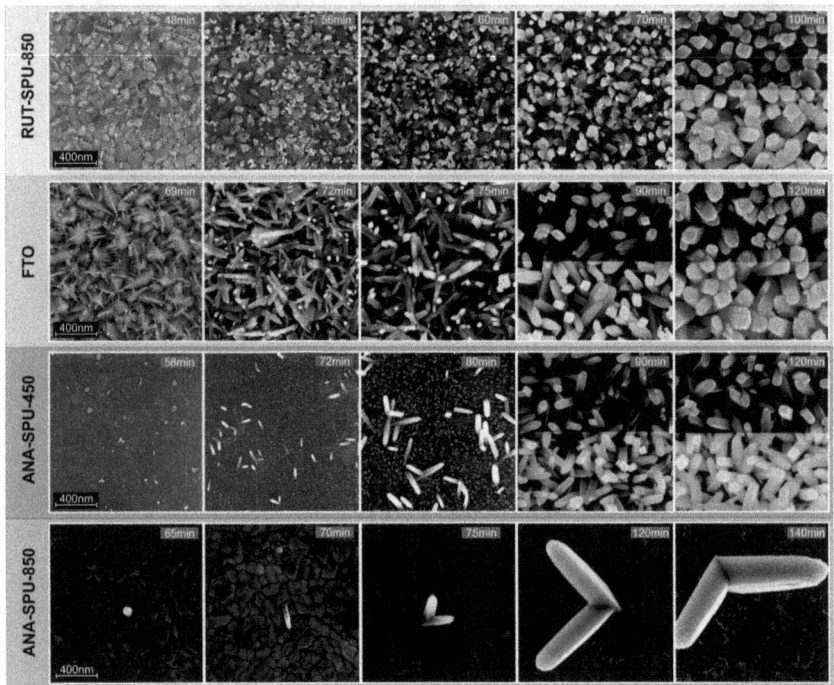

Figure 6.7: *SEM images of selected growth steps on RUT-SPU-850, FTO, ANA-SPU-850, and ANA-SPU-450 (growth on anatase seeds is discussed in Section 6.3).*

However, an exception from this rule is observed. Tiny defects on rutile seed facets, which are not promoting the growth out of the substrate plane, cause a locally confined growth of nanorods as seen in the inset of Figure 6.5 F and H. Since the preferred growth direction for these facets is parallel to the surface, this phenomenon reminds strongly on branching of single rutile nanorods as described by Li et al. [159]. Even the occurring angles between the branched rods and the substrate plane match with the observations of Li et al. This is explained by linear ruptures on {110} and {100} wafer facets. Nanorod fences are created along these ruptures as demonstrated in Figure 6.6 A. Breaking the crystal causes randomly distributed, arcuated relief structures at the breaking edge. But below the nanorod fences, there are rectilinear structures running perpendicular toward the surface as presented in Figure 6.6 B. Due to lowered interatomic binding energies across extended crystal defects, ruptures are preferred sites of fracture and thus, ruptures are reflected in the relief structures of breaking edges. If a rupture forms an intersection line with the crystal surface a straight, one-dimensional atomic surface defect such as a crystal step is created. It is reasonable that crystal steps on the surface provide branching and hence, the creation of sticking out nanorods. Nanorods are still distinguishable from each other within the observed fences, although their side walls are in touch with each other giving the chance of growing together. The fact that this does not happen here underlines the thermodynamic stability of their shape drawn in Figure 3.3. The individuality of the rods is confirmed close to breaking edges, where the fence collapses resulting in undamaged, complete and individual nanorods. Yang et al. and Zhou et al. reported about similar fence-like structures grown at the side wall of a primary single freestanding nanorod [166, 167].

Besides line defects resulting in well-defined growth of fences, there are also more irregular and wider line defects as shown in the inset in Figure 6.5 F. These defects might originate from scratches caused by friction with dirt particles during substrate handling. In contrast to the fences, this kind of defect is observed on thin film seed layers as well. Damaging the surface by scratching offers additional crystal facets. However, they are disordered and thus, nanorods grow randomly on the defect including randomly distributed angels with the substrate plane. In order to confirm this idea, a thin seed layer was scratched manually with a scalpel achieving a similar growth activation as observed on the thin linear defects on rutile crystal wafers. This knowledge is useful for controlling growth activation as described in Chapter 8.6.

> Macroscopic single crystalline seed layers provide the growth in certain crystallographic directions as expected. It is demonstrated that the grown layers consist of tiny fingers. Furthermore, defects on single-crystalline seeds provide either random or well-defined additional growth directions.

6.2.2 Hydrothermal Growth on Sputtered, Sol-gel, and Evaporated Rutile TiO_2 Seed Layers

A detailed illustration of growth stages on all used polycrystalline TiO_2 substrates is shown in Figure 6.7. On polycrystalline rutile films nearly all facets promote the growth. The topographies and the resulting hydrothermal growth on RUT-EBE-850, RUT-SG-850, and RUT-SPU-850 resemble each other and are not discussed separately. The first nanofingers appear after 48 min, which is significantly earlier than for solution based nanorods. This difference is expected since a seed layer acts as a catalyst and promotes a heterogeneous nucleation at a lower temperature compared to the homogeneous nucleation in the solution. The growth direction is distributed randomly due to the pyramid-like shapes of the grains as seen in Figure 6.7 A. In the beginning, several small nanocrystals are formed on each facet. Dependent on the crystal orientation of the facet some fingers are sticking out the surface while others grow more or less horizontally. As the growth continues, nanocrystals cluster and form bundles. Soon, bundles grow to some extent vertically on the substrate, cover the more horizontally growing bundles and start to form an array of bundles. As growth continues, these bundles obtain more and

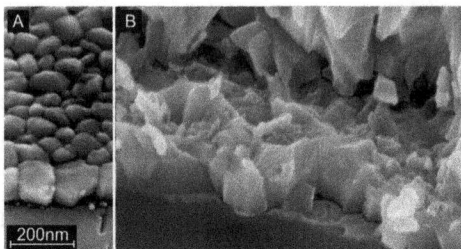

Figure 6.8: *SEM cross-section image of a RUT-SG-850 before (A) and after (B) hydrothermal growth. For clearance, nanorods were removed from the seed layer. In this case, the nanorods were removed simply by the mechanical stress-induced during the breaking process. One can see that the seed layer was not dissolved and acts as an adhesion layer between the silicon substrate and the NRA.*

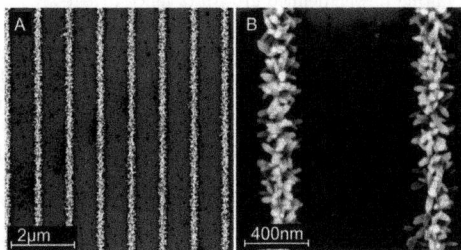

Figure 6.9: *SEM image of NRAs grown on patterned RUT-EBE-850 made by electron-beam lithography.*

more a quadratic cross-section in order to obtain their thermodynamically preferred shape according to the Wulff construction. In doing so, the shape can be controlled by surfactants [156]. The last growth phase is the attachment or ion-by-ion growth in (001) direction resulting in 50–150 nm thick and 1.2–1.6 µm long nanorods within a NRA. The forming nanorods are under pressure of competition concerning the precursor supply. Nanorods including a small angle with the substrate plane will be covered by other rods soon and thus, they are not provided with precursor anymore. As a consequence, vertically growing nanorods have the best requirements to come out on top. This is, why most of the full-grown nanorods face their (001) facet upwards. After a while, the precursor is consumed and the growth stagnates. The diffraction pattern in Figure 6.12 confirms the rutile phase of the grown nanorods.

SEM cross-section images show that rutile TiO2 seed layers do not dissolve significantly during the hydrothermal growth. At the breaking line for the cross-section image, the NRA has spalled partly from the substrate and still a TiO2 layer is observed having about the same thickness as the seed film before the growth process as shown in Figure 6.8.

The advantage of electron-beam evaporation is the well adjustable film thickness and the use in lithography techniques. Hence, this technique is useful to create structured NRAs with optical or electron-beam lithography methods as shown in Figure 6.9.

6.2.3 Hydrothermal Growth on Rutile FTO Seed Layers

FTO is investigated as another polycrystalline rutile seed. The mismatch of lattice parameters between FTO and rutile TiO_2 results in a decreased number of nanofingers per grain facet and the nanofingers are usually not attached to each other as shown in Figure 6.10. However, almost every facet is occupied by small nanocrystals and again, they grow nearly in all directions but within one facet they prefer a common direction. Soon, individual nanofingers are replaced by bunches of nanofingers getting a more and more quadratic cross-section. The following growth process is equal to the one on polycrystalline rutile TiO_2 seed layers. An important consequence of the lower nanofinger density at the beginning of the growth process is the formation of extended channels illustrated in Figure 6.10. In hybrid solar cells, these nanorod-free sites are responsible for shorts, since they allow a contact between the active material and the electrode. [129, 1668, 1669].

For comparison, the bottom of a coiled up NRA-membrane grown on ultrathin TiO_2 seed layers is shown in Figure 6.10 C. Here, the texture is much finer and the film is pinhole-free, because the seed layer is linked much stronger to the NRA due to the better matching of crystal parameters. Furthermore, on rutile TiO_2 seed layers the density of nanofingers is much higher and there is less spacing between the nanofingers. This feature inhibits the creation of voids.

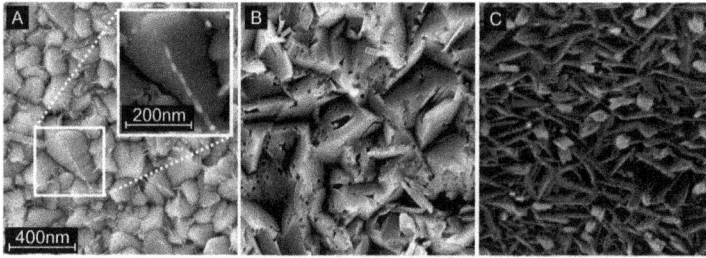

Figure 6.10: *A) SEM image of the very early growth stage of nanorods on FTO after roughly 65 min processing time. Nanocrystals grow neither on every facet nor tight abreast. The consequence is seen in B) SEM image of the bottom of a relieved NRA grown on FTO. The negative "imprint" of FTO is clearly visible. The selective and patchy growth is shown in image (A) results in voids and holes inside the TiO_2 bottom layer formed by horizontally grown rods. C) SEM image of the bottom of a relieved NRA grown on rutile TiO_2.*

The number of nanocrystals appearing on a single FTO grain facet decreases within a few minutes after the first nanocrystals appear. Usually, only one nanocrystal survives on each facet. This development reminds on Ostwald ripening$^\diamond$ and it is explained in more detail in Section 6.4.3.

On rutile TiO_2 seed layers, the growth is provided by each facet. On rutile seed crystals with slightly different lattice parameters, the growth occurs still on most of the facets. Horizontally growing crystals are covered by rather vertically growing nanorods and consequently, the final array consists of almost vertical rods and a dense ground layer. The thickness of as-grown rods correlates with the average grain size in the seed layer.

6.3 Hydrothermal Growth on Anatase Seed Layers

Although not expected from its crystal structure, anatase TiO_2 is a valuable seed layer for rutile TiO_2 nanorods. In contrast to rutile seed layers, the density of nanorods is controlled much easier by film properties as it is demonstrated in this section and Chapter 9.

6.3.1 Hydrothermal Growth on Sputtered Fine-Grained Anatase TiO_2 Seed Layers

As shown in Figure 6.6, single nanorods rise from many positions, but the density of nanorods is slightly lower compared to rutile seed layers. Surface defects and grain boundaries are hot candidates for growth promoting sites. Firstly, because these sites act during the thermally induced ART as nucleation sites as discussed in Section 2.3.1. Secondly, surface defects and grain boundaries show anatase particle-like structures and Li et al. assume that such particles could easily convert into rutile nanocrystallites during the hydrothermal growth [159]. A similar growth triggering process on anatase TiO_2 substrates based on nanoparticles is suggested by Sun et al. [160] even though he does not specify the morphology of these particles. Thus, the large number of grain boundaries observed in TEM images is likely to be the reason for the high density of nanorods.

But having a closer look on the arrays, the difference between rutile films such as RUT-SPU-850 and ANA-SPU-450 becomes clear as illustrated in Figure 6.11. In contrast to RUT-SPU-850, the growth on selected interfaces results in some sites where no nanorods occupy the film.

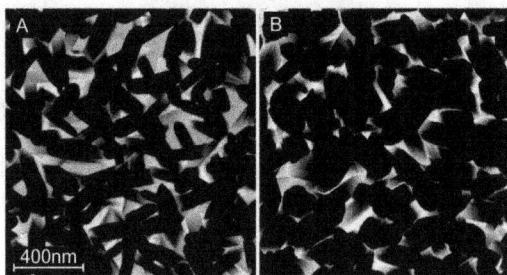

Figure 6.11: *Top view and color inverted SEM images of NRAs grown on ANA-SPU-450 (A) and RUT-SPU-850 (B). On ANA-SPU-450 there are a few sites that are not covered by nanorods (marked reddish), since the growth on grain boundaries and surface defects on fine-grained anatase films is less efficient as the growth on polycrystalline rutile TiO_2 seed layers.*

Using ANA-SPU-450 or ANA-PYR-450 as seed layers offers the advantage that much more inorganic substrates are available for annealing temperatures of 450 °C compared to 850 °C applied for rutile seed layers.

For a thickness of 1 nm and fewer, NRAs are still created but they peel off. The reason behind this effect is found in the inhomogeneous formation of ultrathin films by sputter deposition [1670] and dewetting of TiO_2 on silicon during the annealing process. If the interface between TiO_2 and silicon decreases with decreasing film thickness due to island-like growth, the average adhesion of the grown NRA and the substrates shrinks as well. This effect cannot be compensated by the few horizontally grown nanorods, which are touching the substrate. Firstly, the nanorods are touching the substrate in most cases with an edge only and hence, the interface between rods and substrate is negligible. Secondly, the bonding between TiO_2 films and silicon

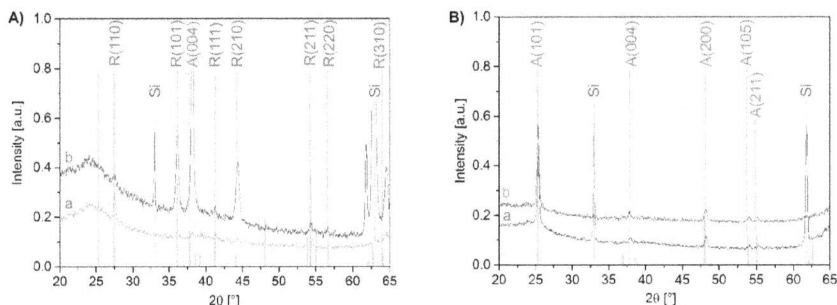

Figure 6.12: *Diffraction pattern of RUT-SPU-850 and ANA-SPU-850 before (a) and after (b) hydrothermal growth process. A) Since the hydrothermal growth results in rutile nanorods the peaks before and after the growth process on rutile seed layers correspond to the rutile morphology. There is only one peak at $2\theta = 38°$ which does not match to one of the three TiO_2 phases, although it is close to the anatase (004) peak. B) The amount of rutile nanorods on ANA-SPU-850 is too low to result in an observable rutile signal. The diffraction pattern indicates that the rearrangement of TiO_2 material on anatase films during the exposure to the hydrothermal solution does not result in significant amounts of rutile TiO_2, but remains anatase instead.*

substrates is formed at temperatures that are much higher than provided during the hydrothermal growth process. Thus, the intrinsic stress of NRAs exceeds the remaining adhesion force. As a consequence, the NRA is coiling up in order to relieve stress. This effect can be used to fabricate levitating NRAs for mechanical, electronic and optical characterization experiments. Horizontally grown nanorods cover areas of bare silicon and form a dense undercoating. Nevertheless, vertically grown nanorods are the shaping structures for the final NRA as it was described for the formation of NRAs on thick seed layers already. As discussed in Section 5.1.1, non-annealed TiO_2 films peel off after the hydrothermal bath likely. This observation indicates that the bonds between the seed film and the nanorods are much stronger than between the seed layer and the substrate. By annealing at 850 °C, the adhesion is improved and sufficiently thick seed layers (thicker than 2 nm) do not peel off anymore.

6.3.2 Hydrothermal Growth on Coarse Grain Sputtered and Sprayed Anatase TiO_2 Seed Layers

Topographies and resulting hydrothermal growth behavior on ANA-SPU-850 and ANA-PYR-850 resemble each other and are not discussed separately. They consist of large grains and thus fewer grain boundaries. As a consequence, less nanorods are expected on these films, which is in good agreement with the experimental results. The density is so low that nanorods are not touching each other. Since branching appears at the very beginning of the growth process, branched nanorods often look like forks as shown in Figure 6.7.

Besides single individual rutile TiO_2 nanorods, an additional crystalline layer is formed on anatase films as shown in Figure 6.13 A3. This additional layer makes the film a few nanometers thicker, increases the roughness, and stabilizes the anatase film in the hydrothermal growth solution [573]. Beyond that, also the roughness is increasing. However, the cross-section images in Figure 6.13 A1-3 indicate that the internal structure of the film remains unchanged and the

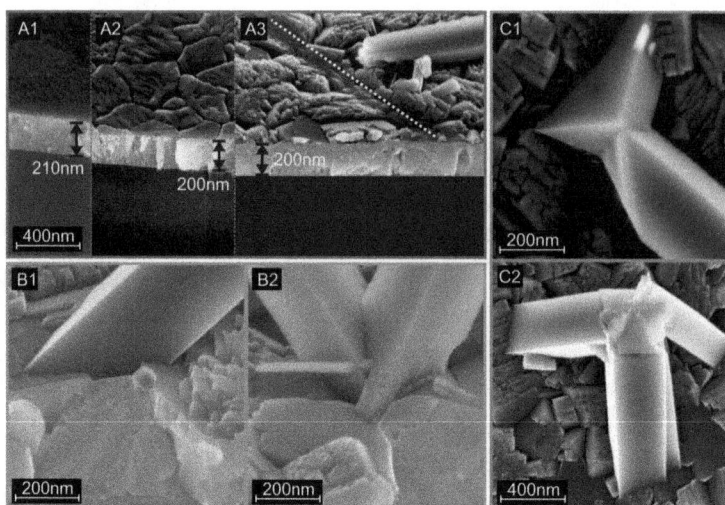

Figure 6.13: *A) SEM cross-section image of a sprayed film (A1), ANA-PYR-850 (A2), and ANA-PYR-850 after the hydrothermal growth process and sonication (A3). The film thickness is shrinking slightly from (A1) to (A2) due to the higher density after crystallization. The film thickness is increasing slightly from (A2) to (A3) due to the formation of a thin and rough surface layer. This layer is less developed in places where nanorods were located on the substrate (marked with dotted line). The cross-section area of the film is colored bluish in (A) and (B). B) SEM cross-section image of ANA-PYR-850 directly below sticking out nanorods. It is seen clearly that the film is not penetrated by the rods. C) Sticking out rods with removed branches on ANA-PYR-850. The former grain boundaries between the branches are colored bluish.*

visible changes appear within a thin surface layer only. According to XRD measurements shown in Figure 6.12 B the film remains anatase and there is no indication that the structural changes observed on the surface correspond to any other phase but anatase.

In Figure 6.13 A3 one lying nanorod was removed by sonication. Its former position is marked with the yellow dotted line. Since its long axis was parallel to the seed layer it is reasonable that the rod was created in solution and settled out on the seed layer resulting in a weak bond between the seed layer and the substrate. The surface below the removed rod is deepened. Since there is no sharp imprint of the square-edged rod, the film was never touching the rod tightly. One explanation for the depression is a reduced precursor supply inside gaps that are even broader than the precursor diameter. This in good agreement with the observation that the hydrothermal growth is suppressed by covering the seed layer with a silicon plate – even if a gap between the seed layer and the silicon is seen with the naked eye. Another reason might be the increased consumption of precursor material close to the rutile nanorod. Due to the low Gibbs free energy solved precursor molecules. Dissolved titanium oxide from the seed layer crystallize more likely as rutile TiO$_2$ on the nanorod than as anatase TiO$_2$ on the seed layer.

It is excluded that all nanorods on ANA-PYR-850 and ANA-APU-850 settle out from solution, because they are sticking out the seed film and cannot be removed easily by sonication. Nanorods

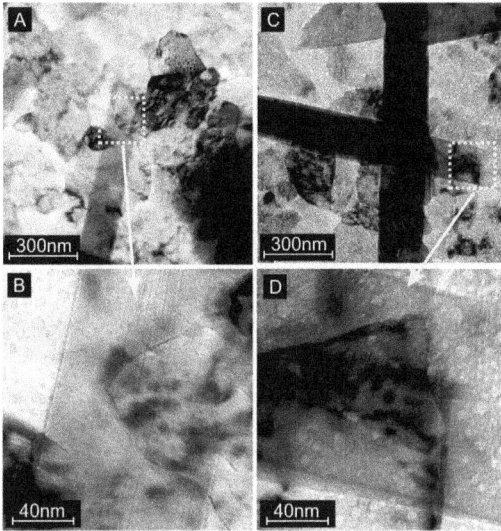

Figure 6.14: *TEM image of two nanorods grown on ANA-SPU-850 with different magnifications. Firstly, no penetration of the rod into the seed layer is observed. Secondly, a clear nucleation site is not detectable, but in the center of the interface between the rods and the seed there are always grain boundaries inside the seed layer acting eventually as nucleation sites.*

condensed from solution on blank silicon substrates never stick out. The central part of the rod growing on a seed ensures a strong bonding between the rod and the substrate. Hence, the rods are hard to remove from the substrate by sonication as demonstrated in Figure 6.13 C1 and 2. But some branches break during sonication. Their breaking areas are marked bluish in the SEM image. It came out that the breaking of branches is much more likely during sonication than interrupting the whole nanorod from the seed layer. Thus, there is a quite strong adhesion between the polycrystalline substrate and the lower tip of the rod. A reasonable explanation is that rods sticking out the substrate grow directly on a small seed crystal belonging to the polycrystalline seed layer.

The strong adhesion does not originate from deep anchoring of the rods in the anatase film, since SEM cross-section images do not show any penetration of the seed layer below the nanorods as demonstrated in Figure 6.13 B1,2. Also, the TEM images in Figure 6.14 confirm the absence of any penetration of the seed layer by the rutile nanorods. That is, why it is more likely that the adhesion force is assembled by the thin interface between the seed and the nanorod. Anatase nanoparticles, grain boundaries, and crystal defects are supposed to be growth promoters. No candidate triggering the growth is detected clearly in Figure 6.14, but grain boundaries seem to be always present in the area where nanorods rise from the seed layer. The TEM images confirm the existence of voids in ANA-SPU-850. If such a void sits on the surface, it could act as a growth promoting site due to an increased surface roughness and defect density.

As they grow slowly in the ⟨110⟩ direction, sticking out nanorods begin to cover the neighboring grains. The growth of individual nanorods on anatase films is faster compared to nanorods in

Figure 6.15: *Schematic illustration of hydrothermal growth on rutile and anatase seed layers. On rutile films, nearly all facets act as a seed forming dense NRAs. On anatase films nanorods have to nucleate either on small rutile inclusions, or more likely, on small anatase particles that convert into rutile nanocrystals as explained in Section 2.3.1. Rods grown in solution drop on the substrate and show no strong adhesion to it.*

NRAs on rutile films. A reasonable explanation is the high precursor consumption of dense NRAs. In addition, the resulting nanorods on ANA-SPU-850 and ANA-PYR-850 are slightly larger than their relatives on RUT-SPU-850.

Of course, many lying nanorods might arise from solution and settled out on the anatase films. Nevertheless, the total number of nanorods on ANA-SPU-850 and ANA-PYR-850 is significantly larger than on blank silicon.

On anatase films, the density of rods is correlated with the grain boundary density. It is suggested that grain boundaries provide crystal structures on the seed surface which are transformed into rutile TiO$_2$ during the hydrothermal growth process. Their adhesion to the seed layer is as strong as for nanorods grown on rutile films, but their diameter does not depend on the grain size. A graphical comparison between the growth on rutile and anatase films is given in Figure 6.15.

6.4 Stages of the Hydrothermal Growth Process

The relatively slow and continuous temperature rise shown in Figure 4.1 is responsible for permanently changing environmental conditions. Thus, the growth process is separated into different stages. The overlap of these stages is given by the heating rate and composition of the hydrothermal growth solution. In this section, more common and substrate independent features are addressed. A schematic overview is given in Figure 6.16.

6.4.1 Nucleation

As soon as a certain temperature is exceeded, the chemical reaction turning the primary precursor titanium(IV) butoxide into the final precursor TiCl$_4$ begins. The nucleation starts at a critical temperature and concentration. In this period, the transformation of diluted TiCl$_4$ into

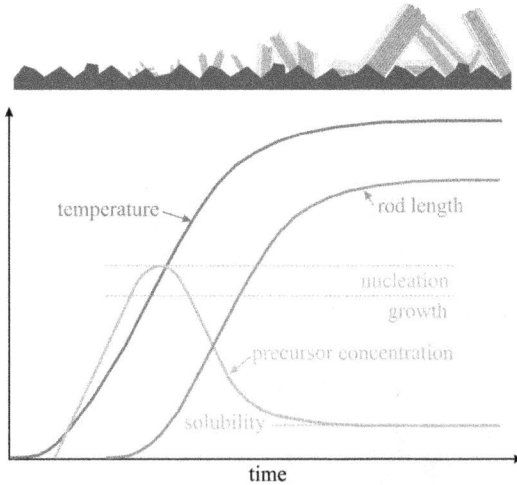

Figure 6.16: *Schematic overview on the stages of the hydrothermal growth process following a La Mer nucleation [1671, 1672]. At a certain temperature, titanium(IV) butoxide converts continuously into the final precursor TiCl₄. Above a critical precursor concentration, the nucleation begins and the precursor concentration is reduced again. As a consequence, it drops below the minimal concentration necessary for nucleation. In the following, no further nuclei are formed and the previously created nuclei grow consuming the remaining precursor until an equilibrium between growth and dissolution has set. Hence, the as-grown nanorods are mostly monodisperse. After the precursor is consumed the nanorods are affected by dissolving and recrystallization processes.*

solid TiO_2 results in nuclei. Nucleation centers are found in aggregates (homogeneous nucleation) and dirt particles as well as on seeds (heterogeneous nucleation). The most important thermodynamic driving force for nucleation is the low Gibbs free energy of rutile bulk TiO_2. However, due to the relatively high surface energy of rutile TiO_2, an energy barrier has to be overcome before the nucleation works efficiently. And even rutile TiO_2 dissolves in hot hydrochloric acid and especially tiny crystallites with a large surface-to-volume ratio are unstable. As a consequence, an equilibrium between nucleation and dissolving is established.

6.4.2 Period of Crystal Growth

Since precursor is consumed during the nucleation the concentration drops and below a critical concentration the nucleation stops and nuclei continue growing. Within this stage, nanorods achieve their final size and shape. With increasing size they become more and more stable due to the low Gibbs free energy of the rutile TiO_2. From now on, it is unfavorable to create new nucleation sites and the crystallization takes place at already existing nanorod surfaces only. This property results in nearly monodisperse full-grown nanorods. Since the growth is preferred in the [001] direction, the resulting nanocrystals have a (001) top facets and {110} sidewalls as discussed in Section 3.2. This aspired shape is not given in the early growth stage. But the primary crystal structures are transformed during the saturation stage more and more into nanorods with a quadratic cross-section. This is seen impressively for the growth on rutile seeds as shown in Figure 6.6.

The presence of rutile facets lowers the energy barrier for nucleation and decrease the surface area that is exposed to hydrochloric acid. As a consequence, the growth process appears at lower temperatures and accordingly shorter dwell times in the oven. For instance, the first nanocrystals are observed after 44(4) min on rutile TiO₂ and after 65(4) min in solution corresponding to 115(5) °C and 140(5) °C, respectively.

Although rutile crystals grow in the [001] direction, the seed must not consist of a (001) facet necessarily. Crystallites are stabilized at any rutile facet. But only facets with a projection on the (001) facet such as the {111} facet promote the growth of sticking out nanorods as shown impressively on macroscopic rutile single crystals in Figure 6.5.

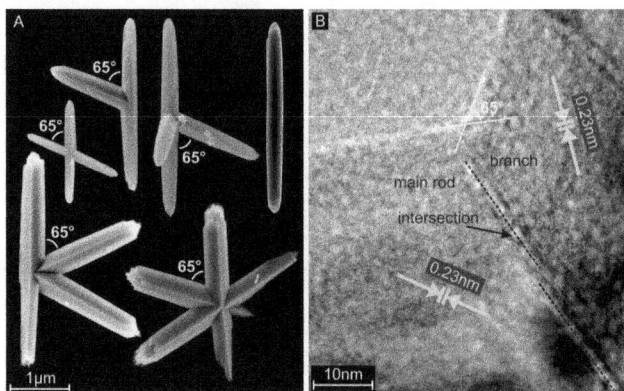

Figure 6.17: *A) SEM image of single and branched rutile nanorods grown on ANA-SPU-850 and ANA-PYR-850. There are very defined angles between the main rod and the branches. Thus, the origin of branching is not a random crystal defect. It is more likely that certain crystal planes are preferred for creating twin planes. Branching appears only in the early growth stage. That is why all branches have roughly the same length. B) TEM image of branching site. The intersection between the main rod and the branch is marked with a dotted purple line. The rutile {200} plane with its d-spacing of 0.23 nm is seen clearly.*

6.4.3 Ostwald Ripening

The number might still be changed by Ostwald ripening° [155, 165, 1673]. An observation that might originate from Ostwald ripening is the decreasing number of nanocrystallites on FTO in the early growth stage. Often only one nanocrystal per facet is left. Ostwald ripening was reported by Yang et al. for acid vapor oxidation growth of rutile TiO₂ nanorods [165], but according to literature it was not documented for hydrothermal growth of rutile TiO₂ nanorods in aqueous solution so far. There is no clear evidence for Ostwald ripening on other seed films used in this thesis. It is known that Ostwald ripening appears during the growth of very similar systems such as fibrous WO₃ nanorods [1674]. The appearance of Ostwald ripening would emphasize the interplay of crystallization, dissolving, and recrystallization during the growth process.

6.4.4 One- and Multi-directional Growth (Branching)

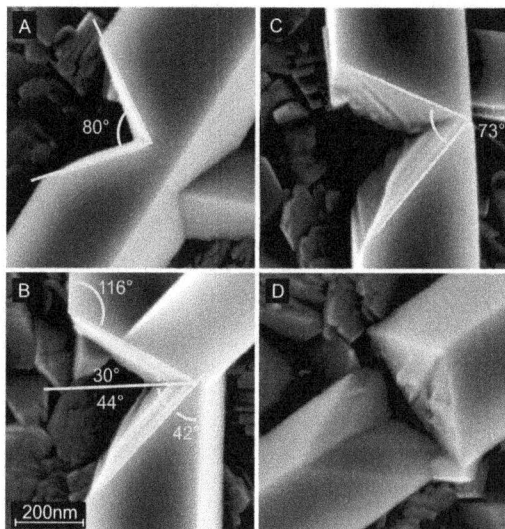

Figure 6.18: *SEM image of rods with removed branches. The rods were detached by sonication. The lower plane in (C) and lower left plane in (D) are significantly smoother than their complement. Due to its inclination, it is supposed to be the {101} plane. This plane is reported to be one of the possible intersections between main rods and branches [159].*

Many nanorods form branches in well-defined directions. Beyond that, all branches of an individual nanorod complex show nearly the same length. This symmetry indicates that the branching process takes place during or right after the initial nucleation. Only for low hydrochloric acid concentration branching appears for a longer period within the hydrothermal growth process. For the commonly applied hydrochloric acid concentrations branching becomes energetically unlikely because the formation of new branches would create a further high energetic (001) facet on the low energetic {110} sidewalls. As discussed in Section 3.2.2, this results in an energy barrier for the formation of branches. Due to the experimental observation, the energy barrier seems to increase with increasing crystal size. Besides the crystal size, also the increasing temperature changes the thermodynamic requirements for branching. But applying the hydrothermal growth process on large nanorods used as seed shows that no further branches are formed. Thus, the temperature change plays a minor role in suppressing the formation of branches.

The most frequent branched nanorods have one branch and are documented in detail by Guang-Lai et al. [154], but also more branches are observed as shown in Figure 6.17. The branches include defined angles that are discussed in Section 3.2.3.

Another very common structure are nanohedgehogs. These structures might have their origin in larger aggregates consisting of molecules or dirt particles in the solution and thus, they cannot be traced back to individual specific crystal defects. As a consequence, no primary rod can be identified and angles between branches are randomly distributed. Nanohedgehogs are formed in solution/ on substrates resulting in spherical/ hemispherical structures. Since these structures

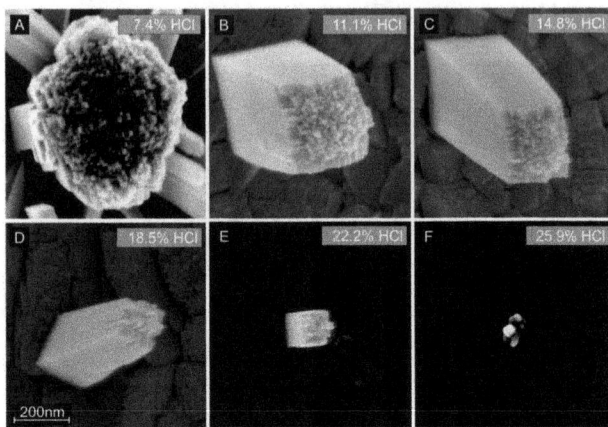

Figure 6.19: *SEM images of nanorods grown in different hydrochloric acid concentrations. Oven temperature and growth time is the same for all samples. The substrate is a polycrystalline anatase TiO₂ film made by a sprayed film annealed at 850 °C for 2 h.*

can reach several microns in diameter, they cause damages in hybrid solar cells but yield a high surface area for photocatalysis application. Their density correlates with the purity grade of the growth solution.

Crystal defects usually reduce the mechanical stability of a crystal and hence, they act often as breaking points under mechanical stress. Thus, it is easy to separate branches from primary rods by sonication. As demonstrated in Figure 6.18, a flat interface is expected where there is only a small lattice mismatch and a rough surface is expected for the remaining side. And indeed, a closer look on the uncovered interfaces in Figure 6.18 confirms that the two opposing facets show a different roughness.

Using this method, a huge number of bare {101} facets is gained. SEM images show that these planes are flat within the resolution of the electron microscope. Due to the model describing {101} twin planes as branching centers these planes are assumed to be atomically flat [159]. The existence of such flat defect planes in crystals gives rise for specific electronic applications. Such a defect layer represents a local change of rutile stoichiometry and hence, a shift of the conduction band. DFT calculations by Morgan and Watson indicate a low formation energy of oxygen vacancies on {101} facets [1590]. Since oxygen vacancies are electron donors [198, 292, 915, 933, 1675, 1676], the density of mobile electrons should be increased at {101} twin planes and thus, a n-n^--n junction is expected for electrons moving perpendicular to the defect layer. But Morgan and Watson also calculated bandgap states close to the valence band, which would give rise to a n-p-n type junction [1590]. For the sake of completeness, it should be noted that bare and embedded {101} facets belonging to twin plane defects are supposed to differ due to relaxation and hence, the results from Morgan and Watson might not be completely transferable to the atomic constellation and band structure in {101} twin planes.

The crystal morphology of single branched nanorods is investigated with TEM. Inside a branch d-spacings and angles between crystal planes fit very well to rutile TiO₂ as shown in Figure 6.17. The situation in the branching point is less clear. In some locations, Moiré° give rise to the assumption that more than one crystal orientation or even other morphologies exists. For example, the d-spacing of 2.23 Å meshes with the {200}-plane of rutile, but also to the {112}-plane of anatase and the {121}- or {400}-plane of brookite. The d-spacing of 2.48 Å meshes with the {011}-plane of rutile, but also to the {102}-plane of brookite. But none of the observed angles meshes with the calculated angles between the mentioned crystal planes for each morphology. But it could be that two morphologies lying on top of each other so that the mismatch of the angles does not exclude any of the morphologies necessarily. This perception meshes with the appearance of Moiré fringes as well.

Influence of Hydrochloric Acid Concentration and Processing Temperature

The resulting shape which is formed under defined experimental condition depends strongly on growth conditions such as the pH value and reaction temperature [61, 129, 162]. A low pH value slows down the hydrolysis of the precursor [153, 159]. Nanorods grown in different hydrochloric acid concentrations ranging from $c_{HCl} = 7.4\%$ to 25.9% at 180 °C oven temperature on ANA-SPU-850 are compared. The result is shown in Figure 6.19. Obviously, the size is decreasing with increasing hydrochloric acid concentration. But also the diameter of the fingers forming the substructure is decreasing with increasing hydrochloric acid concentration. Since the number of nanorods is almost independent from the precursor concentration the nucleation period is mostly unaffected by the titanium(IV) butoxide concentration. But the growth rate depends dramatically on this concentration. This is reasonable since the surface energy and hence the crystallization rate depends strongly on the chemical environment. And it seems that branching is very likely for low hydrochloric acid concentrations as shown in Figure 6.19 A.

Nucleation and branching appear only within a small time period in the beginning. Consequently, the resulting rods and branches are almost monodisperse. There are smooth interfaces between branches and these interfaces do not show the finger-related structure which is seen on the tip of each nanorod. This suggests strongly an inhomogeneous distribution of crystal defects and grain boundaries inside the nanorods. The dimensions of rods are controlled with the concentration of the hydrochloric acid.

7 Non-Equilibrium Growth Model for Fibrous Rutile TiO$_2$ Nanorods

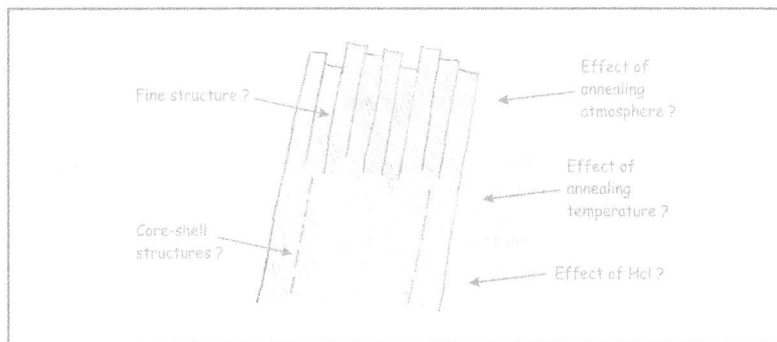

Figure 7.1: *Scientific objectives investigated in this chapter. In which way are fine structure and core-shell structures affected by post-annealing employing different temperatures and atmospheres? Does long exposure to hot hydrochloric acid influence the fine structure?*

The growth process of nanocrystals has a great impact on chemical, electronic, or optical properties of the resulting structures. Amongst others, hydrothermal methods are popular techniques to fabricate such nanocrystals. A typical feature of hydrothermally grown rutile TiO$_2$ nanorods is a substructure consisting of a multitude of small nanofingers as described in Section 3.2.4. This substructure is observed for all investigated nanorods independent from the kind of seed. Thus, these nanocrystals are not single-crystalline, which is disadvantageous for high charge carrier mobilities. Interestingly, this feature is also found in some other nanocrystalline materials listed in Section 3.2.4, which are valuable for various applications. Consequently, the understanding and control of this substructure are supposed to be an essential key to tune and improve device performances. Motivated by this fact and based on fundamental observations during the hydrothermal growth process of rutile TiO$_2$ nanorods on different TiO$_2$ seed layers, a material independent thermodynamic model is introduced in this chapter, which is able to describe the formation of such substructures in rutile TiO$_2$ nanocrystals. Thereby, a non-equilibrium state between the crystallization and surface energy minimization is suggested. Finally, experimentally observable consequences of this model are demonstrated and methods to control the substructure are presented. The content of this chapter is also dealt with in the article [1677].

7.1 Model for the Origin of the Typical Fine Structure in Rutile TiO$_2$ Nanorods

By investigating the hydrothermal growth and its results on different seed films an extensive view on the origin of typical fine structures is achieved. Besides homogeneous nucleation, there is heterogeneous nucleation at suitable rutile surfaces, anatase particles, or precursor aggregates. Since temperature and pressure are increasing while the concentration of the final precursor is decreasing due to the creation of all the nuclei, the chemical environment changes as well. When the precursor concentration drops below a critical value, nucleation stops and the nuclei continue growing. Along this time, the crystal size is increasing rapidly and a defined shape is formed. It is assumed that the fine structure is developed during the growth stage.

7.1.1 Facet-Dependent Crystallization Speeds

Within the growth stage, rutile TiO$_2$ nanorods gain their final size, while two physical processes are competing. On the one hand, there is the chemical reaction converting TiCl$_4$ molecules into solid and crystalline TiO$_2$. On the other hand, there is the Wulff construction representing the shape with the lowest surface energy for a given volume [1678].

First, a closer look on the chemical reaction is made. Crystallization is enabled by lowering the Gibbs free energy $G = H - TS$, with the enthalpy H, temperature T, and entropy S. Since TiO$_2$ is an anisotropic crystal, the surface Gibbs free energies of different crystal facets are not equal [1679]. As a consequence, the reaction speeds and thus, the crystal growth speeds $v_{I,(001)}$ and $v_{I,\{110\}}$ on the (001) and {110} facets differ (Figure 7.2).

7.1.2 Facet-Dependent Growth Speeds Corresponding to Wulff Construction

As another effect, surface energy minimization has to be considered. The facet-dependent atomic arrangement of the (001) and {110} facets results in a differently strong interaction with the growth solution and thus in different surface energies. Calculations based on density functional theory (DFT) show that {110} facets have a lower surface energy than (001) facets [298, 561, 1586–1590]. To keep the {110} facets large compared to the (001) facets the growth speed $v_{II,(001)}$ in the [001] direction has to be faster than the growth speed $v_{II,\{110\}}$ in $\langle 110 \rangle$ direction. The index I for the speed v is related to the real chemical reaction speed, while the index II corresponds to the optimal growth speed which is needed to satisfy the Wulff construction. For the sake of completeness, it is mentioned that chemical reactions appear mostly in forward and backward direction at the same time. Thus, the mentioned speeds v_I are effective growth speeds.

7.1.3 Non-Equilibrium Growth Model Describing the Origin of the Fine Structure

While the speed v_I depends on the reaction probability, v_{II} results simply from interaction forces between the crystal and the solution. Only for a slow chemical reaction, v_I equals v_{II} and the surface energy is minimized at all times. And only v_I is observable since it represents the actual value of the growth speed, while v_{II} is the desired value determined by the thermodynamic equilibrium. If v_I becomes large compared to v_{II}, the surface energy is not minimized and the system is not in thermodynamic equilibrium anymore. The lower the temperature the more significant is this effect, since statistical processes keeping the crystal's surface energy minimized work less efficient at lower temperatures due to lower particle mobilities.

Non-equilibrium growth conditions are not rare in nature as demonstrated by De Yoreo and Vekilov in the case of biomineralization [1680]. But this model has never been applied to describe the hydrothermal growth of metal oxide nanostructures.

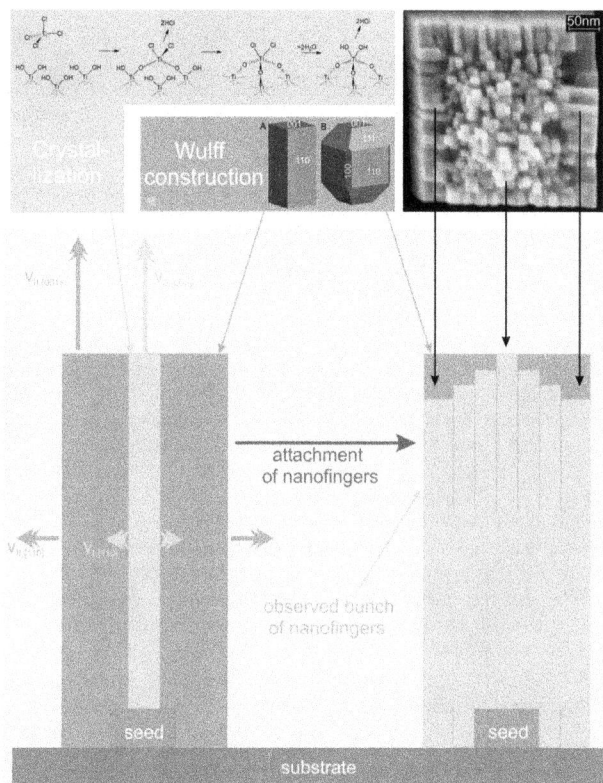

Figure 7.2: *Schematic representation of the non-equilibrium growth model: Due to a fast chemical reaction converting $TiCl_4$ into TiO_2 on the (001) facet of nanofingers (green), the dimensions of the nanofingers differ from the Wulff construction (violet). Based on the fast growth in [001] direction, the fingers would have to grow much faster in the $\langle 110 \rangle$ direction to fit the dimensions of the Wulff construction (left side). In order to reduce the surface energy, additional fingers are attached to the central one. The bunch of fingers is in equilibrium with the Wulff construction. The outermost fingers are a little bit expanded due to better precursor support (right side). The SEM image shows that the model is in line with the experimental results. To simplify matters, the Wulff construction is drawn for the (001) and {110} facets only (Wulff construction (A)). Other facets might appear under certain conditions as well [168] (Wulff construction (B)). The Wulff constructions (A) and (B) were drafted using WulffMaker [1581].*

It is suggested that nature attaches many crystals to each other in order to reduce the ratio between {110} and (001) facet areas and to push the system back to the state of minimized surface energy. Thus, the competition between the chemical reaction and the ambition to minimize the surface energy is the origin of the fine structure in TiO_2 rutile nanorods. A possible mechanism for the attachment process is introduced later.

In addition, such a fine structure increases the low energetic {110} and decreases the high energetic (001) facet area. Thus, the creation of such a fine structure is thermodynamically supported. In extreme cases, this effect results in hollow nanorods or nanotubes. And the occurrence of {110} facets as grain boundaries of all mentioned facets in Figure 7.2 is not astonishing since facets with low surface energies appear more frequently than ones with higher energies [1681]. In the lower part of the primary rod, the fine structure is energetically not advantageous anymore and thus, it might heal up in the remaining growth process time. Indeed, TEM studies by Wisnet et al. proved that the fine structure is missing in the lower part of the rods [168]. This growth model is illustrated in Figure 7.2.

During the nucleation stage, the nuclei could gather by Ostwald ripening, where some rods dissolve and other that are slightly larger or already attached to each other remain. During the growth stage, the attachment of fingers is most likely driven by a heterogeneous nucleation at the sidewalls of existing rods similar to the growth on macroscopic {110} facets (Figure 6.5). A quite similar explanation for the fine structure in WO_3 nanorods was suggested by Chi et al. [1626] and is based on a model for the growth of ZnO nanorods by Ye et al. [1682]. The resemblance to different materials is not unexpected since this model depends rather on growth conditions than on material compositions. Singh et al. considered besides chemical bonds additional interaction forces between precursor particles and the crystal to predict the shape of the resulting crystal. They observed also deviations from the Wulff construction since their predicted shape is given by the relative attachment/surface energy instead of the absolute surface energy [1683]. Thus, it is reasonable that the fine structure is observed in all solution and seed layer based TiO_2 nanorods as well as in rutile films grown on rutile single crystal wafers with {001} and {111} facets investigated in this work.

7.1.4 From Polycrystalline Fine Structure Towards Single Crystal

In the final growth stage, the interplay between dissolving and recrystallization becomes important again. The interfaces between fingers are energetically unfavorable compared to a single crystal. Thus, there is a tendency of the system to vanish these interfaces to form a perfect single crystal. Driven by atomic diffusion, the extended defects disappear from bottom to top. The top region retains the fine structure for the following reasons: First, the tip region of a rod is younger than the bottom and thus, it was not exposed to the growth solution as long as the bottom part and there was less time for atomic rearrangements. Second, the fingers have grown over one micron side by side from the bottom region to the top. Since the growth of neighboring rods is not perfectly synchronized, lattice displacements become stronger towards the tip and it becomes harder to replace grain boundaries by single-crystalline material. Close to the tip, even small voids appear between the fingers suppressing the merging of the fingers. In addition, lattice mismatch is enhanced by the increasing chemical reaction speed resulting from the increasing temperature during the hydrothermal growth process. And it has to be considered that a preexisting roughness on the top surface of the rod is even increased, since the crystallization is preferred on heights compared to depressions due to different precursor supply. Third, the fingers have slightly different lengths. As a consequence, the very upper part of the {110} sidewall facets of many fingers are laid open and share a common interface with the solution. Since these facets have a lower surface energy than the (001) facet and other facets on the tip, the formation of rutile single-crystalline bulk TiO_2 has to bring up enough energy

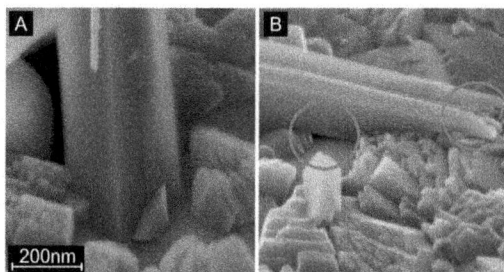

Figure 7.3: *A) SEM image of the bottom of a nanorod (red) grown on small anatase TiO$_2$ islands (green) on silicon (blue). B) SEM image of the bottom of a nanorod grown tilted on ANA-SPU-850.*

to overcome the increase of surface energy, which is necessary to reduce the low energetic {110} facets [1681].

The introduced model assumes a high crystallization rate along the long principal axis of the nanorod. Consequently, the surface energy of the whole crystal is not minimized. To reduce the surface energy, several crystals are attached to each other, which changes the ratio of the exposed facets. Finally, a nanorod consists of a bundle of such crystals and the crystals become the fingers that form the substructure in each rod. Several experimental findings promote this model as presented in the following.

7.2 Observable Consequences of the Non-Equilibrium Growth Model

The finger structure increases the electrical resistance of the whole nanorod and thus, it is unfavorable for applications such as DSSCs and photocatalysis. This motivates procedures to decrease or vanish the fine structure. Here, predictions of the model about features of the fine structure are presented and methods are depicted to reduce the number of fingers in each nanorod.

7.2.1 Attachment of Fingers

(1) Nanorods grown on small seed grains on silicon share a common interface with the silicon surface in their outer region as shown in Figure 7.3 A. Since the surface of the grains is several nanometers above the silicon surface the outer parts of the nanorod must have been attached later and grown towards the silicon and the tip, respectively. The blank reference silicon substrate yields no standing nanorods at all. Thus, it is concluded that the outer nanofingers of a nanorod are not growing on the substrate, but are attached to primary central fingers. Close to the bottom of these rods, cracks in parallel to the [001] direction appear, which are similar to the cracks in the sidewalls close to the tip region rising from interfaces between different fingers.

(2) A similar effect is observed for nanorods including a small angle between their long axis and the substrate plane. As shown in Figure 7.3 B, fingers attached to the upper part of the rod were not able to reach the substrate (right red circle). From the way the rod has been grown across

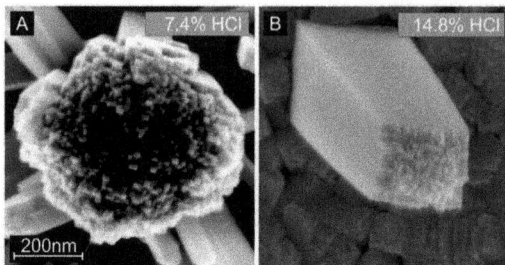

Figure 7.4: *SEM images of single rutile TiO₂ nanorods grown on ANA-SPU-850 employing different hydrochloric acid concentrations. For the low concentration, branching is promoted and the cross-section is not quadratic anymore. The steps on the sidewalls are supposed to be related to the fingers forming the nanorod. Since these steps continue from the tip to the bottom it is concluded that also the fingers inside the rod range from the tip to the bottom.*

heights on the substrate (left red circle), it is concluded that the rod was grown from the left to the right side (towards the substrate). This fact excludes that the rod is grown on the seed layer with its complete bottom side and thus, it provides strongly the assumption that fingers were attached subsequently. As shown in Figure 7.3 A, a nanorod sticking out the surface covered with tiny seed crystallites has a common interface with the blank silicon. It is believed that there was a central bunch of nanofingers grown on an anatase particle. To conserve the Wulff construction additional nanofingers were attached to the sidewalls. These nanofingers have been growing towards the tip and the bottom of the nanorod closing the gap between the nanorod and the silicon. For clarity, the SEM image was colored subsequently.

(3) The encapsulation of the inner bunch of fingers by the outermost fingers restrict the precursor support for the inner fingers. Thus, the growth trends to result in a lot of defects. High defect densities decrease the chemical stability of the core of nanorods. It is observed that after a long exposure to hydrochloric acid the inner part dissolves and hollow nanorods remain as shown in Figure 7.7 D.

(4) If a finger is in touch with another finger on a certain sidewall, it cannot grow in the ⟨110⟩ direction easily anymore. Only the outermost fingers have sidewalls, which are not in contact with any other finger. Hence, these fingers grow unhindered outwards, but their growth is restricted in any other direction. As a result, many outer fingers are supposed to have rectangular instead of quadratic cross-sections. This is in good agreement with the experiments as shown in Figure 7.2.

(5) Nanorods grown using a low hydrochloric acid concentration have no quadratic cross-section, but more a round and ribbed surface as shown in Figure 7.4 A. Therefore, there are fine steps with dimensions corresponding to the width of the nanofingers. It is reasonable that these steps arise from attached nanofingers. These steps range over the whole length of the rod. Hence, they must have had enough time to grow from the bottom to the top of the rod. This assumption is supported by item (1) and (2) additionally. From this it follows that the creation of new fingers happens at an early point in time during the growth process, which is in good agreement with item (3) and (4).

Figure 7.5: *HR TEM image of the fine structure in rutile TiO$_2$ nanorods grown at 220 °C for 3 h on FTO. The d-spacing of 2.8 Å belongs to the {110} facet of rutile TiO$_2$. The inset shows the fast Fourier transform (FFT) of the nanofingers. The white lines indicate the traces of nanofingers.*

7.2.2 Dependence of the Finger's Diameter on the Process Temperature

A system remains in thermodynamic equilibrium if exchange processes such as energy or particle transfer work efficiently. Here, the Wulff construction is obtained only, if dissolving, recrystallization, and particle transport on the surface and in the growth solution is fast enough to compensate the fast crystallization process at the (001) facet. In the model, it is assumed that the Wulff construction is not achieved by a single nanofinger during the hydrothermal growth process. For higher temperatures, the particle movements in the solution are faster and thus, the shape of the fingers should approach the Wulff construction for higher process temperatures. Since the supposed Wulff construction equals approximately the shape of the resulting nanorods, the fingers should become thicker in order to adapt the Wulff construction. Or said in other words, the aspect ratio of the fingers is supposed to decrease with increasing process temperature. Since the diameter of the nanorod is not changing significantly with temperature, the number of fingers per nanorod should decrease with increasing temperature as well. This is observed experimentally as illustrated in Figure 7.5 and 7.8.

As expected and demonstrated in Figure 7.5 and 7.6, the shape and internal structure of nanorods grown on FTO are affected by the hydrochloric acid concentration and process temperature. The diameter of the nanofingers is increasing with increasing process temperature. At 150 °C, the diameter is a few nanometers, at 180 °C, it is roughly 5–10 nm, and at 220 °C, it is sometimes more than 30 nm. Applying a process temperature of 220 °C can result in nanorods consisting of a single nanofinger only. Thus, these rods are supposed to be really single-crystalline.

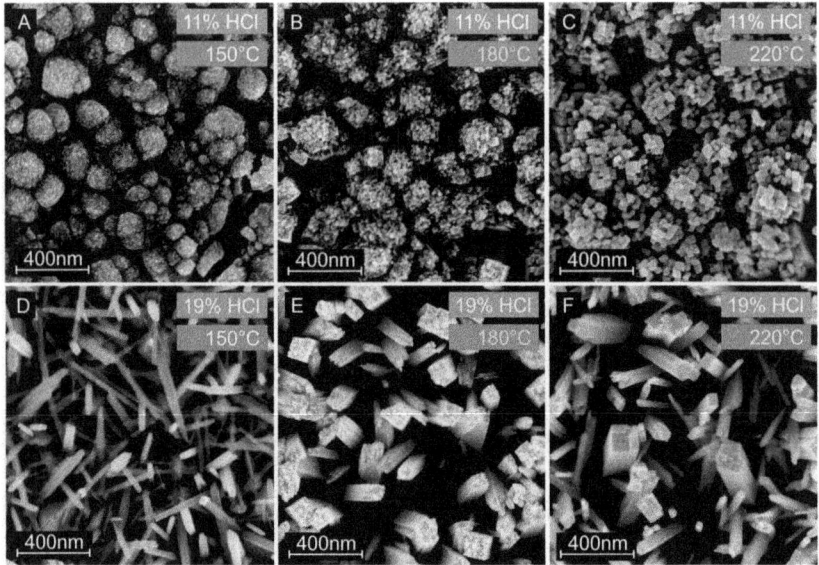

Figure 7.6: *SEM images of NRAs grown on FTO employing different hydrochloric acid concentrations and temperatures. The cross-section of the fingers is increasing with increasing temperature. For 220 °C and 19% HCl some nanorods are composed of a single nanofinger only.*

The diameter of nanorods is increasing with decreasing hydrochloric acid concentration as the chemical reaction runs faster and finally, one gains a dense film, where the shape of single rods disappear and only the fine structure remains visible. This is in good agreement with growth experiments using titanium tetrachloride instead of titanium(IV) butoxide as a precursor, where a faster reaction is expected resulting in polycrystalline films [61, 1684]. For an HCl concentration of $c_{HCl} = 22.2\%$, the majority of nanorods grow nearly in parallel to the substrate. Finally, no more nanorods grow for a concentration of $c_{HCl} \geq 25.9\%$. The growth behavior is slightly shifted for an oven temperature of 150 °C. Here, already for $c_{HCl} \geq 22.2\%$, the growth is suppressed completely. Even at an oven temperature of 120 °C, nanorods are able to grow albeit very slow.

7.2.3 Post-Growth Atomic Rearrangement in Hydrochloric Acid

The model is based on the assumption that the envelope shape of the fingers forming the nanorod is not the thermodynamic equilibrium shape corresponding to the Wulff construction. This means the resulting fine structure is energetically unfavorable. To optimize the total surface energy $\gamma = \sum (\partial G / \partial A_i)_{T,p,n}$, the ratio of different facet areas A_i on nanofingers has to be changed. If no precursor is available, this could be realized by dissolving and recrystallization processes only. The exposure of nanorods to hydrochloric acid after the hydrothermal growth process shows that the fine structure changes towards less but thicker fingers as demonstrated in

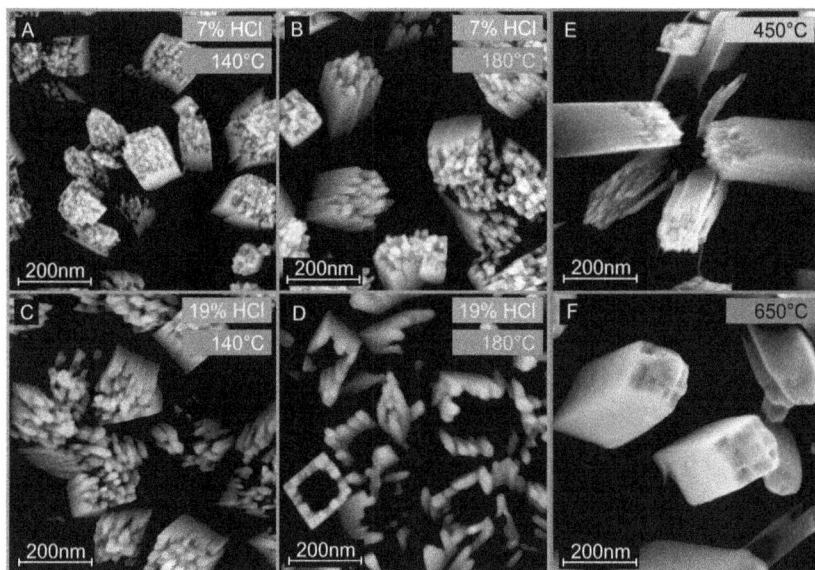

Figure 7.7: *SEM images of nanorods treated in hydrochloric acid (A-D) at different temperatures and concentrations for 4 h. The cross-section of fingers increases for sufficiently high temperatures and hydrochloric acid concentrations. For even higher temperatures and HCl concentrations, the inner part of the rods dissolve. For comparison, nanorods grown on ANA-SPU-450 were exposed to 450 °C and 650 °C in oxygen rich atmosphere for 2 h. The fine structure is reduced for 650 °C only.*

Figure 7.7. If temperature and hydrochloric acid concentration are too high, the inner part of the rod is dissolved due to high crystal defect densities lowering the chemical stability [767, 1607]. A similar effect is observed for IrO_2 nanorods [1685]. It is assumed that defects arise from an incomplete crystallization process in the inner part of the rod due to precursor limitation by the outer nanofingers. Increasing the exposure time to 6 h results mostly in the same shapes as for 4 h. Hence, the major atomic rearrangement takes place within in the first 4 h and the resulting thicker fingers are represented in Figure 7.8. This figure summarizes processes that affect the fine structure. No post-treatment was done on the nanorods in Panel A and B, which have been grown at 180 and 220 °C respectively. An additional hydrochloric acid treatment (4 h at 140 °C in 19% HCl) was performed on the nanorod presented in Panel C. The nanorod in Panel D was grown at 180 °C and annealed at 1050 °C in vacuum. An enlarged image detail of Panel 1 can be seen in Panel 2 and the SAED pattern is displayed in Panel 3. The SAED pattern are arbitrary rotated against the bright field images.

7.2.4 Post-Growth Annealing in Different Atmospheres

For this study, nanorods are annealed at 450 °C, 650 °C, 850 °C, and 1050 °C in vacuum, nitrogen rich, oxygen rich, and forming gas (95% N_2, 5% H_2) atmosphere, respectively. The results are shown partly in Figure 7.9.

179

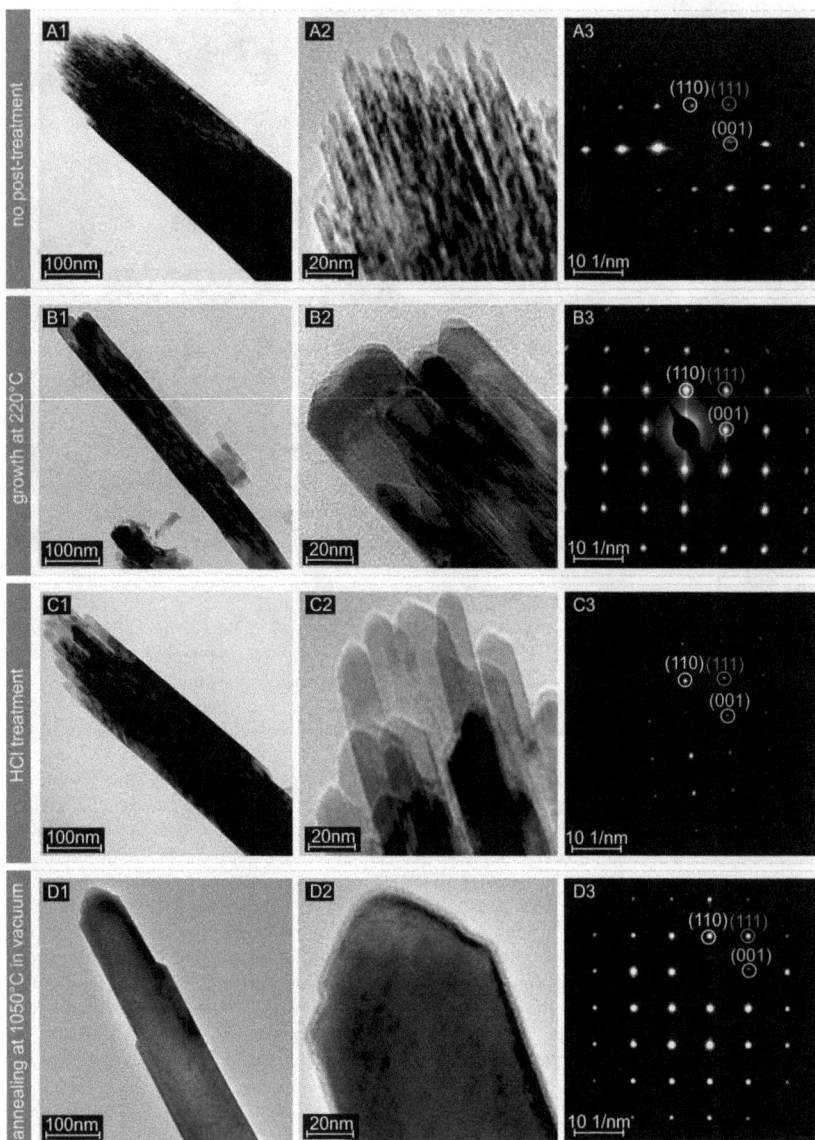

Figure 7.8: *TEM images summarizing techniques that influence the fine structure.*

For an annealing temperature of 450 °C, the fine structure keeps unchanged in any atmosphere. At 650 °C, the fine structure is reduced significantly. While hydrochloric acid supports atomic rearrangement by dissolving and recrystallization, elevated temperatures promote the atomic rearrangement by higher atomic mobilities. Anyway, the results of both treatments show clearly that the crystal shape of nanorods including the fine structure is not the energetically most favorable shape.

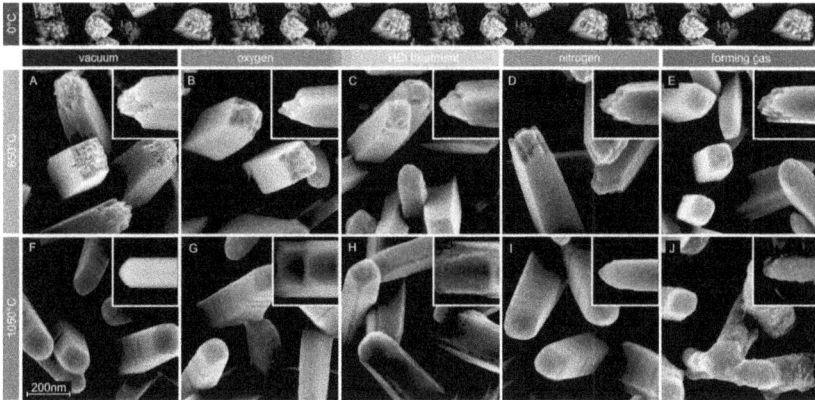

Figure 7.9: *Effect of annealing in different atmospheres and temperatures for* 2 h *and* 5 °C/s *temperature slew rate. The main images show NRAs on sputtered ANA-SPU-450 and the insets show lying rods on sputtered ANA-SPU-850.*

Oxygen atmosphere stabilizes the stoichiometric ratio of TiO_2 by avoiding the formation of oxygen vacancies. Nitrogen atmosphere prevents the annihilation of oxygen vacancies, but a nitrogen doping in the rutile TiO_2 nanorods was not observed [214]. In contrast to oxygen, hydrogen reduces TiO_2. But for annealing at 650 °C, the macroscopic result of the atomic rearrangement does not seem to depend on the annealing atmosphere. Only in the case of vacuum, the finger structure is less reduced compared to the gases. The reason for this behavior might be the low thermal conductivity of vacuum resulting in a temperature gradient between the TiO_2 nanorods and the thermometer used for controlling the temperature inside the oven. After annealing at 1050 °C, there is no difference between vacuum and gas atmosphere anymore. The fine structure disappeared completely as demonstrated in Figure 7.8. Steps are formed on the sidewalls of the rods and the cross-section is decreasing towards the tip of the rods. On edges, the {100} facets become significantly larger. Many rods lying on sputtered ANA-SPU-850 seem to merge with the film partly as shown in the inset of Figure 7.9 G. This points up the high atomic mobility and indicates an efficient defect healing. Thus, it is assumed that nanorods lose their mesocrystalline structure and become more single-crystalline during annealing. A simple test probing the crystal quality is to test the chemical stability with concentrated hydrochloric acid. Thereby, rods without post-treatment and rods annealed in oxygen rich atmosphere were exposed to 18.5% HCl at 180 °C for 4 h. The results are shown in Figure 7.9 C for 650 °C and in Figure 7.9 H for 1050 °C annealing. In contrast to the non-annealed rods, the inner part of rods annealed at 650 °C is not dissolved during the hydrochloric acid exposure. Only a few slacks appear on the sidewalls. For rods annealed at 1050 °C, the {110} facets and the tip facets remain unchanged, while {100} facets are etched slightly. This indicates the low defect density in these nanocrystals. The etching of the {100} facets is expected, since these facets are not developed during the hydrothermal growth using the same hydrochloric acid concentration as for the post-deposition hydrochloric acid exposure. Hence, these facets are energetically unfavorable and relatively vulnerable in the acid.

The shape of the rods annealed in hydrogen rich atmosphere at 1050 °C is standing out certainly. The surface is roughened indicating a strong effect of the hydrogen on the TiO₂ rods as shown in Figure 7.9 J. If the reduction of TiO₂ is too advanced, the rutile morphology becomes unstable [1686]. As a consequence, the material changes its atomic arrangement and thus, the rods start changes its shape. Possible morphologies are cubic TiO or orthorhombic Ti₂O₃ [1655, 1656]. But also fractions of amorphous oxide or pure titanium might appear. These rods are etched by exposure to 18.5% concentrated hydrochloric acid solution easily, since they have not the outstanding chemical resistivity like single-crystalline rutile TiO₂.

The thickness of fingers is increased at elevated growth temperatures. However, the dimensions of the fingers are also enhanced by a post-treatment in hot hydrochloric acid. The finestructure is reduced or even vanished by post-annealing above 600 °C. All these features are in good agreement with the introduced model at the beginning of this chapter.

8 Expanding the Horizon of Position-Controlled Hydrothermal Growth Methods

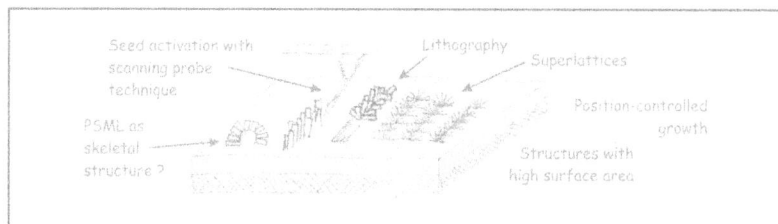

Figure 8.1: *Scientific objectives investigated in this chapter. In which way can lithography, PSML, and scanning probe lithography be used for position-controlled hydrothermal growth? Which methods are suitable for the creation of superlattices?*

Controlling the position of nanostructures or adding an additional superlattice provides significantly more application compared to homogeneous NRAs, for instance. A precise control over the position is necessary for applications demanding a high spatial resolution such as (photo)transistors, data storage devices, gas detectors, or medical implants. Additionally, also basic research projects need nanostructures in well-defined locations in order to investigate their interaction with heat, light, electrical fields, chemicals, or biological material. Hence, position-controlled fabrication techniques are nothing less than the backbone in the advancement of nanotechnologies. In order to provide as many developers and researchers as possible with these techniques, they must be precise and inexpensive. In this chapter, some ways are shown to realize these requirements for rutile TiO_2 nanorods.

A nanostructure in combination with a superlattice provides further advantages. First, it provides the possibility to enhance the surface which is important for surface-dependent applications such as photocatalysis, solar cells, or energy storage. Hence, there is a huge demand for hierarchically structured large area TiO_2 composites [1686–1691]. Second, the periodicity of a superlattice might be used to improve the interaction with waves such as the absorbance of light, which is, again, essential for applications such as photocatalysis or photovoltaics. Thirdly, the surface structure affects the interaction with liquids and thus, it can be used to adjust the wetting properties of a nanostructured surface which might be interesting for chemical or biological applications. In this chapter, the fabrication of superlattices for rutile TiO_2 nanostructures is demonstrated as well.

8.1 From Standard Methods to New Advanced Lithography Techniques

Conventional lithography is presented as a first position-controlled fabrication technique for rutile TiO_2 nanorod arrangements as drafted in Figure 8.2. Optical and electron-beam lithography are applied to deposit rutile TiO_2 seed layers selectively on a polished silicon substrate. Optical contact lithography provides large structures and is suitable for large area patterning. In this work, the technique is refined resulting in structures with efficient light scattering. Additionally, a monolayer of polystyrene spheres is employed as a skeleton for hemispherical seed layers providing an extra-large surface and additional light scattering. Comparably, a concept for highly light scattering structures made by electron-beam lithography is introduced.

Masks made of developed photoresists and monolayers of polystyrene spheres (PSML) are taken since these methods are quite common and well-established. For instance, Norasetthekul et al. and Sinha et al. applied optical (UV and deep-UV) lithography to fabricate patterned TiO_2 films [1692, 1693]. Park et al. employed light stamping lithography (LSL) [1694]. Thereby, a poly(dimethylsiloxane) (PDMS) stamp is placed on the sample and illuminated with UV light (254 nm wavelength) light. Wherever the PDMS is touching the substrate, bonds are created between the PDMS and the substrate. During lift-off, PDMS bonded to the substrate remains and is employed as a mask for the following ALD of TiO_2.

Another approach employs the deposition of a seed layer on pre-structured surfaces as shown in this chapter. Wang and Shi showed that selective growth of TiO_2 nanorods on silicon nanowires works well [1695]. In this work, FIB milling is used to produce very small and individual structures made of silicon as a substrate for TiO_2 nanorods.

But there are other interesting position-controlled deposition methods for thin TiO_2 films as well. Besides LSL, direct nanoimprint lithography (NIL) is used for various one and two-dimensional patterns with sub-nanometer spacing [1696]. For NIL, a stamp with a single or periodic nanostructure is pressed in the sample. Shan et al. were even able to create a large pattern by controlling the air flow used for drying a drop of TiO_2 sol on a flat substrate [1697]. Nikoobakht et al. deposited catalysts promoting the growth of ZnO nanorods on selected positions on the sample [1698]. This method is related to the techniques introduced in the Chapters 8.6 and 9.

8.2 Confined TiO_2 NRAs Made by Optical Contact Lithography

Experimental details of optical lithography are reported in Chapter 4. For the conventional and extended optical lithography, 8 nm and 20 nm thin RUT-EBE-850 films are used, respectively. The titanium is deposited after the development of the photoresist and the film is oxidized after lift-off. TiO_2 nanorods are made with the hydrothermal method described in previous chapters. Details and standard values of the film and nanorod fabrication are found in Chapter 4.

8.2.1 Standard Structures

Using optical contact lithography seed stripes with 2 µm width and 2 µm pitch are fabricated. The result after the hydrothermal growth is shown in Figure 8.3 A. This method is well established in micro-manufacturing and serves as a reference and initial point for the presented nano-manufacturing techniques.

8.2.2 Levitating TiO_2 NRAs

To increase both, scattering and surface area of subwavelength rutile TiO_2 nanorods, the conventional optical lithography is expanded as drafted in Figure 8.3 A and B. For this approach,

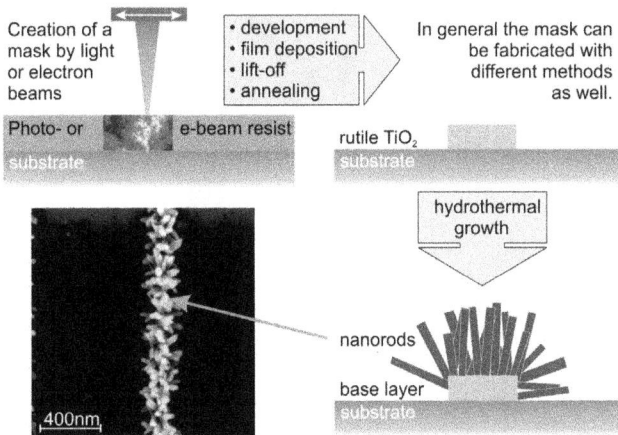

Figure 8.2: *Sketch showing the fabrication of confined NRAs by lithography. First, a mask is created. Second, the seed layer is deposited on the mask. Third, the hydrothermal growth is applied. The example shown in the image was fabricated by e-beam lithography.*

the seed layer is deposited via sputter deposition instead of electron-beam evaporation. Due to the higher processing pressure, atoms approach the substrate with a broad distribution of incident angles. Thus, the photoresist is covered with a conformal layer including the vertical side walls of the structure boundaries. For sufficiently thick films, this results in a stable bond between the seed layer on the substrate and the photoresist. For small pitches, the seed layer on the photoresist remains as a levitating film after the lift-off as shown in Figure 8.3 D. The grain size of the levitating film is larger than the fixed film. This might result from the much larger surface of levitating films promoting efficient diffusion of atoms at the surface [1699]. After the hydrothermal growth, the whole seed layer is occupied by nanorods as shown in 8.3 C. The levitating seeds cause curvatures on the surface and the nanorods become omnidirectionally oriented. Due to the larger grains in the levitating film the diameter of the resulting nanorods is increased as well.

8.3 Double Superlattice by Optical Lithography and PSML

PSML as a tool for structuring TiO_2 films in order to improve light absorption and to enlarge the surface was employed by several groups before [1700, 1701]. In particular, this technique is used to improve the efficiency of photocatalysis devices [1702].

In this work, three different kinds of structures fabricated with PSML are presented. The fabrication of the polystyrene spheres and monolayer is described in Chapter 4. Further processing steps are shown in Figure 8.4. For all structures, 20 nm titanium is deposited by sputter deposition on the PSML. The structure shown in Figure 8.4 A is created by removing the spheres in oxygen plasma and via sonication. The remaining titanium is distributed in highly ordered and triangular shaped layers with lateral dimensions between approximately 10 and 50 nm. During annealing at 850 °C in an oxygen rich atmosphere, the triangular titanium shapes turn into more circular shaped particles consisting mainly of rutile TiO_2. It is assumed that the circular shape

185

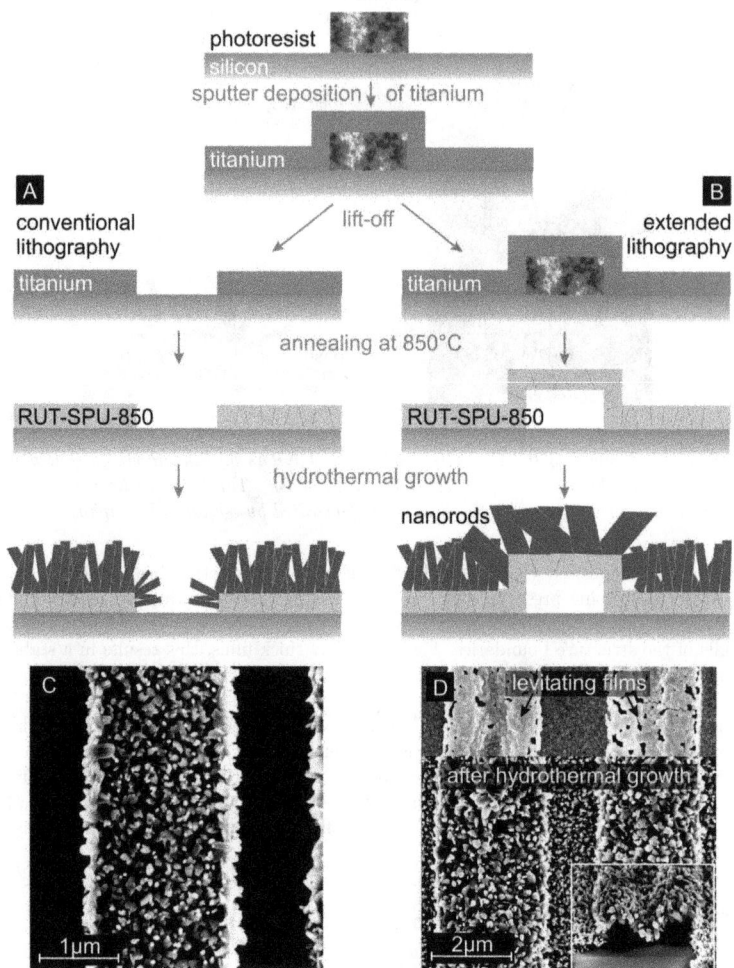

Figure 8.3: *NRA with superlattice made by optical lithography: A) Structured NRA made by complete lift-off. The achieved structure is shown in the SEM image in Figure (C). B) During careful lift-off the titanium film remains on narrow gaps and is turned into a levitating rutile film after annealing. Since the grains in levitating films are slightly larger than elsewhere, the resulting nanorods have larger diameters in these regions as shown in the SEM image in Figure (D). The inset in Figure (D) is a cross-section image of the structure shown around the inset.*

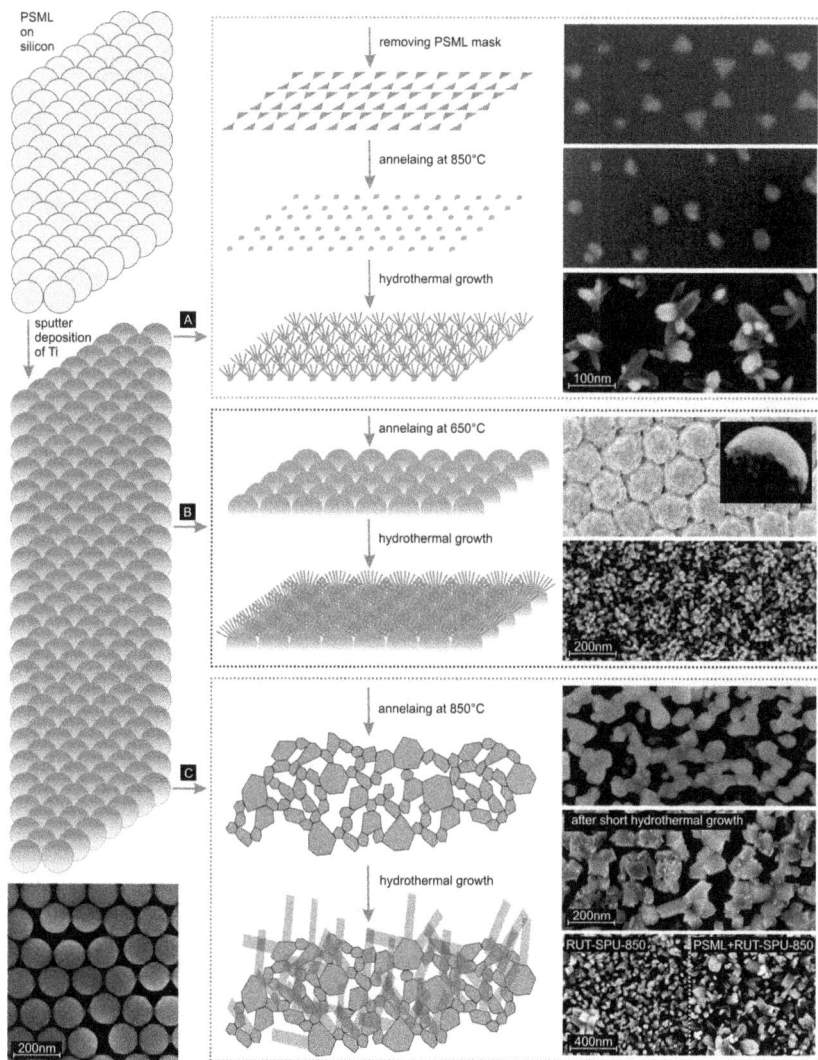

Figure 8.4: *Structures made via PSML masking. The diameter of the spheres used in this figure is 160 nm. Details are described in the text.*

results from minimizing the surface or TiO_2/Si-interface energy of each grain. The hydrothermal growth was applied for 60 min resulting in ordered nanorod flowers.

For the other two structures, the PSML works rather as a skeleton than a mask. The PSML was not removed after the deposition of titanium and hence, the polystyrene was combusted during the annealing in an oxygen rich atmosphere. Annealing at 600 °C results in hollow hemispheres consisting of polycrystalline rutile TiO_2 as demonstrated in Figure 8.4 B. Applying the hydrothermal growth for 60 min, a well-ordered hedgehog structure with a very large TiO_2 surface was created. Hence, this structure satisfies an important requirement for an efficient photocatalysis.

Annealing the sample at 850 °C results in rutile TiO_2 grains as described in Chapter 5, but the diameter of the grains is larger than in a dense polycrystalline film manufactured with the same deposition and annealing parameters. The grains are linked with each other but there are spacings between many grains as well. This is most likely a result of the porous structure caused by the hollow polystyrene spheres. As presented in Figure 8.4 C, the large and less densely packed grains promote nanorods having larger diameters compared to their relatives on dense polycrystalline seed films. This outcome is reasonable since usually a single facet of a grain provides a single nanorod. Hence, if the facets are larger the diameter of the rods becomes enlarged as well. A similar explanation is that the density of facets is reduced if the facets are larger and separated by pores. As discussed in Chapter 6, the growth of nanorods is disturbed and partly hindered by neighboring rods which limits their width. By increasing the average distance between neighboring rods, their interference with each other is reduced and the average width of the rods is increased. Consequently, this structure provides NRAs with larger spacings and thus, the steric hindrance for larger molecules and agglomerates is reduced which might be useful for photocatalytic or energy storage applications.

Optical lithography is a simple and inexpensive tool to grain large structured NRAs. An extension of the optical lithography is introduced that generates locally levitating NRAs. Employing seed layer fabrication on PSML in combination with different annealing temperatures provides a wide range of structures such as nanorod bushes, hollow hedgehogs, and nanorods on a mesoporous seed film.

8.4 Confined TiO_2 NRAs Made by Electron-Beam Lithography

Experimental details of electron-beam lithography are reported in Section 4.1.4. The film deposition and hydrothermal growth procedure is the same as for optical lithography using 8 nm thin RUT-EBE-850 films. The results are presented in Figure 8.5.

The motivation to use electron-beam lithography is to gain smaller structures that are advantageous for an enhanced light interaction as shown in Chapter 10 and describe a further step towards single nanorods devices such as transistors, gas sensors, or UV-photodetectors. The limits of the fast and inexpensive optical lithography originate from the long wavelengths of visible light. Since a simple mask aligner in combination with a mercury lamp is used for exposure, the resolution for the applied optical lithography is limited to a few hundred nanometers. This limit can be extended with more advanced methods such as excimer laser lithography [1703]. Electron-beam lithography is used as an alternative to fabricate narrower stripes as shown in Figure 8.5. The resolution limit is given by the electron energy and the resist properties. Stripe widths of only 3 nm are possible [175].

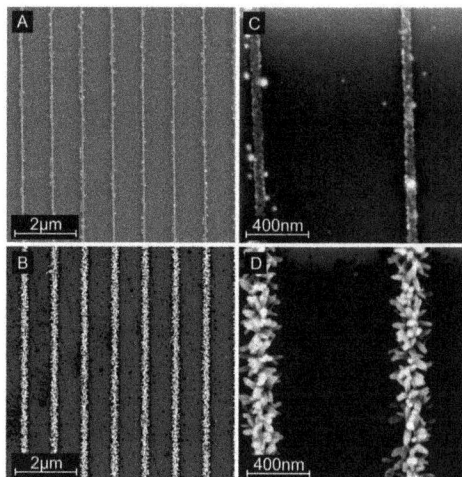

Figure 8.5: *SEM images of a superlattice made by electron-beam lithography: A,C) seed layers before hydrothermal growth using different magnifications; B,D) nanorods grown on structured seed layer with different magnifications. The stripes are* 60 nm *wide, the pitch is* 1 µm *and the rods are roughly* 200 nm *long.*

Electron-beam lithography provides narrow lines covered with randomly orientated nanorods. However, nanorods directing along the line are very rare. The width of the lines is in the order of the rod length.

8.5 Focused Ion Beam Triggered Growth of TiO₂ Nanorods

In this section, the position-controlled fabrication of rutile TiO₂ NRAs using focused ion beam (FIB) milling is demonstrated. This technique, which displayed in Figure 8.6, results in seed layers with confined geometries. The method is based on focused ion beam milling in combination with sputter deposition and generates nanorods oriented mainly in parallel to the substrate plane. There are no published fabrication techniques promoting only the horizontal growth, although horizontally grown nanorods are of special interest for integrated circuit and lab-on-a-chip applications. Similar to lithography techniques, the focused ion beam triggered hydrothermal growth is transferable to nanostructures made of different materials as well. The effect of orientation of the presented NRA geometries on visible light scattering and the interaction between visible light and subwavelength-diameter TiO₂ nanorods are discussed in Chapter 10.

8.5.1 Focussed Ion Beam Milling

Two approaches are used for this technique. Firstly, 40 nm thick RUT-SPU-850 films on silicon are removed by milling and thus, the growth was locally suppressed. Secondly, a 20 nm thin ANA-SPU-450 film was deposited on a silicon substrate with prefabricated trenches made by

189

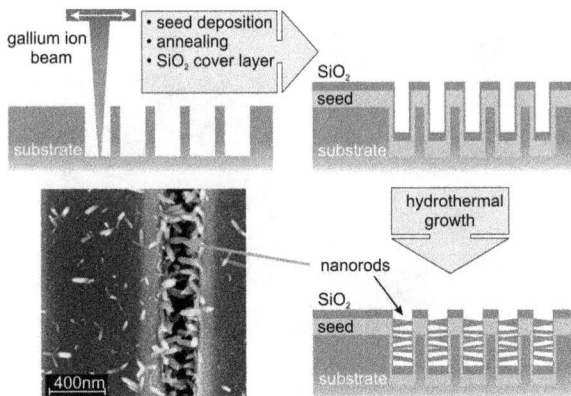

Figure 8.6: *Sketch of FIB triggered growth of confined TiO_2 NRA. In narrow trenches, the layers deposited by sputter deposition become thinner compared to the unaffected surface due to shadow effects. The TiO_2 layer is thick enough to provide the growth even inside the trenches. The SiO_2 layer thickness is adjusted to a value which is large enough to prevent the growth of TiO_2 nanorods on the unaffected sample surface, but which is too thin to prevent the growth inside the trenches.*

FIB milling. To prevent the growth in untreated regions a thin SiO_2 layer is deposited via RF sputter deposition. Here, the anatase film is preferred as a seed layer, because SiO_2 layers block the growth of nanorods much better on these films than on rutile films. TiO_2 nanorods are made by a hydrothermal method in an acid aqueous solution containing hydrochloric acid (14.8%) and 350 µL titanium(IV) butoxide. The samples are autoclaved at 180 °C for 75 min. Details and standard values of the film and nanorod fabrication are found in Chapter 4.

8.5.2 Confined NRAs Made by Focused Ion Beam Milling

An active seed layer on silicon is milled using a FIB in order to gain trenches as shown in Figure 8.7 A. Inside the trenches, the seed is removed. After the hydrothermal growth process, trenches remain as nanorod free areas. Since the ion beam has no sharp limits the substrate is milled close to the desired beam impact as well. This fact results in soft edges of the milled structures and thus limits the resolution. To reduce this effect, a 200 nm thick protecting titanium layer is deposited on the seed layer before milling and removed afterward using nitrohydrochloric acid. The improvement can be seen when comparing the orange marked stripe widths affected by an ion beam in Figure 8.7 A and C. The resulting structures covered by rutile TiO_2 nanorods are presented in Figure 8.7 B and D, respectively.

To gain a significant amount of horizontally oriented nanorods, a seed layer was deposited on an already existing structure made by FIB milling. The seed layer was deposited by sputter deposition in order to cover every niche of the structured substrate. In a next step, a thin SiO_2 layer was deposited on the seed layer in order to suppress the formation of nanorods.

The important parameter is the thickness of both films, since it is thinner on the walls inside the narrow trench compared to the unaffected surface due to shadow effects during the sputter deposition. The TiO_2 layer is so thick that there is a continuous seed layer in- and outside the

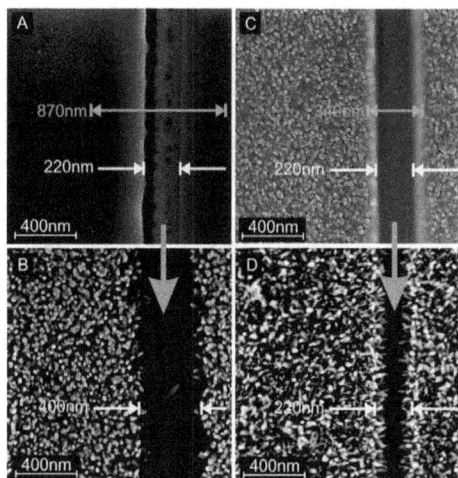

Figure 8.7: *SEM images of trenches made by FIB milling on a substrate covered with a seed layer, before (A,C) and after (B,D) hydrothermal growth. By applying an additional titanium cover layer (C,D), the resulting stripe width is nearly halved. The yellow arrows mark the average width of the desired trenches and the orange arrows mark the average width of the stripes affected by the ion beam due to its non sharp-edged profile. In Figure A, the transition between the area which is unaffected and the region affected by the ion beam is seen hardly. Tiny grains which are clearly seen in the unaffected region become less distinct and finally disappear in the affected region. This effect influences the growth (density and orientation) of the nanorods slightly and can be seen in Figure B by comparing the NRA in- and outside the affected area which is marked by the 870 nm bar in Figure A.*

trenches. The additional SiO_2 layer is thick enough to prevent the growth of nanorods on the unaffected sample surface, but inside the trenches it is too thin to have this property. Consequently, the growth of rutile TiO_2 nanorods is promoted inside the trenches only as displayed in Figure 8.8. To demonstrate that this effect does not originate from trapping precursor aggregates inside the narrow trenches, the same technique was applied on more freestanding vertical walls as shown in Figure 8.9. Even on tilted planes the growth of nanorods is still reduced significantly. Inside the narrow trenches, the density of nanorods should increase in deeper regions since the SiO_2 is expected to be thinner in larger distances from the surface.

Focused ion beam milling provides narrow trenches. Adding a thick seed and a thin SiO_2 layer to this trench structure, the growth of nanorods becomes restricted to the inner walls of the trenches. This effect is related to the reduced deposition rate of SiO_2 inside the trenches. Furthermore, the rods are oriented mostly in parallel to the substrate surface.

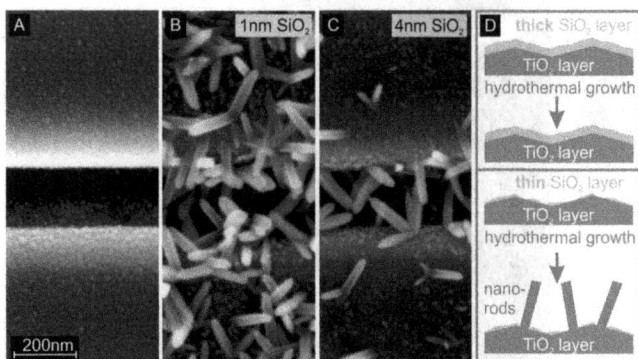

Figure 8.8: *SEM images of nanorods grown in a trench fabricated by FIB milling. Image (A) shows the trench covered by a seed layer before hydrothermal growth. (B) and (C) show similar trenches after hydrothermal growth process. The samples in image (B) and (C) were covered with 1 nm and 4 nm SiO_2 before the growth, respectively. The effect of the SiO_2 layer thickness on the growth is illustrated in Figure D. An ultra-thin SiO_2 does not cover the TiO_2 film completely due to a thickness inhomogeneity caused by the sputter deposition process. Uncovered areas promote the growth of nanorods. Thick SiO_2 do not show any holes and prevent the generation of nanorods.*

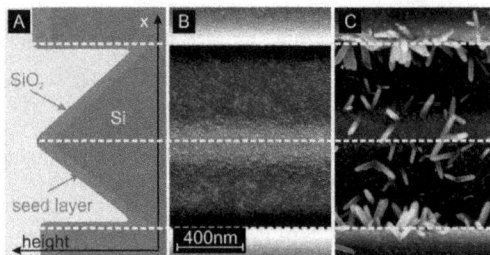

Figure 8.9: *SEM images demonstrating the results of hydrothermal growth on differently tilted substrate planes created by FIB milling. The structures are covered with a seed layer and 4 nm SiO_2. The profile is shown in Figure A. Figure B shows the trench before and Figure C after the hydrothermal growth process. The density of nanorods is clearly reduced in flat and 45° tilted areas, while the density is relatively high on vertical planes. The different layer thicknesses on differently tilted planes result from shadow effects and the angle distribution of the approaching atoms during the sputter deposition. On vertical planes, the amount of deposited SiO_2 is negligible.*

8.6 Advanced Scanning Probe Lithography Using Anatase-to-Rutile Transition of Localized Nanoparticles

In this chapter, the position-controlled hydrothermal growth of rutile TiO_2 nanorods using a new and advanced scanning probe lithography method is demonstrated. The method is displayed in Figure 8.10. Thereby, a silicon tip as it is used for conventional AFM was pulled across an ANA-SPU-850 film. During this process, dust is created that contains tiny anatase TiO_2 nanoparticles likely. Due to the lattice mismatch between anatase and rutile as well as the few grain boundaries, the growth of rutile nanorods is not provided on ANA-SPU-850 in general. The work of Li et al. forms the basis of the development of the presented technique. They suppose the transformation of anatase into rutile nanoparticles as a major process during the growth and branching of rutile TiO_2 nanorods [159]. The content of this chapter is also dealt with in the article [1704].

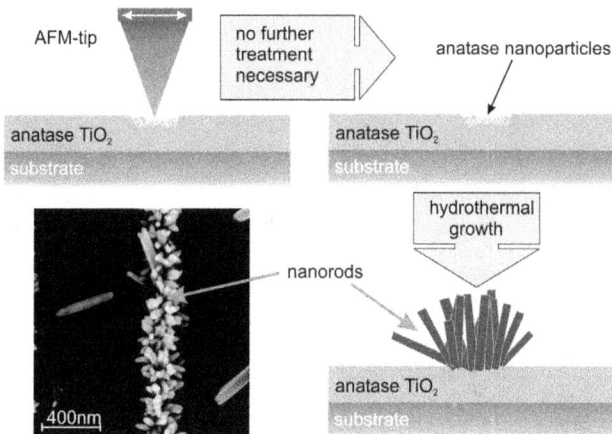

Figure 8.10: *Sketch of the fabrication of confined NRAs by the presented advanced scanning probe lithography.*

Later it was observed by different groups that rutile TiO_2 nanorods grow on anatase particles with a diameter of approximately 25 nm [1705, 1706]. And even if rutile TiO_2 nanocrystals are already present, it is assumed that anatase clusters diffuse along rutile TiO_2 surfaces towards low energetic facets and perform a solid state phase transformation into rutile TiO_2 on these facets [1578]. Thus, it is an obvious possibility to generate anatase nanoparticles locally by scratching across an anatase film using a conventional AFM tip in order to promote a position-controlled hydrothermal growth. It should be noted that the applied force during scratching is much higher than used for conventional AFM surface characterization in contact mode. Consequently, the AFM tip is worn out faster than during a standard surface characterization. In this work, there is no critical force determined which is needed for activating the hydrothermal growth. It is reasonable that it depends on the exact kind and shape of tip as well.

Nanorods grown on these structures are omnidirectional. As it will be demonstrated in Chapter 10, this property results in high light scattering efficiencies. This is because nanorods oriented perpendicular to the incident light show an "antenna" effect enhancing light scattering [713, 714].

193

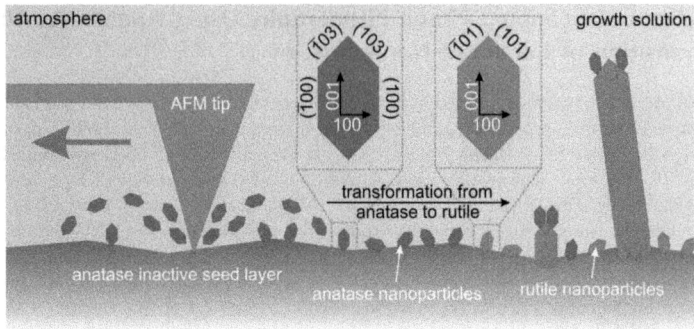

Figure 8.11: *Schematic drawing of activating inactive anatase TiO$_2$ films: Scratching with an AFM tip causes anatase nanoparticles. In growth solution, these anatase particles are transformed into rutile particles and promote the growth of rutile TiO$_2$ nanorods.*

And there are a lot of nanorods within the restricted NRA which are oriented perpendicular to the incident light. Locally enhanced visible light scattering is determined with optical dark field images. Due to the small tip radius, the resolution of this method is excellent and it is inexpensive and faster compared to electron-beam lithography and similar methods providing a position-controlled growth of semiconducting TiO$_2$ nanostructures. The high light scattering efficiency of the fabricated structure is demonstrated and explained in Chapter 10.

8.6.1 Advanced Scanning Probe Lithography

The advanced scanning probe lithography is applied on 40 nm thin ANA-SPU-850 films using an Innova AFM from Bruker (former Vecco) in contact mode. The applied force is significantly higher than typically chosen for topography scanning. OTESPA-R3 (from Bruker AFM probes, Camarillo, US) silicon tips with a spring constant of approximately 26 N/m are used. The tip radius is less than 10 nm before scratching but it is enlarged during the lithography process due to abrasion. Each line is scratched 256 times (128 times in each direction) in order to generate a sufficient amount of seed sites. The technique is performed under ambient conditions with a writing speed of 10 µm/s (scan frequency: 0.5 Hz; line length: 10 µm). The samples are placed top-side down in a Teflon liner. The growth solution consisted of hydrochloric acid (14.8%) and 350 µL titanium(IV) butoxide. The samples are autoclaved in the growth solution for 75 min at 180 °C. Details and standard values of the film and nanorod fabrication are found in Chapter 4.

Scratching with an AFM tip across an anatase film is supposed to generate anatase nanoparticles. Anatase nanoparticles transform into rutile nanoparticles during the hydrothermal growth process [159, 1707]. These rutile nanoparticles serve as seeds for rutile nanorods as drawn in Figure 8.11. The experimental success is presented in Figure 8.12.

8.6.2 Excluding Competing Processes Resulting in Localized Growth of Nanorods

Before specific properties of this technique are discussed, it should be excluded that AFM tip-induced growth is not simply a surface roughening effect. In Figure 8.13, the difference between a scratched film (8.13 A) and a mechanically broken film (8.13 B) is demonstrated. The scratched

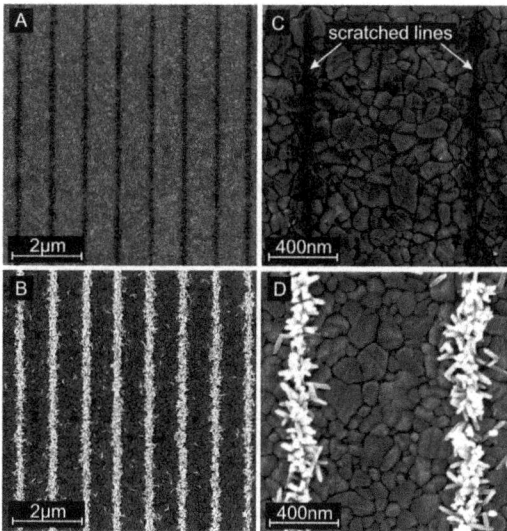

Figure 8.12: *SEM image after scratching and before hydrothermal growth with different magnifications (A and C); superlattice consisting of nanorods gained after hydrothermal growth process (B and D). The stripe width is less than 100 nm, the pitch is 1 µm and the rod length is approximately 100 nm.*

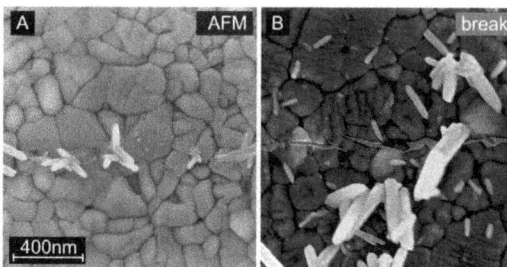

Figure 8.13: *A) SEM image after short hydrothermal growth on scratched line. The line is written only a few times in order to generate fewer growth sites compared to Figure 8.12. For clarity, nanorods are colored bluish and the scarified surface is colored brownish. B) SEM image after applying several times the hydrothermal growth to a stress-induced line of breakage. For clarity, the breaking line is colored purple.*

line in Figure 8.13 A is marked brownish, the breaking line in Figure 8.13 B is marked purple. Such a breaking line is induced by mechanical stress generated by scratching the sample with a diamond writer manually. Hereby, the mechanical stress is guided through the sample from the center of the scratch to surrounding regions. Usually, the film is removed close to the scratch while tiny cracks such as the one shown in the figure appear in a distance of a few microns. The fracture appears as defined straight lines through grains or on grain boundaries indicating preferred breaking lines along specific crystal planes. Nanorods grow only in the scratched area, while there is no preferred growth observable at the breaking line in Figure 8.13 B. Nanorods growing around the breaking line might originate from anatase particles released during the breaking process.

Besides surface roughening surface, charging is another candidate that could promote the growth on scratched regions. The tip causes a lot of friction and charges the film locally. The charges might be trapped in defect states generated by the scratching process. Furthermore, the native silicon oxide layer prevents a quick charge transport into the conductive doped silicon substrate. A charged film might attract the precursor more likely and the density of rods is increased. But charging should also increase the nucleation probability on different seeds promoting the hydrothermal growth of rutile TiO_2 nanorods as well. As a reference system, reduced WO_3 is chosen because rutile WO_2 is harder than the silicon tip and therefore, film roughening becomes unlikely as a competing effect. Even for the hardest applied scratching force, the growth is not enhanced by the presented scanning probe lithography method on reduced WO_3. The density of grown nanorods on the scratched sites is the same than elsewhere. This outcome is not expected if charging would be the major driving force for the locally enhanced hydrothermal growth of rutile TiO_2 nanorods.

Nanorods grown on ANA-SPU-850 films resist even long sonication, which is more reasonable for nanorods being linked to nanoparticles bound to the anatase film rather than nanorods bound to the seed layer by Coulomb interaction only.

8.6.3 Special Features of the Presented Advanced Scanning Probe Lithography

Growth on activated seeds can be suppressed by an electron bombardment as shown in Figure 8.14. Inside the yellow rectangle, the sample was scratched the same way as elsewhere but afterward, it was exposed to an intense electron-beam. The exposed area stays free of nanorods during the hydrothermal growth process. Since this effect becomes significantly stronger with increasing beam intensity and remains for many days in humid environment it results more likely from electron-beam-induced surface smoothing as described by Strauss et al. [1708] and Uno et al. [1709] rather than from surface charging. A reasonable reason for surface charging is the trapping of electrons from the electron-beam in neutral traps such as OH groups [1710]. A detailed discussion of this effect is found in the end of Chapter 9. Figure 8.14 C shows the effect of scratching an area for several times. The density of lines scratched several times is slightly higher compared to lines scratched only once. However, the difference is marginal. Figure 8.14 D shows that light scattering is enhanced by TiO_2 nanorod lines locally.

Last but not least, a look on the penetration depth of this method is taken. For this purpose, a 3 nm thin SiO_2 layer was deposited on the anatase film. Even after scratching hundreds of times at the same position with the hardest employable force, the growth was triggered only in a few places as demonstrated in Figure 8.15. First, anatase TiO_2 (5.5 to 6 Mohs) is a little bit softer than SiO_2 (7 Mohs) [1711]. Second, significantly less force was applied during scratching the uncovered anatase film. Consequently, it is reasonable that the AFM tip technique penetrates the upper 1–3 nm of the anatase film only. In addition, silicon nanoparticles might act as nucleation sites such as other dirt particles appearing in the growth solution. But the experiment on SiO_2

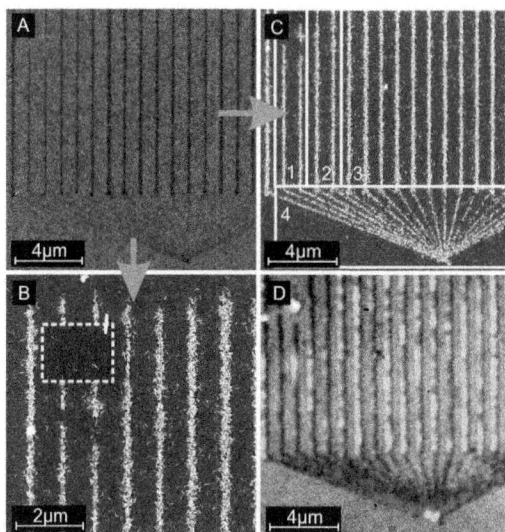

Figure 8.14: *Figure A to C show SEM images. Figure A is the selectively scratched area before hydrothermal growth on an ANA-SPU-850 layer. Figure B shows scratched lines after the hydrothermal growth process. Inside the yellow rectangle, the substrate is scratched as elsewhere, but the growth is suppressed by intensive electron-beam exposure. Figure C displays the area shown in Figure A after the hydrothermal growth process. The lines in area 1 are scratched 32 times per line, in area 2 64 times per line, in area 3 128 times per line and in area 4 only once per line. Figure D is a BF image taken with an optical microscope of the same area as shown in Figure C. The colors in Figure D are amplified subsequently.*

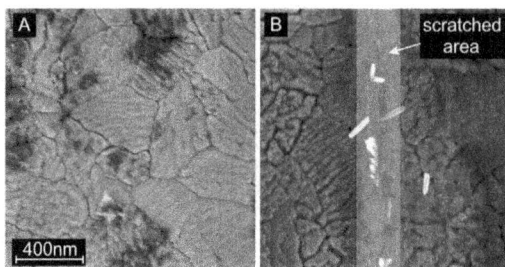

Figure 8.15: *SEM images of a sample where the ANA-SPU-850 layer is covered with a 3 nm thin SiO$_2$ layer. The scratched area is shown before (A) and after (B) the hydrothermal growth process was applied. The density of nanorods is much less compared to Figure 8.14. The yellow area marks roughly the region where the AFM tip scratched the sample.*

showed that the growth of rutile nanorods is not triggered by silicon nanoparticles being scraped off the AFM tip.

Advantages of the Scratching Method

Although writing speed is slow compared to lithography and one has to exchange the tip from time to time, there are some clear advantages of this method. First, it works under ambient conditions and the setup is inexpensive compared to electron-beam lithography. Still, the resolution is roughly of the same order. Second, scratching influences a very thin regime at the sample surface only and the majority of the seed film remains unchanged. As a consequence, this method can be applied to ultra-thin films of a few nanometers thickness only. Thirdly, this technique does not need any pre- and post-processing such as the resist handling for lithography. This enables the opportunity to apply this method on very uneven or pre-structured substrates. Furthermore, it does not matter if the seed is a film or bulk material and by which technique the inactive seed is deposited.

This scanning probe technique is based on the principle that anatase nanocrystals are converted into rutile crystals during the hydrothermal growth process. The transformation is a consequence of the different surface and bulk energies of anatase and rutile. Above a critical diameter, the surface energy of anatase becomes so large that a phase transition into rutile is precipitated. Hence, locally distributed seeds can be obtained by generating anatase nanocrystals during scratching with a hard tip across a flat anatase film. The growth will be provided in areas with anatase nanoparticles only, but not on the unaffected and clean film.

9 LASER-Induced Position-Controlled Hydrothermal Growth

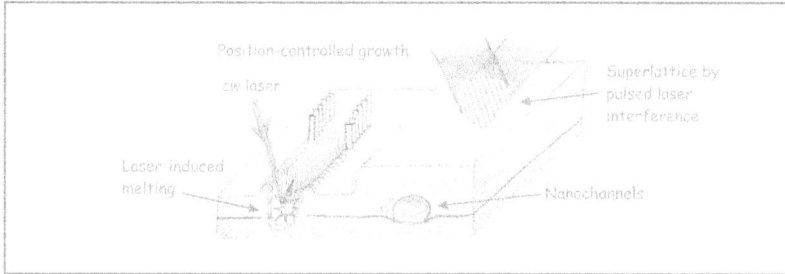

Figure 9.1: *Scientific objectives investigated in this chapter. In which way can cw and pulsed laser radiation used for position-controlled hydrothermal growth? How does the laser radiation interact with the samples?*

In this chapter, a completely different approach to generate local rutile seeds is presented. A thin anatase TiO_2 layer on silicon is melted by a laser beam. During the cooling and solidification process, rutile TiO_2 is formed partly. Due to the Gaussian intensity profile of the laser beam, only the central part of the beam delivers enough energy to exceed the melting temperature. It is demonstrated that laser-induced melting works for cw and pulsed laser mode. It is inexpensive, fast, and applicable in ambient atmosphere. The content of this chapter is also dealt with in the article [1712].

9.1 Basics about Laser Lithography Techniques

A lot of structural, thermodynamic, electronic, and optical properties have to be considered in order to describe the physical processes properly. A general introduction of these properties is given in Chapter 2. Here, the interplay between important effects and properties contributing to the light-induced melting process are discussed in detail.

It should be noted that the temperature dependence of the reflectivity and the absorption is not discussed in this section since it is not clear so far where the absorption, that results in the initial temperature increase, takes place. This effect will be discussed in Section 9.2.5.

9.1.1 Formation of Interlayer between TiO_2 and Silicon Substrates

Laser-induced melting calls for absorption of the applied wavelength. This requirement is satisfied if the material shows either a strong photon-phonon coupling, a sufficiently small bandgap, or a sufficiently high density of filled intra-bandgap trap states.

The photon-phonon interaction in materials consisting of nanoparticles such as powders or poly-crystalline thin films differs often from their bulk properties. These discrepancies might originate from field enhancements near anisotropic nanoparticles, a decrease of the photon group velocity, localized photons, and further effects causing nonlinear behavior [1713]. Kudryavtseva et al. reported about such nanostructure related effects for ITO and SiO_2 nanoparticles with diameters between 200–320 nm excited by 20 ns laser pulses at 694.3 nm [1713]. Ren et al. found that in anatase TiO_2 nanoparticles with a diameter of 11 nm the direct excitation of TO phonons by light with a wavelength of 532 nm is rather unlikely [1714]. But the incident light excites electrons into the conduction band and by relaxation to the conduction band minimum energy is transferred from electrons to phonons [1714]. And phonon confinement in anatase TiO_2 nanoparticles support local heating significantly [1715].

Due to its large bandgap energy, the absorption of pure anatase TiO_2 by electronic excitations is limited to UV light. But defects and impurities shifts the absorption into the visible regime. Generally, the absorption coefficient of silicon is much higher than that of anatase TiO_2 in the visible regime [1716, 1717]. Thus, one might expect that most of the beam energy is absorbed in the silicon close to the TiO_2/silicon interface.

Furthermore, one has to consider the formation of an additional layer consisting of a Ti–Si–O blend between the silicon substrate and the thin anatase TiO_2 film [585, 1718]. The thermal energy supplied by annealing at 850 °C during the fabrication of the polycrystalline TiO_2 films used in this work is supposed to be high enough to stimulate the diffusion of oxygen into silicon and silicon into TiO_2 [1225]. Nemanich et al. reported that silicon starts to diffuse into titanium at 300 °C already [1719].

Diffusion works best if pure titanium is annealed on silicon as it is performed for the fabrication of RUT-SPU-850. The diffusivity of titanium atoms into silicon is given by

$$D_{ti} = 1.45 \cdot 10^{-6} \exp\left(-1.73 \cdot 10^5 / R_g T\right) \text{m}^2\text{s}^{-1}. \tag{9.1}$$

where R_g is the gas constant and T the temperature [1227, 1720]. This results in a diffusivity of roughly 3×10^{-17} m^{2}s^{-1} at 850 °C. The diffusivity of oxygen in silicon is roughly one order of magnitude smaller [1227]. Here, titanium-silicon phases are formed as shown in Figure 9.4. Morgan et al. documented the formation of different titanium-silicon phases at different annealing temperatures on silicon and Si_3N_4 substrates [1721]. For instance, Corcoran et al. have observed thin titanium silicide layers after annealing a titanium film on silicon at 750 °C [1722]. In particular, $TiSi_2$ seems to be created quite easily [1723]. Additional phases such as $TiSi$, $TiSi_2$, and TiO_x are created already after an annealing time of 15 min at 700 °C and $TiSi_4$ is observed after 85 min annealing at 600 °C below 6.7×10^{-7} mbar [1724]. Ti_5Si_3, $TiSi$, $TiSi_2$, and Ti_5Si_4 was detected between titanium and silicon after annealing at 500 °C for 20 min only [1725]. The composition of such an interface layer is shown in Figure 9.2 A [1725]. In conclusion, annealing titanium films on silicon at 850 °C provides the formation of titanium-silicon phases easily. But does diffusion and intermixing still work if titanium is exchanged by TiO_2?

The presence of silicon during annealing causes a degradation of the dielectric properties of TiO_2 [1219]. This degradation is supposed to originate from retarded crystallization [1726, 1727] and intermixing with SiO_2 [1728] forming a SiO_2 layer between the silicon substrate and TiO_2 layer [1219]. Mosaddeq-ur Rahman et al. reported about the formation of an additional SiO_2 layer between TiO_2 and silicon after annealing the sample at 800 °C in air for 6 h [1729]. Hauf et al. assumed that silicon acts as a reducing agent by forming SiO_2 during annealing at 1000 °C in an inert atmosphere [1730]. The diffusivity of silicon in TiO_2 films is given by

$$D_{si} = 5.3 \cdot 10^{-19} \exp\left(-32.6 / R_g T\right) \text{m}^2\text{s}^{-1}. \tag{9.2}$$

(the authors of the original publication use 32.6 instead of 32,600 in the argument of the exponential function but the value of 32.6 does not fit their reported experimental data very well; it is not clear if this correction displays the expression for the diffusivity correctly). This results in roughly 5×10^{-21} m^{2}s^{-1} at 850 °C [1731]. Thus, the calculated penetration depth $2\sqrt{Dt}$ for a processing time of 2 h is roughly 12 nm. The relatively small activation energy for silicon diffusion of 32 kJ/mol is unlikely for bulk diffusion and supposed to be a hint for grain boundary diffusion [1731].

But not only silicon is diffusing into TiO$_2$. Hocine and Mathiot reported about the diffusion of titanium in silicon between 950 °C and 1200 °C [1732]. Besides absorption, the melting temperature is a crucial property for laser-induced melting. Silicon melts at 1410 °C while anatase TiO$_2$ melts at 1802 °C (assuming that anatase TiO$_2$ is heated up so quickly that there is no time for the ART). One has to consider that the melting temperature of a thin grainy anatase film is lower than the melting temperature of bulk TiO$_2$. In addition, the melting temperature of a Ti–Si–O blend differs from the melting temperature of the silicon substrate and the anatase TiO$_2$ film as well. The phase diagram reveals an eutectic temperature of 1330 °C for a titanium:silicon ratio of 1:3 (16% silicon in titanium) and 1:1, which is lower than the melting temperature $T_{m,Si} = 1410$ °C and $T_{m,\beta Ti} = 1670$ °C of pure silicon and β-titanium, respectively. In an analogous manner, Ti–Si–O phases are formed if TiO$_2$ is annealed on silicon substrates. The phase diagram of these phases is shown in Figure 9.2. The most energetically preferred diffusion at elevated temperatures is that of titanium and oxygen into silicon [1225]. Due to the resolution limit of EDX, the thickness of such blends is not determined in this work.

Last but not least, there might be TiO$_{2-x}$ phases close to the interface in rutile films resulting from incomplete oxidation of titanium [971].

As displayed in Figure 9.3, titanium–oxygen phases have relatively high melting temperatures between 1700 °C and 1885 °C which are in the same range than the melting temperature of anatase $T_{m,ana} = 1802$ °C [1737]. The silicon–oxygen phases have slightly lower melting temperature of about $T_{m,SiO} = 1713$ °C [1738].

The absorption properties of mixed layers might be completely different. What is known so far is that silicon doped TiO$_2$ absorbs visible light efficiently [1739, 1740].

> At elevated temperatures, atoms begin to diffuse across the interface between the substrate and the covering TiO$_2$. This results in mixed phases with different physical properties.

9.1.2 Light In-Coupling in Anatase and Rutile TiO$_2$ Films

An experimental analysis of the optical properties of ANA-SPU-850 and RUT-SPU-850 was performed by ellipsometry and the results are represented in Figure 9.5. The shown refractive index and extinction coefficient are fitted assuming that there is a homogeneous TiO$_2$ layer on silicon with a clear interface. Effects rising from mixed phases, surface roughness, and grain boundaries are neglected. Hence, the data represent an effective medium and might differ from properties of perfectly homogeneous and flat TiO$_2$ layers on silicon. The refractive index at 532 nm is slightly smaller compared to Jellison et al. and Miao et al.. And although absorption of visible light in polycrystalline rutile films is known, it is notable that the fitting of the extinction coefficient results in a local maximum around 460 nm. Absorption results from defect states in the bandgap and usually it decreases with decreasing photon energy [748]. Since the value of the absorption coefficient at 460 nm is very low compared to values for UV light the presented extinction coefficient might be affected by discrepancies between the fitting model and the investigated layer system.

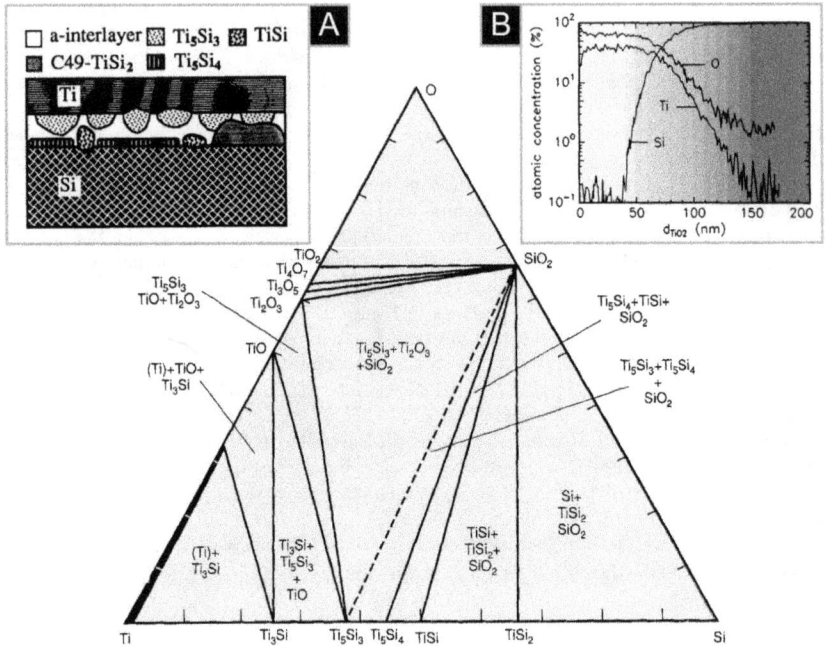

Figure 9.2: *An isothermal section of the Ti–Si–O phase diagram at 800 °C. Reprinted from [1733] with permission of Springer. Inset A) Formation of different Ti–Si phases at the titanium/silicon interface indicating that there is an exchange of atoms between both layers at elevated temperatures. Reprinted from [1725], with permission of Springer. Inset B) secondary neutral mass spectroscopy (SNMS) measurement displays the penetration of titanium and oxygen into silicon. The 50 nm TiO$_2$ film was made by metal organic vapor deposition (MOCVD) on a (100) silicon wafer followed by annealing at 900 °C for 6 h in oxygen atmosphere resulting in a polycrystalline anatase film. Reprinted from [1225], Copyright 1992, with permission from Elsevier.*

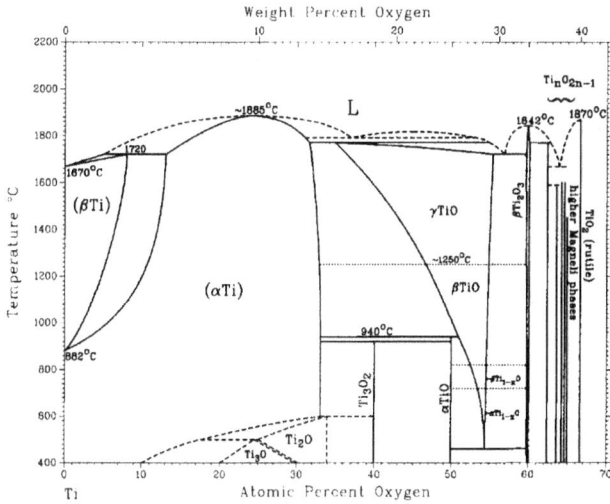

Figure 9.3: *Phase diagram of the Ti–O system. Reprinted from [1734], with permission of Springer. A similar diagram is presented by Wahlbeck and Gilles [1735].*

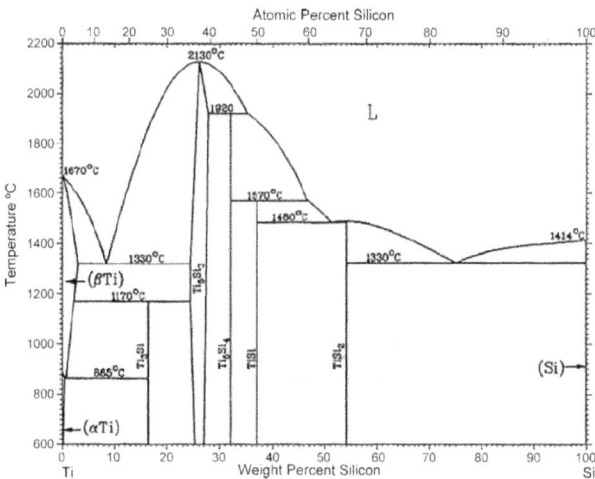

Figure 9.4: *Phase diagram of the Ti–Si system. Reprinted from [1736], Copyright 2006, with permission from Elsevier.*

And because of the small values of the extinction coefficient at 532 nm, these discrepancies do not affect the calculated reflectance represented in Figure 9.5 C and D significantly. Similar to the fitting model for the optical properties the calculated reflectance is based on a homogeneous TiO_2 layer and interflections are taken into account. Surface reflection limits the total absorbed energy. It is possible to reduce this loss mechanism by adding an additional SiO_2 layer on top of the TiO_2. The refractive index of SiO_2 is in between the refractive indices of air and TiO_2 and consequently, a cascade of refractive indices with relative low steps is formed reducing the reflection and increasing the absorbed energy. In this work, SiO_2 layers with a thickness of 5, 20, and 80 nm are applied. Besides the cascade of refractive indices, the reflection is influenced by interference effects inside the layer system as demonstrated in Figure 9.5 D.

It is reasonable that ANA- and RUT-SPU-850 systems do not consist of homogeneous layers as described in the last section. There might be local gradients of the refractive index and extinction coefficient due to mixed phases and because of tilted facets of the grains the assumption of normal incidence might be too imprecise as well. In the following, light in-coupling in ANA- and RUT-SPU-850 systems is discussed beginning from the simple assumption that light which is transmitted through the surface can be absorbed in principle. Step by step, more details are discussed in order to receive a complete idea where light is reflected, transmitted, or potentially absorbed. The physical origin of absorption is discussed in detail in the next sections.

Light has to be transmitted through the surface of the sample in order to be absorbed inside the material. Thus, the reflectivity of the sample is almost as important as the absorption. Reflection occurs on any interface and it increases with an increasing difference of the refractive index of both adjacent materials. In this work, monochromatic light with wavelengths of 532 nm and 266 nm is applied. Optical properties such as the refractive index n, the extinction coefficient κ, and the transmittance° Ξ^T collected from different reports are shown in Table 9.1. By the experimentally determined values for transmittance one gets a rough idea how much energy is available for absorption inside the layers below the surface. The absorption itself is discussed at a later date. The transmittance Ξ^T of a flat interface for normal incidence is given by

$$\Xi_i^T = 1 - \left(\frac{(n_1 - n_{air})^2 + \kappa_1^2}{(n_1 + n_{air})^2 + \kappa_1^2} \right), \tag{9.3}$$

where $n_{air} \approx 1$ is the refractive index of air, n_1 is the refractive index of the dielectric material and κ_1 is the absorption coefficient. The transmittance through a non-absorbing material with two interfaces is given by the product $\Xi_1^T \cdot \Xi_2^T$.

Due to its high refractive index, blank and polished silicon is supposed to reflect the incident light very well. The transmittance of the air/silicon interface is roughly 0.63. In combination with its relatively low absorption coefficient it is not expected that melting is induced for low beam energies at 532 nm laser radiation.

The transmittance of the air/TiO_2 interface is about 0.75 to 0.83 as represented in Table 9.1. Since TiO_2 is a bad visible light absorber most of the transmitted photons arrive the TiO_2/silicon interface. By adding a TiO_2 layer on top of the silicon the total transmittance from air to silicon is increased from 0.63 up to 0.81 as displayed in Table 9.2. Related to the low absorption coefficient of silicon at 532 nm, most of the transmitted light is not absorbed close to the TiO_2/silicon interface. But related to the large penetration depth in silicon the absorbing volume is quite large and in combination with the good thermal conductivity of silicon one needs a lot of laser power to induce a melting process. As a consequence, the melting would appear rather in a large volume than in distinct positions for sufficiently high laser powers.

For thin films, another effect has to be discussed. The high reflectivity of the polished silicon substrate and the covering titanium–oxide film cause thin film interference. A standing electromagnetic wave is formed in the covering film. Assuming perpendicular incidence the field

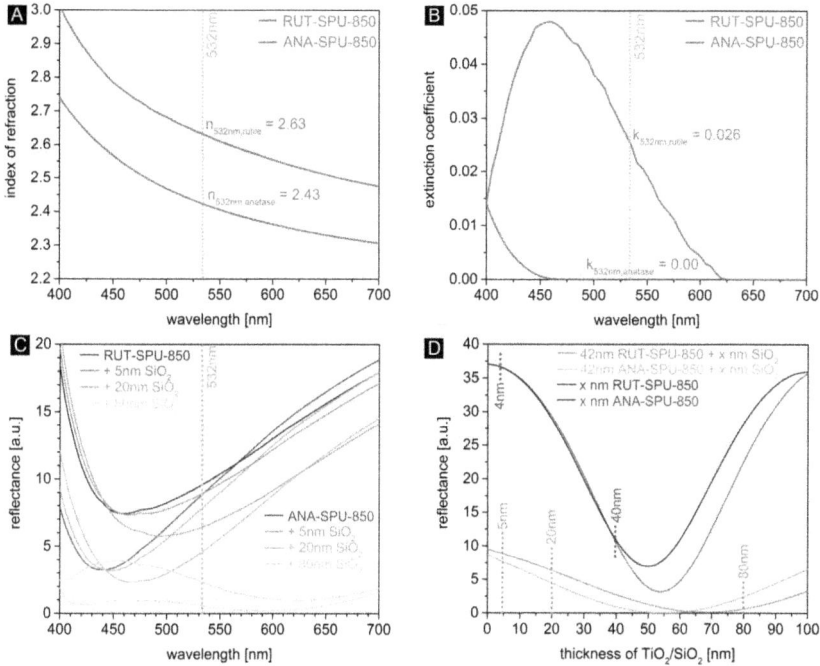

Figure 9.5: *Refractive index n (A) and extinction coefficient (B) of ANA-SPU-850 and RUT-SPU-850 obtained by ellipsometry (details in Chapter 4). The behavior of the extinction coefficient of rutile TiO_2 is not expected. In particular, the higher absorption compared to anatase, the peak at 460 nm, and the decreasing absorption towards smaller wavelengths is unphysical. Most likely, this is related to discrepancies of the real system from the basic conditions that are assumed for the employed Tauc-Lorenzt dispersion formula. The calculated reflectance as a function of wavelength for differently thick SiO_2 cover layers (C) and as a function of the thickness of the TiO_2 film and the covering SiO_2 layer on 42 nm thick TiO_2 films (D) using the optical properties represented in Panel A and B. The simulations are based on a relatively accurate and commonly applied model using a transfer matrix formalism [1743]. This formalism considers the interference of coherent transmitted and reflected waves at each interface in the multi-layered system [1744, 1745].*

	n_{532nm}	n_{266nm}	k_{532nm}	k_{266nm}	Ξ^T_{532nm}	Ξ^T_{266nm}	references
silicon	4.1420*	1.9974*	0.0324*	4.4428*	0.627	0.056	[1716]
SiO$_2$	1.4799*	1.5180*	0.0000*	0.0009*	0.963	0.958	[1746]
anatase TiO$_2$	2.98	1.7	0.000	3.3	0.753	0.099	[1739]
anatase TiO$_2$, bulk	2.5o/2.4e	3.2o/2.6e	0.0o/0.0e	1.6o/1.4e	0.826o/0.830e	0.561o/0.606e	[1741]
anatase TiO$_2$, film	2.6	3.5	0.0	1.5	0.802	0.569	[1741]
anatase TiO$_2$, film	2.6	2.5	0.0	1.8	0.802	0.506	[1742]
anatase TiO$_2$, film	2.43		0.00		0.826		this work
rutile TiO$_2$	3.04	2.3	0.036	1.36	0.745	0.611	[1739]
rutile TiO$_2$, bulk	2.6o/2.9e	2.4o/2.5e	0.0o/0.0e	1.7o/2.9e	0.802o/0.763e	0.525o/0.271e	[1741]
rutile TiO$_2$, film	2.8	2.9	0.0	2.0	0.776	0.482	[1742]
rutile TiO$_2$, film	2.63		0.026		0.798		this work
TiSi$_2$	2.27	1.62	2.70	2.22			[1739]
Ti$_5$Si$_4$, film	1.8		2.2				[1740]
titanium, bulk	1.8237*	0.21178*	3.0106*	1.5768*	0.151	0.001	[1747]
titanium, film	1.9		2.9		0.180		[1740]

Table 9.1: *Refractive indices* n, *absorption coefficients* κ, *and transmittance* Ξ^T *(for vacuum* $\rightarrow$ *material transition) of layer materials used in this work.* *Values are calculated using the tool on http://refractiveindex.info/.* "o" *and* "e" *label the ordinary and extraordinary values. The films made by Jellison et al. and Miao et al. are fabricated by reactive sputter deposition of titanium [1741, 1742]. All values are determined at room temperature using Equation 9.3 and neglecting interflection.*

is maximized for an optical film thickness of $\lambda/2$ in the middle of the film and the reflection is maximized. In contrast, the transmittance is maximized for an optical film thickness of $\lambda/4$ which acts like an antireflection coating. Due to the high refractive index of TiO$_2$, interference effects appear already for blue light in films with an effective thickness of less than 40 nm. The optical cavity increases the total absorbed energy in the TiO$_2$ layer and graduates the low absorption coefficient of TiO$_2$ partly. If intra-bandgap states are present, the absorption becomes even more significant.

It is conspicuous that the absorption of titanium–silicon phases is increased dramatically towards pure TiO$_2$. Since the absorption coefficients of pure silicon and pure TiO$_2$ are very low for the applied 532 nm light, these mixed phases might have an important role for light absorption. As discussed above, the penetration depth of silicon in TiO$_2$ is at least 12 nm. Considering that the absorption coefficient for mixed phases is in the order of pure titanium it becomes reasonable that these phases work as an efficient light absorber, especially if the film thickness satisfies the above mentioned $\lambda/4$ which corresponds to a thickness of roughly 51 nm in rutile and 55 nm in anatase TiO$_2$. In addition, one should remember that a lot of these phases have a lower melting temperature than pure silicon or TiO$_2$. As a consequence, one expects the melting effect to appear at the silicon/TiO$_2$ interface.

Light In-Coupling through Air/RUT-SPU-850 Interface

Above, more general requirements for the absorption of energy from incident light are discussed. But in addition, one has to consider the individual structure of the used TiO$_2$ layers.

And because of the higher refractive index of rutile compared to anatase TiO$_2$ the transmittance of the air/rutile and air/rutile/silicon interface is only 0.75 and 0.73, respectively. Both values are lower than those for anatase instead of rutile TiO$_2$ layers as shown in Table 9.2. Thus, less energy arrives at the TiO$_2$/silicon interface, where the melting process is supposed to be induced.

But another effect compensates the enhanced reflection. So far, the TiO$_2$ films were treated as a flat effective medium neglecting their topography. In case of RUT-SPU-850, the facets have

	Ξ^T
air/silicon	0.63
air/rutile	0.75
air/rutile/silicon	0.73
air/rutile/anatase	0.74
air/rutile/anatase/silicon	0.70
air/anatase	0.81
air/anatase/silicon	0.77
air/anatase/rutile	0.81
air/anatase/rutile/silicon	0.79

Table 9.2: *Transmittance Ξ^T of the applied single and double layer constellations used in this work neglecting the absorption in TiO_2 and assuming flat surfaces. All values are determined at room temperature using Equation 9.3.*

a diameter between 10 nm and 50 nm and differ significantly from grains in ANA-SPU-850 that are discussed in the next section. Due to the pyramid-like grain structure of rutile the microscopic surface is tilted towards normal incidence. The transmittance through a rough surface is a complex problem, but in general the transmittance is enhanced for rough TiO_2 surfaces on silicon [795, 1748, 1749]. Basically, this behavior results from two effects. First, a rough surface is seen by light with wavelengths, which are significantly larger than the surface structures, as a mesoporous material having a lower effective refractive index than bulk material. As a consequence, a refractive index cascade is created reducing the reflectance. Second, multi-reflections appear on grainy surfaces guiding the light into the sample [1748]. This effect decreases the reflectance of RUT-SPU-850 compared to ANA-SPU-850.

Finally, one has to consider the effect of different silicon diffusion constants in rutile and anatase resulting in a thinner or differently mixed Ti–Si–O phase. But a lower silicon diffusion rate during the fabrication of RUT-SPU-850 is unlikely since the raw material for these films is pure titanium and it was mentioned before that silicon diffusion appears already for relatively low annealing temperatures and short annealing times. The diffusivity of silicon in TiO_2, which is the raw material for ANA-SPU-850, is supposed to be lower compared to pure titanium. According to this feature, the mixed phase should be thinner between ANA-SPU-850 and silicon than between RUT-SPU-850 and silicon. But one has to consider that rutile TiO_2 is formed quickly during the annealing of titanium at 850 °C in oxygen-rich atmosphere. And due to the dense packaging of atoms in rutile crystals the diffusion rate is supposed to drop significantly in the beginning of the annealing process. Furthermore, it cannot be excluded that the composition of the mixed phase is different for both cases resulting by chance in a thermally more stable mixed phase for the RUT-SPU-850/silicon system compared to the ANA-SPU-850/silicon system. Further investigations need high resolution element mapping of the interface region between the TiO_2 layers and the silicon substrate which goes beyond the scope of this work.

Light In-Coupling through Air/ANA-SPU-850 Interface

In contrast to rutile films, anatase grains have almost flat surfaces and thus, the micro- and macroscopic angles of incidence are approximately equal. As a consequence, topographic effects are supposed to be less significant on ANA-SPU-850 compared to RUT-SPU-850. The transmittance of air/ANA-SPU-850 and air/ANA-SPU-850/silicon is 0.81 and 0.77, respectively and, in particular, higher than for rutile films. And in contrast to rutile TiO_2, the refractive index of anatase TiO_2 is more similar to the refractive indices of many Ti–Si–O phases as shown in Table 9.1 or reported by Pulker et al. [1750]. So light passes the air/anatase interface is transmitted more likely into a highly absorbing mixed phase at the TiO_2/silicon interface.

As discussed above, the Ti–Si–O interlayer should be thinner for ANA-SPU-850 compared to RUT-SPU-850. But besides the thickness, the composition and element distribution might differ between both films. So it cannot be claimed that the supposed thinner Ti–Si–O interlayer results in a lower total absorption in general.

Light In-Coupling through Air/RUT-/ANA-SPU-850 Double Layer Interfaces

Although it goes beyond the scope of this work to apply high resolution element mapping in order to find out the exact element distribution at the TiO_2/silicon interface, an additional experiment could help to shed light on the effects contributing to the laser-induced melting at the TiO_2/silicon interface. For this experiment, RUT-SPU-850/ANA-SPU-850/silicon and ANA-SPU-850/RUT-SPU-850/silicon double layers are applied.

The calculated transmittance is 0.74 and 0.81 for the air/rutile/anatase and air/anatase/rutile system, respectively. Thus, the latter constellation is expected to be preferred for laser-induced melting, especially because it has the same transmittance as the air/anatase interface. But the preparation of the RUT-SPU-850/ANA-SPU-850 is limited because the deposited TiO_2 which is the raw material for ANA-SPU-850 films converts likely into rutile TiO_2 due to the presence of the subjacent rutile film. This emphasizes the role of grain boundary and surface-induced growth of rutile TiO_2 as discussed in Chapter 2.3. Thus, the RUT-SPU-850/ANA-SPU-850 film is more like a thick rutile film. To prevent this effect a 4 nm thin silicon layer was introduced between the different TiO_2 layers always.

The transmittance of TiO_2 layers depends on their refractive index and surface roughness. Since the surface of rutile consists of pyramidal structures, the transmittance is expected to be higher than for anatase TiO_2. However, the refractive index of rutile is larger which provides an enhanced reflectivity.

9.1.3 Absorption of Incident Light

As demonstrated above, light-incoupling is a complex process. Therefore, it is reasonable to measure the absorbance of the applied samples in an integrating sphere. The results are represented in Figure 9.6. The absorption of ANA-SPU-850 is almost fitting with the reflectance calculated from ellipsometry data in Figure 9.5: low reflectance is related to high absorbance. For RUT-SPU-850 the trend is similar but the measured absorbance maximum is blue-shifted compared to the calculated reflectance minimum. This might be related to effects that are not considered in the fitting model applied for the ellipsometry data as listed above. Here, the origin of absorption is discussed in detail.

Absorption of visible light occurs in Ti–Si–O mixed phases, TiO_{2-x} resulting from incomplete oxidation of titanium in rutile films, and the subjacent silicon. Optical properties of some typical mixed phases are represented in Table 9.1. In this section, the physical origins of this absorption should be discussed.

The typical light absorption in semiconductors originates from the excitation of single electrons from the valence to the conduction band. Although the bandgap of anatase TiO_2 is 3.2 eV, absorption in the visible light regime is easily introduced by defects, interface and surface states [613]. Diffusion of silicon into TiO_2 causes additional intra-bandgap states especially at the TiO_2/silicon interface. As a consequence, many mixed Ti–Si–O phases show a significantly increased visible light absorption compared to single-crystalline TiO_2. $TiSi_2$ is one of the most popular titanium silicides and it is known as a broad range absorber between 360 nm and 800 nm [754, 1751, 1752].

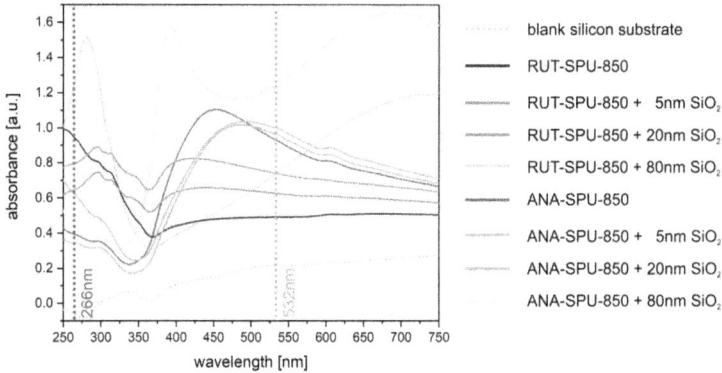

Figure 9.6: *Absorbance of ANA- and RUT-SPU-850 samples without and with different SiO_2 cover layers measured in an integrating sphere.*

In addition, such mixed phases are often highly conductive indicating that the density of free charge carriers is quite high. Hence, for phases such as $TiSi_2$ [1753] and Ti_5Si_4 [1754] light might be absorbed by photon-plasmon coupling as well.

In titanium-doped silicon, deep electronic levels rise within the bandgap. In particular, Chen et al. determined two donor states at $E_d = E_C - 260\,meV$ and $E_d = E_C - 290\,meV$, and an acceptor level at $E_a = E_V + 100\,meV$ [1755]. Mathiot and Hocine reported about similar states at $E_d = E_C - 90\,meV$, $E_d = E_C - 300\,meV$, and $E_a = E_V + 260\,meV$ [1756]. Olea et al. achieved densities of up to $1 \times 10^{18}\,cm^{-3}$ substitutional titanium atoms [1757]. In addition, free electrons in the highly conductive titanium-doped silicon result in efficient light absorption at 530 nm as well [1758]. Light absorption is increased for anatase TiO_2 films on SiO_2 that were annealed at 1000 °C in oxygen rich atmosphere [660]. On the one hand, the transmitted energy density is decreased by scattering due to the increased surface roughness after annealing [588, 592]. But the reported transmittance is also decreased due to additional absorbing states created by the increased level of Ti^{3+} interstitials [136, 588, 589].

The bandgap of pure anatase and rutile TiO_2 suggests a negligible absorption in the visible regime. Consequently, the absorption is caused by the subjacent silicon substrate or Ti–Si–O phases close to the interface.

9.1.4 Heat Dissipation in the Silicon/TiO_2 Interface Region

Fast heat dissipation due to high thermal conductivities is inappropriate for efficient local laser-induced melting. Unfortunately, the thermal conductivities of bulk anatase (8.5 W/mK [1759, 1760]), rutile (6.531 W/mK at 373 K [295]), and silicon (156 W/mK at 300 K [1761]) are relatively high at room temperature.

But fortunately, there are effects that decrease the thermal conductivity. Firstly, grain boundaries, interfaces, and crystal defects reduce the thermal conductivity by phonon scattering and low thermal conductivity in certain crystal directions as it was already discussed by Xia et al.

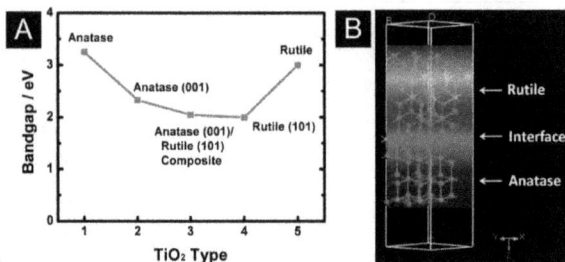

Figure 9.7: *A) Bandgap energies of anatase (001) and rutile (101) surfaces compared to bulk anatase and rutile. B) Schematic drawing of the temperature amplitude in different regions close to the rutile-anatase interface. Panel A and B reprinted with permission from [613]. Copyright (2013) American Chemical Society.*

and mentioned in context with the ART in Chapter 2.3.1 [613]. Furthermore, in anatase TiO_2 nanocrystals with diameters below 20 nm, phonon confinement localizes induced heat [554]. Secondly, phonon-phonon scattering increases with increasing temperature in TiO_2. Thus, heat dissipation during laser-induced heating is a self-limiting process [1762].

> The heat dissipation is limited by phonon-phonon and phonon-grain boundary scattering.

9.1.5 Melting Parameters

Finally, one has to consider the melting points of the different components inside the TiO_2 films, silicon substrate, and mixed phases. All experiments were performed at ambient pressure and thus, only the melting temperature is discussed. During a slow temperature rise in the order of 100 K/s or less as it is typically applied in resistive and RTP ovens, anatase and brookite TiO_2 convert into rutile before their melting temperature is exceeded. But the conversion takes time and laser-induced heating by non-radiative electron relaxations or plasmon decay happens within the order of nanoseconds or even faster [1117, 1489]. Assuming that the melting temperature of roughly 2000 K is reached after 20 ns the corresponding temperature rise is 1×10^{11} K/s! Thus, it is reasonable to take the melting temperatures of all involved crystals structures into account. The calculated melting temperature $T_{m,\infty}$ of bulk rutile, anatase, and brookite TiO_2 are 2143 K, 2075 K, and 2098 K, respectively [1737]. But similar to the ART (Chapter 2.3.1), for particles below 20 nm surface tension becomes important and the melting temperature becomes size- and shape-dependent. According to Guisbiers et al., the ratio of the shape-dependent melting temperature T_m is given by

$$\frac{T_m}{T_{m,\infty}} = 1 - \frac{\alpha_{sh}}{2L}, \qquad (9.4)$$

where $\alpha_{sh} = 2AL(\gamma_s - \gamma_l)/(V \Delta H_{m,\infty})$ is the shape parameter. L, A, and V is the shortest dimension, the surface area, and the volume of the nanoparticle, respectively. $\Delta H_{m,\infty}$ is the melting enthalpy for the bulk material and γ_s and γ_l are the surface energies of the liquid and solid phases [1737].

The standard thickness of RUT-SPU-850 and ANA-SPU-850 films is 40 nm, which corresponds to L in almost all grains in both kind of films. Only in RUT-SPU-850, there are a few smaller grains having diameters down to 20 nm. Because the thickness of both films is nearly the same, A and V are proportional to the squared diameter d of the grains. Consequently, T_m depends only on the thickness d^* and is given by $T_m/T_{m,\infty} \approx 1 - C^*/d^*$, where C^* is a constant depending on intrinsic properties. The equality in Equation 9.4 has to be replaced by an approximated relation since the topography of grains in RUT-SPU-850 and ANA-SPU-850 films differ. Hence, only the melting temperature of RUT-SPU-850 films might be lowered slightly by surface energy effects and it seems to play a minor role for laser-induced melting in this work. For films with smaller grains the size dependency of the films might be more important.

Similar to the ART, impurities are able to stabilize or destabilize the TiO_2 films and the silicon substrate. Ti_nO_{2n-1} melts at a lower temperature than single-crystalline rutile TiO_2. But since the lowest reported melting temperature is 1700 °C, the effect of oxygen vacancies is not significant [1734]. Si^{4+} doping supported by the nearby silicon substrate does not create any oxygen vacancies in TiO_2 and thus, it stabilizes the TiO_2 film. But for sufficiently high silicon doping titanium–silicon phases with different crystal structures and hence, different melting temperatures appear. An overview on the specific melting temperatures is given in the phase diagram in Figure 9.4. There are many phases with lower melting temperatures than anatase TiO_2 and even than pure silicon. If high melting temperature phases such as Ti_5Si_3 with $T_{m,\infty} = 2403\,K$ are present, they do not prevent laser-induced melting necessarily as long as there are enough other mixed phases with low melting temperatures. Most of these mixed phases have melting temperatures of less than 1873 K and melt at similar temperatures as the subjacent silicon [1736].

> The melting behavior of the TiO_2/silicon system is affected by the appearance of mixed Ti–Si–O phases and grain boundaries.

9.1.6 Recrystallization Process

The composition and behavior of the TiO_2 melt gained from laser-induced heating of thin polycrystalline TiO_2 films on silicon is unknown so far. But there are some ideas what happens when the melt cools down and goes below the melting temperature. At first sight, one might suggest that the solidification results in rutile TiO_2 because no anatase phase is observed above 1900 K when heating up a TiO_2 film. But one has to consider that the nucleation kinetics in a TiO_2 melt is driven by different forces compared to the ART. First of all, the solidification depends on the difference of the Gibbs free energy ΔG_V^L between the liquid and solid TiO_2 phases. As discussed in Chapter 2.3, rutile is thermodynamically more stable than anatase due to its lower total Gibbs free energy. Hence, considering Gibbs free energies only the nucleation of rutile TiO_2 should be formed from the melt. But similar to previous discussions, the interface energy between the solid phases and the melt affects the solidification process as well and give rise for the creation of thermodynamically less stable phases as it was reported for α- and γ-Al_2O_3 by McPherson or β-$MoSi_2$ by Fan and Ishigaki [1763, 1764] already.

According to calculations from Li and Ishigaki, the interface energy between liquid TiO_2 and solid rutile or anatase TiO_2 is $\gamma^{L,rutile} = 931.5\,\mathrm{ergs/cm^3}$ or $\gamma^{L,anatase} = 379.2\,\mathrm{ergs/cm^3}$, respectively. Considering Gibbs free energies and interface energies, the extended ratio of Gibbs free energies is given by

$$\frac{\Delta G^{*,rutile}}{\Delta G^{*,anatase}} = \left(\frac{\gamma^{L,rutile}}{\gamma^{L,anatase}} \right)^3 \cdot \left(\frac{\Delta G_V^{L/anatase}}{\Delta G_V^{L/rutile}} \right)^2 . \tag{9.5}$$

211

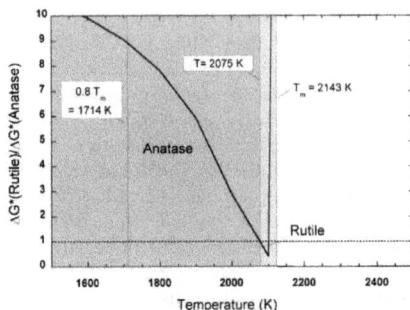

Figure 9.8: *Temperature-dependent $\Delta G^*(rutile)/\Delta G^*(anatase)$ ratio showing the preferred morphology after solidification. This ratio includes the bulk Gibbs free energies ($\Delta G(rutile/anatase)$) and the surface energies between the rutile/anatase phase and liquid TiO_2. Reprinted from [1765], Copyright 2002, with permission from Elsevier.*

This ratio is displayed in Figure 9.8. The steep increase around the melting temperature of rutile TiO_2 ($T_{m,\infty} = 2143\,\text{K}$) is caused by the convergence of the Gibbs free energy of the liquid phase and rutile TiO_2. The minimum at 2143 K is the pseudo-melting temperature of anatase. Thus, anatase could not be formed above this temperature although $\Delta G^{*,rutile}/\Delta G^{*,anatase} < 1$. In summary, anatase nucleates below 2075 K and rutile between 2075 K and 2143 K.

A direct consequence is that the formed morphology depends strongly on the cooling rate and the degree of supercooling. While rutile is formed preferentially under equilibrium conditions slightly below its melting temperature, anatase is formed as a result of high cooling rates and hence, significant supercooling [1766]. McPherson reported that usually a supercooling temperature of $0.82T_m = 1714\,\text{K}$ is needed for solidification even for slower cooling rates. As a consequence, anatase is the most likely outcome from the nucleation process of small TiO_2 droplets. But nevertheless, rutile is found as well [419, 1767–1769]. Li and Ishigaki suggest that rutile is formed due to insufficient cooling mechanisms or heat released from fusion during the nucleation [1765]. In this case, the temperatures could be sufficiently high for such a long time that the nucleation process takes place in equilibrium resulting in rutile or already formed anatase is converted in rutile by an ART. Thus, spherical particles have often rutile cores and anatase shells because the cooling rate is much lower in the core than close to the surface [1765].

The morphology of resolidified TiO_2 depends mainly on the cooling rate. Only at very high temperatures, the formation of rutile is provided. If the melt is cooled down too quickly, there is no time for recrystallization in the rutile morphology and anatase is obtained.

9.2 Results and Discussion of Laser Lithography Techniques

The previous discussion motivates two different laser based lithography techniques: laser-induced oxidation lithography (LIOX) and laser-induced melting lithography (LiMeL).

9.2.1 Laser-Induced Oxidation Lithography

A 532 nm cw laser is used in combination with a controllable xy-stage to heat up a 40 nm thin titanium film on silicon locally. As a metal, titanium absorbs light with the applied wavelength efficiently as shown in Figure 9.9. The locally induced heat increases the oxygen diffusion rate and increases the oxidation rate [1641, 1642]. The heating rate is orders of magnitude higher compared to any oven and the formation of rutile TiO$_2$ is provided in the same way as described for the formation of polycrystalline rutile films using the RTP oven in Chapter 5. But even if anatase particles are created as described by Nakamura et al., they might act as nucleation seeds for rutile TiO$_2$ nanorods as described in Chapter 8.6 or they become unstable at higher temperatures [1770, 1771].

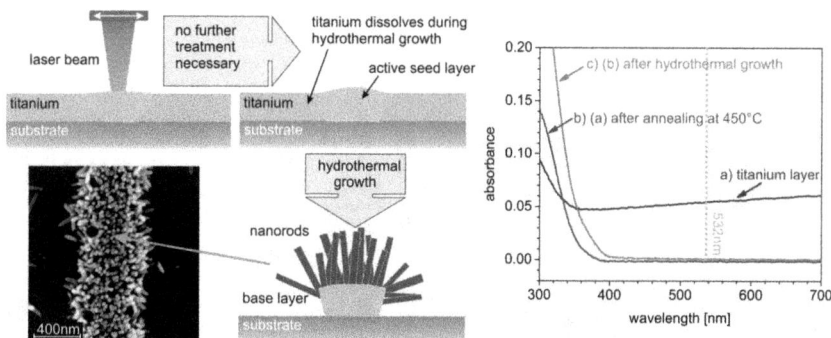

Figure 9.9: *Schematic drawing of generating confined NRAs by LIOX. The titanium film is heated locally by the laser beam in air. This results in an enhanced oxygen diffusion and the film becomes oxidized locally. Since oxidation is accompanied by a crystallization process a part of the created TiO$_2$ crystallizes with the rutile morphology giving rise to the growth of nanorods during the hydrothermal growth process. On the right side, the absorption of the initial titanium layer, RUT-SPU-450 (anatase), and RUT-SPU-450 after hydrothermal growth is displayed.*

During the hydrothermal growth, the titanium film is dissolved or desorbs from the silicon substrate [1772]. Only the oxidized parts remain as a patterned seed layer for rutile nanorods as demonstrated in Figure 9.9 and 9.10 A-D.

Since oxidation is a top-down process, this technique supports nearly any kind of flat and solid substrate. The heat is generated at the surface of the titanium film in particular. Therefore, also heat sensitive substrates such as organic solids are suitable. In this work, the method was applied to polyether ether ketone (PEEK) for instance. In comparison to conventional optical or electron-beam lithography, this method needs less fabrication steps. Besides rutile TiO$_2$ nanorods, this technique is applicable for any material that can be oxidized and provides an epitaxial growth of nanocrystals. This includes at least a few metal oxides and their doped variations.

213

Unfortunately, the simplicity of this method provides some disadvantages as well. If the samples are stored under ambient conditions or any other oxygen or water rich atmosphere a global oxide layer is formed that becomes resistant in the hydrochloric acid and, even worse, provides the growth of nanorods anywhere as demonstrated in Figure 9.10 E and F. Heat generation and oxygen supply is maximized at the titanium surface and therefore, a gradient of the oxygen density is established between the surface and the titanium/substrate interface. As a consequence, the oxidation process remains incomplete at the interface and the degradation of the titanium film in this region leads to the ablation of the whole NRA grown on the surface as shown in Figure 9.10 G. In addition, this technique does not provide an intermixing of silicon and TiO$_2$ at the interface by an annealing process as it is applied for RUT-SPU-850 films for instance. In the following sections, the function of such mixed phases as an adhesion layer will be emphasized. So this method should work better for thin films where a continuous oxide region between the surface and interface is created. But one has to consider that the absorbed energy in the titanium film decreases for decreasing film thicknesses. The absorption coefficient for the subjacent silicon is much lower than for titanium. Hence, in case of ultrathin titanium films, most of the heat is generated in a relatively large volume inside the silicon far away from titanium film. This fact indicates the existence of an optimal film thickness and laser power. Besides the applied wavelength, the resolution is limited by the lateral heat dissipation which cannot be neglected for a metal such as titanium. Thus, the minimal line width that was achieved in this work was approximately equal to the applied wavelength.

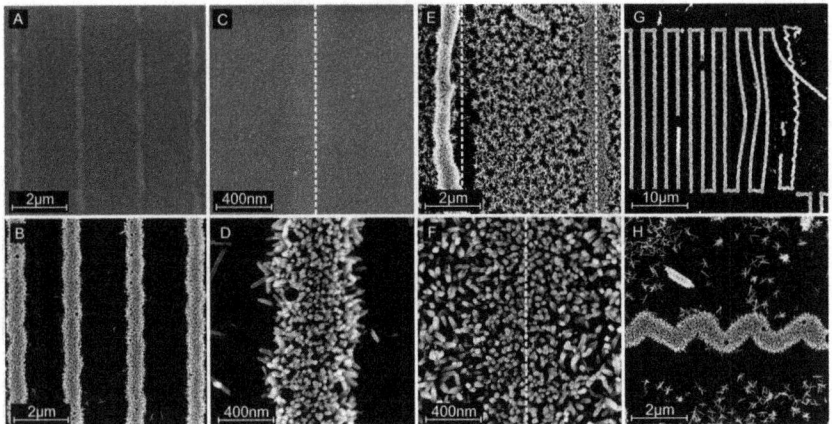

Figure 9.10: *SEM images of confined NRAs created by laser oxidation on a* 40 nm *thick titanium film. The gained structure is shown before (A,C) and after (B,D) hydrothermal growth with different magnifications. Remains of the titanium in the unaffected regions oxidize during storing or in the growth solution and act as nucleation seeds for nanorods as well (E,F). The adhesion of as-deposited titanium on silicon is relatively weak and the gained NRA pattern lines are easily removed by sonication or even rinsing (G). A general problem are oscillations of the setup resulting in sinusoidal instead of linear pattern lines (H). The yellow dashed lines are a guide to the eye and mark the path of the laser beam.*

A more general and purely technical problem is the stability of the stage. If vibrations of the underground are not damped sufficiently one gets a line shape as shown in Figure 9.10 H, for instance. But this problem has to be solved for all presented techniques in this work. In particular, for laser-based lithography methods the technical solution for a stable focusing is found in CD- and DVD-writer devices. A related technique to LIOX is oxidation by electrolysis by applying a sufficient potential between a CAFM tip and the sample covered with a thick water film [1773]. Due to the small tip radius of AFM tips, the resolution is better compared with LIOX. But since there is no heat the oxidation is restricted much more to the film surface as it is the case for LIOX. Thus, electrolysis-induced oxidation works for ultrathin films only and might be a good complementary technique for LIOX. Because LIOX is a contact free method, there is no tip consumption and the sampling rate might be higher.

A titanium film is heated locally with a cw laser. The heat provides the oxidation process and TiO_2 is created in defined positions. During the hydrothermal growth, the remaining titanium dissolves and the oxidized parts provide the hydrothermal growth of rutile nanorods.

9.2.2 cw Laser-Induced Melting Lithography (CiMeL)

To avoid some disadvantages of the oxidation technique, another laser-based technique was developed in this work. The basic idea is to exchange the titanium film by an anatase TiO_2 film and exceed the melting temperature by laser-induced heating. For specific cooling rates, the melt solidifies as rutile TiO_2 and provides an epitaxial growth of rutile TiO_2 nanorods. This process is called cw laser-induced melting lithography (CiMeL) and is accompanied by many interesting physical effects that will be discussed in the following.

9.2.3 High Energy cw Laser-Induced Melting Lithography (HELM)

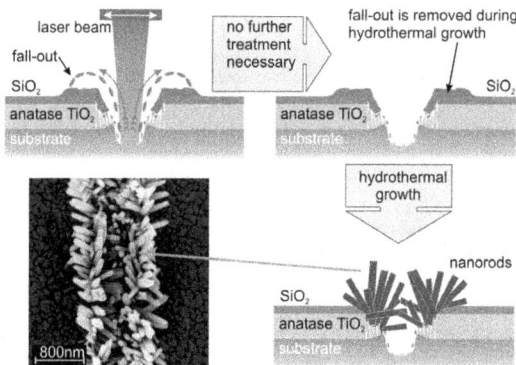

Figure 9.11: *Schematic drawing of generating confined NRAs by HELM. Silicon melts due to laser heating and the covering ANA-SPU-850 films is supposed to be converted into rutile particles acting as seeds during the hydrothermal growth. The fall-out might originate from inert gas condensation of evaporated material or splashes of liquefied matter.*

For relatively high injected light powers, the absorbed energy in the mixed Ti–Si–O phases or the silicon substrate is high enough to trigger melting. The solidified material contains a lot of

215

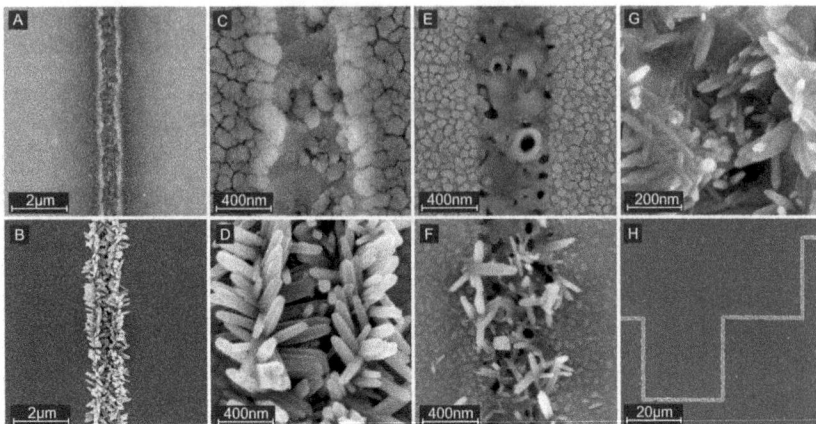

Figure 9.12: *SEM images of confined NRAs created by HELM. The gained structure on* 60 nm *thick films is shown before (A,C) and after (B,D) hydrothermal growth with different magnifications. The structure gained on* 5 nm *thin films is shown before (E) and after (F) hydrothermal growth. Panel G shows a zoom-in of an early growth stage inside a pattern line for* 60 nm *thick ANA-SPU-850. A zoom-out of a confined NRA on a* 60 nm *thick ANA-SPU-850 film is shown in Panel H. The dark region besides the trench in Panel A results from the fall-out mentioned in Figure 9.11 and displays not a topographical feature.*

seed crystals, especially at the edges of the fused regions. The hydrothermal growth results in inhomogeneously oriented nanorods as shown in Figure 9.11. The technique is applied to films between 5 and 60 nm and the density of seeds is lower for 5 nm thin films. In contrast to thick films, where the growth of nanorods is preferred at the edges of the pattern line (Figure 9.12 D), nanorods are distributed homogeneously inside the pattern line for thin films (Figure 9.12 F). For thick ANA-SPU-850 films, there are many nucleation sites at the left and right side. By branching effects, the nanorods start growing in any possible direction and occupy the center of the pattern line as well. In Figure 9.12 E, typical features of solidified silicon are seen, namely bubbles and tubes [1774]. These structures indicate that the silicon is heated up above its boiling temperature. Additionally, these deposits might contain TiO_2 particles from the anatase TiO_2 film and the Ti–Si–O phases above the evaporated silicon. Fortunately, these deposits disappear during the hydrothermal growth. It is not completely clear, if they dissolved or are washed off. Deposits from laser evaporation are known to stick tightly at surfaces and therefore, washing off is rather unlikely. But it is reasonable that tiny particles with a high surface-to-volume ratio are less stable in hot hydrochloric acid compared to macroscopic particles or films. The roughness resulting from these structures is responsible for the disorder of nanorods grown in these areas. For the fabrication of structures having a high contrast of the photocatalytic activity, it is necessary to keep the areas around the highly photoactive rutile nanorods inert. This is realized with a thin SiO_2 film for instance. Such additional films do not disturb the function of this technique significantly. In case of SiO_2, it seems that the melting process is even enhanced as discussed in the following sections.

9.2.4 Low Energy cw Laser-Induced Melting Lithography (LELM)

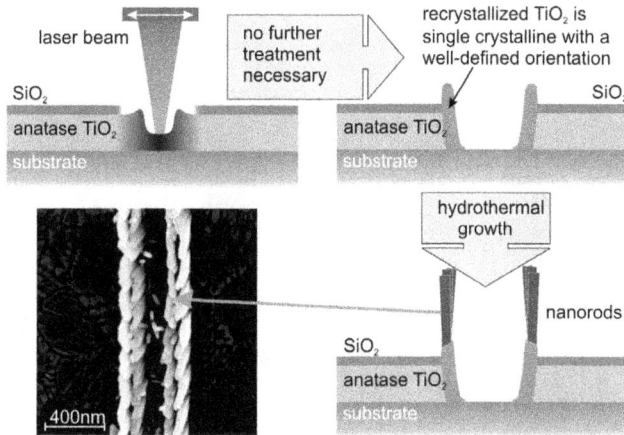

Figure 9.13: *Sketch of generating confined NRAs by LELM. A mixed Ti–Si–O phase is molten by laser heating. The solidified material crystallizes partly in the rutile morphology forming the seed for rutile nanorods. A closer look on the shown double wall (SEM image, bottom left) shows that each wall consists of two nanorod rows with a tiny spacing of roughly* 20 nm. *This is due to the particular crystal structure of the solidified* TiO_2 *bulges.*

For a certain range of injected energies, the situation differs from the above described laser-induced silicon melting. The absorption coefficient of Ti–Si–O phases is higher than that of TiO_2 or silicon. If these mixed phases are sufficiently thick, a significant part of the beam energy is absorbed in these mixed phases. So above a critical laser power, mixed phases at the TiO_2/silicon interface begin to melt. But if the laser power is not too high, the pure silicon below remains mostly solid and unaffected. Dewetting forces the melt to form two bulges at the edges of the molten area as drawn in Figure 9.13. Since the nearby environment of the melt remains hot after the laser spot has continued to the next position, the thermal conductivity of the subjacent silicon is lower compared to silicon at room temperature. As a consequence, the cooling rate is relatively slow and according to Figure 9.8 the melt is supposed to solidify mostly as rutile TiO_2.

The liquid-solid phase boundary moves behind the laser spot along the writing direction. As a consequence, the melt crystallizes almost with the same crystal orientation as the melt that solidified a moment before. Thus, a memory effect of the crystal orientation appears which is seen in the oriented growth of nanorods on the solidified bulges as demonstrated impressively in Figure 9.13 and 9.14. If the melt is interrupted by tiny gaps, spontaneous jumps in the crystal orientation are observed as shown in Figure 9.14 G. The cross-sections in Figure 9.14 F and H prove that the silicon substrate (dark grey) is not affected by the low energy cw laser-induced melting process. The vertical stripes in Panel H originate from FIB milling used to create the cross-sections. The width and shape of the generated nanostructure is controlled by the laser power as shown in Figure 9.14 J. For very low laser powers, the bulge-forming behavior does not appear and a single instead of a double nanorod-wall (shown in Figure 9.13 for the first time) is formed with a width below 100 nm, which is around five times smaller than the applied

Figure 9.14: *SEM images of double nanowalls made by laser-induced melting. Bulges gained from laser-induced melting before (A,C) and after (B,D) hydrothermal growth. Tilted and upright standing nanowalls are shown in Panel E and G, respectively. The cross-sections of these nanowalls (F,H) indicate that laser melting does not affect regions deep inside the silicon substrate. The bright layer is the ANA-SPU-850 film and the darker region below is the silicon substrate. The vertical linear structures in the lower part (bulk silicon) result from FIB milling that was applied to create these cross-sections. Panel J demonstrate the opportunity of available structures by varying the beam energy density. In the right part of the image, the energy density is reduced by defocusing the laser beam. In the left part of the image, it is demonstrated that double walls could be generated for a length of many microns. The length of these highly ordered structures is limited by defects in the recrystallized seed or beam instabilities.*

wavelength. The outstanding resolution originates from the fact that only the central part of the laser beam supplies enough energy to exceed the melting temperature. In particular, the resolution is affected much less by the lateral heat dissipation as it is the case for LIOX. The solidified material is linked strongly to the substrate which is expressed by the high mechanical stability of the grown rods during sonication.

The effect of the beam energy on the melting behavior of ANA-SPU-850 on silicon is represented in Figure 9.15.

An anatase film on silicon is melted with a cw laser (532 nm). A part of the resolidified material provides the growth of rutile TiO_2 nanorods. For low laser powers, the material crystallizes with a defined crystal orientation which is conserved for a few microns along the created line. Consequently, the nanorods resulting from this technique are oriented as well and form long walls. According to the current state of affairs, the presented technique is the only method for the fabrication of position-controlled rutile TiO_2 nanorods resulting in highly ordered nanorods.

9.2.5 Pulsed Laser-Induced Pattern Generation

The cw-laser is useful for individual and complex structures since controlling the xy-stage is quite easy. With an additional control of the z-axis, the focus of the laser beam is withdrawn and approached to the sample and thus, the linewidth becomes adjustable or the structure is even interrupted without blocking the laser. But with increasing size the writing time becomes longer as well. For large area patterning, a technique manipulating the sample just with a single flash – as it is applied for photo-lithography – would be much faster. For such an application, the building plan of the required structure has to be transferred from the real space expressed by the movements of an xyz-stage into the phase space as it is realized for holograms. A hologram provides a 3D distribution of enhanced light intensities that might be used to induce melting of an anatase TiO_2 film for instance. The simplest kind of a hologram is a two beam interference. The result is a 1D pattern consisting of parallel lines. Here, the laser is not focused anymore, but the cross-section of the interfering beams equals the whole area on the sample that has to be patterned. Thus, the beam power has to be increased by a few orders of magnitude which exceed the limits of cw-lasers and calls for the use of pulsed lasers. In this work, a 12 ns pulsed laser beam at 532 nm and 266 nm is used. Details are described in Chapter 4. The effect of laser-induced melting for pulsed lasers is similar in many respects to cw laser-induced melting. But here is no memory effect of the crystal orientation expected since melting, forming and solidifying happen simultaneously on the whole sample providing cooling rate-dependent short distance correlations only. Some interesting dependencies of the melting behavior are investigated and discussed in the following. An overview of the involved effects is given in Figure 9.26.

Effect of the Wavelength, Morphology, and an Additional Thin SiO_2 Layer

The first effect on incident light is the reflection and transmission on the surface. Since the difference between the refractive indices of air and TiO_2 is quite large, the reflection is high as well. Hence, a simple method to inject more light energy into the system is to add an additional layer with a refractive index between the ones of air and TiO_2. A good candidate is SiO_2 with a refractive index of roughly 1.5 in the visible light regime. SiO_2 is a suitable core shell material because it is inert, chemically stable, insulating, and does not absorb low energetic UV radiation and visible light. The simulated reflectance and measured absorption of ANA-SPU-850 and RUT-SPU-850 films was discussed above and is represented in Figure 9.5 and 9.6. Figure

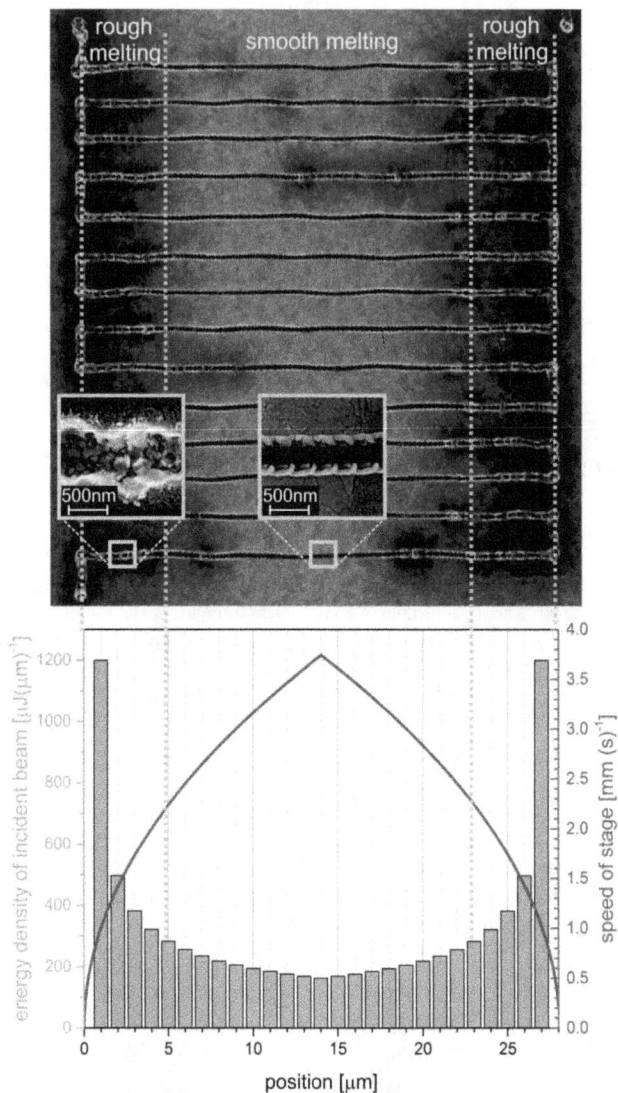

Figure 9.15: *Effect of the beam energy on cw laser-induced melting: The incident beam energy depends on the position since the stage is continuously accelerating or decelerating. The melting behavior for lower energies is much more defined as shown in the SEM image. For high energies, a significant part of the subjacent silicon melts as well. The diagram shows the position depend stage speed and incoming energy per $1\,\mu m^2$ pixel. Calculations are made in Section 9.2.2.*

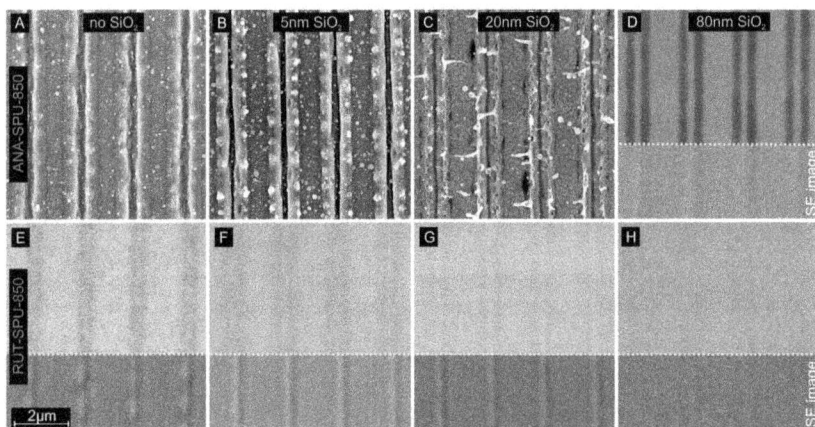

Figure 9.16: *SEM images showing the effect of differently thick SiO_2 cover layers on the melting behavior of the subjacent TiO_2 and silicon substrate. To illustrate the surface structure more clearly secondary electron (SE) images are added for selected samples.*

9.16 demonstrates the consequences for laser-induced melting of ANA- and RUT-SPU-850 films. The molten structures for both kind of films look very similar and the calculated differences of the reflectance displayed in Figure 9.5 are not mirrored by the experiment. The results are more consistent with the measured absorbance shown in Figure 9.6 which does not show significant differences except for 80 nm thick SiO_2.

The slightly different shapes of the molten structures gained by increasing the thickness of the covering SiO_2 layer up to 20 nm originate most likely from thermodynamic and mechanic properties of the layer system. Liquid TiO_2 or silicon own specific wetting angels on TiO_2 films. On the one hand, Figure 9.16 A and B demonstrate that the melt on blank anatase films covers a larger area compared to films covered with SiO_2. On the other hand, the melt forms much more distinct bulges on SiO_2 which is seen very clearly in the same figure as well as in Figure 9.18 A/C and in Figure 9.19 E/F. This observation indicates strongly that the wetting angle of liquid TiO_2 is lower on anatase than on SiO_2. Regarding the high polarizability of solid TiO_2 this behavior is quite reasonable. The splashes in Figure 9.16 C might rise from the fact that the covering SiO_2 forms a mechanical barrier for the increasing amount of melt during the light exposure. At the certain point in time, the pressure of the melt becomes such high that the SiO_2 bursts and the melt is catapulted in the air and forms the observed splashes instead of bulges. The completely different behavior of 80 nm thick SiO_2 layers has to be explained in another way. Figure 9.17 D demonstrates clearly that the subjacent anatase TiO_2 and silicon is not changed in any kind. This fact implies that most of the photons are absorbed by the SiO_2 layer. Usually SiO_2 is transparent in the visible regime but samples with an 80 nm thick SiO_2 layer have an important feature which is different from all other investigated samples. As on all other samples, the light is reflected and a standing wave is formed by the overlap of the incident and reflected light. But in contrast to all other samples, simulations represented in Figure 9.17 A-C point out that the first maximum of the electrical field amplitude of this standing wave is inside the 80 nm thick SiO_2 layer. As a consequence, the electrical field is almost doubled in a relatively large

221

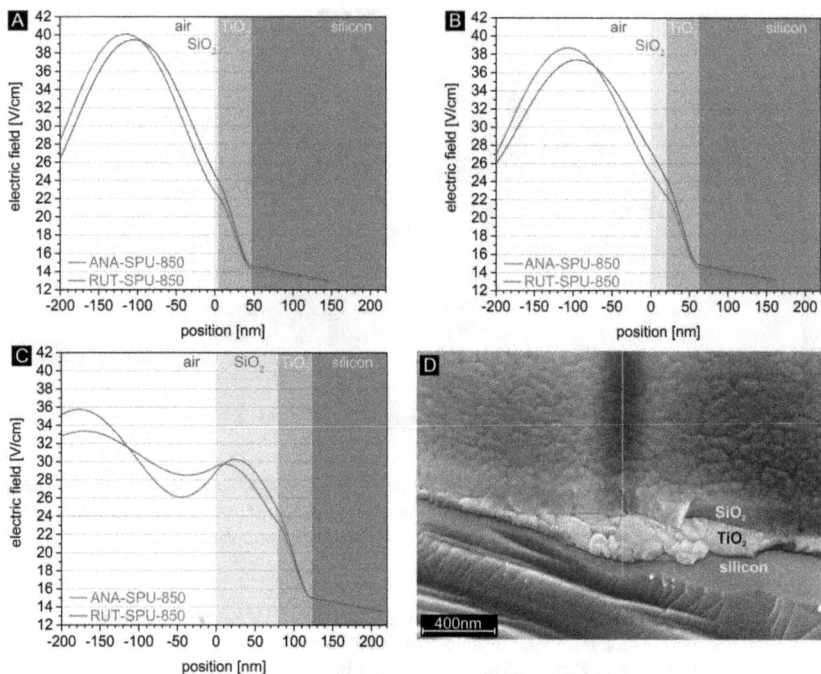

Figure 9.17: *Field distribution in ANA- and RUT-SPU-850 covered by 5 (A), 20 (B), and* 80 nm *(C) thick SiO$_2$ applying light with a wavelength of* 532 nm. *The calculations are based on the same transfer matrix formalism which is also employed in Figure 9.5 [1754]. D) SEM cross-section image of supposed non-linear effect in an* 80 nm *thick SiO$_2$ shell. As a guide to the eye the yellow dotted line marks the center of the exposed pattern stripe.*

volume inside the SiO$_2$. It is known that sufficiently high beam energies induce a self-absorption [1775]. Due to the high energy density of the pulsed laser, a huge number of electrons might be excited from the valence band into the conduction band of the SiO$_2$. The amorphous and defect-rich structure of the sputtered SiO$_2$ is supposed to keep the bandgap small and support the excitation of electrons over and above. The energy of the excited electrons is transferred to the lattice partly and phonons are created. It appears reasonable that the induced heat changes the inner structure of the amorphous SiO$_2$. The induced heat is able to change the structure of the amorphous SiO$_2$. This is a comprehensible explanation for the dark color of the exposed SiO$_2$ layer in Figure 9.17 D. In conclusion, the energy of the incident light beam is supposed to be used for the structure change in the SiO$_2$ layer and cannot be used for liquefying the ANA-SPU-850 anymore. In addition, the reflectance might be increased while the conduction band is filled with excited electrons.

As an intermediate result, the calculated differences of the reflectance for ANA- and RUT-SPU-850 samples seem to play a minor role for the melting behavior and the search for more important factors promoting the melting goes on.

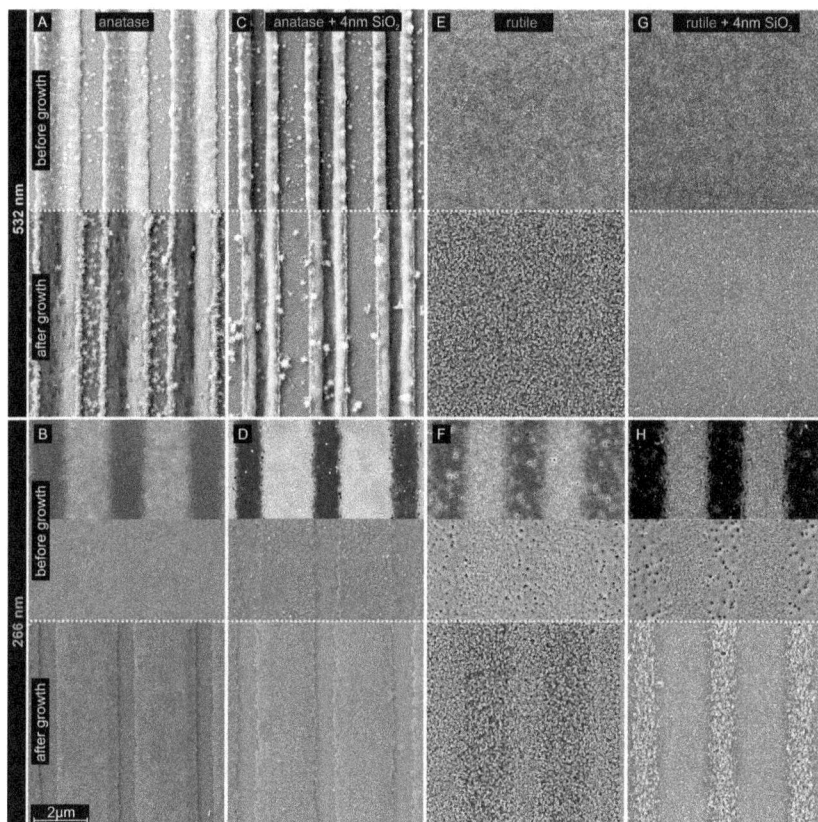

Figure 9.18: *Effect of the incident wavelength and an additional thin SiO_2 layer on the laser-induced structures gained by pulsed laser interference. Panel A-D and E-H show anatase and rutile films, respectively. The films in Panel C, D, G, and H are covered with a 4 nm thin SiO_2 layer. The samples in Panel A, C, E, and G were exposed to 532 nm light while the others were exposed to 266 nm light. The rutile films are not affected by 532 nm light and for 266 nm light, they show almost the same kind of molten structure. For anatase films, pattern lines are broader and more homogeneous in the absence of SiO_2. The topography is affected most strongly by the 532 nm light exposure on anatase films and there is a clear bulge-forming behavior for the samples covered with a SiO_2 layer (C). The thickness of the TiO_2 film is 40 nm for all samples.*

223

The wavelength of the incident light has a great effect on melting as shown impressively in Figure 9.18. The melting behavior of ANA-SPU-850 for 532 nm light with and without a covering SiO_2 layer was described above. The hydrothermal growth gives another hint for the crystal structure of the solidified material. Assuming that a lot of silicon melts within the pattern lines the solidified material in the created structures is supposed to consist mainly of silicon with minor inclusion of titanium and oxygen. This assumption is consistent with the EDX measurements shown in Figure 9.22 A. But also the anatase TiO_2 layer melted and blended with the molten silicon since the surface structure of the ANA-SPU-850 disappeared in the center of the pattern lines completely. And TiO_2 flakes that might have been blown away from the pattern lines by the laser-induced blast are found nowhere on the whole sample. The results from the hydrothermal growth shown in Figure 9.18 A and C indicate that inclusions of titanium and oxygen in silicon are not able to form nucleation sights for rutile nanorods. But at the edges of the molten structures, nanorods grow very likely. Regarding the fact that the melting temperature of the subjacent silicon is around 1400 °C, it is reasonable that anatase TiO_2 which is in contact with the hot melt converts into rutile TiO_2 within a solid-solid or even solid-liquid-solid phase transition. Due to the geometry of the solidified material and the covering SiO_2 layer, the stripe of hydrothermally grown rutile nanorods is more restricted on samples having a thin SiO_2 layer on top.

The effect of 266 nm light on ANA-SPU-850 is almost the same for samples with and without a thin covering SiO_2 layer as displayed in Figure 9.18 B and D. Anatase TiO_2 absorbs UV light very well and melts. In contrast to the unaffected ANA-SPU-850 film, the solidified oxide has a smooth and flat surface. It is reasonable that recrystallization happens so fast that grains are so small that they cannot be detected within the resolution limit of the SEM anymore. And it cannot be excluded that the rearranged TiO_2 might be even amorphous. In fact, the affected material dissolves during the hydrothermal treatment. This observation fits to the feature that amorphous TiO_2 and grain boundaries of anatase TiO_2 are chemically rather unstable. Another important finding is that the subjacent silicon remains unaffected. Hence, exposing sufficiently thick anatase layers to 266 nm light results in melting of TiO_2 only.

Interestingly, RUT-SPU-850 remains completely unaffected by 532 nm light exposure. Again, the calculated reflectance does not explain this observation as the reflectance of ANA- and RUT-SPU-850 are almost the same as displayed in Figure 9.5. This is another evidence that the reflectance of the samples plays a minor role for the melting behavior. The measured absorbance represented in Figure 9.6 shows clearly that RUT-SPU-850 absorbs much less energy compared to ANA-SPU-850. But this fact does not tell us where the absorption takes place. According to literature listed in Section 9.1.3 silicon and titanium, the raw materials of RUT-SPU-850, are intermixing with each other likely even at temperatures below the applied annealing temperature. These mixed phases are supposed to be rather good absorbers in the visible light regime. Furthermore, most of these intermixed phases melt at similar temperatures as silicon and below the melting temperature of anatase TiO_2.

Nevertheless, melting of RUT-SPU-850 does not appear for the same laser power as applied for the ANA-SPU-850 samples. So either the supposed absorbing mixed phase is less pronounced in RUT-SPU-850 or the assumptions used to calculate the reflectance are imprecise. And in deed, an important difference between ANA- and RUT-SPU-850 films was not considered so far. The surface of RUT-SPU-850 films is very rough with a lot of tilted facets. Reflection and transmission of rough surfaces is a complex topic and is not discussed in detail here. Often surface roughness is used to reduce the reflection in solar cells [225]. Hereby, light incoupling is improved by scattering or interflection on several facets of a grainy film [696]. So it seems that the suggested effect of surface roughness is opposing the experimental results. But one has to consider that scattering and interflection expand the beam cross-section inside the material and consequently, the incident beam energy is distributed across a larger area on the TiO_2/silicon interface. This might be an alternative explanation besides the eventually less pronounced absorbing mixed layer for the missing melting. But since rutile grains in RUT-SPU-850 films are one order of magnitude smaller than the wavelength of the incident light, one does not expect a significant contribution from the roughness. The film should behave more like an effective medium. But also the assumption of an effective medium raises questions regarding the experimental results. The effective refractive index of the rough surface is supposed to be smaller than the refractive index of the subjacent dense TiO_2. Hence, a cascade of the effective refractive index is expected which is supposed to reduce the reflectance and increase the injected beam power.

The heat capacity and thermal conductivity of bulk TiO_2 are similar and cannot be used as an explanation for efficient heat localization in anatase films. Since the grains in RUT-SPU-850 films are smaller than in ANA-SPU-850 films the heat transport should be reduced in rutile films. Of course, the propagation of phonons through rutile/rutile interfaces might be more efficient than through anatase/anatase interfaces as discussed in Chapter 2. But since the thermal conductivity of the subjacent silicon is relative high, these thermodynamic properties are not expected to play a major role anyway.

In case of RUT-SPU-850, the hydrothermal growth does not deliver any new insights into the film structure. On blank RUT-SPU-850, a dense NRA is growing and on RUT-SPU-850 covered with a thin SiO_2 layer, almost no rods are growing. Only in very narrow stripes, the growth is slightly enhanced. It is reasonable that the SiO_2 degraded in these stripes due to the self-absorption effect described above.

Applying light with a wavelength of 266 nm to RUT-SPU-850 films the layer melts similarly to ANA-SPU-850 films as demonstrated in Figure 9.18 F and H. Pinholes indicate a contraction of the TiO_2 due to a heat-induced change of the density. The hydrothermal growth results in slightly different rod geometries on exposed and unaffected areas which is directly related to the different geometries of the seeds in both areas. For samples with a thin SiO_2 cover layer, the growth is promoted only on the recrystallized stripes where the SiO_2 was cracked by the rearrangement process of the subjacent TiO_2. The most important finding from Panel F and H is that liquefied RUT-SPU-850 films recrystallize mainly in the rutile morphology. Another clear hint for the missing melting process of RUT-SPU-850 films is not given from these experiments.

So far, different reasons for the creation of the observed seed for rutile nanorods are investigated. Using a wavelength of 532 nm and a defined power density, only the anatase film melts. An additional SiO_2 layer changes the reflectivity, but it is not able to change the melting significantly or induce melting of the rutile films. For a wavelength of 266 nm, both films melt due to the high absorption in the UV regime, but the resolidification generates no seeds. This suggests that the melting process delivering rutile seeds in anatase films does not happen on the surface.

Effect of the Injected Energy on the Melting Behavior

The energy-dependent melting behavior of ANA-SPU-850 is demonstrated in Figure 9.19. At very low energy densities (Figure 9.19 A and 9.21 A), only a few cracks appear on the surface indicating that the melting process at the TiO_2/silicon interface caused strain inside the ANA-SPU-850 film.

At higher beam densities (Figure 9.19 B and 9.20 A), the relief of the grains disappear indicating that the melting process effects the whole ANA-SPU-850 film between the film/silicon interface

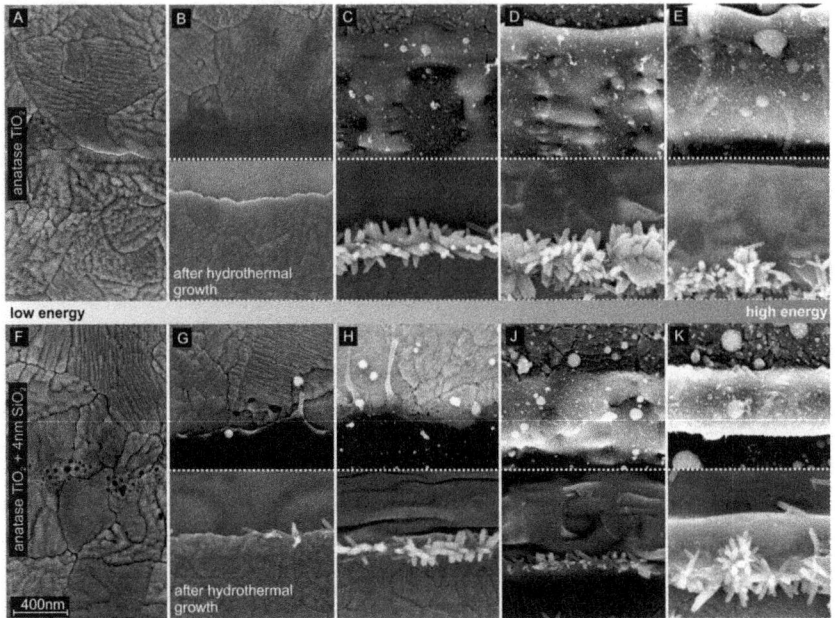

Figure 9.19: *Effect of (pulsed) laser beam energy density and additional SiO$_2$ cover layer on the melting effect on ANA-SPU-850 films. The energy density is increased from Panel A to E and F to K, respectively.*

and the surface. This behavior reminds on the melting of ANA-SPU-850 using UV light as shown in Figure 9.18 B. Hence, this is a clear evidence that ANA-SPU-850 films melt under the exposure of 532 nm light. This means that either the whole ANA-SPU-850 consists of a mixed phase with a high absorption coefficient and relatively low melting temperature or the film heats up likely due to phonon confinement as discussed in Section 2.3.1. In contrast to the melting behavior shown in Figure 9.18 A and C, the subjacent silicon is almost not affected in Figure 9.19 B. Hence, the low thermal conductivity within the anatase TiO$_2$ becomes important. At the same time, the high thermal conductivity in silicon plays a minor role, because if phonons are confined within anatase grains the thermal transport through the anatase/silicon interface is supposed to be bad as well. The missing nanorods after the hydrothermal growth indicate that the temperature was too low to trigger the ART. The sharp edges of the patterned structure in Figure 9.19 B are a strong hint for a mainly solid-solid phase transition. The cross-sections of pattern lines for different applied energy densities presented in Figure 9.20 reveals further interesting details. For rather low energy densities (Figure 9.20 A), the ANA-SPU-850 is clearly distinguishable from the silicon substrate. For slightly higher energies, the silicon substrate is domed with a maximum in the center of the pattern line as demonstrated in Figure 9.20 B and 9.21 B. In particular, the height profile in the latter Panel shows shallow trenches besides the domed region. This effect was described by Wysocki et al. and is related to a melting and reso-lidification process in the subjacent silicon [1777]. The volume expansion might be enhanced by

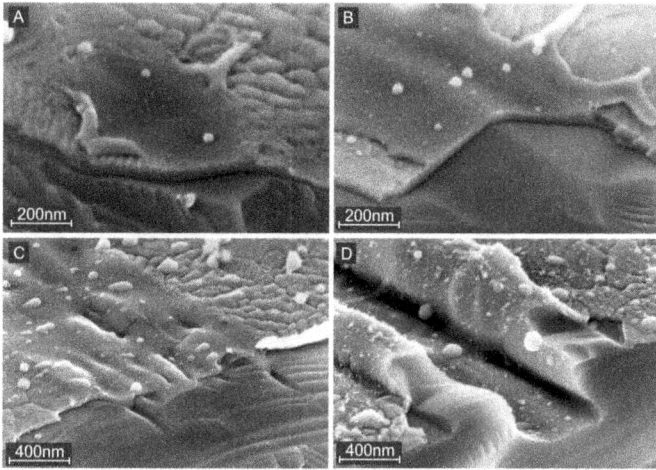

Figure 9.20: *Effect of energy density on pulsed laser-induced melting. All figures show SEM cross-section images. The applied energy density is increasing from Panel A to D.*

the contamination of silicon with titanium and oxygen atoms near the interface, which results in a mixed phase with different physical properties. But due to a lack of spatially resolved element analysis data, this assumption cannot be proved. Although silicon is known to increase its density at high temperatures, a volume expansion cannot be excluded completely for lower temperatures [1778].

For even higher injected energies (Figure 9.20 C), the silicon is liquefied resulting in a volume contraction [1779] and a homogeneous material distribution across the pattern line. The surface becomes rougher, but a top layer is still recognizable on the silicon substrate as displayed in Figure 9.21 C.

Increasing the energy density further, the viscosity of the melt is low enough for a strong dewetting effect leaving a deep trench in the center of the pattern line and bulges besides the trench behind as demonstrated in Figure 9.19 E, 9.20 D, and 9.21 D. The trench has a depth of more than 250 nm which is a clear proof that a huge amount of silicon is melting besides the TiO_2. There is no top layer observable anymore. But an interesting feature is magnified in Figure 9.22 B. The yellow arrow marks an intact part of the ANA-SPU-850 covered by the bulge. This observation emphasizes that the laser-induced melting happens in the area of the trench only and the melt flows across the neighboring ANA-SPU-850 film afterward. This observation proves that in some positions the bulges are created by dewetting and not directly from the melting process. But it could also happen that the emerging bulge bends the film up as demonstrated in Figure 9.22 C. It seems that there are no hard criteria separating these two events. If the film is bent up the bulge and the silicon substrate form a homogeneous solid. Hence, the bulge consists mainly of silicon.

From now on, the width of the pattern line does not increase with increasing energy density, but the melting effect seems to penetrate the silicon substrate deeper and deeper. The sample is covered with droplets and dust consisting most likely of low crystalline anatase TiO_2. This assumption is supported by the EDX measurements displayed in Figure 9.22 A and the fact that

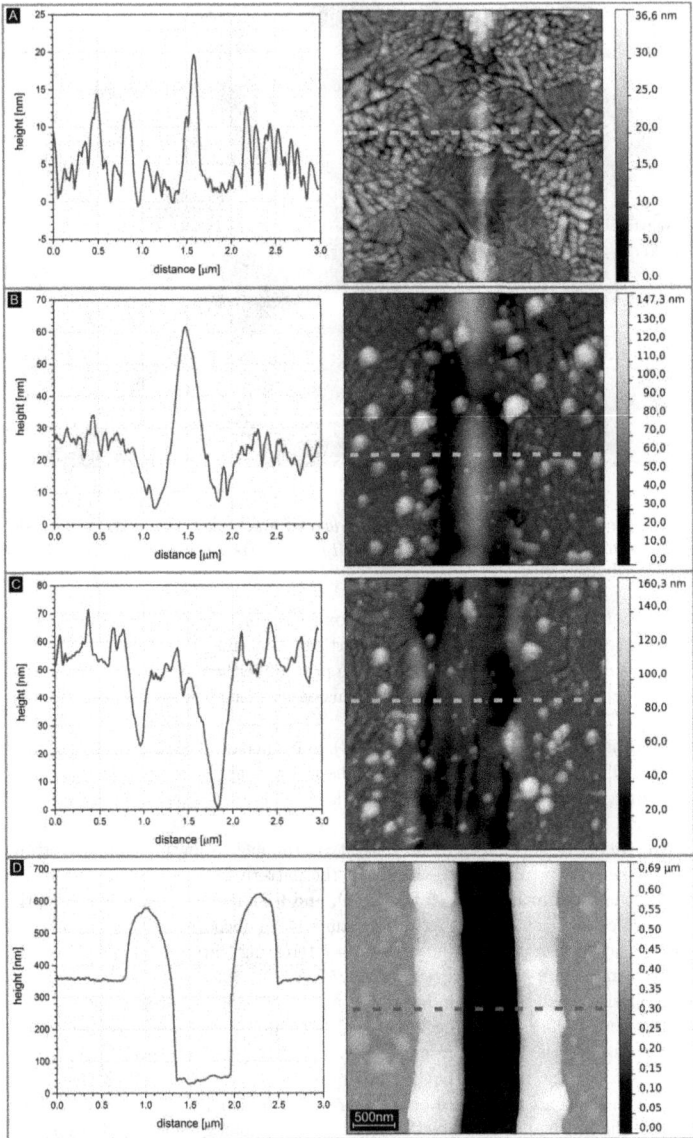

Figure 9.21: *AFM images of structures made by laser-induced melting using increasing energies from Image A to D. The dashed lines mark the positions of the height profiles shown on the left side. The displayed images are related to the cross-section images shown in Figure 9.20. A similar structuring behavior was observed by Baumgart et al. [1776].*

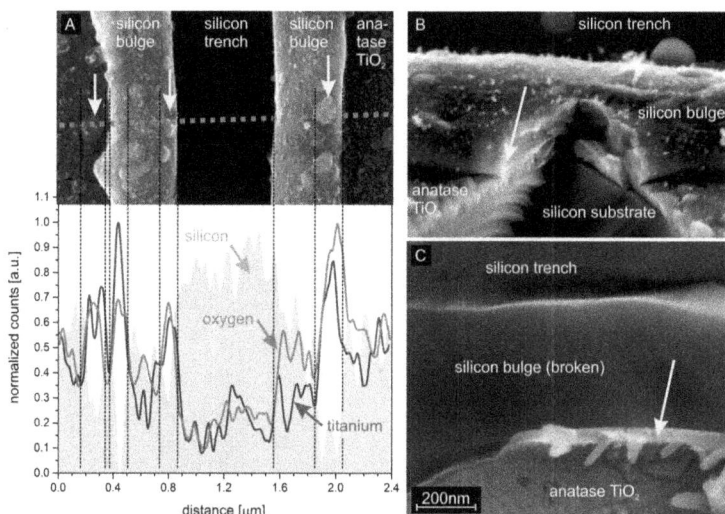

Figure 9.22: *Structure and composition of a bulge investigated by SEM and EDX: A) SEM top view and EDX line scan of bulges made with high beam powers. Between the bulges, almost no titanium and oxygen is covering the silicon substrate. Similar to the unaffected TiO$_2$ film on silicon, the bulges seem to consist of similar amounts of titanium, oxygen and silicon. But there are some drops which consist clearly of titanium and oxygen. B) Cross-section of a bulge showing that the anatase film (yellow arrow) remains below the outer part of the bulge indicating that the molten material moves across the neighboring anatase film during the bulge-forming process. C) In contrast to Panel B, the top view on a tilted cross-section in another position demonstrates that here, the anatase film was bent up (yellow arrow) while the bulge was formed and the bulge is homogeneously connected with the subjacent silicon substrate.*

small droplets of liquid TiO$_2$ crystallize as anatase TiO$_2$ due to supercooling as discussed in connection with Figure 9.8. Since these particles are chemically unstable, they disappear during the hydrothermal growth process. The generation of droplets and bulges are a clear evidence that the material was liquefied during the light exposure. The induced heat is already high enough to provide the ART and generate seeds for rutile nanorods at the edges of the affected stripes as discussed above in connection with Figure 9.18 A.

Another effect that might play a role in the creation of rutile seed layers is the cooling speed of the liquefied TiO$_2$ as discussed in the connection with Figure 9.8. If a solid-liquid-solid phase transition is responsible for the creation of rutile seeds it needs a slow cooling rate to avoid supercooling. So to predict the location with the highest number of rutile seeds one has to look for the location with the slowest cooling rate. The SEM cross-section image in Figure 9.20 shows that close to the center of the molten stripes, there is no difference between the molten regions and the subjacent silicon anymore. Thus, it is reasonable that the thermal contact between this region and the silicon substrate is relatively good and the cooling rate is quite high. Hence, the formation of anatase is expected here. But it might also happen that the mixture of silicon and TiO$_2$ prevents the formation of any TiO$_2$ crystal structure. Because of their strong surface bend-

229

ing, the edges of the molten bulges have a quite large surface exposed to air which is a bad heat conductor. The outermost region is separated partly from the subjacent silicon by the 40 nm thick ANA-SPU-850 film. Besides the increased heat resistance related to the additional interfaces, the thermal conductivity of ANA-SPU-850 is more than one order of magnitude smaller than that of silicon. In other words, the ANA-SPU-850 film works as a thermal insulator and thus, the cooling rate is minimal in the outer regions of the molten stripes. So the region where rutile is formed most likely is at the rim of the pattern lines.

Samples being covered by an additional SiO_2 layer show a similar trend. Dewetting seems to be improved since the bulges are more pronounced compared to the blank ANA-SPU-850 film as demonstrated in Figure 9.19 E and K.

In conclusion, these experiments give further hints that phonon confinement is the driving force for the melting process in ANA-SPU-850 films. Especially because it is proofed now that the anatase TiO_2 film is able to melt independently from the melting behavior of the subjacent silicon.

At lower power densities and before the melting affects the surface, the volume of the substrate increases which suggest that the incident light is absorbed somewhere close to the interface below the TiO_2 film. At higher laser powers, a huge amount of material is liquefied and forms two parallel bulges. The rutile seeds are found at the boundary between the solidified material and the unaffected TiO_2 film.

Effect of Annealing Temperature on Melting Behavior

Another important hint for mixed phases at the TiO_2/silicon is given by laser beam exposure of differently annealed films. Sputtered amorphous TiO_2 films without any post-annealing are ablated by the laser-induced blast easily (Figure 9.23 A). This indicates that the adhesion between the film and the substrate is relatively low. Nevertheless, a lot of heat must have been induced into the ablated TiO_2 since the ablated stripes form stable seeds for the hydrothermal growth while the unaffected TiO_2 dissolves during the growth process. ANA-SPU-450 films trigger the melting of some interface material but the covering film is blown away by the laser-induced blast as well (Figure 9.23 C). Since blank silicon does not melt for similar laser powers at all, this observation indicates that atoms diffused from the TiO_2 film into the silicon substrate changing its absorption or melting temperature. Only ANA-SPU-850 films melt completely and run through a dewetting process. This comes along with the generation of SiO_2 nanoparticles appearing as bright nanoparticles on the SEM image (Figure 9.23 E). Rutile TiO_2 films are made of titanium. The as-deposited titanium melts because of its metallic absorption properties and shows a typical bulge-forming behavior (Figure 9.23 B). The hot titanium oxidizes easily and acts as a seed layer for rutile nanorods as discussed in detail in Section 9.2.1. After annealing the titanium film at 450 °C in an oxygen-rich atmosphere, it consists of anatase TiO_2. The melting behavior (Figure 9.23 D) is similar to annealed TiO_2 at 450 °C (Figure 9.23 C): The TiO_2 film cracks and the sample surface is partly liquefied. And finally, RUT-SPU-850 films do not show any melting effect as reported previously (Figure 9.23 F).

In conclusion, these experiments deliver the final proof that the role of the refractive index and the reflectance play a minor role. All layers have very similar refractive indices but their melting behavior is completely different. And the deposition technique is also much less important than the morphology. The subjacent silicon melts easier for higher annealing temperatures of the covering TiO_2 film. Again, this is a strong hint for the role of an absorbing mixed phase with a relatively low melting temperature.

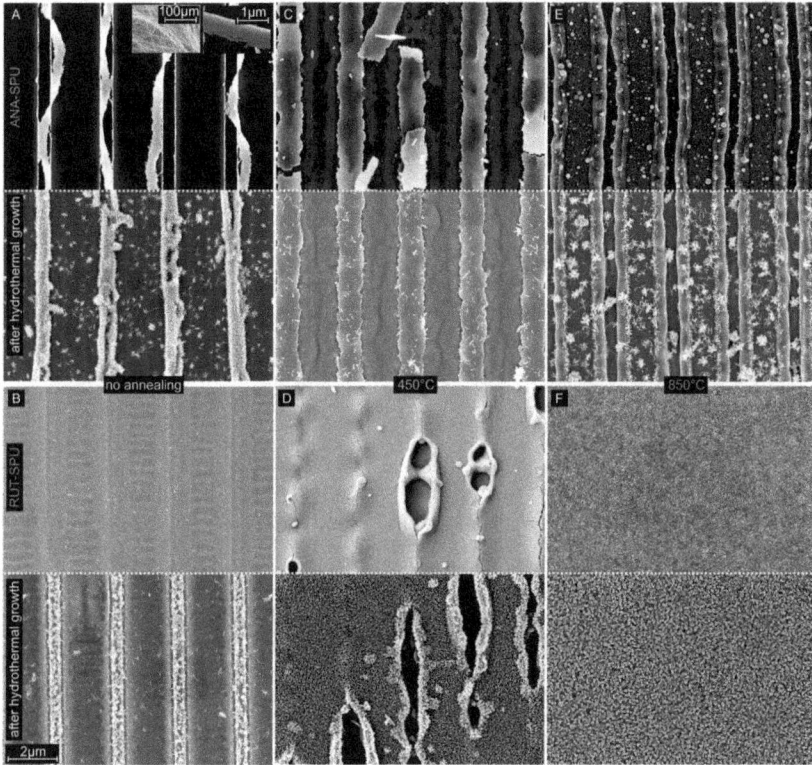

Figure 9.23: *Effect of annealing temperature on pulsed laser-induced melting. The insets in Figure A are zoom-ins and -outs of the same sample. Panel A, C, and E show sputtered TiO_2 films without annealing, annealed at 450 °C (ANA-SPU-450), and annealed at 850 °C (ANA-SPU-850), respectively. Panel B, D, and F show sputtered titanium films as deposited, annealed at 450 °C (anatase, but different from ANA-SPU-450), and annealed at 850 °C (RUT-SPU-850), respectively.*

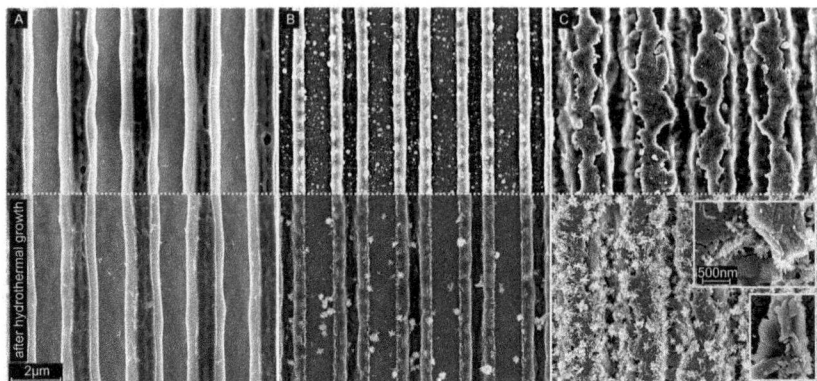

Figure 9.24: *Effect of film thickness on melting behavior: A) 4 nm, B) 40 nm, and C) 400 nm ANA-SPU-850 on silicon before (top) and after hydrothermal growth (bottom). The insets demonstrate the activated growth of nanorods on the molten side of anatase flakes while the opposing silicon substrate remains blank during the hydrothermal growth.*

Film Thickness-Dependent Melting Behavior

To investigate the role of Fabry-Pérot cavities° and localize the origin of the melting process laser-induced melting was applied to films with different thicknesses.

Blank silicon does not melt for the applied laser power. But already a 4 nm thick ANA-SPU-850 layer results in almost the same melting behavior as it is observed for 40 nm thick layers as shown in Figure 9.24 A and B. Interestingly, the growth of nanorods is not enhanced at the outer part of the bulges as it is common for thicker ANA-SPU-850 films. The reason might be the low amount of TiO_2 close to the bulges or the rapid cooling of the thin film. As discussed in connection with Figure 9.8, rapid cooling of TiO_2 melts results likely in supercooling bearing anatase TiO_2. According to Figure 9.5, the calculated reflectance for 0 and 4 nm TiO_2 is almost the same for 532 nm light. Hence, this experiment is another proof that reflectance and Fabry-Pérot cavities are not an important driving factor for laser-induced melting.

Exposing 400 nm thick ANA-SPU-850 films to 532 nm light pulses, the melting behavior is still very similar to thinner films. The only difference is that the bulge is not as distinct as on thinner films since the ANA-SPU-850 film is so thick that the melt is not able to flow across the surface of the unaffected layer as demonstrated in Figure 9.24 C. Since thick layers provide a lot of TiO_2 the growth of nanorods is enhanced significantly at the sidewalls of the created stripes. On the bottom of the trenches, no seed layer is formed. Only a minor part of the ANA-SPU-850 layer melted during the light exposure. The layer which was on top of the area melted by the incident light is blown away by the created laser blast and settles down as flakes in other parts on the sample. An example of such a flake is represented in the inset of Figure 9.25 A. It is seen clearly that the top side which was the former surface is not affected and the bottom side is smoothed by a solid-liquid-solid phase transition. Two of these flakes are shown after the hydrothermal growth in the insets of Figure 9.24 C. The upper inset displays a flake facing with its upper side upwards. The upper side was the former layer surface. The top side is not covered by nanorods but there is an intensive growth of rods on the bottom side which was the former interface between the TiO_2 and the silicon. The lower inset offers a

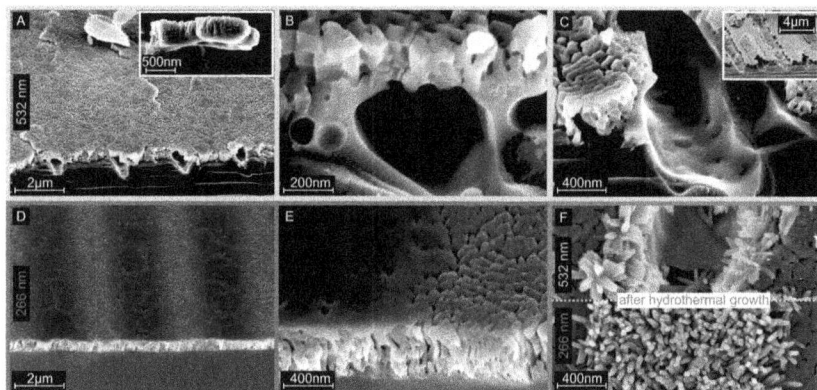

Figure 9.25: *Effect of film thickness on the structure made by pulsed laser-induced melting. A 532 nm laser beam was applied to 4 nm (A), 40 nm (B), and 400 nm (C-F) thick ANA-SPU-850 films on silicon. The inset in Panel C shows a piece of the laser treated film that was removed from the film by the laser-induced blast. Higher laser powers result in ablation of the ANA-SPU-850 film (D). Nanochannels are surrounded by a layer that is clearly distinguishable from the environmental TiO_2 and silicon (E,F).*

direct view on the bottom side of such a flake. The grown nanostructures seem to be oriented in a view specific directions only indicating that the crystallinity on the bottom of the flake is quite high. This feature implies that the cooling rate was rather slow in this area. By applying lower laser powers the ANA-SPU-850 is not blown away due to the pronounced adhesion and nanochannels are generated at the TiO_2/silicon interface as demonstrated in Figure 9.25 A-C. The inner surface of the nanochannels is covered with a thin layer which is clearly separated from the silicon substrate and unaffected TiO_2 film. The composition of this layer remains unknown but some statements can be made anyway: 1) The system contains silicon, titanium, and oxygen only. Hence, the layer consists of these elements or a fraction of them only. 2) Due to the clear interface between the layer and the silicon substrate, it is unlikely that the layer consists of pure silicon. This thesis is supported by the observation in Figure 9.22 C where bulges which emerge from liquefied silicon are *not* separated from the subjacent silicon. 3) The layer does not provide the hydrothermal growth of rutile nanorods. Thus, the layer does not consist of rutile TiO_2. 4) The layer is chemically stable and remains unchanged during the hydrothermal growth. Hence, the layer does not seem to consist of amorphous TiO_2 which dissolves easily under hydrothermal growth conditions. Considering all these statements, the layer could consist of anatase TiO_2 which would be reasonable because of the fast cooling rate resulting from the good thermal contact to the subjacent silicon. But it could consist of a highly silicon-doped TiO_2 material with a different morphology as well. Only the layer which is directly linked to the unaffected TiO_2 provides the growth of nanorods as it is shown in the insets in Figure 9.24 C (the bottom of the flakes is equal with the ceiling of the nanochannels). Another interesting feature is the appearance of spherical holes besides the nanochannels. These holes indicate the temporary presence of vaporized TiO_2. Interestingly, the solid material, which was situated in the volume of the nanochannels and holes before their creation, cannot be ablated or evaporated since the surface of the sample remains totally unaffected during the laser treatment. Hence, the density of the surrounding material, in particular, the molten and solidified matter, must be extremely high or a significant amount of atoms must be diffused into the nearby substrate similar to a eutectic system.

But the most important finding of the experiments on thick ANA-SPU-850 films is that the melting process appears at the TiO_2/silicon surface obviously. Furthermore, it is assured that both, the TiO_2 and the silicon melted during light exposure.

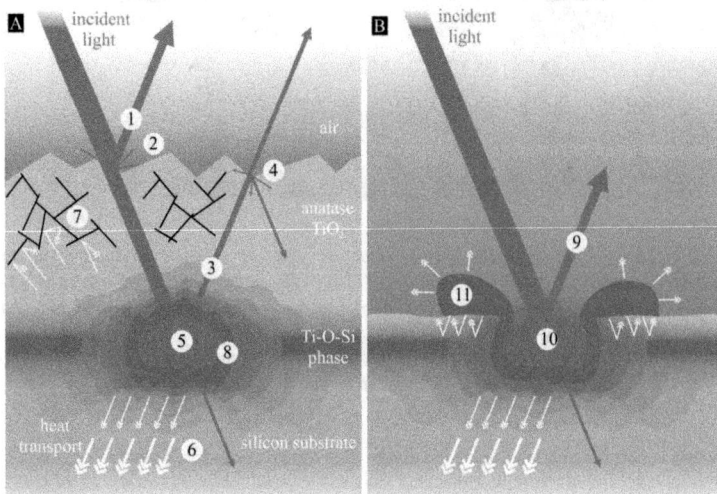

Figure 9.26: *Schematic drawing of laser-induced melting in thick (A) and thin (B) ANA-SPU-850 films. The reflection of incident light depends on the difference of the refractive indices of the interface materials (1,9). In addition, the roughness (2) of the film affects the reflectance as well. There is also reflection (3) at the ANA-SPU-850/silicon interface that could be reduced by an additional Ti–Si–O mixed phase. Usually in thin films interflection events (4) take place that are influenced by the film roughness as well. Due to the high absorption coefficients of some of the mixed phases, light is absorbed (5,10) strongly in this layer. The absorption induces heat dissipating into the silicon and ANA-SPU-850 film. But the thermal resistance is increased by phonon-phonon scattering (6) in hot regions and by phonon-grain boundary scattering (7) in the ANA-SPU-850 film. Thus, the heat is concentrated in the ANA-SPU-850/silicon interface promoting the melting (8) of this phase. In case of thin films, the melt strikes the surface and forms bulges (11) due to dewetting.*

The situation for using 266 nm light is completely different and more simple. Due to the high absorption coefficient of ANA-SPU-850 films for UV light the beam energy is absorbed directly at the surface of the film as demonstrated in Figure 9.25 D and E. The TiO_2 melts and solidifies mostly as rutile TiO_2 providing the growth of rutile TiO_2 nanorods (Figure 9.25 F). According to Figure 9.25 E, the melting process does not affect more than the upper 20 nm of the ANA-SPU-850 film. The TiO_2/silicon interface remains completely unaffected.

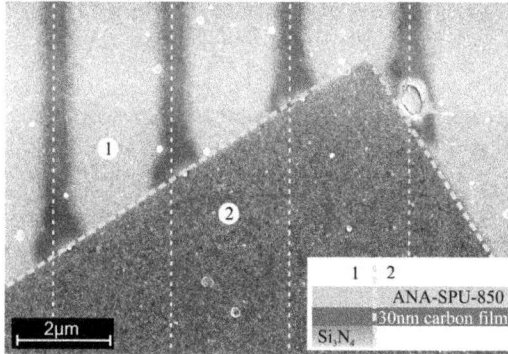

Figure 9.27: *ANA-SPU-850 film deposited on a Si_3N_4 TEM grid covered with a thin amorphous carbon layer. In area (2) the subjacent Si_3N_4 is missing.*

> Experiments on thick films deliver the final proof that the melting process in anatase TiO_2/silicon systems appears at the TiO_2/silicon interface. It is suggested that a mixed Ti–Si–O phase is responsible for an efficient light absorption and a low melting temperature.

Effect of the Temperature-Dependent Absorption Coefficient

One has to consider for both wavelengths that the absorption changes likely as soon as the solid starts to melt since the absorption coefficient for liquid TiO_2 and silicon differs from the corresponding values in the solid phases. For silicon, it was observed that the absorption coefficient increases dramatically with temperature and is significantly higher at the melting temperature compared to room temperature [1780, 1781]. The absorption is related mainly to the direct bandgap of silicon [1781]. For TiO_2, an increase of the absorption coefficient is also conceivable, since the decay of the crystal structure at the melting point results in local oxygen-deficient regions with high mobile electron densities similar to reduced black TiO_2. This feature is an important issue for describing the melting process for 532 nm light. As displayed in Figure 9.20 A, melting for low laser powers appears in the anatase TiO_2 layer at first. Close to the interface, melting might be triggered by an increased absorption and lowered melting temperature due to silicon dopants. Heating might be enhanced by phonon confinement as well. As soon as the TiO_2 is liquefied the absorption coefficient of that layer is supposed to increase and a self-absorption is initiated. The efficient heating triggered by such a self-absorption is able to initialize the melting process in the nearby silicon as well. Since the absorption coefficient of liquid silicon is much higher than for solid silicon for light with a wavelength of 532 nm, a self-absorption effect is expected in silicon as well [1782]. In conclusion, a minor melting event inside the anatase TiO_2 is supposed to start a chain reaction resulting in the presented trenches, bulges, and channels.

Effect of the Substrate on Melting Behavior

The meaning of the substrate for the laser-induced melting process is emphasized in Figure 9.27. The ANA-SPU-850 film was deposited on a Si_3N_4 TEM grid covered with a thin amorphous carbon layer. It is assumed that the exchange of atoms between the TiO_2 and Si_3N_4 is not prohibited by the carbon film completely. Melting appears on the substrate (1) only and there is absolutely no influence on the ANA-SPU-850 film deposited on the levitating carbon film (2). The broadening of the molten lines at the edges of the subjacent Si_3N_4 substrate might result from reflection effects on the edges or reduced heat dissipation due to the missing substrate in one direction.

The missing melting in RUT-SPU-850 films might be a consequence of several effects: Firstly, the silicon diffusion rate is supposed to be relatively low in the densely packed rutile crystals as described in Section 9.1.1 and 9.1.2. As a consequence, the mixed phase is thinner. Secondly, in contrast to rutile TiO_2, the absorption of annealed TiO_2 is reported to be significantly increased as mentioned in Section 9.1.3. Thirdly, due to the reduced silicon diffusion in rutile TiO_2, the melting temperature is not lowered as much as in mixed phases created between anatase TiO_2 and silicon. And fourthly, phonon confinement is expected in anatase rather than in rutile TiO_2 as discussed in Section 2.3.2.

10 Optical Properties of Rutile TiO$_2$ Nanostructures

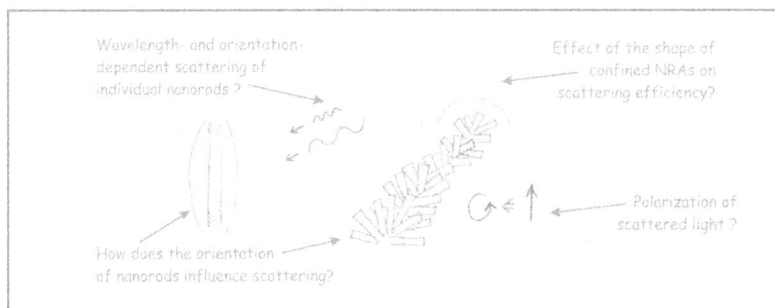

Figure 10.1: *Scientific objectives investigated in this chapter. In which way differs dielectric scattering of an individual nanorod from scattering of a confined NRA? How does the orientation of the nanorods influence scattering? Does scattering show polarization effects? Does the scattering efficiency depend on the shape of confined NRAs?*

Optical properties include a lot of effects in semiconducting TiO$_2$ nanostructures such as dielectric functions, dielectric scattering, photon-phonon interaction, photon-electron interaction, and photon-plasmon interaction. Based on the definition of optical, electronic, and optoelectronic properties in Section 2.5, only optical properties are discussed in this chapter. Hence, dielectric scattering comes into focus and absorption is not investigated here. Although there is no exact theory describing dielectric scattering on non-spherical particles, it is commonly known that strongly anisotropic particles such as the investigated TiO$_2$ nanorods show polarization-dependent light scattering. It was shown in previous chapters that the orientation of single nanorods as well as the creation of superlattices of nanorods are controllable by the presented nanofabrication techniques. Thus, also scattering properties of the prepared samples become designable. This is demonstrated for selected structures in the following. Because the dimensions of the presented structures are smaller, equal, or larger than the incident light wavelength, the scattering behavior is complex and not only described by one effect as drafted in Figure 10.2.

In the beginning and before discussing the results for particular structures, general assumptions about the expected behavior of different nanorod constellations are introduced which are based on the known optical properties of semiconducting nanostructures described in Section 2.5.

The images are taken with a digital camera and it was insured that their colors correspond to the perception by the human eye. Thus the colors indicate wavelength-dependent scattering and reflection effects. It should be noted that these images show only tendencies and are not as informative as a spectrometer (e.g. as used for micro-PL).

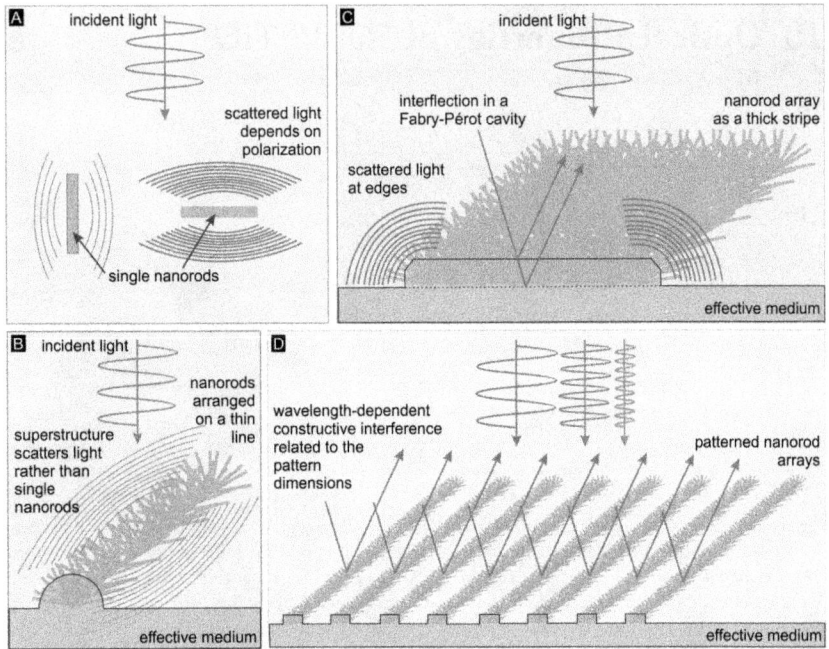

Figure 10.2: *Size-dependent scattering on TiO₂ nanostructures: λ is the wavelength of the incident light, l_{rod} the length of a nanorod, l_{line}/w_{line} the length/width of linear nanorod arrays, and d_{line} the distance between parallel NRAs. A) $\lambda \leq l_{rod}$: Scattering on single nanorods behaving as a dielectric antenna is very sensitive to the polarization of the incident light. B) $\lambda > l_{rod}$, $\lambda \leq l_{line}$, and $\lambda > w_{line}$: Single nanorods and the spacing between them are described as an effective medium. The anisotropic shape of the effective medium is correlated to the linear NRA which scatters light as a dielectric antenna. C) $\lambda > l_{rod}$, $\lambda \leq l_{line}$, and $\lambda < w_{line}$: The structure separates in an inner part with a homogeneous NRA and an outer part including the edges of the NRA. Scattering on the inner part is low since the homogeneous NRA forms a rather flat effective medium. Here, thin film interference is expected since the effective medium creates a Fabry-Pérot cavity. In contrast, on the edges light is scattered very well since the curved surface of the effective medium offers a lot of reflection angles. The polarization of the incident light plays almost no role. D) $\lambda > l_{rod}$, $\lambda \leq l_{line}$, $\lambda > w_{line}$, and $\lambda \approx d_{line}$: The parallel and line-shaped NRAs form a grid of dielectric antennae promoting the reflection of specific wavelengths in well-defined directions. The reflected light is enhanced for incident light polarized in parallel to the lines.*

Rutile TiO$_2$ NRAs as an Effective Medium

As discussed in Chapter 2, dense NRAs behave more like an effective medium° than assemblies of individual dielectric nanorods. Here, reflection and transmission of light are described predominantly by the rules of geometric optics. Because a lot of reflection angles are offered in curved areas, light is reflected on round edges very well. Typical wavelength-dependent effects in effective media are thin film interference, refraction at interfaces, and dispersion in the bulk. Polarization-dependent reflection is described rather by the Fresnel equations than with the anisotropic shape of the scattering particles.

In Figure 10.3, the effective refractive index is plotted as a function of the volume fraction of TiO$_2$ in the NRA based on Equation 2.12. Typically, one-third of the array consists of TiO$_2$. The major part of the volume consists of the spacing between the rods. $\epsilon_{TiO2,c} = 8.427$ and $\epsilon_{TiO2,a} = 6.843$ are taken as dielectric constant parallel to the c- and a-axis, respectively (see Section 2.2.4).

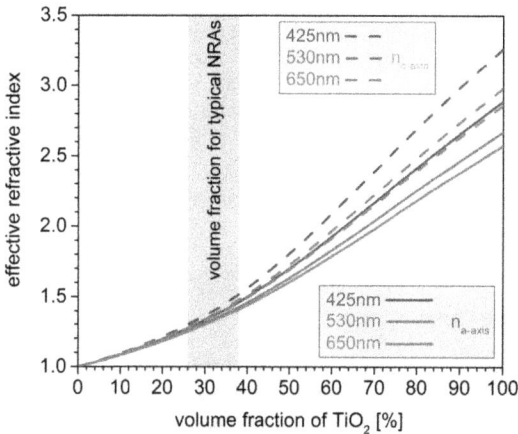

Figure 10.3: *Effective refractive index as a function of the volume fraction of TiO$_2$ in 1.5 µm thick NRAs in air ($n_{air} \approx 1$). The simulations are based on the Bruggeman effective-medium approximation (EMA) model and Equation 2.12. The wavelength-dependent refractive index is taken for 425 nm, 532 nm, and 650 nm from Figure 2.9.*

The fact that the rods are not aligned exactly in parallel results in disparities between the real system and the model. This becomes significant for low rod densities and at edges of NRAs grown in confined geometries. But this effect is compensated by another feature of the NRA: For normal incidence and nanorods growing strictly perpendicular to the substrate, the electrical field of the incident light is oscillating perpendicularly to the (001) direction of the rutile nanocrystal (c-axis). Along the a/b-axis the refractive index is significantly lower than along the c-axis. For tilted nanorods or besides normal incidence, the field of the incident light is oscillating along the c-axis partly resulting in a larger effective refractive index. In addition, the dense polycrystalline bottom layer with a thickness of 40 nm increases the effective refractive index further and causes a gradient of the refractive index. For small nanorods, the influence of the bottom layer might become dominant.

Anisotropic Nanorod Assemblies as Polarizers

In the second regime, the fabricated superlattices are large enough to interact with the incident light, but the nanorods as their basic elements are too small for individual light scattering and behave still like an effective medium. Now, the anisotropy of the superlattice becomes relevant. Here, mostly narrow and long, rectangular shaped seed areas are applied. Usually, their long side is larger and their short side is smaller than the incident wavelength. Thus, light polarized in parallel to their long side is supposed to be scattered more efficiently than differently polarized light (see also Section 2.5). Since scattering depends on the relation between structure dimensions and the wavelength, scattering of blue light should be less polarization-dependent than green or red light.

Individual Nanorods as Polarizers

The third regime describes scattering of light on nanorods having their length in the same order as the incident wavelength. Due to the dielectric antenna effect described in Section 2.5 scattering is supposed to depend strongly on the polarization. In addition, scattering might show a length-dependent wavelength selectivity. One should note that the orientation of the long axis of the nanorods differs in some structures from the long axis of the superlattice elements. For these structures, one can distinguish between the contribution from single nanorods and patterned NRAs by characterizing the polarization of the scattered light.

10.1 Scattering of Individual TiO₂ Nanorods

Scattering on individual rutile TiO₂ nanorods depends on the orientation of the nanorod in relation to the polarization of the incident light, the length, and the diameter of the rod.

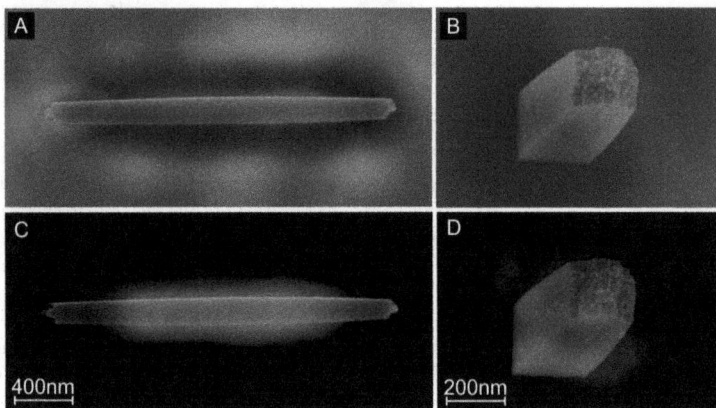

Figure 10.4: *Single nanorods imaged by overlapping pictures from an optical microscope and SEM. Reflection of white LED light on a lying (A) and sticking out (B) nanorod imaged by using the bright field (BF) mode of the optical microscope. Scattered light on a lying (C) and an upright nanorod (D) from the same light source imaged in dark field (DF) mode. Details about the imaging setup are found in Section 4.2.*

10.1.1 Polarization-Dependent Scattering on Individual Nanorods

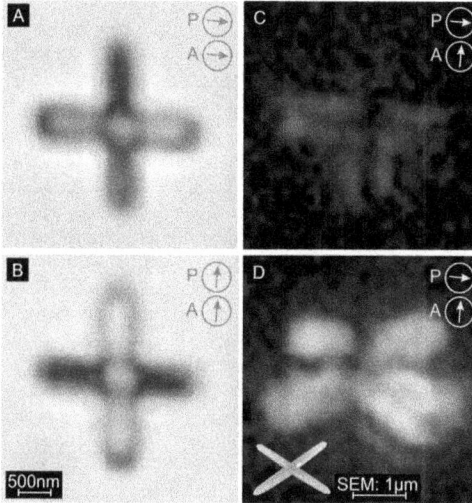

Figure 10.5: *Optical image of branched nanorods. The orientation of the polarizer **P** and analyzer **A** are shown in the upper right corner of each image. A,B) The branch appears bright if the light is polarized in parallel to its long axis. C) The incident light is polarized in parallel to one of the axes. But due to the position of the analyzer, only light polarized in parallel to the other axis is detectable. In this case, the branched rod becomes invisible. D) A neighboring branched rod appears bright under the same polarizer constellation. The branches are tilted by roughly 45° compared to Panel C and the polarization of the scattered light is tilted towards the incident light. The crossed nanorods in the lower left corner of Panel D are imaged with SEM and serve as an example for the structures imaged in Panel A to D with the optical microscope.*

As demonstrated in Figure 10.4, the nanorod scatters light more efficiently if the incident light is polarized in parallel to the c-axis. This is primarily explained by the antenna effect. Vertically grown nanorods scatter light as well because of two reasons. Firstly, the diameter of the rods is roughly one order of magnitude larger than the minimum diameter of TiO₂ particles that show still a detectable scattering of visible light [674]. However, their diameter is much less than $\lambda/2$ and hence, the scattering efficiency is reduced compared to horizontally oriented rods which have a significant larger equivalent spherical radius as described in Section 2.5. Secondly, focused light contains polarization directions perpendicular to the substrate surface as demonstrated in Figure 4.5. For these components of the incident light, scattering on vertically aligned nanostructures is more efficient due to the antenna effect.

The nanorod appears smudgy in the optical image since its diameter is nearly one order of magnitude smaller than the wavelength of the incident light. In Figure 10.4 C , the "optical" diameter appears to be about 500 nm which is more than three times larger than its actual diameter determined by SEM. The length of the nanorod appears in optical images shorter than in SEM images. Hence, it is excluded that the large width of nanorods observed in optical images results from an imaging artifact. The reduced length in Figure 10.4 C is explained by a decrease of the scattered light intensity towards the poles of an antenna as described by Paniagua-Dominguez et al. [1783]. The surface region close to the nanorod is supposed to show an effective refractive index that is lower than that of rutile TiO_2 and silicon but still significantly larger than 1. The characteristics of such an effective medium result in interference effects especially close to the poles of the rod where less light is scattered as shown in Figure 10.4 A. In addition, more light is transmitted into the silicon due to the cascade of the refractive indices. Scattering and transmission reduce the intensity of the directly reflected light which appears as dark areas in the optical microscope images in Figure 10.5 A and B.

The polarization-dependent scattering of rutile TiO_2 nanorods becomes even more significant for branched rods as demonstrated in Figure 10.5. Panel C shows that there is no scattered light polarized perpendicularly to the incident direction of polarization if the incident light is polarized in parallel to one of the two branches. The polarization plane of the incident light in Figure D includes an angle of roughly 45° with the branches. Thus, the direction of polarization of scattered light is tilted by 45° towards the incident polarization. Hence, the scattered light is able to pass the analyzer and the branches appear bright.

The observed colors in Figure 10.5 originate most likely from interference effects. In summary, the findings in Figure 10.5 emphasize that scattering on single nanorods depends strongly on the polarization of the incident light. Consequently, nanorods act as polarizers.

10.1.2 Size Effects on Light Scattering

Rutile TiO_2 nanorods grown on patterned seed layers are usually much smaller than 3 μm. Figure 10.6 shows the size-dependent scattering of individual rods and those embedded in a dense array. This figure indicates that the scattering efficiency drops with decreasing rod size. Hence, for individual rods shorter than 500 nm the contribution to scattering is negligible. This is not related to physical limitation, since 500 nm particles should still be able to scatter visible light. However, the scattering cross-section becomes smaller than the resolution limit of the microscope. This behavior is described by the estimated linear volume dependence of the polarizability of anisotropic particles according to Equation 2.10. As a consequence, sufficiently large nanorods appear very bright in DF images, even if they are vertically oriented or lying. For dense NRAs, the image becomes grainier with increasing size of the nanorods. Diffuse scattering and an increased light transmission related to the cascade of the effective refractive index lower the intensity of the backscattered light. Thus, the sample with long rods appears darker in the optical BF image. But the distribution of nanorods in an array is not homogeneous. Often, the spacing between nanorods is less than the average diameter of the rods and the array has to be considered as an effective medium. Its surface is rough due to the inhomogeneous distribution of the rods. This roughness is assumed to cause the grainy appearance of the NRA in Figure 10.6 F1. The different colors in Panel D1, E1, and F1 originate from interference effects since the thickness of the effective medium increases with the length of the nanorods.

DF imaging is less sensitive to such interference effects which becomes evident from the panels H, K, and M. In contrast to individual rods, the DF image shows no scattering, even for large rods. This is in good agreement with the assumption that the NRA forms an effective medium. The roughness of the effective medium is much lower compared to individual nanorods on a polished silicon substrate. In particular, the surface roughness of the effective medium formed by an NRA is too low for efficient light scattering. This is why even the DF image of an NRA consisting of nanorods with a length of roughly 1.5 μm appears dark while individual nanorods appear bright.

The DF images in Figure 10.6 indicate that the surface roughness of the subjacent seed layer has a minor contribution visible light scattering and reflection significantly [696].

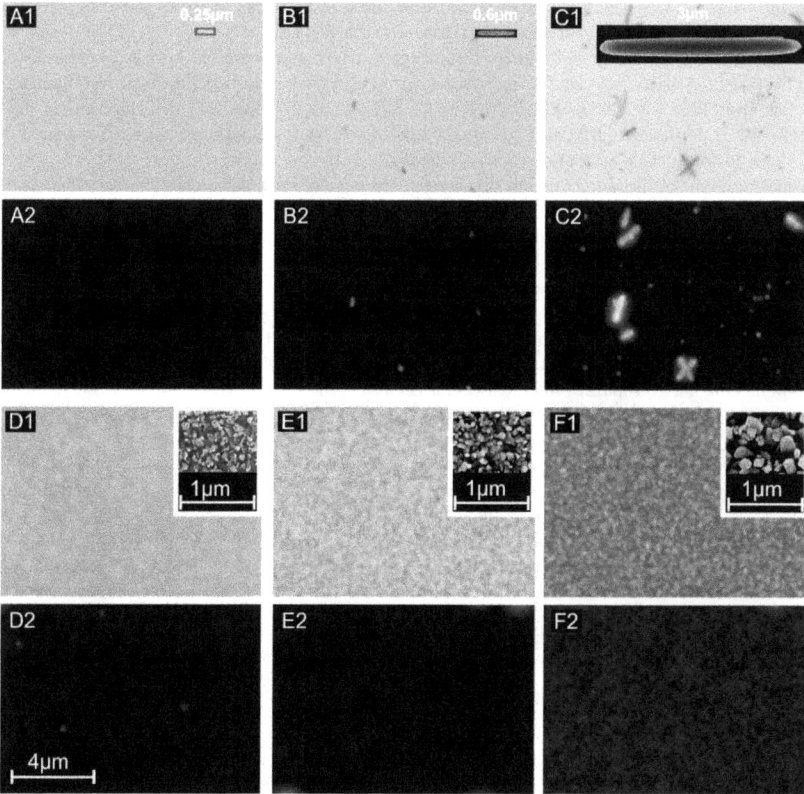

Figure 10.6: *A1,B1,C1) BF images of individual nanorods with various sizes on blank silicon. A2,B2,C2) DF images of the same areas as imaged with BF mode. D1,E1,F1) BF images of NRAs with increasing nanorod size. D2,E2,F2) DF images of the same areas as imaged with BF mode. The inset shows SEM images of the various samples. All SEM images have a four times larger magnification compared to the optical image.*

10.1.3 Increasing Light Scattering with the Orientation of Nanorods

Since the cross-section for Rayleigh scattering drops with the sixth power of the particle diameter and the fourth power of the incident light frequency, tiny nanorods are rather weak scatterers for visible light.

Furthermore, the equivalent spherical radius is much smaller for light approaching in parallel to the long rod axis, compared to normal incidence. Zhang et al. calculated the dependence of scattering efficiency on particle size and wavelength for TiO$_2$ spheres in air [717]. Based on these calculations and the findings from Rayleigh scattering on small TiO$_2$ nanoparticles listed in Section 2.5, visible light scattering becomes strongly wavelength-dependent for rod diameters below approximately 40 nm and disappears for rod diameters below 10 nm if the incident light is polarized perpendicularly to the long rod axis. Thus, methods gaining more tilted nanorods without losing surface area are favored for strongly scattering samples.

A huge amount of tilted nanorods is observed at the edge of seed layers, for instance. Increasing the density of such edges would be a reasonable approach on the way to enhanced visible light scattering. But implementing edge zones enlarges the area without nanorods and this decreases the TiO$_2$ surface area. Thus, it is more advantageous to avoid blank areas by using waved substrates in order to gain more tilted nanorods.

Besides the orientation of single nanorods, another aspect on structured substrates has to be considered. Assemblies of nanorods behave often like an effective medium which is curved at edges and hence, they offer a lot of reflection angles that increase back reflection of focused light as well. One has to separate these effects for each sample.

Individual TiO$_2$ nanorods behave as dielectric antennas in the visible regime as expected. Consequently, the scattering becomes negligible for rods shorter than half a micron and thinner than roughly 20 nm. In addition, nanorods that are parallel to the incident light scatter hardly light. Rods in an array scatter light only if the spacing between the rods are large enough. Otherwise, they behave like an effective medium and thin film interference effects occur. The border zone between individual and collective scattering behavior is illuminated in the following.

10.2 Scattering Superlattices Fabricated with Optical Contact Lithography

Optical contact lithography is a fast and low-cost method to fabricate structures with sizes of roughly 1 μm or larger. This section demonstrates how to design structures with high scattering efficiencies and which effects contribute to scattering.

10.2.1 Scattering at Levitating NRA Membranes

Adding nanorods to shaped surfaces increases scattering. On the one hand, nanorods offer different orientations on curved substrates and hence, there are always some nanorods exhibiting a large scattering cross-section independent of the incident angle. On the other hand, rutile TiO$_2$ nanorod assemblies might act as an effective medium with a relatively high refractive index. As a consequence, the effective refractive index of the subjacent structured surface and thus light scattering is enhanced. This is similar to the optimization of scattering using core-shell structures [710, 711]. Liu et al. reported about enhanced scattering from their spherical nanoparticles after they attached roughly 10 nm wide and 150 nm long TiO$_2$ nanorods [1784].

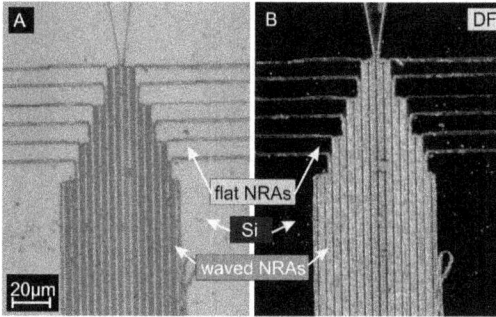

Figure 10.7: *Light scattering at structures made with optical contact lithography. The image was taken with an optical microscope using BF (A) and DF (B) mode. The blue (A) and black (B) background results from the blank silicon substrate. All other parts are covered with an NRA. The horizontally oriented wide stripes show enhanced scattering only at the edges while the remaining array scatters hardly any light. The levitating, vertically oriented stripes scatter more light and appear bright in the DF image.*

Scattering on NRAs with Different Geometries

A comparison between wide, narrow, and levitating stripes and the effect of edges of a very similar structure is given in Figure 10.7. The image was taken using the BF and DF mode of an optical microscope. Blank silicon appears blueish, homogeneous NRAs on broad and narrow stripes reddish, and edges as well as levitating films dark. In the image taken by DF mode, blank silicon appears black, homogeneous NRAs on broad and narrow stripes dark blue, and edges as well as levitating films bright. The mostly vertically oriented nanorods on flat seed layers have a small equivalent spherical radius for normal incidence and hence, light with short wavelengths is scattered predominantly causing a reddish color in BF mode and a blue color in the DF mode image. In addition, the color of the flat NRAs is affected by thin film interference in the BF image. In DF images, such interference effects do not appear and the blueish color is caused only by scattering. Nanorods on the edges of the structure and on levitating seed layers are omnidirectional. Hence, they offer differently sized cross-sections including the largest possible cross-section which is achieved for horizontally oriented nanorods. As a consequence, light with all wavelengths is scattered, tinting the areas in the BF mode image dark and in the DF image bright. As expected, visible light scattering is significantly enhanced on levitating seed layers. This is demonstrated impressively in Figure 10.8 where the DF mode in an optical microscope is applied in combination with a white LED light source.

Scattering at Levitating NRAs

Figure 10.8 offers detailed insight into light scattering properties of the patterned NRA fabricated by optical lithography. The BF images show no polarization-dependent light scattering effects. The grainy appearance known from the full-grown NRAs in Figure 10.6 occurs in Figure 10.8 on levitating films exclusively. Areas, where the film is attached to the silicon substrate, appear homogeneously reddish. Here, one has to consider that the rods have a length of only 500 nm in order to avoid an overgrowth of the superlattice. Thus, the NRA fixed on the silicon

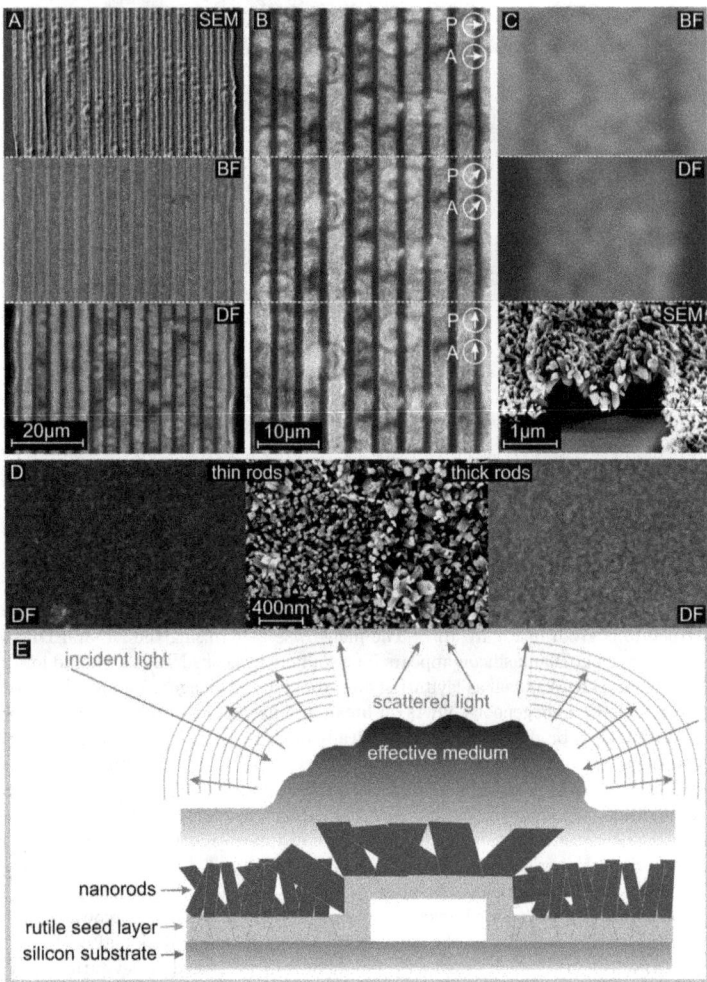

Figure 10.8: *Scattering at levitating NRAs as presented in Figure 10.7: A) SEM (top), optical BF (middle), and optical DF (bottom) image of the investigated structure. B) Magnified DF image of the structure shown in Panel A detecting light with different polarization directions. Here, **P** (polarizer) indicates the polarization of the incident light and **A** (analyzer) the polarization of the scattered light. C) A single stripe of a levitating NRA imaged by optical BF and DF microscopy. The SEM image shows a cross-section of a similar part of the sample. D) Scattered light from flat NRAs consisting of thin (left) and thick (right) nanorods. The SEM image shows both kinds of NRAs. The intersection between both kinds of NRAs is marked with a yellow dotted line.*

substrate behaves certainly as an effective medium and the reddish color results from thin film interference. The levitating films imbed voids between the bottom of the seed film and the silicon. One might expect that these voids are the reason for the enhanced scattering behavior of levitating NRAs. In the optical images, one can see clearly that the grainy character of the levitating film is almost homogeneously spread on the levitating NRA. This is in contrast to the size distribution of the voids underneath the NRA as shown on the SEM cross-section image in Figure 10.8 C. In some places, the levitating film touches the silicon surface whereas in some other places, there is a distance of more than half a micron between the underneath of the NRA and the silicon surface. That is why the void structure is not supposed to contribute to the scattering properties of the levitating NRA. It is more reasonable that the grainy appearance of the levitating NRA is caused by the much larger grains of the levitating rutile seed films. An increasing facet size yields larger rod diameters. As discussed above, it is not the length of the nanorods but their diameter and the spacing between the rods that causes their grainy appearance in the optical images. The size effect of the grains and nanorods on visible light scattering is demonstrated in Figure 10.8 D and E. Since the thickness of 500 nm long nanorods grown on levitating films equals approximately the thickness of 1.5 μm long nanorods grown on fixed seed film, both kinds of NRAs are supposed to show a similar grainy structure in the optical images.

In some areas, DF images show an enhanced scattering for light polarized in parallel to the line pattern in some areas. Since this is not the preferred growth direction of the nanorods, the polarization dependence is caused rather by the superlattice than by individual nanorods. In DF images, the distribution and size of the voids underneath the NRA play a more important role as for BF images. Structures that are elevated further up show an enhanced scattering efficiency due to their larger cross-section area for light at large incident angles. Thus, scattering efficiency is reduced if levitating films do not form a single bow but touch the silicon in between. For light that is polarized in parallel to the line pattern, the scattering efficiency is enhanced by the dielectric antenna effect.

Standard NRAs that are attached to the substrate and are composed of small and thin rods behave as an effective medium. In contrast to that, nanorods with the same length but with a slightly larger diameter and a marginally lower density on a levitating ground layer behave more like a powder of intensively scattering dielectric particles.

10.3 Scattering on Narrow Linear NRA

Anisotropic and narrow NRAs show strong polarization-dependent scattering. In this work, the narrow structures are made with electron-beam lithography, scanning probe lithography, and laser-induced melting. The smallest dimension of the resulting structures is smaller than the incident wavelength.

10.3.1 Scattering at Superlattices Fabricated with Electron-Beam Lithography

In the beginning, scattering on NRAs grown on 60 nm and 1 μm wide TiO_2 seed layers made with electron-beam lithography on polished silicon as shown in Figure 10.9 is discussed. For broad lines, significant scattering appears on the edges, only as demonstrated in Figure 10.9. Apart from the edges, the NRA behaves mainly like a flat effective medium similar to the attached NRAs created by optical lithography shown in Figure 10.7 B. Mostly blue light is scattered by vertically aligned nanorods according to the description of single particle scattering with small

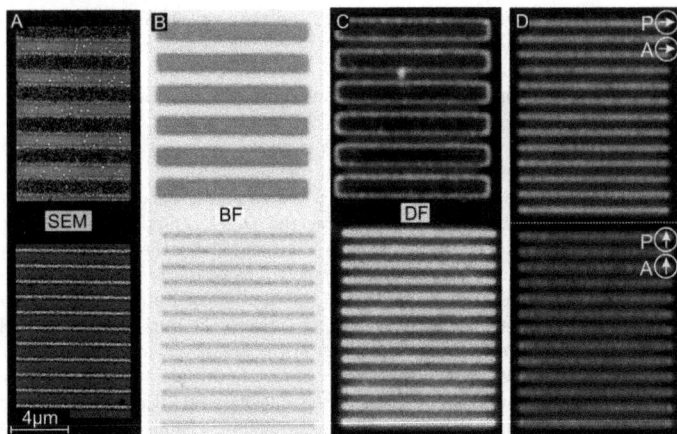

Figure 10.9: *White light scattering on structures made by electron-beam lithography: The image was made by SEM (A) as well as an optical microscope using BF (B) and DF (C,D) mode. The line-shaped NRAs have a width of 1 μm (top) and 60 nm (bottom). The pitch of the applied pattern is 1 μm in both cases. The rods are roughly 200 nm long and 20 nm wide. The DF image demonstrates clearly that only the edges of the broad NRAs (top) scatter light efficiently. In contrast, scattering at thin NRAs (bottom) is more efficient. D) DF images of polarization-dependent scattering display the increased scattering efficiency for incident light polarized in parallel to the line pattern.*

equivalent spherical radii in Figure 2.10. The length of the rods is roughly 200 nm in order to prevent an overgrowth of the superlattice. Since there are many nanorods that are grown perpendicularly to the line pattern and horizontally on the silicon substrate, the average width of the rectangular NRA is about 400 nm.

Polarization-Dependent Scattering

The DF image in Figure 10.9 D indicates that scattering is enhanced if the line is in the plane of polarization. This effect is similar to the scattering on structures fabricated by optical lithography and emphasizes the dominance of the superlattice over the contribution of single rods. Hence, the narrow NRAs behave like an anisotropically shaped effective medium for visible light and act as dielectric antennae.

Figure 10.10 shows the capability of the patterned NRA to change the polarization of incident light. The optical images were taken in the BF mode. For incident light polarized exactly parallel to the line pattern, the center of the lines appears dark. This might be related to the strong scattering efficiency of the antenna-like array which emits light with the same polarization only. By tilting the incident polarization just by a few degrees the polarization of the scattered light differs from the incident polarization in a few spots on the line pattern. In Figure 10.11, three effects that might be involved in affecting the polarization of the scattered light on narrow NRAs are drafted. The SEM images covered by the BF and DF images in Figure 10.10 do not allow to prove the dominance or exclude one of the suggested effects. As expected for structures acting as linear polarizers, the highest light intensity is detected at a tilt angle of 45°. For larger angles,

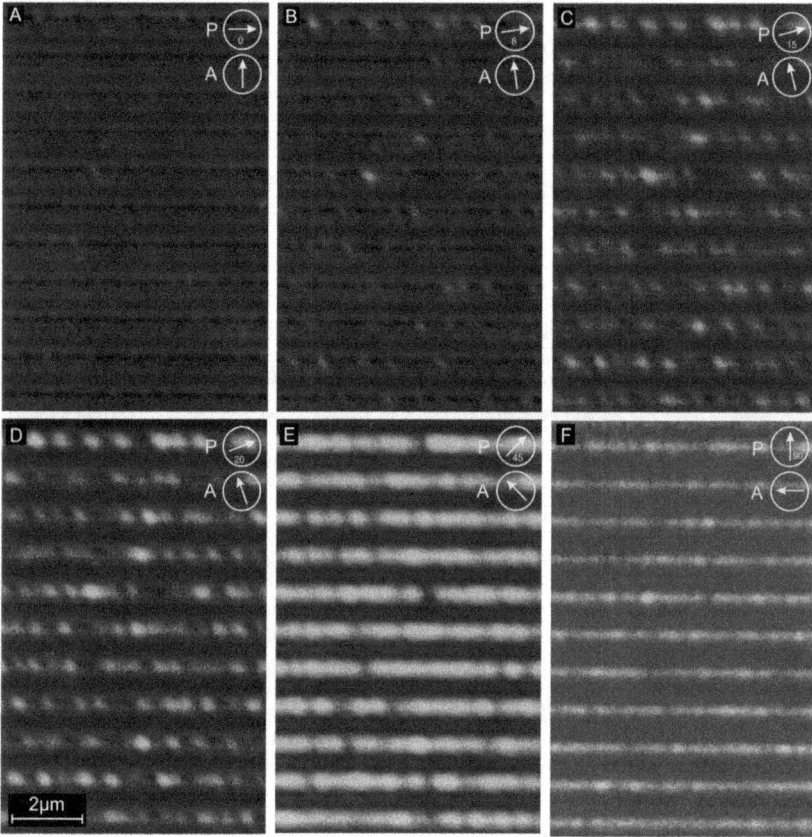

Figure 10.10: *Polarization-dependent light scattering on narrow NRAs made by electron-beam lithography. Here, an overlay of an optical BF image and a SEM image demonstrates the ability of the structure to change the polarization of incident light. Hence, polarizer P and analyzer A are aligned orthogonally. Obviously, the effect on scattering by changing the polarization shows a very sensitive angle dependency. The tilt angle of P and A in relation to Panel A is given in each frame in degrees. Since the superstructure works as a polarizer, the influence on the polarization is maximized for an angle of 45°. But the effect is not homogeneously distributed across the lines. At first sight, there is no obvious relation between darker areas and the subjacent structure of the NRA. These differences are most likely caused by tiny density changes in the NRA that are not observable in the SEM image.*

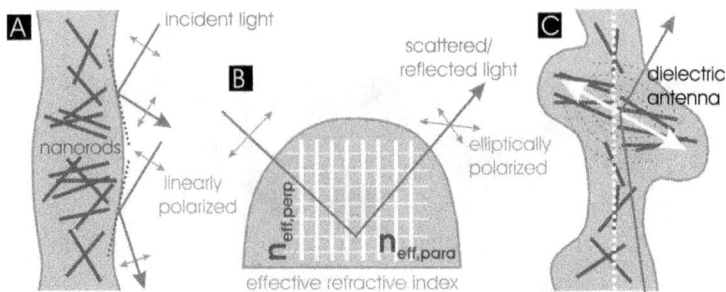

Figure 10.11: *Processes affecting the polarization direction of the reflected and scattered light in real patterns: A) Inhomogeneities of the nanorod density and the orientation distribution cause a warped surface of the effective medium. Consequently, photons which are reflected in dissimilar positions on a line pattern are polarized slightly different. B) If these inhomogeneities cause gradients of the effective refractive index within the order of the wavelength or shorter the refractive index depends on the polarization orientation. Hence, most of the photons become elliptically polarized and might not pass the analyzer. C) Another effect of these inhomogeneities could be the formation of antenna-like structures which are not oriented in parallel to the main direction of the line pattern and act as additional polarizers.*

the intensity of light changing its polarization during scattering is decreasing and distributed more homogeneously along the line pattern. But even for incident light polarized perpendicularly to the line pattern, there is scattered light with a changed polarization state.

Astonishingly, DF images do not show the dark stripe in the center of the linear NRAs for incident light polarized in parallel to the line pattern. Applying the DF mode, light approaches with large incident angles only. According to Figure 4.5, a significant part of this light is polarized perpendicularly to the sample surface. Hence, the light is composed of two components (in and out of plane) interacting differently with the NRAs and the light might become elliptically polarized more likely than in case of steep incidence.

A narrow line of randomly oriented nanorods with a width in the order of the rod length scatters light which is polarized in parallel to the line more efficient than perpendicularly polarized light. Hence, the line acts rather than an effective medium and the scattering properties of the single rods disappear.

10.3.2 Scattering Superlattices Fabricated by Scanning Probe Lithography

In contrast to electron-beam lithography, scanning probe lithography does not require any additional seed layer. Here, the film is affected within a few nanometers close to the surface and scattering can originate from the nanorods only. In particular, any contribution to scattering from the seed layer is excluded for these structures. The similarity of white light scattering on electron-beam and scanning probe lithography generated structures indicates that scattering on seed grains is insignificant for the structures made by electron-beam lithography as well.

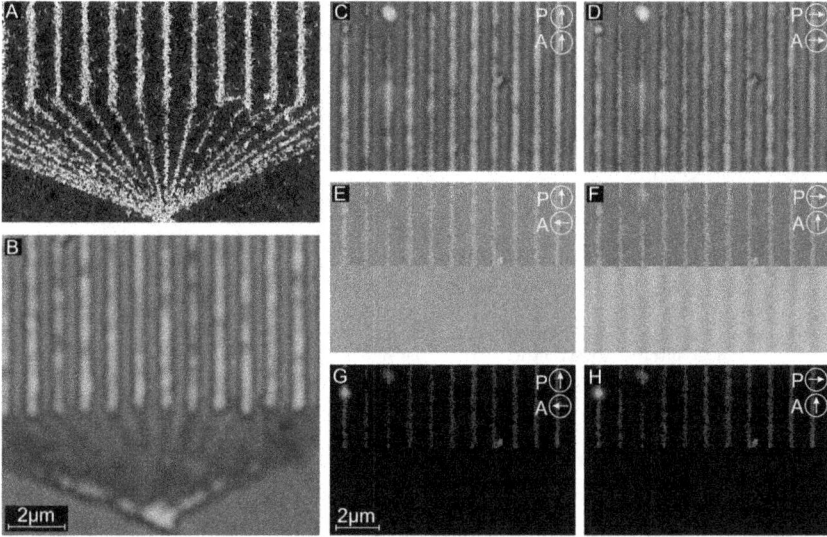

Figure 10.12: *Scattering on superlattices fabricated by scanning probe lithography. A) SEM image of nanorods on line patterns. B) Optical BF image of the same structure as shown in Panel A. C,D) Polarization-dependent scattering observed in BF mode. E,F) The ability to change the polarization probed in BF mode. G,H) The ability to change the polarization probed in DF mode. The upper part of the Panel E to H displays an overlay of the SEM and optical image. The lower part displays the optical image only.*

Polarization-Dependent Scattering

The scattering behavior of structures made by scanning probe lithography is demonstrated in Figure 10.12. Polarization-dependent effects are shown in Figure 10.12 C to H. Panel C and D show that visible light scattering is enhanced if the light is polarized in parallel to the line pattern. Hence, these structures function as a dielectric antenna consisting of an anisotropically shaped effective medium similar to the line pattern created by the electron-beam lithography. The SEM images covered by optical images in Figure 10.12 demonstrate that the brightness of scattered light is very sensitive to the local density and orientation of nanorods. Here, the correlation between rod density and scattering efficiency is clearer compared to Figure 10.10. And, similar to the structures created by electron-beam lithography, there is no light scattered which is polarized perpendicularly to the line pattern if the incident light is polarized in parallel to it. This indicates that this effect is rather a general scattering feature of such narrow nanorod assemblies than something which depends on a specific fabrication technique. The changed polarization of scattered light is much less distinct in the DF compared to the BF mode (Figure G and H), however, it is still observable. This is similar to the scattering behavior of structures created by electron-beam lithography.

In some areas of the linear NRAs, the polarization of the scattered light is changed in a different way than in most of the other parts. For structures, fabricated with electron-beam lithography, the correlation between the position-dependent polarization effect and the exact local density and orientation of the nanorods is hardly seen. The linear NRAs formed by the scanning probe method are not grown on an anisotropic seed layer but on a homogeneous TiO$_2$ film. Consequently, no effect from the seed layer is expected. The observations on these structures suggest that the space-resolved polarization-dependent scattering originates from local nanorod accumulations with a high density and a preferred orientation as illustrated in Figure 10.11.

10.3.3 Scattering at Laser-Fabricated Superlattices

The rutile TiO$_2$ nanostructures created by laser-induced melting presented in Chapter 9 are one of the most interesting structures for visible light scattering due to its high order. Light scattering appears in nearly all positions of the double wall system, no matter if the wall is upright or strongly tilted.

Nanowalls as an Effective Medium

Due to the small dimensions of the structure, an effective medium is supposed to describe the interaction with visible light sufficiently. Thus, the variety of colors seen in the BF images in Figure 10.13 C to E results most likely from thin film interference effects. Strongly tilted walls are equivalent to films with a thickness of roughly 30 nm which corresponds to the width of a single rod. In these areas, violet light is reflected efficiently. Upright walls appear as a much thicker effective medium. Related to their tilt angle they reflect blue, green, yellow, or red light preferentially. Similar to the previously presented superstructures, the intensity of the reflected light is maximized for incident light which is polarized in parallel to the line pattern.

Polarization-Dependent Scattering

The line patterns presented in Figure 10.13 acts as a polarizer similar to the previously presented patterns. An interesting feature is revealed in Panel C to E. Obviously, the interference effect is coupled with the polarization effect of the line pattern. It can be seen that the same region of a nanowall enhances different wavelengths for different incident angles and detected polarization directions. This effect is explained by the highly anisotropic dielectric nanowalls which have direction-dependent refractive indices. This is in good agreement with the model presented in Figure 10.11 B and the generation of elliptically polarized light in the previously presented structures.

The DF images depend much less on the polarization than the BF images. This effect is expected since it is known from Figure 4.5 that light with large incident angles has a large and polarization-dependent field amplitude perpendicular to the surface plane. Thus, the height of the nanowalls is not negligible and contributes to scattering as well.

The wavelengths that are reflected from linear nanowalls depend on the polarization of the incident light. This suggests that the direction-dependent refractive index of this structure plays an important role.

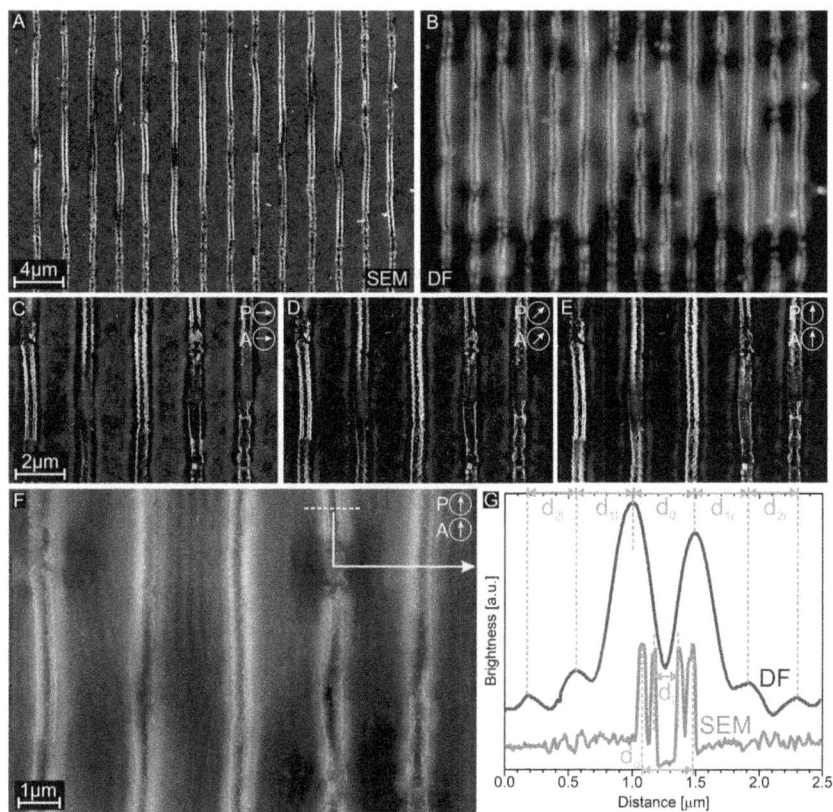

Figure 10.13: *Wavelength-dependent scattering, interference effects, and direction-dependent scattering on TiO_2 nanostructures created by laser-induced seed melting. A) SEM image of the applied structure. B) Color enhanced optical DF image of the same structure as shown in Panel A. There are intensity patterns between the structures in certain regions. A comment to these patterns is given in the appendix (Section 13.5). C to E) Zoom of the SEM image in Panel A overlaid with the BF image for different incident polarization angles showing that parts of the structure that act as polarizers. P and A are given as described in Figure 10.5. The different colors rise from interference effects. F) Same area as shown in Panel C to E probed by DF mode with light that is polarized in parallel to the nanowalls. G) The brightness of the cross-section marked with a blue dotted line in Panel F for the SEM and DF image. The lateral distance in the SEM image represents the real dimensions of the structure while the DF image shows the dimensions detected by an optical microscope in the DF mode.*

11 Time Evolution of the Transient Current in TiO$_{2-x}$ Nanocrystals

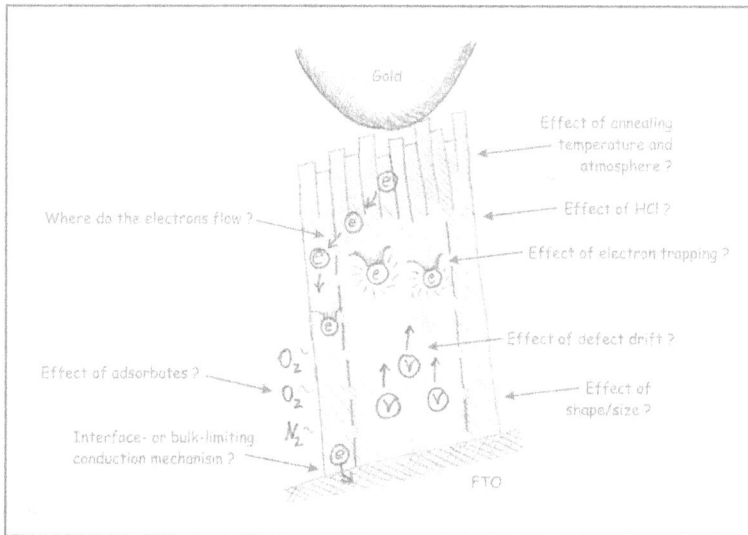

Figure 11.1: *Scientific objectives investigated in this chapter. Where do the electrons flow? In which way is the electron transport affected by trapping, defect drift, electrode materials, post-annealing temperature, post-annealing atmosphere, post-growth hydrochloric acid exposure, structure shape, structure size, and adsorbates?*

The electronic properties of rutile TiO$_2$ nanorods are quite complex. Dependent on its stoichiometry, it has properties of an insulating dielectric material, an n-type semiconductor, or a metallic material. An elaborate behavior of the transient current appears when the oxide switches between different properties during voltage or current stress. In particular, nanorods without post-annealing offer a huge number of defects such as oxygen vacancies and extended defects that might change in terms of density and distribution while applying electronic stress. A time-dependent transient current originates from surface reactions [1785], charging [1346], a shift of charges into another region of the nanocrystal [1343, 1344], the creation or the shift of defects [1343, 1344], or a Mott transition [216]. Up to the present, there is a lack of detailed reports describing the relation between the internal structure of TiO$_2$ nanorods and their electronic properties [1574]. Here, current at constant voltage stress (CVS) was applied to differently composed TiO$_2$ nanostructures. Induced changes of the electronic properties affect the $I - V$ characteristics. Possible bulk- and interface-limited conduction mechanisms and effects causing a time-dependent current at CVS conditions are listed in Chapter 2.

In some cases, many effects are overlapping and an analytical description of the transient current behavior is not possible. To learn more about the complex behavior, the CVS behavior of the investigated rutile TiO_2 nanocrystals is demonstrated and discussed firstly. Subsequently, an interpretation of the $I - V$ characteristics based on the results of the CVS measurements is given. Typically, time-dependent effects cause hysteresis behavior in $I - V$ characteristics. A summarizing image of the supposed effects describing the overall behavior of the measured transient currents is drawn. The investigated nanocrystals are listed in Figure 11.2. It is noted that the shown results are just a minor but very interesting excerpt from several intensive studies employing large-area and micro-electrodes on TiO_2 films and PtIr instead of gold tips on NRAs. A part of these results is published elsewhere [214, 1786].

To keep this chapter short, experimental characteristics concerning measurements with metal tips as top electrodes, capacity effects, the relation between displacement current and measured current, effect of light on the measurements, the approximation of Joule heating at high current densities, the idea behind the measurement protocol, and possible reasons for asymmetric $I - V$ characteristics are found in the appendix (Section 13.6.1).

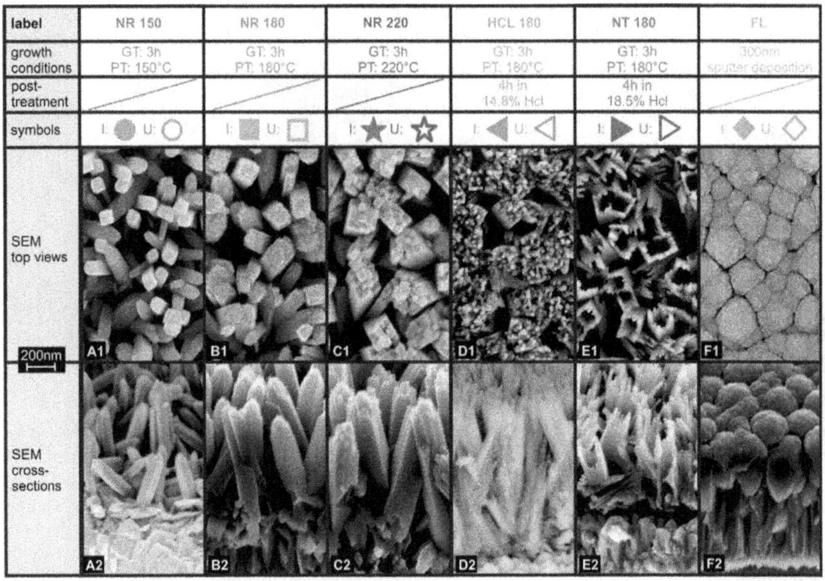

label	NR 150	NR 180	NR 220	HCL 180	NT 180	FL
growth conditions	GT: 3h PT: 150°C	GT: 3h PT: 180°C	GT: 3h PT: 220°C	GT: 3h PT: 180°C	GT: 3h PT: 180°C	300nm sputter deposition
post-treatment				4h in 14.8% Hcl	4h in 18.5% Hcl	
symbols	I: ● U: ○	I: ■ U: □	I: ★ U: ☆	I: ◀ U: ◁	I: ▶ U: ▷	I: ◆ U: ◇
SEM top views	A1	B1	C1	D1	E1	F1
SEM cross-sections	A2	B2	C2	D2	E2	F2

Figure 11.2: *Samples employed for electronic characterization. All hydrothermally grown nanocrystals were grown in 14.8% HCl with 350 µl titanium(IV) butoxide for a growth time (GT) of 3 h at various process temperature (PT). Two chemical post-treatments were applied to obtain larger fingers (HCL) and nanotubes (NT). The shown symbols for the transient current (I) and applied voltage (U) are used in the following figures. SEM top views (A1-F1) and cross-sections (A2-F2) are displayed as well. Since the SEM images show no differences for as-grown and annealed nanorods, the images are not displayed here. Post-annealing was performed at 550 °C in vacuum and oxygen atmosphere.*

11.1 General Model for the Electron Transport in As-Grown Rutile TiO_{2-x} Nanorods

In this section, a model describing the most outstanding observed features of the transient current in the CVS and $I - V$ characteristics of as-grown rutile TiO_{2-x} NRAs is introduced. Based on the drift and diffusion of electrons and ionized donors and the creation of charged regions by trapping, this model describes mechanisms that affect the transient current. The model is an extension of a model suggested by Miao et al. for planar TiO_2 memristive devices [1434]. Besides reports in the literature describing similar systems, the two components in the model are motivated by the following experimental observation. A short sequence of CVS was performed on as-grown nanorods as shown in Figure 11.3. The two linear slopes indicate two time constants which suggest strongly the occurrence of at least two time-dependent mechanisms. Obviously, one of the effects is relaxing completely and consequently, the initial current amplitude is constant for all pulses while the final current amplitude is increasing. In general, the fast and slow processes are correlated with subsurface charging and ion migration [915, 916, 1207, 1326, 1354, 1355, 1434, 1492].

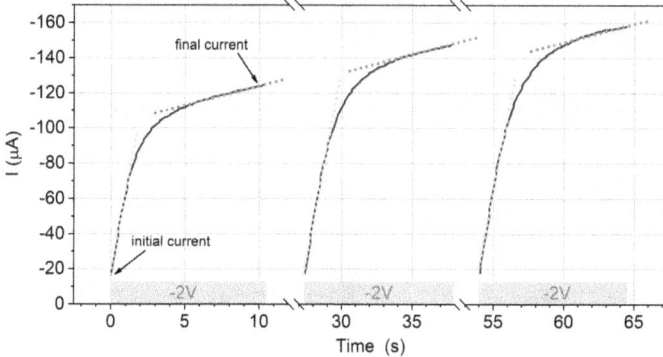

Figure 11.3: *Transient current through a $PtIr$-$tip/TiO_2NRA/FTO$ device with nanorods grown at 180 °C without post-annealing. Three consecutive pulses with the same voltage were applied. The transient current in each pulse is affected by two time constants which are suggested by the linear lines. The initial current in each pulse is the same for all pulses while the final current is increasing. This indicates that the system is affected by at least two mechanisms with different rise and relaxation times.*

11.1.1 Separating the Rod into an Upper and a Bottom Part

It is reasonable that the electronic properties of the nanorods are linked with their atomic structure [1206, 1787]. In Chapter 7, it was shown that as-grown nanorods consist of fingers that are pronounced mainly in the upper part of the rods [168]. Hence, the nanorod is separated into the surface including surface states, a defect-rich top part, and a defect-poor bottom part. The defects are mainly oxygen vacancies with thermally excitable electron donor levels. The oxygen vacancy is a neutral defect and it becomes positively charged when electrons are excited from their donor states. Surface traps are most likely neutral hydroxyl adsorbates which become negatively charged when they trap an electron [1232].

The bottom part is linked to FTO and approaches the properties of highly crystalline rutile TiO_2 with a low number of intrinsic mobile charge carriers. Hence, the response to an electrical field is expected to be limited to a fast and completely reversible displacement of core electrons close to atomic nuclei. This generates a typical capacity-related time-dependence. If electrons are injected into this part of the rod, they exceed the number of intrinsic charge carriers already at relatively low applied electrical fields. In this situation, the current will be space-charge-limited. The upper part is connected with the gold tip and is dominated by crystal defects such as grain boundaries (mainly parallel to the long rod axis) and oxygen vacancies. At room temperature, these defects are ionized partly and deliver mobile electrons. As a consequence, the conduction of electrons is supposed to be affected by Poole-Frenkel emission rather than SCLC [1213]. An ionized oxygen vacancy is positively charged and attracted from the cathode [1788]. Therefore, oxygen vacancies accumulate at the side of the top part facing towards the cathode.

In hydrothermally grown nanorods, there are at least two origins for the time dependence of transient currents: electron trapping/detrapping and ion migration.

11.1.2 The Effect of CVS on the Electronic Landscape

Thermal equilibrium: The work function of gold and FTO are 5.1 [789–791] and 4.8 eV [769, 777–780], respectively and the chemical potential of nanocrystalline rutile TiO_2 is 4.9 eV [766]. The latter value is assumed for the bottom part of the nanorod which has the slightest deviation from stoichiometric TiO_2 crystals. Hence, an accumulation layer is formed at the FTO/TiO_2 interface and a depletion layer is found at the Au/TiO_2 interface suggesting an ohmic contact towards the FTO and a Schottky barrier at the gold interface [79, 1789]. However, it has to be considered that electrons tunnel from the FTO into TiO_2 interface states and create an interface dipole that annihilates the accumulation layer and might even cause a depletion layer. Due to the small energy between the Fermi level and CBM in TiO_2 as well as the large contact area, the FTO/TiO_2 interface is supposed to behave as a quasi-ohmic contact. An out-diffusion of oxygen ions to the FTO related to its higher work function compared to titanium is suggested to lower the FTO/TiO_2 barrier additionally [1790–1792]. Related to the high donor density in the upper part, the Fermi level is raised and thus, the conduction band is lowered. A schematic drawing of this model is given in Figure 11.4. This results in an internal interface (abbreviated with "T/T"). It is noted that the nanorod has not only inhomogeneities along its long axis but also perpendicular to it. This is considered later when the electronic characteristics of each structure are discussed in detail.

Positively biased gold tip: Injected and thermally activated electrons move from the FTO to the gold tip and are faced with two energy barriers at the FTO/TiO_2 and Au/TiO_2 interface as illustrated in Figure 11.5. Electrons might pass the Schottky barrier at the Au/TiO_2 interface via Schottky emission, thermionic field emission, Fowler-Nordheim (FN) tunneling, or trap-assisted tunneling (TAT). At the T/T interface, there is a superior number of ionized oxygen vacancies due to the higher slope of the conduction band. Consequently, a positive space-charge is created [1788]. On the one hand, the positive charge increases the internal electrical field towards the FTO interface and hence lowers this interface barrier [1347]. On the other hand, the positive charge lowers the electrical field towards the gold electrode and decreases the current in the upper part of the nanorod. In summary, the positive charge is able to promote or attenuate the transient current and might even switch the conduction mechanism from interface- into bulk-limited conduction. Since it is assumed that the FTO contact is quasi-ohmic, the switching is not

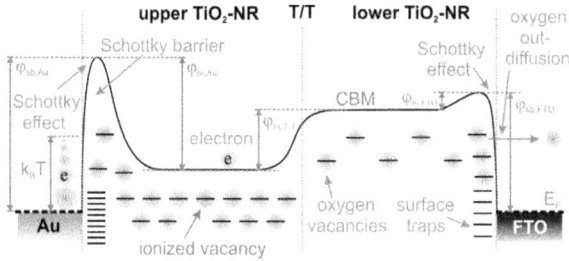

Figure 11.4: *Schematic drawing of the assumed electronic landscape in hydrothermally grown rutile TiO_2 nanorods in the virgin state. The energetic distribution of electrons above the Fermi level at a temperature T is displayed with a vertical sequence of blue spheres. The interface between the upper and lower part is labeled with T/T. The conduction band minimum is labeled with CBM.*

supposed to be caused by this effect in the investigated nanocrystals. The effect of the positive charge is enhanced by mobile ionized oxygen vacancies that drift towards the T/T interface and accumulate there. The enhanced density of donors pulls the CBM further downwards. Furthermore, the vacancy drift affects the Schottky barrier thickness at the Au/TiO$_2$ interface. The drifting oxygen vacancies lower the donor density close to the Schottky barrier and the CBM is raising. Consequently, the Schottky barrier height decreases slightly [79]. Vacancies that are situated inside the depletion area are ionized partly as a result of the internal field near the gold electrode. Close to the maximum of the Schottky barrier, the internal field becomes strong and ionized vacancies might be able to drift towards the maximum of the Schottky barrier. However, this drift appears immediately after contacting the gold tip with the TiO$_2$ nanorods and since the applied field does not increase the edge steepness of the Schottky barrier no further drift of vacancies towards the TiO$_2$/Au interface is expected.

Relaxation of effects caused by positively biased gold tip: Injected electrons remain in the lowered conduction band of the defect-rich TiO$_2$ since they are encapsulated by the Schottky barrier and the T/T interface. This electron bubble is a non-equilibrium state of the system and it tends to relax via electron diffusion towards the electrodes [1793]. The positive charge at the T/T interface is eliminated by the diffusing electrons which lift the CBM locally and pushes electrons towards the FTO. The depletion region and Schottky barrier are still extended and electrons cannot diffuse to the gold electrode efficiently. Neutral defects diffuse along density gradients and consequently, the thickness of the Schottky barrier is reduced. On the one hand, this requires low binding energies that are not found in highly crystalline TiO$_2$ and it is not confirmed that neutral oxygen vacancies can diffuse along grain boundaries and other extended defects at room temperature easily. On the other hand, the thickness of the depletion layer in TiO$_2$ is supposed to be only a few nanometers [777], and in highly doped TiO$_2$ even less. Hence, assuming a diffusion of oxygen vacancies of less than roughly 2 nm is already reasonable to explain measurable relaxation effects.

Negatively biased gold tip: As soon as a negative potential is applied to the gold tip, the electron bubble is pulled to the FTO and generates a (non-exponentially) decaying current, which is overlaid by the transient current from the gold tip to the FTO. Besides that, oxygen vacancies are ionized and drift towards the Au/TiO$_2$ interface as illustrated in Figure 11.6 [79]. The enhanced vacancy density lowers and narrows the Schottky barrier and the interface current

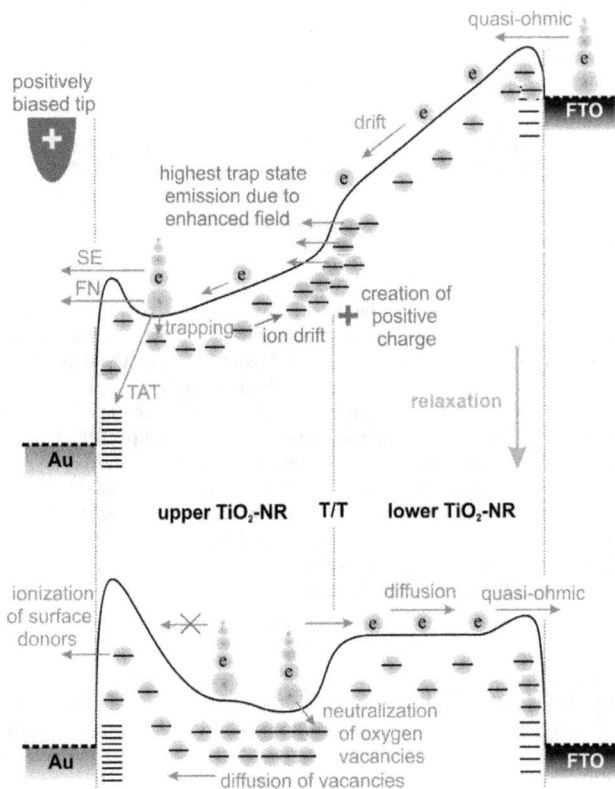

Figure 11.5: *Schematic drawing of the assumed electronic landscape in hydrothermally grown rutile TiO$_2$ nanorods in case of a positively biased gold tip potential and during relaxation after CVS (Schottky emission (SE), Fowler-Nordheim tunneling (FN), trap-assisted tunneling (TAT)).*

is increased [1794]. Furthermore, the edge steepness of the Schottky barrier is increased and oxygen vacancies situated near the Schottky barrier maximum are ionized preferentially and a positive space-charge is created. This space-charge reduces the Schottky barrier additionally and in combination with the image charge in the gold tip a strong interface dipole is formed. The positive charge lowers the slope of the conduction band in the bulk and thus, a switching from interface- to bulk-limited conduction mechanism might appear. Independent from the polarity of the tip, this model follows the received opinion that the resistive switching appears mainly in a thin subsurface layer [1207, 1794].

Relaxation of effects caused by negatively biased gold tip: The electron bubble appears again and tends to relax. This time, the Schottky barrier at the Au/TiO_2 interface is still minimized and a bunch of electrons escapes immediately to the gold tip causing a decaying discharge current. As time goes by, the accumulated ionized oxygen vacancies close to the interface are neutralized by diffusing electrons and diffuse towards the bulk driven by the density gradient. Consequently, the initial shape of the Schottky barrier is established and the remaining electrons in the electron bubble diffuse mainly towards the FTO.

The presented model is used to describe all investigated nanostructures and is based on structural properties which are known from previous chapters or other scientific reports. The main elements of this model are an inhomogeneous defect distribution in the rod, the formation of charged volumes inside the rod, ion drift and relaxation close to the Schottky barrier, and trapping and detrapping of electrons. The model is presented in the beginning as a support for the reader. It should provide an easier understanding of the complex experimental results in the following.

11.1.3 A Simplified Model for I–V Measurements

The driving direction from negative to positive applied bias is called "forward" direction while the opposite direction is called the "backward" direction. According to the context, these terms have to be distinguished from "forwardly" and "reversely" biased Schottky barriers. The latter terms define the polarity of the applied field for a single Schottky barrier. The presented $I - V$ measurements begin with a negatively biased gold tip and continue in the forward direction.

High negative applied voltage: Ionized oxygen vacancies drift towards the Au/TiO_2 interface and additionally, vacancies close to the Schottky barrier are ionized preferentially causing a positive space-charge. Hence, the Schottky barrier is lowered or the contact is even metalized and the conduction mechanism might switch from interface- to bulk-limited conduction.

Low negative applied voltage in forward direction: The migration of ionized oxygen vacancies is decreased and the Au/TiO_2 contact keeps the properties created at higher bias.

Low positive applied voltage in forward direction: The reduced Schottky barrier is conserved when the polarity of the applied field is switched and the conductivity of the interface remains high. However, even for ohmic conduction deviations from the linear slope around 0 V might appear which originate from the discharging of the electron bubble. The accumulated oxygen vacancies begin to diffuse towards the bulk, and ionized vacancies close to the Schottky barrier are neutralized and the barrier becomes larger. The conduction mechanism might switch from a bulk- to an interface-limited conduction again.

High positive applied voltage: Oxygen vacancies become ionized again and drift towards the T/T interface. The accumulation of positively charged vacancies lowers the barrier at the

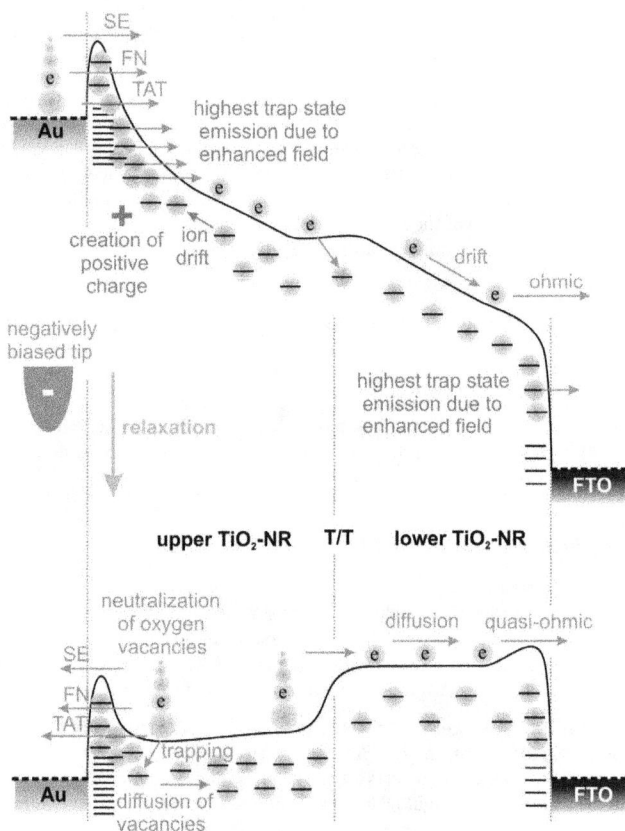

Figure 11.6: *Schematic drawing of the assumed electronic landscape in hydrothermally grown rutile* TiO_2 *nanorods in case of a negatively biased gold tip potential and during relaxation after CVS (Schottky emission (SE), Fowler-Nordheim tunneling (FN), trap-assisted tunneling (TAT)).*

FTO/TiO_2 interface. The conduction band near the Schottky barrier is flat and many ionized donors become neutralized which lifts the CBM and hence, the effective Schottky barrier is lowered. Both effects enhance the interface currents and the conduction mechanism might switch from an interface- to a bulk-limited mechanism again. Nevertheless, ionized donors move from the Schottky barrier towards the bulk and support an increase of the barrier.

Low positive applied voltage in backward direction: Since the edge steepness of the Schottky barrier is affected much less by positive bias, the barrier might not become as low as for negative bias. Consequently, a lower slope at $0\,V$ is expected for the backward direction.

Low negative applied voltage in backward direction: The less pronounced effect on the Schottky barrier should also be seen for negative bias, when the barrier is reversely biased. Since the electrons have to overcome a higher barrier from the gold to the TiO_2 compared to the opposite direction, the conduction mechanism might become interface-limited below $0\,V$.

11.2 General Model for the Electron Transport in Annealed Rutile $TiO_{2\text{-}x}$ Nanorods

In this section, the expected differences between as-grown and annealed nanocrystals are discussed. Post-annealing changes the structure including the distribution of electronic states and hence, the shape of the $I - V$ curve is affected significantly. The investigated CVS changes rather the transient current amplitude than the shape of the $I - V$ curve.

11.2.1 The Effect of Annealing in Vacuum on the Electronic Landscape

The nanorod consists of small crystalline fingers separated by grain boundaries. These boundaries are too large to be annihilated during annealing at $550\,°C$. At elevated temperatures, crystal defects diffuse likely and accumulate at grain boundaries [287] and the surface of the rods [214]. Consequently, the surface and the grain boundaries become very conductive. Inside the fingers, the conduction band rises and the depletion layer width increases. Inside the grain boundaries, both values decrease due to the high donor density. Hence, the conduction is focused on the grain boundaries and surface. The mobility in the grain boundaries is decreased by the higher momentum scattering frequency [1788], but the number of mobile charge carriers is increased by the high dopant concentration. Nevertheless, the conductivity of the whole rod decreases due to the low cross-section of the grain boundaries compared to the total rod cross-section. More important, the conductive surface connects the grain boundaries with the Au/TiO_2 and FTO/TiO_2 interface. In particular, the quasi-ohmic contacts at the latter interface prevent the creation of an electron bubble with a long time and a large number of electrons. Since trapping in metallic volumes is not a dominant process for the charge transport, the creation of a positive space-charge during CVS is expected to be reduced drastically.

Also, the drift of ions is focused on grain boundaries and the surface since their binding energies are significantly lower compared to the highly crystalline fingers [1788].

Surface traps such as neutral hydroxyl groups become negatively charged when electrons are trapped [1232]. Consequently, the canalized electrons at the surface interact with the negatively charged surface traps and the current is reduced by this self-controlled transistor effect.

11.2.2 The Effect of Annealing in Oxygen on the Electronic Landscape

Oxygen annealing reduces the density of oxygen vacancies at the surface significantly [214]. However, the grain boundaries are much less affected by the oxygen annealing and hence, the donor density in grain boundaries is most likely higher than on the surface. Thus, structures with grain boundaries that are completely encapsulated by surrounding TiO$_2$ material have similar features as related structures annealed in vacuum. In contrast, structures such as nanotubes without encapsulated grain boundaries are significantly less conductive. Related to the defect-poor shell, there is a thicker Schottky barrier and a reduced number of oxygen vacancies at the Au/TiO$_2$ interface compared to the rods annealed in vacuum. Aside from interfaces, the effect of surface traps is less important since the surface has a low conductivity anyway.

Annealing affects the density, distribution, and composition of point defects. Oxygen vacancies diffuse towards interfaces and the surface forming a defective and conductive shell. Annealing in oxygen reduced the number of surface defects.

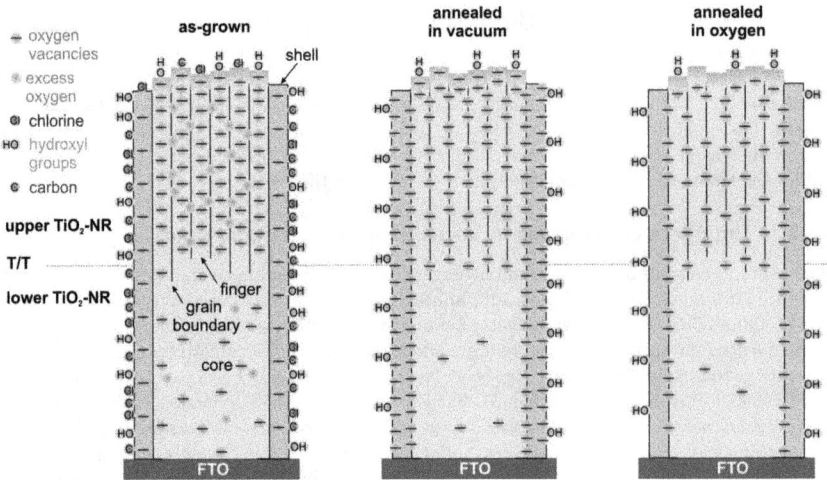

Figure 11.7: *Schematic drawing of defect distribution in as-grown and annealed nanorods. During annealing, some defects are healed and other defects move towards grain boundaries or the surface where they accumulate and form thin conductive pathways. Annealing in oxygen atmosphere reduces the number of surface defects more than annealing in vacuum. The surface defects are mostly oxygen vacancies, hydroxyl groups, and additionally for as-grown nanorods chlorine and carbon residues.*

11.3 Electron Transport in Specific Structures

In this section, the experimental observations for particular nanostructures are discussed and linked with the general model introduced above. The discussed results include polarity-dependent short CVS (Figure 11.8) and long CVS (Figure 11.9, 11.11, 11.13) including their relaxation (Figure 11.10, 11.12, 11.14) as well as $I - V$ characteristics (Figure 11.15 to 11.20). Since the exact contact area is not known, the absolute current instead of the current density is given in the graphs. Hence, quantitative results concerning the charge carrier density or the Schottky barrier height are not presented. The discussion is mainly focused on the physical effects.

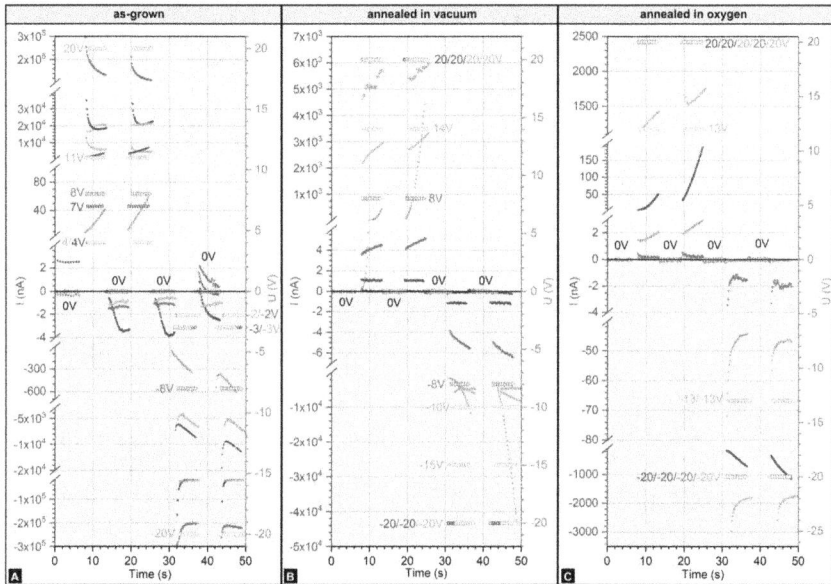

Figure 11.8: *Results for short timescale CVS measurements for all investigated samples. The color and symbol code is given in Figure 11.2. The filled and empty symbols display the measured transient current and the applied bias, respectively. Related to inaccuracies of the employed setup, the pulses of the various samples do not overlap exactly.*

11.3.1 Rutile TiO_{2-x} Nanorods Grown at 150°C (NR150)

As-grown: The quickly declining current observed during the short CVS measurements is related to electron trapping in near-surface trap states such as ionized oxygen vacancies (Figure 11.8). The electron trapping results in a new bias-dependent equilibrium population density. Consequently, the number of free charge carriers and the transient current is varied. It is excluded that the quickly declining current corresponds to a discharge process of a conventional capacitor since the density of stored electrons calculated from the discharge curve would be in the order of 10^{27} electrons/cm^3. This value is about 4 orders of magnitude larger than the electron density in a metal and hence not reasonable.

265

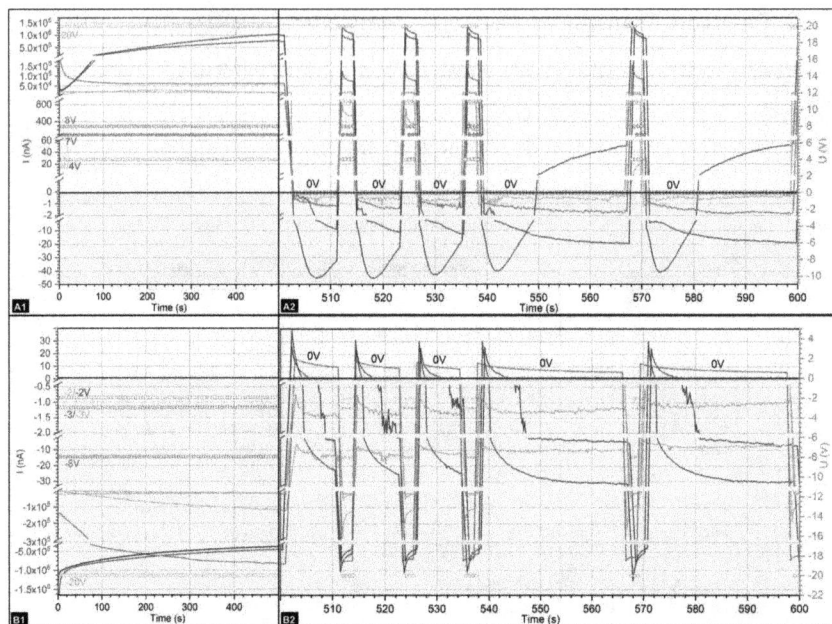

Figure 11.9: *Results for long timescale CVS measurements of as-grown/ as-deposited nanostructures for positive (A1) and negative (B1) bias. Figure A2 and B2 show the first four probing pulses after the long CVS. The color and symbol code are equal to Figure 11.8. The curve for NR150 (light blue) and sputtered films (green) are almost identical for positive CVS.*

The current is increased only to a very small extent during long positive CVS (Figure 11.9) suggesting that ion drift has hardly any effect on the transient current or does not occur at all. Although the effect of positive CVS is small for this kind of NRA, a slow relaxation process is observable (Figure 11.10), which indicates that a few crystal ions are rearranging after the external electrical field is turned off. The relaxation brings the systems in an almost insulating state with a lower transient current as measured at the beginning of the long CVS. Hence, the high transient current at the beginning of the long positive CVS is a consequence of the current stress which was applied during the $I-V-t$ measurements directly before the CVS (the $I-V-t$ measurement is mentioned in Section 13.6.1 and has no further meaning here).

For long negative CVS, the current is increasing and a clear and slow relaxation is observed after the induced stress. This is related to the efficient suction process of ionized oxygen vacancies by the negatively charged tip. Consequently, the current promoting effects succeed and the transient current is rising. The current amplitude in the beginning of the first probe pulses is significantly higher than anytime during CVS. This emphasizes the current blocking property of trapped charges. After the CVS, ionized donors are neutralized by diffusing electrons. Hence, the trapping rate at the beginning of a probe pulse is quite low and the electrons can pass the nanorod without interacting with charged trap species considerably. However, as soon as the donors become ionized in the applied field, the current is affected significantly by trapping and detrapping processes.

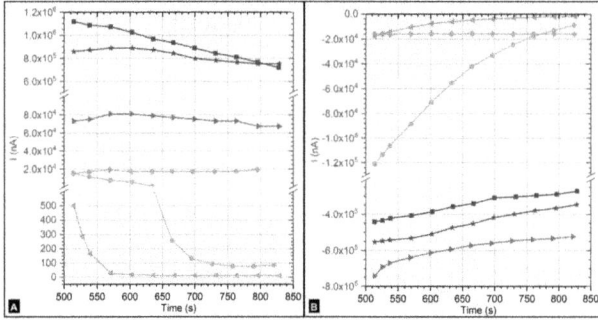

Figure 11.10: *Relaxation after long positive (A) and negative (B) CVS extracted from the probe pulses (the last current value of each pulse is taken) for as-grown nanostructures.*

The current at CVS conditions for the negatively biased tip is about one order of magnitude higher compared to the positively biased tip although a lower voltage is applied. This is related to the different energy barriers for electrons. However, it is notable that also the discharge current at $0\,V$ is also about one order of magnitude larger compared to the positively biased tip. The discharging current is a measure of the number of electrons in the electron bubble. Based on the discharge current, the calculated number of stored charges is in the order 10^{22} electrons/cm^3. Hence, the system behaves similarly to a series connection of a large capacity and a huge resistance. Thus, for short periods of less than a minute, the system behaves like a battery. From that assumption, the capacity of a single nanorod can be estimated. Since the current during CVS is a superposition of a time-dependent transient and a time-dependent charging current it is not clear if the charging process was completed when the bias was turned off after CVS. Hence, the following approximation gives a lower limit of the capacity. The discharge current of roughly $10\,nA$ decays abruptly after $200\,s$. Thus, the emitted charge is in the order of $1\,\mu C$ ($\approx$ 10^{13} electrons). Since the order of emitted charges does not depend on the polarity of the CVS, it is assumed that the capacity is almost voltage-independent and hence given by $C = Q/U$. For an applied voltage of $-2\,V$, the capacity is in the order of $1\,\mu F$. Assuming that roughly 1,000 to 10,000 nanorods were connected with the tip, the capacity of a single nanorod is roughly 0.1 to $1\,nF$. The slope at the beginning of the discharging process is roughly $50\,pA/s$ and hence, the time constant τ is about $200\,s$ and $R = \tau/C$, which is the effective average resistance of a single nanorod at $0\,V$, is in the order of $100\,M\Omega$. It will be demonstrated that a resistance in the same order results from the $I - V$ characteristics of these rods as well. It has to be considered that this capacity is not a conventional geometric capacity and the applied bias of $2\,V$ during CVS and charging have no physical meaning during discharging. Furthermore, the approximated resistance of $100\,M\Omega$ is only valid for a certain electron density in the electron bubble causing a well-defined internal field. Due to the presence of at least one Schottky barrier, the resistance increases for a decreasing electron density in the bubble. This might also explain the abrupt drop of the discharge current after $200\,s$.

The discharge current points for both polarities towards the FTO indicating that the potential barrier at the T/T interface is much lower than the Schottky barrier at the Au/TiO$_2$ interface. After each probing pulse, the discharge current needs some seconds to build up which is most likely a measure of the electron mobility in the defect-poor part between the electron bubble

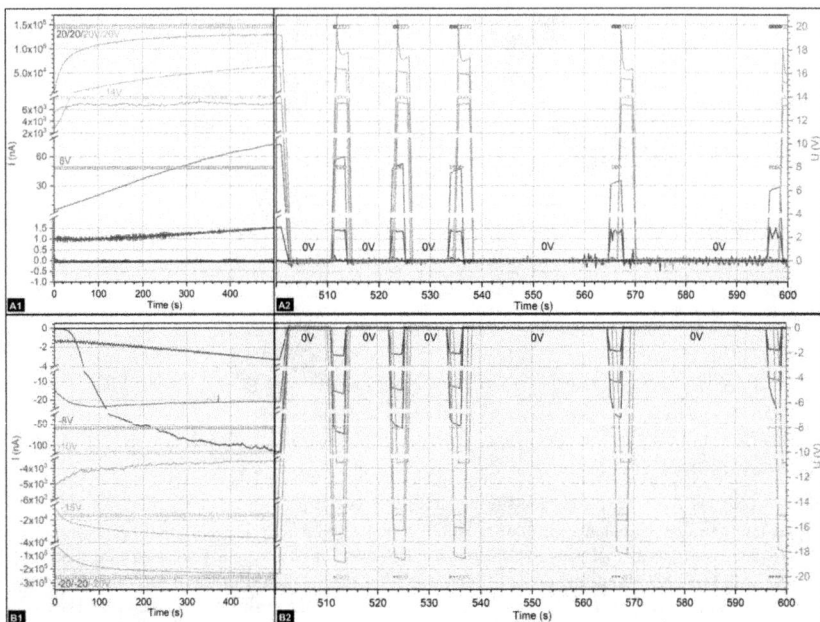

Figure 11.11: *Results for long timescale CVS measurements of nanostructures annealed in vacuum for positive (A1) and negative (B1) bias. Figure A2 and B2 show the first four probing pulses after the long CVS. The color and symbol code are equal to Figure 11.8.*

and the FTO/TiO_2 interface. The time-dependent transient current reminds strongly on the charge extraction with a linearly increasing voltage sweep (CELIV) [1196]. This similarity is not unexpected since the stored charges generate an electrical field between the center of charge and the grounded electrodes which pull the electrons in a defined direction. However, because the exact charge distribution is not known, the self-induced voltage remains unknown. Thus, the correlation between the electron mobility and the time-dependent discharge current is complex and not discussed further here.

The ion drift and relaxation as well as trapping and detrapping of electrons result in a time-dependent current. When the external field is turned off, some electrons remain in the upper part of the nanorod and find their way towards one of the electrodes. The electron density of the charged nanorod is approximately 10^{22} electrons/cm^3.

The $I - V$ characteristic (Figure 11.15) shows clearly, that an ohmic conduction is established at negative bias that results in a linear slope around 0 V. The slope corresponds to a resistance of roughly 500 MΩ, which is in the same order as the resistance estimated from the capacity. Above

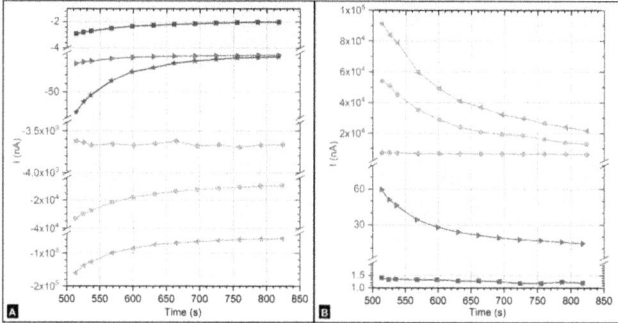

Figure 11.12: *Relaxation after long positive (A) and negative (B) CVS extracted from the probe pulses (the last current value of each pulse is taken) for nanostructures annealed in vacuum.*

0.4 V, the high conductivity is reduced. This is related either to oxygen vacancy migration or electron trapping in ionized oxygen vacancies near the surface. Due to the small space-charge region, oxygen vacancies have to move only a few nanometers to cause a significant effect on the transient current. However, the neutralization of ionized oxygen vacancies increases the Schottky barrier and is also a reasonable explanation for the decline of the current with increasing positive bias. At higher positive voltages, the current amplitude recovers partly in coincidence with the model. Nevertheless, the conduction behavior is clearly not ohmic at 0 V. This observation emphasizes the assumption that the transient current is affected by a positively biased tip much less than by the opposite field polarity.

In the low resistance state, the low-bias resistance is in the order of $10^8 \, \Omega$. This follows from the time constant of the discharge current as well as from the $I - V$ characteristics. Additionally, a negatively biased tip affects the time dependence more than a positively biased tip.

Annealed in vacuum: There is no fast decreasing current in the CVS measurements for both polarities anymore. It is assumed that there are charge transport pathways between the highly conductive grain boundaries, the surface, and the electrodes.

The current is increasing during long CVS measurements for both polarities, but for the negatively biased tip, the increase is faster. This observation fits to the assumption that the effect of the external field on oxygen vacancies (drift and ionization) is enhanced for an electrical field pointing towards the tip. The reduced resistivity relaxes for both polarities with roughly the same rate (Figure 11.12). Compared to non-annealed nanorods, the current enhancement is much stronger and faster since the oxygen vacancy density is higher at the surface.

The Fowler-Nordheim plot in Figure 11.15 indicates that the shape of the Schottky barrier at the Au/TiO_2 interface has changed during annealing. In contrast to non-annealed rods, where no Fowler-Nordheim tunneling is observed, a linear behavior up to ± 1 V appears for the annealed rods. At the same time, there is no hint for a thermal emission at low bias. Hence, the Schottky barrier is still significantly higher than $k_B T$ but the thickness must be in the order of 1 nm to allow efficient tunneling. Besides the Fowler-Nordheim tunneling, there are ohmic pathways

269

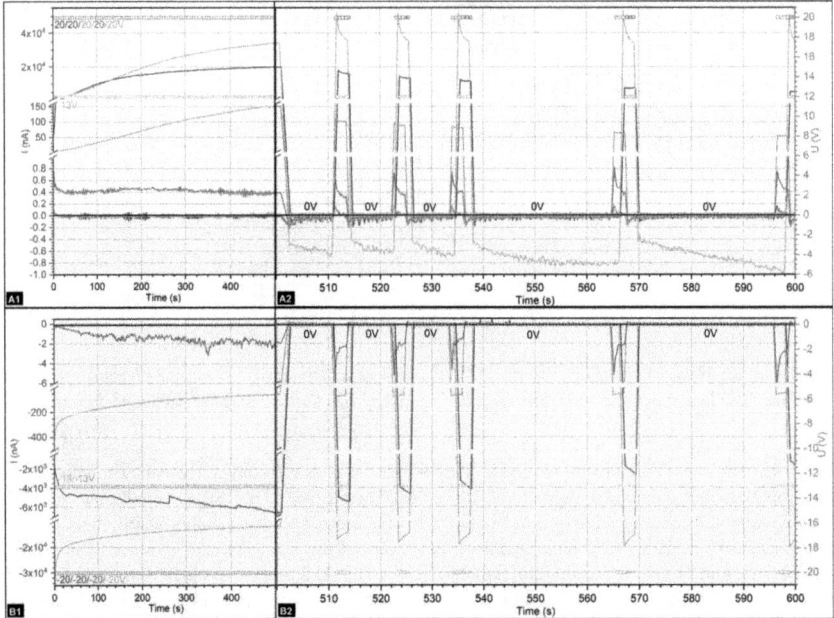

Figure 11.13: *Results for long timescale CVS measurements of nanostructures annealed in oxygen for positive (A1) and negative (B1) bias. Figure A2 and B2 show the first four probing pulses after the long CVS. The color and symbol code are equal to Figure 11.8.*

with a resistance in the order of 10 (after CVS) and 100 MΩ (before CVS) (Figure 11.15 B1). This corresponds to an extremely high resistance between 1 and 10^3 GΩ per single rod which indicates the minor role of these ohmic pathways.

In forward direction, the conduction mechanism is affected by a switching process for both polarities and cannot be determined clearly. The backward direction seems to be almost unaffected by these switching processes and SCLC and Schottky emission is clearly identified for high positive and negative bias, respectively (Figure 11.15 B3 and B4).

Considering the larger bias range applied for annealed rods, the decrease of the total conductivity due to the small cross-section of the grain boundaries becomes apparent. This indicates that the transient current is not limited to grain boundaries in non-annealed nanorods.

Annealing reduces the number of stored charges drastically which might be related to highly conductive pathways between the electrodes. It is suggested that the electrons are restricted to these pathways since the surrounding TiO_2 becomes more crystalline and hence, the energy gap between the grain boundaries and other parts of the rod is enhanced.

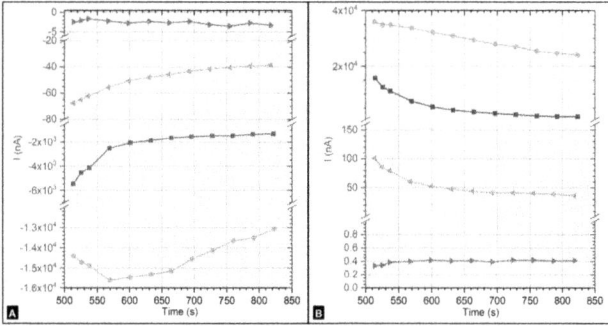

Figure 11.14: *Relaxation after long positive (A) and negative (B) CVS extracted from the probe pulses (the last current value of each pulse is taken) for nanostructures annealed in oxygen.*

Annealed in oxygen: Short CVS measurements show that the fast initial current decline, which is related to the release of trapped electrons, appears only for negative bias (Figure 11.8). This feature is common for almost all investigated structures and allows a deep insight into the electronic landscape of the nanorod. It is reasonable to assume that the crystallinity of the fingers and in the defect-poor bottom part of the nanorod is increasing during annealing. Hence, for a positively biased tip, the electrons can hardly enter the highly crystalline bottom part and prefer grain boundaries and the surface or sub-surface regions as possible pathways. Grain boundaries are rare in the bottom part and the surface is oxidized during annealing in an oxygen atmosphere. Nevertheless, TEM images show that the defective surface shell is a few nanometers thick and hence, not the complete shell might be oxidized. Furthermore, the oxidation of the defective surface layers might not necessarily result in homogeneous stoichiometric insulating TiO_2. Therefore, there is the possibility of conductive pathways between the FTO/TiO_2 and the Au/TiO_2 interface. As a consequence, defects on grain boundaries in the upper part of the rod cannot trap electrons under positive bias because almost no electrons find their way through the highly crystalline bottom part and the partly conductive shell guides the electrons directly toward the Au/TiO_2 interface.

For a negatively biased gold tip, electrons follow the electrical field on any conductive path that is offered close to the Au/TiO_2 interface. Hence, a significant part of the injected electrons flow along the grain boundaries towards the T/T interface. At this interface, they are either injected into the defect-poor part of the rod or they tunnel through the thin fingers towards the conductive sub-surface. The presence of a high number of electrons in the grain boundaries provides trapping in case of a negatively biased tip. Nevertheless, the suggested tunnel current through the fingers towards the sub-surface results in local oxide degradation and conductive filaments might be generated between the grain boundaries and the shell. This is a reasonable explanation why the initial fast current decay is recovered during long CVS (compare Figure 11.8 C and 11.13 A2).

The discharge current points towards the FTO interface and is building up very slowly (Figure 11.14). This indicates that the conductivity through the highly crystalline bottom part or the conductive sub-surface layer is low which is in good agreement with the assumption that the momentum scattering rate in highly defective parts is high and the electron density in the defect-poor bottom part is low.

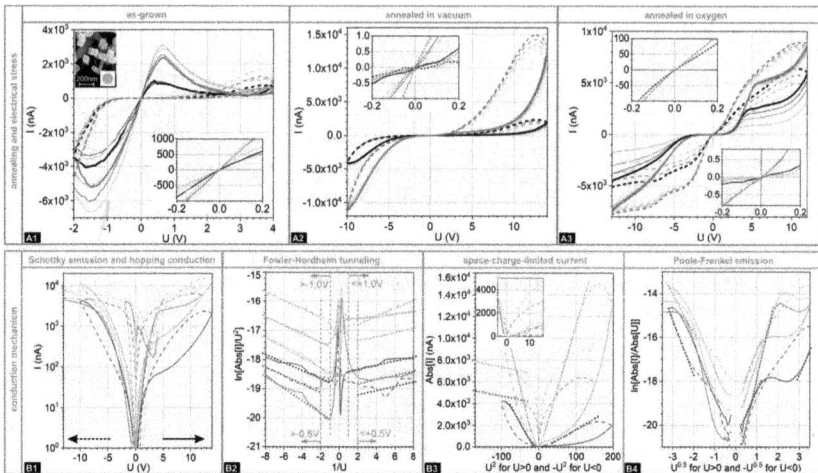

Figure 11.15: $I - V$ characteristics of as-grown (A1), in vacuum annealed (A2), and in oxygen annealed (A3) TiO$_2$ nanorods grown at 150 °C. Five $I - V$ cycles (including forward (solid lines) and backward (dashed lines) direction) are displayed. The order from the first to the last measurement is given by the color of the lines. For measurements before CVS, the first measurement is colored in light gray and the color becomes darker and finally black for the last measurement. For measurements after CVS, the first measurement is colored in light blue and the color becomes darker and finally violet for the last measurement. Additionally, the final measurement in each sequence is drawn with a thick line and used for the plots in B to derive the conduction limiting mechanism. The insets are magnifications of the $I - V$ characteristics near 0 V and have the same axis as the large diagrams. In B, the $I - V$ characteristics of as-grown, vacuum-, and oxygen-annealed nanorods is drawn in green, blue, and red, respectively. Dark and light colors display measurements before and after CVS. Dotted lines emphasize linear parts indicating a particular limiting conduction mechanism. In the semilogarithmic plot, all measurements are displayed, but in the other plots (B2 to B4) only relevant measurements are shown. The as-grown nanorods have an ohmic behavior at 0 V (Panel A1), but Fowler-Nordheim tunneling suggests the existence of a thin barrier for the rods annealed in vacuum and oxygen (Panel B2). The SCLC (Panel B3) and Poole-Frenkel plot (Panel B4) indicate a bulk-limited conduction for high bias. The SCLC mechanism is favored for higher bias or currents over the Poole-Frenkel emission. More details are found in the text.

The induced current enhancement during long positive CVS relaxes slowly (Figure 11.14) which is explained by the lower oxygen vacancy mobility close to the oxidized interface. The low conductivity of electrons in grain boundaries and the low mobility of oxygen vacancies in the interface layer cause another effect. The positive space-charge created at the T/T interface during positive CVS is annihilated only slowly by diffusing electrons and remains until the negative CVS. As soon as the gold tip is biased negatively, the positive space-charge is annihilated quickly due to the large excess of electrons and the CBM is raised. Consequently, the edge steepness of the Schottky barrier at the Au/TiO_2 interface is decreased, which broadens and increases the barrier and reduces the transient interface current. Furthermore, the low density of oxygen vacancies at the Au/TiO_2 interface diminishes the creation of a positive charge close to the Schottky barrier by ionization of vacancies and the leak of mobile vacancies in this volume limits the accumulation of electron donors near the interface. Hence, both current promoting effects are small and slow. In total, the current limiting effects become stronger during the negative CVS (Figure 11.13 B1). Since the filling of ionized traps is irreversible when the applied field is turned off, the induced effect during long CVS is hardly relaxing (Figure 11.14).

Annealing in oxygen does not turn the nanorod in a perfect insulator. Not all defects can be healed during oxygen annealing and there might be additional kinds of defects besides oxygen vacancies. However, oxygen vacancies are the most important electron donors in TiO_2 which are reduced on the surface during annealing. Consequently, the high transient current after annealing is supposed to restricted in a subsurface layer.

11.3.2 Rutile TiO_{2-x} Nanorods Grown at 180°C (NR180)

These nanorods have a similar structure as the rods grown at 150 °C, but the fingers are thicker and the grain boundary density is reduced. Effects that are similar to the ones described for NR150 samples above are mentioned but not discussed in detail once more. The focus is rather on the differences between the various investigated structures.

As-grown: Similar to NR150, the short CVS measurements (Figure 11.8) indicate the presence of trapped electrons for both polarities. Long CVS with a positively biased tip results in a significant increase of the current and there is a relatively large discharge current (Figure 11.9 A). The increase of the discharge current at 0 V is most likely related to the limited electron diffusion in the bottom part of the nanorod. The following decrease of the discharge current is not only due to the reduction of electrons in the electron bubble but also induced by an increasing discharge current in the opposite direction across the Au/TiO_2 interface. The current towards the gold is increasing much slower since the Schottky barrier height needs to relax via the diffusion of oxygen vacancies close to the interface and annihilation of the positive charge created by ionized oxygen vacancies at the T/T interface. Interestingly, the system recovers within the short probe pulses and hence, the time-dependence of the discharge current is the same after each probe pulse (Figure 11.9 A2). Since the peak height of the probing pulses does not change over time, it is reasonable that oxygen vacancies are attracted from the bulk material to the interface region during long CVS and remain in this region after the bias is turned off. Thus, the sequence of probing pulses and 0 V periods is supposed to shift these accumulated vacancies only a little bit forward and backward on a nanometer scale. To approximate the capacity of these nanorods, a much longer time-dependent measurement is necessary. However, the obtained data allow the conclusion that the capacity of a single rod is significantly larger than the one of the rods grown at 150 °C (1 nF). After negative CVS, the electrons move directly through the reduced Schottky barrier at the Au/TiO_2 interface until the barrier relaxes, which

results in a quickly decreasing discharge current pointing towards the gold tip. When the barrier is relaxed, the discharge current changes its sign and the electrons leave the nanorod via the FTO/TiO$_2$ interface (Figure 11.9 A2).
CVS does not always promote the current flow as shown in Figure 11.16 B1. Long CVS at a negatively biased tip results in a decrease of the current similar to the NR150 sample annealed in oxygen atmosphere. The failure of current promoting effects at the interface is most likely caused by dominant current blocking effects in the bulk. Since trap generation is usually used to describe an increase of the conductivity [1405], a charging process is supposed to reduce the conductivity. Negatively charged traps are rare in the bulk and hence, the trapping-induced annihilation of the positive charge created during positive CVS is used to describe the time-dependent behavior. The lapse of this charge lowers the conduction band significantly which provides bulk-limiting conduction mechanisms. However, due to the continuous switching, the dominant conduction mechanism is hardly determinable from the $I - V$ characteristics. It is noted that the effect of the positive space-charge must be really strong which suggests that the central position of the positive charge is close to the Au/TiO$_2$ interface and the density of ionized oxygen vacancies is extremely high. Since higher growth temperature provide larger fingers and the hydrochloric acid treatment even enlarges existing fingers, the defective upper part might be smaller in rods grown at 180 °C than in rods grown at 150 °C. This explanation is supported by the data of nanorods grown at 220 °C which show roughly the same behavior for long positive and negative CVS as the rods grown at 180 °C.

The reduction of the Schottky barrier and the generation of an internal charge can turn a bulk-limited into an interface-limited conduction behavior. The relaxation of the Schottky barrier can switch the sign of the discharge current.

The $I - V$ characteristics, displayed in Figure 11.16, show clearly that CVS increases the conductivity of the whole sample significantly and stabilizes the properties in a way that all five demonstrated $I - V$ curves overlap after CVS. This means that all effects induced within an $I - V$ cycle are able to relax on a shorter time scale than the cycle is passed through. Hence, it is assumed that the ionization-induced positive charge is quite low and constant. This is reasonable since the $I - V$ characteristics are measured after the long negative CVS which is supposed to annihilate the positive charge. For a better understanding, only the stabilized $I - V$ characteristics after CVS are discussed (Figure 11.16 A2). At a negatively biased tip, oxygen vacancies drift towards the Au/TiO$_2$ interface and oxygen vacancies at the Schottky barrier are ionized due to the increased edge steepness of the barrier. This process metallizes the barrier and this property is "frozen" when the bias is lowered. Hence the conduction becomes ohmic with an effective resistance in the order of 1 to 10 MΩ per rod at 0 V. At 0.4 V, the ions are drawn back from the interface and become neutralized partly recovering the Schottky barrier. It is noted that the field amplitude for the current peak is the same as for nanorods grown at 150 °C. The clear negative peak at around −0.5 V is most likely induced by the detrapping and ionization of oxygen vacancies close to the surface. The generated positive charge induces a negative image charge in the gold tip and hence, a current builds up. As soon as most of the donor states are excited, the changing current declines quickly, but subsequently, the transient current through the reduced Schottky barrier rises with increasing negative field amplitude. Consequently, the observed peak for both polarities depends on the voltage sweep rate. However, the exact nature of the two peaks needs to be investigated in further experiments.

The resistive switching is based on trapping and detrapping of electrons and ion migration.

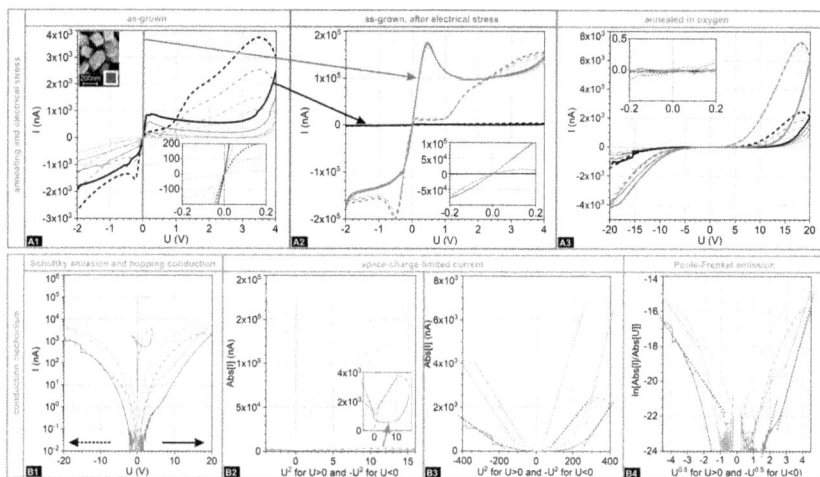

Figure 11.16: $I - V$ characteristics of TiO_2 nanorods grown at $180\,°C$. This figure is assembled as Figure 11.15. For as-grown nanorods, the $I - V$ curves before (A1) and after (A2) CVS are shown. The overlapping lines in Figure A2 indicate that the system becomes stabilizes during CVS. The data from nanorods annealed in vacuum are not shown. Most likely a large TiO_2 particle was between the gold tip and the NRA during the measurement suppressing the transient current almost completely. Sometimes, the particles are created during the hydrothermal growth process and settle down on the NRA. After oxygen annealing, the features at low bias disappear but at high bias, the behavior is partly conserved (Panel A3). The shape of the $I - V$ curve for as-grown nanorods does not allow to define a specific limiting conduction process (Panel B1 and B2). For the rods annealed in oxygen, Poole-Frenkel emission appears at lower bias and SCLC at higher bias. One has to consider that the SCLC plot (Panel B3) opens the view rather on high applied voltages and the Poole-Frenkel plot brings out the features at lower voltages (Panel B4).

Annealed in vacuum: There is almost no transient current in the NR180 sample. This does not fit to any expectation. However, a reasonable explanation for the low current are particles that settle down on the sample during the hydrothermal growth process. These particles consist of TiO_2 nanorods and are shaped like sea urchins. If they are situated between the tip and the NRA, the distance between the two electrodes is enhanced and hence, the electrical field across the TiO_2 nanocrystals becomes very low. Additionally, the number of grain boundaries is increased since the electrons need to pass several nanorods. In summary, the data allow no deeper insight into the electronic properties of nanorods and are neither discussed nor shown in the $I - V$ characteristics. However, this measurement proves indirectly that all other shown data are reliable and not affected by these particles.

Annealed in oxygen: According to the short CVS pulses, there is no distinct and fast effect from electron trapping. This suggests that the grain boundaries and surfaces are connected with each other. In particular, there are most likely direct conductive paths between the T/T interface and the conductive sub-surface layer in order to avoid bag-like potential shapes when a bias is applied.

For long CVS with a positively biased tip, the current is increased similarly as the nanorods grown at 150 °C (Figure 11.13). However, the induced effect is building up and relaxing faster (Figure 11.14). This indicates that either the oxygen vacancy diffusion or the annihilation of the positive charge appears on a faster timescale. Since there are no electron bubbles there is no discharging current for both polarities (Figure 11.13). For negative long CVS, the transient current is fluctuating and soft breakdowns appear which are clear indicators of ion migration and oxide degradation. Also, this effect is relaxing quickly after the CVS. Hence, the time-dependence is rather attributed to drifting oxygen vacancies than to ionization-induced charging. However, the latter effect might also take place.

Compared to non-annealed nanorods, the current is significantly lower which is related to the small cross-section of the residual conductive pathways and the increased Schottky barrier at the Au/TiO$_2$ interface after annealing in oxygen atmosphere. Similar to non-annealed nanorods, the $I - V$ curves are stabilized after CVS (Figure 11.16 A2). Due to the continuously changing properties, the conduction mechanism is not clearly identified for any bias value. For negative bias, the Schottky barrier is reduced and an SCLC is observed in forward direction as expected [79]. For positive bias and in forward direction, the conduction mechanism is dominated by Poole-Frenkel emission for lower bias and SCLC above 17 V (Figure 11.16 B3 and B4). This is in good agreement with previous studies. Considering a rod length of about 1.5 μm, the average electrical field between the electrodes is up to 200 kV/cm, which is enough for a significant current contribution from the Poole-Frenkel emission [283, 1233, 1795]. Furthermore, it is reported that Poole-Frenkel emission appears usually as long as there is a significant amount of unfilled traps. However, at a critical injection rate all traps are filled and the bulk-limited conduction mechanism is dominated by SCLC [1203, 1213].

> The bulk-limited conduction behavior is dominated by Poole-Frenkel conduction at low bias or low currents and SCLC for high bias or high currents.

Here, an interesting feature is observed. A similar kink as observed at 2 V for oxygen-annealed nanorods grown at 150 °C appears also for the nanorods grown at 180 °C. Combining the results from both kinds of nanorods, it seems that Poole-Frenkel emission switches between a low trap level below 4 V and a deeper trap level above 9 V. This transition depends on the applied CVS for both, the NR150 and NR180 sample. The appearance of two distinguishable trap levels is in contrast to the assumption that trap levels are distributed densely in bulk TiO$_2$. However, discrete trap levels are reported for the surface of TiO$_2$ nanocrystals and a conduction based on two trap levels was suggested by Schwarzburg and Willig [1354]. Hence, this transition suggests that a significant part of the transient current is guided through a sub-surface conductive path. However, a similar result is expected for a single trap level depth and a charge transport via two conduction bands, which was reported by Becker and Hosler employing a second conduction band 0.05 eV above the CBM [820]. Slightly different slopes of the linear behavior in Poole-Frenkel plots before and after the transition denote different effective electron masses, which is reasonable if two conduction bands are involved.

At high positive bias, the barrier is lowered again and the complete backward direction is dominated by SCLC. Hence, for oxygen annealed nanorods grown at 180 °C, the conduction is mostly bulk-limited although a thin insulating layer is supposed to cover the surface. This feature indicates that the conductivity of the electron pathways is relatively low compared to non-annealed rods.

The Poole-Frenkel plots suggest that electrons are emitted from two different trap levels into the conduction band or from a single trap level into two conduction bands in different voltage regimes. This effect is distinct in some of the presented defect-rich nanostructures which have been annealed in an oxygen atmosphere. The appearance of various distinct energy levels concludes that the electrons traverse the structure close to their surface.

11.3.3 Rutile TiO$_{2-x}$ Nanorods Grown at 220°C (NR220)

The structure of these nanorods is similar to rods grown at 150 and 180 °C, but they have the thickest fingers and consequently the lowest grain boundary density. This feature suggests slightly different electronic properties.

As-grown: In the beginning, there is no fast relaxation process for positive short CVS (Figure 11.8). During long positive CVS, the current is increasing in the same way as for NR180 (Figure 11.9 A1). After long CVS, the relaxation is slower than for the NR180 sample (Figure 11.10). From this observation, it follows that during the CVS, conductive paths between the FTO/TiO$_2$ and the T/T interface are formed. It might be related to these paths or the low height of the T/T barrier that the major part of the electrons diffuse from the electron bubble towards the FTO/TiO$_2$ interface during the discharge process at 0 V (Figure 11.9 A2). The build-up of the discharge current is much slower (half a minute) than for NR180 (a few seconds). This indicates a very slow diffusion constant for electrons. Since momentum scattering is unlikely due to the high crystallinity of the NR220, the high resistivity results from the low donor density [1203, 1354, 1355]. Additionally, the small cross-section of the few grain boundaries between the fingers is also responsible for the reduced charge transport.

For long negative CVS, the behavior of NR180 and NR220 is the same, but the nanorods grown at 220 °C have the highest discharging current among all investigated structures after negative CVS (Figure 11.9). On the one hand, these rods could have the largest electron bubbles. On the other hand, the barrier of the T/T interface might be relatively small due to the lower defect density compared to rods grown at lower temperatures. Consequently, the electrons can escape easily providing a high discharging current.

The continuous change of the Schottky barrier is expressed by the various slopes appearing in the $I - V$ curves at 0 V (Figure 11.17 A1). The $I - V$ curves are not stabilized after CVS, which means that the time constants for the emergence and relaxation of conduction affecting processes are larger than the time needed for recording a single $I - V$ curve. This is in good agreement with the slow electron and vacancy diffusion expected in the highly crystalline nanorods. The forward direction is dominated intensively by the continuous property changes and hence, the conduction mechanism cannot be determined. The backward direction for positive bias is marked by SCLC for strong external fields and Schottky emission for low fields. In the forward direction, the conduction at 0 V is completely ohmic indicating an effective lowering of the Schottky barrier. In backward direction, the conduction is ohmic for high electrical fields and it is limited by Schottky emission when the Schottky barrier recovers. Hence, only for the backward direction, Fowler-Nordheim tunneling is identified at 0 V (Figure 11.17 B2).

Annealed in vacuum: During annealing in vacuum the nanorods become almost insulating (Figure 11.8). This suggests that there are not enough surface defects to provide a measurable current in the applied voltage regime for both polarities.

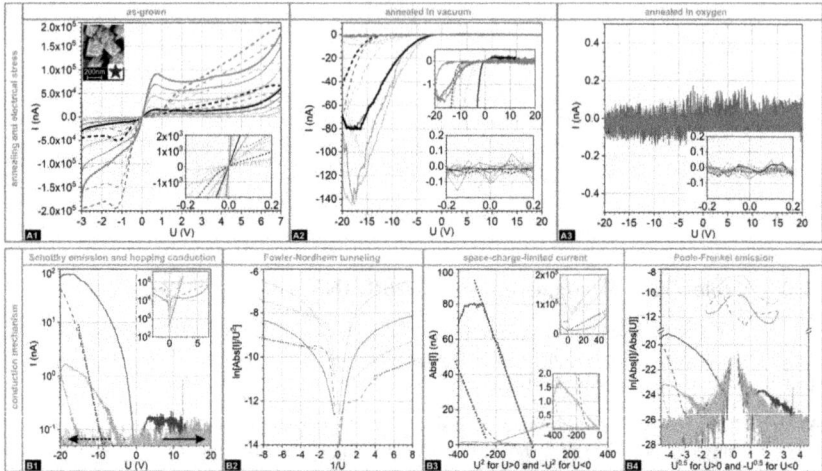

Figure 11.17: $I - V$ characteristics of TiO_2 nanorods grown at 220 °C. This figure is assembled as Figure 11.15. The as-grown nanorods have similar low-bias features as the rods grown at 180 °C but even after CVS, they are not stabilized (Panel A1). After annealing in vacuum, a conductivity changing process is only observed for negative bias (Panel A2). Here, the current is dominated by SCLC (Panel B3) which indicates that the current is focused in tiny filaments which are highly conductive and the effect of the Schottky barrier and Poole-Frenkel emission becomes negligible. The resistive switching process at negative bias is very clear. The transient current is suppressed after annealing the rod in oxygen.

The contrast between the high conductivity of non-annealed nanorods and rods annealed in vacuum originates most likely from the high stoichiometry of nanorods grown at 220 °C. Although there are many defects in non-annealed rods, the total ratio between titanium and oxygen atoms in a single nanorod is close to 1:2. During annealing, the atoms rearrange and oxygen vacancies recombine partly with oxygen atoms, which are diffusing inside the oxide. The grain boundaries and a certain number of surface defects resist the annealing process and provide still conductive pathways between the electrodes. However, these pathways have to be created by employing strong electrical fields which is demonstrated impressively for the long negative CVS displayed in Figure 11.17 B1. The noisy behavior of the current is a sign of significant ion migration [1348, 1796].

The low grain boundary density and the relatively broad length distribution of the fingers provide tiny contact areas between the rod and the gold at the tip of the tallest finger of each nanorod. Hence, there is a distance of a few nanometers between the gold contact and the nearest grain boundary. Therefore, the created filaments are situated most likely on the surface of the tallest finger and continue on the surface or in a grain boundary close to the surface towards the FTO. Hence, no potential bag is created and the formation of an electron bubble is hindered. Consequently, no quickly relaxing current is observed. The outcome that the filament is created for negative bias is probably linked with the high electrical field at the gold tip close to the contact and the relatively high number of defects in the upper part of the nanorod compared to the bottom part. After long negative CVS, the induced conductivity relaxes due to surface ion diffusion which is more efficient than bulk diffusion.

The $I - V$ characteristics in Figure 11.17 A2 demonstrates that the filament creation and annihilation results in a bipolar resistive switching. This observation suggests that oxygen vacancies are withdrawn from the gold contact if the tip is positively biased. The current through the formed filaments is dominated by SCLC. During their creation, the current is limited by Schottky emission. According to Section 2.8, this $I - V$ behavior is expected for conductive filaments.

Annealed in oxygen: The combination of low defect generation at a growth temperature of 220 °C and a further surface defect reduction during annealing in oxygen limits the number of oxygen vacancies dramatically. Hence, there are not enough vacancies for building up conductive paths and the electrical field is too weak to create new oxygen vacancies. Consequently, the creation of conductive paths is prohibited and no transient current is observed.

The transient current is already quite low before annealing due to the high crystallinity and the low grain boundary density of this kind of rods. Vacuum annealing reduces the conductivity further and finally, annealing in oxygen results in a completely insulating nanorod within the applied voltage range. This suggests that the defect density is so low that no defective and conductive subsurface layer remains after the oxidation. This observation is in good agreement with the structure observed in TEM images.

11.3.4 Rutile TiO$_{2-x}$ Nanorods Grown at 180°C and treated with hot hydrochloric acid (HCL180)

These rods have a similar basic structure as rods without any post-treatment. However, the fingers are slightly enlarged by dissolving and recrystallization processes. These processes might affect the electronic properties of the surface and grain boundaries.

As-grown: In contrast to rods without chemical post-treatment (NR180), quickly declining currents do not occur during the short CVS measurements (Figure 11.8 A). The small current drop in the second negative short CVS pulse is most likely related to the trapping of electrons which means that a few trap states were formed along the conductive paths.
For long positive CVS, the current is increased significantly and this effect relaxes quickly after the CVS [1] (Figure 11.9). A short decaying current appears in each probe pulse, but the low total current suggests a small number of encapsulated electrons and thus, no discharge current is expected and observed at 0 V. For long negative CVS, the current increases only a little bit indicating that the ion drift already saturated or trapping and detrapping rates are almost equal. Also here, the effect is relaxing (Figure 11.10) and a small discharge peak occurs in each probe pulse. However, the total current is about one order of magnitude larger than for positive CVS and hence, a discharge current occurs. This current builds up quickly indicating a relatively high electron diffusion rate. The time constant of the discharge current is $\tau \approx$ 200 s, the estimated capacity of a single rod is between 1 and 100 pF, and hence, the effective resistance according to the RC-circuit at the beginning of the discharge process is in order of 10 to 1000 GΩ. Such a large RC value limits frequency-dependent electronic measurements such as impedance spectroscopy significantly [1207]. This RC-nanorod-electron bubble combination

---

[1] Due to a technical malfunction, not the complete data set is available. The behavior of transient current during positive CVS is assumed from the relaxation behavior. For all other investigated samples, a distinct relaxation appears only if there was an enhancement of the current during CVS before. Hence, it is reasonable to assume a current ascent during positive long CVS.

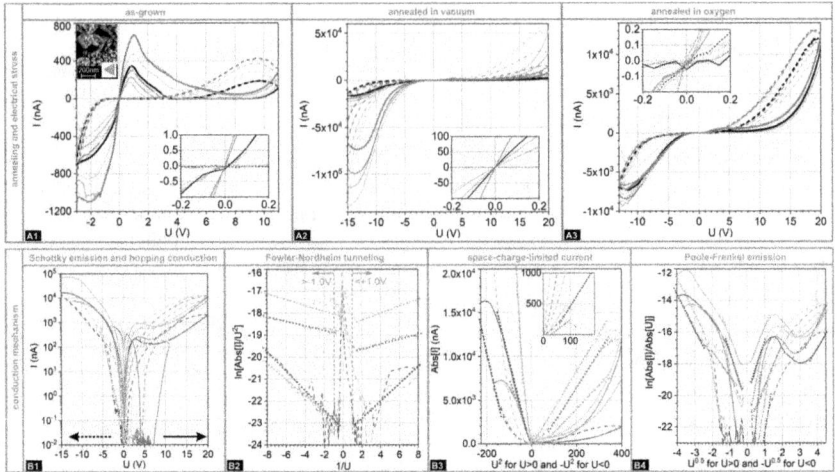

Figure 11.18: $I - V$ characteristics of TiO$_2$ nanorods grown at 180 °C after hydrochloric acid post-treatment. This figure is assembled as Figure 11.15. The low-bias features (Panel A1, B1, and B4) are similar to the $I - V$ characteristics of the nanorods without hydrochloric acid treatment. For the annealed nanorods, Poole-Frenkel emission appears for low bias (Panel B4) and SCLC for high bias and currents (Panel B3). The similarity of the nanorods annealed in vacuum and oxygen in Panel B4 indicates that the charge transport does not take place directly on the surface but in a subsurface layer or in grain boundaries where the environmental oxygen cannot annihilate oxygen vacancies.

offered the lowest stable current (in the order of 0.1 to 1 pA per nanorod [2]) in all presented structures.

The added line in Figure 11.18 demonstrates that the electron bubble is not charged significantly by the probe pulse. Hence, this specific measurement supports the general validity of the CVS-probe technique.

The $I-V$ characteristics of as-grown HCL180 and NR150 are very similar (compare Figure 11.15 and 11.18). This suggests that the defect density, which is relatively high in nanorods grown at low process temperatures, is increased and controlled by the hydrochloric acid treatment. The creation of defects is most likely a consequence of the dissolving and recrystallization process. In contrast, the finger diameter is slightly enlarged during the treatment. Thus, the grain boundary density as the only parameter is not enough to predict the properties of such a nanostructure. The total current in HCL180 is slightly lower than in NR150. On the one hand, this might be linked with the larger length of the rods resulting in a lower electrical field. On the other hand, this could be a consequence of the smaller contact area caused by the broader length distribution of the fingers in each nanorod.

Annealed in vacuum: Annealing in vacuum increases the defect density on the surface and grain boundaries further. This enhances ion diffusion on the surface and along grain boundaries,

---

[2] The values have to be given in a range since the number of nanorods which are connected with the gold tip is approximated between 1,000 and 10,000 (based on SEM images).

which results in the fastest observed current increase during CVS measurements for both polarities among all investigated samples. Consequently, the current-promoting effect relaxes quickly after the bias is turned off (Figure 11.8 B).
It is noteworthy that long positive CVS provokes a quickly declining current which might be related to the formation of a potential bag in the upper part of the nanorod (Figure 11.11). It is reasonable that a high ion mobility supports such a formation. Since this effect is not observed for a negatively biased tip, it is reasonable that oxygen vacancies are pulled into the grain boundaries towards the T/T interface where they accumulate and build up a positive charge causing a local potential drop. However, no discharge current occurs when the bias is turned off. Hence, the electron bubble might be not so large compared to the structures discussed before or the potential barriers in the relaxing system might be higher.
The $I - V$ characteristics are similar to the NR150 sample (compare Figure 11.15 and 11.18). However, the peak above 0 V observed for non-annealed samples is still slightly expressed. For negative bias in the forward direction, the current is dominated by SCLC since the Schottky barrier is reduced (Figure 11.18 B3). After the peak at 4 V, the Schottky barrier is recovered and the current is dominated by thermionic field emission. This conduction mechanism suggests a narrow Schottky barrier which results from the low CBM and small depletion depth caused by the huge defect density. A positively biased tip reduced the Schottky barrier and an SCLC appears in the backward direction. After the polarity is switched again, the Schottky barrier recovers and a Schottky emission appears. Below -5 V, the Schottky barrier is reduced significantly and SCLC dominates (Figure 11.18 B1).

Annealed in oxygen: Due to the high initial defect density, annealing in an oxygen atmosphere is not able to annihilate all defects and hence, the CVS and $I - V$ characteristics after annealing in oxygen atmosphere are very similar to the ones for nanorods grown at 150 °C and annealed in oxygen atmosphere. This includes the conduction through two trap levels with a transition between 4 and 9 V. From this comparison, it follows that the conduction is limited by Poole-Frenkel emission in the forward direction for positive bias although the shown $I - V$ plots suggest rather a Schottky emission (Figure 11.18 B1 and B4). For the negatively biased tip, the $I - V$ curve is similar to the one for rods without chemical post-treatment (but also grown at 180 °C). This indicates that not all features are changed by the exposure to hot hydrochloric acid.

The similarity between the transient current through nanorods grown at 150 °C and nanorods grown at 180 °C and treated with hot hydrochloric acid indicates that the chemical post-treatment increases the number of defects. Since the crystalline fingers inside the acid-treated rods are significantly larger, it is suggested that the grain boundary and surface defect density is extremely high.

11.3.5 Rutile TiO$_{2-x}$ Nanotubes Grown at 180°C and etched with hot hydrochloric acid (NT180)

Nanotubes do not have completely embedded grain boundaries and the etching with hydrochloric acid is supposed to affect the electronic properties more significantly as the previously discussed hydrochloric post-treatment (HCL180). The size of the fingers is similar to the nanorods grown at 180 °C. The inner upper part of the nanorods is removed by etching.

As-grown: A quickly declining current is observed for both polarities in the short CVS measurements (Figure 11.8). This is correlated with the internal structure of nanorods. In Chapter 7, it is suggested that the inner part is more defective and hence chemically less stable than the outer fingers. Supported by TEM studies, the introduced model assumes that the upper part is more defective than the bottom part. As a consequence, the inner upper part of the rod is removed first. In the next step, the acid increases the defect density on the inner and outer walls of the tube. Since the surface-to-volume ratio for nanotubes is much larger than for nanorods, the regions affected by defects in the etched tubes are more dominant than in nanorods. The etching process might create a potential bag in the inner part of the tube which can be filled with electrons.

In case of long positive CVS, the current is decreasing (Figure 11.9). This is in contrast to all other structures. It has to be considered that oxygen vacancies are not the only defects but there are also neutral traps such as hydroxyl groups on the surface. These neutral traps become negatively charged after they have trapped an electron. In nanotubes, all conductive paths are situated close to a surface and hence, enhanced Coulomb scattering appears between the mobile and the trapped electrons. If the mechanisms, which reduce the Schottky barrier are not strong enough, the mentioned repulsive Coulomb interaction becomes dominant and the current is reduced. Consequently, the conduction mechanism is expected to become bulk-limited which is in good agreement with the SCLC shown in the right inset of Figure 11.19 (Panel B3, backward direction). Nevertheless, the Fowler-Nordheim plot (Panel B2) indicates a thin Schottky barrier at the Au/TiO$_2$ interface which is reasonable due to the high donor density in the contact region. This effect relaxes due to the emission of electrons from their trap sites when trapping is reduced by the lack of electrons from the applied transient current.

For long CVS with a negatively biased tip, the current is increasing. As discussed previously, the reduction of the Schottky barrier is more efficient for negative CVS and hence, a distinct lowering of the Schottky barrier is reasonable. In addition, the Fowler-Nordheim plot shows, that the conduction is limited by tunneling for much lower voltages (higher $1/U$) in the forward than in the backward direction. This indicates that the Schottky barrier is narrowed more efficiently when the system was exposed to a negatively biased gold tip before. Additionally, the discharging current pointing towards the gold tip at $0\,\mathrm{V}$ after negative CVS is about one order of magnitude larger than the discharging current pointing towards the FTO after positive CVS. Besides the amplitude, the direction of the discharge current is a strong hint for an efficient reduction of the Schottky barrier during negative CVS. Furthermore, the discharge current towards the gold tip has no measurable build-up period since the electron bubble is supposed to adjoin the Schottky barrier. The discharging current in the other direction needs at least half a minute to reach the maximum current amplitude which is a hint for intensive electron momentum scattering on the conductive paths between the electron bubble and the FTO/TiO$_2$ interface.

The current reduction for a positively biased tip and the current enhancement for a negatively biased tip affect also the $I - V$ characteristics (Figure 11.19). This feature keeps the transient current in forward direction always above the transient current in backward direction.

Annealed in vacuum: The total current is reduced by three to five orders of magnitude. According to the lack of a quickly decaying current during the probe pulses, the number of trapped electrons is very low. This feature is described already for the NR150 and HCL180 samples.

Since the defects are supposed to accumulate at the surface and grain boundaries during annealing, the transient current is restricted to these parts of the tube. Hence, the cross-section of the conductive paths is very small and the momentum scattering rate of electrons is high. In this situation, the impact of negatively charged hydroxyl groups is distinct. Furthermore, the effective mass of electrons might be changed in the highly restricted and defective regions. For the nanorods treated with hydrochloric acid (HCL180), it is assumed that the ratio between titanium and oxygen atoms is close to 1:2 and hence, annealing results in highly stoichiometric TiO$_2$ crystals. This explanation is applicable also for the nanotubes.

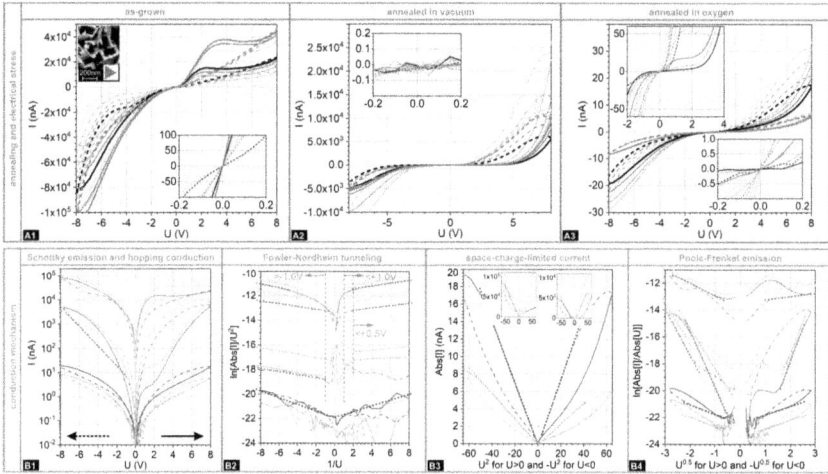

Figure 11.19: $I - V$ characteristics of TiO_2 nanotubes. This figure is assembled as Figure 11.15. Also for the tubes, the low-bias effect which is correlated to near-surface ion migration and charging is vanished during annealing (Panel A1 to A3 and B1). In particular, the $I - V$ characteristics become highly symmetric (Panel B1 and B3). This is another strong sign for a bulk-limited conduction mechanism since the two interfaces are still dissimilar and should not provide a symmetric $I - V$ characteristics. Consequently, SCLC and Poole-Frenkel (Panel B3 and B4) emission dominate the charge transport at higher voltages and a tunneling barrier is observed at low bias (Panel B2).

The increasing current for positive CVS is likely caused by the ionization-induced formation of a positive charge close to the FTO/TiO_2 interface (Figure 11.11). Hence, the fast relaxation is related to the trapping of diffusing electrons after the bias is turned off. The enhancement of the current for negative CVS is much faster, which is expected for a negatively biased gold tip as discussed previously. The small current decrease might be related to electron trapping in neutral hydroxyl groups.

Due to the effective Schottky barrier reduction at negative bias, the $I - V$ curve is dominated by SCLC in forward direction (Figure 11.19 B3). As the polarity switches, the barrier relaxes and the current is limited by thermionic field emission indicating that the barrier is still thin enough to allow tunneling. At larger bias, the positive ionization-induced charge enhances the current across the FTO/TiO_2 interface and hence, SCLC occurs in the backward direction. The high symmetry of the $I - V$ characteristics is another evidence for the dominance of bulk-limiting processes during an $I - V$-cycle. For negative bias, the Schottky barrier relaxes and the conduction is limited by thermionic field emission again.

Annealed in oxygen: Related to the restriction of the current to paths near the surface, the conductivity of nanotubes is reduced further and almost no transient current is measured. The creation of a highly resistive oxide shell allows the formation of an electron bubble as described for the nanorods that have been exposed to hot hydrochloric acid (HCL180).

For long CVS, a very low and noisy current appears that probes the continuous creation and degradation of conductive paths as well as trapping and detrapping processes (Figure 11.13). A relaxation after CVS is not observed which suggests the creation of a stable defect distribution (Figure 11.14). The observed current is most likely guided by the few grain boundaries and the noisy behavior results from regions where the electrons need to pass the surface. A hint for a conductive sub-surface layer is given by the Fowler-Nordheim plots. Fowler-Nordheim tunneling is only allowed for barrier thicknesses of a few nanometers. Hence, the edge steepness of the barrier must be quite large which is only fulfilled if there is a highly defective region directly below the surface. However, the structure of such an ultra-thin insulating shell needs to be analyzed more deeply with TEM in further studies.

The reduction of the transient current during continuous $I - V$ mapping before CVS might be related to electron trapping in neutral hydroxyl groups and a subsequently increased Coulomb scattering rate. After CVS, the $I - V$ curves are stabilized. At negative bias and in forward direction, the current is dominated by SCLC for larger bias and Poole-Frenkel emission for lower bias. For positive bias, the current is limited by Poole-Frenkel emission and the previously described transition between two trap levels appears here, too. However, the transition takes place at a much lower bias (between 1 and 2 V).

In nanotubes, all electron pathways are close to the surface and hence, these structure react very sensitive on changes of the surface. In particular, charged surface traps limit the transient current via Coulomb interaction. Annealing in vacuum reduces the transient current significantly. This suggests that the post-treatment with the hot hydrochloric acid generates a lot of defects, but does not necessarily change the stoichiometry of the rod. Hence, the defects can be healed even in the absence of environmental oxygen during annealing.

11.3.6 TiO_{2-x} Films Deposited via Sputter Deposition (FI)

The TiO_2 film is used as a reference sample without a T/T interface, fingers, highly oriented grain boundaries, or remains from the hydrothermal growth process.

As-deposited: During CVS, the transient current through amorphous thin TiO_2 films is almost constant, although the applied field is much higher (Figure 11.8). In nanorods grown at 180 °C without post-annealing, the same current was obtained for a twenty times lower electrical field. This property suggests that hydrothermal growth provides a large number of oxygen vacancies which enhance the electron transport. Additionally, the oriented grain boundaries between the fingers promote a directed ion drift and diffusion and allow an efficient and controllable switching behavior. It is concluded that the binding energy of ionized oxygen vacancies is significantly higher in amorphous TiO_2 films compared to hydrothermally grown TiO_2 nanorods.

Also, the amorphous film has extended defects similar to the grain boundaries in nanorods. However, their distance is quite large and they are situated at the lowest positions on the rough film surface. Hence, a conduction through these extended defects is only possible if long filaments are formed across the surface from the metal tip towards the extended defects. The measurements suggest that this does not happen.

Nevertheless, $I - V$ cycles are only stabilized after CVS (Figure 11.20 A1). According to the inset in Figure 11.20 A1 and the Fowler-Nordheim plot in Figure 11.20 B2, the Schottky barrier is reduced during CVS. However, the Schottky barrier seems to be quite low since an ohmic conduction dominates the current at higher voltages. After the CVS and despite the ohmic contact resistance, the conduction is mainly limited by Poole-Frenkel emission for both polarities. This behavior might be explained by the annihilation of grain boundary leakage currents, e.g.

by negative charging. After the CVS, the electrons have to pass the defective bulk TiO_2 that provides a higher Schottky barrier and more interaction with traps.

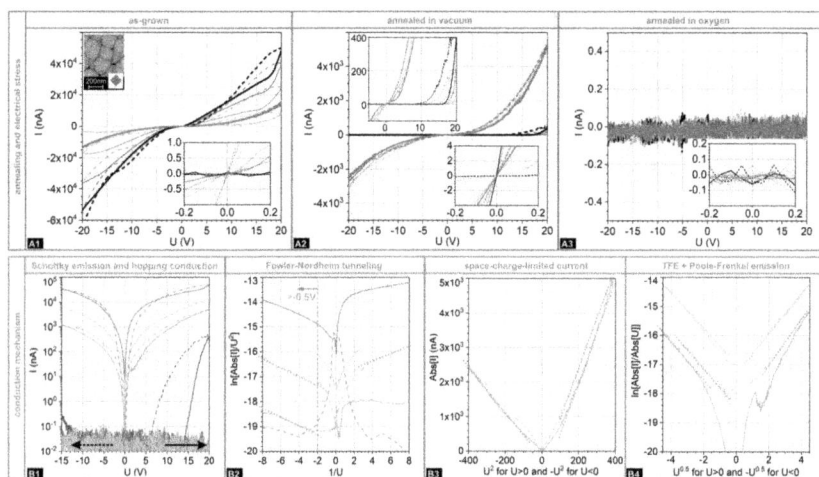

Figure 11.20: $I - V$ characteristics of sputtered TiO_2 films. This figure is assembled as Figure 11.15. In particular after CVS, the film shows a very symmetric and bulk-limited conduction mechanism, even for the as-deposited film (Panel A, B1, B3, and B4). The latter observation indicates that the film has a homogeneous structure as expected from sputter deposition. Furthermore, a homogeneous material should show a polarity- and electrode material-independent $I - V$ characteristic if the conduction is limited by a bulk process. Consequently, this observation indicates that the structure of the nanorods is not homogeneous as assumed in the introduced model.

Annealed in vacuum: Annealing in vacuum increases the crystallinity in the bulk and might generate a defective surface layer. Particularly after CVS, the contact becomes ohmic as shown in the lower inset of Figure 11.20 A2. Hence, defects at the interfaces are rearranged during CVS and as a consequence, the electron transport is promoted. Therefore, the electrons are injected into the TiO_2 easier, but their transport is still limited by bulk properties as demonstrated by the $I - V$ plots in Figure 11.20 B3 and B4. However, the exact conduction mechanism (SCLC or Poole-Frenkel emission) cannot be determined from the presented data. For negative CVS, the current is even decreasing which might be related to charging effects. The noisy behavior suggests continuous fluctuations of defects and trapped charges.

Annealed in oxygen: After annealing in oxygen, no transient current is measured within the applied voltage range since there are no oxygen vacancies that can drift and the electrical field ($\approx 50\,kV/cm$) is too weak for trap generation in crystalline TiO_2 [1212]. This result emphasizes the meaning of sub-surface of grain boundary-induced conductivity in hydrothermally grown rutile TiO_2 naorods.

> The films act as a reference system with a rather homogeneous defect distribution. This homogeneous distribution is expressed in symmetric $I - V$ curves. The time-dependent effects are less distinct than in nanorods. This might be related to a lower defect density and the missing grain boundaries in films. In particular, the grain boundaries provide directed pathways for mobile oxygen vacancies and electrons towards the Schottky barrier at the gold tip. In conclusion, grain boundaries are supposed to play an important role for the resistive switching behavior in rutile TiO$_2$ nanorods.

11.3.7 The Effect of Adsorbed Oxygen on The Transient Current

A nanorod which was grown at 180 °C and post-annealed at 550 °C in vacuum was exposed to oxygen while the transient current was measured at a constant bias. As soon as the oxygen is filled in the evacuated measuring chamber, the transient current is extinguished immediately. Oxygen adsorbates are attached preferentially to oxygen vacancies and function as an electron scavenger. Consequently, the number of mobile electrons and the conductivity are reduced [8, 739, 928, 1078, 1090]. The small number of mobile electrons increases the depletion layer and broadens the Schottky barrier which lowers the transient current even more [1512]. As displayed in Figure 11.21, the current decays quickly which indicates that most of the electrons involved in the charge transport are situated near the surface. A similar conclusion was made for TiO$_2$ nanoparticles elsewhere [739, 747, 1090]. Furthermore, this result is in good agreement with the core-shell structure analysis made by TEM presented in Chapter 7. The influence by further effects on the time-dependent charge transport behavior, which are most likely not relevant for the presented data, are discussed in the appendix (Section 13.6.2).

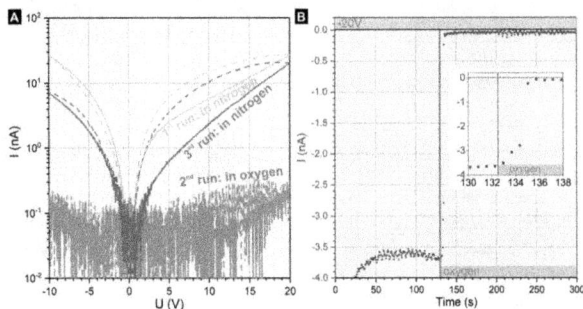

Figure 11.21: *Effect of an oxygen environment on the conductivity of nanorods grown at 180 °C and annealed in vacuum. In Panel A, three measurements in forward (solid lines) and backward (dashed lines) are displayed: the first in nitrogen (light blue), the second in oxygen (pink), and the third in nitrogen (dark blue) atmosphere. The figure demonstrates clearly that the current blocking effect in oxygen atmosphere recovers partly in nitrogen atmosphere. Panel B shows that the current is immediately blocked when the evacuated chamber is flooded with oxygen. The inset demonstrates that it takes less than 2 s to block the current which is attributed to gas diffusion in the chamber and the adsorption process on the TiO$_2$ surface.*

When oxygen is filled into the evacuated measuring cell, the transient current in an annealed rod disappears immediately. The oxygen is adsorbed on the surface of nanorods and acts as an electron scavenger. In agreement with other results shown in this chapter, it is suggested that the quick and drastic response to gaseous oxygen originates from a conduction near the surface of the nanorod.

12 Conclusion and Outlook

Conclusion

This work contains important guidelines for handling, designing, and controlling the growth and properties of rutile TiO_2 nanorods. In the beginning, a detailed summary about the structural, crystallographic, thermodynamic, optical, electronic, and optoelectronic properties of TiO_2 was given. Additionally, the state of research of hydrothermally grown rutile TiO_2 nanorods was presented.

One of the first research topics of this thesis was to compare rutile nanorods grown on polycrystalline rutile and anatase TiO_2 films. At first glance, the growth on anatase TiO_2 was unexpected, since it was assumed that the creation of rutile nanorods occurs most likely on seeds with the same crystal shape. The growth of rutile nanorods on anatase films was explained by the existence of tiny splints or nanoparticles on the surface of anatase films. During the hydrothermal treatment, these particles convert into rutile nanoparticles more easily than large grains in anatase films. In addition, the density of nanorods on seed films was controlled by the fabrication technique of the seed layer. On polycrystalline anatase films, the density of nanorods increases with the density of grain boundaries. Another important part of this work was the documentation of the entire growth process including nucleation, early growth stages, and the development of full-grown structures such as individual nanorods, dense NRAs, or anything in between.

A closer look into single nanorods was performed, ending up with a model that contributes to the understanding of the universal fine structure of each rod. The model assumes that the crystallization of nanorods happens so quickly that the surface of the growing nanorod and its environment are not in thermodynamic equilibrium and hence, the Wulff construction is not satisfied. By attaching several nanorods to each other, the overall shape is able to match the requirements of the Wulff construction. Hence, the typical mesocrystalline fine structure of nanorods is supposed to be a result of the association of many tiny high-aspect ratio nanocrystals. After a certain exposure time in the hydrothermal growth solution, these nanocrystals become merged due to dissolving and recrystallization processes. As a consequence, the full-grown nanorods become single-crystalline in the bottom (older) part. Due to higher particle mobilities, the imbalance between both competing processes is decreasing with increasing temperature. Thus, the aspect ratio and the number of fingers per nanorod is decreasing with increasing process temperature.

This thesis answered not only questions concerning the fine structure of single rods but offered routes to control the superlattice consisting of many nanorods as well. Several methods for position-controlled nanorod fabrication techniques were presented in this work. As a starting point, optical and electron-beam lithography were chosen as two conventional techniques to deposit seed films in confined geometries. Applying electron-beam lithography resulted in very narrow seed layers having a width of a few nanometers only. Huge patterns with wider seed layers were realized by applying the much cheaper and faster optical lithography.

To create a large patterned seed layer with a homogeneous superlattice, the resist was replaced by a monolayer of polystyrene spheres acting as a mask for resistive or electron-beam driven physical vapor deposition (PVD). The characteristics of the superlattice were controlled by the size of the spheres and the post-deposition treatment. Removing the PSML before annealing results in tiny seed islands with well-defined distances. Annealing the seed film on the polystyrene spheres at 600 °C results in hollow TiO_2 hemispheres. Increasing the annealing temperature to 850 °C, a porous film with large columnar rutile grains is formed.

Using patterned masks is not the only technique to achieve a position-controlled growth. The substrate itself might also be patterned by creating deep trenches using FIB milling for instance. It was demonstrated that a SiO_2 cover layer on top of the seed layer reduces the density of grown nanorods with increasing SiO_2 film thickness. Inside of trenches with a width of less than roughly 200 nm, the sputter deposition rate of SiO_2 is reduced. For optimized sputter para-meters, the layer becomes sufficiently thin inside the trenches and thick enough elsewhere, so that nanorods grow inside the trenches only. Surprisingly, this method promotes the growth parallel to the main substrate plane. Hence, this is the first presented technique providing a control of position *and* direction of the nanorods.

Besides position-controlled methods based on well-established structuring techniques, new pathways were discovered, developed, and described in this work. A scanning probe method using an atomic force microscope (AFM) equipped with a thin silicon tip was applied to scratch the surface of an anatase TiO_2 film. Along the narrow scratch, anatase splints and nanoparticles were created that converted into rutile seed particles during the hydrothermal growth process. The anatase-to-rutile transformation (ART) is supported by the fact that growing anatase particles become unstable and convert into rutile particles. The reason for this behavior is the low bulk energy of rutile compared to anatase which becomes dominant for increasing particle dimensions. As a consequence, each scratch provides the growth of TiO_2 nanorods while the untreated anatase TiO_2 remains mainly empty. The resolution of this technique is given by the radius of the silicon tip and the roughness of the film. In this work, the resolution was very close to electron-beam lithography. Interestingly, smearing the substrate only once activated the growth in a lot of sites along the scratch already. This method appeared to be very surface sensitive and affects less than the upper 4 nm of the film. It provides a higher resolution and is more flexible (no mask needed) compared to optical lithography and it is much less expensive than electron-beam lithography. Thus, it complements the versatile toolkit of advanced scanning probe lithography methods [1797].

As another technique, a laser-induced melting method was developed. It was demonstrated that similar results were achieved by cw- and pulsed laser radiation. While cw-laser-induced melting is more suitable for individual structures, interference patterning using pulsed lasers is preferential for large area superlattices. Applying wavelengths with 266 nm and 532 nm showed two amazing results. First, the investigated anatase film on silicon melts much easier for 532 nm laser radiation compared to blank silicon or levitating anatase films. Second, the rutile films are not affected in any way by this wavelength. Taking diffusion, multiple reflection, phonon dispersion, surface roughness, dielectric functions, and other properties of the complex TiO_2 system into account, it appears that the most reasonable explanation for the strange melting behavior of anatase films is based on the light absorption of mixed Ti–O–Si phases as well as on a significant rise of absorption for liquefied material. At the beginning, light is absorbed by a Ti–O–Si phase at the TiO_2/silicon interface which was created by a diffusion process during the high-temperature fabrication of the film. It was assumed that this layer absorbs more light and melts at lower temperatures than pure and almost transparent anatase. The whole layer system reflects less light than polished blank silicon. As soon as the mixed phase is liquefied partly, the absorption increases rapidly. Nearby TiO_2 and silicon start melting due to the increasing amount of absorbed energy. Hence, their absorption coefficient rises as well. In the case of rutile TiO_2 films, the mixed phase is supposed to have a much smaller volume or different composition and thus, light absorption might be relatively low or the decrease of the melting temperature is much less distinct compared to anatase films. The growth of rutile TiO_2 nanorods is provided close to the interface between the anatase film and the solidified melt. In the case of cw-laser radiation, the liquefied Ti–O–Si phase crystallizes mostly as rutile TiO_2 with a homogeneous crystal orientation over several microns along the linear structure written by the laser. As a consequence, the nanorods arrange themselves in parallel to each other over a long distance as well. So, this is another fast and inexpensive method providing control over the position *and* orientation of rutile TiO_2 nanorods.

In general, the approach was to create seeds in selected areas on a substrate, which does not provide the growth of nanorods elsewhere. Although large area patterning techniques are advantageous for some applications, TiO_2 NRAs need to be accessible for applications which need precise, flexible, and individual structuring as well. Examples are lab-on-a-chip devices where rutile TiO_2 nanorods could provide locally increased surface roughening, microchannels, photocatalytic molecule degradation, visible light scattering, superhydrophobic/hydrophilic surface gradients, or gas sensing [86–88]. In addition, a controlled orientation of nanorods is necessary for device integration, optical, and optoelectronic properties.

The large refractive index of rutile TiO_2 and the high anisotropic shape of the nanorods and their superlattices turn these systems into interesting light scatterers. It was demonstrated that light scattering on single nanorods and NRAs shaped as narrow lines depends on the polarization of the incident light. The model of an equivalent spherical radius describes the strongly wavelength-dependent scattering behavior and rather a low scattering efficiency for light polarized perpendicular to the long axis and the strong white light scattering for light polarized in parallel to the long axis. Scattering on anisotropic conglomerations of nanorods is dominated by the orientation of the superstructure rather than by the orientation of individual nanorods within the conglomeration. Dense and homogeneous arrays apart from edges were described best with an effective medium approximation.

This work contained fundamental research on semiconducting nanocrystals having various applications in optoelectronic devices. Hence, optoelectronic properties are of great interest. However, a detailed look at all these properties would have needed further time-consuming investigations, since the optoelectronic properties are probably even more complex than any other property of TiO_2. Thus, this work dealt with the electrical properties of rutile TiO_2 NRAs only and some outstanding effects were demonstrated and described. Chemical and heat treatment change the distribution and density of donors and traps. It turned out that oxygen vacancies form a thin metallic shell if the nanorod is annealed in a nitrogen atmosphere. This was explained by the missing extinction of oxygen vacancies that diffuse from the bulk to the surface during post-annealing. On the one hand, this shell converts the former rectifying contact into an ohmic contact. On the other hand, the shell has a relatively large series resistance due to its small cross-section. In general, the I-V characteristics contained different limiting conduction mechanisms for specific bias ranges, growth conditions, and post-treatments. Besides the structure-dependent conduction mechanisms, time-dependent transient current effects were investigated. It was found that reversible defect drifting and electron trapping events are most likely responsible for the hysteresis observed in the I-V characteristics.

Outlook

The results presented in this thesis form the basis of various new research projects. In the following section, a selection of the most interesting projects is listed.

In Chapter 5, the properties of various potential seed layers were discussed. For applications requiring flexible substrates such as indium tin oxide/polyethylene terephthalate (ITO/PET) and indium tin oxide/polyethylene naphthalate (ITO/PEN) [1798], it is of great interest to deposit **seed layers on plastic foils**. In the case of electronic applications such as solar cells or lithium batteries, the foil needs to be conductive such as indium tin oxide (ITO) coated polyethylene terephthalate (PET). A challenge is to reduce the process temperature for seed fabrication below the melting temperature of the applied plastic material. Another approach could be the deposition of the seed layer on a temperature-resistant substrate, covered with a chemically and mechanically stable film carrier, and transfer it to the flexible substrate after annealing. For this method, promising seed carriers are $Ca_2Nb_3O_{10}$-TiO_2 nanosheets [1799].

In Chapter 6, the growth of rutile TiO_2 nanorods on rutile and anatase layers was investigated. In this study, the growth steps between nucleation and several hundred nanometers long nanorods were documented. However, it is not clear which is the **smallest stable nanocrystal** that could be created with the hydrothermal method on specific seeds. The answer to this question is important for decreasing the dimensions of devices such as rutile TiO_2 nanorod-based transistors. An impressive example of anatase TiO_2 nanorods with a thickness of only one unit cell and their potential application in field effect transistors is presented by Liu and Yang and Deák et al. [71, 1800].

In Chapter 7, the creation of the inner structure of rutile TiO_2 nanorods was investigated. The nanorods were grown on silicon or FTO substrates and hence, diffusion of atoms from and into the substrate took place during high-temperature annealing. Furthermore, the existence of a substrate below the nanorods limits the annealing temperature. To avoid the diffusion of substrate atoms and a limitation of the annealing temperature, the development of levitating **NRA membranes** is important. NRA membranes provide the cultivation of high-purity single nanocrystals via high-temperature post-annealing. These single crystals can be refined by a further annealing step in a specific atmosphere or after depositing another solid material on the rods before the second annealing process. This second post-treatment results likely in well-defined core-shell structures or doping gradients which are valuable for specific electronic applications such as field effect transistors and energy or data storage devices. An example of the formation of a high-temperature-induced SiO_2-shell/TiO_2-core structure is reported by Folger et al. [1801]. However, the research on NRA membranes and the inner structure of high-temperature annealed nanorods is still in the fledgling stages.

In Chapter 8, existing methods to control the confined growth of NRAs on a substrate precisely were extended and new inexpensive and fast methods were introduced. Nanorods on primary and secondary superstructures such as fabricated by PSML on photoresist patterns offer a high surface area and are applicable for **photonic crystals** enhancing the coupling of incident light, for instance, in photocatalytic devices. PSML structuring on flat substrates compete with laser interference lithography which is usually faster and more accurate. However, on structured substrates (see Figure 13.7) and if the spherical structure of the polystyrene spheres should be conserved in the completed device, the PSML based structuring outperforms laser interference lithography.

Besides the position-controlled growth, the **orientation-controlled growth** is needed for certain device integrations. It is shown that focused ion beam (FIB) milled substrates and laser-induced melting already offer a limited control over the orientation of the nanorods. The achieved control of the orientation has to be expanded for larger nanorod assemblies in order to use them in complex optoelectronic devices. The functional principles of the demonstrated position-controlled fabrication techniques are not restricted to rutile TiO_2 nanorods necessarily and they are transferable to many different materials. Mainly polymorphology and epitaxial growth are required for the presented techniques. These requirements are satisfied by many further metal oxides and other hydrothermally grown nanocrystals. Some of them are listed in the introduction and their potential applications in integrated circuits and lab-on-a-chip devices are highlighted.

In Chapter 9, a laser-induced melting method was introduced as a part of a position-controlled growth method which is simple to apply. However, the working principle is rather complex. Obviously, a light absorbing layer below the anatase TiO_2, which consists of a TiO_2/silicon blend and is created during the initial annealing process, plays a key role for the melting process. Electronic circuits often require insulating substrates and hence, it is useful to apply the laser melting on such a substrate. So far, it has not been possible to create a melting triggering layer on an **insulating substrate** such as silica (SiO_2) or sapphire (Al_2O_3). Nevertheless, to apply this method on **metal or organic electrodes** is valuable as well, for instance to investigate the electronic properties of small NRAs in confined geometries. In general, on organic or thin metallic electrodes, it is difficult to deposit seed layers due to the high temperatures applied during the seed layer generation. This problem can be solved by using low-temperature seed fabrication methods such as spatial atomic layer deposition (SALD).

In Chapter 10, light scattering on NRAs with confined geometries and superlattices were investigated. In order to optimize light coupling, the parameters of these structured NRAs need to be improved further. **Wavelength-dependent scattering** and reflection as a function of the specific structure and angle of incidence are central issues. The presented model for polarization-dependent scattering needs to be refined in order to control light scattering reliably. The successful optimization of light coupling can be demonstrated by fabricating efficient photocatalytic devices and solar cells as well as sensitive UV- and gas detectors.

In Chapter 11, the influence of post-treatment processes on the electronic properties of rutile TiO_2 NRAs was investigated. This research topic offers a huge amount of scientific questions that are waiting to be answered.
One of the most promising candidates to gain further insight into the electronic properties are the NRA membranes which have been mentioned above. These membranes offer the whole range of defect doping densities and make various kinds of impurity doping accessible. At the same time, uncontrolled diffusion from the substrate is suppressed. As a consequence, the membranes are supposed to be the new workhorse for determining the energy landscape and the mobilities of the electrons in TiO_2 nanorods, in particular, if individual nanorods are extracted from the membranes for characterizing single nanorods.

In general, the electron mobility is an intensively discussed property of semiconductors. Comparing the mobilities determined by terahertz and impedance spectroscopy, the effect of grain boundaries and crystallographic direction are examined. Gradients of electron mobilities appear likely if gradients of defect and doping densities are present. In such systems, an applied current could be influenced by space charge effects and cause position-dependent chemical properties on the nanorod surface such as photocatalytic activity or wetting.
For electronic applications, **surface properties** are often more important than bulk properties. And the surface is affected by doping in particular (Schottky barrier, band bending, surface traps). Hence, it is interesting to investigate the spatial and energetic trap and donor state distribution as well as the electronic landscape inside the nanostructure and close to the TiO_2/electrode interface. In doing so, the presented model for charge conduction can be refined

by employing different electrode material (different work functions/ Schottky barriers), temperatures (temperature-dependent conduction mechanisms such as thermionic emissions), and wavelengths in the UV-vis-NIR regime (photodoping, trap level analysis).
It was shown in this work, that the surface properties influence not only the TiO_2/electrode interface, but also contribute to the electron conduction between the electrodes significantly. Especially, high vacancy densities in shell layers created by post-annealing deserve a special mentioning as documented in the last chapter. The presented results raise new scientific questions. What is the critical vacancy density for band conduction in nanorods and the vacancy rich shell layers? Could the thickness of this highly conductive shell layer be decreased sufficiently enough to achieve two dimensional electron systems (2DES)? Is it possible to fabricate nanorods showing plasmonic effects in their metallic shells and pure dielectric behavior in the core? This combination gives rise for the coupling of magnetic and electric resonances applicable for nanostructured metamaterials.

Besides the thermally driven diffusion of oxygen vacancies, charge defects can be shifted by external electrical fields as well. The effect of these fields and the injected charges need to be investigated further in order to distinguish between ion migration and electron trapping **resistive switching** mechanisms. These experiments also cover the creation and healing of extended defects, but intrinsic extended defects should be considered, too. For instance, are 2DES created in extremely flat grain boundaries inside branched nanorods and could their properties be tuned by the vacancy or doping density of the surrounding TiO_2 branches? Do these grain boundaries behave like a n-n^--n junction?

It was shown that, charging and discharging effects result in a time-dependent transient current behavior. Thus, the nanorod acts as a self-controlled **field effect transistor**. In order to use this property for technical applications, a floating gate can be placed close to a single nanorod separated by a thin oxide layer. What would be even more elegant, to use the nanorod itself as a gate by filling and emptying traps similar to a silicon oxide-nitride oxide-silicon (SONOS) memory device [1802, 1803].

Besides the composition and crystallinity, the electronic and optoelectronic properties are additionally influenced by the close neighborhood of the nanorods. Attached molecules, atomic clusters, quantum dots, and other nanoobjects are able to change surface states and to inject or extract electrons, which is useful for gas sensing and photon detection. In particular, gold nano-dots and anisotropic nanoobjects absorb light by the excitation of plasmons. These plasmons decay partly into hot electrons which are able to enter the conduction band of TiO_2 and increase the density of mobile electrons significantly. Such **plasmon-driven phototransistors** would exceed the sensitivity of directly excited TiO_2 and expand the working range into the visible light spectrum.

In addition to the investigated research topics, further scientific problems dealing with rutile TiO_2 nanostructures can be handled more easily with the findings in this thesis. For instance, dopants might not only be used to adjust electronic but also **magnetic properties**. The anisotropic shape of the rods could be transferred to the magnetic properties of each nanocrystal. Decreasing structure sizes cause **quantum confinement** and the related changes of the electronic band structure have to be investigated and tentatively controlled by defect or impurity doping. For applications in **medical and biological research** such as lab-on-a-chip devices, the interaction between the presented rutile TiO_2 nanostructures and living cells in dependence on the incident photon energy and intensity is of interest. It does not require an exceeding amount of creativity to extend the mentioned list of research projects further.

Final Comment

In conclusion, the growth process of rutile TiO_2 nanorods was elucidated in detail, various new position-controlled fabrication methods for these nanorods were presented, and a brief insight into selected optical, electronic, and optoelectronic properties of differently structured and cured nanorod conglomerations was given. The findings in this thesis were a substantial and fruitful base for further research on and development of rutile TiO_2 and metal oxide nanostructure arrangements employable for basic research as well as advanced applications. Only the full power over nanostructure fabrication provides an extensive understanding and control of its physical and chemical properties – or as Richard Feynman said: "What I cannot create, I do not understand."

13 Appendix

13.1 Chapter 5

13.1.1 Thin Seed Layers Made of Different Rutile Materials

For some electronic applications and to confirm the assignability of the growth mechanism to different rutile metal oxides two further rutile seed films made up of different metal oxides are employed.

Transparent Conductive FTO Seed Layers

Commercially available FTO as described in Section 4.1.1 was used as a seed layer. SnO_2 is one of the other rutile metal oxides with slightly different lattice parameters ($a = 4.7552\,\text{Å}$ and $c = 3.1992\,\text{Å}$ [1804]). This mismatch influences the growth of rutile TiO_2 nanorods slightly and will be discussed in Chapter 6.2. The fluorine doping does not influence the lattice parameters of SnO_2 significantly [1805], but increases the electronic conductivity extensively without decreasing the transparency too much [1806, 1807]. In addition, FTO is chemically stable and thus, FTO is usable as a transparent, conductive seed layer for rutile TiO_2 nanorods [1808–1810]. The topography of FTO is very similar to that of rutile TiO_2 as demonstrated in Figure 13.1. The XRD measurements in Figure 13.2 show the crystal similarity of FTO and rutile TiO_2 layers clearly. Since the film thickness of FTO is one order of magnitude thicker, the peaks for FTO are much better expressed.

Tungsten Dioxide (WO_2)

The WO_3 layer was deposited by resistive evaporation and is mainly amorphous. After annealing for 10 min at 700 °C in oxygen or hydrogen rich atmosphere, polycrystalline films were formed. AFM and SEM images in Figure 13.3 demonstrate clearly the different topographies for different annealing atmospheres. Amazingly, the discrepancy in their diffraction patterns shown in Figure 13.2 is much less significant as it is expected from AFM and SEM images. The XRD

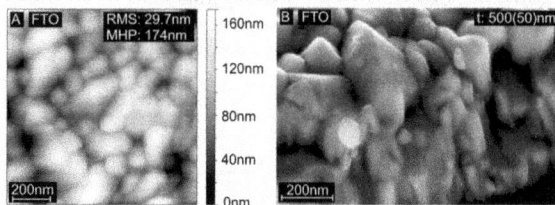

Figure 13.1: *AFM image of a commercially available FTO film (A) and its cross-section imaged by SEM (B). The RMS roughness and (MHP) are calculated from AFM-measurements and the film thickness was determined from SEM cross-section images.*

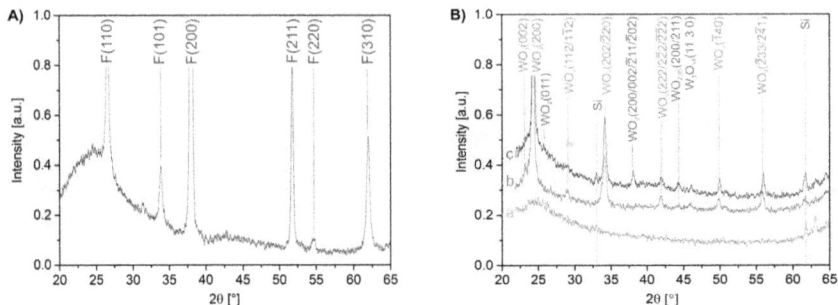

Figure 13.2: *A) XRD pattern of FTO. For peak determination, the XRD pattern published by Lee and Lu [1805] is used. B) XRD pattern of the as-deposited amorphous tungsten oxide film (a), after annealing at 700 °C in oxygen (b) and forming gas (95% N_2, 5%H_2) (c) rich atmosphere for 10 min, respectively. Details are found in the text. The reference diffraction data for WO_3, WO_2, $WO_{2.90}$, and W_5O_{14} are 20-1323 [1811], 32-1393 [1812] (primary reference: Calculated from ICSD using POWD-12++, (1997)), 18-1417 [1813], and 41-0745 [1812], respectively.*

measurement confirms that tungsten oxide annealed in oxygen rich atmosphere converts into the triclinic WO_3 crystal structure. Annealing in a hydrogen atmosphere is supposed to reduce the WO_3 resulting in WO_{3-x} forms such as WO_2. The structure of WO_2 is monoclinic (M), but it could be converted into pseudorutile (R) by a lattice vector transformation: $a_M \approx 2c_R$, $b_M \approx a_R$, and $c_M \approx b_R - c_R$ [1814]. The lattice parameters are $a_M = 5.56$ Å, $b_M = 4.90$ Å, $c_M = 5.56$ Å, and $\beta = 120.47°$ [1815]. Due to the diffraction pattern in Figure 13.2, there is still a lot of WO_3 in the reduced film. Nevertheless, additional peaks such as the $2\theta \approx 38°$ indicate that there is at least a small amount of WO_2 present. An interesting feature is also the peak at $2\theta \approx 44.5°$ that does neither belong to WO_3 nor WO_2, but it is observed for intermediate forms such as $WO_{2.90}$. Furthermore, the WO_3 peak at $2\theta \approx 29°$ disappears for the reduced WO_3 completely. A longer annealing time in hydrogen atmosphere results in extensive dewetting and formation of nanoparticles which is not an aspired structure in this work. Nevertheless, these nanoparticles might contain larger amounts of pseudorutile WO_2.

In conclusion, most of the film crystallizes as WO_3 even in reducing hydrogen atmosphere. However, there are also clear hints for the existence of reduced WO_{3-x} that might even form tiny pseudorutile nanocrystals. Hence, tow effects on the hydrothermal growth are expected. Firstly, pseudorutile parts should stabilize the film under hydrothermal growth conditions and secondly, a few nanorods should grow on the pseudorutile nanocrystals that are exposed on the surface of the reduced oxide film.

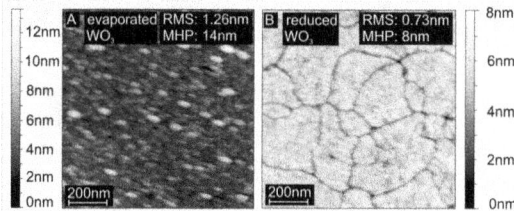

Figure 13.3: *Topography of a tungsten oxide film from AFM measurements before (A) and after (B) annealing in forming gas.*

13.2 Chapter 6

13.2.1 Hydrothermal Growth on TiCl$_4$ Treatment Processed Seed Layers

Since these films are not used in this work extensively a detailed investigation of the film properties is left out in Chapter 5.1.
In contrast to films annealed at 450 °C, an increased density of nanorods is observed on TiCl$_4$ processed films annealed at 850 °C as demonstrated in Figure 13.4 A to D. Compared to blank silicon substrates the density is increased significantly and there are some nanorods sticking out indicating a strong adhesion between the nanorods and the substrate. Obviously, some growth promoting seeds are formed.
TiO$_2$ layers made from TiCl$_4$ treatment behave similarly to sol-gel films. Before annealing, the layer consists of an organic compound and needs time to rearrange and crystallize as TiO$_2$. This process is slower than the temperature rise to 850 °C during annealing and hence, an anatase phase is never created. But also ART promoting impurities might explain the presence of rutile TiO$_2$ after annealing at 850 °C.
For the observed result, it does not matter if rutile results from the ART or was formed without anatase as an interstage product. This technique offers a simple and inexpensive way to get a lot of individual rods on a substrate.

13.2.2 Hydrothermal Growth on Reduced WO$_3$ Seed Layers

Although there is a mismatch of the lattice constants between TiO$_2$ and WO$_2$, growth of TiO$_2$ nanorods is still expected as observed on FTO. Indeed, applying the hydrothermal growth method on the reduced WO$_3$ films results in a few TiO$_2$ nanorods, while there are no rods on polycrystalline WO$_3$ films annealed in oxygen as shown in Figure 13.4 F to H. The diffraction pattern of WO$_3$ layers annealed in oxygen and hydrogen atmosphere are very similar as shown in Figure 13.2. So, only a small part of the film is expected to be crystallized in the pseudorutile morphology. Consequently, the density of nanorods is relatively low.
Rods grown on reduced WO$_3$ have a slightly different shape compared to rods grown on TiO$_2$ layers. This could either result from heavy tungsten doping or from changed growth conditions when the growth solution is mixed with a high amount of tungsten ions generated by the dissolving WO$_3$ film. Furthermore, reduced films have a higher chemical resistivity compared to non-reduced WO$_3$ films that might be a result of the new crystal structure as well.

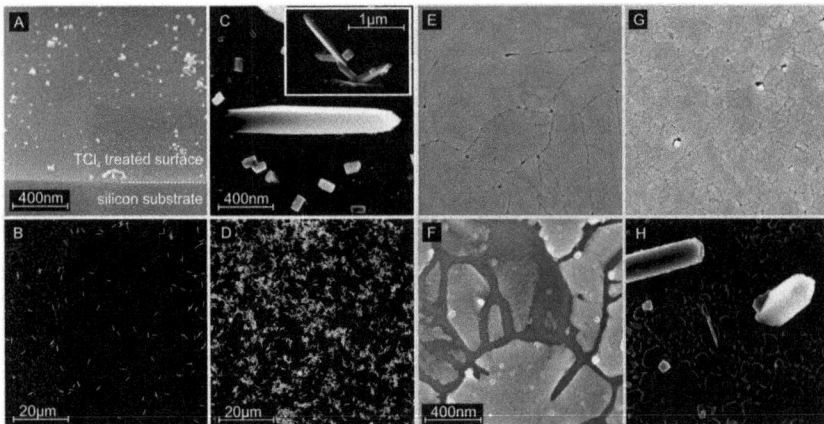

Figure 13.4: *SEM cross-section image of a TiCl₄ treated silicon wafer without annealing (A), after hydrothermal growth on a film annealed at 450 °C (B) and 850 °C (C and D). and hydrothermal growth on reduced WO₃ layers: SEM image of an as-deposited WO₃ film (E), and the same film after reduction (G). (F) and (H) show the samples from (E) and (G) after the hydrothermal process, respectively.*

13.2.3 Hydrothermal Growth on ALD and SALD Seed Layers

The hydrothermal growth process was applied to ALD- and SALD films as well. The results are presented in Figure 13.5. Dependent on the parameters applied for the film fabrication, different film morphologies are expected [1335, 1437, 1816]. The roughly 9 nm thick ALD film appears to be mainly amorphous. At least, no grains were observed in the SEM and there was no clear morphology observable in the XRD data. However, a small fraction of crystalline material cannot be excluded with XRD due to the low film thickness. After the hydrothermal growth, dense NRAs were observed on both, as-deposited and annealed films employing an annealing temperature of 450 °C. Since amorphous TiO₂ films usually dissolve during the hydrothermal growth process, this outcome indicates that a certain amount of crystalline TiO₂ was created within the ALD process. The density of grown nanorods is reduced significantly for films that were annealed at 850 °C after the ALD process. This behavior is similar to sputtered TiO₂ films and might indicate an increase of anatase material which is reasonable for higher annealing temperatures. And even more important, this result suggests strongly that the ALD process did not produce rutile TiO₂. First, rutile would not convert into anatase. Second, if there were rutile particles in the as-deposited ALD film, the amorphous TiO₂ around these particles would crystallize in the rutile morphology and promote the hydrothermal growth of rutile TiO₂ nanorods. This was shown with the annealing of amorphous TiO₂ on a rutile substrate in Chapter 13.2.5. Hence, the ALD films annealed at 450 °C might consist of fine grains promoting a dense growth of rutile nanorods as described for ANA-SPU-450 films.

The films fabricated by the SALD process behave differently. Already the topography and morphology shown in Figure 5.3 and 5.4 differ strongly from the ALD films. After the hydrothermal growth, only a view nanorods appear on the samples similar to ANA-SPU-850 or ANA-PYR-850. Hence, the SALD film affects the growth in the same way as other films with large anatase grains. This is another experimental indicator that the hydrothermal growth on anatase films is linked with grain boundaries: the more grain boundaries, the higher the nanorod density.

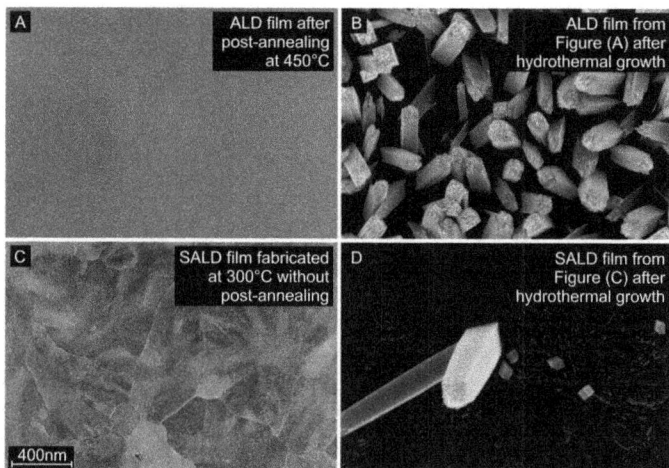

Figure 13.5: *SEM images of ALD (A) and SALD (C) films created with different process parameters as described in Section 4.1.2. The film properties were analyzed in Section 5.1. The film displayed in Panel A was annealing at 450 °C in oxygen atmosphere for 2 hours after deposition. Panel B and D show the results after applying the hydrothermal growth process for rutile TiO$_2$ nanorods employing 14.8% hydrochloric acid and 350 µL titanium(IV) butoxide at 180 °C for 3 hours. The SALD film with its significantly larger grains prevents the growth of dense NRAs similar to ANA-SPU-850.*

13.2.4 Probing the Rutile Phase with Hydrothermal Growth

There are just a few techniques to determine the local crystal structure such as TEM. But preparing samples for TEM studies is time-consuming. Here, a simple and efficient method is introduced for finding regions exposing rutile TiO$_2$ facets. As discussed in the previous chapter, the hydrothermal growth of rutile TiO$_2$ nanorods works only on rutile seed crystal efficiently. Of course, there are always some nanorods formed on statistically distributed dirt particles, but the density of nanorods is increased drastically on rutile seeds. This effect can be used for detecting rutile structures that are exposed on the surface of samples as shown below.

13.2.5 Crystal Structure and Hydrothermal Growth on Rutile/Anatase Sandwiches

Due to their important meaning in photocatalysis, anatase/rutile bilayer structures are discussed intensively in literature [766, 1817, 1818]. Wen et al. demonstrated the use of rutile/anatase core-shell nanorod arrays for photoelectrochemical water splitting [1819]. The VBM of anatase is below that of rutile [1817] and hence, the hole transport becomes at the anatase/rutile blocks holes while electrons can diffuse easily across this interface at room temperature.

The observation made within this work indicate that the crystal orientation of silicon substrates have no influence on the morphology of TiO$_2$ films. As described in Chapter 5, the decision if a film becomes anatase or rutile during annealing at the same annealing temperature is driven mainly by internal effects. But by depositing TiO$_2$ on already existing anatase or rutile films changes the initial condition for crystallization completely. These experiments are important to learn about the crystallization of TiO$_2$ close to anatase and rutile crystal facets. The morphology of the resulting TiO$_2$ films is probed by the hydrothermal method.

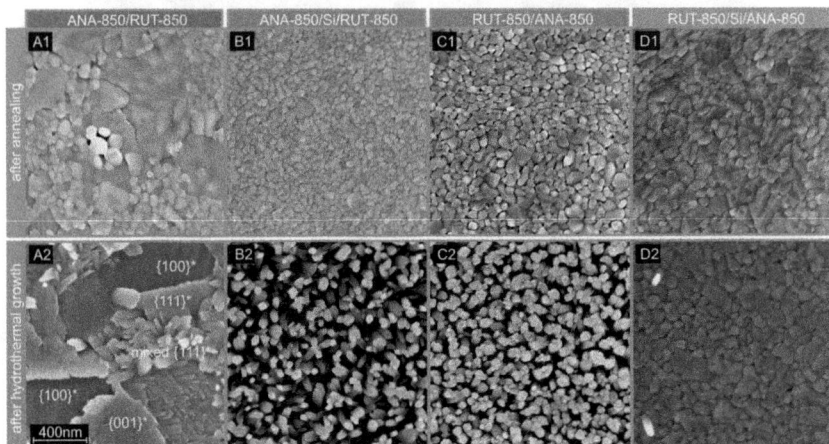

Figure 13.6: *Top views of different rutile/anatase sandwiches made by SEM: A1) silicon/40 nm ANA-SPU-850/40 nm RUT-SPU-850, B1) silicon/40 nm ANA-SPU-850/4 nm silicon/40 nm RUT-SPU-850, C1) silicon/40 nm RUT-SPU-850/40 nm ANA-SPU-850, D1) silicon/40 nm RUT-SPU-850/4 nm silicon/40 nm ANA-SPU-850. The images in the lower row show the samples after applying hydrothermal growth. In this case, the labels RUT and ANA do not mean that the correspondent layers are rutile or anatase necessarily. The label RUT-SPU-850 and ANA-SPU-850 is related to the fabrication procedure only. The (*) points out that the normal vectors of the listed crystal facets must not stick out perpendicular from the surface of the substrate. Details are discussed in the text.*

Firstly, an ANA-SPU-850 was deposited on silicon. Then, a titanium layer was placed on the anatase film and annealed at 850 °C applying the same recipe as used for the fabrication of RUT-SPU-850. The resulting film shown in Figure 13.6 A1 consists of both, large and small grains in equal measure. Small grains are usually observed in rutile films and large grains are related more to anatase films. Hence, based on SEM images only the new film could be a mixture of anatase and rutile grains which would be interesting for photocatalytic applications. To shed light on the morphology of the new film, the hydrothermal growth method was applied on the sample and the result is shown in Figure 13.6 A2. Obviously, the growth of rutile nanostructures is supported on every facet. Small grains provide the growth of rutile nanorods as known from RUT-SPU-850 layers. On large grains, there is a uniform growth direction indicating that these grains consist of a single crystal orientation. The mentioned orientations are assumed from comparing the gained nanostructures with the growth on rutile single crystals as discussed in the beginning of this chapter. The (*) points out that the normal vectors of the listed crystal facets must not stick out perpendicular from the surface of the substrate. This method is useful for generating rutile films with a high content of large grains. And it demonstrates, that the formation of rutile TiO_2 is energetically preferred even in the presence of an anatase crystal. But nevertheless, there is a clear influence of the subjacent anatase film on the crystallization of rutile TiO_2. As discussed in Section 2.3.1, the $(100)_{rutil}/(100)_{anatase}$ interface is expected to be very stable [614]. The low energy of these interfaces provide their generation during the annealing process and hence, large rutile grains are formed on selected anatase grains. For comparison, a 4 nm thick silicon layer is placed between the ANA-SPU-850 and the titanium layer before annealing using sputter deposition. The thin silicon layer does not alter the topography of the ANA-SPU-850 but it covers the anatase facets efficiently. Indeed, a RUT-SPU-850 film consisting of small grains only is formed as it is observed on flat silicon substrates comparably. As expected, a typical dense NRA is grown on this seed layer, which is represented in Figure 13.6 B2.

Now let's reverse the order and deposit a RUT-SPU-850 film as the first layer. In presence of a rutile film, the formation of the thermodynamically less stable anatase is supposed to be suppressed. And as a matter of fact, annealing a sputtered amorphous TiO_2 layer on top of RUT-SPU-850 with the same parameters as used for ANA-SPU-850 films results in a film consisting of small grains similar to the primary RUT-SPU-850 (Figure 13.6 C1). But the grains are slightly larger which is reasonable since small grains might grow together during the secondary annealing step and additional TiO_2 from the covering amorphous TiO_2 layer supports a further growth of grains as well. As the diameter of rods is increasing with increasing grain size rods embedded in the resulting NRA are slightly thicker than rods on a standard RUT-SPU-850 film (Figure 13.6 C2). An additional 4 nm thin silicon layer prevents the catalyst effect of the subjacent rutile films and an anatase film is formed as a top layer which does not promote the growth of rutile nanorods as demonstrated in Figure 13.6 D1 and D2.

13.3 Chapter 8

13.3.1 Double Superlattice by Optical Lithography and PSML

For the latter applications, an even larger surface might be interesting. If needed, the PSML can be combined with the optical lithography as demonstrated in Figure 13.7. In the presented examples, a superlattice was fabricated by the optical lithography. By adjusting the lattice parameters the light coupling is enhanced which is important for light-driven applications such as photocatalysis or photovoltaics.

13.3.2 Applying Scratching on Reduced WO₃ Films

As mentioned in the introduction, rutile TiO_2 is a model system and some techniques developed in this thesis might be transferable to other rutile materials. In Chapter 13.1, reduced WO_3 with pseudorutile parts are presented. In Figure 13.8, it is observed that rutile TiO_2 nanorods grow on scratched and reduced WO_3 but not on scratched and untreated WO_3 films as demonstrated in Figure 13.8. For scratching, a silicon AFM tip, a polyester wipe, plastic tweezers, stainless steel tweezers, and a diamond writer tip are applied. Since WO_2 is harder than silicon and the applied force using the AFM is small, the film could not be affected by the AFM tip. But the situation is different if one rubs with a polyester wipe across the reduced WO_3 film. The density of nanorods is increased in the smeared region significantly. Scratching using plastic tweezers and diamond writers were not successful since the tungsten oxide film was removed completely. Hence, it is assumed that scratching performed with a diamond AFM tip might be an adequate technique to gain locally grown TiO_2 NRAs on reduced WO_3 films.

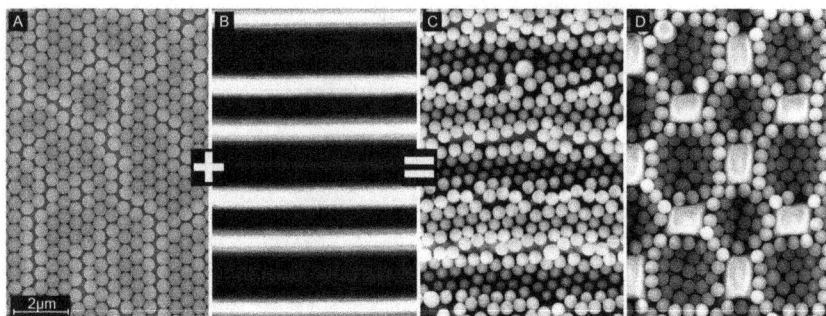

Figure 13.7: *Applying double masks: A) PSML on a flat substrate; B) Mask made by lithography on a flat substrate; C) PSML on the mask shown in Panel B; D) PSML on a mask with a two-dimensional pattern. The diameter of the spheres used in this Figure is* 350 nm.

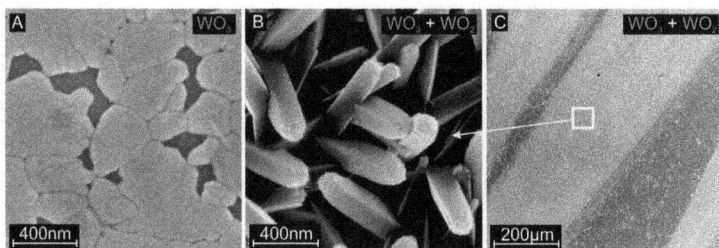

Figure 13.8: *SEM images taken after the hydrothermal growth on scratched areas: A) almost no rods on an untreated WO$_3$ film; B) dense NRA on a reduced film; C) Zoom-out of (B). Scratching was performed by rubbing the film with a polyester tissue.*

13.4 Chapter 9

13.4.1 Correlation Between Provided Beam Energy and Melting Behavior for cw Laser Melting

The injected power is not controlled by changing the beam power during writing but by changing the shift velocity of the stage. In order to achieve rectangular kinks within the applied meander line, the stage accelerates and decelerates in a defined way before and after the kinks. The time t-dependent acceleration is given by $\ddot{x}(t) = a^* \sin^2(\beta t)$ [1820]. Since the acceleration has to be vanished after reaching the maximum shift velocity v_m, (βt) has to be equal π and thus, $\beta = \ddot{x}_a/v_m \cdot \pi$ using the given average acceleration of $\ddot{x}_a = 500\,\mathrm{mm/s^2}$. According to the primitive, the velocity is given by $\dot{x}(t) = a^* \left[\frac{1}{2}t - \frac{1}{4\beta}\sin(2\beta t)\right]$. v_m is achieved after the total acceleration time of $t_a = v_m/\ddot{x}_a$. From this it follows that

$$a^* = \frac{1}{\dfrac{t_a}{2v_m} - \dfrac{1}{4\beta v_m}\sin(2\beta t_a)} \tag{13.1}$$

and since the sine becomes zero for $t = t_a$ as discussed above, it follows that $a^* = 2v_m/t_a = 2\ddot{x}_a$. Thus, the equations of motion are given by

$$\ddot{x}(t) = 2\ddot{x}_a \sin^2(\beta t), \tag{13.2}$$

$$\dot{x}(t) = 2\ddot{x}_a \left[\frac{1}{2}t - \frac{1}{4\beta}\sin(2\beta t)\right], \tag{13.3}$$

$$x(t) = 2\ddot{x}_a \left[\frac{1}{4}t^2 + \frac{1}{8\beta^2}\left(\cos(2\beta t) - 1\right)\right]. \tag{13.4}$$

For high and low energy laser-induced melting, the shift velocity v_m is 0.05 mm/s and 10 mm/s, respectively. For high energy-induced melting, the distance of acceleration is 2 nm and thus negligible. In case of low energy-induced melting, the total distance for acceleration is 100 µm and thus, twice as long as the longest line in the presented meander pattern. As a consequence, v_m is never achieved for the 30 µm long lines. The applied laser power for high and low energy laser-induced melting are 560–600 mW and 590–610 mW, respectively. It is assumed that the focused laser beam irradiates an area of 1 µm². (The diameter of the beam focused by a 50X objective is roughly 1.4 µm and hence, the total exposed area is about 1.5 µm². But due to the Gaussian beam profile, most of the energy is delivered in the center of the beam and the simplification that the total beam energy is concentrated within 1 µm² is reasonable [1821].) To estimate the energy attaining the surface of the sample, the path of the laser beam is sectioned in a row of quadratic 1 µm² pixels. The energy delivered in each pixel is calculated from the product of the beam power P and the time Δt a single pixel is traversed by the beam. Δt is determined numerically using Equation 13.4. As a further approximation, reflection losses inside the objective are neglected and a beam power of 0.6 W is taken for all samples. For high energy laser-induced melting lithography, the maximum speed is so slow that the acceleration period does not play any role and the energy per pixel is 12 mJ for all pixels. For low energy laser-induced melting lithography, the energy per pixel is continuously changing between 0.2 and 1.2 mJ as demonstrated in Figure 9.15.

13.4.2 Suppressing Hydrothermal Growth by Intensive Electron-Beam Exposure

As demonstrated in Figure 13.9, intense electron-beam exposure works as an eraser for locally created seed structures. To explain this feature one has to consider different effects caused by high energetic electrons striking the sample surface.

The incident electron-beam causes heat and the more focused the beam is, the higher is the achieved temperature [1822, 1823]. The temperature rise is amplified by a decrease of the thermal conductivity at high temperatures due to phonon scattering. As shown for the formation of anatase TiO_2 at 450 °C atom rearrangement appears already at a temperature far below the melting temperature. So anatase nanoparticles might merge with the subjacent anatase grains and work no longer as seed crystals. This explains the observation on samples with TiO_2 anatase nanoparticles made by the AFM-tip based scanning probe lithography. But it is not likely that the thermodynamically stable rutile seeds created by laser-induced melting decompose and prevent the hydrothermal growth of rutile nanorods.

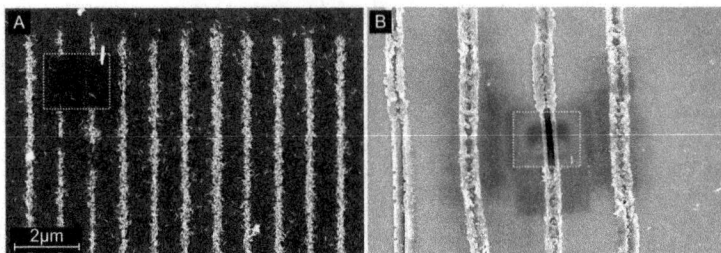

Figure 13.9: *SEM images showing the effect of intensive electron currents on the growth behavior: The marked rectangular area was exposed to intensive electron radiation before hydrothermal growth but after activating the growth locally by scanning probe lithography (introduced in Chapter 8.6) (A) and cw laser-induced melting (introduced in this Chapter) (B).*

Electron scattering causes atomic displacements, so-called knock-on damages [1822, 1823]. Due to the high threshold displacement energy in bulk TiO_2, they play a role on the TiO_2 surface only [1822, 1823]. But the atomic structure on the surface is important for the epitaxial growth of nanorods.

Radiolysis is another effect that occurs during the exposure by high energetic electrons. Bremsstrahlung provides X-rays that are able to crack chemical bonds inside the TiO_2. It is known that already UV radiation changes the density of adsorbed oxygen inducing superhydrophilicity of the TiO_2 surface [32, 34]. So it is reasonable that broadband x-ray radiation causes the same effect. However, an increased hydrophilicity is not supposed to prevent the approach of $TiCl_4$ molecules to the surface [1824].

Hydrocarbon contamination results in an electron-beam-induced deposition of carbon at the exposed areas [1822]. A thin carbon layer is likely to resist the growth conditions and covers the seeds during hydrothermal growth. However, the chamber of the SEM is almost hydrocarbon free and the sputtered film is not supposed to carry a significant amount of hydrocarbon molecules. So this explanation is not completely satisfying.

As described by the induced-electrical-field model electrons are trapped in subsurface layers causing high electrical fields inside the TiO_2 [1822, 1823, 1825]. It is reasonable that a negatively charged surface prevents the approach of $TiCl_4$ with its electronegative chlorine-ions. This general explanation can be applied to any technique providing local seeds for the hydrothermal growth. Since the electrons are fixed in subsurface layers, they are not desorbed by water molecules under ambient conditions and remain as growth-prohibiting elements during the whole hydrothermal growth process. However, another possible reason is that growth-promoting surfaces are damaged by the intensive electron-beam so extensively that they lose their growth promoting property. This explanation is supported by the observation that the samples could be stored for many days under ambient conditions without losing the growth suppression feature. In contrast, charges trapped on surfaces are transferred to water molecules in the gas phase within a few days, even if it is a slow process as assumed above.

In conclusion, the hydrothermal growth of nanorods is controllable by electron-beam exposure on samples treated by scanning probe lithography and laser-induced melting technique precisely!

13.5 Chapter 10

13.5.1 Scattering on Superstructures Made by PSML

The advantage of superstructures are their additional resonances with electromagnetic waves and large surface area as reported by Evlyukhin et al., Lu et al., and Baek et al., for instance [677, 724, 1826]. Here, it is demonstrated that a primary superstructure of the seed layer made with a PSML can be placed on an additional, secondary superstructure made with optical lithography. The structures shown in Figure 13.10 are covered by a 20 nm thick titanium layer.

On flat substrates, the pattern structure is displayed by an inhomogeneous visible light scattering. Similar to NRAs, scattering is enhanced on edges and crystal defects in the hexagonal ordered PSML. These mostly 1D defects appear as smudge bright lines in the DF image (Figure 13.10 A bottom). The secondary superstructure displayed in Figure 13.10 B works as a grating for visible light which can be seen easily with the naked eye. The PSML on top of the lithography pattern forms a hexagonal 2D lattice similar to flat substrates but with more crystal defects. Due to scattering events, the nature of a grating for visible light is reduced. But Figure 13.10 C indicates that the reflected light displays both the PSML and the patterned photoresist. A similar feature is observed if the 1D pattern of the photoresist is replaced by a 2D superlattice (Figure 13.10 D).

The preparation of NRAs on these structures works similar on the other presented superstructures and is not discussed in this thesis. Applying these structures to TiO_2 nanorods and gold particles are valuable for photocatalysis devices, for instance [1099].

13.5.2 Intensity Pattern Occurring Near the Nanowalls

Finally, this report allocates phenomena observed for light scattering at double walls in particular. In Figure 10.13 F, stripes parallel to the double walls appear which are pronounced clearly beside upright walls. Their most dominant color is blue. Especially for the inner stripes, the whole visible spectrum is observed while blue seems to be the most inner and red the most outer color. Such an order is known for diffraction effects and might be already a hint for the origin of the observed stripes.

First of all, it should be verified if the requirements for the appearance of diffraction patterns are satisfied in the applied setup. Such quantum mechanic effects require sufficient coherence. Interference patterns caused by diffraction occur if the coherence length is larger than the difference of the path length of interfering waves. In the shown experiments, the illumination source is a white light LED and typical coherence lengths of LEDs are in the order of 10 μm [1827]. The distance of the double walls is about 2 μm. Thus, the difference of the path length for small angles is assumed to be more than two orders of magnitude smaller than the coherence length of LEDs. For this relation, the appearance of diffraction patterns is reasonable.

Of course, a diffraction effect might result from the two walls within the same double wall complex as well. Due to the smaller distance between the diffraction elements, the diffraction pattern is expected to be more stretched compared to the diffraction pattern of neighboring

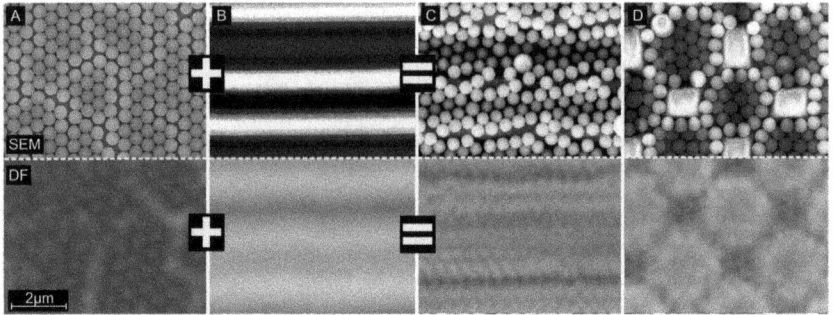

Figure 13.10: *Scattering at superstructures made with a monolayer of polystyrene spheres (PSML) and optical lithography demonstrated using SEM (top) and optical DF (bottom) images: A) Flat PSML consisting of spheres with a diameter of 380 nm. B) Blank and developed photoresist shaped as parallel lines with a width and pitch of 2 μm. The thickness of the photoresist is roughly 1 μm. The colors result from interference effects. The wavelength of the reflected light depends on the reflection angle and thus, it behaves like a diffraction grating as drafted in Figure 10.2 D. C) Photoresists shown in Panel B covered with the PSML shown in Panel A. The intensity of the reflected light is decreased and the properties of a diffraction grating are reduced. D) Similar sample as shown in Panel C, but applying a 2D instead of a 1D pattern. All structures are covered with a 20 nm thin titanium film fabricated with sputter deposition to obtain sharper SEM images.*

double walls. In this case, the diffraction pattern is expected on both sides of all double walls with the same intensity as elsewhere. This is not observed for white light illumination.

However, there is another feature indicating that the intensity pattern is generated by diffraction. Figure 10.13 G shows the brightness profile of the scattered light in the neighborhood of a double wall in comparison to the actual structure dimension measured with SEM. The width of the gap within a double wall is $d_i = 150$ nm. The total width of a double wall complex is $d_o = 400$ nm. In the optical image, the distance between the two bright lines corresponds to a geometrical distance of $d_0 = 500$ nm. The average distances between the first and second intensity maxima on both sides of the diffraction pattern are $d_1 = 430$ nm and $d_2 = 380$–390 nm, respectively. It should be noted that the corresponding distance on the optical image between the center of the dark line in the middle of the double wall and the brightness maximum of the bright line is 250 nm only. This is smaller than the applied wavelength and should not be detected by an optical microscope in combination with the applied objective so clearly. Only interference effects from diffraction could result in such tiny intensity patterns.

There are mainly two possible origins for such patterns: the structure on the sample and the optical elements in the microscope. At first sight and especially for white LED illumination as shown in Figure 10.13 B and F, there are some hints indicating that the observed intensity pattern results from diffraction on nanowalls. For instance, the pattern is not continued outside of the outermost double walls as demonstrated in Figure 10.13 B. And there is no visible pattern between double walls where only one of the double walls is upright.

But close to the resolution limit of the optical microscope, diffraction effects might be also induced by the focusing optics and pinholes used in the microscope. These so-called Airy patterns° would appear symmetrically besides a scattering particle which is not observed here.

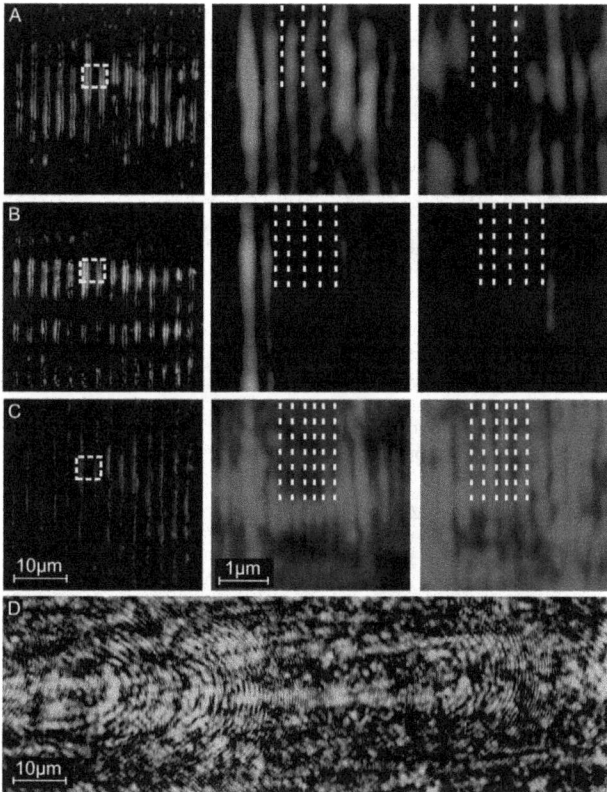

Figure 13.11: *Wavelength-dependent reflection on TiO₂ nanochannels using a* 633 nm *(A),* 532 nm *(B), and* 405 nm *(C) diode laser, respectively. For all samples, the laser power is in the same order, but since the effects are compared qualitatively only, the laser power is not measured in this study. The middle and right images are magnifications of the yellow squares in the left images. The average distances of the intensity minima are* 433(20) nm *(A),* 304(20) nm *(B), and* 223(20) nm *(C), respectively. These distances do not change with the working distance of the microscope as determined from comparing images recorded in (middle) and out of (right) focus. D) Out of focus image of an area besides the patterned surface shown in Image B. Here, equidistant intensity maxima and minima appear as well having the same distance as determined from Image B and elliptical intensity patterns appear around dirt particles as seen in the left part of the image. These observations indicate that the intensity patterns are rather Airy fringes than interference patterns of the generated nanostructure grating.*

In order to find more details about the origin of the intensity pattern, the sample is illuminated with laser diodes emitting at 633 nm, 532 nm, and 405 nm. As demonstrated in Figure 13.11 A to C, the distance between the intensity minima depends strongly on the applied wavelength and decreases with decreasing wavelength which is expected for diffraction on gratings or slits. However, the bright laser light illuminates features which suggest that the interference pattern is caused by the nanowalls. First, in relation to the applied wavelengths and the distance of the nanowalls, only maxima up to the fifth order should be visible. But the pattern is observed on the whole sample and in particular in areas without any nanostructure. Second, the maxima are equidistant which is not expected for diffraction on gratings and slits. Third, the distance of the intensity pattern does not depend on the working distance and appears also in the neighborhood of round dirt particles as demonstrated in Figure 13.11 D. All these observations indicate that Airy fringes are the dominating origin for the observed intensity pattern. However, an overlap of Airy fringes with diffraction patterns of the applied nanostructures cannot be excluded.

13.6 Chapter 11

13.6.1 Experimental Characteristics

Some peculiarities of the setup and the measurements are presented which are useful for the understanding of the presented experimental outcomes.

Electronic Measurements Using Metal Tips as Top Electrodes

Using metal tips as top contacts is a widely spread experimental method to obtain quickly and locally resolved information about the electronic properties of TiO_2 films and nanostructures [220, 1207, 1788, 1828]. A round apex helps to contact roughly the same number of nanocrystals in each measurement. Similar to extended planar electrodes, measurements with large tips average the properties of several nanocrystals. Furthermore, pinholes in a film or an array which are smaller than the dimensions of the apex do not cause shorts. Capacitive effects are less distinct for tip electrodes and the sheet resistance of thin film electrodes does not need to be considered. Thus, there are significant differences between measurements on flat TiO_2 films which are sandwiched between extended electrodes and TiO_2 nanostructures measured with a metal tip as top electrode [1207].

Effect of adsorbates: The interface between the tip and the TiO_2 is exposed to environmental conditions such as humidity, dust, or the partial pressure of a specific gas. Hence, there are adsorbates that result in additional conducting paths, electron donors and traps, insulating and injection promoting layers, or electrochemical reactions [1207, 1829–1831]. The effects of the most important adsorbates are discussed in Section 2.6 to 2.8. Early experiments employing a conductive atomic force microscope (CAFM) under ambient conditions showed tremendous current fluctuations and unreasonably high current densities. Most likely, this behavior was related to electrochemical reaction generating negatively charged oxygen ions (O^-) and hydroxyl groups (OH^-) from the ubiquitous water film [1078, 1207, 1829, 1830]. This assumption was reinforced by the observation that the $I - V$ characteristics depended rather on the day than on the sample preparation. Besides affecting the transient current directly, these charge fluctuations change the electrostatic forces between the tip and the rods and hence promote uncontrollable oscillations of the tip. As a consequence, the measurements were transferred into a sealed chamber that was evacuated, purged, and filled with nitrogen or oxygen. The tip was exchanged by a macroscopic probe without any feedback-loop control and placed manually on the array. The gas pressure was kept above the atmospheric pressure to avoid a contamination with air.

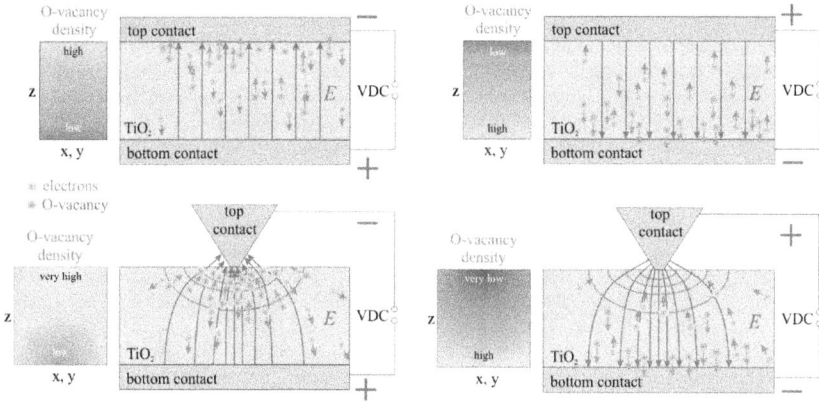

Figure 13.12: *The electrical field close to the tip is much higher (dense electrical flux lines) than near a planar electrode. As a consequence, an asymmetric I−V characteristic is not only given by different electrode materials and oxide surfaces but also by the geometry of the electrode. Additionally, the assumed effect on the distribution of oxygen vacancies is drawn.*

To reduce the water film without changing the electronic properties of the TiO_2 each sample was heated to 120 °C on a hotplate before it was placed in the chamber. However, the hydroxyl groups remained but their effect is supposed to be much less considerable than the effect of a water film [1207]. The pressurized gas provided dry and defined environmental conditions. Gold as a noble tip material prevents the creation of an additional insulating layer between the tip and the TiO_2. Pre-existing impurities on the gold tip were covered with an additional gold layer which was deposited by resistive evaporation shortly before the tip was installed in the experimental setup.

Distribution of the electrical field: The geometry of the tip affects strongly the distribution of the electrical field at the interface and inside the TiO_2, which is illustrated in Figure 13.12 [1207]. The electrical field close to a round tip is similar to the field around a charged sphere. Hence, the highest field amplitude is found near the surface of the tip and it decreases towards the flat bottom electrode. As a consequence, the assumption that the electrical field is equal to the applied bias divided by the oxide thickness is not valid in this case. The electrical field on both electrode/TiO_2 interfaces differs and hence, a precise quantitative analysis of electronic properties based on models for bulk- and interface-limited conduction mechanisms is hardly possible. An additional local field enhancement is caused by the roughness of the tip surface. Tiny dips result in a large field enhancement and field-dependent effects such as ion migration, energy barriers, and trapping/detrapping rates are amplified locally [1832].

Furthermore, the enhanced field facilitates electrochemical reactions and field-dependent effects, such as resistive switching, preferentially in the oxide volume close to the tip [1207]. To reduce field inhomogeneities a tip with a large tip radius in the order of a millimeter is employed.

Effect of the contact area: The contact area between the tip and the nanorods depends on the tip radius and the roughness of the tip and the NRA. Typically, TiO_2 NRAs have a very rough "surface". First of all, the rods fill not the whole volume of the array but there are large material-free spacings. Furthermore, according to the fine structure of each nanorod, their

311

surface is far from being flat and not all nanorods have the same length. As a consequence, the contact area is hard to determine and can only be approximated. Hence, the absolute value of the measured current is given rather than the current density for the $I-V$ characteristics shown in this work. This is another reason why quantitative analysis on a nanostructured surface is almost impossible with tip methods. Nevertheless, tip-based measurements are a powerful tool to determine qualitative differences for a large number of samples in a short time. Thus, this method is ideal for field studies to get a first impression about relative differences of electronic properties of various fabrication and post-treatment parameters.

In addition, there is another advantage using flat bottom and tip top electrodes. Related to the restricted contact area and the inhomogeneous electrical field, the current is distributed across a much larger area at the bottom interface. As a consequence, this interface does hardly limit the total current even if the Schottky barrier is slightly higher at the top interface. This feature results in a quasi-ohmic contact at the bottom interface [1207, 1788].

The high resistance of the rather small tip/TiO$_2$ contact area has to be considered for impedance spectroscopy since the RC term is large and hence, the interface capacity limits the accessible frequency bandwidth [1207, 1833]. Consequently, no impedance spectroscopy was made on the investigated nanostructures. Additionally, the high resistance of the tip contact generates heat at the tip apex and the gold might melt for high current densities [1207, 1834]. Therefore, the current was kept below roughly 1 mA in all measurements.

Capacity Effects

In general, the capacity is defined as the number of additional charges that are stored in a device when the voltage is increased by a certain value. In TiO$_2$, there are various effects contributing to the storage of charges such as the polarization of ions and bonds, ionization of donors, trapping of electrons, creation of interface dipoles, and ion migration. Hence, the investigated system has a complex equivalent circuit of capacities and resistors which are connected in parallel and series. It is distinguished between the pure dielectric capacity that involves only local charge displacements and a non-dielectric capacity which involves the separation of free charge carriers. Put simply and in more technical terms, the system is a combination of a capacity and a battery. The battery effects are discussed in detail later.

Here, it is demonstrated that the geometric dielectric capacity does not play an important role in the investigated samples. Effects resulting from the drift or diffusion of mobile charge carriers are often not symmetric for positive and negative applied bias. Nevertheless, the charging and discharging current of a pure dielectric rutile TiO$_2$ capacity does not depend on the sign of the applied field. The reason is the crystal symmetry of rutile along the principal axis. Furthermore, the dielectric capacity is very low. This is related to the small contact area at the tip/NRA interface and the relatively large length of the NRA compared to the thickness of TiO$_2$ films. The dielectric capacity of the tip/NRA/FTO system is extracted from samples that have been annealed in an oxygen atmosphere which show no transient current but a symmetric charging and discharging current. The corresponding data are shown in Figure 13.13. It is assumed that the dielectric capacity is completely charged when the bias is switched off. This assumption is reasonable since the observed discharge curve for charging times of 3 and 500 s is completely the same. The sputtered TiO$_2$ film and the NRA have the same capacity since the NRA is a factor of 3 thicker than the film, but only less than one third of the NRA's volume is filled with TiO$_2$. After the sample was charged at 20 V, an initial discharge current of roughly 200 pA occurred. The measurable discharge process takes about 2 s and hence, the stored charge is roughly 200 pC and the capacity is in the order of 10 pF. This value can be used to approximate the equivalent electrode area which is the geometrical electrode area of a planar electrode/film/electrode system with the same capacity. From $A_{eq} = \frac{Cd_{ox}}{\epsilon \epsilon_s}$ with $d_{ox} = 0.35\,\mu\text{m}$ and $\epsilon_s \approx 100$ it follows that the equivalent area is in the order of 10^{-9} m^2 (about 10×10 µm). It has to be considered that this area is much larger than the real contact area because the electrical field beside the contact area is almost as strong as below the contact area and induces a polarization of rods that are not connected with the tip. Further deviations originate from the inhomogeneous field between the tip and the bottom electrode. However, related to the large tip radius ($r_t \approx 1$ mm) the inhomogeneity is less important compared to a CAFM tip ($r_t \approx 10$ nm), for instance. If the depletion layer contributes significantly to the capacity, the oxide thickness has to be replaced by the depletion layer thickness. For a depletion layer thickness of a few nanometers, the approximated area shrinks to roughly 10×10 µm. Nevertheless, the combination of a low donor density in the annealed samples and the similar Fermi levels of gold, TiO$_2$, and FTO prohibit the creation of a pronounced narrow depletion layer.

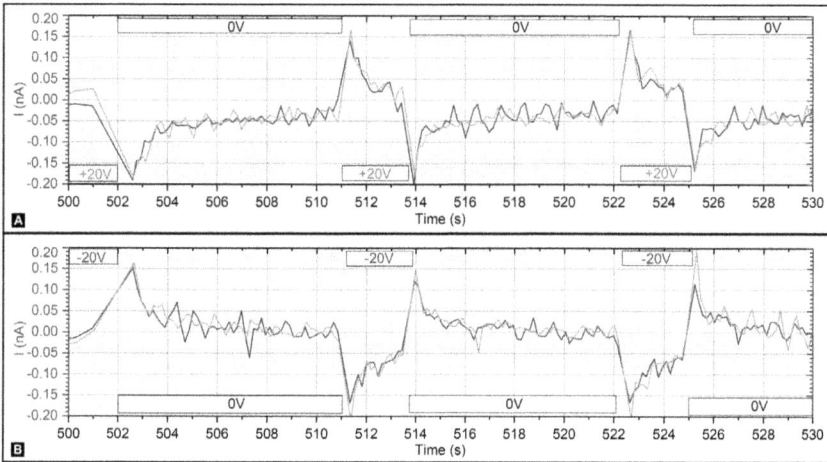

Figure 13.13: *The response of short voltage pulses on the current signal of NR220 (purple) and sputtered films (green) after annealing in an oxygen atmosphere. The color code is the same as introduced in Figure 11.2. These samples were insulating in the applied bias regime and consequently, no disturbing transient current appears. The first pulse is 500 s long and only the last seconds of this pulse are displayed. Then, two voltage pulses (+(A) and -(B) 20 V) with a duration of 3 s and a time lag of 8 s are applied and the current signal is recorded. The crystal structures of the two samples are different and they show unsimilar transient current behaviors as discussed later. Hence, a completely symmetric response to an external field as shown in this figure is not expected if moving electrons or ions are involved. Therefore, it is suggested that the seen signal originates from the pure dielectric capacity of the sample.*

The Relation Between Displacement Current and Measured Current

After an electron is injected into the nanorod, a positive image charge is created in the cathode (Schottky effect). Since a positive charge is created at the same time when the electron leaves the cathode, no displacement current is generated. As the electron departs from the interface its charge becomes shielded by the polarization charges in the TiO$_2$ and the image charge is annihilated. Hence, a displacement current is created pulling an electron from the cathode supply line into the cathode. When the drifting electron approaches the anode, a positive image charge is induced and consequently, an electron has to leave the anode towards the anode supply line. Again, a displacement current is created pointing in the same direction as the displacement current at the cathode. When the electron is injected into the anode, it recombines with its image charge and no displacement current is generated. In conclusion, the measured current is nothing else than a displacement current.

For a low number of trapped or mobile charges, the polarization of TiO_2 is efficient enough to neutralize each charge within a certain radius in bulk TiO_2. For high charge densities, the field of the charges cannot be compensated by polarization charges completely and a steady image charge is situated on each electrode. If the center of the total charge shifts, the image charges change as well and a displacement current is measured.

Electronic Measurements in the Dark

Electronic properties are not only affected by the internal structure, electrode materials, and adsorbates but also by incident light. In Section 2.9, the optoelectronic properties are discussed only briefly since it would go beyond the scope of this thesis. However, it should be noted that even the exposure to visible light might generate mobile electrons and holes and hence, the measurements were performed in a completely dark chamber.

Approximation of Joule Heating at High Current Densities

The electronic properties are influenced by several temperature-dependent mechanisms in the investigated metal-insulator-metal (MIM) structures above room temperature. In TiO_2, the mobility of ions, the density of mobile electrons, and the momentum scattering frequency of electrons is increasing while the conductivity of the metallic electrodes and the density of adsorbates is decreasing with increasing temperature. Hence, it is necessary to approximate the effect of Joule heating.

Although several temperature-dependent effects are mentioned above, not all of them influence the electronic properties to the same extent. The density of adsorbates is most likely stable since the water film was already reduced by baking and the applied vacuum before the measurements, the remaining OH-groups are supposed to desorb not below 450 °C anyway, nitrogen molecules do not have a significant influence on the electronic properties at all, and oxygen molecules reduce the current and subsequently the Joule heating to zero. The temperature coefficient of resistance of gold is 0.0034 /°C at room temperature [1835]. Thus, an increase of 200 K would increase the resistance of the gold tip by almost 70%. However, the total resistance of the gold-tip/FTO system (without TiO_2) is in the order of 10 Ω, which is orders of magnitudes smaller than the resistance of any TiO_2 nanostructure. Additionally, only a small part of the tip and the FTO are warmed up by the dissipating heat from the TiO_2 resistor. Consequently, the effect of Joule heating on the metallic electrodes is negligible for stable currents. Nevertheless, if the resistance of the TiO_2 is changing considerably due to a resistive switching effect, the current might become so high that the tip is melting at the interface area. The most important temperature-dependent effects in TiO_2 are thermally activated ion and electron drift. The diffusion of crystal defects or the detrapping rate are proportional to $\exp\left(\frac{E_a}{k_b T}\right)$ where E_a is the activation energy of ion movement or electron detrapping, respectively.

In order to find an upper limit of the induced Joule heat, the sample was externally heated to roughly 200 °C with a soldering tip close to the measuring tip. A constant voltage (here $-2\,\mathrm{V}$) over time measurement was performed before, during, and after the external heating which is shown in Figure 13.14. During the external heat stress, the current over time behavior is similar

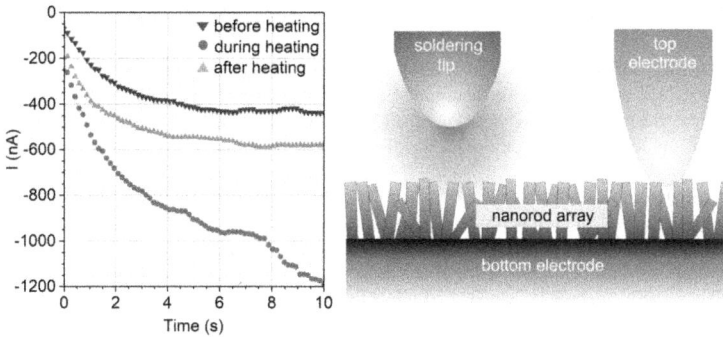

Figure 13.14: *Effect of temperature on stress-induced transient current: transient current before* (▼), *while* (●), *and after* (▲) *heat exposure.*

to the non-heated sample, but the current becomes two times larger and noisy as without heating. After some seconds, it shows fluctuations around a certain level. This might be explained by the increased mobility of atoms, which results in charge transport by percolation pathways changing their structure at elevated temperatures frequently. The observed relaxation time after the external heating is much longer than for applying bias at room temperature. Furthermore, the current does not relax to the initial value and hence, the CVS measurement after external heating shows slightly higher values as the measurement before heating. This feature is also an indication for an alteration of the atomic structure in the TiO_2 film. However, the current after heating is still much lower than the current during heating. The soldering tip might also reduce the water film on the sample but this effect is supposed to decrease the current [1078, 1207, 1829, 1830]. It is concluded that the sample temperature does not even increase approximately by 200 K for currents below 1 µA (see Figure 13.14).

A similar conclusion is obtained by calculating the injected electrical energy. It is assumed that a bulk-limited conduction mechanism is dominant and that the whole electrical energy $W_{el} = U \cdot \int_{\Delta t} I(t)dt$ is transformed into heat energy $W_{th} = \frac{3}{2}k_b(T_{rt} + \Delta T)$ within the volume of the connected nanocrystals (T_{rt} = room temperature). Hence, the temperature is roughly proportional to the number of injected charges. However, a heat flux conducts the heat energy to neighboring materials and the assumed proportionality is not valid in case of metallic electrodes having a high thermal conductivity and heat capacitance. Furthermore, it is assumed from previous measurements that the resistance of a 1 µm long and 50 nm wide single nanorod is roughly 1 MΩ and 10,000 rods are connected by the gold tip. Hence, the total resistance of the 10,000 parallel nanorods is $R_r = 100\,\Omega$ and their volume is in the order of $10^{-17}\,m^3$. The electrical power is given by $P = R_r I^2$. For transient currents of 1 µA and 1 mA, power densities of $10\,mW/mm^3$ and $10\,kW/mm^3$ are obtained, respectively. Compared to a standard macroscopic $100\,\Omega$ resistor with a volume of $5\,mm^3$, the same power density is achieved with currents in the order of 20 mA and 20 A, respectively. This comparison shows that currents below 1 µA are not expected to heat up the sample significantly. However, transient currents of roughly 1 mA are either heating the sample likely above the melting temperature of gold or the average resistance of the connected nanorods was decreased dramatically by resistive switching effects before.

315

The Measurement Protocol

Many investigated samples changed their electronic properties during the characterization on a long timescale. As soon as a measurement is performed at a certain position, the properties are changed. Additionally, the long timescales for some effects made the experiments very time-consuming. As a consequence, an efficient measurement protocol (Figure 13.15) was developed that considered several features at once.

①: In the first part of the measurement, the voltage range resulting in current amplitudes below $1\,\mu A$ was determined. The limitation of the current should avoid a quick oxide degradation.

②: Within the second step, five $I - V$ measurements employing the voltage range obtained in the first step were performed starting from the lowest voltage towards the highest voltage and back. These measurements provided the typical hysteresis behavior. The revision of $I - V$ cycles demonstrates the stability of the system. The sweep rate was fixed at $0.08\,V/s$.

③: Accordingly, an $I - V - t$ measurements was applied in the same voltage range but with much larger voltage steps. For each voltage step, the voltage was kept constant for 30 s. These measurements, which were performed twice with the same driving direction as the $I - V$ measurements, give first insights into the voltage-dependent evolution of the transient current. These results are not discussed in this thesis as they do not offer important additional information about the electronic properties. However, it should be considered that $I - V - t$ measurements cause already significant voltage and current stress to the sample.

④: The fourth step was a cohered sequence of constant voltage over time $(I - t)$ measurements that were performed with the highest voltage for both polarities obtained from step one. Each sequence started with two positive and two negative pulses (pulse length and time lag: 6 s) followed by a 500 s lasting voltage stress and short voltage pulses in the end. The idea is similar as in a pump-probe experiment where the long CVS corresponds to the pump pulse and the short pulses correspond to the probe pulses. Their duration is so short (3 s) that they affect the material much less than the long (500 s) pulse. From this measurement, the long-time change of the oxide under voltage and current stress, relaxation times, and discharging properties were obtained.

⑤: In the final step, the $I - V$ measurements from step two were repeated in order to gain the $I - V$ characteristics of stressed nanocrystals.

Reasons for Asymmetric I–V Characteristics

The barrier height for electrons between the electrodes and the TiO_2 nanorods is mainly given by the Fermi levels of gold, FTO, and TiO_2 and the electron affinity of TiO_2. The slightly different work functions of gold and FTO, as well as the inhomogeneous structure of TiO_2 nanocrystals, cause an asymmetric $I - V$ characteristic. The asymmetry is expressed with the rectification factor which is defined as $Z = \frac{I(U)}{I(-U)}$ [1077].

However, besides the electrode material, there are a few additional effects that influence the asymmetry of such a MIM device. As discussed above, it is important if the top electrode is an extended flat film or a metal tip with a certain tip radius. Further differences could appear due to different interface roughness, inhomogeneous impurity distribution, and morphology or defect gradients between the two interfaces [1206, 1787]. Oxygen ions and vacancies are pulled in an external electrical field from one electrode to the other. While they accumulate at the sealed bottom electrode, they desorb partly at the open tip electrode. The latter effect depends on the environmental oxygen pressure [1206, 1787]. These additional effects could even switch the rectification factor from $Z < 1$ to $Z > 1$.

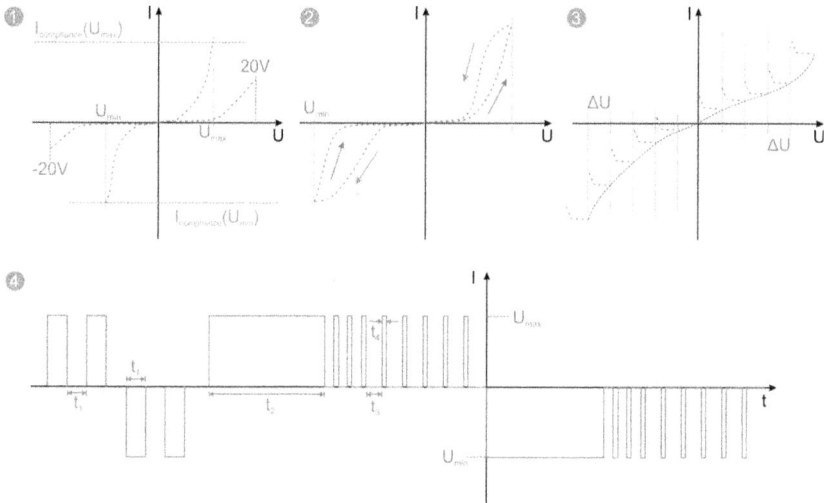

Figure 13.15: *Measurement protocol for electronic properties:* ① *determining the maximal voltage range that results in currents not larger than* $1\,\mu A$, ② $I - V$ *measurements,* ③ $I - V - t$ *measurements* $(\Delta U = 0.5\,V)$, ④ $I - t$ *measurements,* ⑤ *equal to step* ③.

13.6.2 The Influence of Further Effects on the Time-Dependent Charge Transport Behavior

Dipole Relaxation

The polarized Ti–O bonds reduce the total internal field and cause surface dipoles that increase the band bending at the electrode/TiO$_2$ interface. After the external field is turned off, the induced dipoles do not relax immediately as discussed in Section 2.8. However, the discussed data do not give any hint that the time-dependent transient current is affected by a slow dipole relaxation. First, switching the sign of the applied field is expected to push the system back into its initial state. But often, the current amplitude is increasing further. In addition, as soon as the current amplitude is relaxed after changing the polarity of the applied field, the dipole polarization is expected to be built up in the opposite direction due to the symmetry of the rutile crystal structure along a principal axis. This was not observed in any sample. There is also no feasible link between dipole relaxation and the short declining currents which occur when the bias is turned on. Dipole relaxation does not explain, why the current amplitude during the probe pulses exceeds the current amplitude at the end of the long CVS. In summary, it is suggested that dipole relaxation is not a dominant effect in the investigated samples.

Oxide Degradation

The introduced model ignores the creation of traps because the applied fields are too weak to break the bonds in crystalline TiO$_2$. Particularly, in nanorods annealed at 550 °C, the trap generation is supposed to be very low [70]. Ion migration is suggested only in defective regions where the binding energy is reduced. This assumption is supported by the observation that the transient current shows a smooth behavior, in particular for high current amplitudes. In case of serious oxide degradation, the current becomes noisy. Additionally, oxide degradation is irreversible while some of the observed effects relax to the initial state. Furthermore, famous degradation mechanisms such as the hydrogen release or anode hole injection model deal with voltages which are one and two orders of magnitude higher than the highest applied voltage in this study [1347].

Charge Trapping Ignoring the Built-Up of a Space-Charge and Oxygen Vacancy Migration

The possibility that the observed changes of the current amplitude are only caused by electrons that are "lost" and "gained" by trapping and detrapping is not very likely. This would mean that the integral of the current change ΔI is correlated with the number of trapped electrons and hence give a lower limit of the defect density N_t. For non-annealed nanorods grown at 180 °C, this integral would correspond to roughly 10^{30} electrons/cm^3 which is seven orders of magnitude higher than the atom density in a metal. The short effect is explained by the detrapping of electrons. Also here, the detrapping must cause secondary effects such as charging or barrier lowering which affect the transient current much more than the "lost" and "gained" electrons by trapping and detrapping. E.g., the integral for ΔI observed during the discharge effect in non-annealed nanorods grown at 180 °C would correspond to a lower trap density limit of 10^{26} electrons/cm^3. In conclusion, the system behaves rather like a self-controlled transistor than a pure charge storage device. Hence, the impact of trapped electrons and shifted ions must be considered much more than their number.

Glossary

Abrasiveness is a material employed for polishing and grinding such as pumice, sandpaper, or emery.

Absorbance is the ratio of the incident light power and the sum of reflected, scattered, and transmitted light power. The relation between absorbance and transmittance is given in the glossary entry for "transmittance".

Acousto-optic materials are applied in Bragg cells to generate diffraction and frequency shifts of light with sound waves. Typically, they are induced by a piezoelectric transducer which is attached to the acousto-optic material.

Airy patterns are caused by the diffraction of a light beam at an aperture. For homogeneous illumination of the aperture, circular rings with an intensity maximum in the center are observed. The pattern is defined by the illuminating wavelength and the size of the aperture. Airy patterns can appear if objects with dimensions in the same order as the wavelength of the illuminating light are imaged with an optical microscope.

Amphiphilic substances are both, hydrophilic and oleophilic.

Arrhenius equation is typically employed to describe the temperature-dependence of a reaction rate and determine the energy of activation. In detail, the empirical equation is used to calculate the temperature-dependence of diffusion, creep rates, and population of crystal vacancies, for instance.

Asbestosis is a disease affecting the tissue of the lungs. It is caused by rod-shaped particles such as asbestos fibers. Typical consequences of asbestosis are severe shortness of breath and certain cancers.

Biodiesel is composed of long-chain alkyl esters made from vegetable oil or animal fat. Biodiesel is used in standard diesel engines.

Bloch waves are usually employed to describe electrons as wave packets in a periodically-repeating potential. Hence, a Bloch wave is nothing else than the product of a plane wave and a periodic function. They describe the energy eigenstates of electrons in a crystal.

Bohr radius is a physical constant defined as $a_0 = \hbar/m_e c\alpha = 5.291\,772\,106\,7(12) \times 10^{-11}$ m, where m_e is the rest mass of the electron, c is the speed of light, and α is the fine structure constant. The Bohr radius equals roughly the radius of a hydrogen atom.

Brillouin scattering is an inelastic scattering and appears when energy of an incident electro-magnetic wave is used to generate phonons, polarons, or magnons in a solid. Since Brillouin scattering depends on the crystallographic direction of the incident light, it is a tool to determine a part of the dispersion relation of the listed quasiparticles.

Cicatrization is the formation of scar in a healing wound.

Curie temperature is defined as the material-dependent temperature at which permanent magnetic properties are replaced by induced magnetism in materials. Similarly, the Curie temperature defines the phase transition between ferroelectricity and paraelectricity.

Deacon process was used to the oxidation of hydrogen chloride and employed for the generation of alkalis. Later the Deacon process was replaced mainly by electrolytic processes.

Debye temperature is the temperature at which all phonon modes are exited and hence, it is correlated with the specific heat capacity of a material. Usually, the Debye temperature is temperature- and pressure-dependent.

Drude-model is based on the kinetic gas theory and treats electrons in solids as a free electron gas. The model describes the charge transport in mainly metallic materials assuming that there are free electrons bouncing and rebounding off heavy and fixed positive ions..

Dye-sensitized solar cells are solar cells containing dyes in order to optimize light absorption. If free charge carriers are not generated inside the dye molecule, the absorbed energy has to be transferred to another molecule that takes over the generation of free charge carriers.

Effective mediums are composed materials with an inhomogeneous distribution of the refractive index. And the average distance of boundaries between volumes with different refractive indices is smaller than the wavelength of the incident light. Hence, light is not sensitive for such densely packed boundaries and the interaction between light and matter such as reflection is based on a homogeneous refractive index which is built up from the refractive indices of the individual components. A typical example are nanostructures on a solid-air interface.

Electrophysiology covers the science of electric characteristics of biological cells and tissues.

Fabry-Pérot cavities are optical cavities consisting of two parallel mirrors.

Glan-Taylor polarizer is typically used as a polarizer or polarizing beam splitter. The polarizer consists of two right-angled prisms made of a birefringent material. The prisms are separated on their long facet with a narrow gap. Employing total reflexion and the Brewster's angle, s-polarized light is reflected and the p-polarized light is transmitted.

Gloss is the property of an surface describing the ability for light reflection in a specular direction. Typical parameters influencing the gloss are the refractive index, the incident angle, and the surface topography of an object.

Hybrid solar cells are mostly heterojunction solar cells consisting of several organic and inorganic layers. Typically, these layers are separated in photoactive, electron/hole blocking, and electrode layers.

Hydrophobic/superhydrophilic molecules are repulsed/attracted by water molecules. In general, hydrophobic molecules tend to be badly solved in polar solvents. The contact angle between a superhydrophilic substance and water is almost zero.

Kapton® is a polyimide developed by *DuPont*™ with an outstanding thermal and chemical stability.

Lab-on-a-chip devices are laboratories scaled down to structure sizes of a few microns or below. They provide the automatic analysis of microfluids and work similar as macroscopic laboratories. Usually, they are integrated in chips for controlling and data readouts.

Lorentz oscillators are used to model photon-electron interactions. Hereby, an electromagnetic wave excites harmonic oscillations of an electron which is bound to an atomic core.

Mössbauer spectroscopy employs the Doppler effect to generate gamma quantums with a defined energy distribution by oscillating the gamma source mechanically. This narrow energy spectrum is used for an absorption measurement of the sample to determine the hyperfine structure of the absorbing atomic cores. The hyperfine structure results from an electron-core interaction and hence, it depends on the environment of the absorbing core such as oxidation states, spin states, magnetic properties, electronegativity of ligands, and other chemical properties..

Magnéli phases are crystal structures of non-stoichiometric crystalline materials. Theses phases could be defined as highly ordered defect structures and represent the opposite of randomly distributed crystal defects.

Mie scattering describes the exact solution for scattering of electromagnetic plane waves by homogeneous spheres based on the Maxwell's equations. Typically, it is used for particles with a diameter in the same range as the wavelength of the incident light. It can be transferred to any other system where separated equations for radial and angular-dependent solutions exist. This is not the case for nanorods. Hence, only an approximation of Mie scattering can be employed to describe scattering at dielectric rods.

Moiré fringes or patterns are cause by the interference of at least two lattices lying on top of each other. To result in Moiré fringes the patterns must be displaced or rotated. For slightly different lattice parameters, Moiré fringes appear as well. In this work, Moiré fringes appear on TEM images showing a stack of crystal planes.

Mott transition is the transition between metal and nonmetal behavior in condensed matter. As the temperature in semiconductors such as transition metal oxides with a lot of intra-bandgap defect states rises, more and more electrons are released into the conduction band and the material might become metallic at a certain temperature. However, the mobile electrons interact with each other and thus, the energy of the system is increasing. At some point the energy is so high that the systems lowers its energy by localizing the electrons and the material becomes insulting again.

Nanorods are rectangular nanocrystals with two short and one long dimension. Typically, their aspect ration is between 1:5 and 1:100. Structures with smaller/larger aspect ratio are called nanopillars/nanowires. Structures with two long and one short dimension are nanoflakes. Hollow nanorods are called nanotubes.

Oleophilic substances show an affinity for oil rather than for water.

Osteoblasts are somatic cells building up bones. They rise from mesenchymal stem cells and attach on bones. They segregate crosslinked collagen and other specialized proteins such as osteocalcin and osteopontin which are the building blocks for the bone matrix. This matrix is mineralized with hydroxyapatite, which is a calcium and phosphate-based mineral.

Ostwald ripening is a process where two equally sized particles in the same solution obtain slightly different sizes by dissolving and solidification processes. As a consequence, the surface-to-volume ratio becomes slightly different resulting in a different total energy for both particles and the solidification on the surface of the larger particle becomes more likely. Thus, in a solution the larger particles obtain size while smaller particles dissolve.

Oxophilicity is the tendency of atoms or molecules to create oxides by hydrolysis.

Peptization is the process employed for the generation of stable dispersions of colloidal particles from precipitates by shaking them with the dispersion medium and small amounts of electrolyte used as peptization agent. The peptization agent reduces the electric repulsion of the precipitates. The opposite effect is used to generate nanoparticles. Here, peptization is employed to split agglomerates into primary particles by adjusting the surface energy, adding a surfactant, or by applying a charge. For instance, ammonium cations are deposited on a titania surface in order to charge it positively and by electrostatic repulsion the agglomerated titania decays into nanoparticles.

Polarons are quasiparticles including a charged particle such as an electron or hole inside a solid and a deformation of the atomic arrangement around the particle. In strongly polarizable crystals such as ionic crystals, positive/negative ions are attracted/repelled by mobile charged particles resulting in a Coulomb interaction between the atomic lattice and the charged particle. Hence, polarons result from the electron-phonon-interaction.

Polymorph materials can adopt more than one crystal structure without changing the stoichiometry. For instance, brookite, rutile, and anatase are polymorphs of TiO_2..

Raman scattering is an inelastic scattering of electromagnetic waves on matter. Typically, Raman scattering describes the excitation and relaxation of phonons or vibrational modes in crystals or molecules.

Raman spectroscopy is mainly used to detect certain molecules or crystal structures via their typical bonds which have well-defined vibrational, rotational, and further low-frequency modes. It is based on Raman scattering described in the glossary entry "Raman scattering". Often, this kind of spectroscopy is used complementary to IR-spectroscopy.

Rayleigh scattering describes the scattering of electromagnetic waves by particles that are much smaller than the wavelength of incident radiation. Similar as Mie scattering, Rayleigh scattering is an elastic scattering mechanism based on microscopic dipole scattering and hence, it depends on the dielectric function of the particles.

Redispersibility describes the ability for separating basic particles or molecules from agglomerates and aggregates in solvents.

Reflectance is the ratio of the incident light power and the power which is reflected at an interface..

Resistive switching is the capability of a material to change its resistance reversibly. Usually, voltage or current stress is employed to induce resistive switching.

Rheology is the study of flowing liquids and soft matter.

Stoichiometry is the ratio of elements in a defect free chemical compound or crystal.

Supercapacitors (SC) are electric double-layer capacitors (EDLC) having a significantly higher energy density than electrolytic capacitors, accepting and delivering charges faster than batteries, and performing more charge and discharge cycles than rechargeable batteries. To improve the capacity supercapacitors employ either electrostatic double-layer, electrochemical pseudocapacitors, or a mixture of both (hybrid capacitors).

Superlattices are well-defined structures of an already patterned or otherwise structured surface. An example are NRAs on spherical seed layers that are arranged with optical lithography. Here, the nanorods are the primary structure, the spherical seed layers are the secondary structure and primary superlattice, and the arrangement made by the optical lithography represents the tertiary structure and secondary superlattice..

Tinting strength is a measurement of the covering power of a color pigment. It depends on the type of pigment, the fineness of the grinding, and the amount of pigment. With increasing tinting strength the color of a paint is conserved longer when mixing it with white pigments.

Transmittance is the ratio of the incident light power and the power which is transmitted at an interface. For samples with a negligible scattering, the transmittance and absorbance are correlated with $T = \exp -\tau = 10^{-\tilde{A}}$, where τ is the optical depth and $\tilde{A}$ is the absorbance.

Wulff construction is a theoretical tool to gain the shape of a crystal for a given volume in a defined environment in thermodynamic equilibrium. This tool compares the surface energy of different crystal facets in order to find the shape with the lowest total surface energy.

Young's modulus or also called the elastic modulus, describes the stiffness of a material. It pulls together stress (force per unit area) and strain (proportional deformation) in a solid.

List of Symbols

a_B Bohr radius
$\mathring{A}$ absorbance
A_{abs} absorption
a_p polaron hopping distance
A area
α_F Fröhlich coupling constant
α_h localization radius for hopping
α_{sh} shape parameter

B magnetic field

c vacuum speed of light ($2.99793 \times 10^8\,\mathrm{m/s}$)
C capacity

d thickness or diameter

e charge of an electron $= -q$
$\mathcal{E}$ electrical field
e index for "electron"
E energy
E_g bandgap energy
E_c conduction band minimum energy
E_d depth of a donor state (in relation to CBM)
E_F Fermi level
E_v valence band maximum energy
ϵ dielectric constant of vacuum ($= 1/\mu_{vac}c^2$, $8.8541878176 \times 10^{-12}\,\mathrm{As/Vm}$)
ϵ_d dynamic dielectric constant
ϵ_i image force dielectric constant
ϵ_∞ high frequency dielectric constant
ϵ_o optical dielectric constant
ϵ_s static dielectric constant
η number or density (of particles)
η_e density of thermally activated mobile electrons
η_t density of tunneling electrons

G Gibbs free energy
γ surface tensions
γ^* scattering rate

H enthalpy
$\hbar$ Planck constant ($1.05459 \times 10^{-34}\,(\mathrm{J \cdot s})$)

I electrical current
J electrical current density
I_n intensity (in counts per time, power or a.u.)
I_s reverse saturation current

k momentum
k_b Boltzmann constant ($1.38062 \times 10^{-23}\,\mathrm{J/K}$)
$\mathbf{k}$ wave vector
κ absorption coefficient

l path length
λ wavelength
Λ probability

M_c number of equivalent minima in the CBM
m_e electron rest mass ($9.10956 \times 10^{-31}\,\mathrm{kg}$)
m_e^* effective electron mass (conduction band mass)
m_{de} density-of-state effective electron mass
m_{dh} density-of-state effective hole mass
m_h^* effective hole mass (valence band mass)
m^{**} electron polaron mass
m_t^* tunneling effective mass of charge carriers
μ chemical potential
μ_e drift electron mobility
μ_e^* band structure-derived electron mobility
μ_e^{**} electron polaron mobility
μ_H Hall mobility
μ_T Thz mobility

N_c effective DOS in conduction band
N_d donor density
N_F DOS at Fermi level
N_t trap density
N_v effective DOS in the valence band
n_{id} ideality factor of a diode
n refractive index
$\hat{n}$ complex refractive index
ν_d deposition rate
ν_h hopping rate

ω_D angular Debye frequency
ω_o angular optical phonon mode frequency

ω_{LO} angular LO phonon mode frequency

P (excess) pressure

φ_{bi} effective build-in potential

$\Delta\varphi_{bi}$ lowering of build-in potential due to image charge effect

φ_{fb} flat-band potential

φ_{sb} height of Schottky barrier

φ_{dt} depth of a deep trap

φ_{st} depth of a shallow trap

Φ_W work function

q elementary charge ($1.602\,19 \times 10^{-19}$ C)

r radius

R resistance

R_g gas constant ($8.314\,459\,8(48)$ J/(molK))

R_H Hall coefficient

R^* Richardson constant

R^{**} effective Richardson constant

ρ resistivity

S entropy

s_{gr} specific gravity

σ conductivity

T temperature

T_C Curie temperature

t time

t_b time-to-breakdown

t_{in} tinting strength

τ time constant

τ_c collision time

τ_{ch} (dis-)charging time

τ_{def} momentum scattering time for defect scattering

τ_{ph} momentum scattering time for phonon scattering

τ_s momentum scattering time

τ_t tunneling time

Θ_D Debye temperature

U_i inner energy

U voltage or bias

v velocity

v_D drift velocity

v_F Fermi velocity

v_s scattering-limited velocity

V volume

W_a activation or binding energy

W_D disorder energy

W_H hopping energy

x position

χ electron affinity

Ξ^R reflectance

Ξ^T transmittance

ξ_i ionization potential

List of Acronyms

2DES two dimensional electron system

AALD atmospheric atomic layer deposition
AcAc acetylacetone
AFM atomic force microscope
AGG abnormal grain growth
ALD atomic layer deposition
AMO amorphous
ANA anatase
AP-SALD atmospheric pressure spatial atomic layer deposition
APT anatase-to-α-PbO$_2$ transformation
ART anatase-to-rutile transformation

BF bright field

CAFM conductive atomic force microscope
CB conduction band
CBM conduction band minimum
CCD charge-coupled device
CELIV charge extraction with a linearly increasing voltage sweep
CiMeL cw laser-induced melting lithography
CNOS complementary metal-oxide-semiconductor
CVD chemical vapor deposition
cw continuous wave

DC direct current
DEA diethanolamine
DF dark field
DFT density functional theory
DLS diffusive light scattering
DLTS deep level transient spectroscopy
DOS density of states
DOS$_{ph}$ density of phonon states
DSSC dye-sensitized solar cell
DTA differential thermal analysis

EBE electron-beam evaporation
EBID electron-beam-induced deposition
EDX energy-dispersive X-ray spectroscopy
EIS electrochemical impedance spectroscopy

EMA Bruggeman effective-medium approximation
ESAVD electrostatic spray assisted vapor deposition
ESR electron spin resonance
EUV extreme ultraviolet lithography

FESEM field emission scanning electron microscope
FET field-effect transistor
FIB focused ion beam
FN Fowler-Nordheim
FTO fluorine tin oxide
FZ float zone

GRG generalized Rayleigh-Gans

HCl hydrochloric acid
HELM high-energy cw Laser-induced melting lithography
HF hydrofluoric acid
HREELS high-resolution electron energy loss spectroscopy
HRS high resistance state
HRTEM high-resolution transmission electron microscope

I-t current over time
I-V current over voltage
ICPB intimately coupled photobiocatalysis
IR infrared
IS impedance spectroscopy
ITO indium tin oxide

LED light emitting diode
LELM low-energy cw laser-induced melting lithography
LiMeL laser-induced melting lithography
LIOX laser-induced oxidation lithography
LIXPS low intensity x-ray spectroscopy
LL Lorentz-Lorenz
LO longitudinal optical phonon
LRS low resistance state

LSL light stamping lithography

MD molecular dynamics
MEA monoethanolamine
MG Maxwell-Garnett
MHP maximum height of the profile
MIBK methylisobutylketon
MIM metal-insulator-metal
MOCVD metal-organic chemical-vapor deposition
MOSFET metal-oxide-semiconductor field-effect transistor
MSE mean squared error
MT multiple-trapping (model)

n-MOS nanostructured metal oxide semiconductors
NDR negative differential resistance
NIL nanoimprint lithography
NIR near infrared
NRA nanorod array

ODPA octadecylphosphonic acid
OLED organic light-emitting diodes

P-F Poole-Frenkel emission
PDMS poly(dimethylsiloxane)
PECVD plasma enhanced chemical vapor deposition
PEEK polyether ether ketone
PEN polyethylene naphthalate
PET polyethylene terephthalate
PMMA poly(methyl methacrylate)
PSML monolayer of polystyrene spheres
PVD physical vapor deposition
PXRD powder X-ray diffraction
PYR spray pyrolysis

REM Rasterelektronenmikroskop
RF radio frequency
RMS root-mean-squared roughness
RRAM resistive random access memories
RTP rapid temperature processing
RUT rutile

RWNS random walk numerical simulations

SAD selected area diffraction
SALD spatial atomic layer deposition
SCLC space-charge-limited current
SDS sodium dodecyl sulfate
SEM scanning electron microscope
SG sol-gel method
SILC stress-induced leakage current
SNMS secondary neutral mass spectroscopy
SONOS silicon-oxide-nitride-oxide-silicon
SPU sputter deposition
STM scanning tunneling microscope
SWCNT single wall carbon nanotube

TAT trap-assisted tunneling
TCO transparent conductive oxide
TDDB time-dependent dielectric breakdown
TEM transmission electron microscope
Ti–Si–O titanium–silicon–oxygen
TiO$_2$ titanium dioxide
TO transversal optical phonon
TOF time-of-flight
TSC thermally stimulated currents
TTIP titanium tetraisopropoxide

UGS upgraded slag
UPS ultraviolet photoemission spectroscopy
UV ultra-violet
UV-NIL ultraviolet-assisted nanoimprint lithography
UV-Vis-IR ultraviolet-visible-infrared

VASE variable angle spectroscopic ellipsometry
VB valence band
VBM valence band maximum
VRH variable range hopping

XPS X-ray photoelectron spectroscopy
XRD X-ray diffraction

ZnO zinc oxide

List of Figures

List of Tables

References

[1] Zhengyang Weng, Huan Guo, Xiangmei Liu, Shuilin Wu, K. W. K. Yeung, and Paul. K. Chu. Nanostructured tio$_2$ for energy conversion and storage. *RSC Adv.*, 3:24758–24775, 2013. doi: 10.1039/C3RA44031A. URL http://dx.doi.org/10.1039/C3RA44031A.

[2] Xiaobo Chen and Samuel S. Mao. Titanium dioxide nanomaterials: Synthesis, properties, modifications, and applications. *Chemical Reviews*, 107(7):2891–2959, 2007. doi: 10.1021/cr0500535. URL http://dx.doi.org/10.1021/cr0500535. PMID: 17590053.

[3] Michael R. Hoffmann, Scot T. Martin, Wonyong. Choi, and Detlef W. Bahnemann. Environmental applications of semiconductor photocatalysis. *Chemical Reviews*, 95(1):69–96, 1995. doi: 10.1021/cr00033a004. URL http://dx.doi.org/10.1021/cr00033a004.

[4] L. Rideh, A. Wehrer, D. Ronze, and A. Zoulalian. Photocatalytic degradation of 2-chlorophenol in Tio$_2$ aqueous suspension: Modeling of reaction rate. *Industrial & Engineering Chemistry Research*, 36(11):4712–4718, 1997. doi: 10.1021/ie970100m. URL http://dx.doi.org/10.1021/ie970100m.

[5] D Shchukin, S Poznyak, A Kulak, and P Pichat. Tio$_2$-In$_2$o$_3$ photocatalysts: preparation, characterisations and activity for 2-chlorophenol degradation in water. *Journal of Photochemistry and Photobiology A: Chemistry*, 162(2–3):423 – 430, 2004. ISSN 1010-6030. doi: http://dx.doi.org/10.1016/S1010-6030(03)00386-1. URL http://www.sciencedirect.com/science/article/pii/S1010603003003861.

[6] Akira Fujishima, Tata N. Rao, and Donald A. Tryk. Titanium dioxide photocatalysis. *Journal of Photochemistry and Photobiology C: Photochemistry Reviews*, 1(1):1 – 21, 2000. ISSN 1389-5567. doi: 10.1016/S1389-5567(00)00002-2. URL http://www.sciencedirect.com/science/article/pii/S1389556700000022.

[7] Yijing Wu, Haoquan Yan, and Peidong Yang. Semiconductor nanowire array: Potential substrates for photocatalysis and photovoltaics. *Topics in Catalysis*, 19(2):197–202, 2002. ISSN 1022-5528. doi: 10.1023/A:1015260008046. URL http://dx.doi.org/10.1023/A%3A1015260008046.

[8] Hubert Gnaser, Bernd Huber, and Christiane Ziegler. Nanocrystalline tio$_2$ for photocatalysis. In *Encyclopedia of Nanoscience and Nanotechnology*, volume 6, pages 505–535. American Scientific Publishers, 2004.

[9] Kazuhito Hashimoto, Hiroshi Irie, and Akira Fujishima. Tio$_2$ photocatalysis: A historical overview and future prospects. *Japanese Journal of Applied Physics*, 44(12R):8269, 2005. URL http://stacks.iop.org/1347-4065/44/i=12R/a=8269.

[10] Jaturong Jitputti, Yoshikazu Suzuki, and Susumu Yoshikawa. Synthesis of tio$_2$ nanowires and their photocatalytic activity for hydrogen evolution. *Catalysis Communications*, 9(6):1265 – 1271, 2008. ISSN 1566-7367. doi: 10.1016/j.catcom.2007.11.016. URL http://www.sciencedirect.com/science/article/pii/S1566736707004980.

[11] Shahed U. M. Khan, Mofareh Al-Shahry, and William B. Ingler. Efficient photochemical water splitting by a chemically modified n-tio$_2$. *Science*, 297(5590):2243–2245, 2002. ISSN 0036-8075. doi: 10.1126/science.1075035. URL http://science.sciencemag.org/content/297/5590/2243.

[12] Johanes P.H. Sukamto, Christopher S. McMillan, and William Smyrl. Photoelectrochemical investigations of thin metal-oxide films: Tio$_2$, al$_2$o$_3$, and hfo$_2$ on the parent metals. *Electrochimica Acta*, 38(1):15 – 27, 1993. ISSN 0013-4686. doi: https://doi.org/10.1016/0013-4686(93)80005-K. URL http://www.sciencedirect.com/science/article/pii/001346869380005K.

[13] Jiajie Cen, Qiyuan Wu, Mingzhao Liu, and Alexander Orlov. Developing new understanding of photoelectrochemical water splitting via in-situ techniques: A review on recent progress. *Green Energy & Environment*, 2(2):100 – 111, 2017. ISSN 2468-0257. doi: https://doi.org/10.1016/j.gee.2017.03.001. URL http://www.sciencedirect.com/science/article/pii/S2468025717300476. Special Issue on New Energy Catalysis.

[14] Chengzhi Luo, Xiaohui Ren, Zhigao Dai, Yupeng Zhang, Xiang Qi, and Chunxu Pan. Present perspectives of advanced characterization techniques in tio$_2$-based photocatalysts. *ACS Applied Materials & Interfaces*, 9(28):23265–23286, 2017. doi: 10.1021/acsami.7b00496. URL http://dx.doi.org/10.1021/acsami.7b00496. PMID: 28628307.

[15] Yongcai Qiu, Siu-Fung Leung, Zhanhua Wei, Qingfeng Lin, Xiaoli Zheng, Yuegang Zhang, Zhiyong Fan, and Shihe Yang. Enhanced charge collection for splitting of water enabled by an engineered three-dimensional nanospike array. *The Journal of Physical Chemistry C*, 118(39):22465–22472, 2014. doi: 10.1021/jp507800t. URL http://dx.doi.org/10.1021/jp507800t.

[16] Xinrui Cao, Guohui Tian, Yajie Chen, Jian Zhou, Wei Zhou, Chungui Tian, and Honggang Fu. Hierarchical composites of tio$_2$ nanowire arrays on reduced graphene oxide nanosheets with enhanced photocatalytic hydrogen evolution performance. *J. Mater. Chem. A*, 2: 4366–4374, 2014. doi: 10.1039/C3TA14272H. URL http://dx.doi.org/10.1039/C3TA14272H.

[17] Jianming Zhang, Mitra Vasei, Yuanhua Sang, Hong Liu, and Jerome P. Claverie. Tio$_2$ at carbon photocatalysts: The effect of carbon thickness on catalysis. *ACS Applied Materials & Interfaces*, 8(3):1903–1912, 2016. doi: 10.1021/acsami.5b10025. URL http://dx.doi.org/10.1021/acsami.5b10025. PMID: 26716463.

[18] Nan Zhang and Yi-Jun Xu. The endeavour to advance graphene-semiconductor composite-based photocatalysis. *CrystEngComm*, 18: 24–37, 2016. doi: 10.1039/C5CE01712B. URL http://dx.doi.org/10.1039/C5CE01712B.

[19] Songtian Li, Pengwei Du, Xiao Yang, Lu Yao, and Kesheng Cao. Enhanced photocatalytic and photoelectrochemical activity via sensitization and doping of novel tio$_2$ nanowire/nanoleaf arrays: dual synergistic effects between tio$_2$-n and cds-mn. *RSC Adv.*, 6: 13670–13679, 2016. doi: 10.1039/C5RA24189H. URL http://dx.doi.org/10.1039/C5RA24189H.

[20] Qiang Zhou, Shanqing Zhang, Xianfeng Yang, Qili Wu, Huijun Zhao, and Mingmei Wu. Rutile nanowire array electrodes for photoelectrochemical determination of organic compounds. *Sensors and Actuators B: Chemical*, 186:132 – 139, 2013. ISSN 0925-4005. doi: 10.1016/j.snb.2013.05.060. URL http://www.sciencedirect.com/science/article/pii/S0925400513006333.

[21] Chong Liu, Neil P. Dasgupta, and Peidong Yang. Semiconductor nanowires for artificial photosynthesis. *Chemistry of Materials*, 26(1): 415–422, 2014. doi: 10.1021/cm4023198. URL http://dx.doi.org/10.1021/cm4023198.

[22] Xin Yu, Xun Han, Zhenhuan Zhao, Jian Zhang, Weibo Guo, Caofeng Pan, Aixue Li, Hong Liu, and Zhong Lin Wang. Hierarchical tio$_2$ nanowire/graphite fiber photoelectrocatalysis setup powered by a wind-driven nanogenerator: A highly efficient photoelectrocatalytic device entirely based on renewable energy. *Nano Energy*, 11:19 – 27, 2015. ISSN 2211-2855. doi: http://dx.doi.org/10.1016/j.nanoen.2014.09.024. URL http://www.sciencedirect.com/science/article/pii/S221128551420231X.

References

[23] Siqi Liu, Chuang Han, Zi-Rong Tang, and Yi-Jun Xu. Heterostructured semiconductor nanowire arrays for artificial photosynthesis. *Mater. Horiz.*, 3:270–282, 2016. doi: 10.1039/C6MH00063K. URL http://dx.doi.org/10.1039/C6MH00063K.

[24] Yoshihisa Ohko, Yosuke Utsumi, Chisa Niwa, Tetsu Tatsuma, Koichi Kobayakawa, Yuichi Satoh, Yoshinobu Kubota, and Akira Fujishima. Self-sterilizing and self-cleaning of silicone catheters coated with TiO_2 photocatalyst thin films: A preclinical work. *Journal of Biomedical Materials Research*, 58(1):97–101, 2001. ISSN 1097-4636. doi: 10.1002/1097-4636(2001)58:1<97::AID-JBM140>3.0.CO;2-8. URL http://dx.doi.org/10.1002/1097-4636(2001)58:1<97::AID-JBM140>3.0.CO;2-8.

[25] M. Kalbacova, J. M. Macak, F. Schmidt-Stein, C. T. Mierke, and P. Schmuki. TiO_2 nanotubes: photocatalyst for cancer cell killing. *physica status solidi (RRL) - Rapid Research Letters*, 2(4):194–196, 2008. ISSN 1862-6270. doi: 10.1002/pssr.200802080. URL http://dx.doi.org/10.1002/pssr.200802080.

[26] Kazuya Nakata, Munetoshi Sakai, Tsuyoshi Ochiai, Taketoshi Murakami, Katsuhiko Takagi, and Akira Fujishima. Antireflection and self-cleaning properties of a moth-eye-like surface coated with TiO_2 particles. *Langmuir*, 27(7):3275–3278, 2011. doi: 10.1021/la200438p. URL http://dx.doi.org/10.1021/la200438p. PMID: 21391643.

[27] Lei Liu, Zhaoyang Liu, Hongwei Bai, and Darren Delai Sun. Concurrent filtration and solar photocatalytic disinfection/degradation using high-performance ag/TiO_2 nanofiber membrane. *Water Research*, 46(4):1101 – 1112, 2012. ISSN 0043-1354. doi: http://dx.doi.org/10.1016/j.watres.2011.12.009. URL http://www.sciencedirect.com/science/article/pii/S0043135411007810.

[28] Hongwei Bai, Zhaoyang Liu, Lei Liu, and Darren Delai Sun. Large-scale production of hierarchical TiO_2 nanorod spheres for photocatalytic elimination of contaminants and killing bacteria. *Chemistry - A European Journal*, 19(9):3061–3070, 2013. ISSN 1521-3765. doi: 10.1002/chem.201204013. URL http://dx.doi.org/10.1002/chem.201204013.

[29] Dandan Zhou, Shuangshi Dong, Junlong Shi, Xiaochun Cui, Dongwon Ki, Cesar I. Torres, and Bruce E. Rittmann. Intimate coupling of an n-doped TiO_2 photocatalyst and anode respiring bacteria for enhancing 4-chlorophenol degradation and current generation. *Chemical Engineering Journal*, 317(Supplement C):882 – 889, 2017. ISSN 1385-8947. doi: https://doi.org/10.1016/j.cej.2017.02.128. URL http://www.sciencedirect.com/science/article/pii/S1385894717302930.

[30] Rong Wang, Kazuhito Hashimoto, Akira Fujishima, Makota Chikuni, Eiichi Kojima, Atsushi Kitamura, Mitsuhide Shimohigoshi, and Toshiya Watanabe. Light-induced amphiphilic surfaces. *Nature*, 388(6641):431–432, 1997.

[31] Masato Takeuchi, Kenji Sakamoto, Gianmario Martra, Salvatore Coluccia, and Masakazu Anpo. Mechanism of photoinduced superhydrophilicity on the TiO_2 photocatalyst surface. *The Journal of Physical Chemistry B*, 109(32):15422–15428, 2005. doi: 10.1021/jp058075i. URL http://dx.doi.org/10.1021/jp058075i. PMID: 16852956.

[32] Xinjian Feng, Jin Zhai, and Lei Jiang. The fabrication and switchable superhydrophobicity of TiO_2 nanorod films. *Angewandte Chemie International Edition*, 44(32):5115–5118, 2005. ISSN 1521-3773. doi: 10.1002/anie.200501337. URL http://dx.doi.org/10.1002/anie.200501337.

[33] Eiji Hosono, Hirofumi Matsuda, Itaru Honma, Masaki Ichihara, and Haoshen Zhou. Synthesis of a perpendicular TiO_2 nanosheet film with the superhydrophilic property without uv irradiation. *Langmuir*, 23(14):7447–7450, 2007. doi: 10.1021/la701117a. URL http://dx.doi.org/10.1021/la701117a. PMID: 17552549.

[34] Yue Li, Takeshi Sasaki, Yoshiki Shimizu, and Naoto Koshizaki. Hexagonal-close-packed, hierarchical amorphous TiO_2 nanocolumn arrays: Transferability, enhanced photocatalytic activity, and superamphiphilicity without uv irradiation. *Journal of the American Chemical Society*, 130(44):14755–14762, 2008. doi: 10.1021/ja805077q. URL http://dx.doi.org/10.1021/ja805077q. PMID: 18844352.

[35] Masahiro Miyauchi, Zhifu Liu, Zhi-Gang Zhao, Srinivasan Anandan, and Hiromasa Tokudome. Visible-light-driven superhydrophilicity by interfacial charge transfer between metal ions and metal oxide nanostructures. *Langmuir*, 26(2):796–801, 2010. doi: 10.1021/la902177x.

[36] Makoto Chikuni, Eiichi Kojima, Atsushi Kitamura, Mitsuhide Shimohigoshi, and Toshiya Watanabe. Light-induced amphiphilic surfaces. *Nature*, 388:431–432, 1997.

[37] G. Krabbes, R. Krausze, and Vuong Hoanh. Untersuchungen zum isothermen schnitt des systems ti-si-o bei 1273 k. *Zeitschrift für anorganische und allgemeine Chemie*, 558(1):41–45, 1988. ISSN 1521-3749. doi: 10.1002/zaac.19885580105. URL http://dx.doi.org/10.1002/zaac.19885580105.

[38] Andrei Kolmakov and Martin Moskovits. Chemical sensing and catalysis by one-dimensional metal-oxide nanostructures. *Annual Review of Materials Research*, 34(1):151–180, 2004. doi: 10.1146/annurev.matsci.34.040203.112141. URL http://dx.doi.org/10.1146/annurev.matsci.34.040203.112141.

[39] L. Francioso, A.M. Taurino, A. Forleo, and P. Siciliano. TiO_2 nanowires array fabrication and gas sensing properties. *Sensors and Actuators B: Chemical*, 130(1):70 – 76, 2008. ISSN 0925-4005. doi: 10.1016/j.snb.2007.07.074. URL http://www.sciencedirect.com/science/article/pii/S0925400507004497. Proceedings of the Eleventh International Meeting on Chemical Sensors IMCS-11IMCS 2006IMCS 11.

[40] Dang Thi Thanh Le, Dang Duc Vuong, and Nguyen Duc Chien. Synthesis and lpg-sensing properties of TiO_2 nanowires. *Journal of Physics: Conference Series*, 187(1):012086, 2009. doi: 10.1088/1742-6596/187/1/012086. URL http://stacks.iop.org/1742-6596/187/i=1/a=012086.

[41] Wei-Cheng Tian, Yu-Hsuan Ho, Chao-Hao Chen, and Chun-Yen Kuo. Sensing performance of precisely ordered TiO_2 nanowire gas sensors fabricated by electron-beam lithography. *Sensors*, 13(1):865, 2013. ISSN 1424-8220. doi: 10.3390/s130100865. URL http://www.mdpi.com/1424-8220/13/1/865.

[42] Yong Yang, Yan Liang, Guozhong Wang, Liangliang Liu, Cailei Yuan, Ting Yu, Qinliang Li, Fanyan Zeng, and Gang Gu. Enhanced gas-sensing properties of the hierarchical TiO_2 hollow microspheres with exposed high-energy 001 crystal facets. *ACS Applied Materials & Interfaces*, 7(44):24902–24908, 2015. doi: 10.1021/acsami.5b08372. URL http://dx.doi.org/10.1021/acsami.5b08372. PMID: 26497199.

[43] O.K. Varghese, D. Gong, M. Paulose, K.G. Ong, E.C. Dickey, and C.A. Grimes. Extreme changes in the electrical resistance of titania nanotubes with hydrogen exposure. *Advanced Materials*, 15(7-8):624–627, 2003. ISSN 1521-4095. doi: 10.1002/adma.200304586. URL http://dx.doi.org/10.1002/adma.200304586.

[44] Gopal K. Mor, Karthik Shankar, Oomman K. Varghese, and Craig A. Grimes. Photoelectrochemical properties of titania nanotubes. *Journal of Materials Research*, 19(10):2989–2996, 2004. doi: 10.1557/JMR.2004.0370.

[45] André Decroly, Arnaud Krumpmann, Marc Debliquy, and Driss Lahem. Nanostructured TiO_2 layers for photovoltaic and gas sensing applications. In Marcelo L. Larramendy and Sonia Soloneski, editors, *Green Nanotechnology - Overview and Further Prospects*, chapter 05. InTech, Rijeka, 2016. doi: 10.5772/62316. URL http://dx.doi.org/10.5772/62316.

[46] Heba Abunahla and Baker Mohammad. *Memristor Technology: Synthesis and Modeling for Sensing and Security Applications.* Springer, 2018.

[47] J. Domaradzki, E. Prociow, D. Kaczmarek, D. Wojcieszak, D. Gatner, and M. Lapinski. Gasochromic effect in nanocrystalline tio$_2$ thin films doped with ta and pd. *Acta Physica Polonica A*, 116:126–128, 2009.

[48] S. Ghayeb Zamharir, M. Ranjbar, and H. Salamati. Excimer laser treatment of tio$_2$/wo$_3$ thin films for self-cleaning gasochromic applications: Preparation and characterization. *Solar Energy Materials and Solar Cells*, 130:27 – 35, 2014. ISSN 0927-0248. doi: https://doi.org/10.1016/j.solmat.2014.06.029. URL http://www.sciencedirect.com/science/article/pii/S0927024814003389.

[49] J. Domaradzki, M. Mazur, D. Wojcieszak, D. Kaczmarek, and T. Jedrzejak. Investigation of optical response of gasochromic thin film structures through modelling of their transmission spectra under presence of organic vapor. *Acta Physica Polonica A*, 127(6):1702–1705, 2015.

[50] V. C. Anitha, Arghya Narayan Banerjee, and Sang Woo Joo. Recent developments in tio$_2$ as n- and p-type transparent semiconductors: synthesis, modification, properties, and energy-related applications. *Journal of Materials Science*, 50(23):7495–7536, 2015. ISSN 1573-4803. doi: 10.1007/s10853-015-9303-7. URL http://dx.doi.org/10.1007/s10853-015-9303-7.

[51] Komathi Shanmugasundaram, Gopalan Sai-Anand, Anantha-Iyengar Gopalan, Hyun-Gyu Lee, Ho Kwon Yeo, Shin-Won Kang, and Kwang-Pill Lee. Direct electrochemistry of cytochrome c with three-dimensional nanoarchitectured multicomponent composite electrode and nitrite biosensing. *Sensors and Actuators B: Chemical*, 228:737 – 747, 2016. ISSN 0925-4005. doi: http://dx.doi.org/10.1016/j.snb.2016.01.109. URL http://www.sciencedirect.com/science/article/pii/S0925400516301095.

[52] Na Bin, Weili Li, Xuehua Yin, Xiaohua Huang, and Qingyun Cai. Electrochemiluminescence aptasensor for tio$_2$/cds:mn hybrids for ultrasensitive detection of cytochrome c. *Talanta*, 160(Supplement C):570 – 576, 2016. ISSN 0039-9140. doi: https://doi.org/10.1016/j.talanta.2016.07.046. URL http://www.sciencedirect.com/science/article/pii/S0039914016305495.

[53] Debabrata Sarkar, Chandan. K. Ghosh, S. Mukherjee, and Kalyan K. Chattopadhyay. Three dimensional ag$_2$o/tio$_2$ type-ii (p–n) nanoheterojunctions for superior photocatalytic activity. *ACS Applied Materials & Interfaces*, 5(2):331–337, 2013. doi: 10.1021/am302136y. URL http://dx.doi.org/10.1021/am302136y. PMID: 23245288.

[54] Dezhong Zhang, Chunyu Liu, Bo Yin, Ruiliang Xu, Jingran Zhou, Xindong Zhang, and Shengping Ruan. Organics filled one-dimensional tio$_2$ nanowires array ultraviolet detector with enhanced photo-conductivity and dark-resistivity. *Nanoscale*, 9:9095–9103, 2017. doi: 10.1039/C7NR03408C. URL http://dx.doi.org/10.1039/C7NR03408C.

[55] Gang Liu, Feng Li, Da-Wei Wang, Dai-Ming Tang, Chang Liu, Xiuliang Ma, Gao Qing Lu, and Hui-Ming Cheng. Electron field emission of a nitrogen-doped tio$_2$ nanotube array. *Nanotechnology*, 19(2):025606, 2008. URL http://stacks.iop.org/0957-4484/19/i=2/a=025606.

[56] Jyh-Ming Wu, Han C. Shih, and Wen-Ti Wu. Electron field emission from the oriented tio$_2$ nanowires prepared by thermal evaporation. *Chemical Physics Letters*, 413(4):490 – 494, 2005. ISSN 0009-2614. doi: https://doi.org/10.1016/j.cplett.2005.07.113. URL http://www.sciencedirect.com/science/article/pii/S000926140501153X.

[57] Akhmad Herman Yuwono, Junmin Xue, John Wang, Hendry Izaac Elim, Wei Ji, Ying Li, and Timothy John White. Transparent nanohybrids of nanocrystalline tio$_2$ in pmma with unique nonlinear optical behavior. *J. Mater. Chem.*, 13:1475–1479, 2003. doi: 10.1039/B211976E. URL http://dx.doi.org/10.1039/B211976E.

[58] Michael Grätzel. Photoelectrochemical cells. *Nature*, 414(6861):338–344, 2001. doi: 10.1038/35104607. URL http://dx.doi.org/10.1038/35104607.

[59] Emil Enache-Pommer, Janice E. Boercker, and Eray S. Aydil. Electron transport and recombination in polycrystalline tio$_2$ nanowire dye-sensitized solar cells. *Applied Physics Letters*, 91(12):123116, 2007. doi: http://dx.doi.org/10.1063/1.2783477. URL http://scitation.aip.org/content/aip/journal/apl/91/12/10.1063/1.2783477.

[60] Emil Enache-Pommer, Bin Liu, and Eray S. Aydil. Electron transport and recombination in dye-sensitized solar cells made from single-crystal rutile tio$_2$ nanowires. *Phys. Chem. Chem. Phys.*, 11:9648–9652, 2009. doi: 10.1039/B915345D. URL http://dx.doi.org/10.1039/B915345D.

[61] Bin Liu and Eray S. Aydil. Growth of oriented single-crystalline rutile tio$_2$ nanorods on transparent conducting substrates for dye-sensitized solar cells. *Journal of the American Chemical Society*, 131(11):3985–3990, 2009. doi: 10.1021/ja8078972. URL http://dx.doi.org/10.1021/ja8078972.

[62] Jin Young Kim, Jun Hong Noh, Kai Zhu, Adam F. Halverson, Nathan R. Neale, Sangbaek Park, Kug Sun Hong, and Arthur J. Frank. General strategy for fabricating transparent tio$_2$ nanotube arrays for dye-sensitized photoelectrodes: Illumination geometry and transport properties. *ACS Nano*, 5(4):2647–2656, 2011. doi: 10.1021/nn200440u. URL http://dx.doi.org/10.1021/nn200440u. PMID: 21395234.

[63] A. Ruiz, A. Cornet, G. Sakai, K. Shimanoe, J. R. Morante, and N. Yamazoe. Preparation of cr-doped tio$_2$ thin film of p-type conduction for gas sensor application. *Chemistry Letters*, 31(9):892–893, 2002. doi: 10.1246/cl.2002.892.

[64] Soumen Das, Jin-Hwan Kim, Yong-Kyu Park, and Yoon-Bong Hahn. Solution processed ni-doped tio$_2$ p-type channel in field effect transistor assembly with <10 nm thin ba$_{0.5}$sr$_{0.5}$tio$_3$ dielectric layer. *Applied Physics Letters*, 98(20):202102, 2011. doi: 10.1063/1.3592736. URL http://dx.doi.org/10.1063/1.3592736.

[65] Y. Sobajima, S. Kato, T. Matsuura, T. Toyama, and H. Okamoto. Study of the light-trapping effects of textured zno:al/glass structure tco for improving photocurrent of a-si:h solar cells. *Journal of Materials Science: Materials in Electronics*, 18(1):159–162, Oct 2007. ISSN 1573-482X. doi: 10.1007/s10854-007-9186-8. URL https://doi.org/10.1007/s10854-007-9186-8.

[66] Soumen Das, Daan Liu, Jong Bae Park, and Yoon-Bong Hahn. Metal-ion doped p-type tio$_2$ thin films and their applications for heterojunction devices. *Journal of Alloys and Compounds*, 553(Supplement C):188 – 193, 2013. ISSN 0925-8388. doi: https://doi.org/10.1016/j.jallcom.2012.11.110. URL http://www.sciencedirect.com/science/article/pii/S0925838812020944.

[67] S. Ait Ali Yahia, L. Hamadou, M.J. Salar-García, A. Kadri, V.M. Ortiz-Martínez, F.J. Hernández-Fernández, A. Pérez de los Rios, and N. Benbrahim. Tio$_2$ nanotubes as alternative cathode in microbial fuel cells: Effect of annealing treatment on its performance. *Applied Surface Science*, 387(Supplement C):1037 – 1045, 2016. ISSN 0169-4332. doi: https://doi.org/10.1016/j.apsusc.2016.07.018. URL http://www.sciencedirect.com/science/article/pii/S0169433216314453.

[68] Matthew D. Pickett, Dmitri B. Strukov, Julien L. Borghetti, J. Joshua Yang, Gregory S. Snider, Duncan R. Stewart, and R. Stanley Williams. Switching dynamics in titanium dioxide memristive devices. *Journal of Applied Physics*, 106(7):074508, 2009. doi: 10.1063/1.3236506. URL http://scitation.aip.org/content/aip/journal/jap/106/7/10.1063/1.3236506.

References

[69] Su-Ting Han, Ye Zhou, and V. A. L. Roy. Towards the development of flexible non-volatile memories. *Advanced Materials*, 25(38): 5425–5449, 2013. ISSN 1521-4095. doi: 10.1002/adma.201301361. URL http://dx.doi.org/10.1002/adma.201301361.

[70] M K Bera and C K Maiti. Charge trapping properties of ultra-thin tio$_2$ films on strained-si. *Semiconductor Science and Technology*, 22 (7):774, 2007. URL http://stacks.iop.org/0268-1242/22/i=7/a=017.

[71] Peter Deák, Bálint Aradi, Alessio Gagliardi, Huynh Anh Huy, Gabriele Penazzi, Binghai Yan, Tim Wehling, and Thomas Frauenheim. Possibility of a field effect transistor based on dirac particles in semiconducting anatase-tio$_2$ nanowires. *Nano Letters*, 13(3):1073–1079, 2013. doi: 10.1021/nl304350v. URL http://dx.doi.org/10.1021/nl304350v. PMID: 23394361.

[72] G Zhang, H Wu, C Chen, T Wang, J Yue, and C Liu. Transparent and flexible capacitors based on nanolaminate al$_2$o$_3$/tio$_2$/al$_2$o$_3$. *Nanoscale Research Letters*, 10(76), 2015. doi: 10.1186/s11671-015-0784-8.

[73] Sarfraj H. Mujawar, Swapnil B. Ambade, T. Battumur, Rohan B. Ambade, and Soo-Hyoung Lee. Electropolymerization of polyaniline on titanium oxide nanotubes for supercapacitor application. *Electrochimica Acta*, 56(12):4462 – 4466, 2011. ISSN 0013-4686. doi: 10.1016/j.electacta.2011.02.043. URL http://www.sciencedirect.com/science/article/pii/S0013468611002581.

[74] C Natarajan, K Setoguchi, and G Nogami. Preparation of a nanocrystalline titanium dioxide negative electrode for the rechargeable lithium ion battery. *Electrochimica Acta*, 43(21–22):3371 – 3374, 1998. ISSN 0013-4686. doi: 10.1016/S0013-4686(97)10140-2. URL http://www.sciencedirect.com/science/article/pii/S0013468697101402.

[75] Junling Guo and Jinping Liu. Topotactic conversion-derived li$_4$ti$_5$o$_{12}$-rutile tio$_2$ hybrid nanowire array for high-performance lithium ion full cells. *RSC Adv.*, 4:12950–12957, 2014. doi: 10.1039/C4RA00111G. URL http://dx.doi.org/10.1039/C4RA00111G.

[76] Jun Jin, Shao-Zhuan Huang, Jing Liu, Yu Li, Li-Hua Chen, Yong Yu, Hong-En Wang, Clare P. Grey, and Bao-Lian Su. Phases hybriding and hierarchical structuring of mesoporous tio$_2$ nanowire bundles for high-rate and high-capacity lithium batteries. *Advanced Science*, pages n/a–n/a, 2015. ISSN 2198-3844. doi: 10.1002/advs.201500070. URL http://dx.doi.org/10.1002/advs.201500070.

[77] Ming-Shan Wang, Wei-Li Song, and Li-Zhen Fan. Interconnected tiox/carbon hybrid framework incorporated silicon for stable lithium ion battery anodes. *J. Mater. Chem. A*, 3:12709–12717, 2015. doi: 10.1039/C5TA00964B. URL http://dx.doi.org/10.1039/C5TA00964B.

[78] Mohammad Arif Ishtiaque Shuvo, Gerardo Rodriguez, Md Tariqul Islam, Hasanul Karim, Navaneet Ramabadran, Juan C. Noveron, and Yirong Lin. Microwave exfoliated graphene oxide/tio$_2$ nanowire hybrid for high performance lithium ion battery. *Journal of Applied Physics*, 118(12):125102, 2015. doi: http://dx.doi.org/10.1063/1.4931380. URL http://scitation.aip.org/content/aip/journal/jap/118/12/10.1063/1.4931380.

[79] Minghao Yu, Zilong Wang, Yi Han, Yexiang Tong, Xihong Lu, and Shihe Yang. Recent progress in the development of anodes for asymmetric supercapacitors. *J. Mater. Chem. A*, 4:4634–4658, 2016. doi: 10.1039/C5TA10542K. URL http://dx.doi.org/10.1039/C5TA10542K.

[80] Taeseup Song and Ungyu Paik. Tio$_2$ as an active or supplemental material for lithium batteries. *J. Mater. Chem. A*, 4:14–31, 2016. doi: 10.1039/C5TA06888F. URL http://dx.doi.org/10.1039/C5TA06888F.

[81] Yu Liu and Yefeng Yang. Recent progress of tio$_2$-based anodes for li ion batteries. *J. Nanomaterials*, 2016:2:2–2:2, January 2016. ISSN 1687-4110. doi: 10.1155/2016/8123652. URL http://dx.doi.org/10.1155/2016/8123652.

[82] Wei Wen, Jin-Ming Wu, Yin-Zhu Jiang, Lu-Lu Lai, and Jian Song. Pseudocapacitance-enhanced li-ion microbatteries derived by a tin@tio$_2$ nanowire anode. *Chem*, 2(3):404 – 416, 2017. ISSN 2451-9294. doi: https://doi.org/10.1016/j.chempr.2017.01.004. URL http://www.sciencedirect.com/science/article/pii/S2451929417300281.

[83] Shaofu Wang, Dandan Qu, Yun Jiang, Wan-Sheng Xiong, Hong-Qian Sang, Rong-Xiang He, Qidong Tai, Bolei Chen, Yumin Liu, and Xing-Zhong Zhao. Three-dimensional branched tio$_2$ architectures in controllable bloom for advanced lithium-ion batteries. *ACS Applied Materials & Interfaces*, 8(31):20040–20047, 2016. doi: 10.1021/acsami.6b05559. URL http://dx.doi.org/10.1021/acsami.6b05559. PMID: 27420343.

[84] Qingqing Sun, Kaixiang Chen, Yubin Liu, Yafeng Li, and Mingdeng Wei. Rutile tio$_2$ mesocrystals as a novel sulfur host for high performance lithium-sulfur batteries. *Chemistry - A European Journal*, pages n/a–n/a, 2017. ISSN 1521-3765. doi: 10.1002/chem.201703130. URL http://dx.doi.org/10.1002/chem.201703130.

[85] Lars M. Bjursten, Lars Rasmusson, Seunghan Oh, Garrett C. Smith, Karla S. Brammer, and Sungho Jin. Titanium dioxide nanotubes enhance bone bonding in vivo. *Journal of Biomedical Materials Research Part A*, 92A(3):1218–1224, 2010. ISSN 1552-4965. doi: 10.1002/jbm.a.32463. URL http://dx.doi.org/10.1002/jbm.a.32463.

[86] Harold Craighead. Future lab-on-a-chip technologies for interrogating individual molecules. *Nature*, 442(7101):387 – 393, 2006. doi: 10.1038/nature05061. URL http://dx.doi.org/10.1038/nature05061.

[87] Minbaek Lee, Ku Youn Baik, Meg Noah, Young-Kyun Kwon, Jeong-O Lee, and Seunghun Hong. Nanowire and nanotube transistors for lab-on-a-chip applications. *Lab Chip*, 9:2267–2280, 2009. doi: 10.1039/B905185F. URL http://dx.doi.org/10.1039/B905185F.

[88] Y. C. Lim, A. Z. Kouzani, and W. Duan. Lab-on-a-chip: a component view. *Microsystem Technologies*, 16(12):1995–2015, 2010. ISSN 1432-1858. doi: 10.1007/s00542-010-1141-6. URL http://dx.doi.org/10.1007/s00542-010-1141-6.

[89] Jae Hyung Park, Jim M. Coy, T. Serkan Kasirga, Chunming Huang, Zaiyao Fei, Scott Hunter, and David H. Cobden. Measurement of a solid-state triple point at the metal-insulator transition in vo$_2$. *Nature*, 500:431–434, August 2013. doi: 10.1038/nature12425. URL http://www.nature.com/nature/journal/v500/n7463/full/nature12425.html.

[90] Xiaojing Zou, Gang Xiao, Sunxiang Huang, Tingyong Chen, and Chia-Ling Chien. Magnetotransport properties of polycrystalline and epitaxial chromium dioxide nanowires. *Journal of Applied Physics*, 103(7):07D710, 2008. doi: 10.1063/1.2836800. URL http://scitation.aip.org/content/aip/journal/jap/103/7/10.1063/1.2836800.

[91] Arnold F. Holleman, Egon Wiberg, and Nils Wiberg. *Lehrbuch der anorganischen Chemie*, volume 102. de Gruyter, 2007.

[92] Ulrich K. Thiele. The current status of catalysis and catalyst development for the industrial process of poly(ethylene terephthalate) polycondensation. *International Journal of Polymeric Materials and Polymeric Biomaterials*, 50(3-4):387–394, 2001. doi: 10.1080/00914030108035115.

[93] Z.G. Bai, D.P. Yu, H.Z. Zhang, Y. Ding, Y.P. Wang, X.Z. Gai, Q.L. Hang, G.C. Xiong, and S.Q. Feng. Nano-scale Geo$_2$ wires synthesized by physical evaporation. *Chemical Physics Letters*, 303(3–4):311 – 314, 1999. ISSN 0009-2614. doi: 10.1016/S0009-2614(99)00066-4. URL http://www.sciencedirect.com/science/article/pii/S0009261499000664.

[94]   Donald B. Rogers, Robert D. Shannon, Arthur W. Sleight, and Joseph L. Gillson. Crystal chemistry of metal dioxides with rutile-related structures. *Inorganic Chemistry*, 8(4):841–849, 1969. doi: 10.1021/ic50074a029. URL http://dx.doi.org/10.1021/ic50074a029.

[95]   B. Hu, L. Mai, W. Chen, and F. Yang. From moo_2 nanobelts to moo_2 nanorods: Structure transformation and electrical transport. *ACS Nano*, 3(2):478–482, 2009. doi: 10.1021/nn800844h.

[96]   J. Zhou, N.-S. Xu, S.-Z. Deng, J. Chen, J.-C. She, and Z.-L. Wang. Large-area nanowire arrays of molybdenum and molybdenum oxides: Synthesis and field emission properties. *Advanced Materials*, 15(21):1835–1840, 2003. ISSN 1521-4095. doi: 10.1002/adma.200305528. URL http://dx.doi.org/10.1002/adma.200305528.

[97]   L.C. Yang, Q.S. Gao, Y.H. Zhang, Y. Tang, and Y.P. Wu. Tremella-like molybdenum dioxide consisting of nanosheets as an anode material for lithium ion battery. *Electrochemistry Communications*, 10(1):118 – 122, 2008. ISSN 1388-2481. doi: 10.1016/j.elecom.2007. 11.009. URL http://www.sciencedirect.com/science/article/pii/S138824810700433X.

[98]   L.C. Yang, Q.S. Gao, Y. Tang, Y.P. Wu, and R. Holze. moo_2 synthesized by reduction of moo_3 with ethanol vapor as an anode material with good rate capability for the lithium ion battery. *Journal of Power Sources*, 179(1):357 – 360, 2008. ISSN 0378-7753. doi: 10.1016/j.jpowsour.2007.12.099. URL http://www.sciencedirect.com/science/article/pii/S0378775308000098.

[99]   Yifeng Shi, Bingkun Guo, Serena A. Corr, Qihui Shi, Yong-Sheng Hu, Kevin R. Heier, Liquan Chen, Ram Seshadri, and Galen D. Stucky. Ordered mesoporous metallic moo_2 materials with highly reversible lithium storage capacity. *Nano Letters*, 9(12):4215–4220, 2009. doi: 10.1021/nl902423a. URL http://dx.doi.org/10.1021/nl902423a.

[100]   Yongming Sun, Xianluo Hu, Wei Luo, and Yunhui Huang. Self-assembled hierarchical moo_2/graphene nanoarchitectures and their application as a high-performance anode material for lithium-ion batteries. *ACS Nano*, 5(9):7100–7107, 2011. doi: 10.1021/nn201802c. URL http://dx.doi.org/10.1021/nn201802c.

[101]   Liang Zhou, Hao Bin Wu, Zhiyu Wang, and Xiong Wen (David) Lou. Interconnected moo_2 nanocrystals with carbon nanocoating as high-capacity anode materials for lithium-ion batteries. *ACS Applied Materials & Interfaces*, 3(12):4853–4857, 2011. doi: 10.1021/am201351z. URL http://dx.doi.org/10.1021/am201351z.

[102]   Yun Xu, Ran Yi, Bin Yuan, Xiaofei Wu, Marco Dunwell, Qianglu Lin, Ling Fei, Shuguang Deng, Paul Andersen, Donghai Wang, and Hongmei Luo. High capacity moo_2/graphite oxide composite anode for lithium-ion batteries. *The Journal of Physical Chemistry Letters*, 3(3):309–314, 2012. doi: 10.1021/jz201619r. URL http://dx.doi.org/10.1021/jz201619r. PMID: 26285844.

[103]   Jingguo Liu, Zhengjun Zhang, Chunyu Pan, Ye Zhao, Xin Su, Ya Zhou, and Dapeng Yu. Enhanced field emission properties of Moo$_2$ nanorods with controllable shape and orientation. *Materials Letters*, 58(29):3812 – 3815, 2004. ISSN 0167-577X. doi: 10.1016/j.matlet. 2004.07.034. URL http://www.sciencedirect.com/science/article/pii/S0167577X04005877.

[104]   K Makise, K Mitsuishi, M Shimojo, and K Furuya. A nanosized photodetector fabricated by electron-beam-induced deposition. *Nanotechnology*, 20(42):425305, 2009. doi: 10.1088/0957-4484/20/42/425305. URL http://stacks.iop.org/0957-4484/20/i=42/a=425305.

[105]   A.A. Balandin and I.D. Rozhdestvenskaya. Some catalytic properties of molybdenum trioxide and dioxide. *Bulletin of the Academy of Sciences of the USSR, Division of chemical science*, 8(11):1804–1810, 1959. ISSN 0568-5230. doi: 10.1007/BF00914749. URL http://dx.doi.org/10.1007/BF00914749.

[106]   A. Katrib, P. Leflaive, L. Hilaire, and G. Maire. Molybdenum based catalysts. i. moo_2 as the active species in the reforming of hydrocarbons. *Catalysis Letters*, 38(1-2):95–99, 1996. ISSN 1011-372X. doi: 10.1007/BF00806906. URL http://dx.doi.org/10.1007/BF00806906.

[107]   Christian Martin Cuba-Torres, Oscar Marin-Flores, Craig D. Owen, Zhouhong Wang, Manuel Garcia-Perez, M. Grant Norton, and Su Ha. Catalytic partial oxidation of a biodiesel surrogate over molybdenum dioxide. *Fuel*, 146:132 – 137, 2015. ISSN 0016-2361. doi: 10.1016/j.fuel.2015.01.003. URL http://www.sciencedirect.com/science/article/pii/S0016236115000083.

[108]   Herbert Over. Atomic-scale understanding of the hcl oxidation over ruo_2, a novel deacon process. *The Journal of Physical Chemistry C*, 116(12):6779, 2012. doi: 10.1021/jp212108b.

[109]   Chi-Chang Hu, Kuo-Hsin Chang, Ming-Champ Lin, , and Yung-Tai Wu. Design and tailoring of the nanotubular arrayed architecture of hydrous ruo_2 for next generation supercapacitors. *Nano Letters*, 6(12):2690–2695, 2006. URL http://dx.doi.org/10.1021/nl061576a.

[110]   Pochiang Chen, Haitian Chen, Jing Qiu, and Chongwu Zhou. Inkjet printing of single-walled carbon nanotube/ruo_2 nanowire supercapacitors on cloth fabrics and flexible substrates. *Nano Research*, 3(8):594–603, 2010. ISSN 1998-0124. doi: 10.1007/s12274-010-0020-x. URL http://dx.doi.org/10.1007/s12274-010-0020-x.

[111]   Yumin Lee, Byeong Uk Ye, Hak Ki Yu, Jong-Lam Lee, Myung Hwa Kim, and Jeong Min Baik. Facile synthesis of single crystalline metallic ruo_2 nanowires and electromigration-induced transport properties. *The Journal of Physical Chemistry C*, 115(11), 2011. doi: 10.1021/jp200426s.

[112]   Hui Xia, Ying Shirley Meng, Guoliang Yuan, Chong Cui, and Li Lu. A symmetric ruo_2/ruo_2 supercapacitor operating at 1.6 v by using a neutral aqueous electrolyte. *Electrochemical and Solid-State Letters*, 15(4):A60–A63, 2012. doi: 10.1149/2.023204esl. URL http://esl.ecsdl.org/content/15/4/A60.abstract.

[113]   D. J. Fixsen, P. G. A. Mirel, A. Kogut, and M. Seiffert. A low noise thermometer readout for ruthenium oxide resistors. *Review of Scientific Instruments*, 73(10):3659–3663, 2002. doi: 10.1063/1.1505108. URL http://scitation.aip.org/content/aip/journal/rsi/73/10/10.1063/1.1505108.

[114]   Y.-L. Chueh, C.-H. Hsieh, M.-T. Chang, L.-J. Chou, C.S. Lao, J.H. Song, J.-Y. Gan, and Z.L. Wang. Ruo$_2$ nanowires and ruo_2/tio_2 core/shell nanowires: From synthesis to mechanical, optical, electrical, and photoconductive properties. *Advanced Materials*, 19(1): 143–149, 2007. ISSN 1521-4095. doi: 10.1002/adma.200601830. URL http://dx.doi.org/10.1002/adma.200601830.

[115]   Yee-Lang Liu, Zong-Yi Wu, Kuei-Jiun Lin, Jr-Jeng Huang, Fu-Rong Chen, Ji-Jung Kai, Yong-Han Lin, Wen-Bin Jian, and Juhn-Jong Lin. Growth of single-crystalline ruo_2 nanowires with one- and two-nanocontact electrical characterizations. *Applied Physics Letters*, 90(1):013105, 2007. doi: 10.1063/1.2428669. URL http://scitation.aip.org/content/aip/journal/apl/90/1/10.1063/1.2428669.

[116]   Y. Wang, X. Jiang, and Y. Xia. A solution-phase, precursor route to polycrystalline Sno_2 nanowires that can be used for gas sensing under ambient conditions. *Journal of the American Chemical Society*, 125(52):16176–16177, 2003. doi: 10.1021/ja037743f.

[117]   A. Kolmakov, DO. Klenov, Y. Lilach, S. Stemmer, and M. Moskovits. Enhanced gas sensing by individual sno_2 nanowires and nanobelts functionalized with pd catalyst particles. *Nano Letters*, 5(4):667–673, 2005. doi: 10.1021/nl050082v.

References

[118] Young-Jin Choi, In-Sung Hwang, Jae-Gwan Park, Kyoung Jin Choi, Jae-Hwan Park, and Jong-Heun Lee. Novel fabrication of an sno_2 nanowire gas sensor with high sensitivity. *Nanotechnology*, 19(9):095508, 2008. doi: 10.1088/0957-4484/19/9/095508. URL http://stacks.iop.org/0957-4484/19/i=9/a=095508.

[119] Suresh Gubbala, Vidhya Chakrapani, Vivekanand Kumar, and Mahendra K. Sunkara. Band-edge engineered hybrid structures for dye-sensitized solar cells based on sno_2 nanowires. *Advanced Functional Materials*, 18(16):2411–2418, 2008. ISSN 1616-3028. doi: 10.1002/adfm.200800099. URL http://dx.doi.org/10.1002/adfm.200800099.

[120] Min-Sik Park, Guo-Xiu Wang, Yong-Mook Kang, David Wexler, Shi-Xue Dou, and Hua-Kun Liu. Preparation and electrochemical properties of sno_2 nanowires for application in lithium-ion batteries. *Angewandte Chemie*, 119(5):764–767, 2007. ISSN 1521-3757. doi: 10.1002/ange.200603309. URL http://dx.doi.org/10.1002/ange.200603309.

[121] Eric N. Dattoli, Qing Wan, Wei Guo, Yanbin Chen, Xiaoqing Pan, and Wei Lu. Fully transparent thin-film transistor devices based on sno_2 nanowires. *Nano Letters*, 7(8):2463–2469, 2007. doi: 10.1021/nl0712217.

[122] A Jha, B D O Richards, G Jose, T Toney Fernandez, C J Hill, J Lousteau, and P Joshi. Review on structural, thermal, optical and spectroscopic properties of tellurium oxide based glasses for fibre optic and waveguide applications. *International Materials Reviews*, 57(6):357–382, 2012. doi: 10.1179/1743280412Y.0000000005. URL http://dx.doi.org/10.1179/1743280412Y.0000000005.

[123] W.A. Bonner, S. Singh, L.G. Van Uitert, and A.W. Warner. High quality tellurium dioxide for acousto-optic and non-linear applications. *Journal of Electronic Materials*, 1(1):154–164, 1972. ISSN 0361-5235. doi: 10.1007/BF02660359. URL http://dx.doi.org/10.1007/BF02660359.

[124] Vitaly B. Voloshinov. Special issue acousto optics/imaging anisotropic light diffraction on ultrasound in a tellurium dioxide single crystal. *Ultrasonics*, 31(5):333 – 338, 1993. ISSN 0041-624X. doi: 10.1016/0041-624X(93)90066-9. URL http://www.sciencedirect.com/science/article/pii/0041624X93900669.

[125] Khu Vu, Kunlun Yan, Zhe Jin, Xin Gai, Duk-Yong Choi, Sukanta Debbarma, Barry Luther-Davies, and Steve Madden. Hybrid waveguide from as_2s_3 and er-doped teo_2 for lossless nonlinear optics. *Optics Letters*, 38(11):1766–1768, Jun 2013. doi: 10.1364/OL.38.001766. URL http://ol.osa.org/abstract.cfm?URI=ol-38-11-1766.

[126] P.C Yen, R.S Chen, C.C Chen, Y.S Huang, and K.K Tiong. Growth and characterization of oso_2 single crystals. *Journal of Crystal Growth*, 262(1–4):271 – 276, 2004. ISSN 0022-0248. doi: 10.1016/j.jcrysgro.2003.10.021. URL http://www.sciencedirect.com/science/article/pii/S0022024803019109.

[127] Stuart F. Cogan. Neural stimulation and recording electrodes. *Annual Review of Biomedical Engineering*, 10(1):275–309, 2008. doi: 10.1146/annurev.bioeng.10.061807.160518. URL http://dx.doi.org/10.1146/annurev.bioeng.10.061807.160518. PMID: 18429704.

[128] J.H. Shim, Y. Lee, M. Kang, J. Lee, J.M. Baik, Y. Lee, C. Lee, and M.H. Kim. Hierarchically driven iro_2 nanowire electrocatalysts for direct sensing of biomolecules. *Analytical Chemistry*, 84(8):3827–3832, 2012. doi: 10.1021/ac300573b.

[129] Akshay Kumar, Anuj R. Madaria, and Chongwu Zhou. Growth of aligned single-crystalline rutile tio_2 nanowires on arbitrary substrates and their application in dye-sensitized solar cells. *The Journal of Physical Chemistry C*, 114(17):7787–7792, 2010. doi: 10.1021/jp100491h. URL http://dx.doi.org/10.1021/jp100491h.

[130] Hua Wang, Yusong Bai, Qiong Wu, Wei Zhou, Hao Zhang, Jinghong Li, and Lin Guo. Rutile tio_2 nano-branched arrays on fto for dye-sensitized solar cells. *Physical Chemistry Chemical Physics*, 13:7008–7013, 2011. doi: 10.1039/C1CP20351G. URL http://dx.doi.org/10.1039/C1CP20351G.

[131] Huang Qing and Gao Lian. A simple route for the synthesis of rutile tio_2 nanorods. *Chemistry Letters*, 32(7):638–639, 2003. doi: 10.1246/cl.2003.638. URL https://doi.org/10.1246/cl.2003.638.

[132] P. Löbl, M. Huppertz, and D. Mergel. Nucleation and growth in tio_2 films prepared by sputtering and evaporation. *Thin Solid Films*, 251(1):72 – 79, 1994. ISSN 0040-6090. doi: http://dx.doi.org/10.1016/0040-6090(94)90843-5. URL http://www.sciencedirect.com/science/article/pii/0040609094908435.

[133] Youl-Moon Sung and Hee-Je Kim. Sputter deposition and surface treatment of tio_2 films for dye-sensitized solar cells using reactive rf plasma. *Thin Solid Films*, 515(12):4996 – 4999, 2007. ISSN 0040-6090. doi: http://dx.doi.org/10.1016/j.tsf.2006.10.079. URL http://www.sciencedirect.com/science/article/pii/S004060900601220X. The Third International Symposium on Dry Process (DPS 2005).

[134] H. Tang, K. Prasad, R. Sanjinès, P. E. Schmid, and F. Lévy. Electrical and optical properties of tio_2 anatase thin films. *Journal of Applied Physics*, 75(4):2042–2047, 1994. doi: http://dx.doi.org/10.1063/1.356306. URL http://scitation.aip.org/content/aip/journal/jap/75/4/10.1063/1.356306.

[135] George Atanassov, Roland Thielsch, and Dimitar Popov. Optical properties of tio_2, y_2o_3 and ceo_2 thin films deposited by electron beam evaporation. *Thin Solid Films*, 223(2):288 – 292, 1993. ISSN 0040-6090. doi: http://dx.doi.org/10.1016/0040-6090(93)90534-V. URL http://www.sciencedirect.com/science/article/pii/004060909390534V.

[136] K.Narasimha Rao, M.A. Murthy, and S. Mohan. Optical properties of electron-beam-evaporated tio_2 films. *Thin Solid Films*, 176(2): 181 – 186, 1989. ISSN 0040-6090. doi: http://dx.doi.org/10.1016/0040-6090(89)90091-6. URL http://www.sciencedirect.com/science/article/pii/0040609089900916.

[137] Jaan Aarik, Aleks Aidla, Teet Uustare, and Väino Sammelselg. Morphology and structure of tio_2 thin films grown by atomic layer deposition. *Journal of Crystal Growth*, 148(3):268–275, 1995. ISSN 0022-0248. doi: http://dx.doi.org/10.1016/0022-0248(94)00874-4. URL http://www.sciencedirect.com/science/article/pii/0022024894008744.

[138] Jaan Aarik, Aleks Aidla, Alma-Asta Kiisler, Teet Uustare, and Väino Sammelselg. Effect of crystal structure on optical properties of tio_2 films grown by atomic layer deposition. *Thin Solid Films*, 305(1):270–273, 1997. ISSN 0040-6090. doi: http://dx.doi.org/10.1016/S0040-6090(97)00135-1. URL http://www.sciencedirect.com/science/article/pii/S0040609097001351.

[139] V. Pore, A. Rahtu, M. Leskelä, M. Ritala, T. Sajavaara, and J. Keinonen. Atomic layer deposition of photocatalytic tio_2 thin films from titanium tetramethoxide and water. *Chemical Vapor Deposition*, 10(3):143–148, 2004. ISSN 1521-3862. doi: 10.1002/cvde.200306289. URL http://dx.doi.org/10.1002/cvde.200306289.

[140] J.S. King, E. Graugnard, and C.J. Summers. Tio_2 inverse opals fabricated using low-temperature atomic layer deposition. *Advanced Materials*, 17(8):1010–1013, 2005. ISSN 1521-4095. doi: 10.1002/adma.200400648. URL http://dx.doi.org/10.1002/adma.200400648.

[141] Stacey D. Standridge, George C. Schatz, and Joseph T. Hupp. Toward plasmonic solar cells: Protection of silver nanoparticles via atomic layer deposition of tio_2. *Langmuir*, 25(5):2596–2600, 2009. doi: 10.1021/la900113e. URL http://dx.doi.org/10.1021/la900113e. PMID: 19199715.

[142] Michael Grätzel. Sol-gel processed tio$_2$ films for photovoltaic applications. *Journal of Sol-Gel Science and Technology*, 22(1):7–13, Sep 2001. ISSN 1573-4846. doi: 10.1023/A:1011273700573. URL https://doi.org/10.1023/A:1011273700573.

[143] D. P. Macwan, Pragnesh N. Dave, and Shalini Chaturvedi. A review on nano-tio$_2$ sol-gel type syntheses and its applications. *Journal of Materials Science*, 46(11):3669–3686, Jun 2011. ISSN 1573-4803. doi: 10.1007/s10853-011-5378-y. URL https://doi.org/10.1007/s10853-011-5378-y.

[144] Wen Wen Xu, Robert Kershaw, Kirby Dwight, and Aaron Wold. Preparation and characterization of tio$_2$ films by a novel spray pyrolysis method. *Materials Research Bulletin*, 25(11):1385 – 1392, 1990. ISSN 0025-5408. doi: http://dx.doi.org/10.1016/0025-5408(90)90221-M. URL http://www.sciencedirect.com/science/article/pii/002554089090221M.

[145] Ladislav Kavan and Michael Grätzel. Highly efficient semiconducting tio$_2$ photoelectrodes prepared by aerosol pyrolysis. *Electrochimica Acta*, 40(5):643 – 652, 1995. ISSN 0013-4686. doi: http://dx.doi.org/10.1016/0013-4686(95)90400-W. URL http://www.sciencedirect.com/science/article/pii/001346869590400W.

[146] M.O Abou-Helal and W.T Seeber. Preparation of tio$_2$ thin films by spray pyrolysis to be used as a photocatalyst. *Applied Surface Science*, 195(1):53 – 62, 2002. ISSN 0169-4332. doi: http://dx.doi.org/10.1016/S0169-4332(02)00533-0. URL http://www.sciencedirect.com/science/article/pii/S0169433202005330.

[147] M.Dj Blešić, Z.V Šaponjić, J.M Nedeljković, and D.P Uskoković. Tio$_2$ films prepared by ultrasonic spray pyrolysis of nanosize precursor. *Materials Letters*, 54(4):298 – 302, 2002. ISSN 0167-577X. doi: http://dx.doi.org/10.1016/S0167-577X(01)00581-X. URL http://www.sciencedirect.com/science/article/pii/S0167577X0100581X.

[148] P. M. Sommeling, B. C. O'Regan, R. R. Haswell, H. J. P. Smit, N. J. Bakker, J. J. T. Smits, J. M. Kroon, and J. A. M. van Roosmalen. Influence of a ticl$_4$ post-treatment on nanocrystalline tio$_2$ films in dye-sensitized solar cells. *The Journal of Physical Chemistry B*, 110 (39):19191–19197, 2006. doi: 10.1021/jp061346k. URL http://dx.doi.org/10.1021/jp061346k. PMID: 17004768.

[149] Fritz J. Knorr, Dongshe Zhang, and Jeanne L. McHale. Influence of ticl$_4$ treatment on surface defect photoluminescence in pure and mixed-phase nanocrystalline tio$_2$. *Langmuir*, 23(17):8686–8690, 2007. doi: 10.1021/la700274k. URL http://dx.doi.org/10.1021/la700274k. PMID: 17658758.

[150] Sang-Wha Lee, Kwang-Soon Ahn, Kai Zhu, Nathan R. Neale, and Arthur J. Frank. Effects of ticl$_4$ treatment of nanoporous tio$_2$ films on morphology, light harvesting, and charge-carrier dynamics in dye-sensitized solar cells. *The Journal of Physical Chemistry C*, 116 (40):21285–21290, 2012. doi: 10.1021/jp3079887. URL http://dx.doi.org/10.1021/jp3079887.

[151] B. Stjerna, E. Olsson, and C. G. Granqvist. Optical and electrical properties of radio frequency sputtered tin oxide films doped with oxygen vacancies, f, sb, or mo. *Journal of Applied Physics*, 76(6):3797–3817, 1994. doi: 10.1063/1.357383. URL http://dx.doi.org/10.1063/1.357383.

[152] E. Elangovan and K. Ramamurthi. Studies on micro-structural and electrical properties of spray-deposited fluorine-doped tin oxide thin films from low-cost precursor. *Thin Solid Films*, 476(2):231 – 236, 2005. ISSN 0040-6090. doi: http://dx.doi.org/10.1016/j.tsf.2004.09.022. URL http://www.sciencedirect.com/science/article/pii/S0040609004013689.

[153] Xinjian Feng, Karthik Shankar, Oomman K. Varghese, Maggie Paulose, Thomas J. Latempa, and Craig A. Grimes. Vertically aligned single crystal tio$_2$ nanowire arrays grown directly on transparent conducting oxide coated glass: Synthesis details and applications. *Nano Letters*, 8(11):3781–3786, 2008. doi: 10.1021/nl802096a. URL http://dx.doi.org/10.1021/nl802096a. PMID: 18954124.

[154] Li Guang-Lai, Wang Guang-Hou, and Hong Jian-Ming. Morphologies of rutile form tio$_2$ twins crystals. *Journal of Materials Science Letters*, 18(15):1243–1246, 1999. ISSN 0261-8028. doi: 10.1023/A:1006627007970. URL http://dx.doi.org/10.1023/A%3A1006627007970.

[155] Takayuki Hirai, Hiroshi Sato, and Isao Komasawa. Mechanism of formation of titanium dioxide ultrafine particles in reverse micelles by hydrolysis of titanium tetrabutoxide. *Industrial & Engineering Chemistry Research*, 32(12):3014–3019, 1993. doi: 10.1021/ie00024a009. URL http://dx.doi.org/10.1021/ie00024a009.

[156] Young-wook Jun, Maria F. Casula, Jae-Hwan Sim, Sang Youl Kim, Jinwoo Cheon, , and A. Paul Alivisatos. Surfactant-assisted elimination of a high energy facet as a means of controlling the shape of Tio$_2$ nanocrystals. *Journal of the American Chemical Society*, 125(51):15981–15985, 2003. doi: 10.1021/ja0369515. URL http://dx.doi.org/10.1021/ja0369515. PMID: 14677990.

[157] Lu-Lu Lai and Jin-Ming Wu. A facile synthesis of hierarchical tio$_2$ for dye adsorption and photocatalysis. *RSC Adv.*, 4:36212–36217, 2014. doi: 10.1039/C4RA04790G. URL http://dx.doi.org/10.1039/C4RA04790G.

[158] Bo Li, Jin-Ming Wu, Tao-Tao Guo, Ming-Zao Tang, and Wei Wen. A facile solution route to deposit tio$_2$ nanowire arrays on arbitrary substrates. *Nanoscale*, 6:3046–3050, 2014. doi: 10.1039/C3NR05786K. URL http://dx.doi.org/10.1039/C3NR05786K.

[159] Dongsheng Li, Frank Soberanis, Jia Fu, Wenting Hou, Jianzhong Wu, and David Kisailus. Growth mechanism of highly branched titanium dioxide nanowires via oriented attachment. *Crystal Growth & Design*, 13(2):422–428, 2013. doi: 10.1021/cg301388e. URL http://dx.doi.org/10.1021/cg301388e.

[160] Chenghua Sun, Nuanxia Wang, Shiyi Zhou, Xiujie Hu, Shuyun Zhou, and Ping Chen. Preparation of self-supporting hierarchical nanostructured anatase/rutile composite tio$_2$ film. *Chem. Commun.*, pages 3293–3295, 2008. doi: 10.1039/B805072D. URL http://dx.doi.org/10.1039/B805072D.

[161] Panpan Sun, Xintong Zhang, Xueping Liu, Lingling Wang, Changhua Wang, Jikai Yang, and Yichun Liu. Growth of single-crystalline rutile tio$_2$ nanowire array on titanate nanosheet film for dye-sensitized solar cells. *J. Mater. Chem.*, 22:6389–6393, 2012. doi: 10.1039/C2JM16695J. URL http://dx.doi.org/10.1039/C2JM16695J.

[162] S. Venkatachalam, H. Hayashi, T. Ebina, and H. Nanjo. Preparation and characterization of nanostructured tio$_2$ thin films by hydrothermal and anodization methods. *Optoelectronics - Advanced Materials and Devices*, 2013. doi: 10.5772/51254. URL http://www.intechopen.com/books/optoelectronics-advanced-materials-and-devices/preparation-and-characterization-of-nanostructured-tio2-thin-films-by-hydrothermal-and-anodization-m.

[163] Jingyu Wang, Xijiang Han, Wei Zhang, Zhike He, Chao Wang, Ruxiu Cai, and Zhihong Liu. Controlled growth of monocrystalline rutile nanoshuttles in anatase tio$_2$ particles under mild conditions. *CrystEngComm*, 11:564–566, 2009. doi: 10.1039/B822117K. URL http://dx.doi.org/10.1039/B822117K.

[164] Puhong Wen, Hiroshi Itoh, Weiping Tang, and Qi Feng. Single nanocrystals of anatase-type tio$_2$ prepared from layered titanate nanosheets: Formation mechanism and characterization of surface properties. *Langmuir*, 23(23):11782–11790, 2007. doi: 10.1021/la701632t. URL http://dx.doi.org/10.1021/la701632t. PMID: 17935363.

341

[165] Xianfeng Yang, Jianle Zhuang, Xiuyan Li, Dihu Chen, Gangfeng Ouyang, Zhongquan Mao, Yaxiong Han, Zhenhui He, Chaolun Liang, Mingmei Wu, and Jimmy C. Yu. Hierarchically nanostructured rutile arrays: Acid vapor oxidation growth and tunable morphologies. *ACS Nano*, 3(5):1212–1218, 2009. doi: 10.1021/nn900084e. URL http://dx.doi.org/10.1021/nn900084e. PMID: 19400581.

[166] Xianfeng Yang, Chongjun Jin, Chaolun Liang, Dihu Chen, Mingmei Wu, and Jimmy C. Yu. Nanoflower arrays of rutile tio$_2$. *Chem. Commun.*, 47:1184–1186, 2011. doi: 10.1039/C0CC04216A. URL http://dx.doi.org/10.1039/C0CC04216A.

[167] Weijia Zhou, Xiaoyan Liu, Jingjie Cui, Duo Liu, Jing Li, Huaidong Jiang, Jiyang Wang, and Hong Liu. Control synthesis of rutile tio$_2$ microspheres, nanoflowers, nanotrees and nanobelts via acid-hydrothermal method and their optical properties. *CrystEngComm*, 13: 4557–4563, 2011. doi: 10.1039/C1CE05186E. URL http://dx.doi.org/10.1039/C1CE05186E.

[168] Andreas Wisnet, Sophia B. Betzler, Rachel V. Zucker, James A. Dorman, Peter Wagatha, Sonja Matich, Eiji Okunishi, Lukas Schmidt-Mende, and Christina Scheu. Model for hydrothermal growth of rutile wires and the associated development of defect structures. *Crystal Growth & Design*, 14(9):4658–4663, 2014. doi: 10.1021/cg500743u. URL http://dx.doi.org/10.1021/cg500743u.

[169] B.W. Smith and K. Suzuki. *Microlithography: Science and Technology, Second Edition*. Optical science and engineering. CRC Press, 2007. ISBN 9781420051537. URL https://books.google.de/books?id=_hTLDCeIYxoC.

[170] W. Ehrfeld and H. Lehr. Deep x-ray lithography for the production of three-dimensional microstructures from metals, polymers and ceramics. *Radiation Physics and Chemistry*, 45(3):349 – 365, 1995. ISSN 0969-806X. doi: http://dx.doi.org/10.1016/0969-806X(93)E0007-R. URL http://www.sciencedirect.com/science/article/pii/0969806X93E0007R. Applications of synchrotron X-radiation.

[171] H. Guckel. High-aspect-ratio micromachining via deep x-ray lithography. *Proceedings of the IEEE*, 86(8):1586–1593, Aug 1998. ISSN 0018-9219. doi: 10.1109/5.704264.

[172] Burn Jeng Lin. Deep uv lithography. *Journal of Vacuum Science and Technology*, 12(6):1317–1320, 1975. doi: 10.1116/1.568527. URL http://dx.doi.org/10.1116/1.568527.

[173] Youngjo Tak and Kijung Yong. Controlled growth of well-aligned zno nanorod array using a novel solution method. *The Journal of Physical Chemistry B*, 109(41):19263–19269, 2005. doi: 10.1021/jp0538767. URL http://dx.doi.org/10.1021/jp0538767. PMID: 16853488.

[174] B. S. Kang, S. J. Pearton, and F. Ren. Low temperature (100c) patterned growth of zno nanorod arrays on si. *Applied Physics Letters*, 90(8):083104, 2007. doi: http://dx.doi.org/10.1063/1.2709631. URL http://scitation.aip.org/content/aip/journal/apl/90/8/10.1063/1.2709631.

[175] C. Vieu, F. Carcenac, A. Pépin, Y. Chen, M. Mejias, A. Lebib, L. Manin-Ferlazzo, L. Couraud, and H. Launois. Electron beam lithography: resolution limits and applications. *Applied Surface Science*, 164(1–4):111 – 117, 2000. ISSN 0169-4332. doi: http://dx.doi.org/10.1016/S0169-4332(00)00352-4. URL http://www.sciencedirect.com/science/article/pii/S0169433200003524.

[176] Jingbiao Cui and Ursula Gibson. Low-temperature fabrication of single-crystal zno nanopillar photonic bandgap structures. *Nanotechnology*, 18(15):155302, 2007. URL http://stacks.iop.org/0957-4484/18/i=15/a=155302.

[177] Benjamin Weintraub, Yulin Deng, and Zhong L. Wang. Position-controlled seedless growth of zno nanorod arrays on a polymer substrate via wet chemical synthesis. *The Journal of Physical Chemistry C*, 111(28):10162–10165, 2007. doi: 10.1021/jp073806v. URL http://dx.doi.org/10.1021/jp073806v.

[178] Q Ahsanulhaq, Jin-Hwan Kim, and Yoon-Bong Hahn. Controlled selective growth of zno nanorod arrays and their field emission properties. *Nanotechnology*, 18(48):485307, 2007. URL http://stacks.iop.org/0957-4484/18/i=48/a=485307.

[179] Vitor R. Manfrinato, Lihua Zhang, Dong Su, Huigao Duan, Richard G. Hobbs, Eric A. Stach, and Karl K. Berggren. Resolution limits of electron-beam lithography toward the atomic scale. *Nano Letters*, 13(4):1555–1558, 2013. doi: 10.1021/nl304715p. URL http://dx.doi.org/10.1021/nl304715p. PMID: 23488936.

[180] Rick Merritt. Euv nudges toward 10nm. *EE Times, News & Analysis*, 2014.

[181] Jung Mi-Hee and Lee Hyoyoung. Selective patterning of zno nanorods on silicon substrates using nanoimprint lithography. *Nanoscale Research Letters*, 6(1):1–11, 2011. doi: 10.1186/1556-276X-6-159.

[182] Hyeong-Ho Park, Xin Zhang, Keun Woo Lee, Ka Hee Kim, Sang Hyun Jung, Deok Soo Park, Young Su Choi, Hyun-Beom Shin, Ho Kun Sung, Kyung Ho Park, Ho Kwan Kang, Hyung-Ho Park, and Chul Ki Ko. Position-controlled hydrothermal growth of zno nanorods on arbitrary substrates with a patterned seed layer via ultraviolet-assisted nanoimprint lithography. *CrystEngComm*, 15:3463–3469, 2013. doi: 10.1039/C3CE27069F. URL http://dx.doi.org/10.1039/C3CE27069F.

[183] Yuekun Lai, Zequan Lin, Jianying Huang, Lan Sun, Zhong Chen, and Changjian Lin. Controllable construction of zno/tio$_2$ patterning nanostructures by superhydrophilic/superhydrophobic templates. *New J. Chem.*, 34:44–51, 2010. doi: 10.1039/B9NJ00325H. URL http://dx.doi.org/10.1039/B9NJ00325H.

[184] Seok-Soon Kim, Chaemin Chun, Jae-Chul Hong, and Dong-Yu Kim. Well-ordered tio$_2$ nanostructures fabricated using surface relief gratings on polymer films. *J. Mater. Chem.*, 16:370–375, 2006. doi: 10.1039/B512104C. URL http://dx.doi.org/10.1039/B512104C.

[185] Deying Xia, Ying-Bing Jiang, Xiang He, and S. R. J. Brueck. Titania nanostructure arrays from lithographically defined templates. *Applied Physics Letters*, 97(22):223106, 2010. doi: http://dx.doi.org/10.1063/1.3521462. URL http://scitation.aip.org/content/aip/journal/apl/97/22/10.1063/1.3521462.

[186] Zhen Xu, Min Yin, Jing Sun, Guqiao Ding, Linfeng Lu, Paichun Chang, Xiaoyuan Chen, and Dongdong Li. 3d periodic multiscale tio$_2$ architecture: a platform decorated with graphene quantum dots for enhanced photoelectrochemical water splitting. *Nanotechnology*, 27(11):115401, 2016. URL http://stacks.iop.org/0957-4484/27/i=11/a=115401.

[187] Alexander E. Krasnok, Andrey E. Miroshnichenko, Pavel A. Belov, and Yuri S. Kivshar. All-dielectric optical nanoantennas. *Opt. Express*, 20(18):20599–20604, Aug 2012. doi: 10.1364/OE.20.020599. URL http://www.opticsexpress.org/abstract.cfm?URI=oe-20-18-20599.

[188] K. G. Lee, X. W. Chen, H. Eghlidi, P. Kukura, R. Lettow, A. Renn, V. Sandoghdar, and S. Gotzinger. A planar dielectric antenna for directional single-photon emission and near-unity collection efficiency. *Nat Photon*, 5(3):166–169, 2011. doi: 10.1038/nphoton.2010.312. URL http://www.nature.com/nphoton/journal/v5/n3/abs/nphoton.2010.312.html#supplementary-information.

[189] W.-Q. Han, L.J. Wu, R.F. Klie, and Y.M. Zhu. Enhanced optical absorption induced by dense nanocavities inside titania nanorods. *Advanced Materials*, 19(18):2525–2529, 2007. ISSN 1521-4095. doi: 10.1002/adma.200700540. URL http://dx.doi.org/10.1002/adma.200700540.

[190] Quanjun Li, Jingwei Zhang, Bingbing Liu, Ming Li, Ran Liu, Xianglin Li, Honglei Ma, Shidan Yu, Lin Wang, Yonggang Zou, Zepeng Li, Bo Zou, Tian Cui, and Guangtian Zou. Synthesis of high-density nanocavities inside tio$_2$-b nanoribbons and their enhanced electrochemical lithium storage properties. *Inorganic Chemistry*, 47(21):9870–9873, 2008. doi: 10.1021/ic800758d. URL http://dx.doi.org/10.1021/ic800758d. PMID: 18837547.

[191] D. Praveen Kumar, N. Lakshmana Reddy, M. Mamatha Kumari, B. Srinivas, V. Durga Kumari, B. Sreedhar, V. Roddatis, O. Bondarchuk, M. Karthik, B. Neppolian, and M.V. Shankar. cu$_2$o-sensitized tio$_2$ nanorods with nanocavities for highly efficient photocatalytic hydrogen production under solar irradiation. *Solar Energy Materials and Solar Cells*, 136:157 – 166, 2015. ISSN 0927-0248. doi: http://dx.doi.org/10.1016/j.solmat.2015.01.009. URL http://www.sciencedirect.com/science/article/pii/S0927024815000136.

[192] Hong Zhu, Jie Tao, and Xiang Dong. Preparation and photoelectrochemical activity of cr-doped tio$_2$ nanorods with nanocavities. *The Journal of Physical Chemistry C*, 114(7):2873–2879, 2010. doi: 10.1021/jp9085987. URL http://dx.doi.org/10.1021/jp9085987.

[193] Zhimin Ren, Chao Chen, Rong Hu, Kaiguang Mai, Guodong Qian, and Zhiyu Wang. Two-step self-assembly and lyotropic liquid crystal behavior of tio$_2$ nanorods. *J. Nanomaterials*, 2012:71:71–71:71, January 2012. ISSN 1687-4110. doi: 10.1155/2012/180989. URL http://dx.doi.org/10.1155/2012/180989.

[194] Arnaud Dessombz, Claude R. Pasquier, Patrick Davidson, and Corinne Chanéac. Evidence for photoconductivity anisotropy in aligned tio$_2$ nanorod films. *The Journal of Physical Chemistry C*, 114(46):19799–19802, 2010. doi: 10.1021/jp1063275. URL http://dx.doi.org/10.1021/jp1063275.

[195] Arnaud Dessombz, David Chiche, Patrick Davidson, Pierre Panine, Corinne Chanéac, and Jean-Pierre Jolivet. Design of liquid-crystalline aqueous suspensions of rutile nanorods: Evidence of anisotropic photocatalytic properties. *Journal of the American Chemical Society*, 129(18):5904–5909, 2007. doi: 10.1021/ja0684491. URL http://dx.doi.org/10.1021/ja0684491. PMID: 17429969.

[196] Terukazu Kosako, Yutaka Kadoya, and Holger F. Hofmann. Directional control of light by a nano-optical yagi-uda antenna. *Nat Photon*, 4(5):312–315, May 2010. ISSN 1749-4885. doi: 10.1038/nphoton.2010.34. URL http://dx.doi.org/10.1038/nphoton.2010.34.

[197] F. A. Grant. Properties of rutile (titanium dioxide). *Rev. Mod. Phys.*, 31:646–674, Jul 1959. doi: 10.1103/RevModPhys.31.646. URL http://link.aps.org/doi/10.1103/RevModPhys.31.646.

[198] C. J. Kevane. Oxygen vacancies and electrical conduction in metal oxides. *Phys. Rev.*, 133:A1431–A1436, Mar 1964. doi: 10.1103/PhysRev.133.A1431. URL http://link.aps.org/doi/10.1103/PhysRev.133.A1431.

[199] P. Kofstad. Note on the defect structure of rutile (tio$_2$). *Journal of the Less Common Metals*, 13(6):635 – 638, 1967. ISSN 0022-5088. doi: http://dx.doi.org/10.1016/0022-5088(67)90111-7. URL http://www.sciencedirect.com/science/article/pii/0022508867901117.

[200] Akira Ohmori, Kyeung-Chae Park, Masayuki Inuzuka, Yoshiaki Arata, Katsunori Inoue, and Nobuya Iwamoto. Electrical conductivity of plasma-sprayed titanium oxide (rutile) coatings. *Thin Solid Films*, 201(1):1 – 8, 1991. ISSN 0040-6090. doi: http://dx.doi.org/10.1016/0040-6090(91)90149-R. URL http://www.sciencedirect.com/science/article/pii/004060909190149R.

[201] Simon Springer. *Free carriers in nanocrystalline titanium dioxide thin films*. PhD thesis, Université de Bâle, 2004.

[202] Eiichi Yagi, Ryukiti R. Hasiguti, and Masakazu Aono. Electronic conduction above 4 k of slightly reduced oxygen-deficient rutile tio$_{2-x}$. *Phys. Rev. B*, 54:7945–7956, Sep 1996. doi: 10.1103/PhysRevB.54.7945. URL http://link.aps.org/doi/10.1103/PhysRevB.54.7945.

[203] Hua-Xi Zhang, Ming Zhao, and Qing Jiang. Effect of oxygen vacancies on electronic structures and field emission properties of tio$_2$ nanotubes: A density-functional theory investigation. *Applied Physics Letters*, 103(2):023111, 2013. doi: 10.1063/1.4813546. URL http://dx.doi.org/10.1063/1.4813546.

[204] S.G. Park, B.M. Clemens, Y. Nishi, and B. Magyari-Köpe. *Resistance Switching Mechanism in TiO$_2$*. Stanford University, Department of Materials Science and Engineering, 2011. URL https://books.google.de/books?id=G3Fbsa9X2gUC.

[205] Y. Lin. *Resistive Switching in TiO$_2$ Thin Films*. Schriften des Forschungszentrums Jülich / Reihe Information: Reihe Information. Forschungszentrum Jülich, Zentralbibliothek, 2011. ISBN 9783893367078. URL https://books.google.de/books?id=0kqkFMvVAG8C.

[206] T. Ihara, M. Miyoshi, M. Ando, S. Sugihara, and Y. Iriyama. Preparation of a visible-light-active tio$_2$ photocatalyst by rf plasma treatment. *Journal of Materials Science*, 36(17):4201–4207, 2001. ISSN 1573-4803. doi: 10.1023/A:1017929207882. URL http://dx.doi.org/10.1023/A:1017929207882.

[207] Daniel S. Choi, Daniel Y. Joh, Thomas Lee, Marissa Milchak, Hebing Zhou, Yongkoo Kang, and Jong in Hahm. Position- and orientation-controlled polarized light interaction of individual indium tin oxide nanorods. *Applied Physics Letters*, 104(8):083112, 2014. doi: 10.1063/1.4866794. URL http://dx.doi.org/10.1063/1.4866794.

[208] Xuqiang Zhang, Chengwei Wang, Jianbiao Chen, Weidong Zhu, Aizhen Liao, Yan Li, Jian Wang, and Li Ma. Enhancement of the field emission from the tio$_2$ nanotube arrays by reducing in a nabh$_4$ solution. *ACS Applied Materials & Interfaces*, 6(23):20625–20633, 2014. doi: 10.1021/am503379y. URL http://dx.doi.org/10.1021/am503379y. PMID: 25408536.

[209] Keu Hong Kim, Eung Ju Oh, and Jae Shi Choi. Electrical conductivity of "hydrogen-reduced" titanium dioxide (rutile). *Journal of Physics and Chemistry of Solids*, 45(11):1265 – 1269, 1984. ISSN 0022-3697. doi: http://dx.doi.org/10.1016/0022-3697(84)90026-X. URL http://www.sciencedirect.com/science/article/pii/002236978490026X.

[210] James G. Highfield and Michael Graetzel. Discovery of reversible photochromism in titanium dioxide using photoacoustic spectroscopy: implications for the investigation of light-induced charge-separation and surface redox processes in titanium dioxide. *The Journal of Physical Chemistry*, 92(2):464–467, 1988. doi: 10.1021/j100313a043. URL http://dx.doi.org/10.1021/j100313a043.

[211] Rong Wang, Nobuyuki Sakai, Akira Fujishima, Toshiya Watanabe, and Kazuhito Hashimoto. Studies of surface wettability conversion on tio$_2$ single-crystal surfaces. *The Journal of Physical Chemistry B*, 103(12):2188–2194, 1999. doi: 10.1021/jp983386x. URL http://dx.doi.org/10.1021/jp983386x.

[212] Dmitrii V. Sidelev, Yuriy N. Yurjev, Valeriy P. Krivobokov, Evgenii V. Erofeev, Olga V. Penkova, and Vadim A. Novikov. The oxygen-deficient tio$_2$ films deposited by a dual magnetron sputtering. *Vacuum*, 134:29 – 32, 2016. ISSN 0042-207X. doi: http://dx.doi.org/10.1016/j.vacuum.2016.09.007. URL http://www.sciencedirect.com/science/article/pii/S0042207X16302317.

[213] U. Diebold. Structure and properties of tio$_2$ surfaces: a brief review. *Applied Physics A*, 76(5):681–687, 2003. ISSN 1432-0630. doi: 10.1007/s00339-002-2004-5. URL http://dx.doi.org/10.1007/s00339-002-2004-5.

[214] Alena Folger, Julian Kalb, Lukas Schmidt-Mende, and Christina Scheu. Tuning the electronic conductivity in hydrothermally grown rutile tio$_2$ nanowires: Effect of heat treatment in different environments. *Nanomaterials*, 7(10):289, 2017. doi: 10.3390/nano7100289. URL

[215] H. Akinaga and H. Shima. Resistive random access memory (reram) based on metal oxides. *Proceedings of the IEEE*, 98(12):2237–2251, Dec 2010. ISSN 0018-9219. doi: 10.1109/JPROC.2010.2070830.

[216] Akihito Sawa. Resistive switching in transition metal oxides. *Materials Today*, 11(6):28 – 36, 2008. ISSN 1369-7021. doi: https://doi.org/10.1016/S1369-7021(08)70119-6. URL http://www.sciencedirect.com/science/article/pii/S1369702108701196.

[217] Rainer Waser, Regina Dittmann, Georgi Staikov, and Kristof Szot. Redox-based resistive switching memories - nanoionic mechanisms, prospects, and challenges. *Advanced Materials*, 21(25-26):2632–2663, 2009. ISSN 1521-4095. doi: 10.1002/adma.200900375. URL http://dx.doi.org/10.1002/adma.200900375.

[218] Fu-Chien Chiu, Peng-Wei Li, and Wen-Yuan Chang. Reliability characteristics and conduction mechanisms in resistive switching memory devices using zno thin films. *Nanoscale Research Letters*, 7(1):178, 2012. ISSN 1556-276X. doi: 10.1186/1556-276X-7-178. URL http://dx.doi.org/10.1186/1556-276X-7-178.

[219] J. Joshua Yang, Matthew D. Pickett, Xuema Li, Ohlberg Douglas A. A., Duncan R. Stewart, and R. Stanley Williams. Memristive switching mechanism for metal/oxide/metal nanodevices. *Nature nanotechnology*, 3(7):429–433, 2008. doi: 10.1038/nnano.2008.160. URL http://dx.doi.org/10.1038/nnano.2008.160.

[220] B. J. Choi, D. S. Jeong, S. K. Kim, C. Rohde, S. Choi, J. H. Oh, H. J. Kim, C. S. Hwang, K. Szot, R. Waser, B. Reichenberg, and S. Tiedke. Resistive switching mechanism of Tio2 thin films grown by atomic-layer deposition. *Journal of Applied Physics*, 98(3):033715, 2005. doi: 10.1063/1.2001146. URL http://dx.doi.org/10.1063/1.2001146.

[221] B. Gao, B. Sun, H. Zhang, L. Liu, X. Liu, R. Han, J. Kang, and B. Yu. Unified physical model of bipolar oxide-based resistive switching memory. *IEEE Electron Device Letters*, 30(12):1326–1328, Dec 2009. ISSN 0741-3106. doi: 10.1109/LED.2009.2032308.

[222] Kyung Min Kim, Byung Joon Choi, Yong Cheol Shin, Seol Choi, and Cheol Seong Hwanga. Anode-interface localized filamentary mechanism in resistive switching of tio2 thin films. *Applied Physics Letters*, 91, 2007. doi: 10.1063/1.2749846.

[223] Deok-Hwang Kwon, Kyung Min Kim, Jae Hyuck Jang, Jong Myeong Jeon, Min Hwan Lee, Gun Hwan Kim, Xiang-Shu Li, Gyeong-Su Park, Bora Lee, Seungwu Han, Miyoung Kim, and Cheol Seong Hwang. Atomic structure of conducting nanofilaments in tio2 resistive switching memory. *Nature Nanotechnology*, 5(2):148–153, 2010. doi: 10.1038/nnano.2009.456. URL http://dx.doi.org/10.1038/nnano.2009.456.

[224] Christina Rohde, Byung Joon Choi, Doo Seok Jeong, Seol Choi, Jin-Shi Zhao, and Cheol Seong Hwang. Identification of a determining parameter for resistive switching of tio2 thin films. *Applied Physics Letters*, 86(26):262907, 2005. doi: 10.1063/1.1968416. URL http://dx.doi.org/10.1063/1.1968416.

[225] Elison Matioli, Carl Neufeld, Michael Iza, Samantha C. Cruz, Ali A. Al-Heji, Xu Chen, Robert M. Farrell, Stacia Keller, Steven DenBaars, Umesh Mishra, Shuji Nakamura, James Speck, and Claude Weisbuch. High internal and external quantum efficiency ingan/gan solar cells. *Applied Physics Letters*, 98(2):021102, 2011. doi: 10.1063/1.3540501. URL http://aip.scitation.org/doi/abs/10.1063/1.3540501.

[226] Jagan Singh Meena, Simon Min Sze, Umesh Chand, and Tseung-Yuen Tseng. Overview of emerging nonvolatile memory technologies. *Nanoscale Res Lett*, 9(1):526–526, Sep 2014. ISSN 1931-7573. doi: 10.1186/1556-276X-9-526. URL http://www.ncbi.nlm.nih.gov/pmc/articles/PMC4182445/. 1556-276X-9-526[PII].

[227] Jeong-Mo Hwang. Bringing non-volatile memory blocks to socs using the sonos process. *Simtek Corporation*, 2007.

[228] Krishnaswamy Ramkumar and Venkatraman Prabhakar. Methods to integrate sonos into cmos flow, December 23 2014. URL https://www.google.com/patents/US8916432. US Patent 8,916,432.

[229] Manuel Jesús Gázquez, Juan Pedro Bolívar, Rafael Garcia-Tenorio, and Federico Vaca. A review of the production cycle of titanium dioxide pigment. *Materials Sciences and Applications*, 5:441–458, 2014. doi: 10.4236/msa.2014.57048.

[230] Donald Knittel. *Titanium and Titanium Alloys*, volume 3rd edition, pages 98–130. John Wiley and Sons, Hoboken, 1983.

[231] R.L. Rudnick and S. Gao. 3.01 - composition of the continental crust. In Heinrich D. Holland and Karl K. Turekian, editors, *Treatise on Geochemistry*, pages 1 – 64. Pergamon, Oxford, 2003. ISBN 978-0-08-043751-4. doi: http://dx.doi.org/10.1016/B0-08-043751-6/03016-4. URL http://www.sciencedirect.com/science/article/pii/B0080437516030164.

[232] A. Stwertka. *Guide to the elements*. revised edition. *Oxford University Press, London*, 1998.

[233] V. A. Williams. *Detrital Heavy Mineral Deposit*, pages 1609–1614. Australasian Institute of Mining and Metallurgy, 1990.

[234] Jelks Barksdale. Titanium, its occurrence, chemistry, and technology. *Soil Science, The Roland Press Company, New York*, 70(2nd edition):0038–075X, 1950.

[235] J. Whitehead. *Encyclopaedia of Chemical Technology*, volume 3rd Edition, chapter Titanium Compounds (Inorganic), pages 131–176. John Wiley and Sons, Hoboken, 1983.

[236] Ch. H. Moore. Formation and properties of single crystals of synthetic rutile. *Transactions of the American Institute of Mining and Metallurgical Engineers*, 184:194, 1949.

[237] L. S. Dubrovinsky, N. A. Dubrovinskaia, V. Swamy, J. Muscat, N. M. Harrison, R. Ahuja, B. Holm, and B. Johansson. Materials science: The hardest known oxide. *Nature*, 410(6829):653–654, 2001. doi: 10.1038/35070650. URL http://dx.doi.org/10.1038/35070650.

[238] R. Adams. The world market for tio2 feedstocks. *Minerals Industry International*, pages 9–14, 1994.

[239] J. Gambogi. *Mineral Commodity Summaries*. US Geological Survey, Government Printing Office, Washington DC, 2011.

[240] J. Gambogi. *Mineral Commodity Summaries*. US Geological Survey, Government Printing Office, Washington DC, 2009.

[241] J. Gambogi. *Mineral Commodity Summaries*. US Geological Survey, Government Printing Office, Washington DC, 2010.

[242] J. Kischekewitz, W.D. Griebler, and M. Liedekerke. *White Pigments*, volume 2nd Edition, pages 43–82. Wiley-VCH, Weinheim, 1998.

[243] Thomas P. Battle, Dat Nguyen, and James W. Reeves. *Queneau International Symposium: Extractive Metallurgy of Copper, Nickel and Cobalt*, volume 1, pages 925–943. TMS, Warrendale, 1993.

[244] CA) Borowiec, Krzysztof (Tracy, CA) Grau, Alfonso E. (St-Bruno, CA) Gueguin, Michel (Tracy, and CA) Turgeon, Jean-fran Cedilla Ois (Tracy. Method to upgrade titania slag and resulting product, 11 1998. URL http://www.freepatentsonline.com/5830420.html.

[245] CA) Borowiec, Krzysztof (Tracy, CA) Grau, Alfonso E. (St-Bruno, CA) Gueguin, Michel (Tracy, Turgeon, and CA) Jean-francois (Tracy. Tio_2 containing product including rutile, pseudo-brookite and ilmenite, 3 2003. URL http://www.freepatentsonline.com/6531110.html.

[246] Dimitrios Filippou and Guillaume Hudon. Iron removal and recovery in the titanium dioxide feedstock and pigment industries. *Journal of the Minerals Metals and Materials Society*, 61(10):36–42, 2009. ISSN 1543-1851. doi: 10.1007/s11837-009-0150-3. URL http://dx.doi.org/10.1007/s11837-009-0150-3.

[247] Jürgen H. Braun, Andrejs Baidins, and Robert E. Marganski. Tio_2 pigment technology: a review. *Progress in Organic Coatings*, 20(2): 105 – 138, 1992. ISSN 0300-9440. doi: http://dx.doi.org/10.1016/0033-0655(92)80001-D. URL http://www.sciencedirect.com/science/article/pii/003306559280001D.

[248] G.S. McNulty. Production of titanium dioxide. In *Plenary Lecture. NORM V International Conference, Sevilla, 19-22 March 2007*, pages 169–188, 3 2007.

[249] ICIS. Titanium dioxide (tio2) uses and market data, icb chemical profile. Technical report, ICB Chemical Profile, 2010.

[250] Robert J. Anderson. The metallurgy of titanium. *Journal of the Franklin Institute*, 184(4):469 – 508, 1917. ISSN 0016-0032. doi: http://dx.doi.org/10.1016/S0016-0032(17)90337-0. URL http://www.sciencedirect.com/science/article/pii/S0016003217903370.

[251] René Marchand, Luc Brohan, and Michel Tournoux. Tio_2(b) a new form of titanium dioxide and the potassium octatitanate $k_2ti_8o_{17}$. *Materials Research Bulletin*, 15(8):1129–1133, 1980. ISSN 0025-5408. doi: http://dx.doi.org/10.1016/0025-5408(80)90076-8. URL http://www.sciencedirect.com/science/article/pii/0025540880900768.

[252] M. Latroche, L. Brohan, R. Marchand, and M. Tournoux. New hollandite oxides: Tio_2(h) and $k^{0.06}tio_2$. *Journal of Solid State Chemistry*, 81(1):78 – 82, 1989. ISSN 0022-4596. doi: http://dx.doi.org/10.1016/0022-4596(89)90204-1. URL http://www.sciencedirect.com/science/article/pii/0022459689902041.

[253] J. Akimoto, Y. Gotoh, Y. Oosawa, N. Nonose, T. Kumagai, K. Aoki, and H. Takei. Topotactic oxidation of ramsdellite-type $li_{0.5}tio_2$, a new polymorph of titanium dioxide: Tio_2 (r). *Journal of Solid State Chemistry*, 113(1):27 – 36, 1994. ISSN 0022-4596. doi: http://dx.doi.org/10.1006/jssc.1994.1337. URL http://www.sciencedirect.com/science/article/pii/S0022459684713375.

[254] P. Y. Simons and F. Dachille. The structure of tio_2ii, a high-pressure phase of tio_2. *Acta Crystallographica*, 23(2):334–336, Aug 1967. doi: 10.1107/S0365110X67002713. URL http://dx.doi.org/10.1107/S0365110X67002713.

[255] H. Sato, S. Endo, M. Sugiyama, T. Kikegawa, O. Shimomura, and K. Kusaba. Baddeleyite-type high-pressure phase of tio_2. *Science*, 251(4995):786–788, 1991. ISSN 0036-8075. doi: 10.1126/science.251.4995.786. URL http://science.sciencemag.org/content/251/4995/786.

[256] Natalia A. Dubrovinskaia, Leonid S. Dubrovinsky, Rajeev Ahuja, Vitaly B. Prokopenko, V. Dmitriev, H.-P. Weber, J. M. Osorio-Guillen, and Börje Johansson. Experimental and theoretical identification of a new high-pressure tio_2 polymorph. *Physical Review Letters*, 87:275501, Dec 2001. doi: 10.1103/PhysRevLett.87.275501. URL http://link.aps.org/doi/10.1103/PhysRevLett.87.275501.

[257] M. Mattesini, J. S. de Almeida, L. Dubrovinsky, N. Dubrovinskaia, B. Johansson, and R. Ahuja. High-pressure and high-temperature synthesis of the cubic Tio_2 polymorph. *Phys. Rev. B*, 70:212101, Dec 2004. doi: 10.1103/PhysRevB.70.212101. URL http://link.aps.org/doi/10.1103/PhysRevB.70.212101.

[258] S. C. Abrahams and J. L. Bernstein. Rutile: Normal probability plot analysis and accurate measurement of crystal structure. *The Journal of Chemical Physics*, 55(7):3206–3211, 1971. doi: http://dx.doi.org/10.1063/1.1676569. URL http://scitation.aip.org/content/aip/journal/jcp/55/7/10.1063/1.1676569.

[259] K. V. Krishna Rao, S. V. Nagender Naidu, and Leela Iyengar. Thermal expansion of rutile and anatase. *Journal of the American Ceramic Society*, 53(3):124–126, 1970. ISSN 1551-2916. doi: 10.1111/j.1151-2916.1970.tb12051.x. URL http://dx.doi.org/10.1111/j.1151-2916.1970.tb12051.x.

[260] Jeremy K. Burdett, Timothy Hughbanks, Gordon J. Miller, James W. Richardson Jr., and Joseph V. Smith. Structural-electronic relationships in inorganic solids: powder neutron diffraction studies of the rutile and anatase polymorphs of titanium dioxide at 15 and 295 k. *Journal of the American Chemical Society*, 109(12):3639–3646, 1987. doi: 10.1021/ja00246a021. URL http://dx.doi.org/10.1021/ja00246a021.

[261] A. Beltrán, L. Gracia, and J. Andrés. Density functional theory study of the brookite surfaces and phase transitions between natural titania polymorphs. *The Journal of Physical Chemistry B*, 110(46):23417–23423, 2006. doi: 10.1021/jp0643000. URL http://dx.doi.org/10.1021/jp0643000. PMID: 17107193.

[262] Joseph Muscat, Varghese Swamy, and Nicholas M. Harrison. First-principles calculations of the phase stability of tio_2. *Phys. Rev. B*, 65:224112, Jun 2002. doi: 10.1103/PhysRevB.65.224112. URL http://link.aps.org/doi/10.1103/PhysRevB.65.224112.

[263] Konstantin I. Hadjiivanov and Dimitar G. Klissurski. Surface chemistry of titania (anatase) and titania-supported catalysts. *Chem. Soc. Rev.*, 25:61–69, 1996. doi: 10.1039/CS9962500061. URL http://dx.doi.org/10.1039/CS9962500061.

[264] P.A. Cox. *Transition Metal Oxides: An Introduction to Their Electronic Structure and Properties*. The International series of monographs on chemistry. OUP Oxford, 2010. ISBN 9780199588947. URL https://books.google.de/books?id=DdcUDAAAQBAJ.

[265] H.H. Kung. *Transition Metal Oxides: Surface Chemistry and Catalysis*. Studies in Surface Science and Catalysis. Elsevier Science, 1989. ISBN 9780080887425. URL https://books.google.de/books?id=S4CBzt6_OSYC.

[266] Oleg V. Krylov. *Catalysis by Nonmetals: Rules for Catalyst Selection*. Academic Press Inc., 1970.

[267] W. H. Baur. Über die verfeinerung der kristallstrukturbestimmung einiger vertreter des rutiltyps: Tio_2, sno_2, geo_2 und mgf_2. *Acta Crystallographica*, 9(6):515–520, Jun 1956. doi: 10.1107/S0365110X56001388. URL http://dx.doi.org/10.1107/S0365110X56001388.

[268] R. Restori, D. Schwarzenbach, and J. R. Schneider. Charge density in rutile, tio_2. *Acta Crystallographica Section B*, 43(3):251–257, Jun 1987. doi: 10.1107/S0108768187097921. URL http://dx.doi.org/10.1107/S0108768187097921.

[269] C. J. Howard, T. M. Sabine, and F. Dickson. Structural and thermal parameters for rutile and anatase. *Acta Crystallographica Section B*, 47(4):462–468, Aug 1991. doi: 10.1107/S010876819100335X. URL http://dx.doi.org/10.1107/S010876819100335X.

345

References

[270] Cristiana Di Valentin, Gianfranco Pacchioni, and Annabella Selloni. Origin of the different photoactivity of N-doped anatase and rutile tio$_2$. *Phys. Rev. B*, 70:085116, Aug 2004. doi: 10.1103/PhysRevB.70.085116. URL http://link.aps.org/doi/10.1103/PhysRevB.70.085116.

[271] Per Kofstad. *Nonstoichiometry, diffusion, and electrical conductivity in binary metal oxides*. Wiley-Interscience New York, 1972.

[272] Hj. Matzke. *Diffusion in nonstoichiometric oxides*, pages 155–230. Academic Press: New York, 1981. doi: 10.1016/B978-0-12-655280-5. X5001-9.

[273] M. K. Nowotny, L. R. Sheppard, T. Bak, and J. Nowotny. Defect chemistry of titanium dioxide. application of defect engineering in processing of tio$_2$-based photocatalysts. *The Journal of Physical Chemistry C*, 112(14):5275–5300, 2008. doi: 10.1021/jp077275m. URL http://dx.doi.org/10.1021/jp077275m.

[274] Soack Dae Yoon, Yajie Chen, Aria Yang, Trevor L Goodrich, Xu Zuo, Dario A Arena, Katherine Ziemer, Carmine Vittoria, and Vincent G Harris. Oxygen-defect-induced magnetism to 880 k in semiconducting anatase tio$_2$-δ films. *Journal of Physics: Condensed Matter*, 18(27):L355, 2006. URL http://stacks.iop.org/0953-8984/18/i=27/a=L01.

[275] L.A. Bursill and M.G. Blanchin. Structure of small oxygen vacancy defects in nonstoichiometric rutile. *Journal of Solid State Chemistry*, 51(3):321 – 335, 1984. ISSN 0022-4596. doi: http://dx.doi.org/10.1016/0022-4596(84)90349-9. URL http://www.sciencedirect.com/science/article/pii/0022459684903499.

[276] J. Nowotny, T. Bak, M.K. Nowotny, and L.R. Sheppard. Titanium dioxide for solar-hydrogen ii. defect chemistry. *International Journal of Hydrogen Energy*, 32(14):2630 – 2643, 2007. ISSN 0360-3199. doi: http://dx.doi.org/10.1016/j.ijhydene.2006.09.005. URL http://www.sciencedirect.com/science/article/pii/S0360319906004319. International Conference on Materials for Hydrogen Energy: Solar Hydrogen (ICMHE 2004).

[277] Bernhard Neumann, Peter Bogdanoff, Helmut Tributsch, Shanmugasundaram Sakthivel, and Horst Kisch. Electrochemical mass spectroscopic and surface photovoltage studies of catalytic water photooxidation by undoped and carbon-doped titania. *The Journal of Physical Chemistry B*, 109(35):16579–16586, 2005. doi: 10.1021/jp051339g. URL http://dx.doi.org/10.1021/jp051339g. PMID: 16853109.

[278] P. Kofstad. Thermogravimetric studies of the defect structure of rutile (tio$_2$). *Journal of Physics and Chemistry of Solids*, 23(11):1579 – 1586, 1962. ISSN 0022-3697. doi: http://dx.doi.org/10.1016/0022-3697(62)90240-8. URL http://www.sciencedirect.com/science/article/pii/0022369762902408.

[279] Katrine Seip Førland. The defect structure of rutile. *Acta Chemica Scandinavica*, 18(5):1267–1275, 1964.

[280] J. B. Moser, R. N. Blumenthal, and D. H. Whitmore. Thermodynamic study of nonstoichiometric rutile (tio$_{2-x}$). *Journal of the American Ceramic Society*, 48(7):384–384, 1965. ISSN 1551-2916. doi: 10.1111/j.1151-2916.1965.tb14769.x. URL http://dx.doi.org/10.1111/j.1151-2916.1965.tb14769.x.

[281] L.M. Atlas and G.J. Schlehman. reported by moser, jb et al. *J. Am. Ceram. Soc*, 48:384, 1965.

[282] C.B. Alcock, S. Zador, and B.C.H. Steele. A thermodynamic study of dilute solutions of defects in the rutile structure tio$_{2-x}$, nbo$_{2-x}$, and tio$_{0.5}$nbo$_{0.5}$o$_{2\pm x}$. In *Proc. Br. Ceram. Soc*, volume 8, page 231, 1967.

[283] Ming-Kwei Lee, Jung-Jie Huang, and Tsung-Shun Wu. Electrical characteristics improvement of oxygen-annealed mocvd-tio$_2$ films. *Semiconductor Science and Technology*, 20(6):519, 2005. URL http://stacks.iop.org/0268-1242/20/i=6/a=007.

[284] M. K. Nowotny, T. Bak, J. Nowotny, and C. C. Sorrell. Titanium vacancies in nonstoichiometric tio$_2$ single crystal. *physica status solidi (b)*, 242(11):R88–R90, 2005. ISSN 1521-3951. doi: 10.1002/pssb.200541186. URL http://dx.doi.org/10.1002/pssb.200541186.

[285] M. K. Nowotny, T. Bak, and J. Nowotny. Electrical properties and defect chemistry of tio$_2$ single crystal. i. electrical conductivity. *The Journal of Physical Chemistry B*, 110(33):16270–16282, 2006. doi: 10.1021/jp0606210. URL http://dx.doi.org/10.1021/jp0606210. PMID: 16913753.

[286] M. K. Nowotny, T. Bak, and J. Nowotny. Electrical properties and defect chemistry of tio$_2$ single crystal. ii. thermoelectric power. *The Journal of Physical Chemistry B*, 110(33):16283–16291, 2006. doi: 10.1021/jp060622s. URL http://dx.doi.org/10.1021/jp060622s. PMID: 16913754.

[287] M. K. Nowotny, T. Bak, and J. Nowotny. Electrical properties and defect chemistry of tio$_2$ single crystal. iii. equilibration kinetics and chemical diffusion. *The Journal of Physical Chemistry B*, 110(33):16292–16301, 2006. doi: 10.1021/jp060623k. URL http://dx.doi.org/10.1021/jp060623k. PMID: 16913755.

[288] M. K. Nowotny, T. Bak, and J. Nowotny. Electrical properties and defect chemistry of tio$_2$ single crystal. iv. prolonged oxidation kinetics and chemical diffusion. *The Journal of Physical Chemistry B*, 110(33):16302–16308, 2006. doi: 10.1021/jp060624c. URL http://dx.doi.org/10.1021/jp060624c. PMID: 16913756.

[289] J. Nowotny, T. Bak, and T. Burg. Electrical properties of polycrystalline tio$_2$ at elevated temperatures. electrical conductivity. *physica status solidi (b)*, 244(6):2037–2054, 2007. ISSN 1521-3951. doi: 10.1002/pssb.200642514. URL http://dx.doi.org/10.1002/pssb.200642514.

[290] J. Nowotny, T. Bak, and T. Burg. Electrical properties of polycrystalline tio$_2$. Equilibration kinetics. *Ionics*, 13(2):71–78, 2007. ISSN 1862-0760. doi: 10.1007/s11581-007-0075-1. URL http://dx.doi.org/10.1007/s11581-007-0075-1.

[291] J. Nowotny, T. Bak, and T. Burg. Electrical properties of polycrystalline tio$_2$. prolonged oxidation kinetics. *Ionics*, 13(2):79–82, 2007. ISSN 1862-0760. doi: 10.1007/s11581-007-0076-0. URL http://dx.doi.org/10.1007/s11581-007-0076-0.

[292] In Sun Cho, Manca Logar, Chi Hwan Lee, Lili Cai, Fritz B. Prinz, and Xiaolin Zheng. Rapid and controllable flame reduction of tio$_2$ nanowires for enhanced solar water-splitting. *Nano Letters*, 14(1):24–31, 2014. doi: 10.1021/nl4026902. URL http://dx.doi.org/10.1021/nl4026902. PMID: 24295287.

[293] J. R. Smith, F. C. Walsh, and R. L. Clarke. Electrodes based on magnéli phase titanium oxides: the properties and applications of ebonex® materials. *Journal of Applied Electrochemistry*, 28(10):1021–1033, 1998. ISSN 1572-8838. doi: 10.1023/A:1003469427858. URL http://dx.doi.org/10.1023/A:1003469427858.

[294] Leandro Liborio and Nicholas Harrison. Thermodynamics of oxygen defective magnéli phases in rutile: A first-principles study. *Phys. Rev. B*, 77:104104, Mar 2008. doi: 10.1103/PhysRevB.77.104104. URL http://link.aps.org/doi/10.1103/PhysRevB.77.104104.

[295] Ulrike Diebold. The surface science of titanium dioxide. *Surface Science Reports*, 48(5–8):53 – 229, 2003. ISSN 0167-5729. doi: http://dx.doi.org/10.1016/S0167-5729(02)00100-0. URL http://www.sciencedirect.com/science/article/pii/S0167572902001000.

[296] Hiroshi Onishi and Yasuhiro Iwasawa. Stm-imaging of formate intermediates adsorbed on a tio$_2$(110) surface. *Chemical Physics Letters*, 226(1):111 – 114, 1994. ISSN 0009-2614. doi: http://dx.doi.org/10.1016/0009-2614(94)00712-8. URL http://www.sciencedirect.com/science/article/pii/0009261494007128.

[297] Ulrike Diebold, J. F. Anderson, Kwok-On Ng, and David Vanderbilt. Evidence for the tunneling site on transition-metal oxides: Tio$_2$(110). *Phys. Rev. Lett.*, 77:1322–1325, Aug 1996. doi: 10.1103/PhysRevLett.77.1322. URL http://link.aps.org/doi/10.1103/PhysRevLett.77.1322.

[298] Madhavan Ramamoorthy, David Vanderbilt, and R. D. King-Smith. First-principles calculations of the energetics of stoichiometric tio$_2$ surfaces. *Phys. Rev. B*, 49:16721–16727, Jun 1994. doi: 10.1103/PhysRevB.49.16721. URL http://link.aps.org/doi/10.1103/PhysRevB.49.16721.

[299] N. M. Harrison, X.-G. Wang, J. Muscat, and M. Scheffler. The influence of soft vibrational modes on our understanding of oxide surface structure. *Faraday Discuss.*, 114:305–312, 1999. doi: 10.1039/A906386B. URL http://dx.doi.org/10.1039/A906386B.

[300] J.-M. Pan, B. L. Maschhoff, U. Diebold, and T. E. Madey. Interaction of water, oxygen, and hydrogen with tio$_2$(110) surfaces having different defect densities. *Journal of Vacuum Science & Technology A*, 10(4):2470–2476, 1992. doi: http://dx.doi.org/10.1116/1.577986. URL http://scitation.aip.org/content/avs/journal/jvsta/10/4/10.1116/1.577986.

[301] Michael A. Henderson. An hreels and tpd study of water on tio$_2$(110): the extent of molecular versus dissociative adsorption. *Surface Science*, 355(1):151 – 166, 1996. ISSN 0039-6028. doi: http://dx.doi.org/10.1016/0039-6028(95)01357-1. URL http://www.sciencedirect.com/science/article/pii/0039602895013571.

[302] R. Schaub, P. Thostrup, N. Lopez, E. Lægsgaard, I. Stensgaard, J. K. Nørskov, and F. Besenbacher. Oxygen vacancies as active sites for water dissociation on rutile tio$_2$(110). *Phys. Rev. Lett.*, 87:266104, Dec 2001. doi: 10.1103/PhysRevLett.87.266104. URL http://link.aps.org/doi/10.1103/PhysRevLett.87.266104.

[303] Shushi Suzuki, Ken-ichi Fukui, Hiroshi Onishi, and Yasuhiro Iwasawa. Hydrogen adatoms on tio$_2$(110)−(1×1) characterized by scanning tunneling microscopy and electron stimulated desorption. *Phys. Rev. Lett.*, 84:2156–2159, Mar 2000. doi: 10.1103/PhysRevLett.84.2156. URL http://link.aps.org/doi/10.1103/PhysRevLett.84.2156.

[304] Philipp Scheiber, Martin Fidler, Olga Dulub, Michael Schmid, Ulrike Diebold, Weiyi Hou, Ulrich Aschauer, and Annabella Selloni. (sub)surface mobility of oxygen vacancies at the tio$_2$ anatase (101) surface. *Phys. Rev. Lett.*, 109:136103, Sep 2012. doi: 10.1103/PhysRevLett.109.136103. URL http://link.aps.org/doi/10.1103/PhysRevLett.109.136103.

[305] M. G. Blanchin, L. A. Bursill, and D. J. Smith. Precipitation phenomena in non-stoichiometric oxides. i. pairs of crystallographic shear planes in reduced rutiles. *Proceedings of the Royal Society of London A: Mathematical, Physical and Engineering Sciences*, 391(1801): 351–372, 1984. ISSN 0080-4630. doi: 10.1098/rspa.1984.0017. URL http://rspa.royalsocietypublishing.org/content/391/1801/351.

[306] Min Li, Wilhelm Hebenstreit, Ulrike Diebold, Alexei M. Tyryshkin, Michael K. Bowman, Glen G. Dunham, and Michael A. Henderson. The influence of the bulk reduction state on the surface structure and morphology of rutile tio$_2$(110) single crystals. *The Journal of Physical Chemistry B*, 104(20):4944–4950, 2000. doi: 10.1021/jp9943272. URL http://dx.doi.org/10.1021/jp9943272.

[307] R. A. Bennett. The re-oxidation of the substoichiometric tio$_2$(110) surface in the presence of crystallographic shear planes. *PhysChemComm*, 3:9–14, 2000. doi: 10.1039/B001938K. URL http://dx.doi.org/10.1039/B001938K.

[308] Hiroshi Onishi and Yasuhiro Iwasawa. Reconstruction of tio$_2$(110) surface: Stm study with atomic-scale resolution. *Surface Science*, 313(1):L783 – L789, 1994. ISSN 0039-6028. doi: http://dx.doi.org/10.1016/0039-6028(94)91146-0. URL http://www.sciencedirect.com/science/article/pii/0039602894911460.

[309] R.E. Tanner, M.R. Castell, and G.A.D. Briggs. High resolution scanning tunnelling microscopy of the rutile tio$_2$(110) surface. *Surface Science*, 412–413:672 – 681, 1998. ISSN 0039-6028. doi: http://dx.doi.org/10.1016/S0039-6028(98)00464-6. URL http://www.sciencedirect.com/science/article/pii/S0039602898004646.

[310] Ulrike Diebold, Jeremiah Lehman, Talib Mahmoud, Markus Kuhn, Georg Leonardelli, Wilhelm Hebenstreit, Michael Schmid, and Peter Varga. Intrinsic defects on a tio$_2$(110)(1×1) surface and their reaction with oxygen: a scanning tunneling microscopy study. *Surface Science*, 411(1-2):137 – 153, 1998. ISSN 0039-6028. doi: http://dx.doi.org/10.1016/S0039-6028(98)00356-2. URL http://www.sciencedirect.com/science/article/pii/S0039602898003562.

[311] Hiroshi Onishi and Yasuhiro Iwasawa. Dynamic visualization of a metal-oxide-surface/gas-phase reaction: Time-resolved observation by scanning tunneling microscopy at 800 k. *Phys. Rev. Lett.*, 76:791–794, Jan 1996. doi: 10.1103/PhysRevLett.76.791. URL http://link.aps.org/doi/10.1103/PhysRevLett.76.791.

[312] Min Li, Wilhelm Hebenstreit, Leo Gross, Ulrike Diebold, M.A. Henderson, D.R. Jennison, P.A. Schultz, and M.P. Sears. Oxygen-induced restructuring of the tio$_2$(110) surface: a comprehensive study. *Surface Science*, 437(1-2):173 – 190, 1999. ISSN 0039-6028. doi: http://dx.doi.org/10.1016/S0039-6028(99)00720-7. URL http://www.sciencedirect.com/science/article/pii/S0039602899007207.

[313] M Li, W Hebenstreit, and U Diebold. Oxygen-induced restructuring of the rutile tio$_2$(110)(1 × 1) surface. *Surface Science*, 414 (1-2):L951 – L956, 1998. ISSN 0039-6028. doi: http://dx.doi.org/10.1016/S0039-6028(98)00549-4. URL http://www.sciencedirect.com/science/article/pii/S0039602898005494.

[314] Min Li, Wilhelm Hebenstreit, Ulrike Diebold, Michael A. Henderson, and Dwight R. Jennison. Oxygen-induced restructuring of rutile tio$_2$(110): formation mechanism, atomic models, and influence on surface chemistry. *Faraday Discuss.*, 114:245–258, 1999. doi: 10.1039/A903598D. URL http://dx.doi.org/10.1039/A903598D.

[315] Min Li, Wilhelm Hebenstreit, Ulrike Diebold, Alexei M. Tyryshkin, Michael K. Bowman, Glen G. Dunham, and Michael A. Henderson. The influence of the bulk reduction state on the surface structure and morphology of rutile tio$_2$(110) single crystals. *The Journal of Physical Chemistry B*, 104(20):4944–4950, 2000. doi: 10.1021/jp9943272. URL http://dx.doi.org/10.1021/jp9943272.

[316] R. Lindsay, B. G. Daniels, and G. Thornton. *Geometry of adsorbates on metal oxide surfaces*, volume 9, chapter 5, page 199. Elsevier Science & Technology, Amsterdam, 2001. ISBN 9780080538310.

[317] A.B. Sherrill and M.A. Barteau. Chapter 10 - principles of reactivity from studies of organic reactions on model oxide surfaces. In D.P. Woodruff, editor, *Oxide Surfaces*, volume 9 of *The Chemical Physics of Solid Surfaces*, pages 409 – 442. Elsevier, 2001. doi: http://dx.doi.org/10.1016/S1571-0785(01)80030-4. URL http://www.sciencedirect.com/science/article/pii/S1571078501800304.

[318] Mark A. Barteau. Organic reactions at well-defined oxide surfaces. *Chemical Reviews*, 96(4):1413–1430, 1996. doi: 10.1021/cr950222t. URL http://dx.doi.org/10.1021/cr950222t. PMID: 11848796.

[319] Mark A Barteau. New catalysis from metal oxide surface science. *Studies in Surface Science and Catalysis*, 130:105–114, 2000.

References

[320] Michael A. Henderson. The interaction of water with solid surfaces: fundamental aspects revisited. *Surface Science Reports*, 46(1–8): 1 – 308, 2002. ISSN 0167-5729. doi: http://dx.doi.org/10.1016/S0167-5729(01)00020-6. URL http://www.sciencedirect.com/science/article/pii/S0167572901000206.

[321] David Brinkley, Michelle Dietrich, Thomas Engel, Paul Farrall, Gerhard Gantner, Adam Schafer, and Amy Szuchmacher. A modulated molecular beam study of the extent of h₂o dissociation on tio₂(110). *Surface Science*, 395(2):292 – 306, 1998. ISSN 0039-6028. doi: http://dx.doi.org/10.1016/S0039-6028(97)00633-X. URL http://www.sciencedirect.com/science/article/pii/S003960289700633X.

[322] I. M. Brookes, C. A. Muryn, and G. Thornton. Imaging water dissociation on tio₂(110). *Phys. Rev. Lett.*, 87:266103, Dec 2001. doi: 10.1103/PhysRevLett.87.266103. URL http://link.aps.org/doi/10.1103/PhysRevLett.87.266103.

[323] Elisa L. Romàn, JoséL. de Segovia, Richard L. Kurtz, Roger Stockbauer, and Theodore E. Madey. Ups synchrotron radiation studies of nh3 adsorption on tio₂(110). *Surface Science*, 273(1):40 – 46, 1992. ISSN 0039-6028. doi: http://dx.doi.org/10.1016/0039-6028(92)90274-A. URL http://www.sciencedirect.com/science/article/pii/003960289290274A.

[324] Guangquan Lu, Amy Linsebigler, and John T. Yates. The photochemical identification of two chemisorption states for molecular oxygen on tio₂(110). *The Journal of Chemical Physics*, 102(7):3005–3008, 1995. doi: http://dx.doi.org/10.1063/1.468609. URL http://scitation.aip.org/content/aip/journal/jcp/102/7/10.1063/1.468609.

[325] Philipp Scheiber, Alexander Riss, Michael Schmid, Peter Varga, and Ulrike Diebold. Observation and destruction of an elusive adsorbate with stm: o₂/tio₂(110). *Phys. Rev. Lett.*, 105:216101, Nov 2010. doi: 10.1103/PhysRevLett.105.216101. URL http://link.aps.org/doi/10.1103/PhysRevLett.105.216101.

[326] Ulrike Diebold, Min Li, Olaga Dulub, Eleonore L. D. Hebenstreit, and Wilhelm Hebenstreit. The relationship between bulk and surface properties of rutile tio₂(110). *Surface Review and Letters*, 07(05n06):613–617, 2000. doi: 10.1142/S0218625X0000052X. URL http://www.worldscientific.com/doi/abs/10.1142/S0218625X0000052X.

[327] E.L.D. Hebenstreit, W. Hebenstreit, H. Geisler, C.A. Ventrice Jr, P.T. Sprunger, and U. Diebold. Bulk-defect dependent adsorption on a metal oxide surface: S/tio₂(110). *Surface Science*, 486(3):467 – 474, 2001. ISSN 0039-6028. doi: http://dx.doi.org/10.1016/S0039-6028(01)01067-6. URL http://www.sciencedirect.com/science/article/pii/S0039602801010676.

[328] H. Nörenberg and J. H. Harding. Ordered structures of calcium oxide on tio₂(110) studied by stm and atomistic simulation. *Phys. Rev. B*, 59:9842–9845, Apr 1999. doi: 10.1103/PhysRevB.59.9842. URL http://link.aps.org/doi/10.1103/PhysRevB.59.9842.

[329] Charles T. Campbell. Ultrathin metal films and particles on oxide surfaces: structural, electronic and chemisorptive properties. *Surface Science Reports*, 27(1):1 – 111, 1997. ISSN 0167-5729. doi: http://dx.doi.org/10.1016/S0167-5729(96)00011-8. URL http://www.sciencedirect.com/science/article/pii/S0167572996000118.

[330] Ulrike Diebold, Jian-Mei Pan, and Theodore E. Madey. Proceedings of the 14th european conference on surface science ultrathin metal film growth on tio₂(110): an overview. *Surface Science*, 331:845 – 854, 1995. ISSN 0039-6028. doi: http://dx.doi.org/10.1016/0039-6028(95)00124-7. URL http://www.sciencedirect.com/science/article/pii/003960289500124Y.

[331] Rajendra Persaud and Theodore E. Madey. *Growth, structure and reactivity of ultrathin metal films on TiO₂ surfaces*, volume 8, pages 407–447. Elsevier, Amsterdam, 1997.

[332] Jian-Mei Pan and Theodore E. Madey. The encapsulation of fe on tio₂(110). *Catalysis Letters*, 20(3):269–274, 1993. ISSN 1572-879X. doi: 10.1007/BF00769299. URL http://dx.doi.org/10.1007/BF00769299.

[333] François Pesty, Hans-Peter Steinrück, and Theodore E. Madey. Thermal stability of pt films on tio₂(110): evidence for encapsulation. *Surface Science*, 339(1):83 – 95, 1995. ISSN 0039-6028. doi: http://dx.doi.org/10.1016/0039-6028(95)00605-2. URL http://www.sciencedirect.com/science/article/pii/0039602895006052.

[334] R.A. Bennett, P. Stone, and M. Bowker. Pd nanoparticle enhanced re-oxidation of non-stoichiometric tio₂: Stm imaging of spillover and a new form of smsi. *Catalysis Letters*, 59(2):99–105, 1999. ISSN 1572-879X. doi: 10.1023/A:1019053512230. URL http://dx.doi.org/10.1023/A:1019053512230.

[335] Olga Dulub, Wilhelm Hebenstreit, and Ulrike Diebold. Imaging cluster surfaces with atomic resolution: The strong metal-support interaction state of pt supported on tio₂(110). *Phys. Rev. Lett.*, 84:3646–3649, Apr 2000. doi: 10.1103/PhysRevLett.84.3646. URL http://link.aps.org/doi/10.1103/PhysRevLett.84.3646.

[336] D.R Jennison, O. Dulub, W. Hebenstreit, and U. Diebold. Structure of an ultrathin tio$_x$ film, formed by the strong metal support interaction (smsi), on pt nanocrystals on tio₂(1 1 0). *Surface Science*, 492(1–2):L677 – L687, 2001. ISSN 0039-6028. doi: http://dx.doi.org/10.1016/S0039-6028(01)01460-1. URL http://www.sciencedirect.com/science/article/pii/S0039602801014601.

[337] M. Valden, X. Lai, and D. W. Goodman. Onset of catalytic activity of gold clusters on titania with the appearance of nonmetallic properties. *Science*, 281(5383):1647–1650, 1998. ISSN 0036-8075. doi: 10.1126/science.281.5383.1647. URL http://science.sciencemag.org/content/281/5383/1647.

[338] Victor A. Bondzie, Stephen C. Parker, and Charles T. Campbell. The kinetics of co oxidation by adsorbed oxygen on well-defined gold particles on tio₂(110). *Catalysis Letters*, 63(3):143–151, 1999. ISSN 1572-879X. doi: 10.1023/A:1019012903936. URL http://dx.doi.org/10.1023/A:1019012903936.

[339] Lei Zhang, Frederic Cosandey, Rajendra Persaud, and Theodore E. Madey. Initial growth and morphology of thin au films on tio₂(110). *Surface Science*, 439(1–3):73 – 85, 1999. ISSN 0039-6028. doi: http://dx.doi.org/10.1016/S0039-6028(99)00734-7. URL http://www.sciencedirect.com/science/article/pii/S0039602899007347.

[340] F. Cosandey and T. E. Madey. Growth, morphology, interfacial effects and catalytic properties of au on tio₂. *Surface Review and Letters*, 08(01n02):73–93, 2001. doi: 10.1142/S0218625X01000884. URL http://www.worldscientific.com/doi/abs/10.1142/S0218625X01000884.

[341] J. G. Traylor, H. G. Smith, R. M. Nicklow, and M. K. Wilkinson. Lattice dynamics of rutile. *Phys. Rev. B*, 3:3457–3472, May 1971. doi: 10.1103/PhysRevB.3.3457. URL http://link.aps.org/doi/10.1103/PhysRevB.3.3457.

[342] P. S. Narayanan. Raman spectrum of rutile (tio₂). *Proceedings of the Indian Academy of Sciences - Section A*, 32(4):279, 1950. ISSN 0370-0089. doi: 10.1007/BF03170832. URL http://dx.doi.org/10.1007/BF03170832.

[343] Zhi-Gang Mei, Yi Wang, Shun-Li Shang, and Zi-Kui Liu. First-principles study of lattice dynamics and thermodynamics of tio₂ polymorphs. *Inorganic Chemistry*, 50(15):6996–7003, 2011. doi: 10.1021/ic200349p. URL http://dx.doi.org/10.1021/ic200349p. PMID: 21714527.

[344] Toshiaki Ohsaka, Fujio Izumi, and Yoshinori Fujiki. Raman spectrum of anatase, tio_2. *Journal of Raman Spectroscopy*, 7(6):321–324, 1978. ISSN 1097-4555. doi: 10.1002/jrs.1250070606. URL http://dx.doi.org/10.1002/jrs.1250070606.

[345] S. P. S. Porto, P. A. Fleury, and T. C. Damen. Raman spectra of tio_2, mgf_2, znf_2, fef_2, and mnf_2. *Phys. Rev.*, 154:522–526, Feb 1967. doi: 10.1103/PhysRev.154.522. URL http://link.aps.org/doi/10.1103/PhysRev.154.522.

[346] D.M. Eagles. Polar modes of lattice vibration and polaron coupling constants in rutile (tio_2). *Journal of Physics and Chemistry of Solids*, 25(11):1243 – 1251, 1964. ISSN 0022-3697. doi: http://dx.doi.org/10.1016/0022-3697(64)90022-8. URL http://www.sciencedirect.com/science/article/pii/0022369764900228.

[347] W. Cochran. Dielectric constants and lattice vibrations of cubic ionic crystals. *Zeitschrift für Kristallographie - Crystalline Materials*, 112(1-6):465–471, 1959. doi: 10.1524/zkri.1959.112.jg.465. URL //www.degruyter.com/view/j/zkri.1959.112.issue-1-6/zkri.1959.112.jg.465/zkri.1959.112.jg.465.xml.

[348] D.C. Cronemeyer. in massachusetts institute of technology laboratory for insulation research report no. 46. p. 19, 1951.

[349] Rebecca A. Parker. Static dielectric constant of rutile (tio_2), 1.6–1060 k. *Phys. Rev.*, 124:1719–1722, Dec 1961. doi: 10.1103/PhysRev.124.1719. URL http://link.aps.org/doi/10.1103/PhysRev.124.1719.

[350] W. Cochran. Crystal stability and the theory of ferroelectricity. *Advances in Physics*, 9(36):387–423, 1960. doi: 10.1080/00018736000101229. URL http://dx.doi.org/10.1080/00018736000101229.

[351] B Montanari and N.M Harrison. Lattice dynamics of tio_2 rutile: influence of gradient corrections in density functional calculations. *Chemical Physics Letters*, 364(5–6):528 – 534, 2002. ISSN 0009-2614. doi: http://dx.doi.org/10.1016/S0009-2614(02)01401-X. URL http://www.sciencedirect.com/science/article/pii/S000926140201401X.

[352] Pavlin D. Mitev, Kersti Hermansson, Barbara Montanari, and Keith Refson. Soft modes in strained and unstrained rutile tio_2. *Phys. Rev. B*, 81:134303, Apr 2010. doi: 10.1103/PhysRevB.81.134303. URL http://link.aps.org/doi/10.1103/PhysRevB.81.134303.

[353] Weiguang Yang, Yanhao Yu, Matthew B. Starr, Xin Yin, Zhaodong Li, Alexander Kvit, Shifa Wang, Ping Zhao, and Xudong Wang. Ferroelectric polarization-enhanced photoelectrochemical water splitting in tio_2–$batio_3$ core–shell nanowire photoanodes. *Nano Letters*, 15(11):7574–7580, 2015. doi: 10.1021/acs.nanolett.5b03988. URL http://dx.doi.org/10.1021/acs.nanolett.5b03988. PMID: 26492362.

[354] S. Schöche, T. Hofmann, R. Korlacki, T. E. Tiwald, and M. Schubert. Infrared dielectric anisotropy and phonon modes of rutile tio_2. *Journal of Applied Physics*, 113(16):164102, 2013. doi: http://dx.doi.org/10.1063/1.4802715. URL http://scitation.aip.org/content/aip/journal/jap/113/16/10.1063/1.4802715.

[355] G. A. Samara and P. S. Peercy. Pressure and temperature dependence of the static dielectric constants and raman spectra of tio_2 (rutile). *Phys. Rev. B*, 7:1131–1148, Feb 1973. doi: 10.1103/PhysRevB.7.1131. URL http://link.aps.org/doi/10.1103/PhysRevB.7.1131.

[356] R. J. Gonzalez, R. Zallen, and H. Berger. Infrared reflectivity and lattice fundamentals in anatase tio_2s. *Phys. Rev. B*, 55:7014–7017, Mar 1997. doi: 10.1103/PhysRevB.55.7014. URL http://link.aps.org/doi/10.1103/PhysRevB.55.7014.

[357] D. R. Lide. Crc handbook of chemistry and physics. *CRC Press, Boca Raton*, 2000.

[358] Masayoshi Mikami, Shinichiro Nakamura, Osamu Kitao, and Hironori Arakawa. Lattice dynamics and dielectric properties of tio_2 anatase: A first-principles study. *Phys. Rev. B*, 66:155213, Oct 2002. doi: 10.1103/PhysRevB.66.155213. URL http://link.aps.org/doi/10.1103/PhysRevB.66.155213.

[359] Roman Sikora. Ab initio study of phonons in the rutile structure of tio_2. *Journal of Physics and Chemistry of Solids*, 66(6):1069 – 1073, 2005. ISSN 0022-3697. doi: http://dx.doi.org/10.1016/j.jpcs.2005.01.007. URL http://www.sciencedirect.com/science/article/pii/S0022369705000430.

[360] H. N. Pandey. The theoretical elastic constants and specific heats of rutile. *physica status solidi (b)*, 11(2):743–751, 1965. ISSN 1521-3951. doi: 10.1002/pssb.19650110226. URL http://dx.doi.org/10.1002/pssb.19650110226.

[361] M M Elcombe and A W Pryor. The lattice dynamics of calcium fluoride. *Journal of Physics C: Solid State Physics*, 3(3):492, 1970. URL http://stacks.iop.org/0022-3719/3/i=3/a=002.

[362] B.T.M. Willis and A.W. Pryor. Thermal vibrations in crystallography. *Cambridge University Press*, page 168, 1975.

[363] Jr. Chase, M.W. Nist-janaf themochemical tables, fourth edition. *J. Phys. Chem. Ref. Data, Monograph 9*, 1998.

[364] D. de Ligny, P. Richet, F. E. Westrum Jr, and J. Roux. Heat capacity and entropy of rutile (tio_2) and nepheline ($naalsio_4$). *Physics and Chemistry of Minerals*, 29(4):267–272, 2002. ISSN 1432-2021. doi: 10.1007/s00269-001-0229-z. URL http://dx.doi.org/10.1007/s00269-001-0229-z.

[365] Masaki Akaogi, Hitoshi Yusa, Kimiko Shiraishi, and Toshihiro Suzuki. Thermodynamic properties of α-quartz, coesite, and stishovite and equilibrium phase relations at high pressures and high temperatures. *Journal of Geophysical Research: Solid Earth*, 100(B11):22337–22347, 1995. ISSN 2156-2202. doi: 10.1029/95JB02395. URL http://dx.doi.org/10.1029/95JB02395.

[366] D.A. Ditmars, S. Ishihara, S.S. Chang, G. Bernstein, and E.D. West. Enthalpy and heat-capacity standard reference material: Synthetic sapphire (alpha-al_2o_3) from 10 to 2250 k. *The Journal of Research of the National Institute of Standards and Technology*, 87(2):159, 1982. URL http://www.archive.org/details/jresv87n2p159_A1b.

[367] C. Howard Shomate. Heat capacities at low temperatures of titanium dioxide (rutile and anatase). *Journal of the American Chemical Society*, 69(2):218–219, 1947. doi: 10.1021/ja01194a008. URL http://dx.doi.org/10.1021/ja01194a008.

[368] E Shojaee and M R Mohammadizadeh. First-principles elastic and thermal properties of tio 2 : a phonon approach. *Journal of Physics: Condensed Matter*, 22(1):015401, 2010. URL http://stacks.iop.org/0953-8984/22/i=1/a=015401.

[369] Vilius Palenskis. Drift mobility, diffusion coefficient of randomly moving charge carriers in metals and other materials with degenerated electron gas. *World Journal of Condensed Matter Physics*, 3(1):28333, 2013.

[370] M. Gurfinkel, Hao D. Xiong, K.P. Cheung, J.S. Suehle, J.B. Bernstein, Yoram Shapira, A.J. Lelis, D. Habersat, and N. Goldsman. Characterization of transient gate oxide trapping in sic mosfets using fast i-v techniques. *IEEE Transactions on Electron Devices*, 55(8):2004–2012, August 2008. ISSN 0018-9383. doi: 10.1109/TED.2008.926626.

[371] P N Murgatroyd. Theory of space-charge-limited current enhanced by frenkel effect. *Journal of Physics D: Applied Physics*, 3(2):151, 1970. URL http://stacks.iop.org/0022-3727/3/i=2/a=308.

[372] Jacob M. Schliesser, Stacey J. Smith, Guangshi Li, Liping Li, Trent F. Walker, Thomas Parry, Juliana Boerio-Goates, and Brian F. Woodfield. Heat capacity and thermodynamic functions of nano- Tio$_2$ rutile in relation to bulk- Tio$_2$ rutile. *The Journal of Chemical Thermodynamics*, 81(Supplement C):311 – 322, 2015. ISSN 0021-9614. doi: https://doi.org/10.1016/j.jct.2014.08.002. URL http://www.sciencedirect.com/science/article/pii/S0021961414002481.

[373] S.-M. Lee, David G. Cahill, and Thomas H. Allen. Thermal conductivity of sputtered oxide films. *Phys. Rev. B*, 52:253–257, Jul 1995. doi: 10.1103/PhysRevB.52.253. URL http://link.aps.org/doi/10.1103/PhysRevB.52.253.

[374] YS Touloukian, RW Powell, CY Ho, and MC Nicolaou. Thermophysical properties of matter-the tprc data series. volume 10. thermal diffusivity. Technical report, Thermophysical and electronic properties information analysis center lafayette in, 1974.

[375] C. Y. Ho, R. W. Powell, and P. E. Liley. Thermal conductivity of the elements. *Journal of Physical and Chemical Reference Data*, 1 (2):279–421, 1972. doi: http://dx.doi.org/10.1063/1.3253100. URL http://scitation.aip.org/content/aip/journal/jpcrd/1/2/10.1063/1.3253100.

[376] W. D. Kingery, J. Francl, R. L. Coble, and T. Vasilos. Thermal conductivity: X, data for several pure oxide materials corrected to zero porosity. *Journal of the American Ceramic Society*, 37(2):107–110, 1954. ISSN 1551-2916. doi: 10.1111/j.1551-2916.1954.tb20109.x. URL http://dx.doi.org/10.1111/j.1551-2916.1954.tb20109.x.

[377] Glen A. Slack. Thermal conductivity of pure and impure silicon, silicon carbide, and diamond. *Journal of Applied Physics*, 35(12):3460–3466, 1964. doi: http://dx.doi.org/10.1063/1.1713251. URL http://scitation.aip.org/content/aip/journal/jap/35/12/10.1063/1.1713251.

[378] Charles Kittel. Interpretation of the thermal conductivity of glasses. *Phys. Rev.*, 75:972–974, Mar 1949. doi: 10.1103/PhysRev.75.972. URL http://link.aps.org/doi/10.1103/PhysRev.75.972.

[379] J. W. Anthony, R. A. Bideaux, K. W. Bladh, and M. C. Nichols. *Handbook of Mineralogy, Volume III, Halides, Hydroxides, Oxides*. Mineral Data Publishing, Arizona, 1997.

[380] D. J. Kim, D. S. Kim, S. Cho, S. W. Kim, S. H. Lee, and J. C. Kim. Measurement of thermal conductivity of tio$_2$ thin films using 3ω method. *International Journal of Thermophysics*, 25(1):281–289, Jan 2004. ISSN 1572-9567. doi: 10.1023/B:IJOT.0000022340.65615.22. URL http://doi.org/10.1023/B:IJOT.0000022340.65615.22.

[381] D.J Benford, T.J Powers, and S.H Moseley. Thermal conductivity of kapton tape. *Cryogenics*, 39(1):93 – 95, 1999. ISSN 0011-2275. doi: http://dx.doi.org/10.1016/S0011-2275(98)00125-8. URL http://www.sciencedirect.com/science/article/pii/S0011227598001258.

[382] S A Ishmael, M Slomski, H Luo, M White, A Hunt, N Mandzy, J F Muth, R Nesbit, T Paskova, W Straka, and J Schwartz. Thermal conductivity and dielectric properties of a tio$_2$ -based electrical insulator for use with high temperature superconductor-based magnets. *Superconductor Science and Technology*, 27(9):095018, 2014. URL http://stacks.iop.org/0953-2048/27/i=9/a=095018.

[383] F. Gervais and J.F. Baumard. Lo phonon-plasmon coupling in non-stoichiometric rutile tio$_2$. *Solid State Communications*, 21(8):861 – 865, 1977. ISSN 0038-1098. doi: http://dx.doi.org/10.1016/0038-1098(77)91172-3. URL http://www.sciencedirect.com/science/article/pii/003810987791172a.

[384] Regis J. Betsch, Hong Lee Park, and William B. White. Raman spectra of stoichiometric and defect rutile. *Materials Research Bulletin*, 26(7):613 – 622, 1991. ISSN 0025-5408. doi: http://dx.doi.org/10.1016/0025-5408(91)90104-T. URL http://www.sciencedirect.com/science/article/pii/002554089190104T.

[385] E. A. Akhadov, S. A. Safron, J. G. Skofronick, D. H. Van Winkle, F. A. Flaherty, and Rifat Fatema. Surface lattice dynamics of rutile tio$_2$(110) using helium atom surface scattering. *Phys. Rev. B*, 68:035409, Jul 2003. doi: 10.1103/PhysRevB.68.035409. URL http://link.aps.org/doi/10.1103/PhysRevB.68.035409.

[386] D Wang, J Zhao, B Chen, and C Zhu. Lattice vibration fundamentals in nanocrystalline anatase investigated with raman scattering. *Journal of Physics: Condensed Matter*, 20(8):085212, 2008. URL http://stacks.iop.org/0953-8984/20/i=8/a=085212.

[387] Jonathan E. Spanier, Richard D. Robinson, Feng Zhang, Siu-Wai Chan, and Irving P. Herman. Size-dependent properties of ceo$_{2-y}$ nanoparticles as studied by raman scattering. *Phys. Rev. B*, 64:245407, Nov 2001. doi: 10.1103/PhysRevB.64.245407. URL http://link.aps.org/doi/10.1103/PhysRevB.64.245407.

[388] Satyaprakash Sahoo, A. K. Arora, and V. Sridharan. Raman line shapes of optical phonons of different symmetries in anatase tio$_2$ nanocrystals. *The Journal of Physical Chemistry C*, 113(39):16927–16933, 2009. doi: 10.1021/jp9046193. URL http://dx.doi.org/10.1021/jp9046193.

[389] A. Li Bassi, D. Cattaneo, V. Russo, C. E. Bottani, E. Barborini, T. Mazza, P. Piseri, P. Milani, F. O. Ernst, K. Wegner, and S. E. Pratsinis. Raman spectroscopy characterization of titania nanoparticles produced by flame pyrolysis: The influence of size and stoichiometry. *Journal of Applied Physics*, 98(7):074305, 2005. doi: http://dx.doi.org/10.1063/1.2061894. URL http://scitation.aip.org/content/aip/journal/jap/98/7/10.1063/1.2061894.

[390] A. Diéguez, A. Romano-Rodríguez, A. Vilà, and J. R. Morante. The complete raman spectrum of nanometric sno$_2$ particles. *Journal of Applied Physics*, 90(3):1550–1557, 2001. doi: http://dx.doi.org/10.1063/1.1385573. URL http://scitation.aip.org/content/aip/journal/jap/90/3/10.1063/1.1385573.

[391] H. Richter, Z.P. Wang, and L. Ley. The one phonon raman spectrum in microcrystalline silicon. *Solid State Communications*, 39(5): 625 – 629, 1981. ISSN 0038-1098. doi: http://dx.doi.org/10.1016/0038-1098(81)90337-9. URL http://www.sciencedirect.com/science/article/pii/0038109881903379.

[392] I.H. Campbell and P.M. Fauchet. The effects of microcrystal size and shape on the one phonon raman spectra of crystalline semiconductors. *Solid State Communications*, 58(10):739 – 741, 1986. ISSN 0038-1098. doi: http://dx.doi.org/10.1016/0038-1098(86)90513-2. URL http://www.sciencedirect.com/science/article/pii/0038109886905132.

[393] Akhilesh K. Arora, M. Rajalakshmi, T. R. Ravindran, and V. Sivasubramanian. Raman spectroscopy of optical phonon confinement in nanostructured materials. *Journal of Raman Spectroscopy*, 38(6):604–617, 2007. ISSN 1097-4555. doi: 10.1002/jrs.1684. URL http://dx.doi.org/10.1002/jrs.1684.

[394] Dumitru Georgescu, Lucian Baia, Ovidiu Ersen, Monica Baia, and Simion Simon. Experimental assessment of the phonon confinement in tio$_2$ anatase nanocrystallites by raman spectroscopy. *Journal of Raman Spectroscopy*, 43(7):876–883, 2012. ISSN 1097-4555. doi: 10.1002/jrs.3103. URL http://dx.doi.org/10.1002/jrs.3103.

[395] Subodh K. Gautam, Fouran Singh, I. Sulania, R. G. Singh, P. K. Kulriya, and E. Pippel. Micro-raman study on the softening and stiffening of phonons in rutile titanium dioxide film: Competing effects of structural defects, crystallite size, and lattice strain. *Journal of Applied Physics*, 115(14):143504, 2014. doi: http://dx.doi.org/10.1063/1.4868079. URL http://scitation.aip.org/content/aip/journal/jap/115/14/10.1063/1.4868079.

[396] Toshiaki Ohsaka. Temperature dependence of the raman spectrum in anatase tio₂. *Journal of the Physical Society of Japan*, 48(5): 1661–1668, 1980. doi: 10.1143/JPSJ.48.1661. URL http://dx.doi.org/10.1143/JPSJ.48.1661.

[397] E. Barborini, I. N. Kholmanov, A. M. Conti, P. Piseri, S. Vinati, P. Milani, and C. Ducati. Supersonic cluster beam depositionof nanostructured titania. *The European Physical Journal D - Atomic, Molecular, Optical and Plasma Physics*, 24(1):277–282, 2003. ISSN 1434-6079. doi: 10.1140/epjd/e2003-00189-2. URL http://dx.doi.org/10.1140/epjd/e2003-00189-2.

[398] Y. Djaoued, S. Badilescu, P.V. Ashrit, D. Bersani, P.P. Lottici, and J. Robichaud. Study of anatase to rutile phase transition in nanocrystalline titania films. *Journal of Sol-Gel Science and Technology*, 24(3):255–264, 2002. ISSN 1573-4846. doi: 10.1023/A: 1015357313003. URL http://dx.doi.org/10.1023/A:1015357313003.

[399] Dorian A. H. Hanaor and Charles C. Sorrell. Review of the anatase to rutile phase transformation. *Journal of Materials Science*, 46 (4):855–874, 2011. ISSN 1573-4803. doi: 10.1007/s10853-010-5113-0. URL http://dx.doi.org/10.1007/s10853-010-5113-0.

[400] Hyun Jeong Nam, Kiminori Itoh, and Masayuki Murabayashi. Photocatalytic activity of tio₂ thin film - effect of substrate. *Electrochemistry*, 70(6):429–431, 6 2002. ISSN 1344-3542.

[401] K. N. P. Kumar, K. Keizer, and A. J. Burggraaf. Stabilization of the porous texture of nanostructured titania by avoiding a phase transformation. *Journal of Materials Science Letters*, 13(1):59–61, 1994. ISSN 1573-4811. doi: 10.1007/BF00680274. URL http://dx.doi.org/10.1007/BF00680274.

[402] R. Lee Penn and Jillian F. Banfield. Formation of rutile nuclei at anatase {112} twin interfaces and the phase transformation mechanism in nanocrystalline titania. *American Mineralogist*, 84(5-6):871–876, 1999. doi: 10.2138/am-1999-5-622. URL //www.degruyter.com/view/j/ammin.1999.84.issue-5-6/am-1999-5-622/am-1999-5-622.xml.

[403] A. Rothschild, A. Levakov, Y. Shapira, N. Ashkenasy, and Y. Komem. Surface photovoltage spectroscopy study of reduced and oxidized nanocrystalline tio₂ films. *Surface Science*, 532–535:456 – 460, 2003. ISSN 0039-6028. doi: http://dx.doi.org/10.1016/S0039-6028(03) 00154-7. URL http://www.sciencedirect.com/science/article/pii/S0039602803001547. Proceedings of the 7th International Conference on Nanometer-Scale Science and Technology and the 21st European Conference on Surface Science.

[404] Olinger B Jamieson J. Pressure temperature studies of anatase, brookite, rutile and tio₂ ii: A discussion. *Mineralogical Notes*, 54:1477, 1969.

[405] Stacey J. Smith, Rebecca Stevens, Shengfeng Liu, Guangshe Li, Alexandra Navrotsky, Juliana Boerio-Goates, and Brian F. Woodfield. Heat capacities and thermodynamic functions of tio₂ anatase and rutile: Analysis of phase stability. *American Mineralogist*, 2009.

[406] Hengzhong Zhang and Jillian F. Banfield. Thermodynamic analysis of phase stability of nanocrystalline titania. *J. Mater. Chem.*, 8: 2073–2076, 1998. doi: 10.1039/A802619J. URL http://dx.doi.org/10.1039/A802619J.

[407] A. Navrotsky and O.J. Kleppa. Enthalpy of the anatase-rutile transformation. *Journal of the American Ceramic Society*, 50(11):626–626, 1967. ISSN 1551-2916. doi: 10.1111/j.1151-2916.1967.tb15013.x. URL http://dx.doi.org/10.1111/j.1151-2916.1967.tb15013.x.

[408] T. Mitsuhashi and O.J. Kleppa. Transformation enthalpies of the tio₂ polymorphs. *Journal of the American Ceramic Society*, 62(7-8): 356–357, 1979. ISSN 1551-2916. doi: 10.1111/j.1151-2916.1979.tb19077.x. URL http://dx.doi.org/10.1111/j.1151-2916.1979.tb19077.x.

[409] Juan Yang, Sen Mei, and José M. F. Ferreira. Hydrothermal synthesis of nanosized titania powders: Influence of peptization and peptizing agents on the crystalline phases and phase transitions. *Journal of the American Ceramic Society*, 83(6):1361–1368, 2000. ISSN 1551-2916. doi: 10.1111/j.1151-2916.2000.tb01394.x. URL http://dx.doi.org/10.1111/j.1151-2916.2000.tb01394.x.

[410] Hyun Suk Jung, Hyunho Shin, Jeong-Ryeol Kim, Jin Young Kim, Kug Sun Hong, and Jung-Kun Lee. In situ observation of the stability of anatase nanoparticles and their transformation to rutile in an acidic solution. *Langmuir*, 20(26):11732–11737, 2004. doi: 10.1021/la048425c. URL http://dx.doi.org/10.1021/la048425c. PMID: 15595805.

[411] L. Gerward and J. Staun Olsen. Post-rutile high-pressure phases in tio₂. *Journal of Applied Crystallography*, 30(3):259–264, Jun 1997. doi: 10.1107/S0021889896011454. URL http://dx.doi.org/10.1107/S0021889896011454.

[412] H. Rath, P. Dash, T. Som, P. V. Satyam, U. P. Singh, P. K. Kulriya, D. Kanjilal, D. K. Avasthi, and N. C. Mishra. Structural evolution of tio₂ nanocrystalline thin films by thermal annealing and swift heavy ion irradiation. *Journal of Applied Physics*, 105(7):074311, 2009. doi: http://dx.doi.org/10.1063/1.3103333. URL http://scitation.aip.org/content/aip/journal/jap/105/7/10.1063/1.3103333.

[413] S. A. Borkar and S. R. Dharwadkar. Temperatures and kinetics of anatase to rutile transformation in doped tio₂ heated in microwave field. *Journal of Thermal Analysis and Calorimetry*, 78(3):761–767, 2004. doi: 10.1007/s10973-005-0443-0. URL http://www.akademiai.com/doi/abs/10.1007/s10973-005-0443-0.

[414] G.M Neelgund, S.A. Shivashankar, B.K. Chethana, P.P. Sahoo, and K.J. Rao. Nanocrystalline tio₂ preparation by microwave route and nature of anatase-rutile phase transition in nano tio₂. *Bulletin of Materials Science*, 34(6):1163–1171, 2011. ISSN 0973-7669. doi: 10.1007/s12034-011-0165-6. URL http://dx.doi.org/10.1007/s12034-011-0165-6.

[415] P. Mitrev, G. Benvenuti, P. Hofman, A. Smirnov, N. Kaliteevskaya, and R. Seisyan. Phase transitions in thin titanium oxide films under the action of excimer laser radiation. *Technical Physics Letters*, 31(11):908–911, 2005. ISSN 1090-6533. doi: 10.1134/1.2136949. URL http://dx.doi.org/10.1134/1.2136949.

[416] X. Z. Ding, X. H. Liu, and Y. Z. He. Grain size dependence of anatase-to-rutile structural transformation in gel-derived nanocrystalline titania powders. *Journal of Materials Science Letters*, 15(20):1789–1791, 1996. ISSN 1573-4811. doi: 10.1007/BF00275343. URL http://dx.doi.org/10.1007/BF00275343.

[417] Teruhisa Ohno, Kojiro Tokieda, Suguru Higashida, and Michio Matsumura. Synergism between rutile and anatase tio₂ particles in photocatalytic oxidation of naphthalene. *Applied Catalysis A: General*, 244(2):383 – 391, 2003. ISSN 0926-860X. doi: http: //dx.doi.org/10.1016/S0926-860X(02)00610-5. URL http://www.sciencedirect.com/science/article/pii/S0926860X02006105.

[418] Teruhisa Ohno, Daisuke Haga, Kan Fujihara, Kaoru Kaizaki, and Michio Matsumura. Unique effects of iron(iii) ions on photocatalytic and photoelectrochemical properties of titanium dioxide. *The Journal of Physical Chemistry B*, 101(33):6415–6419, 1997. doi: 10.1021/jp971093i. URL http://dx.doi.org/10.1021/jp971093i.

References

[419] M. Kamal Akhtar, Sotiris E. Pratsinis, and Sebastian V. R. Mastrangelo. Dopants in vapor-phase synthesis of titania powders. *Journal of the American Ceramic Society*, 75(12):3408–3416, 1992. ISSN 1551-2916. doi: 10.1111/j.1151-2916.1992.tb04442.x. URL http://dx.doi.org/10.1111/j.1151-2916.1992.tb04442.x.

[420] Yi Hu, H.-L. Tsai, and C.-L. Huang. Effect of brookite phase on the anatase-rutile transition in titania nanoparticles. *Journal of the European Ceramic Society*, 23(5):691 – 696, 2003. ISSN 0955-2219. doi: http://dx.doi.org/10.1016/S0955-2219(02)00194-2. URL http://www.sciencedirect.com/science/article/pii/S0955221902001942.

[421] Shu Yin, Hiroshi Yamaki, Masakazu Komatsu, Qiwu Zhang, Jinshu Wang, Qing Tang, Fumio Saito, and Tsugio Sato. Preparation of nitrogen-doped titania with high visible light induced photocatalytic activity by mechanochemical reaction of titania and hexamethylenetetramine. *J. Mater. Chem.*, 13:2996–3001, 2003. doi: 10.1039/B309217H. URL http://dx.doi.org/10.1039/B309217H.

[422] Krishnankutty-Nair P. Kumar. Growth of rutile crystallites during the initial stage of anatase-to-rutile transformation in pure titania and in titania-alumina nanocomposites. *Scripta Metallurgica et Materialia*, 32(6):873 – 877, 1995. ISSN 0956-716X. doi: http://dx.doi.org/10.1016/0956-716X(95)93217-R. URL http://www.sciencedirect.com/science/article/pii/0956716X9593217R.

[423] J. Arbiol, J. Cerdà, G. Dezanneau, A. Cirera, F. Peiró, A. Cornet, and J. R. Morante. Effects of nb doping on the tio2 anatase-to-rutile phase transition. *Journal of Applied Physics*, 92(2):853–861, 2002. doi: http://dx.doi.org/10.1063/1.1487915. URL http://scitation.aip.org/content/aip/journal/jap/92/2/10.1063/1.1487915.

[424] Vittorio Loddo, Giuseppe Marcí, Leonardo Palmisano, and Antonino Sclafani. Preparation and characterization of al2o3 supported tio2 catalysts employed for 4-nitrophenol photodegradation in aqueous medium. *Materials Chemistry and Physics*, 53(3):217 – 224, 1998. ISSN 0254-0584. doi: http://dx.doi.org/10.1016/S0254-0584(98)00041-8. URL http://www.sciencedirect.com/science/article/pii/S0254058498000418.

[425] R. A. Spurr and Howard Myers. Quantitative analysis of anatase-rutile mixtures with an x-ray diffractometer. *Analytical Chemistry*, 29(5):760–762, 1957. doi: 10.1021/ac60125a006. URL http://dx.doi.org/10.1021/ac60125a006.

[426] Roger I. Bickley, Teresita Gonzalez-Carreno, John S. Lees, Leonardo Palmisano, and Richard J.D. Tilley. A structural investigation of titanium dioxide photocatalysts. *Journal of Solid State Chemistry*, 92(1):178 – 190, 1991. ISSN 0022-4596. doi: http://dx.doi.org/10.1016/0022-4596(91)90255-G. URL http://www.sciencedirect.com/science/article/pii/002245969190255G.

[427] Z Jing, X Qian, F Zhaochi, L Meijun, and L Can. Uv raman spectroscopic study on tio2. ii. effect of nanoparticle size on the outer/inner phase transformations. *Angewandte Chemie International Edition*, 47:1766, 2008.

[428] Yoshio Iida and Shunro Ozaki. Grain growth and phase transformation of titanium oxide during calcination. *Journal of the American Ceramic Society*, 44(3):120–127, 1961. ISSN 1551-2916. doi: 10.1111/j.1151-2916.1961.tb13725.x. URL http://dx.doi.org/10.1111/j.1151-2916.1961.tb13725.x.

[429] Yu-Hong Zhang and Armin Reller. Phase transformation and grain growth of doped nanosized titania. *Materials Science and Engineering: C*, 19(1–2):323 – 326, 2002. ISSN 0928-4931. doi: http://dx.doi.org/10.1016/S0928-4931(01)00409-X. URL http://www.sciencedirect.com/science/article/pii/S092849310100409X. Current Trends in Nanotechnologies: From Materials to Systems, Proceedings of Symposium S, {EMRS} Spring Meeting 2001, Strasbourg France,.

[430] D. Hanaor. *Thesis*. PhD thesis, The School of Materials Science and Engineering, 2007.

[431] Jing Sun, Lian Gao, and Qinghong Zhang. Synthesizing and comparing the photocatalytic properties of high surface area rutile and anatase titania nanoparticles. *Journal of the American Ceramic Society*, 86(10):1677–1682, 2003. ISSN 1551-2916. doi: 10.1111/j.1151-2916.2003.tb03539.x. URL http://dx.doi.org/10.1111/j.1151-2916.2003.tb03539.x.

[432] Manuel Ocaña, Jose V. Garcia-Ramos, and Carlos J. Serna. Low-temperature nucleation of rutile observed by raman spectroscopy during crystallization of tio2. *Journal of the American Ceramic Society*, 75(7):2010–2012, 1992. ISSN 1551-2916. doi: 10.1111/j.1151-2916.1992.tb07237.x. URL http://dx.doi.org/10.1111/j.1151-2916.1992.tb07237.x.

[433] Maria Suzana P. Francisco and Valmor R. Mastelaro. Inhibition of the anatase-rutile phase transformation with addition of ceo2 to cuo-tio2 system: Raman spectroscopy, x-ray diffraction, and textural studies. *Chemistry of Materials*, 14(6):2514–2518, 2002. doi: 10.1021/cm011520b. URL http://dx.doi.org/10.1021/cm011520b.

[434] Guido Busca, Gianguido Ramis, Jose M. Gallardo Amores, Vicente Sanchez Escribano, and Paolo Piaggio. Ft raman and ftir studies of titanias and metatitanate powders. *J. Chem. Soc., Faraday Trans.*, 90:3181–3190, 1994. doi: 10.1039/FT9949003181. URL http://dx.doi.org/10.1039/FT9949003181.

[435] Hua Chang and Pei Jane Huang. Thermo-raman studies on anatase and rutile. *Journal of Raman Spectroscopy*, 29(2):97–102, 1998. ISSN 1097-4555. doi: 10.1002/(SICI)1097-4555(199802)29:2<97::AID-JRS198>3.0.CO;2-E. URL http://dx.doi.org/10.1002/(SICI)1097-4555(199802)29:2<97::AID-JRS198>3.0.CO;2-E.

[436] Renishaw plc In. Renishaw minerals & inorganic materials database, 2010.

[437] ICDD. Icdd, powder diffraction file, newtown square, pennsylvania, usa, 2016.

[438] Teruhisa Ohno, Koji Sarukawa, and Michio Matsumura. Photocatalytic activities of pure rutile particles isolated from tio2 powder by dissolving the anatase component in hf solution. *The Journal of Physical Chemistry B*, 105(12):2417–2420, 2001. doi: 10.1021/jp003211z. URL http://dx.doi.org/10.1021/jp003211z.

[439] Teruhisa Ohno, Koji Sarukawa, Kojiro Tokieda, and Michio Matsumura. Morphology of a Tio2 photocatalyst (degussa, p-25) consisting of anatase and rutile crystalline phases. *Journal of Catalysis*, 203(1):82 – 86, 2001. ISSN 0021-9517. doi: 10.1006/jcat.2001.3316. URL http://www.sciencedirect.com/science/article/pii/S0021951701933160.

[440] C Suresh, V Biju, P Mukundan, and K.G.K Warrier. Anatase to rutile transformation in sol-gel titania by modification of precursor. *Polyhedron*, 17(18):3131 – 3135, 1998. ISSN 0277-5387. doi: http://dx.doi.org/10.1016/S0277-5387(98)00077-1. URL http://www.sciencedirect.com/science/article/pii/S0277538798000771.

[441] S Rajesh Kumar, C Suresh, Asha K Vasudevan, N.R Suja, P Mukundan, and K.G.K Warrier. Phase transformation in sol-gel titania containing silica. *Materials Letters*, 38(3):161 – 166, 1999. ISSN 0167-577X. doi: http://dx.doi.org/10.1016/S0167-577X(98)00152-9. URL http://www.sciencedirect.com/science/article/pii/S0167577X98001529.

[442] Kiyoshi Okada, Nobuo Yamamoto, Yoshikazu Kameshima, Atsuo Yasumori, and Kenneth J. D. MacKenzie. Effect of silica additive on the anatase-to-rutile phase transition. *Journal of the American Ceramic Society*, 84(7):1591–1596, 2001. ISSN 1551-2916. doi: 10.1111/j.1151-2916.2001.tb00882.x. URL http://dx.doi.org/10.1111/j.1151-2916.2001.tb00882.x.

[443] Alan Matthews. The crystallization of anatase and rutile from amorphous titanium dioxide under hydrothermal conditions. *American Mineralogist*, 61:419–424, 1976.

[444] T. B. Ghosh, Sampa Dhabal, and A. K. Datta. On crystallite size dependence of phase stability of nanocrystalline tio$_2$. *Journal of Applied Physics*, 94(7):4577–4582, 2003. doi: http://dx.doi.org/10.1063/1.1604966. URL http://scitation.aip.org/content/aip/journal/jap/94/7/10.1063/1.1604966.

[445] R.R Bacsa and J Kiwi. Effect of rutile phase on the photocatalytic properties of nanocrystalline titania during the degradation of p-coumaric acid. *Applied Catalysis B: Environmental*, 16(1):19 – 29, 1998. ISSN 0926-3373. doi: http://dx.doi.org/10.1016/S0926-3373(97) 00058-1. URL http://www.sciencedirect.com/science/article/pii/S0926337397000581.

[446] Hyunho Shin, Hyun Suk Jung, Kug Sun Hong, and Jung-Kun Lee. Crystal phase evolution of tio$_2$ nanoparticles with reaction time in acidic solutions studied via freeze-drying method. *Journal of Solid State Chemistry*, 178(1):15 – 21, 2005. ISSN 0022-4596. doi: http://dx.doi.org/10.1016/j.jssc.2004.09.035. URL http://www.sciencedirect.com/science/article/pii/S0022459604005079.

[447] Guangshe Li, Liping Li, Juliana Boerio-Goates, and Brian F. Woodfield. Grain-growth kinetics of rutile tio$_2$ nanocrystals under hydrothermal conditions. *Journal of Materials Research*, 18:2664–2669, 11 2003. ISSN 2044-5326. doi: 10.1017/S0884291400065936. URL http://journals.cambridge.org/article_S0884291400065936.

[448] Guangshe Li, Juliana B. Goates, Brian F. Woodfield, and Liping Li. Evidence of linear lattice expansion and covalency enhancement in rutile tio$_2$ nanocrystals. *Applied Physics Letters*, 85(24):259–261, 2004.

[449] Mingmei Wu, Gang Lin, Dihu Chen, Guangguo Wang, Dian He, Shouhua Feng, , and Ruren Xu. Sol-hydrothermal synthesis and hydrothermally structural evolution of nanocrystal titanium dioxide. *Chemistry of Materials*, 14(5):1974–1980, 2002. doi: 10.1021/cm0102739. URL http://dx.doi.org/10.1021/cm0102739.

[450] Chen-Chi Wang and Jackie Y. Ying. Sol-gel synthesis and hydrothermal processing of anatase and rutile titania nanocrystals. *Chemistry of Materials*, 11(11):3113–3120, 1999. doi: 10.1021/cm990180f. URL http://dx.doi.org/10.1021/cm990180f.

[451] Humin Cheng, Jiming Ma, Zhenguo Zhao, and Limin Qi. Hydrothermal preparation of uniform nanosize rutile and anatase particles. *Chemistry of Materials*, 7(4):663–671, 1995. doi: 10.1021/cm00052a010. URL http://dx.doi.org/10.1021/cm00052a010.

[452] S. T. Aruna, S. Tirosh, and A. Zaban. Nanosize rutile titania particle synthesis a hydrothermal method without mineralizers. *J. Mater. Chem.*, 10:2388–2391, 2000. doi: 10.1039/B001718N. URL http://dx.doi.org/10.1039/B001718N.

[453] M. Gopal, W. J. Moberly Chan, and L. C. De Jonghe. Room temperature synthesis of crystalline metal oxides. *Journal of Materials Science*, 32(22):6001–6008, 1997. ISSN 1573-4803. doi: 10.1023/A:1018671212890. URL http://dx.doi.org/10.1023/A:1018671212890.

[454] Yuanyuan Li, Jinping Liu, and Zhijie Jia. Morphological control and photodegradation behavior of rutile tio$_2$ prepared by a low-temperature process. *Materials Letters*, 60(13–14):1753 – 1757, 2006. ISSN 0167-577X. doi: http://dx.doi.org/10.1016/j.matlet.2005. 12.012. URL http://www.sciencedirect.com/science/article/pii/S0167577X05012565.

[455] Shu Yin, Hitoshi Hasegawa, Daisaku Maeda, Masayuki Ishitsuka, and Tsugio Sato. Synthesis of visible-light-active nanosize rutile titania photocatalyst by low temperature dissolution-reprecipitation process. *Journal of Photochemistry and Photobiology A: Chemistry*, 163(1–2):1 – 8, 2004. ISSN 1010-6030. doi: http://dx.doi.org/10.1016/S1010-6030(03)00289-2. URL http://www.sciencedirect.com/science/article/pii/S1010603003002892.

[456] Andrew Mills, Nicholas Elliott, George Hill, David Fallis, James R. Durrant, and Richard L. Willis. Preparation and characterisation of novel thick sol-gel titania film photocatalysts. *Photochem. Photobiol. Sci.*, 2:591–596, 2003. doi: 10.1039/B212865A. URL http://dx.doi.org/10.1039/B212865A.

[457] X. G. Ma, P. Liang, L. Miao, S. W. Bie, C. K. Zhang, L. Xu, and J. J. Jiang. Pressure-induced phase transition and elastic properties of tio$_2$ polymorphs. *physica status solidi (b)*, 246(9):2132–2139, 2009. ISSN 1521-3951. doi: 10.1002/pssb.200945111. URL http://dx.doi.org/10.1002/pssb.200945111.

[458] Giridhar Madras, Benjamin J. McCoy, and Alexandra Navrotsky. Kinetic model for tio$_2$ polymorphic transformation from anatase to rutile. *Journal of the American Ceramic Society*, 90(1):250–255, 2007. ISSN 1551-2916. doi: 10.1111/j.1551-2916.2006.01369.x. URL http://dx.doi.org/10.1111/j.1551-2916.2006.01369.x.

[459] Masanori Hirano, Chiaki Nakahara, Keisuke Ota, Osamu Tanaike, and Michio Inagaki. Photoactivity and phase stability of zro$_2$-doped anatase-type tio$_2$ directly formed as nanometer-sized particles by hydrolysis under hydrothermal conditions. *Journal of Solid State Chemistry*, 170(1):39 – 47, 2003. ISSN 0022-4596. doi: http://dx.doi.org/10.1016/S0022-4596(02)00013-0. URL http://www.sciencedirect.com/science/article/pii/S0022459602000130.

[460] Robert D. Shannon and Joseph A. Pask. Kinetics of the anatase-rutile transformation. *Journal of the American Ceramic Society*, 48(8): 391–398, 1965. ISSN 1551-2916. doi: 10.1111/j.1151-2916.1965.tb14774.x. URL http://dx.doi.org/10.1111/j.1151-2916.1965.tb14774.x.

[461] C. N. R. Rao. Kinetics and thermodynamics of the crystal structure transformation of spectroscopically pure anatase to rutile. *Canadian Journal of Chemistry*, 39(3):498–500, 1961. doi: 10.1139/v61-059. URL http://dx.doi.org/10.1139/v61-059.

[462] Matthias Batzill, Erie H. Morales, and Ulrike Diebold. Influence of nitrogen doping on the defect formation and surface properties of tio$_2$ rutile and anatase. *Phys. Rev. Lett.*, 96:026103, Jan 2006. doi: 10.1103/PhysRevLett.96.026103. URL http://link.aps.org/doi/10. 1103/PhysRevLett.96.026103.

[463] Gwan Hyoung Lee and Jian-Min Zuo. Growth and phase transformation of nanometer-sized titanium oxide powders produced by the precipitation method. *Journal of the American Ceramic Society*, 87(3):473–479, 2004. ISSN 1551-2916. doi: 10.1111/j.1551-2916.2004. 00473.x. URL http://dx.doi.org/10.1111/j.1551-2916.2004.00473.x.

[464] Jose Criado and Concha Real. Mechanism of the inhibiting effect of phosphate on the anatase → rutile transformation induced by thermal and mechanical treatment of tio$_2$. *J. Chem. Soc., Faraday Trans. 1*, 79:2765–2771, 1983. doi: 10.1039/F19837902765. URL http://dx.doi.org/10.1039/F19837902765.

[465] C.N.R. Rao, A. Turner, and J.M. Honig. Some observations concerning the effect of impurities on the anatase-rutile transition. *Journal of Physics and Chemistry of Solids*, 11(1):173 – 175, 1959. ISSN 0022-3697. doi: http://dx.doi.org/10.1016/0022-3697(59)90056-3. URL http://www.sciencedirect.com/science/article/pii/0022369759900563.

[466] D.R. Stull and H. Prophet. Janaf thermochemical tables. *National Standard Reference Data System*, 2(37):1141, 1971.

References

[467] O. Carp, C.L. Huisman, and A. Reller. Photoinduced reactivity of titanium dioxide. *Progress in Solid State Chemistry*, 32(1–2):33 – 177, 2004. ISSN 0079-6786. doi: http://dx.doi.org/10.1016/j.progsolidstchem.2004.08.001. URL http://www.sciencedirect.com/science/article/pii/S0079678604000123.

[468] Jinsoo Kim, Ki Chang Song, Sandra Foncillas, and Sotiris E. Pratsinis. Dopants for synthesis of stable bimodally porous titania. *Journal of the European Ceramic Society*, 21(16):2863 – 2872, 2001. ISSN 0955-2219. doi: http://dx.doi.org/10.1016/S0955-2219(01)00222-9. URL http://www.sciencedirect.com/science/article/pii/S0955221901002229.

[469] Hengzhong Zhang and Jillian F. Banfield. Phase transformation of nanocrystalline anatase-to-rutile via combined interface and surface nucleation. *Journal of Materials Research*, 15:437–448, 2 2000. ISSN 2044-5326. doi: 10.1557/JMR.2000.0067. URL http://journals.cambridge.org/article_S0884291400054091.

[470] Pelagia I. Gouma and Michael J. Mills. Anatase-to-rutile transformation in titania powders. *Journal of the American Ceramic Society*, 84 (3):619–622, 2001. ISSN 1551-2916. doi: 10.1111/j.1151-2916.2001.tb00709.x. URL http://dx.doi.org/10.1111/j.1151-2916.2001.tb00709.x.

[471] Y. Bessekhouad, D. Robert, and J. V. Weber. Preparation of tio2 nanoparticles by sol-gel route. *International Journal of Photoenergy*, 5(3), 2003. doi: 10.1155/S1110662X03000278.

[472] Qinghong Zhang, Lian Gao, and Jingkun Guo. Effects of calcination on the photocatalytic properties of nanosized tio2 powders prepared by ticl4 hydrolysis. *Applied Catalysis B: Environmental*, 26(3):207 – 215, 2000. ISSN 0926-3373. doi: http://dx.doi.org/10.1016/S0926-3373(00)00122-3. URL http://www.sciencedirect.com/science/article/pii/S0926337300001223.

[473] Z.M. Shi, L. Yan, L.N. Jin, X.M. Lu, and G. Zhao. The phase transformation behaviors of sn2+-doped titania gels. *Journal of Non-Crystalline Solids*, 353(22–23):2171 – 2178, 2007. ISSN 0022-3093. doi: http://dx.doi.org/10.1016/j.jnoncrysol.2007.02.048. URL http://www.sciencedirect.com/science/article/pii/S0022309307003067.

[474] Zhibo Zhang, Chen-Chi Wang, Rama Zakaria, and Jackie Y. Ying. Role of particle size in nanocrystalline tio2-based photocatalysts. *The Journal of Physical Chemistry B*, 102(52):10871–10878, 1998. doi: 10.1021/jp982948+. URL http://dx.doi.org/10.1021/jp982948+.

[475] Lili Hu, Toshinobu Yoko, Hiromitsu Kozuka, and Sumio Sakka. Effects of solvent on properties of sol-gel-derived tio2 coating films. *Thin Solid Films*, 219(1):18 – 23, 1992. ISSN 0040-6090. doi: http://dx.doi.org/10.1016/0040-6090(92)90718-Q. URL http://www.sciencedirect.com/science/article/pii/004060909290718Q.

[476] S. Riyas, G. Krishnan, and P. N. Mohan Das. Anatase-rutile transformation in doped titania under argon and hydrogen atmospheres. *Advances in Applied Ceramics*, 106(5):255–264, 2007. doi: 10.1179/174367607X202645. URL http://dx.doi.org/10.1179/174367607X202645.

[477] S. Vargas, R. Arroyo, E. Haro, and R. Rodríguez. Effects of cationic dopants on the phase transition temperature of titania prepared by the sol-gel method. *Journal of Materials Research*, 14:3932–3937, 10 1999. ISSN 2044-5326. doi: 10.1557/JMR.1999.0532. URL http://journals.cambridge.org/article_S0884291400052353.

[478] T. Ihara, M. Miyoshi, Y. Iriyama, O. Matsumoto, and S. Sugihara. Visible-light-active titanium oxide photocatalyst realized by an oxygen-deficient structure and by nitrogen doping. *Applied Catalysis B: Environmental*, 42(4):403 – 409, 2003. ISSN 0926-3373. doi: http://dx.doi.org/10.1016/S0926-3373(02)00269-2. URL http://www.sciencedirect.com/science/article/pii/S0926337302002692.

[479] J. Nowotny, T. Bak, L. R. Sheppard, and C. C. Sorrell. *Solar-Hydrogen: A Solid-State Chemistry Perspective*, volume 17, pages 169–215. Routledge, 2007. URL https://books.google.de/books?id=iOoeCwAAQBAJ.

[480] KJD Mackenzie. Calcination of titania: V, kinetics and mechanism of the anatase-rutile transformation in the presence of additives. *Transactions and Journal of the British Ceramic Society*, 74(3):77–84, 1975.

[481] KJD Mackenzie. The calcination of titania. iv: the effect of additives on the anatase-rutile transformation. *Transactions and Journal of the British Ceramic Society*, 74:29–34, 1975.

[482] Julio Andrade Gamboa and Daniel M. Pasquevich. Effect of chlorine atmosphere on the anatase-rutile transformation. *Journal of the American Ceramic Society*, 75(11):2934–2938, 1992. ISSN 1551-2916. doi: 10.1111/j.1151-2916.1992.tb04367.x. URL http://dx.doi.org/10.1111/j.1151-2916.1992.tb04367.x.

[483] Dani Gustaman Syarif, Atsumi Miyashita, Tetsuya Yamaki, Taishi Sumita, Yeongsoo Choi, and Hisayoshi Itoh. Preparation of anatase and rutile thin films by controlling oxygen partial pressure. *Applied Surface Science*, 193(1–4):287 – 292, 2002. ISSN 0169-4332. doi: http://dx.doi.org/10.1016/S0169-4332(02)00532-9. URL http://www.sciencedirect.com/science/article/pii/S0169433202005329.

[484] D.J. Reidy, J.D. Holmes, and M.A. Morris. The critical size mechanism for the anatase to rutile transformation in tio2 and doped-tio2. *Journal of the European Ceramic Society*, 26(9):1527 – 1534, 2006. ISSN 0955-2219. doi: http://dx.doi.org/10.1016/j.jeurceramsoc.2005.03.246. URL http://www.sciencedirect.com/science/article/pii/S0955221905000422X.

[485] A.M. Venezia, L. Palmisano, and M. Schiavello. Structural changes of titanium oxide induced by chromium addition as determined by an x-ray diffraction study. *Journal of Solid State Chemistry*, 114(2):364 – 368, 1995. ISSN 0022-4596. doi: http://dx.doi.org/10.1006/jssc.1995.1056. URL http://www.sciencedirect.com/science/article/pii/S0022459685751056.

[486] Ying Yang, Xin jun Li, Jun tao Chen, and Liang yan Wang. Effect of doping mode on the photocatalytic activities of mo/tio2. *Journal of Photochemistry and Photobiology A: Chemistry*, 163(3):517 – 522, 2004. ISSN 1010-6030. doi: http://dx.doi.org/10.1016/j.jphotochem.2004.02.008. URL http://www.sciencedirect.com/science/article/pii/S1010603004000736.

[487] Emanuele Finazzi, Cristiana Di Valentin, and Gianfranco Pacchioni. Nature of ti interstitials in reduced bulk anatase and rutile tio2. *The Journal of Physical Chemistry C*, 113(9):3382–3385, 2009. doi: 10.1021/jp8111793. URL http://dx.doi.org/10.1021/jp8111793.

[488] R. D. Shannon. Revised effective ionic radii and systematic studies of interatomic distances in halides and chalcogenides. *Acta Crystallographica Section A*, 32(5):751–767, Sep 1976. doi: 10.1107/S0567739476001551. URL http://dx.doi.org/10.1107/S0567739476001551.

[489] J. Yang, Y.X. Huang, and Jose Maria F. Ferreira. Inhibitory effect of alumina additive on the titania phase transformation of a sol-gel-derived powder. *Journal of Materials Science Letters*, 16(23):1933–1935, 1997. ISSN 1573-4811. doi: 10.1023/A:1018590701831. URL http://dx.doi.org/10.1023/A:1018590701831.

[490] C.H. Chen, E.M. Kelder, and J. Schoonman. Electrostatic sol-spray deposition (essd) and characterisation of nanostructured tio2 thin films. *Thin Solid Films*, 342(1–2):35 – 41, 1999. ISSN 0040-6090. doi: http://dx.doi.org/10.1016/S0040-6090(98)01160-2. URL http://www.sciencedirect.com/science/article/pii/S0040609098011602.

[491] J Yang and J.M.F Ferreira. On the titania phase transition by zirconia additive in a sol-gel-derived powder. *Materials Research Bulletin*, 33(3):389 – 394, 1998. ISSN 0025-5408. doi: http://dx.doi.org/10.1016/S0025-5408(97)00249-3. URL http://www.sciencedirect.com/science/article/pii/S0025540897002493.

[492] W. Hume-Rothery, R.E. Smallman, and C.W. Haworth. The structure of metals and alloys. Technical report, The Institute of Metals, 1 Carlton House Terrace, London SW1Y 5DB, UK, 1988.

[493] Jimmy C. Yu, Jiaguo Yu, Wingkei Ho, Zitao Jiang, and Lizhi Zhang. Effects of f- doping on the photocatalytic activity and microstructures of nanocrystalline tio$_2$ powders. *Chemistry of Materials*, 14(9):3808–3816, 2002. doi: 10.1021/cm020027c. URL http://dx.doi.org/10.1021/cm020027c.

[494] Yasutaka Takahashi and Yoshihiro Matsuoka. Dip-coating of tio$_2$ films using a sol derived from ti(o – i – pr)$_4$-diethanolamine-h$_2$o-i-proh system. *Journal of Materials Science*, 23(6):2259–2266, 1988. ISSN 1573-4803. doi: 10.1007/BF01115798. URL http://dx.doi.org/10.1007/BF01115798.

[495] Zhong L. Wang, Jin S. Yin, Wei Dong Mo, and Z. John Zhang. In-situ analysis of valence conversion in transition metal oxides using electron energy-loss spectroscopy. *The Journal of Physical Chemistry B*, 101(35):6793–6798, 1997. doi: 10.1021/jp971593b. URL http://dx.doi.org/10.1021/jp971593b.

[496] Eunnsox F. Heald and Clair W. Weiss. Kinetics and mechanism of the anatase/rutile transformation, as catalyzed by ferric oxide and reducing conditions. *American Mineralogist*, 57:10–23, 1972.

[497] R. Janes, L.J. Knightley, and C.J. Harding. Structural and spectroscopic studies of iron (iii) doped titania powders prepared by sol-gel synthesis and hydrothermal processing. *Dyes and Pigments*, 62(3):199 – 212, 2004. ISSN 0143-7208. doi: http://dx.doi.org/10.1016/j.dyepig.2003.12.003. URL http://www.sciencedirect.com/science/article/pii/S0143720803002535.

[498] F. C. Gennari and D. M. Pasquevich. Kinetics of the anatase-rutile transformation in tio$_2$ in the presence of fe$_2$o$_3$ fe$_2$o$_3$. *Journal of Materials Science*, 1998.

[499] D.J. Reidy, J.D. Holmes, and M.A. Morris. Preparation of a highly thermally stable titania anatase phase by addition of mixed zirconia and silica dopants. *Ceramics International*, 32(3):235 – 239, 2006. ISSN 0272-8842. doi: http://dx.doi.org/10.1016/j.ceramint.2005.02.009. URL http://www.sciencedirect.com/science/article/pii/S0272884205000647.

[500] U. Gesenhues. Calcination of metatitanic acid to titanium dioxide white pigments. *Chemical Engineering & Technology*, 24(7):685–694, 2001. ISSN 1521-4125. doi: 10.1002/1521-4125(200107)24:7<685::AID-CEAT685>3.0.CO;2-1. URL http://dx.doi.org/10.1002/1521-4125(200107)24:7<685::AID-CEAT685>3.0.CO;2-1.

[501] Krishnankutty-Nair P. Kumar, Derek J. Fray, Jalajakumari Nair, Fujio Mizukami, and Tatsuya Okubo. Enhanced anatase-to-rutile phase transformation without exaggerated particle growth in nanostructured titania-tin oxide composites. *Scripta Materialia*, 57(8): 771 – 774, 2007. ISSN 1359-6462. doi: http://dx.doi.org/10.1016/j.scriptamat.2007.06.039. URL http://www.sciencedirect.com/science/article/pii/S1359646207004708.

[502] B. Grzmil, B. Kic, and M. Rabe. Inhibition of the anatase-rutile phase transformation with addition of k$_2$o, p$_2$o$_5$, and li$_2$o. *Chemical Papers*, 58(6):410–414, 2004.

[503] M. Bettinelli, V. Dallacasa, D. Falcomer, P. Fornasiero, V. Gombac, T. Montini, L. Romanò, and A. Speghini. Photocatalytic activity of tio$_2$ doped with boron and vanadium. *Journal of Hazardous Materials*, 146(3):529 – 534, 2007. ISSN 0304-3894. doi: http://dx.doi.org/10.1016/j.jhazmat.2007.04.053. URL http://www.sciencedirect.com/science/article/pii/S0304389407005389. Environmental Applications of Advanced Oxidation Processes.

[504] P. P. Ahonen, E. I. Kauppinen, J. C. Joubert, J. L. Deschanvres, and G. Van Tendeloo. Preparation of nanocrystalline titania powder via aerosol pyrolysis of titanium tetrabutoxide. *Journal of Materials Research*, 14:3938–3948, 10 1999. ISSN 2044-5326. doi: 10.1557/JMR.1999.0533. URL http://journals.cambridge.org/article_S0884291400052365.

[505] S Rajesh Kumar, Suresh C Pillai, U.S Hareesh, P Mukundan, and K.G.K Warrier. Synthesis of thermally stable, high surface area anatase-alumina mixed oxides. *Materials Letters*, 43(5–6):286 – 290, 2000. ISSN 0167-577X. doi: http://dx.doi.org/10.1016/S0167-577X(99)00275-X. URL http://www.sciencedirect.com/science/article/pii/S0167577X9900275X.

[506] Sujith Perera and Edward G. Gillan. High-temperature stabilized anatase tio$_2$ from an aluminum-doped ticl$_3$ precursor. *Chem. Commun.*, pages 5988–5990, 2005. doi: 10.1039/B512148E. URL http://dx.doi.org/10.1039/B512148E.

[507] Chunzhong Li, Liyi Shi, Dongmei Xie, and Hailiang Du. Morphology and crystal structure of al-doped tio$_2$ nanoparticles synthesized by vapor phase oxidation of titanium tetrachloride. *Journal of Non-Crystalline Solids*, 352(38–39):4128 – 4135, 2006. ISSN 0022-3093. doi: http://dx.doi.org/10.1016/j.jnoncrysol.2006.06.036. URL http://www.sciencedirect.com/science/article/pii/S0022309306009628.

[508] L. E. Depero, A. Marino, B. Allieri, E. Bontempi, L. Sangaletti, C. Casale, and M. Notaro. Morphology and microstructural properties of tio$_2$ nanopowders doped with trivalent al and ga cations. *Journal of Materials Research*, 15:2080–2086, 10 2000. ISSN 2044-5326. doi: 10.1557/JMR.2000.0299. URL http://journals.cambridge.org/article_S0884291400060298.

[509] Masanori Hirano, Keisuke Ota, and Hiroyuki Iwata. Direct formation of anatase (tio$_2$)/silica (sio$_2$) composite nanoparticles with high phase stability of 1300°c from acidic solution by hydrolysis under hydrothermal condition. *Chemistry of Materials*, 16(19):3725–3732, 2004. doi: 10.1021/cm040055q. URL http://dx.doi.org/10.1021/cm040055q.

[510] Masaru Yoshinaka, Ken Hirota, and Osamu Yamaguchi. Formation and sintering of tio$_2$ (anatase) solid solution in the system tio$_2$-sio$_2$. *Journal of the American Ceramic Society*, 80(10):2749–2753, 1997. ISSN 1551-2916. doi: 10.1111/j.1151-2916.1997.tb03190.x. URL http://dx.doi.org/10.1111/j.1151-2916.1997.tb03190.x.

[511] Y Suyama and A Kato. The inhibitory effect of sio$_2$ on the anatase-rutile transition of tio$_2$. *J. Ceram. Soc. Jpn*, 86(3):119–125, 1978.

[512] Yoshiro Ohtsuka, Yoshinori Fujiki, and Yoshio Suzuki. Impurity effects on anatase-rutile transformation. *The Journal of the Japanese Association of Mineralogists, Petrologists and Economic Geologists*, 77(4):117–124, 1982. doi: 10.2465/ganko1941 77 117.

[513] Konstantin I. Hadjiivanov, Dimitar G. Klissurski, and Anatoly A. Davydov. Study of phosphate-modified tio$_2$ (anatase). *Journal of Catalysis*, 116(2):498 – 505, 1989. ISSN 0021-9517. doi: http://dx.doi.org/10.1016/0021-9517(89)90116-4. URL http://www.sciencedirect.com/science/article/pii/0021951789901164.

[514] Geoffrey C. Bond, Antal J. Sárkány, and Geoffrey D. Parfitt. The vanadium pentoxide-titanium dioxide system. *Journal of Catalysis*, 57(3):476 – 493, 1979. ISSN 0021-9517. doi: http://dx.doi.org/10.1016/0021-9517(79)90013-7. URL http://www.sciencedirect.com/science/article/pii/0021951779900137.

References

[515] F. Bregani, C. Casale, L.E. Depero, I. Natali-Sora, D. Robba, L. Sangaletti, and G.P. Toledo. Materials for sensors: Functional nanoscaled structures of the 1995 e-mrs spring conference temperature effects on the size of anatase crystallites in mo-tio2 and w-tio2 powders. *Sensors and Actuators B: Chemical*, 31(1):25 – 28, 1996. ISSN 0925-4005. doi: http://dx.doi.org/10.1016/0925-4005(96) 80011-6. URL http://www.sciencedirect.com/science/article/pii/0925400596800116.

[516] Laura E. Depero, Paolo Bonzi, Mirella Musci, and Cristina Casale. Microstructural study of vanadium-titanium oxide powders obtained by laser-induced synthesis. *Journal of Solid State Chemistry*, 111(2):247 – 252, 1994. ISSN 0022-4596. doi: http://dx.doi.org/10.1006/ jssc.1994.1224. URL http://www.sciencedirect.com/science/article/pii/S0022459684712242.

[517] L.E. Depero. Influence of vanadium and tungsten substitution on the stability of anatase. *Journal of Solid State Chemistry*, 104(2): 470 – 475, 1993. ISSN 0022-4596. doi: http://dx.doi.org/10.1006/jssc.1993.1184. URL http://www.sciencedirect.com/science/article/ pii/S0022459683711849.

[518] Saila Karvinen. The effects of trace elements on the crystal properties of tio2. *Solid State Sciences*, 5(5):811 – 819, 2003. ISSN 1293-2558. doi: http://dx.doi.org/10.1016/S1293-2558(03)00082-7. URL http://www.sciencedirect.com/science/article/pii/S1293255803000827.

[519] Kyong Chan Heo, Chi Il Ok, Jang Whan Kim, and Byong Kee Moon. The effects of manganese ions and their magnetic properties on the anatase-rutile phase transition of nanocrystalline tio2 : Mn prepared by using the solvothermal method. *Journal of Korean Physical Society*, 47(5):861–865, November 2005.

[520] R Arroyo, G Córdoba, J Padilla, and V.H Lara. Influence of manganese ions on the anatase-rutile phase transition of tio2 prepared by the sol-gel process. *Materials Letters*, 54(5–6):397 – 402, 2002. ISSN 0167-577X. doi: http://dx.doi.org/10.1016/S0167-577X(01)00600-0. URL http://www.sciencedirect.com/science/article/pii/S0167577X01006000.

[521] J.O. Carneiro, V. Teixeira, A. Portinha, A. Magalh nes, P. Coutinho, C.J. Tavares, and R. Newton. Iron-doped photocatalytic tio2 sputtered coatings on plastics for self-cleaning applications. *Materials Science and Engineering: B*, 138(2):144 – 150, 2007. ISSN 0921-5107. doi: http://dx.doi.org/10.1016/j.mseb.2005.08.130. URL http://www.sciencedirect.com/science/article/pii/S092151070600300X. Society of Vacuum Coaters 2nd Symposium on Smart Materials.

[522] Fabiana C. Gennari and Daniel M. Pasquevich. Enhancing effect of iron chlorides on the anatase-rutile transition in titanium dioxide. *Journal of the American Ceramic Society*, 82(7):1915–1921, 1999. ISSN 1551-2916. doi: 10.1111/j.1151-2916.1999.tb02016.x. URL http://dx.doi.org/10.1111/j.1151-2916.1999.tb02016.x.

[523] G. Sankar, K. R. Kannan, and C. N. R. Rao. Anatase-rutile transformation in fe/tio2, th/tio2 and cu/tio2 catalysts and its possible role in metal-support interaction. *Catalysis Letters*, 8(1):27–36, 1991. ISSN 1572-879X. doi: 10.1007/BF00764380. URL http: //dx.doi.org/10.1007/BF00764380.

[524] James L. Gole, Sharka M. Prokes, and Orest J. Glembocki. Efficient room-temperature conversion of anatase to rutile tio2 induced by high-spin ion doping. *The Journal of Physical Chemistry C*, 112(6):1782–1788, 2008. doi: 10.1021/jp075557g. URL http://dx.doi.org/ 10.1021/jp075557g.

[525] Jalajakumari Nair, Padmakumar Nair, Fujio Mizukami, Yoshinao Oosawa, and Tatsuya Okubo. Microstructure and phase transformation behavior of doped nanostructured titania. *Materials Research Bulletin*, 34(8):1275 – 1290, 1999. ISSN 0025-5408. doi: http://dx.doi.org/10.1016/S0025-5408(99)00113-0. URL http://www.sciencedirect.com/science/article/pii/S0025540809900130.

[526] S. Hishita, I. Mutoh, K. Koumoto, and H. Yanagida. Inhibition mechanism of the anatase-rutile phase transformation by rare earth oxides. *Ceramics International*, 9(2):61 – 67, 1983. ISSN 0272-8842. doi: http://dx.doi.org/10.1016/0272-8842(83)90025-1. URL http://www.sciencedirect.com/science/article/pii/0272884283900251.

[527] Jun Lin and Jimmy C. Yu. An investigation on photocatalytic activities of mixed tio2-rare earth oxides for the oxidation of acetone in air. *Journal of Photochemistry and Photobiology A: Chemistry*, 116(1):63 – 67, 1998. ISSN 1010-6030. doi: http://dx.doi.org/10.1016/ S1010-6030(98)00289-5. URL http://www.sciencedirect.com/science/article/pii/S1010603098002895.

[528] Ron Stevens, Weeraw Sunsaneeyametha, and Angkhana Jaroenworaluck. Surface characteristics of zirconia-coated tio2 and its phase transformation. In *Advances in Composite Materials and Structures*, volume 334 of *Key Engineering Materials*, pages 1101–1104. Trans Tech Publications, 3 2007. doi: 10.4028/www.scientific.net/KEM.334-335.1101.

[529] Masanori Hirano, Chiaki Nakahara, Keisuke Ota, and Michio Inagaki. Direct formation of zirconia-doped titania with stable anatasetype structure by thermal hydrolysis. *Journal of the American Ceramic Society*, 85(5):1333–1335, 2002. ISSN 1551-2916. doi: 10.1111/ j.1151-2916.2002.tb00274.x. URL http://dx.doi.org/10.1111/j.1151-2916.2002.tb00274.x.

[530] Laura E. Depero, Luigi Sangaletti, Brigida Allieri, Elza Bontempi, Roberto Salari, Marcello Zocchi, Cristina Casale, and Maurizio Notaro. Niobium-titanium oxide powders obtained by laser-induced synthesis: Microstructure and structure evolution from diffraction data. *Journal of Materials Research*, 13:1644–1649, 6 1998. ISSN 2044-5326. doi: 10.1557/JMR.1998.0226. URL http://journals. cambridge.org/article_S0884291400044629.

[531] Shunichi Hishita, Masasuke Takata, and Hiroaki Yangida. Inhibition of anatase-rutile transformation due to nb2o5 addition. *Journal of the Ceramic Association, Japan*, 86(1000):631–632, 1978. doi: 10.2109/jcersj1950.86.1000_631.

[532] Yongqiang Cao, Tao He, Lusong Zhao, Enjun Wang, Wensheng Yang, and Yaan Cao. Structure and phase transition behavior of sn4+-doped tio2 nanoparticles. *The Journal of Physical Chemistry C*, 113(42):18121–18124, 2009. doi: 10.1021/jp9069288. URL http://dx.doi.org/10.1021/jp9069288.

[533] Richard A. Eppler. Effect of antimony oxide on the anatase-rutile transformation in titanium dioxide. *Journal of the American Ceramic Society*, 70(4):C–64–C–66, 1987. ISSN 1551-2916. doi: 10.1111/j.1151-2916.1987.tb04985.x. URL http://dx.doi.org/10.1111/j.1151-2916. 1987.tb04985.x.

[534] LE Depero, L Sangaletti, B Allieri, F Pioselli, C Casale, and M Notaro. Microstructural properties of ta-doped tio2 powders obtained by laser pyrolysis. In *Materials Science Forum*, volume 278, pages 654–659. Trans Tech Publ, 1998.

[535] Mahlaba A. Debeila, Mpfuzeni C. Raphulu, Emma Mokoena, Miguel Avalos, Vitalii Petranovskii, Neil J. Coville, and Mike S. Scurrell. The effect of gold on the phase transitions of titania. *Materials Science and Engineering: A*, 396(1-2):61 – 69, 2005. ISSN 0921-5093. doi: http://dx.doi.org/10.1016/j.msea.2004.12.047. URL http://www.sciencedirect.com/science/article/pii/S0921509305000213.

[536] K.V. Baiju, C.P. Sibu, K. Rajesh, P. Krishna Pillai, P. Mukundan, K.G.K. Warrier, and W. Wunderlich. An aqueous sol-gel route to synthesize nanosized lanthana-doped titania having an increased anatase phase stability for photocatalytic application. *Materials Chemistry and Physics*, 90(1):123 – 127, 2005. ISSN 0254-0584. doi: http://dx.doi.org/10.1016/j.matchemphys.2004.10.024. URL http://www.sciencedirect.com/science/article/pii/S025405840400536X.

[537] Charles A. LeDuc, Jeffrey M. Campbell, and Joseph A. Rossin. Effect of lanthana as a stabilizing agent in titanium dioxide support. *Industrial & Engineering Chemistry Research*, 35(7):2473–2476, 1996. doi: 10.1021/ie960112s. URL http://dx.doi.org/10.1021/ie960112s.

[538] Yan Qingzhi, Su Xintai, Zhou Yanping, and Ge Changchun. Influence of cerium ions on the anatase-rutile phase transition of tio2 prepared by sol-gel auto-igniting synthesis. *Rare Metals*, 24(2), 2005.

[539] E. Setiawati and K. Kawano. Stabilization of anatase phase in the rare earth; eu and sm ion doped nanoparticle tio2. *Journal of Alloys and Compounds*, 451(1–2):293 – 296, 2008. ISSN 0925-8388. doi: http://dx.doi.org/10.1016/j.jallcom.2007.04.059. URL http://www.sciencedirect.com/science/article/pii/S0925838807008729. The 6th International Conference on f-Elements (ICFE-6)The 6th International Conference on f-Elements (ICFE-6).

[540] K. T. Ranjit, H. Cohen, I. Willner, S. Bossmann, and A. M. Braun. Lanthanide oxide-doped titanium dioxide: Effective photocatalysts for the degradation of organic pollutants. *Journal of Materials Science*, 34(21):5273–5280, 1999. ISSN 1573-4803. doi: 10.1023/A: 1004780401030. URL http://dx.doi.org/10.1023/A:1004780401030.

[541] Yuhong Zhang, Hailiang Xu, Yongxi Xu, Huaxing Zhang, and Yanguang Wang. The effect of lanthanide on the degradation of {RB} in nanocrystalline ln/tio2 aqueous solution. *Journal of Photochemistry and Photobiology A: Chemistry*, 170(3):279 – 285, 2005. ISSN 1010-6030. doi: http://dx.doi.org/10.1016/j.jphotochem.2004.09.001. URL http://www.sciencedirect.com/science/article/pii/S1010603004004034.

[542] A. Turković, M. Ivanda, A. Drašner, V. Vraneša, and M. Peršin. Raman spectroscopy of thermally annealed tio2 thin films. *Thin Solid Films*, 198(1):199 – 205, 1991. ISSN 0040-6090. doi: http://dx.doi.org/10.1016/0040-6090(91)90338-X. URL http://www.sciencedirect.com/science/article/pii/004060909190338X.

[543] Robert D. Shannon. Phase transformation studies in tio2 supporting different defect mechanisms in vacuum-reduced and hydrogen-reduced rutile. *Journal of Applied Physics*, 35(11):3414–3416, 1964. doi: http://dx.doi.org/10.1063/1.1713231. URL http://scitation.aip.org/content/aip/journal/jap/35/11/10.1063/1.1713231.

[544] Zhang Zhang, Anke Weidenkaff, and Amin Reller. Mesoporous structure and phase transition of nanocrystalline tio2. *Materials Letters*, 54(5–6):375 – 381, 2002. ISSN 0167-577X. doi: http://dx.doi.org/10.1016/S0167-577X(01)00597-3. URL http://www.sciencedirect.com/science/article/pii/S0167577X01005973.

[545] V. Chhabra, V. Pillai, B. K. Mishra, A. Morrone, and D. O. Shah. Synthesis, characterization, and properties of microemulsion-mediated nanophase tio2 particles. *Langmuir*, 11(9):3307–3311, 1995. doi: 10.1021/la00009a007. URL http://dx.doi.org/10.1021/la00009a007.

[546] M.Manjari Lal, Vishal Chhabra, Pushan Ayyub, and Amarnath Maitra. Preparation and characterization of ultrafine tio2 particles in reverse micelles by hydrolysis of titanium di-ethylhexyl sulfosuccinate. *Journal of Materials Research*, 13:1249–1254, 5 1998. ISSN 2044-5326. doi: 10.1557/JMR.1998.0178. URL http://journals.cambridge.org/article_S0884291400004149.

[547] L. K. Campbell, B. K. Na, and E. I. Ko. Synthesis and characterization of titania aerogels. *Chemistry of Materials*, 4(6):1329–1333, 1992. doi: 10.1021/cm00024a037. URL http://dx.doi.org/10.1021/cm00024a037.

[548] Ulagaraj Selvaraj, Alamanda V. Prasadarao, Sridhar Komarneni, and Rustum Roy. Sol-gel fabrication of epitaxial and oriented tio2 thin films. *Journal of the American Ceramic Society*, 75(5):1167–1170, 1992. ISSN 1551-2916. doi: 10.1111/j.1151-2916.1992.tb05554.x. URL http://dx.doi.org/10.1111/j.1151-2916.1992.tb05554.x.

[549] M. R. Ranade, A. Navrotsky, H. Z. Zhang, J. F. Banfield, S. H. Elder, A. Zaban, P. H. Borse, S. K. Kulkarni, G. S. Doran, and H. J. Whitfield. Energetics of nanocrystalline tio2. *Proceedings of the National Academy of Sciences*, 99(suppl 2):6476–6481, 2002. doi: 10.1073/pnas.251534898. URL http://www.pnas.org/content/99/suppl_2/6476.abstract.

[550] Amy A. Gribb and Jillian F. Banfield. Particle size effects on transformation kinetics and phase stability in nanocrystalline tio2. *American Mineralogist*, 82(7-8):717–728, 1997. ISSN 0003-004X. doi: 10.2138/am-1997-7-809. URL http://ammin.geoscienceworld.org/content/82/7-8/717.

[551] A. S. Barnard and L. A. Curtiss. Prediction of tio2 nanoparticle phase and shape transitions controlled by surface chemistry. *Nano Letters*, 5(7):1261–1266, 2005. doi: 10.1021/nl050355m. URL http://dx.doi.org/10.1021/nl050355m. PMID: 16178221.

[552] A. S. Barnard and P. Zapol. Predicting the energetics, phase stability, and morphology evolution of faceted and spherical anatase nanocrystals. *The Journal of Physical Chemistry B*, 108(48):18435–18440, 2004. doi: 10.1021/jp0472459. URL http://dx.doi.org/10.1021/jp0472459.

[553] W. Ma, Z. Lu, and M. Zhang. Investigation of structural transformations in nanophase titanium dioxide by raman spectroscopy. *Applied Physics A*, 66(6):621–627, 1998. ISSN 1432-0630. doi: 10.1007/s003390050723. URL http://dx.doi.org/10.1007/s003390050723.

[554] W F Zhang, Y L He, M S Zhang, Z Yin, and Q Chen. Raman scattering study on anatase Tio2 nanocrystals. *Journal of Physics D: Applied Physics*, 33(8):912, 2000. doi: 10.1088/0022-3727/33/8/305. URL http://iopscience.iop.org/0022-3727/33/8/305.

[555] Guillaume Sudant, Emmanuel Baudrin, Dominique Larcher, and Jean-Marie Tarascon. Electrochemical lithium reactivity with nan-otextured anatase-type tio2. *J. Mater. Chem.*, 15:1263–1269, 2005. doi: 10.1039/B416176A. URL http://dx.doi.org/10.1039/B416176A.

[556] L.E. Depero, L. Sangaletti, B. Allieri, E. Bontempi, A. Marino, and M. Zocchi. Correlation between crystallite sizes and microstrains in tio2 anatase. *Journal of Crystal Growth*, 198–199, Part 1:516 – 520, 1999. ISSN 0022-0248. doi: http://dx.doi.org/10.1016/S0022-0248(98)01086-0. URL http://www.sciencedirect.com/science/article/pii/S0022024898010860.

[557] Yun-Hong Zhang, Chak K. Chan, John F. Porter, and Wei Guo. Micro-raman spectroscopic characterization of nanosized tio2 powders prepared by vapor hydrolysis. *Journal of Materials Research*, 13:2602–2609, 9 1998. ISSN 2044-5326. doi: 10.1557/JMR.1998.0363. URL http://journals.cambridge.org/article_S0884291400045994.

[558] J. G. dos Santos, T. Ogasawara, and R. A. Corrêa. Synthesis of mesoporous titania in rutile phase with pore-stable structure. *Brazilian Journal of Chemical Engineering*, 26:555 – 561, 09 2009. ISSN 0104-6632. doi: 10.1590/S0104-66322009000300011. URL http://www.scielo.br/scielo.php?script=sci_arttext&pid=S0104-66322009000300011&nrm=iso.

[559] W. Li, C. Ni, H. Lin, C. P. Huang, and S. Ismat Shah. Size dependence of thermal stability of tio2 nanoparticles. *Journal of Applied Physics*, 96(11):6663–6668, 2004. doi: http://dx.doi.org/10.1063/1.1807520. URL http://scitation.aip.org/content/aip/journal/jap/96/11/10.1063/1.1807520.

[560] A. S. Barnard and P. Zapol. A model for the phase stability of arbitrary nanoparticles as a function of size and shape. *The Journal of Chemical Physics*, 121(9):4276–4283, 2004. doi: http://dx.doi.org/10.1063/1.1775770. URL http://scitation.aip.org/content/aip/journal/jcp/121/9/10.1063/1.1775770.

[561] A. S. Barnard and P. Zapol. Effects of particle morphology and surface hydrogenation on the phase stability of Tio₂. *Phys. Rev. B*, 70:235403, Dec 2004. doi: 10.1103/PhysRevB.70.235403. URL http://link.aps.org/doi/10.1103/PhysRevB.70.235403.

[562] Zhongwu Wang, S K Saxena, V Pischedda, H P Liermann, and C S Zha. X-ray diffraction study on pressure-induced phase transformations in nanocrystalline anatase/rutile (tio₂). *Journal of Physics: Condensed Matter*, 13(36):8317, 2001. URL http://stacks.iop.org/0953-8984/13/i=36/a=307.

[563] G. R. Hearne, J. Zhao, A. M. Dawe, V. Pischedda, M. Maaza, M. K. Nieuwoudt, P. Kibasomba, O. Nemraoui, J. D. Comins, and M. J. Witcomb. Effect of grain size on structural transitions in anatase Tio₂: A raman spectroscopy study at high pressure. *Phys. Rev. B*, 70:134102, Oct 2004. doi: 10.1103/PhysRevB.70.134102. URL http://link.aps.org/doi/10.1103/PhysRevB.70.134102.

[564] Malcolm Nicol and Mei Y. Fong. Raman spectrum and polymorphism of titanium dioxide at high pressures. *The Journal of Chemical Physics*, 54(7):3167–3170, 1971. doi: http://dx.doi.org/10.1063/1.1675305. URL http://scitation.aip.org/content/aip/journal/jcp/54/7/10.1063/1.1675305.

[565] T. Arlt, M. Bermejo, M. A. Blanco, L. Gerward, J. Z. Jiang, J. Staun Olsen, and J. M. Recio. High-pressure polymorphs of anatase tio₂. *Phys. Rev. B*, 61:14414–14419, Jun 2000. doi: 10.1103/PhysRevB.61.14414. URL http://link.aps.org/doi/10.1103/PhysRevB.61.14414.

[566] J. K. Dewhurst and J. E. Lowther. High-pressure structural phases of titanium dioxide. *Phys. Rev. B*, 54:R3673–R3675, Aug 1996. doi: 10.1103/PhysRevB.54.R3673. URL http://link.aps.org/doi/10.1103/PhysRevB.54.R3673.

[567] J. Z. Jiang, J. S. Olsen, L. Gerward, D. Frost, D. Rubie, and J. Peyronneau. Structural stability in nanocrystalline zno. *EPL (Europhysics Letters)*, 50(1):48, 2000. URL http://stacks.iop.org/0295-5075/50/i=1/a=048.

[568] T. Ohsaka, S. Yamaoka, and O. Shimomura. Effect of hydrostatic pressure on the raman spectrum of anatase (tio₂). *Solid State Communications*, 30(6):345 – 347, 1979. ISSN 0038-1098. doi: http://dx.doi.org/10.1016/0038-1098(79)90648-3. URL http://www.sciencedirect.com/science/article/pii/0038109879906483.

[569] J.F. Mammone, S.K. Sharma, and M. Nicol. Raman study of rutile (tio₂) at high pressures. *Solid State Communications*, 34(10):799 – 802, 1980. ISSN 0038-1098. doi: http://dx.doi.org/10.1016/0038-1098(80)91055-8. URL http://www.sciencedirect.com/science/article/pii/0038109880910558.

[570] J.Staun Olsen, L Gerward, and J.Z Jiang. On the rutile/α-pbo₂-type phase boundary of tio₂. *Journal of Physics and Chemistry of Solids*, 60(2):229 – 233, 1999. ISSN 0022-3697. doi: http://dx.doi.org/10.1016/S0022-3697(98)00274-1. URL http://www.sciencedirect.com/science/article/pii/S0022369798002741.

[571] Varghese Swamy, Leonid S. Dubrovinsky, Natalia A. Dubrovinskaia, Alexandre S. Simionovici, Michael Drakopoulos, Vladimir Dmitriev, and Hans-Peter Weber. Compression behavior of nanocrystalline anatase tio₂. *Solid State Communications*, 125(2):111 – 115, 2003. ISSN 0038-1098. doi: http://dx.doi.org/10.1016/S0038-1098(02)00601-4. URL http://www.sciencedirect.com/science/article/pii/S0038109802006014.

[572] F. Di Fonzo, C.S. Casari, V. Russo, M.F. Brunella, A. Li Bassi, and C.E. Bottani. Hierarchically organized nanostructured tio₂ for photocatalysis applications. *Nanotechnology*, 20(1):015604, 2009. URL http://stacks.iop.org/0957-4484/20/i=1/a=015604.

[573] James Ovenstone and Kazumichi Yanagisawa. Effect of hydrothermal treatment of amorphous titania on the phase change from anatase to rutile during calcination. *Chemistry of Materials*, 11(10):2770–2774, 1999. doi: 10.1021/cm990172z. URL http://dx.doi.org/10.1021/cm990172z.

[574] Yun Jeong Hwang, Chris Hahn, Bin Liu, and Peidong Yang. Photoelectrochemical properties of tio₂ nanowire arrays: A study of the dependence on length and atomic layer deposition coating. *ACS Nano*, 6(6):5060–5069, 2012. doi: 10.1021/nn300679d. URL http://dx.doi.org/10.1021/nn300679d. PMID: 22621345.

[575] Y. Zhang, T. Ichihashi, E. Landree, F. Nihey, and S. Iijima. Heterostructures of single-walled carbon nanotubes and carbide nanorods. *Science*, 285(5434):1719–1722, 1999. ISSN 0036-8075. doi: 10.1126/science.285.5434.1719. URL http://science.sciencemag.org/content/285/5434/1719.

[576] R. Debnath and J. Chaudhuri. Inhibiting effect of alpo₄ and sio₂ on the anatase → rutile transformation reaction: An x-ray and laser raman study. *Journal of Materials Research*, 7:3348–3351, 12 1992. ISSN 2044-5326. doi: 10.1557/JMR.1992.3348. URL http://journals.cambridge.org/article_S088429140001921X.

[577] Zhang Yu-Hong and Armin Reller. Investigation of mesoporous and microporous nanocrystalline silicon-doped titania. *Materials Letters*, 57(24–25):4108 – 4113, 2003. ISSN 0167-577X. doi: http://dx.doi.org/10.1016/S0167-577X(03)00345-8. URL http://www.sciencedirect.com/science/article/pii/S0167577X03003458.

[578] Dongchen Wu, Fuwei Mao, Zhihui Yang, Shumei Wang, and Zhufa Zhou. Silicon and aluminum co-doping of titania nanoparticles: Effect on thermal stability, particle size and photocatalytic activity. *Materials Science in Semiconductor Processing*, 23:72 – 77, 2014. ISSN 1369-8001. doi: http://dx.doi.org/10.1016/j.mssp.2014.02.040. URL http://www.sciencedirect.com/science/article/pii/S1369800114001115.

[579] L. S. Armstrong, M. J. Walter, J. R. Tuff, O. T. Lord, A. R. Lennie, A. K. Kleppe, and S. M. Clark. Perovskite phase relations in the system cao–mgo–tio₂–sio₂ and implications for deep mantle lithologies. *Journal of Petrology*, 53(3):611–635, 2012. doi: 10.1093/petrology/egr073. URL http://petrology.oxfordjournals.org/content/53/3/611.abstract.

[580] Weimei Shi, Qifeng Chen, Yao Xu, Dong Wu, and Chun fang Huo. Investigation of the silicon concentration effect on si-doped anatase tio₂ by first-principles calculation. *Journal of Solid State Chemistry*, 184(8):1983 – 1988, 2011. ISSN 0022-4596. doi: http://dx.doi.org/10.1016/j.jssc.2011.05.056. URL http://www.sciencedirect.com/science/article/pii/S0022459611003057.

[581] Yan Su, Jiansheng Wu, Xie Quan, and Shuo Chen. Electrochemically assisted photocatalytic degradation of phenol using silicon-doped tio₂ nanofilm electrode. *Desalination*, 252(1–3):143 – 148, 2010. ISSN 0011-9164. doi: http://dx.doi.org/10.1016/j.desal.2009.10.011. URL http://www.sciencedirect.com/science/article/pii/S001191640901220X.

[582] Dorian A.H. Hanaor, Wanqiang Xu, Michael Ferry, and Charles C. Sorrell. Abnormal grain growth of rutile tio₂ induced by zrsio₄. *Journal of Crystal Growth*, 359:83 – 91, 2012. ISSN 0022-0248. doi: http://dx.doi.org/10.1016/j.jcrysgro.2012.08.015. URL http://www.sciencedirect.com/science/article/pii/S0022024812005672.

[583] Yoshitoshi Saito, Takahiro Takei, Shigeo Hayashi, Atsuo Yasumori, and Kiyoshi Okada. Effects of amorphous and crystalline sio₂ additives on γ-al₂o₃-to-α-al₂o₃ phase transitions. *Journal of the American Ceramic Society*, 81(8):2197–2200, 1998. ISSN 1551-2916. doi: 10.1111/j.1151-2916.1998.tb02608.x. URL http://dx.doi.org/10.1111/j.1151-2916.1998.tb02608.x.

[584] Giovanni A. Battiston, Rosalba Gerbasi, Marina Porchia, and Antonio Marigo. Influence of substrate on structural properties of tio₂ thin films obtained via mocvd. *Thin Solid Films*, 239(2):186 – 191, 1994. ISSN 0040-6090. doi: http://dx.doi.org/10.1016/0040-6090(94) 90849-4. URL http://www.sciencedirect.com/science/article/pii/0040609094908494.

[585] Jong-Wha Kim, Do-Oh Kim, and Yoon-Bong Hahn. Effect of rapid thermal annealing on the structural and electrical properties of Tio₂ thin films prepared by plasma enhanced cvd. *Korean Journal of Chemical Engineering*, 15(2):217–222, 1998. ISSN 0256-1115. doi: 10.1007/BF02707075. URL http://dx.doi.org/10.1007/BF02707075.

[586] Dwi Wicaksana, Akihiko Kobayashi, and Akira Kinbara. Process effects on structural properties of tio₂ thin films by reactive sputtering. *Journal of Vacuum Science & Technology A*, 10(4):1479–1482, 1992. doi: http://dx.doi.org/10.1116/1.578269. URL http://scitation. aip.org/content/avs/journal/jvsta/10/4/10.1116/1.578269.

[587] M. D. Wiggins, M. C. Nelson, and C. R. Aita. Phase development in sputter deposited titanium dioxide. *Journal of Vacuum Science & Technology A*, 14(3):772–776, 1996. doi: http://dx.doi.org/10.1116/1.580387. URL http://scitation.aip.org/content/avs/journal/jvsta/ 14/3/10.1116/1.580387.

[588] M. H. Suhail, G. Mohan Rao, and S. Mohan. dc reactive magnetron sputtering of titanium-structural and optical characterization of tio₂ films. *Journal of Applied Physics*, 71(3):1421–1427, 1992. doi: http://dx.doi.org/10.1063/1.351264. URL http://scitation.aip.org/ content/aip/journal/jap/71/3/10.1063/1.351264.

[589] K. Narasimha Rao and S. Mohan. Optical properties of electron-beam evaporated tio₂ films deposited in an ionized oxygen medium. *Journal of Vacuum Science & Technology A*, 8(4):3260–3264, 1990. doi: http://dx.doi.org/10.1116/1.576575. URL http://scitation.aip. org/content/avs/journal/jvsta/8/4/10.1116/1.576575.

[590] Hae-Yong Lee and Ho-Gin Kim. The role of gas-phase nucleation in the preparation of tio₂ films by chemical vapor deposition. *Thin Solid Films*, 229(2):187 – 191, 1993. ISSN 0040-6090. doi: http://dx.doi.org/10.1016/0040-6090(93)90362-S. URL http://www. sciencedirect.com/science/article/pii/004060909390362S.

[591] L. M. Williams and D. W. Hess. Structural properties of titanium dioxide films deposited in an rf glow discharge. *Journal of Vacuum Science & Technology A*, 1(4):1810–1819, 1983. doi: http://dx.doi.org/10.1116/1.572220. URL http://scitation.aip.org/content/avs/ journal/jvsta/1/4/10.1116/1.572220.

[592] Nicolas Martin, Christophe Rousselot, Daniel Rondot, Franck Palmino, and René Mercier. Microstructure modification of amorphous titanium oxide thin films during annealing treatment. *Thin Solid Films*, 300(1):113 – 121, 1997. ISSN 0040-6090. doi: http://dx.doi. org/10.1016/S0040-6090(96)09510-7. URL http://www.sciencedirect.com/science/article/pii/S0040609096095107.

[593] Nicholas T. Nolan, Michael K. Seery, and Suresh C. Pillai. Spectroscopic investigation of the anatase-to-rutile transformation of sol-gel-synthesized tio₂ photocatalysts. *The Journal of Physical Chemistry C*, 113(36):16151–16157, 2009. doi: 10.1021/jp904358g. URL http://dx.doi.org/10.1021/jp904358g.

[594] Yao-Hsuan Tseng, Hong-Ying Lin, Chien-Sheng Kuo, Yuan-Yao Li, and Chin-Pao Huang. Thermostability of nano-tio₂ and its photo-catalytic activity. *Reaction Kinetics and Catalysis Letters*, 89(1):63–69, 2006. ISSN 1588-2837. doi: 10.1007/s11144-006-0087-2. URL http://dx.doi.org/10.1007/s11144-006-0087-2.

[595] Harold H. Kung and Edmond I. Ko. Preparation of oxide catalysts and catalyst supports - a review of recent advances. *The Chemical Engineering Journal and the Biochemical Engineering Journal*, 64(2):203 – 214, 1996. ISSN 0923-0467. doi: http://dx.doi.org/10.1016/ S0923-0467(96)03139-9. URL http://www.sciencedirect.com/science/article/pii/S0923046796031399.

[596] Suresh C. Pillai, Pradeepan Periyat, Reenamole George, Declan E. McCormack, Michael K. Seery, Hugh Hayden, John Colreavy, David Corr, and Steven J. Hinder. Synthesis of high-temperature stable anatase tio₂ photocatalyst. *The Journal of Physical Chemistry C*, 111(4):1605–1611, 2007. doi: 10.1021/jp065933h. URL http://dx.doi.org/10.1021/jp065933h.

[597] I N Kholmanov, E Barborini, S Vinati, P Piseri, A Podestà, C Ducati, C Lenardi, and P Milani. The influence of the precursor clusters on the structural and morphological evolution of nanostructured tio₂ under thermal annealing. *Nanotechnology*, 14(11):1168, 2003. URL http://stacks.iop.org/0957-4484/14/i=11/a=002.

[598] C. Byun, J.W. Jang, I.T. Kim, K.S. Hong, and B.-W. Lee. Anatase-to-rutile transition of titania thin films prepared by mocvd. *Materials Research Bulletin*, 32(4):431 – 440, 1997. ISSN 0025-5408. doi: http://dx.doi.org/10.1016/S0025-5408(96)00203-6. URL http://www.sciencedirect.com/science/article/pii/S0025540896002036.

[599] H. Y. Zhu, Y. Lan, X. P. Gao, S. P. Ringer, Z. F. Zheng, D. Y. Song, and J. C. Zhao. Phase transition between nanostructures of titanate and titanium dioxides via simple wet-chemical reactions. *Journal of the American Chemical Society*, 127(18):6730–6736, 2005. doi: 10.1021/ja044689+. URL http://dx.doi.org/10.1021/ja044689+. PMID: 15869295.

[600] Teruichiro Kubo, Masanori Kato, Yukuaki Mitarai, Junkichi Takahashi, and Ken Ohkura. Structural change of tio₂ and zno by means of mechanical grinding. *The Journal of the Society of Chemical Industry, Japan*, 66(3):318–321, 1963. doi: 10.1246/nikkashi1898.66.3_318.

[601] T Mohanty, A Pradhan, S Gupta, and D Kanjilal. Nanoprecipitation in transparent matrices using an energetic ion beam. *Nanotechnology*, 15(11):1620, 2004. URL http://stacks.iop.org/0957-4484/15/i=11/a=042.

[602] Kenneth J. Klabunde, Jane Stark, Olga Koper, Cathy Mohs, Dong G. Park, Shawn Decker, Yan Jiang, Isabelle Lagadic, and Dajie Zhang. Nanocrystals as stoichiometric reagents with unique surface chemistry. *The Journal of Physical Chemistry*, 100(30):12142–12153, 1996. doi: 10.1021/jp960224x. URL http://dx.doi.org/10.1021/jp960224x.

[603] Liang-Bin Xiong, Jia-Lin Li, Bo Yang, and Ying Yu. Ti³⁺ in the surface of titanium dioxide: Generation, properties and photocatalytic application. *J. Nanomaterials*, 2012:9:9–9:9, jan 2012. ISSN 1687-4110. doi: 10.1155/2012/831524. URL http://dx.doi.org/10.1155/ 2012/831524.

[604] Pier Carlo Ricci, Alberto Casu, Marcello Salis, Riccardo Corpino, and Alberto Anedda. Optically controlled phase variation of tio₂ nanoparticles. *The Journal of Physical Chemistry C*, 114(34):14441–14445, 2010. doi: 10.1021/jp105091d. URL http://dx.doi.org/10. 1021/jp105091d.

[605] Pier Carlo Ricci, Carlo Maria Carbonaro, Luigi Stagi, Marcello Salis, Alberto Casu, Stefano Enzo, and Francesco Delogu. Anatase-to-rutile phase transition in tio₂ nanoparticles irradiated by visible light. *The Journal of Physical Chemistry C*, 117(15):7850–7857, 2013. doi: 10.1021/jp312325h. URL http://dx.doi.org/10.1021/jp312325h.

[606] Yu-Hong Zhang and Armin Reller. Nanocrystalline iron-doped mesoporous titania and its phase transition. *J. Mater. Chem.*, 11: 2537–2541, 2001. doi: 10.1039/B103818B. URL http://dx.doi.org/10.1039/B103818B.

References

[607] André L. A. Parussulo, Manuel F. G. Huila, Koiti Araki, and Henrique E. Toma. N3-dye-induced visible laser anatase-to-rutile phase transition on mesoporous tio$_2$ films. *Langmuir*, 27(15):9094–9099, 2011. doi: 10.1021/la201838z. URL http://dx.doi.org/10.1021/la201838z. PMID: 21707061.

[608] Jing Zhang, Qian Xu, Meijun Li, Zhaochi Feng, and Can Li. Uv raman spectroscopic study on tio$_2$. ii. effect of nanoparticle size on the outer/inner phase transformations. *The Journal of Physical Chemistry C*, 113(5):1698–1704, 2009. doi: 10.1021/jp808013k. URL http://dx.doi.org/10.1021/jp808013k.

[609] A. Chatterjee, S.-B. Wu, P.-W. Chou, M. S. Wong, and C.-L. Cheng. Observation of carbon-facilitated phase transformation of titanium dioxide forming mixed-phase titania by confocal raman microscopy. *Journal of Raman Spectroscopy*, 42(5):1075–1080, 2011. ISSN 1097-4555. doi: 10.1002/jrs.2806. URL http://dx.doi.org/10.1002/jrs.2806.

[610] R. Alexandrescu, I. Morjan, M. Scarisoreanu, R. Birjega, E. Popovici, I. Soare, L. Gavrila-Florescu, I. Voicu, I. Sandu, F. Dumitrache, G. Prodan, E. Vasile, and E. Figgemeier. Structural investigations on tio$_2$ and fe-doped tio$_2$ nanoparticles synthesized by laser pyrolysis. *Thin Solid Films*, 515(24):8438 – 8445, 2007. ISSN 0040-6090. doi: http://dx.doi.org/10.1016/j.tsf.2007.03.106. URL http://www.sciencedirect.com/science/article/pii/S0040609007004348. First International Symposium on Transparent Conducting Oxides.

[611] H-Y Lee, W-L Lan, T Y Tseng, D Hsu, Y-M Chang, and J G Lin. Optical control of phase transformation in fe-doped tio$_2$ nanoparticles. *Nanotechnology*, 20(31):315702, 2009. URL http://stacks.iop.org/0957-4484/20/i=31/a=315702.

[612] Deanna C. Hurum, Alexander G. Agrios, Kimberly A. Gray, Tijana Rajh, and Marion C. Thurnauer. Explaining the enhanced photocatalytic activity of degussa p25 mixed-phase tio$_2$ using epr. *The Journal of Physical Chemistry B*, 107(19):4545–4549, 2003. doi: 10.1021/jp0273934. URL http://dx.doi.org/10.1021/jp0273934.

[613] Ting Xia, Neng Li, Yuliang Zhang, Michael B. Kruger, James Murowchick, Annabella Selloni, and Xiaobo Chen. Directional heat dissipation across the interface in anatase-rutile nanocomposites. *ACS Applied Materials & Interfaces*, 5(20):9883–9890, 2013. doi: 10.1021/am402983k. URL http://dx.doi.org/10.1021/am402983k. PMID: 24090213.

[614] N. Aaron Deskins, Sebastien Kerisit, Kevin M. Rosso, and Michel Dupuis. Molecular dynamics characterization of rutile-anatase interfaces. *The Journal of Physical Chemistry C*, 111(26):9290–9298, 2007. doi: 10.1021/jp0713211. URL http://dx.doi.org/10.1021/jp0713211.

[615] Xiaobo Chen, Lei Liu, Peter Y. Yu, and Samuel S. Mao. Increasing solar absorption for photocatalysis with black hydrogenated titanium dioxide nanocrystals. *Science*, 331(6018):746–750, 2011. doi: 10.1126/science.1200448. URL http://www.sciencemag.org/content/331/6018/746.abstract.

[616] Ke-Rong Zhu, Ming-Sheng Zhang, Qiang Chen, and Zhen Yin. Size and phonon-confinement effects on low-frequency raman mode of anatase tio$_2$ nanocrystal. *Physics Letters A*, 340(1–4):220 – 227, 2005. ISSN 0375-9601. doi: http://dx.doi.org/10.1016/j.physleta.2005.04.008. URL http://www.sciencedirect.com/science/article/pii/S037596010500530X.

[617] Tian Lan, Xiaoli Tang, and Brent Fultz. Phonon anharmonicity of rutile tio$_2$ studied by raman spectrometry and molecular dynamics simulations. *Phys. Rev. B*, 85:094305, Mar 2012. doi: 10.1103/PhysRevB.85.094305. URL http://link.aps.org/doi/10.1103/PhysRevB.85.094305.

[618] Sean Kelly, Fred H. Pollak, and Micha Tomkiewicz. Raman spectroscopy as a morphological probe for tio$_2$ aerogels. *The Journal of Physical Chemistry B*, 101(14):2730–2734, 1997. doi: 10.1021/jp962747a. URL http://dx.doi.org/10.1021/jp962747a.

[619] D. Bersani, P. P. Lottici, and Xing-Zhao Ding. Phonon confinement effects in the raman scattering by tio$_2$ nanocrystals. *Applied Physics Letters*, 72(1):73–75, 1998. doi: http://dx.doi.org/10.1063/1.120648. URL http://scitation.aip.org/content/aip/journal/apl/72/1/10.1063/1.120648.

[620] M. Ivanda, S. Musić, M. Gotić, A. Turković, A.M. Tonejc, and O. Gamulin. The effects of crystal size on the raman spectra of nanophase tio$_2$. *Journal of Molecular Structure*, 480–481:641 – 644, 1999. ISSN 0022-2860. doi: http://dx.doi.org/10.1016/S0022-2860(98)00921-1. URL http://www.sciencedirect.com/science/article/pii/S0022286098009211.

[621] Fang Tian, Yupeng Zhang, Jun Zhang, and Chunxu Pan. Raman spectroscopy: A new approach to measure the percentage of anatase tio$_2$ exposed (001) facets. *The Journal of Physical Chemistry C*, 116(13):7515–7519, 2012. doi: 10.1021/jp301256h. URL http://dx.doi.org/10.1021/jp301256h.

[622] Jing Zhang, Meijun Li, Zhaochi Feng, Jun Chen, and Can Li. Uv raman spectroscopic study on tio$_2$. i. phase transformation at the surface and in the bulk. *The Journal of Physical Chemistry B*, 110(2):927–935, 2006. doi: 10.1021/jp0552473. URL http://dx.doi.org/10.1021/jp0552473. PMID: 16471625.

[623] Norifusa Satoh, Toshio Nakashima, and Kimihisa Yamamoto. Metastability of anatase: size dependent and irreversible anatase-rutile phase transition in atomic-level precise titania. *Scientific Reports*, 3, 2013.

[624] Hengzhong Zhang and Jillian F. Banfield. Understanding polymorphic phase transformation behavior during growth of nanocrystalline aggregates: Insights from tio$_2$. *The Journal of Physical Chemistry B*, 104(15):3481–3487, 2000. doi: 10.1021/jp000499j. URL http://dx.doi.org/10.1021/jp000499j.

[625] Y. Hwu, Y.D. Yao, N.F. Cheng, C.Y. Tung, and H.M. Lin. X-ray absorption of nanocrystal tio$_2$. *Nanostructured Materials*, 9(1):355 – 358, 1997. ISSN 0965-9773. doi: http://dx.doi.org/10.1016/S0965-9773(97)00082-2. URL http://www.sciencedirect.com/science/article/pii/S0965977397000822.

[626] Meng-Hsiu Tsai, Pouyan Shen, and Shuei-Yuan Chen. Defect generation of anatase nanocondensates via coalescence and transformation from dense fluorite-type tio$_2$. *Journal of Applied Physics*, 100(11):114313, 2006. doi: http://dx.doi.org/10.1063/1.2399889. URL http://scitation.aip.org/content/aip/journal/jap/100/11/10.1063/1.2399889.

[627] Ya Zhou and Kristen A. Fichthorn. Microscopic view of nucleation in the anatase-to-rutile transformation. *The Journal of Physical Chemistry C*, 116(14):8314–8321, 2012. doi: 10.1021/jp301228x. URL http://dx.doi.org/10.1021/jp301228x.

[628] R. Lee Penn and Jillian F. Banfield. Formation of rutile nuclei at anatase {112} twin interfaces and the phase transformation mechanism in nanocrystalline titania. *American Mineralogist*, 84(5–6):871–876, 2015. doi: 10.2138/am-1999-5-622.

[629] Chandana Rath, P Mohanty, A C Pandey, and N C Mishra. Oxygen vacancy induced structural phase transformation in tio$_2$ nanoparticles. *Journal of Physics D: Applied Physics*, 42(20):205101, 2009. URL http://stacks.iop.org/0022-3727/42/i=20/a=205101.

[630] Xinyong Li, Xie Quan, and Charles Kutal. Synthesis and photocatalytic properties of quantum confined titanium dioxide nanoparticle. *Scripta Materialia*, 50(4):499 – 505, 2004. ISSN 1359-6462. doi: http://dx.doi.org/10.1016/j.scriptamat.2003.10.031. URL http://www.sciencedirect.com/science/article/pii/S1359646203007188.

[631] P.I Gouma, P.K Dutta, and M.J Mills. Structural stability of titania thin films. *Nanostructured Materials*, 11(8):1231 – 1237, 1999. ISSN 0965-9773. doi: http://dx.doi.org/10.1016/S0965-9773(99)00413-4. URL http://www.sciencedirect.com/science/article/pii/S0965977399004134.

[632] R. Jalilian, G. U. Sumanasekera, H. Chandrasekharan, and M. K. Sunkara. Phonon confinement and laser heating effects in germanium nanowires. *Phys. Rev. B*, 74:155421, Oct 2006. doi: 10.1103/PhysRevB.74.155421. URL http://link.aps.org/doi/10.1103/PhysRevB.74.155421.

[633] Hua Tang and Irving P. Herman. Raman microprobe scattering of solid silicon and germanium at the melting temperature. *Phys. Rev. B*, 43:2299–2304, Jan 1991. doi: 10.1103/PhysRevB.43.2299. URL http://link.aps.org/doi/10.1103/PhysRevB.43.2299.

[634] Hubert H. Burke and Irving P. Herman. Temperature dependence of raman scattering in ge$_{1-x}$si$_x$ alloys. *Phys. Rev. B*, 48:15016–15024, Nov 1993. doi: 10.1103/PhysRevB.48.15016. URL http://link.aps.org/doi/10.1103/PhysRevB.48.15016.

[635] Junwei Wang, Ashish Kumar Mishra, Qing Zhao, and Liping Huang. Size effect on thermal stability of nanocrystalline anatase tio2. *Journal of Physics D: Applied Physics*, 46(25):255303, 2013. URL http://stacks.iop.org/0022-3727/46/i=25/a=255303.

[636] Antonio Tilocca and Annabella Selloni. Reaction pathway and free energy barrier for defect-induced water dissociation on the (101) surface of tio2-anatase. *The Journal of Chemical Physics*, 119(14):7445–7450, 2003. doi: http://dx.doi.org/10.1063/1.1607306. URL http://scitation.aip.org/content/aip/journal/jcp/119/14/10.1063/1.1607306.

[637] Antonio Tilocca and Annabella Selloni. Vertical and lateral order in adsorbed water layers on anatase tio2(101). *Langmuir*, 20(19):8379–8384, 2004. doi: 10.1021/la048937r. URL http://dx.doi.org/10.1021/la048937r. PMID: 15350117.

[638] Antonio Tilocca and Annabella Selloni. Structure and reactivity of water layers on defect-free and defective anatase tio2(101) surfaces. *The Journal of Physical Chemistry B*, 108(15):4743–4751, 2004. doi: 10.1021/jp037685k. URL http://dx.doi.org/10.1021/jp037685k.

[639] Xue-Qing Gong, Annabella Selloni, Matthias Batzill, and Ulrike Diebold. Steps on anatase tio2(101). *Nature*, 5(8):665–670, 2006. doi: http://dx.doi.org/10.1038/nmat1695. URL http://www.nature.com/nmat/journal/v5/n8/suppinfo/nmat1695_S1.html.

[640] Ting Xia, Joseph W. Otto, Tanmoy Dutta, James Murowchick, Anthony N. Caruso, Zhonghua Peng, and Xiaobo Chen. Formation of tio2 nanomaterials via titanium ethylene glycolide decomposition. *Journal of Materials Research*, 28:326–332, 2 2013. ISSN 2044-5326. doi: 10.1557/jmr.2012.239. URL http://journals.cambridge.org/article_S0884291412002397.

[641] Lukas Thulin and John Guerra. Calculations of strain-modified anatase tio2 band structures. *Phys. Rev. B*, 77:195112, May 2008. doi: 10.1103/PhysRevB.77.195112. URL http://link.aps.org/doi/10.1103/PhysRevB.77.195112.

[642] Wan-Jian Yin, Shiyou Chen, Ji-Hui Yang, Xin-Gao Gong, Yanfa Yan, and Su-Huai Wei. Effective band gap narrowing of anatase tio2 by strain along a soft crystal direction. *Applied Physics Letters*, 96(22):221901, 2010. doi: http://dx.doi.org/10.1063/1.3430005. URL http://scitation.aip.org/content/aip/journal/apl/96/22/10.1063/1.3430005.

[643] R.Lee Penn and Jillian F Banfield. Morphology development and crystal growth in nanocrystalline aggregates under hydrothermal conditions: insights from titania. *Geochimica et Cosmochimica Acta*, 63(10):1549 – 1557, 1999. ISSN 0016-7037. doi: 10.1016/S0016-7037(99)00037-X. URL http://www.sciencedirect.com/science/article/pii/S001670379900037X.

[644] O Chauvet, L Forro, I Kos, and M Miljak. Magnetic properties of the anatase phase of tio2. *Solid State Communications*, 93(8):667 – 669, 1995. ISSN 0038-1098. doi: http://dx.doi.org/10.1016/0038-1098(94)00845-0. URL //www.sciencedirect.com/science/article/pii/0038109894008450.

[645] Xiaohui Wei, Ralph Skomski, B. Balamurugan, Z. G. Sun, Stephen Ducharme, and D. J. Sellmyer. Magnetism of tio and tio2 nanoclusters. *Journal of Applied Physics*, 105(7):07C517, 2009. doi: 10.1063/1.3074509. URL http://dx.doi.org/10.1063/1.3074509.

[646] W. T. Geng and Kwang S. Kim. Structural, electronic, and magnetic properties of a ferromagnetic semiconductor: Co-doped tio2 rutile. *Phys. Rev. B*, 68:125203, Sep 2003. doi: 10.1103/PhysRevB.68.125203. URL http://link.aps.org/doi/10.1103/PhysRevB.68.125203.

[647] Nguyen Ngoc Hai, Nguyen The Khoi, and Pham Van Vinh. Preparation and magnetic properties of tio2 doped with v, mn, co, la. *Journal of Physics: Conference Series*, 187(1):012071, 2009. URL http://stacks.iop.org/1742-6596/187/i=1/a=012071.

[648] Mohamed Bahgat, Ahmed A. Farghali, Ahmed F. Moustafa, Mohamed H. Khedr, and Mohassab Y. Mohassab-Ahmed. Electrical, magnetic, and corrosion resistance properties of tio2 nanotubes filled with nife2o4 quantum dots and ni-fe nanoalloy. *Applied Nanoscience*, 3(3):241–249, 2013. ISSN 2190-5517. doi: 10.1007/s13204-012-0122-8. URL http://dx.doi.org/10.1007/s13204-012-0122-8.

[649] Pedro M. Álvarez, Josefa Jaramillo, Francisco López-Pi nero, and Pawel K. Plucinski. Preparation and characterization of magnetic tio2 nanoparticles and their utilization for the degradation of emerging pollutants in water. *Applied Catalysis B: Environmental*, 100 (1–2):338 – 345, 2010. ISSN 0926-3373. doi: http://dx.doi.org/10.1016/j.apcatb.2010.08.010. URL //www.sciencedirect.com/science/article/pii/S0926337310003565.

[650] A Eucken and A Büchner. Die dielecrizitatskonstante schwach polare kristalle and ihre temperaturabhangigkeit. *Z Phys Chem B*, 27: 321–349, 1935.

[651] Shepard Roberts. Dielectric constants and polarizabilities of ions in simple crystals and barium titanate. *Phys. Rev.*, 76:1215–1220, Oct 1949. doi: 10.1103/PhysRev.76.1215. URL http://link.aps.org/doi/10.1103/PhysRev.76.1215.

[652] J. D'Ans, A. Eucken, G. Joos, and W. A. Roth. *Landolt-Börnstein 6 II/6*. Number 483. Springer Verlag, Berlin, 1959.

[653] J D'Ans, P Ten Bruggengate, A Eucken, G Joos, and WA Roth. *Landolt-Börnstein 6 II/8*. Number 2-145. Springer Verlag, Berlin, 1965.

[654] J. R. DeVore. Refractive indices of rutile and sphalerite. *J. Opt. Soc. Am.*, 41(6):416–419, Jun 1951. doi: 10.1364/JOSA.41.000416. URL http://www.osapublishing.org/abstract.cfm?URI=josa-41-6-416.

[655] C. Bärwald. Mineralogische notizen. *Zeitschrift für Kristallographie*, 7(1-6):167–173, 1883. doi: 10.1524/zkri.1883.7.1.167.

[656] Clas Persson and Antonio Ferreira da Silva. Strong polaronic effects on rutile tio2 electronic band edges. *Applied Physics Letters*, 86(23): 231912, 2005. doi: http://dx.doi.org/10.1063/1.1940739. URL http://scitation.aip.org/content/aip/journal/apl/86/23/10.1063/1.1940739.

[657] Pitak Eiamchai, Pongpan Chindaudom, Artorn Pokaipisit, and Pichet Limsuwan. A spectroscopic ellipsometry study of tio2 thin films prepared by ion-assisted electron-beam evaporation. *Current Applied Physics*, 9(3):707 – 712, 2009. ISSN 1567-1739. doi: http://dx.doi.org/10.1016/j.cap.2008.06.011. URL //www.sciencedirect.com/science/article/pii/S1567173908001314.

References

[658] R. van de Krol, A. Goossens, and J. Schoonman. Mott-schottky analysis of nanometer-scale thin-film anatase tio₂. *Journal of The Electrochemical Society*, 144(5):1723–1727, 1997. doi: 10.1149/1.1837668. URL http://jes.ecsdl.org/content/144/5/1723.abstract.

[659] S. A. Campbell, H. S. Kim, D. C. Gilmer, B. He, T. Ma, and W. L. Gladfelter. Titanium dioxide (tio₂)-based gate insulators. *IBM Journal of Research and Development*, 43(3):383–392, May 1999. ISSN 0018-8646. doi: 10.1147/rd.433.0383.

[660] D.-J. Won, C.-H. Wang, H.-K. Jang, and D.-J. Choi. Effects of thermally induced anatase-to-rutile phase transition in mocvd-grown tio₂ films on structural and optical properties. *Applied Physics A*, 73(5):595–600, 2001. ISSN 1432-0630. doi: 10.1007/s003390100804. URL http://dx.doi.org/10.1007/s003390100804.

[661] Jin Young Kim, Hyun Suk Jung, Jung Hong No, Jeong-Ryeol Kim, and Kug Sun Hong. Influence of anatase-rutile phase transformation on dielectric properties of sol-gel derived tio₂ thin films. *Journal of Electroceramics*, 16(4):447–451, 2006. ISSN 1573-8663. doi: 10.1007/s10832-006-9895-z. URL http://dx.doi.org/10.1007/s10832-006-9895-z.

[662] VV Brus, ZD Kovalyuk, OA Parfenyuk, and ND Vakhnyak. Comparison of optical properties of tio₂ thin films prepared by reactive magnetron sputtering and electron-beam evaporation techniques. *Semiconductor physics, quantum electronics and optoelectronics*, 14(4): 427–432, 2011.

[663] Atefeh Taherniya and Davood Raoufi. The annealing temperature dependence of anatase tio₂ thin films prepared by the electron-beam evaporation method. *Semiconductor Science and Technology*, 31(12):125012, 2016. URL http://stacks.iop.org/0268-1242/31/i=12/a=125012.

[664] M Hemissi and H Amardjia-Adnani. Optical and structural properties of titanium oxide thin films prepared by sol-gel method. *Digest journal of nanomaterials and biostructures*, 2(4):299–305, 2007.

[665] J C Manifacier, J Gasiot, and J P Fillard. A simple method for the determination of the optical constants n, k and the thickness of a weakly absorbing thin film. *Journal of Physics E: Scientific Instruments*, 9(11):1002, 1976. URL http://stacks.iop.org/0022-3735/9/i=11/a=032.

[666] Chien H. Peng and Seshu B. Desu. Metalorganic chemical vapor deposition of ferroelectric pb(zr,ti)o₃ thin films. *Journal of the American Ceramic Society*, 77(7):1799–1812, 1994. ISSN 1551-2916. doi: 10.1111/j.1151-2916.1994.tb07054.x. URL http://dx.doi.org/10.1111/j.1151-2916.1994.tb07054.x.

[667] Han-Yong Joo, Hyeong Joon Kim, Sang June Kim, and Sang Youl Kim. Spectrophotometric analysis of aluminum nitride thin films. *Journal of Vacuum Science & Technology A*, 17(3):862–870, 1999. doi: http://dx.doi.org/10.1116/1.582035. URL http://scitation.aip.org/content/avs/journal/jvsta/17/3/10.1116/1.582035.

[668] Jonas Weickert, Ricky B. Dunbar, Holger C. Hesse, Wolfgang Wiedemann, and Lukas Schmidt-Mende. Nanostructured organic and hybrid solar cells. *Advanced Materials*, 23(16):1810–1828, 2011. ISSN 1521-4095. doi: 10.1002/adma.201003991. URL http://dx.doi.org/10.1002/adma.201003991.

[669] Hua Xu, Xiaoqing Chen, Shuxin Ouyang, Tetsuya Kako, and Jinhua Ye. Size-dependent mie's scattering effect on tio₂ spheres for the superior photoactivity of h₂ evolution. *The Journal of Physical Chemistry C*, 116(5):3833–3839, 2012. doi: 10.1021/jp209378t. URL http://dx.doi.org/10.1021/jp209378t.

[670] Prashant Poudel and Qiquan Qiao. One dimensional nanostructure/nanoparticle composites as photoanodes for dye-sensitized solar cells. *Nanoscale*, 4:2826–2838, 2012. doi: 10.1039/C2NR30347G. URL http://dx.doi.org/10.1039/C2NR30347G.

[671] John Tyndall. On the action of rays of high refrangibility upon gaseous matter. *Philosophical Transactions Royal Society London*, 160: 333–365, 1870.

[672] John Tyndall. On the colour of the lake of geneva and the mediterranean sea. *Nature*, 2:489–490, 1870. doi: 10.1038/002489a0.

[673] Gustav Mie. Beiträge zur optik trüber medien, speziell kolloidaler metallösungen. *Annalen der Physik*, 330(3):377–445, 1908. ISSN 1521-3889. doi: 10.1002/andp.19083300302. URL http://dx.doi.org/10.1002/andp.19083300302.

[674] Z. Said, R. Saidur, and N.A. Rahim. Optical properties of metal oxides based nanofluids. *International Communications in Heat and Mass Transfer*, 59:46 – 54, 2014. ISSN 0735-1933. doi: http://dx.doi.org/10.1016/j.icheatmasstransfer.2014.10.010. URL //www.sciencedirect.com/science/article/pii/S0735193314002450.

[675] Andrew T. Young. Rayleigh scattering. *Appl. Opt.*, 20(4):533–535, Feb 1981. doi: 10.1364/AO.20.000533. URL http://ao.osa.org/abstract.cfm?URI=ao-20-4-533.

[676] C.F. Bohren and D.R. Huffman. *Absorption and Scattering of Light by Small Particles*. Wiley Science Series. Wiley, 2008. ISBN 9783527618163. URL https://books.google.de/books?id=ib3EMXXIRXUC.

[677] Andrey B. Evlyukhin, Carsten Reinhardt, Andreas Seidel, Boris S. Luk'yanchuk, and Boris N. Chichkov. Optical response features of si-nanoparticle arrays. *Phys. Rev. B*, 82:045404, Jul 2010. doi: 10.1103/PhysRevB.82.045404. URL http://link.aps.org/doi/10.1103/PhysRevB.82.045404.

[678] A. García-Etxarri, R. Gómez-Medina, L. S. Froufe-Pérez, C. López, L. Chantada, F. Scheffold, J. Aizpurua, M. Nieto-Vesperinas, and J. J. Sáenz. Strong magnetic response of submicron silicon particles in the infrared. *Opt. Express*, 19(6):4815–4826, Mar 2011. doi: 10.1364/OE.19.004815. URL http://www.opticsexpress.org/abstract.cfm?URI=oe-19-6-4815.

[679] Yuan Hsing Fu, Arseniy I. Kuznetsov, Andrey E. Miroshnichenko, Ye Feng Yu, and Boris Luk'yanchuk. Directional visible light scattering by silicon nanoparticles. *Nature Communications*, 4:1527, 2013. doi: 10.1038/ncomms2538. URL http://dx.doi.org/10.1038/ncomms2538.

[680] D. T. Pierce and W. E. Spicer. Electronic structure of amorphous si from photoemission and optical studies. *Phys. Rev. B*, 5:3017–3029, Apr 1972. doi: 10.1103/PhysRevB.5.3017. URL http://link.aps.org/doi/10.1103/PhysRevB.5.3017.

[681] Hongying Yang, Sukang Zhu, and Ning Pan. Studying the mechanisms of titanium dioxide as ultraviolet-blocking additive for films and fabrics by an improved scheme. *Journal of Applied Polymer Science*, 92(5):3201–3210, 2004. ISSN 1097-4628. doi: 10.1002/app.20327. URL http://dx.doi.org/10.1002/app.20327.

[682] A P Popov, A V Priezzhev, Jürgen Lademann, and Risto Myllylä. Effect of multiple scattering of light by titanium dioxide nanoparticles implanted into a superficial skin layer on radiation transmission in different wavelength ranges. *Quantum Electronics*, 37(1):17, 2007. URL http://stacks.iop.org/1063-7818/37/i=1/a=A03.

[683] Shuang Yang, Yu Hou, Jun Xing, Bo Zhang, Feng Tian, Xiao Hua Yang, and Hua Gui Yang. Ultrathin sno_2 scaffolds for tio_2-based heterojunction photoanodes in dye-sensitized solar cells: Oriented charge transport and improved light scattering. *Chemistry – A European Journal*, 19(28):9366–9370, 2013. ISSN 1521-3765. doi: 10.1002/chem.201300524. URL http://dx.doi.org/10.1002/chem.201300524.

[684] Akira Usami and Hajime Ozaki. Optical modeling of nanocrystalline tio_2 films. *The Journal of Physical Chemistry B*, 109(7):2591–2596, 2005. doi: 10.1021/jp040178y. URL http://dx.doi.org/10.1021/jp040178y. PMID: 16851262.

[685] Michael I. Mishchenko and Larry D. Travis. Light scattering by polydispersions of randomly oriented spheroids with sizes comparable to wavelengths of observation. *Appl. Opt.*, 33(30):7206–7225, Oct 1994. doi: 10.1364/AO.33.007206. URL http://ao.osa.org/abstract.cfm?URI=ao-33-30-7206.

[686] M.I. Mishchenko, J.W. Hovenier, and L.D. Travis. *Light Scattering by Nonspherical Particles: Theory, Measurements, and Applications*. Elsevier Science, 1999. ISBN 9780080510200. URL https://books.google.de/books?id=qT3DwjHXA9cC.

[687] Sami Auvinen, Matti Alatalo, Heikki Haario, Juho-Pertti Jalava, and Ralf-Johan Lamminmäki. Size and shape dependence of the electronic and spectral properties in tio_2 nanoparticles. *The Journal of Physical Chemistry C*, 115(17):8484–8493, 2011. doi: 10.1021/jp112114p. URL http://dx.doi.org/10.1021/jp112114p.

[688] Boris Khlebtsov, Vitaly Khanadeev, Timofey Pylaev, and Nikolai Khlebtsov. A new t-matrix solvable model for nanorods: Tem-based ensemble simulations supported by experiments. *The Journal of Physical Chemistry C*, 115(14):6317–6323, 2011. doi: 10.1021/jp2000078. URL http://dx.doi.org/10.1021/jp2000078.

[689] Shermila Brito Singham. Intrinsic optical activity in light scattering from an arbitrary particle. *Chemical Physics Letters*, 130(1–2):139 – 144, 1986. ISSN 0009-2614. doi: http://dx.doi.org/10.1016/0009-2614(86)80441-9. URL //www.sciencedirect.com/science/article/pii/0009261486804419.

[690] Timo Nousiainen, Michael Kahnert, and Hannakaisa Lindqvist. Can particle shape information be retrieved from light-scattering observations using spheroidal model particles? *Journal of Quantitative Spectroscopy and Radiative Transfer*, 112(13):2213 – 2225, Oct 2011. ISSN 0022-4073. doi: http://dx.doi.org/10.1016/j.jqsrt.2011.05.008. URL //www.sciencedirect.com/science/article/pii/S0022407311001968. Polarimetric Detection, Characterization, and Remote Sensing.

[691] M. R. Vant, R. O. Ramseier, and V. Makios. The complex-dielectric constant of sea ice at frequencies in the range 0.1-40 ghz. *Journal of Applied Physics*, 49(3):1264–1280, 1978. doi: 10.1063/1.325018. URL http://dx.doi.org/10.1063/1.325018.

[692] Clifton E. Dungey and Craig F. Bohren. Light scattering by nonspherical particles: a refinement to the coupled-dipole method. *J. Opt. Soc. Am., A*, 8(1):81–87, Jan 1991. doi: 10.1364/JOSAA.8.000081. URL http://josaa.osa.org/abstract.cfm?URI=josaa-8-1-81.

[693] Alejandro V. Arzola, Petr Jákl, Lukáš Chvátal, and Pavel Zemánek. Rotation, oscillation and hydrodynamic synchronization of optically trapped spheroidal microparticles. *Opt. Express*, 22(13):16207–16221, Jun 2014. doi: 10.1364/OE.22.016207. URL http://www.opticsexpress.org/abstract.cfm?URI=oe-22-13-16207.

[694] Horst W Hoyer and Iris L Doerr. Light scattering studies on solutions of decyltrimethylammonium dodecyl sulfate. *The Journal of Physical Chemistry*, 68(12):3494–3497, 1964.

[695] Hiroshi Kimura. Light-scattering properties of fractal aggregates: numerical calculations by a superposition technique and the discrete-dipole approximation. *Journal of Quantitative Spectroscopy and Radiative Transfer*, 70(4–6):581 – 594, 2001. ISSN 0022-4073. doi: http://dx.doi.org/10.1016/S0022-4073(01)00031-0. URL //www.sciencedirect.com/science/article/pii/S0022407301000310. Light Scattering by Non-Spherical Particles.

[696] Jean-Marie Perrin and Philippe L. Lamy. Light scattering by large rough particles. *Optica Acta: International Journal of Optics*, 30(9): 1223–1244, 1983. doi: 10.1080/713821354. URL http://dx.doi.org/10.1080/713821354.

[697] Piotr J. Flatau, Bruce T. Draine, and Graeme L. Stephens. Light scattering by rectangular solids in the discrete-dipole approximation: a new algorithm exploiting the block–toeplitz structure. *J. Opt. Soc. Am. A*, 7(4):593–600, Apr 1990. doi: 10.1364/JOSAA.7.000593. URL http://josaa.osa.org/abstract.cfm?URI=josaa-7-4-593.

[698] Rosario Vilaplana, Fernando Moreno, and Antonio Molina. Computations of the single scattering properties of an ensemble of compact and inhomogeneous rectangular prisms: implications for cometary dust. *Journal of Quantitative Spectroscopy and Radiative Transfer*, 88(1–3):219 – 231, 2004. ISSN 0022-4073. doi: http://dx.doi.org/10.1016/j.jqsrt.2004.01.010. URL //www.sciencedirect.com/science/article/pii/S0022407304001426. Photopolarimetry in remote sensing.

[699] Nadejda L. Cherkas and Sergey L. Cherkas. *Extinction by the long dielectric needles*. Elsevier, 2016.

[700] R. Schiffer and K. O. Thielheim. Light scattering at dielectric needles and circular disks. In *Astronomische Gesellschaft, Wissenschaftliche Tagung, Basel, Switzerland, Sept. 19-23*. Astronomische Gesellschaft, Mitteilungen, no. 43, 1978, p. 289-293. In German., 1977.

[701] R. Schiffer and K. O. Thielheim. Light scattering by dielectric needles and disks. *Journal of Applied Physics*, 50(4):2476–2483, 1979. doi: 10.1063/1.326257. URL http://dx.doi.org/10.1063/1.326257.

[702] T. Yoshizawa. *Handbook of Optical Metrology: Principles and Applications, Second Edition*. CRC Press, 2015. ISBN 9781466573611. URL https://books.google.de/books?id=1Y-9BwAAQBAJ.

[703] Mingzhao Liu, Philippe Guyot-Sionnest, Tae-Woo Lee, and Stephen K. Gray. Optical properties of rodlike and bipyramidal gold nanoparticles from three-dimensional computations. *Phys. Rev. B*, 76:235428, Dec 2007. doi: 10.1103/PhysRevB.76.235428. URL http://link.aps.org/doi/10.1103/PhysRevB.76.235428.

[704] Markus Grießhammer and Alexander Rohrbach. 5d-tracking of a nanorod in a focused laser beam - a theoretical concept. *Opt. Express*, 22(5):6114–6132, Mar 2014. doi: 10.1364/OE.22.006114. URL http://www.opticsexpress.org/abstract.cfm?URI=oe-22-5-6114.

[705] C.A. Balanis. *Antenna Theory: Analysis and Design*. Wiley, 2016. ISBN 9781118642061. URL https://books.google.de/books?id=1FEBCgAAQBAJ.

[706] Pierre Corfdir, Felix Feix, Johannes K Zettler, Sergio Fernández-Garrido, and Oliver Brandt. Importance of the dielectric contrast for the polarization of excitonic transitions in single gan nanowires. *New Journal of Physics*, 17(3):033040, 2015. URL http://stacks.iop.org/1367-2630/17/i=3/a=033040.

[707] Da-Som Kim and Sun-Kyung Kim. Design rules for core/shell nanowire resonant emitters. *Journal of the Korean Physical Society*, 70 (1):7–11, 2017. ISSN 1976-8524. doi: 10.3938/jkps.70.7. URL http://dx.doi.org/10.3938/jkps.70.7.

References

[708] Yannik Fontana, Pierre Corfdir, Barbara Van Hattem, Eleonora Russo-Averchi, Martin Heiss, Samuel Sonderegger, Cesar Magen, Jordi Arbiol, Richard T. Phillips, and Anna Fontcuberta i Morral. Exciton footprint of self-assembled algaas quantum dots in core-shell nanowires. *Phys. Rev. B*, 90:075307, Aug 2014. doi: 10.1103/PhysRevB.90.075307. URL http://link.aps.org/doi/10.1103/PhysRevB.90.075307.

[709] Grzegorz Grzela, Ramón Paniagua-Domínguez, Tommy Barten, Yannik Fontana, José A. Sánchez-Gil, and Jaime Gómez Rivas. Nanowire antenna emission. *Nano Letters*, 12(11):5481–5486, 2012. doi: 10.1021/nl301907f. URL http://dx.doi.org/10.1021/nl301907f. PMID: 23030698.

[710] Sun-Kyung Kim, Xing Zhang, David J. Hill, Kyung-Deok Song, Jin-Sung Park, Hong-Gyu Park, and James F. Cahoon. Doubling absorption in nanowire solar cells with dielectric shell optical antennas. *Nano Letters*, 15(1):753–758, 2015. doi: 10.1021/nl504462e. URL http://dx.doi.org/10.1021/nl504462e. PMID: 25546325.

[711] Zhiqin Zhong, Ziyuan Li, Qian Gao, Zhe Li, Kun Peng, Li Li, Sudha Mokkapati, Kaushal Vora, Jiang Wu, Guojun Zhang, Zhiming Wang, Lan Fu, Hark Hoe Tan, and Chennupati Jagadish. Efficiency enhancement of axial junction inp single nanowire solar cells by dielectric coating. *Nano Energy*, 28:106 – 114, 2016. ISSN 2211-2855. doi: http://dx.doi.org/10.1016/j.nanoen.2016.08.032. URL //www.sciencedirect.com/science/article/pii/S2211285516303202.

[712] Mitsuhiro Terakawa, Yuto Tanaka, Go Obara, Tatsunori Sakano, and Minoru Obara. Randomly-grown high-dielectric-constant zno nanorods for near-field enhanced raman scattering. *Applied Physics A*, 102(3):661–665, 2011. ISSN 1432-0630. doi: 10.1007/s00339-010-6107-0. URL http://dx.doi.org/10.1007/s00339-010-6107-0.

[713] Vlassis Likodimos, Thomas Stergiopoulos, Polycarpos Falaras, Julia Kunze, and Patrik Schmuki. Phase composition, size, orientation, and antenna effects of self-assembled anodized titania nanotube arrays: A polarized micro-raman investigation. *The Journal of Physical Chemistry C*, 112(33):12687–12696, 2008. doi: 10.1021/jp8027462. URL http://dx.doi.org/10.1021/jp8027462.

[714] Q. Xiong, G. Chen, H.R. Gutierrez, and P.C. Eklund. Raman scattering studies of individual polar semiconducting nanowires: phonon splitting and antenna effects. *Applied Physics A*, 85(3):299–305, 2006. ISSN 1432-0630. doi: 10.1007/s00339-006-3717-7. URL http://dx.doi.org/10.1007/s00339-006-3717-7.

[715] P. M. Rafailov, C. Thomsen, K. Gartsman, I. Kaplan-Ashiri, and R. Tenne. Orientation dependence of the polarizability of an individual WS₂ nanotube by resonant raman spectroscopy. *Phys. Rev. B*, 72:205436, Nov 2005. doi: 10.1103/PhysRevB.72.205436. URL http://link.aps.org/doi/10.1103/PhysRevB.72.205436.

[716] T Yu, X Zhao, Z.X Shen, Y.H Wu, and W.H Su. Investigation of individual cuo nanorods by polarized micro-raman scattering. *Journal of Crystal Growth*, 268(3–4):590 – 595, 2004. ISSN 0022-0248. doi: http://dx.doi.org/10.1016/j.jcrysgro.2004.04.097. URL http://www.sciencedirect.com/science/article/pii/S0022024804005172. ICMAT 2003, Symposium H, Compound Semiconductors in Electronic an d Optoelectronic Applications.

[717] Qifeng Zhang, Daniel Myers, Jolin Lan, Samson A. Jenekhe, and Guozhong Cao. Applications of light scattering in dye-sensitized solar cells. *Phys. Chem. Chem. Phys.*, 14:14982–14998, 2012. doi: 10.1039/C2CP43089D. URL http://dx.doi.org/10.1039/C2CP43089D.

[718] J. M. Elson. Theory of light scattering from a rough surface with an inhomogeneous dielectric permittivity. *Phys. Rev. B*, 30:5460–5480, Nov 1984. doi: 10.1103/PhysRevB.30.5460. URL http://link.aps.org/doi/10.1103/PhysRevB.30.5460.

[719] Mark S. Wheeler, J. Stewart Aitchison, and Mohammad Mojahedi. Three-dimensional array of dielectric spheres with an isotropic negative permeability at infrared frequencies. *Phys. Rev. B*, 72:193103, Nov 2005. doi: 10.1103/PhysRevB.72.193103. URL http://link.aps.org/doi/10.1103/PhysRevB.72.193103.

[720] Vassilios Yannopapas and Alexander Moroz. Negative refractive index metamaterials from inherently non-magnetic materials for deep infrared to terahertz frequency ranges. *Journal of Physics: Condensed Matter*, 17(25):3717, 2005. URL http://stacks.iop.org/0953-8984/17/i=25/a=002.

[721] Akram Ahmadi and Hossein Mosallaei. Physical configuration and performance modeling of all-dielectric metamaterials. *Phys. Rev. B*, 77:045104, Jan 2008. doi: 10.1103/PhysRevB.77.045104. URL http://link.aps.org/doi/10.1103/PhysRevB.77.045104.

[722] Jon A. Schuller, Rashid Zia, Thomas Taubner, and Mark L. Brongersma. Dielectric metamaterials based on electric and magnetic resonances of silicon carbide particles. *Phys. Rev. Lett.*, 99:107401, Sep 2007. doi: 10.1103/PhysRevLett.99.107401. URL http://link.aps.org/doi/10.1103/PhysRevLett.99.107401.

[723] Bogdan-Ioan Popa and Steven A. Cummer. Compact dielectric particles as a building block for low-loss magnetic metamaterials. *Phys. Rev. Lett.*, 100:207401, May 2008. doi: 10.1103/PhysRevLett.100.207401. URL http://link.aps.org/doi/10.1103/PhysRevLett.100.207401.

[724] Y. F. Lu, L. Zhang, W. D. Song, Y. W. Zheng, and B. S. Luk'yanchuk. Laser writing of a subwavelength structure on silicon (100) surfaces with particle-enhanced optical irradiation. *Journal of Experimental and Theoretical Physics Letters*, 72(9):457–459, 2000. ISSN 1090-6487. doi: 10.1134/1.1339899. URL http://dx.doi.org/10.1134/1.1339899.

[725] Junling Song, Hong Bin Yang, Xiu Wang, Si Yun Khoo, C. C. Wong, Xue-Wei Liu, and Chang Ming Li. Improved utilization of photogenerated charge using fluorine-doped tio₂ hollow spheres scattering layer in dye-sensitized solar cells. *ACS Applied Materials & Interfaces*, 4(7):3712–3717, 2012. doi: 10.1021/am300801f. URL http://dx.doi.org/10.1021/am300801f. PMID: 22731936.

[726] Rachel A. Caruso, Andrei Susha, and Frank Caruso. Multilayered titania, silica, and laponite nanoparticle coatings on polystyrene colloidal templates and resulting inorganic hollow spheres. *Chemistry of Materials*, 13(2):400–409, 2001. doi: 10.1021/cm001175a. URL http://dx.doi.org/10.1021/cm001175a.

[727] C. G. Granqvist and O. Hunderi. Optical properties of ultrafine gold particles. *Phys. Rev. B*, 16:3513–3534, Oct 1977. doi: 10.1103/PhysRevB.16.3513. URL http://link.aps.org/doi/10.1103/PhysRevB.16.3513.

[728] Rolf Landauer. Electrical conductivity in inhomogeneous media. *AIP Conference Proceedings*, 40(1):2–45, 1978. doi: 10.1063/1.31150. URL http://aip.scitation.org/doi/abs/10.1063/1.31150.

[729] D. E. Aspnes, J. B. Theeten, and F. Hottier. Investigation of effective-medium models of microscopic surface roughness by spectroscopic ellipsometry. *Phys. Rev. B*, 20:3292–3302, Oct 1979. doi: 10.1103/PhysRevB.20.3292. URL http://link.aps.org/doi/10.1103/PhysRevB.20.3292.

[730] Hyoungwon Park, Kyeong-Jae Byeon, Ki-Yeon Yang, Joong-Yeon Cho, and Heon Lee. The fabrication of a patterned zno nanorod array for high brightness leds. *Nanotechnology*, 21(35):355304, 2010. URL http://stacks.iop.org/0957-4484/21/i=35/a=355304.

[731] D. A. G. Bruggeman. Berechnung verschiedener physikalischer konstanten von heterogenen substanzen. i. dielektrizitätskonstanten und leitfähigkeiten der mischkörper aus isotropen substanzen. *Annalen der Physik*, 416(7):636–664, 1935. ISSN 1521-3889. doi: 10.1002/andp.19354160705. URL http://dx.doi.org/10.1002/andp.19354160705.

[732] Shang-Di Mo and W. Y. Ching. Electronic and optical properties of three phases of titanium dioxide: Rutile, anatase, and brookite. *Phys. Rev. B*, 51:13023–13032, May 1995. doi: 10.1103/PhysRevB.51.13023. URL http://link.aps.org/doi/10.1103/PhysRevB.51.13023.

[733] S.P. Kowalczyk, F.R. McFeely, L. Ley, V.T. Gritsyna, and D.A. Shirley. The electronic structure of srtio3 and some simple related oxides (mgo, al2o3, sro, tio2). *Solid State Communications*, 23(3):161 – 169, 1977. ISSN 0038-1098. doi: http://dx.doi.org/10.1016/0038-1098(77)90101-6. URL http://www.sciencedirect.com/science/article/pii/0038109877901016.

[734] R. Sanjinés, H. Tang, H. Berger, F. Gozzo, G. Margaritondo, and F. Lévy. Electronic structure of anatase tio2 oxide. *Journal of Applied Physics*, 75(6):2945–2951, 1994. doi: 10.1063/1.356190. URL http://dx.doi.org/10.1063/1.356190.

[735] J. Pascual, J. Camassel, and H. Mathieu. Fine structure in the intrinsic absorption edge of tio2. *Phys. Rev. B*, 18:5606–5614, Nov 1978. doi: 10.1103/PhysRevB.18.5606. URL https://link.aps.org/doi/10.1103/PhysRevB.18.5606.

[736] N. Daude, C. Gout, and C. Jouanin. Electronic band structure of titanium dioxide. *Phys. Rev. B*, 15:3229–3235, Mar 1977. doi: 10.1103/PhysRevB.15.3229. URL http://link.aps.org/doi/10.1103/PhysRevB.15.3229.

[737] Damián Monllor-Satoca, Roberto Gómez, Manuel González-Hidalgo, and Pedro Salvador. The "direct-indirect" model: An alternative kinetic approach in heterogeneous photocatalysis based on the degree of interaction of dissolved pollutant species with the semiconductor surface. *Catalysis Today*, 129(1):247 – 255, 2007. ISSN 0920-5861. doi: https://doi.org/10.1016/j.cattod.2007.08.002. URL http://www.sciencedirect.com/science/article/pii/S0920586107005032. Selected Contributions of the 4th European Meeting on Solar Chemistry and Photocatalysis: Environmental Applications (SPEA 4).

[738] Sergio Valencia, Juan Miguel Marín, and Gloria Restrepo. Study of the bandgap of synthesized titanium dioxide nanoparticules using the sol-gel method and a hydrothermal treatment. *The Open Materials Science Journal*, 4:9–14, 2009. doi: 10.2174/1874088X01004010009.

[739] Th. Dittrich. Porous Tio2: Electron transport and application to dye sensitized injection solar cells. *physica status solidi (a)*, 182(1):447–455, 2000. ISSN 1521-396X. doi: 10.1002/1521-396X(200011)182:1<447::AID-PSSA447>3.0.CO;2-G. URL http://dx.doi.org/10.1002/1521-396X(200011)182:1<447::AID-PSSA447>3.0.CO;2-G.

[740] Shafeer Kalathil, Mohammad Mansoob Khan, Arghya Narayan Banerjee, Jintae Lee, and Moo Hwan Cho. A simple biogenic route to rapid synthesis of au@tio2 nanocomposites by electrochemically active biofilms. *Journal of Nanoparticle Research*, 14(8):1051, Jul 2012. ISSN 1572-896X. doi: 10.1007/s11051-012-1051-x. URL https://doi.org/10.1007/s11051-012-1051-x.

[741] K Madhusudan Reddy, Sunkara V Manorama, and A Ramachandra Reddy. Bandgap studies on anatase titanium dioxide nanoparticles. *Materials Chemistry and Physics*, 78(1):239 – 245, 2003. ISSN 0254-0584. doi: https://doi.org/10.1016/S0254-0584(02)00343-7. URL http://www.sciencedirect.com/science/article/pii/S0254058402003437.

[742] R. Zallen and M.P. Moret. The optical absorption edge of brookite tio2. *Solid State Communications*, 137(3):154 – 157, 2006. ISSN 0038-1098. doi: https://doi.org/10.1016/j.ssc.2005.10.024. URL http://www.sciencedirect.com/science/article/pii/S0038109805008896.

[743] M Koelsch, S Cassaignon, J.F Guillemoles, and J.P Jolivet. Comparison of optical and electrochemical properties of anatase and brookite tio2 synthesized by the sol-gel method. *Thin Solid Films*, 403-404(Supplement C):312 – 319, 2002. ISSN 0040-6090. doi: https://doi.org/10.1016/S0040-6090(01)01509-7. URL http://www.sciencedirect.com/science/article/pii/S0040609001015097. Proceedings of Symposium P on Thin Film Materials for Photovoltaics.

[744] Hu Young Jeong, Jeong Yong Lee, and Sung-Yool Choi. Direct observation of microscopic change induced by oxygen vacancy drift in amorphous tio2(110) thin films. *Applied Physics Letters*, 97(4):042109, 2010. doi: 10.1063/1.3467854. URL http://dx.doi.org/10.1063/1.3467854.

[745] Eunae Cho, Seungwu Han, Hyo-Shin Ahn, Kwang-Ryeol Lee, Seong Keun Kim, and Cheol Seong Hwang. First-principles study of point defects in rutile Tio2−x. *Phys. Rev. B*, 73:193202, May 2006. doi: 10.1103/PhysRevB.73.193202. URL https://link.aps.org/doi/10.1103/PhysRevB.73.193202.

[746] John Robertson. Band offsets of wide-band-gap oxides and implications for future electronic devices. *Journal of Vacuum Science & Technology B: Microelectronics and Nanometer Structures Processing, Measurement, and Phenomena*, 18(3):1785–1791, 2000. doi: 10.1116/1.591472. URL http://avs.scitation.org/doi/abs/10.1116/1.591472.

[747] H. Tang, H. Berger, P.E. Schmid, F. Lévy, and G. Burri. Photoluminescence in tio2 anatase single crystals. *Solid State Communications*, 87(9):847 – 850, 1993. ISSN 0038-1098. doi: http://dx.doi.org/10.1016/0038-1098(93)90427-O. URL http://www.sciencedirect.com/science/article/pii/003810989390427O.

[748] H. Tang, F. Lévy, H. Berger, and P. E. Schmid. Urbach tail of anatase tio2. *Phys. Rev. B*, 52:7771–7774, Sep 1995. doi: 10.1103/PhysRevB.52.7771. URL http://link.aps.org/doi/10.1103/PhysRevB.52.7771.

[749] V. E. Henrich, P. A. Cox, and Ulrike Diebold. The surface science of metal oxides. *Physics Today*, 48(2):58, 1995. doi: 10.1063/1.2807916. URL http://dx.doi.org/10.1063/1.2807916.

[750] M. Henzler and W. Göpel. *Oberflächenphysik des Festkörpers*. Teubner Studienbücher Physik. Vieweg+Teubner Verlag, 2013. ISBN 9783322966964. URL https://books.google.de/books?id=pGrwBgAAQBAJ.

[751] Giovanni Cangiani. *Ab-initio study of the properties of TiO2 rutile and anatase polytypes*. PhD thesis, École polytechnique fédérale de Lausanne (EPFL), 2002.

[752] J. R. Bellingham, W. A. Phillips, and C. J. Adkins. Intrinsic performance limits in transparent conducting oxides. *Journal of Materials Science Letters*, 11(5):263–265, Jan 1992. ISSN 1573-4811. doi: 10.1007/BF00729407. URL https://doi.org/10.1007/BF00729407.

[753] Suman Nandy, Arghya Banerjee, Elvira Fortunato, and Rodrigo Martins. A review on cu2o and cu^f-based p-type semiconducting transparent oxide materials: Promising candidates for new generation oxide based electronics. *Reviews in Advanced Sciences and Engineering*, 2(4):273–304, 2013. doi: 10.1166/rase.2013.1045. URL http://www.ingentaconnect.com/content/asp/rase/2013/00000002/00000004/art00004.

[754] Subarna Banerjee, Susanta K. Mohapatra, and Mano Misra. Water photooxidation by tisi2-tio2 nanotubes. *The Journal of Physical Chemistry C*, 115(25):12643–12649, 2011. doi: 10.1021/jp106879p. URL http://dx.doi.org/10.1021/jp106879p.

[755] R. Asahi, Y. Taga, W. Mannstadt, and A. J. Freeman. Electronic and optical properties of anatase tio₂. *Phys. Rev. B*, 61:7459–7465, Mar 2000. doi: 10.1103/PhysRevB.61.7459. URL http://link.aps.org/doi/10.1103/PhysRevB.61.7459.

[756] M Landmann, E Rauls, and W G Schmidt. The electronic structure and optical response of rutile, anatase and brookite tio₂. *Journal of Physics: Condensed Matter*, 24(19):195503, 2012. URL http://stacks.iop.org/0953-8984/24/i=19/a=195503.

[757] Olga Dulub, Cristiana Di Valentin, Annabella Selloni, and Ulrike Diebold. Structure, defects, and impurities at the rutile tio₂ (011)-(2x1) surface: A scanning tunneling microscopy study. *SurfaceScience*, 600:4407–4417, 2006.

[758] J. Ashkenazi and T. Chuchem. Band structure of v2o3, and ti2o3. *Philosophical Magazine*, 32(4):763–785, 1975. doi: 10.1080/14786437508221619. URL http://dx.doi.org/10.1080/14786437508221619.

[759] V. Ern and A. C. Switendick. Electronic band structure of tic, tin, and tio. *Phys. Rev.*, 137:A1927–A1936, Mar 1965. doi: 10.1103/PhysRev.137.A1927. URL http://link.aps.org/doi/10.1103/PhysRev.137.A1927.

[760] L. F. Mattheiss. Electronic structure of the 3d transition-metal monoxides. ii. interpretation. *Phys. Rev. B*, 5:306–315, Jan 1972. doi: 10.1103/PhysRevB.5.306. URL http://link.aps.org/doi/10.1103/PhysRevB.5.306.

[761] Cheng Li, Dan Credgington, Doo-Hyun Ko, Zhuxia Rong, Jianpu Wang, and Neil C. Greenham. Built-in potential shift and schottky-barrier narrowing in organic solar cells with uv-sensitive electron transport layers. *Phys. Chem. Chem. Phys.*, 16:12131–12136, 2014. doi: 10.1039/C4CP01251H. URL http://dx.doi.org/10.1039/C4CP01251H.

[762] David O. Scanlon, Charles W. Dunnill, John Buckeridge, Stephen A. Shevlin, Andrew J. Logsdail, Scott M. Woodley, C. Richard A. Catlow, Michael. J. Powell, Robert G. Palgrave, Ivan P. Parkin, Graeme W. Watson, Thomas W. Keal, Paul Sherwood, Aron Walsh, and Alexey A. Sokol. Band alignment of rutile and anatase tio₂. *Nature Materials*, 12:798–801, 2013. doi: 10.1038/nmat3697.

[763] Michael Grätzel and François P. Rotzinger. The influence of the crystal lattice structure on the conduction band energy of oxides of titanium(iv). *Chemical Physics Letters*, 118(5):474 – 477, 1985. ISSN 0009-2614. doi: http://dx.doi.org/10.1016/0009-2614(85)85335-5. URL http://www.sciencedirect.com/science/article/pii/0009261485853355.

[764] H. Mathieu, J. Pascual, and J. Camassel. Uniaxial stress dependence of the direct-forbidden and indirect-allowed transitions of tio₂. *Phys. Rev. B*, 18:6920–6929, Dec 1978. doi: 10.1103/PhysRevB.18.6920. URL https://link.aps.org/doi/10.1103/PhysRevB.18.6920.

[765] Haipeng Tang. *Electronic properties of anatase TiO₂ investigated by electrical and optical measurements on single crystals and thin films*. PhD thesis, Ecole polytechnique fédérale de Lausanne (EPFL), 1995.

[766] G. Xiong, R. Shao, T.C. Droubay, A.G. Joly, K.M. Beck, S.A. Chambers, and W.P. Hess. Photoemission electron microscopy of tio₂ anatase films embedded with rutile nanocrystals. *Advanced Functional Materials*, 17(13):2133–2138, 2007. ISSN 1616-3028. doi: 10.1002/adfm.200700146. URL http://dx.doi.org/10.1002/adfm.200700146.

[767] Jianjun Tian and Guozhong Cao. Control of nanostructures and interfaces of metal oxide semiconductors for quantum-dots-sensitized solar cells. *The Journal of Physical Chemistry Letters*, 6(10):1859–1869, 2015. doi: 10.1021/acs.jpclett.5b00301. URL http://dx.doi.org/10.1021/acs.jpclett.5b00301. PMID: 26263261.

[768] Yong Xu and Martin A. A. Schoonen. The absolute energy positions of conduction and valence bands of selected semiconducting minerals. *American Mineralogist*, 85(3-4):543–556, 2015. doi: doi:10.2138/am-2000-0416.

[769] A Cultrera, L Boarino, G Amato, and C Lamberti. Band-gap states in unfilled mesoporous nc-tio₂: measurement protocol for electrical characterization. *Journal of Physics D: Applied Physics*, 47(1):015102, 2014. URL http://stacks.iop.org/0022-3727/47/i=1/a=015102.

[770] David Cahen, Gary Hodes, Michael Grätzel, Jean François Guillemoles, and Ilan Riess. Nature of photovoltaic action in dye-sensitized solar cells. *The Journal of Physical Chemistry B*, 104(9):2053–2059, 2000. doi: 10.1021/jp993187t. URL http://dx.doi.org/10.1021/jp993187t.

[771] Jian Cao, Jing-Zhi Sun, Jian Hong, Xin-Guo Yang, Hong-Zheng Chen, and Mang Wang. Direct observation of microscopic photoinduced charge redistribution on tio₂ film sensitized by chloroaluminum phthalocyanine and perylenediimide. *Applied Physics Letters*, 83(9):1896–1898, 2003. doi: 10.1063/1.1608490. URL http://dx.doi.org/10.1063/1.1608490.

[772] J. E. Lyon, M. K. Rayan, M. M. Beerbom, and R. Schlaf. Electronic structure of the indium tin oxide/nanocrystalline anatase (tio₂)/ruthenium-dye interfaces in dye-sensitized solar cells. *Journal of Applied Physics*, 104(7):073714, 2008. doi: 10.1063/1.2963358. URL http://aip.scitation.org/doi/abs/10.1063/1.2963358.

[773] Sebastian Gutmann, Matthäus A. Wolak, Matthew Conrad, Martin M. Beerbom, and Rudy Schlaf. Electronic structure of indium tin oxide/nanocrystalline tio₂ interfaces as used in dye-sensitized solar cell devices. *Journal of Applied Physics*, 109(11):113719, 2011. doi: 10.1063/1.3596544. URL http://dx.doi.org/10.1063/1.3596544.

[774] Shuai Zhong, Sibin Duan, and Yimin Cui. Electrode dependence of resistive switching in au/ni-au nanoparticle devices. *RSC Adv.*, 4: 40924–40929, 2014. doi: 10.1039/C4RA05662K. URL http://dx.doi.org/10.1039/C4RA05662K.

[775] Y. Gassenbauer, R. Schafranek, A. Klein, S. Zafeiratos, M. Hävecker, A. Knop-Gericke, and R. Schlögl. Surface states, surface potentials, and segregation at surfaces of tin-doped in2o3. *Phys. Rev.*, 73:245312, Jun 2006. doi: 10.1103/PhysRevB.73.245312. URL https://link.aps.org/doi/10.1103/PhysRevB.73.245312.

[776] Shujie Wang, Xingtang Zhang, Gang Cheng, Xiaohong Jiang, Yuncai Li, Yabin Huang, and Zuliang Du. Study on electronic transport properties of wo3/tio₂ nanocrystalline thin films by photoassisted scanning electron force microscopy. *Chemical Physics Letters*, 405 (1–3):63 – 67, 2005. ISSN 0009-2614. doi: http://dx.doi.org/10.1016/j.cplett.2005.01.118. URL http://www.sciencedirect.com/science/article/pii/S0009261405001089.

[777] Sven Rühle and David Cahen. Electron tunneling at the tio₂/substrate interface can determine dye-sensitized solar cell performance. *The Journal of Physical Chemistry B*, 108(46):17946–17951, 2004. doi: 10.1021/jp047686s. URL http://dx.doi.org/10.1021/jp047686s.

[778] A K K Kyaw, X W Sun, J L Zhao, J X Wang, D W Zhao, X F Wei, X W Liu, H V Demir, and T Wu. Top-illuminated dye-sensitized solar cells with a room-temperature-processed zno photoanode on metal substrates and a pt-coated ga-doped zno counter electrode. *Journal of Physics D: Applied Physics*, 44(4):045102, 2011. URL http://stacks.iop.org/0022-3727/44/i=4/a=045102.

[779] Xiao Wei, Tengfeng Xie, Dan Xu, Qidong Zhao, Shan Pang, and Dejun Wang. A study of the dynamic properties of photo-induced charge carriers at nanoporous tio 2 / conductive substrate interfaces by the transient photovoltage technique. *Nanotechnology*, 19(27): 275707, 2008. URL http://stacks.iop.org/0957-4484/19/i=27/a=275707.

References

[780] M. Turrión, J. Bisquert, and P. Salvador. Flatband potential of f:sno$_2$ in a tio$_2$ dye-sensitized solar cell: An interference reflection study. *The Journal of Physical Chemistry B*, 107(35):9397–9403, 2003. doi: 10.1021/jp034774o. URL http://dx.doi.org/10.1021/jp034774o.

[781] M. G. Helander, M. T. Greiner, Z. B. Wang, W. M. Tang, and Z. H. Lu. Work function of fluorine doped tin oxide. *Journal of Vacuum Science & Technology A: Vacuum, Surfaces, and Films*, 29(1):011019, 2011. doi: 10.1116/1.3525641. URL https://doi.org/10.1116/1.3525641.

[782] HL Hartnagel, AL Dawar, AK Jain, and C Jagadish. Semiconducting transparent thin films institute of physics publishing. *Bristol, Philadelphia, PA*, 1995.

[783] Shuqin Zhou, Yunqi Liu, Yu Xu, Wenping Hu, Daoben Zhu, Xiaohui Qiu, Chen Wang, and Chunli Bai. Rectifying behaviors of langmuir–blodgett films of an asymmetrically substituted phthalocyanine. *Chemical Physics Letters*, 297(1):77 – 82, 1998. ISSN 0009-2614. doi: https://doi.org/10.1016/S0009-2614(98)01097-5. URL http://www.sciencedirect.com/science/article/pii/S0009261498010975.

[784] Yang Li, Cheng-Yan Xu, and Liang Zhen. Surface potential and interlayer screening effects of few-layer mos$_2$ nanoflakes. *Applied Physics Letters*, 102(14):143110, 2013. doi: 10.1063/1.4801844. URL http://dx.doi.org/10.1063/1.4801844.

[785] Helmut Jungblut, Sheelagh A. Campbell, Michael Giersig, Daniel J. Muller, and Hans Joachim Lewerenz. Scanning tunnelling microscopy observations of biomolecules on layered materials. *Faraday Discuss.*, 94:183–197, 1992. doi: 10.1039/FD9929400183. URL http://dx.doi.org/10.1039/FD9929400183.

[786] Thomas Heim, Dominique Deresmes, and Dominique Vuillaume. Conductivity of dna probed by conducting-atomic force microscopy: Effects of contact electrode, dna structure, and surface interactions. *Journal of Applied Physics*, 96(5):2927–2936, 2004. doi: 10.1063/1.1769606. URL https://doi.org/10.1063/1.1769606.

[787] J. W. G. Wildöer, C. J. P. M. Harmans, and H. van Kempen. Observation of landau levels at the inas(110) surface by scanning tunneling spectroscopy. *Phys. Rev. B*, 55:R16013–R16016, Jun 1997. doi: 10.1103/PhysRevB.55.R16013. URL https://link.aps.org/doi/10.1103/PhysRevB.55.R16013.

[788] Fengmei Gao, Jinju Zheng, Mingfang Wang, Guodong Wei, and Weiyou Yang. Piezoresistance behaviors of p-type 6h-sic nanowires. *Chem. Commun.*, 47:11993–11995, 2011. doi: 10.1039/C1CC14343C. URL http://dx.doi.org/10.1039/C1CC14343C.

[789] Herbert B. Michaelson. The work function of the elements and its periodicity. *Journal of Applied Physics*, 48(11):4729–4733, 1977. doi: http://dx.doi.org/10.1063/1.323539. URL http://scitation.aip.org/content/aip/journal/jap/48/11/10.1063/1.323539.

[790] D. E. Eastman. Photoelectric work functions of transition, rare-earth, and noble metals. *Phys. Rev. B*, 2:1–2, Jul 1970. doi: 10.1103/PhysRevB.2.1. URL https://link.aps.org/doi/10.1103/PhysRevB.2.1.

[791] J.S. Kim, B. Lägel, E. Moons, N. Johansson, I.D. Baikie, W.R. Salaneck, R.H. Friend, and F. Cacialli. Kelvin probe and ultraviolet photoemission measurements of indium tin oxide work function: a comparison. *Synthetic Metals*, 111-112(Supplement C):311 – 314, 2000. ISSN 0379-6779. doi: https://doi.org/10.1016/S0379-6779(99)00354-9. URL http://www.sciencedirect.com/science/article/pii/S0379677999003549.

[792] Ken Onda, Bin Li, and Hrvoje Petek. Two-photon photoemission spectroscopy of Tio$_2$(110) surfaces modified by defects and o$_2$ or h$_2$O adsorbates. *Phys. Rev. B*, 70:045415, Jul 2004. doi: 10.1103/PhysRevB.70.045415. URL https://link.aps.org/doi/10.1103/PhysRevB.70.045415.

[793] Guangming Liu, W. Jaegermann, Jianjun He, Villy Sundström, and Licheng Sun. Xps and ups characterization of the tio$_2$/znpcgly heterointerface: Alignment of energy levels. *The Journal of Physical Chemistry B*, 106(23):5814–5819, 2002. doi: 10.1021/jp014192b. URL http://dx.doi.org/10.1021/jp014192b.

[794] R.A. Smith. *Semiconductors*. Cambridge University Press, London, 1979.

[795] S.M. Sze. *Physics of Semiconductor Devices*. Wiley, 1981. ISBN 9780470068304. URL https://books.google.de/books?id=o4unkmHBHb8C.

[796] J.S. Blackmore. Carrier concentrations and fermi level in semiconductors. *Electronic Communications*, 29(131), 1952.

[797] H.Kordi Ardakani. Electrical and optical properties of in situ "hydrogen-reduced" titanium dioxide thin films deposited by pulsed excimer laser ablation. *Thin Solid Films*, 248(2):234 – 239, 1994. ISSN 0040-6090. doi: http://dx.doi.org/10.1016/0040-6090(94)90017-5. URL http://www.sciencedirect.com/science/article/pii/0040609094900175.

[798] Manuela Jakob, Haim Levanon, and Prashant V. Kamat. Charge distribution between uv-irradiated tio$_2$ and gold nanoparticles: Determination of shift in the fermi level. *Nano Letters*, 3(3):353–358, 2003. doi: 10.1021/nl0340071. URL https://doi.org/10.1021/nl0340071.

[799] David Cahen, Micheal Grätzel, Jean Francois Guillemoles, and Gary Hodes. *Dye-sensitized Solar Cells: Principles of Operation*, chapter Dye Sensitized Solar Cells: Principles of Operation, page 201. Wiley, 2001. ISBN 9783527298365. URL https://books.google.de/books?id=mbVRAAAAMAAJ.

[800] J. Pascual, J. Camassel, and H. Mathieu. Resolved quadrupolar transition in tio$_2$. *Phys. Rev. Lett.*, 39:1490–1493, Dec 1977. doi: 10.1103/PhysRevLett.39.1490. URL https://link.aps.org/doi/10.1103/PhysRevLett.39.1490.

[801] Marius D Stamate. On the non-linear i–v characteristics of dc magnetron sputtered tio$_2$ thin films. *Applied Surface Science*, 205(1):353 – 357, 2003. ISSN 0169-4332. doi: https://doi.org/10.1016/S0169-4332(02)01130-3. URL http://www.sciencedirect.com/science/article/pii/S0169433202011303.

[802] S Monticone, R Tufeu, A.V Kanaev, E Scolan, and C Sanchez. Quantum size effect in tio$_2$ nanoparticles: does it exist? *Applied Surface Science*, 162(Supplement C):565 – 570, 2000. ISSN 0169-4332. doi: https://doi.org/10.1016/S0169-4332(00)00251-8. URL http://www.sciencedirect.com/science/article/pii/S0169433200002518.

[803] K Vos. Reflectance and electroreflectance of tio$_2$ single crystals. ii. assignment to electronic energy levels. *Journal of Physics C: Solid State Physics*, 10(19):3917, 1977. URL http://stacks.iop.org/0022-3719/10/i=19/a=024.

[804] Marks TJ Facchetti A. *Transparent electronics: from synthesis to applications*. Wiley, West Sussex, 2010.

[805] A.N. Banerjee and K.K. Chattopadhyay. Recent developments in the emerging field of crystalline p-type transparent conducting oxide thin films. *Progress in Crystal Growth and Characterisation of Materials*, 50(1-3):52 – 105, 2005. ISSN 0960-8974. doi: http://dx.doi.org/10.1016/j.pcrysgrow.2005.09.001. URL http://www.sciencedirect.com/science/article/pii/S0960897405000409.

References

[806] Hiroshi Kawazoe, Hiroshi Yanagi, Kazushige Ueda, and Hideo Hosono. Transparent p-type conducting oxides: Design and fabrication of p-n heterojunctions. *MRS Bulletin*, 25:28–36, 8 2000. ISSN 1938-1425. doi: 10.1557/mrs2000.148. URL http://journals.cambridge. org/article_S0883769400027147.

[807] K. F. Brennan. *The Physics of Semiconductors: With Applications to Optoelectronic Devices*. Cambridge University Press, 1999. ISBN 9780521596626. URL https://books.google.de/books?id=6JE1XbZpX3IC.

[808] H. T. Grahn. *Introduction to Semiconductor Physics*. 1999. ISBN 9789813105157. URL https://books.google.de/books?id=oWdIDQAAQBAJ.

[809] Mark Lundstrom. *Fundamentals of Carrier Transport*. Cambridge University Press, 2009. ISBN 9780521637244. URL https://books. google.de/books?id=Vrkd_dSC3zwC.

[810] Karlheinz Seeger. *Semiconductor Physics: An Introduction*. Springer-Verlag Berlin Heidelberg, 2004. doi: 10.1007/978-3-662-09855-4.

[811] V.F. Gantmakher and Y.B. Levinson. *Carrier Scattering in Metals and Semiconductors*. Modern Problems in Condensed Matter Sciences. Elsevier Science, 2012. ISBN 9780444598233. URL https://books.google.de/books?id=svWcxGzmqnUC.

[812] B.K. Ridley. *Quantum Processes in Semiconductors*. OUP Oxford, 2013. ISBN 9780199677214. URL https://books.google.de/books?id= 9ZIeAAAAQBAJ.

[813] Włodzimierz Zawadzki. *Mechanisms of electron scattering in semiconductors*. Zakład Narodowy imienia Ossolińskich Wydawnictwo Polskiej Akademii Nauk, 1979.

[814] L. Boltzmann. Weitere studien über das wärmegleichgewicht unter gasmoleküle. *Sitzungsberichte Akad. Wiss. Wien*, 66:275–370, 1872. Translation: Further studies on the thermal equilibrium of gas molecules. In: S. Brush (ed.). Kinetic Theory, Vol. 2. Pergamon Press, Oxford (1966), 88-174.

[815] Carlo Cercignani. *The Boltzmann Equation*, chapter The Boltzmann Equation and Its Applications, pages 40–103. Springer New York, New York, NY, 1988. ISBN 978-1-4612-1039-9. doi: 10.1007/978-1-4612-1039-9_2. URL http://dx.doi.org/10.1007/978-1-4612-1039-9_2.

[816] H. Babovsky. *Die Boltzmann-Gleichung: Modellbildung - Numerik - Anwendungen*. Leitfäden der angewandten Mathematik und Mechanik. Vieweg+Teubner Verlag, 2013. ISBN 9783663120346. URL https://books.google.de/books?id=WCegBwAAQBAJ.

[817] F Bufler and A Schenk. Halbleitertransporttheorie und monte-carlo-bauelementsimulation. *Lecture Notes, ETH Zürich, Switzerland*, 2002.

[818] A. Jüngel. *Transport Equations for Semiconductors*. Lecture Notes in Physics. Springer Berlin Heidelberg, 2009. ISBN 9783540895268. URL https://books.google.de/books?id=C01sCQAAQBAJ.

[819] Debdeep Jena. Charge transport in semiconductors. *E-Publishing, University of Notre Dame*, 2004.

[820] J. H. Becker and W. R. Hosler. Multiple-band conduction in n-type rutile (tio₂). *Phys. Rev.*, 137:A1872–A1877, Mar 1965. doi: 10.1103/PhysRev.137.A1872. URL https://link.aps.org/doi/10.1103/PhysRev.137.A1872.

[821] F. M. Smits. Measurement of sheet resistivities with the four-point probe. *Bell System Technical Journal*, 37(3):711–718, 1958. ISSN 1538-7305. doi: 10.1002/j.1538-7305.1958.tb03883.x. URL http://dx.doi.org/10.1002/j.1538-7305.1958.tb03883.x.

[822] T.C. Tsai W.E. Beadle, R.D. Plummer. Quick reference manual for semiconductor engineers. Technical report, Internal Report, Bell Telephone Laboratories, Incorporated, Reading, Pennsylvania, 1971.

[823] Naoaki Taga, Hidefumi Odaka, Yuzo Shigesato, Itaru Yasui, Masayuki Kamei, and T. E. Haynes. Electrical properties of heteroepitaxial grown tin-doped indium oxide films. *Journal of Applied Physics*, 80(2):978–984, 1996. doi: 10.1063/1.362910. URL http://aip.scitation. org/doi/abs/10.1063/1.362910.

[824] M. Bender, J. Trube, and J. Stollenwerk. Deposition of transparent and conducting indium-tin-oxide films by the r.f.-superimposed dc sputtering technology. *Thin Solid Films*, 354(1):100 – 105, 1999. ISSN 0040-6090. doi: https://doi.org/10.1016/S0040-6090(99)00558-1. URL http://www.sciencedirect.com/science/article/pii/S0040609099005581.

[825] Tadatsugu Minami. New n-type transparent conducting oxides. *MRS Bulletin*, 25(8):38–44, 2000. doi: 10.1557/mrs2000.149.

[826] Naoto Kikuchi, Eiji Kusano, Hidehito Nanto, Akira Kinbara, and Hideo Hosono. Phonon scattering in electron transport phenomena of ito films. *Vacuum*, 59(2):492 – 499, 2000. ISSN 0042-207X. doi: https://doi.org/10.1016/S0042-207X(00)00307-9. URL http: //www.sciencedirect.com/science/article/pii/S0042207X00003079. Proceedings of the Fifth International Symposium on Sputtering and Plasma Processes.

[827] K Ellmer. Resistivity of polycrystalline zinc oxide films: current status and physical limit. *Journal of Physics D: Applied Physics*, 34 (21):3097, 2001. URL http://stacks.iop.org/0022-3727/34/i=21/a=301.

[828] Hidefumi Odaka, Yuzo Shigesato, Takashi Murakami, and Shuichi Iwata. Electronic structure analyses of sn-doped in₂o₃. *Japanese Journal of Applied Physics*, 40(5R):3231, 2001. URL http://stacks.iop.org/1347-4065/40/i=5R/a=3231.

[829] B. Thangaraju. Structural and electrical studies on highly conducting spray deposited fluorine and antimony doped sno₂ thin films from sncl₂ precursor. *Thin Solid Films*, 402(1):71 – 78, 2002. ISSN 0040-6090. doi: https://doi.org/10.1016/S0040-6090(01)01667-4. URL http://www.sciencedirect.com/science/article/pii/S0040609001016674.

[830] Ho-Chul Lee and O. Ok Park. Behaviors of carrier concentrations and mobilities in indium-tin oxide thin films by dc magnetron sputtering at various oxygen flow rates. *Vacuum*, 77(1):69 – 77, 2004. ISSN 0042-207X. doi: https://doi.org/10.1016/j.vacuum.2004. 08.006. URL http://www.sciencedirect.com/science/article/pii/S0042207X04003732.

[831] Ho-Chul Lee and O Ok Park. Electron scattering mechanisms in indium-tin-oxide thin films: grain boundary and ionized impurity scattering. *Vacuum*, 75(3):275 – 282, 2004. ISSN 0042-207X. doi: https://doi.org/10.1016/j.vacuum.2004.03.008. URL http://www. sciencedirect.com/science/article/pii/S0042207X04001952.

[832] Yuzo Shigesato and David C. Paine. Study of the effect of sn doping on the electronic transport properties of thin film indium oxide. *Applied Physics Letters*, 62(11):1268–1270, 1993. doi: 10.1063/1.108703. URL http://dx.doi.org/10.1063/1.108703.

[833] Elias Burstein. Anomalous optical absorption limit in insb. *Phys. Rev.*, 93:632–633, Feb 1954. doi: 10.1103/PhysRev.93.632. URL https://link.aps.org/doi/10.1103/PhysRev.93.632.

[834] Robert G. Breckenridge and William R. Hosler. Electrical properties of titanium dioxide semiconductors. *Phys. Rev.*, 91:793–802, Aug 1953. doi: 10.1103/PhysRev.91.793. URL http://link.aps.org/doi/10.1103/PhysRev.91.793.

[835] R.P. Hübener and Springer-Verlag GmbH. *Leiter, Halbleiter, Supraleiter: Eine kompakte Einführung in Geschichte, Entwicklung und Theorie der Festkörperphysik.* SPRINGER, 2017. ISBN 9783662532812. URL https://books.google.de/books?id=fGe3DQAAQBAJ.

[836] E. Hendry, F. Wang, J. Shan, T. F. Heinz, and M. Bonn. Electron transport in tio2 probed by thz time-domain spectroscopy. *Phys. Rev. B*, 69:081101, Feb 2004. doi: 10.1103/PhysRevB.69.081101. URL http://link.aps.org/doi/10.1103/PhysRevB.69.081101.

[837] J. Bardeen and W. Shockley. Deformation potentials and mobilities in non-polar crystals. *Phys. Rev.*, 80:72–80, Oct 1950. doi: 10.1103/PhysRev.80.72. URL https://link.aps.org/doi/10.1103/PhysRev.80.72.

[838] E. Conwell and V. F. Weisskopf. Theory of impurity scattering in semiconductors. *Phys. Rev.*, 77:388–390, Feb 1950. doi: 10.1103/PhysRev.77.388. URL http://link.aps.org/doi/10.1103/PhysRev.77.388.

[839] H. Ehrenreich. Band structure and electron transport of gaas. *Phys. Rev.*, 120:1951–1963, Dec 1960. URL https://link.aps.org/doi/10.1103/PhysRev.120.1951.

[840] Marshall D. Earle. The electrical conductivity of titanium dioxide. *Phys. Rev.*, 61:56–62, Jan 1942. doi: 10.1103/PhysRev.61.56. URL https://link.aps.org/doi/10.1103/PhysRev.61.56.

[841] D. C. Cronemeyer. Electrical and optical properties of rutile single crystals. *Phys. Rev.*, 87:876–886, Sep 1952. doi: 10.1103/PhysRev.87.876. URL http://link.aps.org/doi/10.1103/PhysRev.87.876.

[842] John G. Simmons. Transition from electrode-limited to bulk-limited conduction processes in metal-insulator-metal systems. *Phys. Rev.*, 166:912–920, Feb 1968. doi: 10.1103/PhysRev.166.912. URL https://link.aps.org/doi/10.1103/PhysRev.166.912.

[843] J. Jaćimović, C. Vaju, A. Magrez, H. Berger, L. Forró, R. Gaál, V. Cerovski, and R. Žikić. Pressure dependence of the large-polaron transport in anatase tio2 single crystals. *EPL (Europhysics Letters)*, 99(5):57005, 2012. URL http://stacks.iop.org/0295-5075/99/i=5/a=57005.

[844] S.I. Gorelik. On electron conductivity in polycrystalline annealed titanium dioxide. *J.Exptl. Theoret. Phys. U.S.S.R.*, 21(826), 1951.

[845] J.T. Devreese. Polarons. *eprint arXiv:cond-mat/0004497*, April 2000.

[846] LD Landau. On the motion of electrons in a crystal lattice. *Phys. Z. Sowjetunion*, 3:664–665, 1933.

[847] A. Kaminski and S. Das Sarma. Polaron percolation in diluted magnetic semiconductors. *Phys. Rev. Lett.*, 88:247202, May 2002. doi: 10.1103/PhysRevLett.88.247202. URL http://link.aps.org/doi/10.1103/PhysRevLett.88.247202.

[848] M. Bonn, F. Wang, J. Shan, T.F. Heinz, and E. Hendry. Ultrafast scattering of electrons in Tio2. *Femtochemistry and Femtobiology*, 2004.

[849] L Jacak, J Krasnyj, and W Jacak. Renormalization of the fröhlich constant for electrons in a quantum dot. *Physics Letters A*, 304(5–6): 168 – 171, 2002. ISSN 0375-9601. doi: https://doi.org/10.1016/S0375-9601(02)01363-4. URL http://www.sciencedirect.com/science/article/pii/S0375960102013634.

[850] Steven JF Byrnes. Basic theory and phenomenology of polarons. *Department of Physics, University of California at Berkeley, Berkeley, CA*, 94720, 2008.

[851] C Kittel. *Quantum theory of solids. John Wiley & Sons, New York*, 1987.

[852] Dwight C. Burnham, Frederick C. Brown, and Robert S. Knox. Electron mobility and scattering processes in agbr at low temperatures. *Phys. Rev.*, 119:1560–1570, Sep 1960. doi: 10.1103/PhysRev.119.1560. URL http://link.aps.org/doi/10.1103/PhysRev.119.1560.

[853] I. G. Austin and N. F. Mott. Metallic and nonmetallic behavior in transition metal oxides. *Science*, 168(3927):71–77, 1970. doi: 10.1126/science.168.3927.71. URL http://www.sciencemag.org/content/168/3927/71.short.

[854] Abdullah Yildiz, Felicia Iacomi, and Diana Mardare. Polaron transport in tio2 thin films. *Journal of Applied Physics*, 108(8):083701, 2010. doi: 10.1063/1.3493742. URL http://dx.doi.org/10.1063/1.3493742.

[855] R.N. Blumenthal, J.C. Kirk, and W.M. Hirthe. Electronic mobility in rutile (tio2) at high temperatures. *Journal of Physics and Chemistry of Solids*, 28(6):1077 – 1079, 1967. ISSN 0022-3697. doi: http://dx.doi.org/10.1016/0022-3697(67)90228-4. URL http://www.sciencedirect.com/science/article/pii/0022369767902284.

[856] I. Bransky and D.S. Tannhauser. Hall mobility of reduced rutile in the temperature range 300–1250°k. *Solid State Communications*, 7 (1):245 – 248, 1969. ISSN 0038-1098. doi: http://dx.doi.org/10.1016/0038-1098(69)90735-2. URL http://www.sciencedirect.com/science/article/pii/0038109869907352.

[857] VN Bogomolov, EK Kudinov, and Yu A Firsov. Polaron nature of current carriers in rutile (tio2). *Soviet Physics Solid State, USSR*, 9 (11):2502, 1968.

[858] V. N. Bogomolov, Yu. A. Firsov, E. K. Kudinov, and D. N. Mirlin. On the experimental observation of small polarons in rutile (tio2). *physica status solidi (b)*, 35(2):555–558, 1969. ISSN 1521-3951. doi: 10.1002/pssb.19690350202. URL http://dx.doi.org/10.1002/pssb.19690350202.

[859] V. N. Bogomolov and D. N. Mirlin. Optical absorption by polarons in rutile (tio2) single crystals. *physica status solidi (b)*, 27(1): 443–453, 1968. ISSN 1521-3951. doi: 10.1002/pssb.19680270144. URL http://dx.doi.org/10.1002/pssb.19680270144.

[860] N. Tsuda, K. Nasu, A. Fujimori, and K. Siratori. *Electronic Conduction in Oxides.* Springer Series in Solid-State Sciences. Springer Berlin Heidelberg, 2013. ISBN 9783662040119. URL https://books.google.de/books?id=SHjtCAAAQBAJ.

[861] Ryukiti R. Hasiguti and Eiichi Yagi. Electrical conductivity below 3 k of slightly reduced oxygen-deficient rutile tio2−x. *Phys. Rev. B*, 49:7251–7256, Mar 1994. doi: 10.1103/PhysRevB.49.7251. URL https://link.aps.org/doi/10.1103/PhysRevB.49.7251.

[862] D. Schrupp, M. Sing, M. Tsunekawa, H. Fujiwara, S. Kasai, A. Sekiyama, S. Suga, T. Muro, V. A. M. Brabers, and R. Claessen. High-energy photoemission on fe3o4 : Small polaron physics and the verwey transition. *EPL (Europhysics Letters)*, 70(6):789, 2005. URL http://stacks.iop.org/0295-5075/70/i=6/a=789.

References

[863] J.T. Devreese. Polarons and bipolarons in oxides. *Zeitschrift für Physik B Condensed Matter*, 104(4):601–604, 1997. ISSN 1431-584X. doi: 10.1007/s002570050495. URL http://dx.doi.org/10.1007/s002570050495.

[864] A. S. Alexandrov and N. Mott. Polarons and bipolarons. World Scientific, Singapore, 1995. pp. 155–157.

[865] Jean-François Baumard and François Gervais. Plasmon and polar optical phonons in reduced rutile TiO_{2-x}. *Phys. Rev. B*, 15:2316–2323, Feb 1977. doi: 10.1103/PhysRevB.15.2316. URL https://link.aps.org/doi/10.1103/PhysRevB.15.2316.

[866] A.E Myasnikova. Temperature dependence of electrical conductivity in systems with large polarons and bipolarons. *Physics Letters A*, 291(6):439 – 446, 2001. ISSN 0375-9601. doi: http://dx.doi.org/10.1016/S0375-9601(01)00740-X. URL http://www.sciencedirect.com/science/article/pii/S037596010100740X.

[867] S. Moser, L. Moreschini, J. Jaćimović, O. S. Barišić, H. Berger, A. Magrez, Y. J. Chang, K. S. Kim, A. Bostwick, E. Rotenberg, L. Forró, and M. Grioni. Tunable polaronic conduction in anatase tio_2. *Phys. Rev. Lett.*, 110:196403, May 2013. doi: 10.1103/PhysRevLett.110. 196403. URL http://link.aps.org/doi/10.1103/PhysRevLett.110.196403.

[868] I.G. Austin and N.F. Mott. Polarons in crystalline and non-crystalline materials. *Advances in Physics*, 18(71):41–102, 1969. doi: 10.1080/00018736900101267. URL http://dx.doi.org/10.1080/00018736900101267.

[869] Michael A. Henderson, William S. Epling, Craig L. Perkins, Charles H. F. Peden, and Ulrike Diebold. Interaction of molecular oxygen with the vacuum-annealed TiO_2(110) surface: Molecular and dissociative channels. *The Journal of Physical Chemistry B*, 103(25): 5328–5337, 1999. doi: 10.1021/jp990655q. URL http://dx.doi.org/10.1021/jp990655q.

[870] W. R. Thurber and A. J. H. Mante. Thermal conductivity and thermoelectric power of rutile (tio_2). *Phys. Rev.*, 139:A1655–A1665, Aug 1965. doi: 10.1103/PhysRev.139.A1655. URL https://link.aps.org/doi/10.1103/PhysRev.139.A1655.

[871] John W. DeFord and Owen W. Johnson. Electron transport properties in rutile from 6 to 40 k. *Journal of Applied Physics*, 54(2): 889–897, 1983. doi: http://dx.doi.org/10.1063/1.332051. URL http://scitation.aip.org/content/aip/journal/jap/54/2/10.1063/1.332051.

[872] T. R. Sandin and P. H. Keesom. Specific heat and paramagnetic susceptibility of stoichiometric and reduced rutile (tio_2) from 0.3 to 20 k. *Phys. Rev.*, 177:1370–1383, Jan 1969. doi: 10.1103/PhysRev.177.1370. URL https://link.aps.org/doi/10.1103/PhysRev.177.1370.

[873] L. A. K. Dominik and R. K. MacCrone. Dielectric relaxations in reduced rutile (TiO_{2-x}) at low temperatures. *Phys. Rev.*, 163:756–768, Nov 1967. doi: 10.1103/PhysRev.163.756. URL http://link.aps.org/doi/10.1103/PhysRev.163.756.

[874] N. Aaron Deskins and Michel Dupuis. Electron transport via polaron hopping in bulk TiO_2: A density functional theory characterization. *Phys. Rev. B*, 75:195212, May 2007. doi: 10.1103/PhysRevB.75.195212. URL http://link.aps.org/doi/10.1103/PhysRevB.75.195212.

[875] J. Schnakenberg. Polaronic impurity hopping conduction. *physica status solidi (b)*, 28(2):623–633, 1968. ISSN 1521-3951. doi: 10.1002/pssb.19680280220. URL http://dx.doi.org/10.1002/pssb.19680280220.

[876] Benjamin J. Morgan, David O. Scanlon, and Graeme W. Watson. Small polarons in nb- and ta-doped rutile and anatase tio_2. *J. Mater. Chem.*, 19:5175–5178, 2009. doi: 10.1039/B905028K. URL http://dx.doi.org/10.1039/B905028K.

[877] Abdullah Yildiz and Diana Mardare. Polaronic transport in tio_2 thin films with increasing nb content. *Philosophical Magazine*, 91(34): 4401–4409, 2011. doi: 10.1080/14786435.2011.623143. URL http://dx.doi.org/10.1080/14786435.2011.623143.

[878] A. Yildiz, S.B. Lisesivdin, M. Kasap, and D. Mardare. Non-adiabatic small polaron hopping conduction in nb-doped tio_2 thin film. *Physica B: Condensed Matter*, 404(8–11):1423 – 1426, 2009. ISSN 0921-4526. doi: http://dx.doi.org/10.1016/j.physb.2008.12.034. URL http://www.sciencedirect.com/science/article/pii/S0921452608007928.

[879] Peter Stallinga. *Theory of Electrical Characterization of (organic) semiconductors*. 2001.

[880] H. P. R. Frederikse. Recent studies on rutile (tio_2). *Journal of Applied Physics*, 32(10):2211–2215, 1961. doi: http://dx.doi.org/10. 1063/1.1777045. URL http://scitation.aip.org/content/aip/journal/jap/32/10/10.1063/1.1777045.

[881] N. Bogomolov and V. P. Zhuze. *Fiz. Tverd. Tela*, 5:3285, 1963. Translation: Soviet Phisics-Solid State, 5 (1963), 2404.

[882] H Fröhlich and NF Mott. The mean free path of electrons in polar crystals. In *Proceedings of the Royal Society of London A: Mathematical, Physical and Engineering Sciences*, volume 171, pages 496–504. The Royal Society, 1939.

[883] H. Fröhlich, H. Pelzer, and S. Zienau. Xx. properties of slow electrons in polar materials. *The London, Edinburgh, and Dublin Philosophical Magazine and Journal of Science*, 41(314):221–242, 1950. doi: 10.1080/14786445008521794. URL http://dx.doi.org/10. 1080/14786445008521794.

[884] L. Forro, O. Chauvet, D. Emin, L. Zuppiroli, H. Berger, and F. Lévy. High mobility n-type charge carriers in large single crystals of anatase (tio_2). *Journal of Applied Physics*, 75(1):633–635, 1994. doi: http://dx.doi.org/10.1063/1.355801. URL http://scitation.aip. org/content/aip/journal/jap/75/1/10.1063/1.355801.

[885] G.A. Acket and J. Volger. On the electron mobility and the donor centres in reduced and lithium-doped rutile (tio_2). *Physica*, 32(10): 1680 – 1692, 1966. ISSN 0031-8914. doi: https://doi.org/10.1016/0031-8914(66)90082-6. URL http://www.sciencedirect.com/science/article/pii/0031891466900826.

[886] P.P. Stallinga. *Electrical Characterization of Organic Electronic Materials and Devices*. Wiley, 2009. ISBN 9780470750179. URL https://books.google.de/books?id=9EYxMIBCZIAC.

[887] Gilles Horowitz, Riadh Hajlaoui, Denis Fichou, and Ahmed El Kassmi. Gate voltage dependent mobility of oligothiophene field-effect transistors. *Journal of Applied Physics*, 85(6):3202–3206, 1999. doi: 10.1063/1.369661. URL http://dx.doi.org/10.1063/1.369661.

[888] M. C. J. M. Vissenberg and M. Matters. Theory of the field-effect mobility in amorphous organic transistors. *Phys. Rev. B*, 57: 12964–12967, May 1998. doi: 10.1103/PhysRevB.57.12964. URL https://link.aps.org/doi/10.1103/PhysRevB.57.12964.

[889] Priti Tiwana, Pablo Docampo, Michael B. Johnston, Henry J. Snaith, and Laura M. Herz. Electron mobility and injection dynamics in mesoporous zno, sno_2, and tio_2 films used in dye-sensitized solar cells. *ACS Nano*, 5(6):5158–5166, 2011. doi: 10.1021/nn201243y. URL http://dx.doi.org/10.1021/nn201243y. PMID: 21595483.

[890] James Lloyd-Hughes and Tae-In Jeon. A review of the terahertz conductivity of bulk and nano-materials. *Journal of Infrared, Millimeter, and Terahertz Waves*, 33(9):871–925, 2012. ISSN 1866-6906. doi: 10.1007/s10762-012-9905-y. URL http://dx.doi.org/10.1007/s10762-012-9905-y.

[891] Neil W Ashcroft and ND Mermin. Solid state. *Physics (New York: Holt, Rinehart and Winston)*, pages 131–151, 1976.

[892] Francis E. Low and David Pines. Mobility of slow electrons in polar crystals. *Phys. Rev.*, 98:414–418, Apr 1955. doi: 10.1103/PhysRev.98.414. URL https://link.aps.org/doi/10.1103/PhysRev.98.414.

[893] R. P. Feynman. Slow electrons in a polar crystal. *Phys. Rev.*, 97:660–665, Feb 1955. doi: 10.1103/PhysRev.97.660. URL https://link.aps.org/doi/10.1103/PhysRev.97.660.

[894] R. P. Feynman, R. W. Hellwarth, C. K. Iddings, and P. M. Platzman. Mobility of slow electrons in a polar crystal. *Phys. Rev.*, 127:1004–1017, Aug 1962. doi: 10.1103/PhysRev.127.1004. URL https://link.aps.org/doi/10.1103/PhysRev.127.1004.

[895] T. D. Schultz. Slow electrons in polar crystals: Self-energy, mass, and mobility. *Phys. Rev.*, 116:526–543, Nov 1959. doi: 10.1103/PhysRev.116.526. URL https://link.aps.org/doi/10.1103/PhysRev.116.526.

[896] Robert W. Hellwarth and Ivan Biaggio. Mobility of an electron in a multimode polar lattice. *Phys. Rev. B*, 60:299–307, Jul 1999. doi: 10.1103/PhysRevB.60.299. URL https://link.aps.org/doi/10.1103/PhysRevB.60.299.

[897] Keith M. Glassford and James R. Chelikowsky. Structural and electronic properties of titanium dioxide. *Phys. Rev. B*, 46:1284–1298, Jul 1992. doi: 10.1103/PhysRevB.46.1284. URL https://link.aps.org/doi/10.1103/PhysRevB.46.1284.

[898] Bernard H. Soffer. Studies of the optical and infrared absorption spectra of rutile single crystals. *The Journal of Chemical Physics*, 35(3):940–945, 1961. doi: http://dx.doi.org/10.1063/1.1701242. URL http://scitation.aip.org/content/aip/journal/jcp/35/3/10.1063/1.1701242.

[899] Hong Lin, Hiromitsu Kozuka, and Toshinobu Yoko. Electrical properties of transparent doped oxide films. *Journal of Sol-Gel Science and Technology*, 19(1):529–532, 2000. ISSN 1573-4846. doi: 10.1023/A:1008784505272. URL http://dx.doi.org/10.1023/A:1008784505272.

[900] Alain Bally. *Electronic properties of nano-crystalline titanium dioxide thin films*. PhD thesis, Laboratory of Thin Film Physics, EPFL, 1999. URL https://infoscience.epfl.ch/record/32570.

[901] Giuseppe Mattioli, Paola Alippi, Francesco Filippone, Ruggero Caminiti, and Aldo Amore Bonapasta. Deep versus shallow behavior of intrinsic defects in rutile and anatase tio$_2$ polymorphs. *The Journal of Physical Chemistry C*, 114(49):21694–21704, 2010. doi: 10.1021/jp1041316. URL http://dx.doi.org/10.1021/jp1041316.

[902] E. J. Verwey, P. W. Haayman, and F. C. Romeijn. Physical properties and cation arrangement of oxides with spinel structures ii. electronic conductivity. *The Journal of Chemical Physics*, 15(4):181–187, 1947. doi: http://dx.doi.org/10.1063/1.1746466. URL http://scitation.aip.org/content/aip/journal/jcp/15/4/10.1063/1.1746466.

[903] N. F. Mott. On the transition to metallic conduction in semiconductors. *Canadian Journal of Physics*, 34(12A):1356–1368, 1956. doi: 10.1139/p56-151. URL http://dx.doi.org/10.1139/p56-151.

[904] C. Zener and R. R. Heikes. Conference on magnetism and magnetic materials, boston, massachusetts, october 16-18. In *Institute of Electrical Engineers, New York, 1957*, 1956.

[905] F. J. Morin. Oxides of the 3d transition metals. *Bell System Technical Journal*, 37(4):1047–1084, 1958. ISSN 1538-7305. doi: 10.1002/j.1538-7305.1958.tb01542.x. URL http://dx.doi.org/10.1002/j.1538-7305.1958.tb01542.x.

[906] R.N. Blumenthal, J. Coburn, J. Baukus, and W.M. Hirthe. Electrical conductivity of nonstoichiometric rutile single crystals from 1000° to 1500°c. *Journal of Physics and Chemistry of Solids*, 27(4):643 – 654, 1966. ISSN 0022-3697. doi: http://dx.doi.org/10.1016/0022-3697(66)90215-0. URL http://www.sciencedirect.com/science/article/pii/0022369766902150.

[907] N. Martin, A.M.E. Santo, R. Sanjinés, and F. Lévy. Energy distribution of ions bombarding tio$_2$ thin films during sputter deposition. *Surface and Coatings Technology*, 138(1):77 – 83, 2001. ISSN 0257-8972. doi: http://dx.doi.org/10.1016/S0257-8972(00)01127-0. URL http://www.sciencedirect.com/science/article/pii/S0257897200011270.

[908] T. Bak, J. Nowotny, M. Rekas, and C.C. Sorrell. Defect chemistry and semiconducting properties of titanium dioxide: Ii. defect diagrams. *Journal of Physics and Chemistry of Solids*, 64(7):1057 – 1067, 2003. ISSN 0022-3697. doi: https://doi.org/10.1016/S0022-3697(02)00480-8. URL http://www.sciencedirect.com/science/article/pii/S0022369702004808.

[909] T Bak, J Nowotny, MK Nowotny, and LR Sheppard. Defect chemistry of titanium dioxide effect of interfaces. *Journal of the Australasian Ceramic Society*, 43(1):49–55, 2007.

[910] J. Van de Lagemaat and A. Frank. Effect of the surface-state distribution on electron transport in dye-sensitized tio$_2$ solar cells: Nonlinear electron-transport kinetics. *J. Phys. Chem. B*, 104:4292, 2000.

[911] Juan Bisquert. Physical electrochemistry of nanostructured devices. *Phys. Chem. Chem. Phys.*, 10:49–72, 2008. doi: 10.1039/B709316K. URL http://dx.doi.org/10.1039/B709316K.

[912] W. H. Leng, Piers R. F. Barnes, Mindaugas Juozapavicius, Brian C. O'Regan, and James R. Durrant. Electron diffusion length in mesoporous nanocrystalline tio$_2$ photoelectrodes during water oxidation. *The Journal of Physical Chemistry Letters*, 1(6):967–972, 2010. doi: 10.1021/jz100051q. URL http://dx.doi.org/10.1021/jz100051q.

[913] Luca Bertoluzzi, Isaac Herraiz-Cardona, Ronen Gottesman, Arie Zaban, and Juan Bisquert. Relaxation of electron carriers in the density of states of nanocrystalline tio$_2$. *The Journal of Physical Chemistry Letters*, 5(4):689–694, 2014. doi: 10.1021/jz4027584. URL http://dx.doi.org/10.1021/jz4027584. PMID: 26270838.

[914] F. Cao, G. Oskam, G. J. Meyer, and P. C. Searson. Electron transport in porous nanocrystalline tio$_2$ photoelectrochemical cells. *J. Phys. Chem.*, 100:17021, 1996.

[915] A. Janotti, J. B. Varley, P. Rinke, N. Umezawa, G. Kresse, and C. G. Van de Walle. Hybrid functional studies of the oxygen vacancy in tio$_2$. *Phys. Rev. B*, 81:085212, Feb 2010. doi: 10.1103/PhysRevB.81.085212. URL http://link.aps.org/doi/10.1103/PhysRevB.81.085212.

[916] A. Janotti, C. Franchini, J. B. Varley, G. Kresse, and C. G. Van de Walle. Dual behavior of excess electrons in rutile tio$_2$. *physica status solidi (RRL) – Rapid Research Letters*, 7(3):199–203, 2013. ISSN 1862-6270. doi: 10.1002/pssr.201206464. URL http://dx.doi.org/10.1002/pssr.201206464.

[917] Elvira Fortunato, David Ginley, Hideo Hosono, and David C Paine. Transparent conducting oxides for photovoltaics. *MRS bulletin*, 32 (03):242–247, 2007.

[918] Gongming Wang, Hanyu Wang, Yichuan Ling, Yuechao Tang, Xunyu Yang, Robert C. Fitzmorris, Changchun Wang, Jin Z. Zhang, and Yat Li. Hydrogen-treated tio2 nanowire arrays for photoelectrochemical water splitting. *Nano Letters*, 11(7):3026–3033, 2011. doi: 10.1021/nl201766h. URL http://dx.doi.org/10.1021/nl201766h. PMID: 21710974.

[919] D. A. Wheeler, Y. C. Ling, R. J. Dillon, R. C. Fitzmorris, C. G. Dudzik, L. Zavodivker, T. Rajh, N. M. Dimitrijevic, G. Millhauser, and C. Bardeen. Probing the nature of bandgap states in hydrogen-treated tio2 nanowires. *J. Phys. Chem. C*, 117:26821, 2013.

[920] Z. Zhang, M. N. Hedhili, H. Zhu, and P. Wang. Electrochemical reduction induced self-doping of tio^{3+}+ for efficient water splitting performance on tio2 based photoelectrodes. *Phys. Chem. Chem. Phys.*, 15:15637, 2013.

[921] S.-E. Lindquist, A. Lindgren, and C. Leygraf. Effects of electrochemical reduction of polycrystalline tio2 photoelectrodes in acidic solutions. *Sol. Energy Mater.*, 15:367, 1987.

[922] Q. Kang, J. Cao, Y. Zhang, L. Liu, H. Xu, and J. Ye. Reduced tio2 nanotube arrays for photoelectrochemical water splitting. *J. Mater. Chem. A*, 1:5766, 2013.

[923] Wei-Dong Zhu, Cheng-Wei Wang, Jian-Biao Chen, and Xu qiang Zhang. Low temperature synthesis of reduced titanium oxide nanotube arrays: Crystal structure transformation and enhanced field emission. *Materials Research Bulletin*, 50:79 – 84, 2014. ISSN 0025-5408. doi: http://doi.org/10.1016/j.materresbull.2013.10.030. URL http://www.sciencedirect.com/science/article/pii/S0025540813008611.

[924] Shelton Fu and Takeshi Egami. Mos and mosfet with transition metal oxides, 1996. URL http://dx.doi.org/10.1117/12.250262.

[925] S. Lakkis, C. Schlenker, B. K. Chakraverty, R. Buder, and M. Marezio. Metal-insulator transitions in ti4o7 single crystals: Crystal characterization, specific heat, and electron paramagnetic resonance. *Phys. Rev. B*, 14:1429–1440, Aug 1976. doi: 10.1103/PhysRevB.14.1429. URL http://link.aps.org/doi/10.1103/PhysRevB.14.1429.

[926] L. R. Sheppard, T. Bak, and J. Nowotny. Electrical properties of niobium-doped titanium dioxide. 1. defect disorder. *The Journal of Physical Chemistry B*, 110(45):22447–22454, 2006. doi: 10.1021/jp0637025. URL http://dx.doi.org/10.1021/jp0637025. PMID: 17091986.

[927] D.M. Smyth. The role of impurities in insultating transition metal oxides. *Progress in Solid State Chemistry*, 15(3):145 – 171, 1984. ISSN 0079-6786. doi: http://dx.doi.org/10.1016/0079-6786(84)90001-3. URL http://www.sciencedirect.com/science/article/pii/0079678684900013.

[928] P. Knauth and H. L. Tuller. Electrical and defect thermodynamic properties of nanocrystalline titanium dioxide. *Journal of Applied Physics*, 85(2):897, 1999. doi: 10.1063/1.369208. URL http://dx.doi.org/10.1063/1.369208.

[929] JJ Baumard, D Panis, and D Ruffier. Electrical conductivity of rutile single crystals at high temperature. *Revue Internationale des Hautes Temperatures et des Refractaires*, 12(4):321–27, 1975.

[930] Jean-Francis Marucco, Jacques Gautron, and Philippe Lemasson. Thermogravimetric and electrical study of non-stoichiometric titanium dioxide tio$_{2-x}$, between 800 and 1100°c. *Journal of Physics and Chemistry of Solids*, 42(5):363 – 367, 1981. ISSN 0022-3697. doi: http://dx.doi.org/10.1016/0022-3697(81)90043-3. URL http://www.sciencedirect.com/science/article/pii/0022369781900433.

[931] Jeri Ann S. Ikeda, Yet-Ming Chiang, Anthony J. Garratt-Reed, and John B. Vander Sande. Space charge segregation at grain boundaries in titanium dioxide: Ii, model experiments. *Journal of the American Ceramic Society*, 76(10):2447–2459, 1993. ISSN 1551-2916. doi: 10.1111/j.1151-2916.1993.tb03965.x. URL http://dx.doi.org/10.1111/j.1151-2916.1993.tb03965.x.

[932] Jeri Ann S. Ikeda and Yet-Ming Chiang. Space charge segregation at grain boundaries in titanium dioxide: I, relationship between lattice defect chemistry and space charge potential. *Journal of the American Ceramic Society*, 76(10):2437–2446, 1993. ISSN 1551-2916. doi: 10.1111/j.1151-2916.1993.tb03964.x. URL http://dx.doi.org/10.1111/j.1151-2916.1993.tb03964.x.

[933] Cristiana Di Valentin, Gianfranco Pacchioni, and Annabella Selloni. Reduced and n-type doped tio2: Nature of ti3+ species. *The Journal of Physical Chemistry C*, 113(48):20543–20552, 2009. doi: 10.1021/jp9061797. URL http://dx.doi.org/10.1021/jp9061797.

[934] AA Samokhvalov and AG Rustamov. Electrical properties of ferrite spinels with a variable content of divalent iron ions. *SOVIET PHYSICS SOLID STATE, USSR*, 7(4):961–+, 1965.

[935] K. Mizushima, M. Tanaka, A. Asai, S. Iida, and John B. Goodenough. Impurity levels of iron-group ions in tio2(ii). *Journal of Physics and Chemistry of Solids*, 40(12):1129 – 1140, 1979. ISSN 0022-3697. doi: http://dx.doi.org/10.1016/0022-3697(79)90148-3. URL http://www.sciencedirect.com/science/article/pii/0022369779901483.

[936] D. C. Cronemeyer. Infrared absorption of reduced rutile tio2 single crystals. *Phys. Rev.*, 113:1222–1226, Mar 1959. doi: 10.1103/PhysRev.113.1222. URL http://link.aps.org/doi/10.1103/PhysRev.113.1222.

[937] Z Klusek, S Pierzgalski, and S Datta. Insulator-metal transition on heavily reduced tio2(1 1 0) surface studied by high temperature-scanning tunnelling spectroscopy (ht-sts). *Applied Surface Science*, 221(1-4):120 – 128, 2004. ISSN 0169-4332. doi: https://doi.org/10.1016/S0169-4332(03)00877-8. URL http://www.sciencedirect.com/science/article/pii/S0169433203008778.

[938] Stefan Wendt, Phillip T. Sprunger, Estephania Lira, Georg K. H. Madsen, Zheshen Li, Jonas Ø. Hansen, Jesper Matthiesen, Asger Blekinge-Rasmussen, Erik Lægsgaard, Bjørk Hammer, and Flemming Besenbacher. The role of interstitial sites in the ti3d defect state in the band gap of titania. *Science*, 320(5884):1755–1759, 2008. ISSN 0036-8075. doi: 10.1126/science.1159846. URL http://science.sciencemag.org/content/320/5884/1755.

[939] Riley E. Rex, Yi Yang, Fritz J. Knorr, Jin Z. Zhang, Yat Li, and Jeanne L. McHale. Spectroelectrochemical photoluminescence of trap states in h-treated rutile tio2 nanowires: Implications for photooxidation of water. *The Journal of Physical Chemistry C*, 120(6):3530–3541, 2016. doi: 10.1021/acs.jpcc.5b11231. URL http://dx.doi.org/10.1021/acs.jpcc.5b11231.

[940] L. Kavan, M. Grätzel, S. E. Gilbert, C. Klemenz, and H. J. Scheel. Electrochemical and photoelectrochemical investigation of single-crystal anatase. *Journal of the American Chemical Society*, 118(28):6716–6723, 1996. doi: 10.1021/ja9541721. URL http://dx.doi.org/10.1021/ja9541721.

[941] A. G. Thomas, W. R. Flavell, A. K. Mallick, A. R. Kumarasinghe, D. Tsoutsou, N. Khan, C. Chatwin, S. Rayner, G. C. Smith, R. L. Stockbauer, S. Warren, T. K. Johal, S. Patel, D. Holland, A. Taleb, and F. Wiame. Comparison of the electronic structure of anatase and rutile tio2 single-crystal surfaces using resonant photoemission and x-ray absorption spectroscopy. *Phys. Rev. B*, 75:035105, Jan 2007. doi: 10.1103/PhysRevB.75.035105. URL http://link.aps.org/doi/10.1103/PhysRevB.75.035105.

[942] Victor E. Henrich, G. Dresselhaus, and H. J. Zeiger. Observation of two-dimensional phases associated with defect states on the surface of tio2. *Phys. Rev. Lett.*, 36:1335–1339, May 1976. doi: 10.1103/PhysRevLett.36.1335. URL https://link.aps.org/doi/10.1103/PhysRevLett.36.1335.

[943] Madhavan Ramamoorthy, R. D. King-Smith, and David Vanderbilt. Defects on tio$_2$ (110) surfaces. *Phys. Rev. B*, 49:7709–7715, Mar 1994. doi: 10.1103/PhysRevB.49.7709. URL https://link.aps.org/doi/10.1103/PhysRevB.49.7709.

[944] Cristiana Di Valentin, Gianfranco Pacchioni, and Annabella Selloni. Electronic structure of defect states in hydroxylated and reduced rutile tio$_2$(110) surfaces. *Phys. Rev. Lett.*, 97:166803, Oct 2006. doi: 10.1103/PhysRevLett.97.166803. URL https://link.aps.org/doi/10.1103/PhysRevLett.97.166803.

[945] V Mikhelashvili and G Eisenstein. Optical and electrical characterization of the electron beam gun evaporated tio$_2$ film. *Microelectronics Reliability*, 41(7):1057 – 1061, 2001. ISSN 0026-2714. doi: https://doi.org/10.1016/S0026-2714(01)00075-0. URL http://www.sciencedirect.com/science/article/pii/S0026271401000750.

[946] Diana Mardare, C. Baban, Raluca Gavrila, M. Modreanu, and G.I. Rusu. On the structure, morphology and electrical conductivities of titanium oxide thin films. *Surface Science*, 507-510:468 – 472, 2002. ISSN 0039-6028. doi: https://doi.org/10.1016/S0039-6028(02)01287-6. URL http://www.sciencedirect.com/science/article/pii/S0039602802012876.

[947] Xiaobo Chen, Lei Liu, Peter Y. Yu, and Samuel S. Mao. Increasing solar absorption for photocatalysis with black hydrogenated titanium dioxide nanocrystals. *Science*, 331(6018):746–750, 2011. ISSN 0036-8075. doi: 10.1126/science.1200448. URL http://science.sciencemag.org/content/331/6018/746.

[948] Jeremy W.J. Hamilton, J. Anthony Byrne, Patrick S.M. Dunlop, Dionysios D. Dionysiou, Miguel Pelaez, Kevin O'Shea, Damian Synnott, and Suresh C. Pillai. Evaluating the mechanism of visible light activity for n,f-tio$_2$ using photoelectrochemistry. *The Journal of Physical Chemistry C*, 118(23):12206–12215, 2014. doi: 10.1021/jp4120964. URL http://dx.doi.org/10.1021/jp4120964.

[949] Isao Nakamura, Nobuaki Negishi, Shuzo Kutsuna, Tatsuhiko Ihara, Shinichi Sugihara, and Koji Takeuchi. Role of oxygen vacancy in the plasma-treated tio$_2$ photocatalyst with visible light activity for no removal. *Journal of Molecular Catalysis A: Chemical*, 161(1):205 – 212, 2000. ISSN 1381-1169. doi: https://doi.org/10.1016/S1381-1169(00)00362-9. URL http://www.sciencedirect.com/science/article/pii/S1381116900003629.

[950] T Ihara, M Ando, and S Sugihara. Preparation of visible light active tio$_2$ photocatalysis using wet method. *Photocatalysis*, 5:19, 2001.

[951] I. Justicia, P. Ordejón, G. Canto, J.L. Mozos, J. Fraxedas, G.A. Battiston, R. Gerbasi, and A. Figueras. Designed self-doped titanium oxide thin films for efficient visible-light photocatalysis. *Advanced Materials*, 14(19):1399–1402, 2002. ISSN 1521-4095. doi: 10.1002/1521-4095(20021002)14:19<1399::AID-ADMA1399>3.0.CO;2-C. URL http://dx.doi.org/10.1002/1521-4095(20021002)14:19<1399::AID-ADMA1399>3.0.CO;2-C.

[952] Jin Wang, De Nyago Tafen, James P. Lewis, Zhanglian Hong, Ayyakkannu Manivannan, Mingjia Zhi, Ming Li, and Nianqiang Wu. Origin of photocatalytic activity of nitrogen-doped tio$_2$ nanobelts. *Journal of the American Chemical Society*, 131(34):12290–12297, 2009. doi: 10.1021/ja903781h. URL http://dx.doi.org/10.1021/ja903781h. PMID: 19705915.

[953] Adriana Zaleska. Doped-tio$_2$: A review. *Recent Patents on Engineering*, 2(3):157–164, 2008. ISSN 1872-2121/2212-4047. doi: 10.2174/187221208786306289. URL http://www.eurekaselect.com/node/92903/article.

[954] Yutaka Furubayashi, Taro Hitosugi, Yukio Yamamoto, Kazuhisa Inaba, Go Kinoda, Yasushi Hirose, Toshihiro Shimada, and Tetsuya Hasegawa. A transparent metal: Nb-doped anatase tio$_2$. *Applied Physics Letters*, 86(25):252101, 2005. doi: http://dx.doi.org/10.1063/1.1949728. URL http://scitation.aip.org/content/aip/journal/apl/86/25/10.1063/1.1949728.

[955] Taro Hitosugi, Yutaka Furubayashi, Atsuki Ueda, Kinnosuke Itabashi, Kazuhisa Inaba, Yasushi Hirose, Go Kinoda, Yukio Yamamoto, Toshihiro Shimada, and Tetsuya Hasegawa. Ta-doped anatase tio$_2$ epitaxial film as transparent conducting oxide. *Japanese Journal of Applied Physics*, 44(8L):L1063, 2005. URL http://stacks.iop.org/1347-4065/44/i=8L/a=L1063.

[956] E Comini, V Guidi, M Ferroni, and G Sberveglieri. Tio$_2$:mo, moo$_3$:ti, tio+wo$_3$ and tio:w layer for landfill produced gases sensing. *Sensors and Actuators B: Chemical*, 100(1):41 – 46, 2004. ISSN 0925-4005. doi: https://doi.org/10.1016/j.snb.2003.12.018. URL http://www.sciencedirect.com/science/article/pii/S0925400503009092. New materials and Technologies in Sensor Applications, Proceedings of the European Materials Research Society 2003 - Symposium N.

[957] E. Comini, G. Sberveglieri, and V. Guidi. Ti-w-o sputtered thin film as n- and p-type gas sensors. *Sensors and Actuators B: Chemical*, 70(1):108 – 114, 2000. ISSN 0925-4005. doi: 10.1016/S0925-4005(00)00571-2. URL http://www.sciencedirect.com/science/article/pii/S0925400500005712. Special Issue in Memory of Professor Wolfgang Gopel.

[958] K. Galatsis, Y.X. Li, W. Wlodarski, E Comini, G Sberveglieri, C Cantalini, S Santucci, and M Passacantando. Comparison of single and binary oxide moo$_3$, tio$_2$ and wo$_3$ sol-gel gas sensors. *Sensors and Actuators B: Chemical*, 83(1):276 – 280, 2002. ISSN 0925-4005. doi: https://doi.org/10.1016/S0925-4005(01)01072-3. URL http://www.sciencedirect.com/science/article/pii/S0925400501010723. Selected Papers from TRANSDUCERS '01 EUROSENSORS XV.

[959] M. Ferroni, V. Guidi, G. Martinelli, E. Comini, G. Sberveglieri, D. Boscarino, and G. Della Mea. Electron microscopy and rutherford backscattering study of nucleation and growth in nanosized w-ti-o thin films. *Journal of Applied Physics*, 88(2):1097–1103, 2000. doi: 10.1063/1.373782. URL http://dx.doi.org/10.1063/1.373782.

[960] M Ferroni, V Guidi, G Martinelli, P Nelli, and G Sberveglieri. Gas-sensing applications of w-ti-o-based nanosized thin films prepared by r.f. reactive sputtering. *Sensors and Actuators B: Chemical*, 44(1):499 – 502, 1997. ISSN 0925-4005. doi: https://doi.org/10.1016/S0925-4005(97)00173-1. URL http://www.sciencedirect.com/science/article/pii/S0925400597001731.

[961] M. Gerlich, S. Kornely, M. Fleischer, H. Meixner, and R. Kassing. Selectivity enhancement of a wo$_3$/tio$_2$ gas sensor by the use of a four-point electrode structure. *Sensors and Actuators B: Chemical*, 93(1):503 – 508, 2003. ISSN 0925-4005. doi: https://doi.org/10.1016/S0925-4005(03)00187-4. URL http://www.sciencedirect.com/science/article/pii/S0925400503001874. Proceedings of the Ninth International Meeting on Chemical Sensors.

[962] Jung-Yup Lee, Jaewon Park, and Jun-Hyung Cho. Electronic properties of n- and c-doped tio$_2$. *Applied Physics Letters*, 87(1):011904, 2005. doi: 10.1063/1.1991982. URL http://dx.doi.org/10.1063/1.1991982.

[963] Ryuhei Nakamura, Tomoaki Tanaka, and Yoshihiro Nakato. Mechanism for visible light responses in anodic photocurrents at n-doped tio$_2$ film electrodes. *The Journal of Physical Chemistry B*, 108(30):10617–10620, 2004. doi: 10.1021/jp048112q. URL http://dx.doi.org/10.1021/jp048112q.

[964] Qingqing Qiu, Shuo Li, Jingjing Jiang, Dejun Wang, Yanhong Lin, and Tengfeng Xie. Improved electron transfer between tio$_2$ and fto interface by n-doped anatase tio$_2$ nanowires and its applications in quantum dot-sensitized solar cells. *The Journal of Physical Chemistry C*, 121(39):21560–21570, 2017. doi: 10.1021/acs.jpcc.7b07795. URL http://dx.doi.org/10.1021/acs.jpcc.7b07795.

[965] N. Umezawa, K. Shiraishi, T. Ohno, H. Watanabe, T. Chikyow, K. Torii, K. Yamabe, K. Yamada, H. Kitajima, and T. Arikado. First-principles studies of the intrinsic effect of nitrogen atoms on reduction in gate leakage current through hf-based high-k dielectrics. *Applied Physics Letters*, 86(14):143507, 2005. doi: 10.1063/1.1899232. URL http://dx.doi.org/10.1063/1.1899232.

References

[966] Zhou Wang, Chongyin Yang, Tianquan Lin, Hao Yin, Ping Chen, Dongyun Wan, Fangfang Xu, Fuqiang Huang, Jianhua Lin, Xiaoming Xie, and Mianheng Jiang. H-doped black titania with very high solar absorption and excellent photocatalysis enhanced by localized surface plasmon resonance. *Advanced Functional Materials*, 23(43):5444–5450, 2013. ISSN 1616-3028. doi: 10.1002/adfm.201300486. URL http://dx.doi.org/10.1002/adfm.201300486.

[967] Zhongbiao Wu, Fan Dong, Weirong Zhao, Haiqiang Wang, Yue Liu, and Baohong Guan. The fabrication and characterization of novel carbon doped tio2 nanotubes, nanowires and nanorods with high visible light photocatalytic activity. *Nanotechnology*, 20:235701, 2009. URL http://iopscience.iop.org/0957-4484/20/23/235701.

[968] Guosheng Wu, Tomohiro Nishikawa, Bunsho Ohtani, and Aicheng Chen. Synthesis and characterization of carbon-doped tio2 nanostructures with enhanced visible light response. *Chemistry of Materials*, 19(18):4530–4537, 2007. doi: 10.1021/cm071244m. URL http://dx.doi.org/10.1021/cm071244m.

[969] Hao Wang and James P Lewis. Effects of dopant states on photoactivity in carbon-doped tio2. *Journal of Physics: Condensed Matter*, 17(21):L209, 2005. URL http://stacks.iop.org/0953-8984/17/i=21/a=L01.

[970] Fan Dong, Sen Guo, Haiqiang Wang, Xiaofang Li, and Zhongbiao Wu. Enhancement of the visible light photocatalytic activity of c-doped tio2 nanomaterials prepared by a green synthetic approach. *The Journal of Physical Chemistry C*, 115(27):13285–13292, 2011. doi: 10.1021/jp111916q. URL http://dx.doi.org/10.1021/jp111916q.

[971] Celine Rüdiger, Filippo Maglia, Silvia Leonardi, Matthias Sachsenhauser, Ian D. Sharp, Odysseas Paschos, and Julia Kunze. Surface analytical study of carbothermally reduced titania films for electrocatalysis application. *Electrochimica Acta*, 71:1 – 9, 2012. ISSN 0013-4686. doi: http://dx.doi.org/10.1016/j.electacta.2012.02.044. URL http://www.sciencedirect.com/science/article/pii/S0013468612002459.

[972] Shigeki Otani, Takaho Tanaka, and Yoshio Ishizawa. Electrical resistivities in single crystals of tic_x and vc_x. *Journal of Materials Science*, 21(3):1011–1014, 1986. ISSN 1573-4803. doi: 10.1007/BF01117387. URL http://dx.doi.org/10.1007/BF01117387.

[973] Vyacheslav N. Kuznetsov and Nick Serpone. On the origin of the spectral bands in the visible absorption spectra of visible-light-active tio2 specimens analysis and assignments. *The Journal of Physical Chemistry C*, 113(34):15110–15123, 2009. doi: 10.1021/jp901034t. URL http://dx.doi.org/10.1021/jp901034t.

[974] Wen Qi Fang, Xue Lu Wang, Haimin Zhang, Yi Jia, Ziyang Huo, Zhen Li, Huijun Zhao, Hua Gui Yang, and Xiangdong Yao. Manipulating solar absorption and electron transport properties of rutile tio2 photocatalysts via highly n-type f-doping. *J. Mater. Chem. A*, 2:3513–3520, 2014. doi: 10.1039/C3TA13917D. URL http://dx.doi.org/10.1039/C3TA13917D.

[975] Run Long, Ying Dai, and Baibiao Huang. Geometric and electronic properties of sn-doped tio2 from first-principles calculations. *The Journal of Physical Chemistry C*, 113(2):650–653, 2009. doi: 10.1021/jp8043708. URL http://dx.doi.org/10.1021/jp8043708.

[976] Yaan Cao, Wensheng Yang, Weifeng Zhang, Guozong Liu, and Polock Yue. Improved photocatalytic activity of sn4+ doped tio2 nanoparticulate films prepared by plasma-enhanced chemical vapor deposition. *New J. Chem.*, 28:218–222, 2004. doi: 10.1039/B306845E. URL http://dx.doi.org/10.1039/B306845E.

[977] Ming Xu, Peimei Da, Haoyu Wu, Dongyuan Zhao, and Gengfeng Zheng. Controlled sn-doping in tio2 nanowire photoanodes with enhanced photoelectrochemical conversion. *Nano Letters*, 12(3):1503–1508, 2012. doi: 10.1021/nl2042968. URL http://dx.doi.org/10.1021/nl2042968. PMID: 22364360.

[978] Bo Sun, Tielin Shi, Zhengchun Peng, Wenjun Sheng, Ting Jiang, and Guanglan Liao. Controlled fabrication of sn/tio2 nanorods for photoelectrochemical water splitting. *Nanoscale Research Letters*, 8(1):462, 2013. ISSN 1556-276X. doi: 10.1186/1556-276X-8-462. URL http://dx.doi.org/10.1186/1556-276X-8-462.

[979] Yandong Duan, Nianqing Fu, Qiuping Liu, Yanyan Fang, Xiaowen Zhou, Jingbo Zhang, and Yuan Lin. Sn-doped tio2 photoanode for dye-sensitized solar cells. *The Journal of Physical Chemistry C*, 116(16):8888–8893, 2012. doi: 10.1021/jp212517k. URL http://dx.doi.org/10.1021/jp212517k.

[980] Kesong Yang, Ying Dai, and Baibiao Huang. First-principles calculations for geometrical structures and electronic properties of si-doped tio2. *Chemical Physics Letters*, 456(1–3):71 – 75, 2008. ISSN 0009-2614. doi: http://doi.org/10.1016/j.cplett.2008.03.018. URL http://www.sciencedirect.com/science/article/pii/S0009261408003643.

[981] Shinji Iwamoto, Seiu Iwamoto, Masashi Inoue, Hisao Yoshida, Tsunehiro Tanaka, and Koji Kagawa. Xanes and xps study of silica-modified titanias prepared by the glycothermal method. *Chemistry of Materials*, 17(3):650–655, 2005. doi: 10.1021/cm040045p. URL http://dx.doi.org/10.1021/cm040045p.

[982] Taro Hitosugi, Naoomi Yamada, Shoichiro Nakao, Yasushi Hirose, and Tetsuya Hasegawa. Properties of tio2-based transparent conducting oxides. *physica status solidi (a)*, 207(7):1529–1537, 2010. ISSN 1862-6319. doi: 10.1002/pssa.200983774. URL http://dx.doi.org/10.1002/pssa.200983774.

[983] Atsushi Hachiya, Shintaro Takata, Yutaro Komuro, and Yuji Matsumoto. Effects of v-ion doping on the photoelectrochemical properties of epitaxial tio2(110) thin films on nb-doped tio2 (110) single crystals. *The Journal of Physical Chemistry C*, 116(32):16951–16956, 2012. doi: 10.1021/jp307185d. URL http://dx.doi.org/10.1021/jp307185d.

[984] Mengjin Yang, Bo Ding, and Jung-Kun Lee. Surface electrochemical properties of niobium-doped titanium dioxide nanorods and their effect on carrier collection efficiency of dye sensitized solar cells. *Journal of Power Sources*, 245:301 – 307, 2014. ISSN 0378-7753. doi: http://doi.org/10.1016/j.jpowsour.2013.06.016. URL http://www.sciencedirect.com/science/article/pii/S0378775313010100.

[985] Andrea Welte, Christoph Waldauf, Christoph Brabec, and Peter J. Wellmann. Application of optical absorbance for the investigation of electronic and structural properties of sol-gel processed tio2 films. *Thin Solid Films*, 516(20):7256 – 7259, 2008. ISSN 0040-6090. doi: https://doi.org/10.1016/j.tsf.2007.12.025. URL http://www.sciencedirect.com/science/article/pii/S0040609007020238. Proceedings on Advanced Materials and Concepts for Photovoltaics EMRS 2007 Conference, Strasbourg, France.

[986] Yasushi Hirose, Naoomi Yamada, Shoichiro Nakao, Taro Hitosugi, Toshihiro Shimada, and Tetsuya Hasegawa. Large electron mass anisotropy in a d-electron-based transparent conducting oxide: Nb-doped anatase tio2 epitaxial films. *Phys. Rev. B*, 79:165108, Apr 2009. doi: 10.1103/PhysRevB.79.165108. URL http://link.aps.org/doi/10.1103/PhysRevB.79.165108.

[987] H Nogawa, A Chikamatsu, Y Hirose, S Nakao, H Kumigashira, M Oshima, and T Hasegawa. Carrier compensation mechanism in heavily nb-doped anatase ti$_{1-x}$nb$_x$o$_{2+\delta}$ epitaxial thin films. *Journal of Physics D: Applied Physics*, 44(36):365404, 2011. URL http://stacks.iop.org/0022-3727/44/i=36/a=365404.

[988] S. X. Zhang, D. C. Kundaliya, W. Yu, S. Dhar, S. Y. Young, L. G. Salamanca-Riba, S. B. Ogale, R. D. Vispute, and T. Venkatesan. Niobium doped tio2: Intrinsic transparent metallic anatase versus highly resistive rutile phase. *Journal of Applied Physics*, 102(1):013701, 2007. doi: http://dx.doi.org/10.1063/1.2750407. URL http://scitation.aip.org/content/aip/journal/jap/102/1/10.1063/1.2750407.

[989] Deependra Das Mulmi, Takao Sekiya, Nozomi Kamiya, Susumu Kurita, Yutaka Murakami, and Tetsuya Kodaira. Optical and electric properties of nb-doped anatase tio$_2$ single crystal. *Journal of Physics and Chemistry of Solids*, 65(6):1181 – 1185, 2004. ISSN 0022-3697. doi: http://dx.doi.org/10.1016/j.jpcs.2003.12.009. URL http://www.sciencedirect.com/science/article/pii/S0022369704000113.

[990] J. Kasai, T. Hitosugi, M. Moriyama, K. Goshonoo, N. L. H. Hoang, S. Nakao, N. Yamada, and T. Hasegawa. Properties of tio$_2$-based transparent conducting oxide thin films on gan(0001) surfaces. *Journal of Applied Physics*, 107(5):053110, 2010. doi: 10.1063/1.3326943. URL http://dx.doi.org/10.1063/1.3326943.

[991] Naoomi Yamada, Taro Hitosugi, Junpei Kasai, Ngoc Lam Huong Hoang, Shoichiro Nakao, Yasushi Hirose, Toshihiro Shimada, and Tetsuya Hasegawa. Transparent conducting nb-doped anatase tio$_2$ (tno) thin films sputtered from various oxide targets. *Thin Solid Films*, 518(11):3101 – 3104, 2010. ISSN 0040-6090. doi: http://dx.doi.org/10.1016/j.tsf.2009.07.205. URL http://www.sciencedirect.com/science/article/pii/S0040609009013868. Transparent Oxides for Electronics and Optiocs (TOEO-6).

[992] Panikar S. Archana, Rajan Jose, Tan Mein Jin, Chellapan Vijila, Mashitah M. Yusoff, and Seeram Ramakrishna. Structural and electrical properties of nb-doped anatase tio$_2$ nanowires. *Journal of the American Ceramic Society*, 93(12):4096–4102, 2010. ISSN 1551-2916. doi: 10.1111/j.1551-2916.2010.04003.x. URL http://dx.doi.org/10.1111/j.1551-2916.2010.04003.x.

[993] Lintao Hou, Pengyi Liu, Yanwu Li, and Chunhong Wu. Enhanced performance in organic light-emitting diodes by sputtering tio$_2$ ultra-thin film as the hole buffer layer. *Thin Solid Films*, 517(17):4926 – 4929, 2009. ISSN 0040-6090. doi: http://dx.doi.org/10.1016/j.tsf.2009.03.017. URL http://www.sciencedirect.com/science/article/pii/S0040609009004581. 4th International Conference on Technological Advances of Thin Films and Surface Coatings.

[994] Tengku Hasnan Tengku Aziz, Muhamad Mat Salleh, and Muhammad Yahaya. Reduction of turn-on voltage in polymer organic light-emitting diode using nanoparticles tio$_2$ thin film as a hole injection layer. *Solid State Sci Technol*, 15:75–83, 2007.

[995] Gordon Thomas. Materials science: Invisible circuits. *Nature*, 389:907–908, 1997. doi: 10.1038/39999.

[996] T. Bak, J. Nowotny, M. Rekas, and C.C. Sorrell. Defect chemistry and semiconducting properties of titanium dioxide: I. intrinsic electronic equilibrium. *Journal of Physics and Chemistry of Solids*, 64(7):1043 – 1056, 2003. ISSN 0022-3697. doi: https://doi.org/10.1016/S0022-3697(02)00479-1. URL http://www.sciencedirect.com/science/article/pii/S0022369702004791.

[997] Soumen Das, Sang-Hoon Kim, Yong-Kyu Park, Cheol-Min Choi, Dae-Young Kim, and Yoon-Bong Hahn. Heterojunction bipolar assembly with cr$_x$ti$_{1-x}$o$_2$ thin films and vertically aligned zno nanorods. *Materials Chemistry and Physics*, 124(1):704 – 708, 2010. ISSN 0254-0584. doi: http://dx.doi.org/10.1016/j.matchemphys.2010.07.040. URL http://www.sciencedirect.com/science/article/pii/S0254058410005870.

[998] Leo Chau-Kuang Liau and Chu-Che Lin. Semiconductor characterization of cr^{3+}-doped titania electrodes with p–n homojunction devices. *Thin Solid Films*, 516(8):1998 – 2002, 2008. ISSN 0040-6090. doi: http://dx.doi.org/10.1016/j.tsf.2007.06.025. URL http://www.sciencedirect.com/science/article/pii/S0040609007009042.

[999] J. Domaradzki and D. Kaczmarek. Electrical and optical properties of tos–s heterojunction devices. *Thin Solid Films*, 516(7):1473 – 1475, 2008. ISSN 0040-6090. doi: http://dx.doi.org/10.1016/j.tsf.2007.05.044. URL http://www.sciencedirect.com/science/article/pii/S0040609007008164. Proceedings of Symposium R on Advances in Transparent Electronics:from materials to devices {EMRS} 2006 Conference, Nice, France.

[1000] D. J. Mowbray, J. I. Martinez, J. M. García Lastra, K. S. Thygesen, and K. W. Jacobsen. Stability and electronic properties of tio$_2$ nanostructures with and without b and n doping. *The Journal of Physical Chemistry C*, 113(28):12301–12308, 2009. doi: 10.1021/jp904672p. URL http://dx.doi.org/10.1021/jp904672p.

[1001] Zhang Zhi-Feng, Deng Zhen-Bo, Liang Chun-Jun, Zhang Meng-Xin, and Xu Deng-Hui. Organic light-emitting diodes with a nanostructured tio$_2$ layer at the interface between {ITO} and {NPB} layers. *Displays*, 24(4–5):231 – 234, 2003. ISSN 0141-9382. doi: http://dx.doi.org/10.1016/j.displa.2004.01.010. URL http://www.sciencedirect.com/science/article/pii/S0141938204000113.

[1002] S.A. Haque, S. Koops, N. Tokmoldin, J.R. Durrant, J. Huang, D.D.C. Bradley, and E. Palomares. A multilayered polymer light-emitting diode using a nanocrystalline metal-oxide film as a charge-injection electrode. *Advanced Materials*, 19(5):683–687, 2007. ISSN 1521-4095. doi: 10.1002/adma.200601619. URL http://dx.doi.org/10.1002/adma.200601619.

[1003] R Könenkamp, Robert C Word, and M Godinez. Electroluminescence in nanoporous tio$_2$ solid-state heterojunctions. *Nanotechnology*, 17(8):1858, 2006. URL http://stacks.iop.org/0957-4484/17/i=8/a=008.

[1004] Xinjian Feng, Karthik Shankar, Maggie Paulose, and Craig A. Grimes. Tantalum-doped titanium dioxide nanowire arrays for dye-sensitized solar cells with high open-circuit voltage. *Angewandte Chemie*, 121(43):8239–8242, 2009. ISSN 1521-3757. doi: 10.1002/ange.200903114. URL http://dx.doi.org/10.1002/ange.200903114.

[1005] Wu Bin-Bin, Pan Feng-Ming, and Yang Yu-E. Annealing effect of pulsed laser deposited transparent conductive ta-doped titanium oxide films. *Chinese Physics Letters*, 28(11):118102, 2011. URL http://stacks.iop.org/0256-307X/28/i=11/a=118102.

[1006] W. Chakhari, J. Ben Naceur, S. Ben Taieb, I. Ben Assaker, and R. Chtourou. Fe-doped tio$_2$ nanorods with enhanced electrochemical properties as efficient photoanode materials. *Journal of Alloys and Compounds*, 708(Supplement C):862 – 870, 2017. ISSN 0925-8388. doi: https://doi.org/10.1016/j.jallcom.2016.12.181. URL http://www.sciencedirect.com/science/article/pii/S0925838816341007.

[1007] Yongcheng Wang, Yue-Yu Zhang, Jing Tang, Haoyu Wu, Ming Xu, Zheng Peng, Xin-Gao Gong, and Gengfeng Zheng. Simultaneous etching and doping of tio$_2$ nanowire arrays for enhanced photoelectrochemical performance. *ACS Nano*, 7(10):9375–9383, 2013. doi: 10.1021/nn4040876. URL http://dx.doi.org/10.1021/nn4040876. PMID: 24047133.

[1008] Li Hao, Jian Wang, Fu-Quan Bai, Mo Xie, and Hong-Xing Zhang. Enhancing electron injection in dye-sensitized solar cells by adopting w6+-doped tio$_2$ nanowires: A theoretical study. *European Journal of Inorganic Chemistry*, 2015(33):5563–5570, 2015. ISSN 1099-0682. doi: 10.1002/ejic.201500813. URL http://dx.doi.org/10.1002/ejic.201500813.

[1009] Juliana Schell, Doru C. Lupascu, João Guilherme Martins Correia, Artur Wilson Carbonari, Manfred Deicher, Marcelo Baptista Barbosa, Ronaldo Domingues Mansano, Karl Johnston, Ibere S. Ribeiro, and ISOLDE collaboration. In and cd as defect traps in titanium dioxide. *Hyperfine Interactions*, 238(1):2, Nov 2016. ISSN 1572-9540. doi: 10.1007/s10751-016-1373-7. URL https://doi.org/10.1007/s10751-016-1373-7.

[1010] B. Qi, Y. Yu, X. He, L. Wu, X. Duan, and J. Zhi. Series of transition metal-doped tio$_2$ transparent aqueous sols with visible-light response. *Mater. Chem. Phys.*, 135:549, 2012.

[1011] N. Serpone. Is the band gap of pristine tio$_2$ narrowed by anion- and cation-doping of titanium dioxide in second-generation photocatalysts? *J. Phys. Chem. B*, 110:24287, 2006.

References

[1012] S. N. R. Inturi, T. Boningari, M. Suidan, and P. G. Smirniotis. Visible-light-induced photodegradation of gas phase acetonitrile using aerosol-made transition metal (v, cr, fe, co, mn, mo, ni, cu, y, ce, and zr) doped tio$_2$. *Appl. Catal., B*, 144:333, 2014.

[1013] Bin Liu, Hao Ming Chen, Chong Liu, Sean C. Andrews, Chris Hahn, and Peidong Yang. Large-scale synthesis of transition-metal-doped tio$_2$ nanowires with controllable overpotential. *Journal of the American Chemical Society*, 135(27):9995–9998, 2013. doi: 10.1021/ja403761s. URL http://dx.doi.org/10.1021/ja403761s. PMID: 23815410.

[1014] Minkyu You, Tae Geun Kim, and Yun-Mo Sung. Synthesis of cu-doped tio$_2$ nanorods with various aspect ratios and dopant concentrations. *Crystal Growth & Design*, 10(2):983–987, 2010. doi: 10.1021/cg9012944. URL http://dx.doi.org/10.1021/cg9012944.

[1015] Rebecca Janisch, Priya Gopal, and Nicola A Spaldin. Transition metal-doped tio$_2$ and zno present status of the field. *Journal of Physics: Condensed Matter*, 17(27):R657, 2005. URL http://stacks.iop.org/0953-8984/17/i=27/a=R01.

[1016] Liqiang Jing, Baifu Xin, Fulong Yuan, Lianpeng Xue, Baiqi Wang, and Honggang Fu. Effects of surface oxygen vacancies on photophysical and photochemical processes of zn-doped tio$_2$ nanoparticles and their relationships. *The Journal of Physical Chemistry B*, 110 (36):17860–17868, 2006. doi: 10.1021/jp063148z. URL http://dx.doi.org/10.1021/jp063148z. PMID: 16956273.

[1017] Libin Yang, Yu Zhang, Weidong Ruan, Bing Zhao, Weiqing Xu, and John R. Lombardi. Improved surface-enhanced raman scattering properties of tio$_2$ nanoparticles by zn dopant. *Journal of Raman Spectroscopy*, 41(7):721–726, 2010. ISSN 1097-4555. doi: 10.1002/jrs.2511. URL http://dx.doi.org/10.1002/jrs.2511.

[1018] Jooho Moon, H. Takagi, Y. Fujishiro, and M. Awano. Preparation and characterization of the sb-doped tio$_2$ photocatalysts. *Journal of Materials Science*, 36(4):949–955, 2001. ISSN 1573-4803. doi: 10.1023/A:1004819706292. URL http://dx.doi.org/10.1023/A:1004819706292.

[1019] J. Xu, Y. Ao, D. Fu, and C. Yuan. A simple route for the preparation of eu, n-codoped tio$_2$ nanoparticles with enhanced visible light-induced photocatalytic activity. *J. Colloid Interface Sci.*, 328:447, 2008.

[1020] Baomei Wen, Chunyan Liu, and Yun Liu. Bamboo-shaped ag-doped tio$_2$ nanowires with heterojunctions. *Inorganic Chemistry*, 44(19): 6503–6505, 2005. doi: 10.1021/ic0505551. URL http://dx.doi.org/10.1021/ic0505551. PMID: 16156602.

[1021] Chao He, Yun Yu, Xingfang Hu, and André Larbot. Influence of silver doping on the photocatalytic activity of titania films. *Applied Surface Science*, 200(1–4):239 – 247, 2002. ISSN 0169-4332. doi: http://doi.org/10.1016/S0169-4332(02)00927-3. URL http://www.sciencedirect.com/science/article/pii/S0169433202009273.

[1022] Laveena P. D'Souza, R. Shwetharani, Vipin Amoli, C.A.N. Fernando, Anil Kumar Sinha, and R. Geetha Balakrishna. Photoexcitation of neodymium doped tio$_2$ for improved performance in dye-sensitized solar cells. *Materials & Design*, 104:346 – 354, 2016. ISSN 0264-1275. doi: http://doi.org/10.1016/j.matdes.2016.05.007. URL http://www.sciencedirect.com/science/article/pii/S0264127516305858.

[1023] Lin Cheng, Xuefeng Xu, Yuyan Fang, Yan Li, Jiaxi Wang, Guojia Wan, Xueying Ge, Liangjie Yuan, Keli Zhang, Lei Liao, and Quan Yuan. Triblock copolymer-assisted construction of 20 nm-sized ytterbium-doped tio$_2$ hollow nanostructures for enhanced solar energy utilization efficiency. *Science China Chemistry*, 58(5):850–857, 2015. ISSN 1869-1870. doi: 10.1007/s11426-014-5237-1. URL http://dx.doi.org/10.1007/s11426-014-5237-1.

[1024] S. I. Shah, W. Li, C.-P. Huang, O. Jung, and C. Ni. Study of nd^{3+}, pd^{2+}, pt^{4+}, and fe^{3+} dopant effect on photoreactivity of tio$_2$ nanoparticles. *Proceedings of the National Academy of Sciences*, 99(suppl 2):6482–6486, 2002. doi: 10.1073/pnas.052518299. URL http://www.pnas.org/content/99/suppl_2/6482.abstract.

[1025] M.M. Rahman, K.M. Krishna, T. Soga, T. Jimbo, and M. Umeno. Optical properties and x-ray photoelectron spectroscopic study of pure and pb-doped tio$_2$ thin films. *Journal of Physics and Chemistry of Solids*, 60(2):201 – 210, 1999. ISSN 0022-3697. doi: http://doi.org/10.1016/S0022-3697(98)00264-9. URL http://www.sciencedirect.com/science/article/pii/S0022369798002549.

[1026] L. G. Devi and S. G. Kumar. Exploring the critical dependence of adsorption of various dyes on the degradation rate using ln^{3+}-tio$_2$ surface under uv/solar light. *Appl. Surf. Sci.*, 261:137, 2012.

[1027] Qi Xiao, Zhichun Si, Zhiming Yu, and Guanzhou Qiu. Sol–gel auto-combustion synthesis of samarium-doped tio$_2$ nanoparticles and their photocatalytic activity under visible light irradiation. *Materials Science and Engineering: B*, 137(1–3):189 – 194, 2007. ISSN 0921-5107. doi: http://doi.org/10.1016/j.mseb.2006.11.011. URL http://www.sciencedirect.com/science/article/pii/S0921510706006702.

[1028] Seokwoo Jeon and Paul V. Braun. Hydrothermal synthesis of er-doped luminescent tio$_2$ nanoparticles. *Chemistry of Materials*, 15(6): 1256–1263, 2003. doi: 10.1021/cm0207402. URL http://dx.doi.org/10.1021/cm0207402.

[1029] R. Asahi, T. Morikawa, T. Ohwaki, K. Aoki, and Y. Taga. Visible-light photocatalysis in nitrogen-doped titanium oxides. *Science*, 293 (5528):269–271, 2001. ISSN 0036-8075. doi: 10.1126/science.1061051. URL http://science.sciencemag.org/content/293/5528/269.

[1030] C. Han, M. Pelaez, V. Likodimos, A. G. Kontos, P. Falaras, K. O'Shea, and D. D. Dionysiou. Innovative visible light-activated sulfur doped tio$_2$ films for water treatment. *Appl. Catal., B*, 107:77, 2011.

[1031] P. Periyat, S. C. Pillai, D. E. McCormack, J. Colreavy, and S. J. Hinder. Improved high-temperature stability and sun-light-driven photocatalytic activity of sulfur-doped anatase tio$_2$. *J. Phys. Chem. C*, 112:7644, 2008.

[1032] M. V. Dozzi and E. Selli. Doping tio$_2$ with p-block elements: Effects on photocatalytic activity. *J. Photochem. Photobiol. C*, 14:13, 2013.

[1033] Aimin Yu, Guangjun Wu, Fuxiang Zhang, Yali Yang, and Naijia Guan. Synthesis and characterization of n-doped tio$_2$ nanowires with visible light response. *Catalysis Letters*, 129(3):507, 2009. ISSN 1572-879X. doi: 10.1007/s10562-008-9832-7. URL http://dx.doi.org/10.1007/s10562-008-9832-7.

[1034] T. Umebayashi, T. Yamaki, H. Itoh, and K. Asai. Band gap narrowing of titanium dioxide by sulfur doping. *Applied Physics Letters*, 81(3):454–456, 2002. doi: 10.1063/1.1493647. URL http://dx.doi.org/10.1063/1.1493647.

[1035] André Schleife, Patrick Rinke, Friedhelm Bechstedt, and Chris G. Van de Walle. Enhanced optical absorption due to symmetry breaking in tio$_{2(1-x)}$s$_{2x}$ alloys. *The Journal of Physical Chemistry C*, 117(8):4189–4193, 2013. doi: 10.1021/jp3106937. URL http://dx.doi.org/10.1021/jp3106937.

[1036] Zhongchun Wang and Xingfang Hu. Fabrication and electrochromic properties of spin-coated tio$_2$ thin films from peroxo-polytitanic acid. *Thin Solid Films*, 352(1):62 – 65, 1999. ISSN 0040-6090. doi: https://doi.org/10.1016/S0040-6090(99)00321-1. URL http://www.sciencedirect.com/science/article/pii/S0040609099003211.

[1037] S. Y. Huang, L. Kavan, I. Exnar, and M. Grätzel. Rocking chair lithium battery based on nanocrystalline tio₂ (anatase). *Journal of The Electrochemical Society*, 142(9):L142–L144, 1995. doi: 10.1149/1.2048726. URL http://jes.ecsdl.org/content/142/9/L142.abstract.

[1038] Anders Hagfeldt, Nicolas Vlachopoulos, and Michael Grätzel. Fast electrochromic switching with nanocrystalline oxide semiconductor films. *Journal of The Electrochemical Society*, 141(7):L82–L84, 1994. doi: 10.1149/1.2055045. URL http://jes.ecsdl.org/content/141/7/L82.abstract.

[1039] E. Prociów, K. Sieradzka, J. Domaradzki, D. Kaczmarek, and M. Mazur. Thin films based on nanocrystalline tio₂ for transparent electronics. *Acta Physica Polonica A*, 116:S72–S74, 2009.

[1040] Vivek Singh, Ignacio J. Castellanos Beltran, Josep Casamada Ribot, and Prashant Nagpal. Photocatalysis deconstructed: Design of a new selective catalyst for artificial photosynthesis. *Nano Letters*, 14(2):597–603, 2014. doi: 10.1021/nl403783d. URL http://dx.doi.org/10.1021/nl403783d. PMID: 24443959.

[1041] Qingsen Meng, Tuo Wang, Enzuo Liu, Xinbin Ma, Qingfeng Ge, and Jinlong Gong. Understanding electronic and optical properties of anatase tio₂ photocatalysts co-doped with nitrogen and transition metals. *Phys. Chem. Chem. Phys.*, 15:9549–9561, 2013. doi: 10.1039/C3CP51476E. URL http://dx.doi.org/10.1039/C3CP51476E.

[1042] Swagata Banerjee, Suresh C. Pillai, Polycarpos Falaras, Kevin E. O'Shea, John A. Byrne, and Dionysios D. Dionysiou. New insights into the mechanism of visible light photocatalysis. *The Journal of Physical Chemistry Letters*, 5(15):2543–2554, 2014. doi: 10.1021/jz501030x. URL http://dx.doi.org/10.1021/jz501030x. PMID: 26277942.

[1043] P.A. Cox. *Transition Metal Oxides: An Introduction to Their Electronic Structure and Properties.* The International series of monographs on chemistry. OUP Oxford, 1992. ISBN 9780199588947. URL https://books.google.de/books?id=DdcUDAAAQBAJ.

[1044] N.F. Mott. Electrons in disordered structures. *Advances in Physics*, 16(61):49–144, 1967. doi: 10.1080/00018736700101265. URL http://dx.doi.org/10.1080/00018736700101265.

[1045] Sir Nevill Francis Mott and Edward Arthur Davis. *Electronic processes in non crystalline materials.* Number pp. 117, 148-150. Clarendon Press, 1971.

[1046] Arghya Narayan Banerjee. The design, fabrication, and photocatalytic utility of nanostructured semiconductors: focus on tio₂-based nanostructures. *Nanotechnology, Science and Applications*, 4, 2011.

[1047] S. Kataoka and T. Suzuki. *Bull. Electrotech. Lab. Tokyo*, 18(732), 1954.

[1048] Alberto Naldoni, Mattia Allieta, Saveria Santangelo, Marcello Marelli, Filippo Fabbri, Serena Cappelli, Claudia L. Bianchi, Rinaldo Psaro, and Vladimiro Dal Santo. Effect of nature and location of defects on bandgap narrowing in black tio₂ nanoparticles. *Journal of the American Chemical Society*, 134(18):7600–7603, 2012. doi: 10.1021/ja3012967. URL http://dx.doi.org/10.1021/ja3012676. PMID: 22519668.

[1049] T S Moss. The interpretation of the properties of indium antimonide. *Proceedings of the Physical Society. Section B*, 67(10):775, 1954. URL http://stacks.iop.org/0370-1301/67/i=10/a=306.

[1050] C. G. Granqvist. *Oxide-Based Electrochromic Materials and Devices Prepared by Magnetron Sputtering*, pages 485–495. Springer Berlin Heidelberg, Berlin, Heidelberg, 2008. ISBN 978-3-540-76664-3. doi: 10.1007/978-3-540-76664-3_13. URL https://doi.org/10.1007/978-3-540-76664-3_13.

[1051] Jiazang Chen, Hua Bing Tao, and Bin Liu. Unraveling the intrinsic structures that influence the transport of charges in tio₂ electrodes. *Advanced Energy Materials*, pages 1700886–n/a, 2017. ISSN 1614-6840. doi: 10.1002/aenm.201700886. URL http://dx.doi.org/10.1002/aenm.201700886. 1700886.

[1052] Juan A. Anta. Electron transport in nanostructured metal-oxide semiconductors. *Current Opinion in Colloid & Interface Science*, 17(3):124–131, June 2012. doi: 10.1016/j.cocis.2012.02.003.

[1053] Anders Hagfeldt, Gerrit Boschloo, Licheng Sun, Lars Kloo, and Henrik Pettersson. Dye-sensitized solar cells. *Chemical Reviews*, 110(11):6595–6663, 2010. doi: 10.1021/cr900356p. URL http://dx.doi.org/10.1021/cr900356p. PMID: 20831177.

[1054] Juan Bisquert, Francisco Fabregat-Santiago, Ivan Mora-Sero, Germa Garcia-Belmonte, and Sixto Gimenez. Electron lifetime in dye-sensitized solar cells: theory and interpretation of measurements. *The Journal of Physical Chemistry C*, 113(40):17278–17290, 2009.

[1055] K. Bernert, C. Oestreich, J. Bollmann, and T. Mikolajick. The influence of bottom oxide thickness on the extraction of the trap energy distribution in sonos (silicon-oxide-nitride-oxide-silicon) structures. *Applied Physics A*, 100(1):249–255, Jul 2010. ISSN 1432-0630. doi: 10.1007/s00339-010-5694-0. URL https://doi.org/10.1007/s00339-010-5694-0.

[1056] José M. Montero and Juan Bisquert. Interpretation of trap-limited mobility in space-charge limited current in organic layers with exponential density of traps. *Journal of Applied Physics*, 110(4):043705, 2011. doi: 10.1063/1.3622615. URL http://dx.doi.org/10.1063/1.3622615.

[1057] Raheleh Mohammadpour, Azam Iraji zad, Anders Hagfeldt, and Gerrit Boschloo. Comparison of trap-state distribution and carrier transport in nanotubular and nanoparticulate tio2 electrodes for dye-sensitized solar cells. *ChemPhysChem*, 11(10):2140–2145, 2010. ISSN 1439-7641. doi: 10.1002/cphc.201000125. URL http://dx.doi.org/10.1002/cphc.201000125.

[1058] B. S. Simpkins, M. A. Mastro, C. R. Eddy Jr., J. K. Hite, and P. E. Pehrsson. Space-charge-limited currents and trap characterization in coaxial algan/gan nanowires. *Journal of Applied Physics*, 110(4):044303, 2011. doi: 10.1063/1.3622145. URL http://dx.doi.org/10.1063/1.3622145.

[1059] Laurence Peter. "sticky electrons" transport and interfacial transfer of electrons in the dye-sensitized solar cell. *Accounts of Chemical Research*, 42(11):1839–1847, 2009. doi: 10.1021/ar900143m. URL http://dx.doi.org/10.1021/ar900143m. PMID: 19637905.

[1060] L.M. Terman. An investigation of surface states at a silicon/silicon oxide interface employing metal-oxide-silicon diodes. *Solid-State Electronics*, 5(5):285 – 299, 1962. ISSN 0038-1101. doi: https://doi.org/10.1016/0038-1101(62)90111-9. URL http://www.sciencedirect.com/science/article/pii/0038110162901119.

[1061] Rajan Jose, Nurbosyn U. Zhanpeisov, Hiroshi Fukumura, Yoshinobu Baba, and Mitsuru Ishikawa. Structure-property correlation of cdse clusters using experimental results and first-principles dft calculations. *Journal of the American Chemical Society*, 128(2):629–636, 2006. doi: 10.1021/ja0565018. URL http://dx.doi.org/10.1021/ja0565018. PMID: 16402851.

References

[1062] Juan Bisquert. Hopping transport of electrons in dye-sensitized solar cells. *The Journal of Physical Chemistry C*, 111(46):17163–17168, 2007. doi: 10.1021/jp077419x. URL http://dx.doi.org/10.1021/jp077419x.

[1063] Don Monroe. Hopping in exponential band tails. *Phys. Rev. Lett.*, 54:146–149, Jan 1985. doi: 10.1103/PhysRevLett.54.146. URL https://link.aps.org/doi/10.1103/PhysRevLett.54.146.

[1064] J Ross Macdonald. *Impedance spectroscopy: emphasizing solid materials and analysis*. John Wiley & Sons, New York, 1987.

[1065] Kan Zhang and Jong Hyeok Park. Surface localization of defects in black tio₂: Enhancing photoactivity or reactivity. *The Journal of Physical Chemistry Letters*, 8(1):199–207, 2017. doi: 10.1021/acs.jpclett.6b02289. URL http://dx.doi.org/10.1021/acs.jpclett.6b02289. PMID: 27991794.

[1066] Haining Chen, Zhanhua Wei, Keyou Yan, Yang Bai, and Shihe Yang. Unveiling two electron-transport modes in oxygen-deficient tio₂ nanowires and their influence on photoelectrochemical operation. *The Journal of Physical Chemistry Letters*, 5(16):2890–2896, 2014. doi: 10.1021/jz5014505. URL http://dx.doi.org/10.1021/jz5014505. PMID: 26278095.

[1067] Xujie Lü, Aiping Chen, Yongkang Luo, Ping Lu, Yaomin Dai, Erik Enriquez, Paul Dowden, Hongwu Xu, Paul G. Kotula, Abul K. Azad, Dmitry A. Yarotski, Rohit P. Prasankumar, Antoinette J. Taylor, Joe D. Thompson, and Quanxi Jia. Conducting interface in oxide homojunction: Understanding of superior properties in black tio₂. *Nano Letters*, 16(9):5751–5755, 2016. doi: 10.1021/acs.nanolett.6b02454. URL http://dx.doi.org/10.1021/acs.nanolett.6b02454. PMID: 27482629.

[1068] Amy L. Linsebigler, Guangquan. Lu, and John T. Yates. Photocatalysis on tio₂ surfaces: Principles, mechanisms, and selected results. *Chemical Reviews*, 95(3):735–758, 1995. doi: 10.1021/cr00035a013. URL http://dx.doi.org/10.1021/cr00035a013.

[1069] Th. Dittrich, J. Weidmann, F. Koch, I. Uhlendorf, and I. Lauermann. Temperature- and oxygen partial pressure-dependent electrical conductivity in nanoporous rutile and anatase. *Applied Physics Letters*, 75(25):3980–3982, 1999. doi: 10.1063/1.125513. URL http://scitation.aip.org/content/aip/journal/apl/75/25/10.1063/1.125513.

[1070] Annabella Selloni. Crystal growth: Anatase shows its reactive side. *Nature Materials*, 7:613 – 615, 2008. doi: 10.1038/nmat2241.

[1071] Amy Linsebigler, Guangquan Lu, and Jr. John T. Yates. Co photooxidation on tio₂(110). *Journal of Physical Chemistry*, 100(16): 6631–6636, 1996.

[1072] W. Göpel, G. Rocker, and R. Feierabend. Intrinsic defects of tio₂(110): Interaction with chemisorbed o₂, h₂, co, and co₂. *Phys. Rev. B*, 28:3427–3438, Sep 1983. doi: 10.1103/PhysRevB.28.3427. URL https://link.aps.org/doi/10.1103/PhysRevB.28.3427.

[1073] H.L. Tuller, J. Claus, and T. Chen. Electronically active grain boundaries: Defects and diffusion. In G. Gusmano and E. Traversa, editors, *Electroceramics-Proc. Fourth Euro Ceramics*, volume 5, pages 443–450, 1995.

[1074] Yoon-Chae Nah, Indhumati Paramasivam, and Patrik Schmuki. Doped tio₂ and tio₂ nanotubes: Synthesis and applications. *ChemPhysChem*, 11(13):2698–2713, 2010. ISSN 1439-7641. doi: 10.1002/cphc.201000276. URL http://dx.doi.org/10.1002/cphc.201000276.

[1075] Reginald M. Penner. Chemical sensing with nanowires. *Annual Review of Analytical Chemistry*, 5(1):461–485, 2012. doi: 10.1146/annurev-anchem-062011-143007. URL http://dx.doi.org/10.1146/annurev-anchem-062011-143007. PMID: 22524224.

[1076] Nicolae Barsan and Udo Weimar. Conduction model of metal oxide gas sensors. *Journal of Electroceramics*, 7(3):143–167, 2001. ISSN 1573-8663. doi: 10.1023/A:1014405811371. URL http://dx.doi.org/10.1023/A:1014405811371.

[1077] Yahya Alivov, Hans H. Funke, Vivek Singh, and Prashant Nagpal. Air-pressure tunable depletion width, rectification behavior, and charge conduction in oxide nanotubes. *ACS Applied Materials & Interfaces*, 7(4):2153–2159, 2015. doi: 10.1021/am5076666. URL http://dx.doi.org/10.1021/am5076666. PMID: 25594471.

[1078] Jenny Nelson, Anuradha M. Eppler, and Ian M. Ballard. Photoconductivity and charge trapping in porous nanocrystalline titanium dioxide. *Journal of Photochemistry and Photobiology A: Chemistry*, 148(1):25 – 31, 2002. ISSN 1010-6030. doi: https://doi.org/10.1016/S1010-6030(02)00035-7. URL http://www.sciencedirect.com/science/article/pii/S1010603002000357. SEMICONDUCTOR PHOTOCHEMISTRY 1 FIRST INTERNATIONAL CONFERENCE ON SEMICONDUCTOR PHOTOCHEMISTRY,UNIVERSITY OF STRATHCLYDE, GLASGOW, JULY 2001.

[1079] Stefano Livraghi, Maria Cristina Paganini, Elio Giamello, Annabella Selloni, Cristiana Di Valentin, and Gianfranco Pacchioni. Origin of photoactivity of nitrogen-doped titanium dioxide under visible light. *Journal of the American Chemical Society*, 128(49):15666–15671, 2006. doi: 10.1021/ja064164c. URL http://dx.doi.org/10.1021/ja064164c. PMID: 17147376.

[1080] Yizheng Jin, Jianpu Wang, Baoquan Sun, James C Blakesley, and Neil C Greenham. Solution-processed ultraviolet photodetectors based on colloidal zno nanoparticles. *Nano letters*, 8(6):1649–1653, 2008.

[1081] Cesare Soci, Arthur Zhang, Bin Xiang, Shadi A Dayeh, DPR Aplin, Jung Park, XY Bao, Yu-Hwa Lo, and Deli Wang. Zno nanowire uv photodetectors with high internal gain. *Nano letters*, 7(4):1003–1009, 2007.

[1082] Jiping Cheng, Yunjin Zhang, and Ruyan Guo. Zno microtube ultraviolet detectors. *Journal of Crystal Growth*, 310(1):57 – 61, 2008. ISSN 0022-0248. doi: https://doi.org/10.1016/j.jcrysgro.2007.08.034. URL http://www.sciencedirect.com/science/article/pii/S0022024807008329.

[1083] Qiang Wang, Jun Jie Li, and Chang Zhi Gu. Enhanced uv response of single anodic tio₂ nanotube: Effect of water-modified microstructures. *The Journal of Physical Chemistry C*, 116(32):16864–16869, 2012. doi: 10.1021/jp304193z. URL http://dx.doi.org/10.1021/jp304193z.

[1084] Katarzyna Zakrzewska, Marta Radecka, and Mieczyslaw Rekas. Effect of nb, cr, sn additions on gas sensing properties of tio₂ thin films. *Thin Solid Films*, 310(1):161 – 166, 1997. ISSN 0040-6090. doi: https://doi.org/10.1016/S0040-6090(97)00401-X. URL http://www.sciencedirect.com/science/article/pii/S004006090700401X.

[1085] V K Khanna and R K Nahar. Carrier-transfer mechanisms and al₂o₃ sensors for low and high humidities. *Journal of Physics D: Applied Physics*, 19(7):L141, 1986. URL http://stacks.iop.org/0022-3727/19/i=7/a=004.

[1086] Ghennady Korotchenkov, Vladimir Brynzari, and Serghei Dmitriev. Electrical behavior of sno₂ thin films in humid atmosphere. *Sensors and Actuators B: Chemical*, 54(3):197 – 201, 1999. ISSN 0925-4005. doi: https://doi.org/10.1016/S0925-4005(99)00016-7. URL http://www.sciencedirect.com/science/article/pii/S0925400599000167.

[1087] Chenghua Sun, Li-Min Liu, Annabella Selloni, Gao Qing (Max) Lu, and Sean C. Smith. Titania-water interactions: a review of theoretical studies. *J. Mater. Chem.*, 20:10319–10334, 2010. doi: 10.1039/C0JM01491E. URL http://dx.doi.org/10.1039/C0JM01491E.

[1088] Zhi Chen and Chi Lu. Humidity sensors: A review of materials and mechanisms. *Sensor Letters*, 3(4):274–295, 2005. doi: doi: 10.1166/sl.2005.045.

[1089] Steven H. Szczepankiewicz, A. J. Colussi, and Michael R. Hoffmann. Infrared spectra of photoinduced species on hydroxylated titania surfaces. *The Journal of Physical Chemistry B*, 104(42):9842–9850, 2000. doi: 10.1021/jp0007890. URL http://dx.doi.org/10.1021/jp0007890.

[1090] E. Konstantinova, J. Weidmann, and Th. Dittrich. Influence of adsorbed water and oxygen on the photoluminescence and epr of por-tio$_2$ (anatase). *Journal of Porous Materials*, 7(1):389–392, Jan 2000. ISSN 1573-4854. doi: 10.1023/A:1009699324013. URL https://doi.org/10.1023/A:1009699324013.

[1091] R. Könenkamp. Carrier transport in nanoporous tio$_2$ films. *Phys. Rev. B*, 61:11057–11064, Apr 2000. doi: 10.1103/PhysRevB.61.11057. URL https://link.aps.org/doi/10.1103/PhysRevB.61.11057.

[1092] Aldo Amore Bonapasta, Francesco Filippone, Giuseppe Mattioli, and Paola Alippi. Oxygen vacancies and {OH} species in rutile and anatase tio$_2$ polymorphs. *Catalysis Today*, 144(1-2):177 – 182, 2009. ISSN 0920-5861. doi: https://doi.org/10.1016/j.cattod.2009.01.047. URL http://www.sciencedirect.com/science/article/pii/S0920586109000807. Selected papers presented at the 5th European Conference on Solar Chemistry and Photocatalysis: Environmental Applications (SPEA 5), 4-8th October 2008.

[1093] Taizo Shibuya, Kenji Yasuoka, Susanne Mirbt, and Biplab Sanyal. A systematic study of polarons due to oxygen vacancy formation at the rutile tio$_2$ (110) surface by gga + u and hse06 methods. *Journal of Physics: Condensed Matter*, 24(43):435504, 2012. URL http://stacks.iop.org/0953-8984/24/i=43/a=435504.

[1094] Gerrit Boschloo and Donald Fitzmaurice. Spectroelectrochemical investigation of surface states in nanostructured tio$_2$ electrodes. *The Journal of Physical Chemistry B*, 103(12):2228–2231, 1999. doi: 10.1021/jp984414e. URL http://dx.doi.org/10.1021/jp984414e.

[1095] P.Madhu Kumar, S. Badrinarayanan, and Murali Sastry. Nanocrystalline tio$_2$ studied by optical, {FTIR} and x-ray photoelectron spectroscopy: correlation to presence of surface states. *Thin Solid Films*, 358(1–2):122 – 130, 2000. ISSN 0040-6090. doi: https://doi.org/10.1016/S0040-6090(99)00722-1. URL http://www.sciencedirect.com/science/article/pii/S0040609099007221.

[1096] Koji Iijima, Masako Goto, Shogo Enomoto, Hideyuki Kunugita, Kazuhiro Ema, Masanori Tsukamoto, Noriya Ichikawa, and Hiroshi Sakama. Influence of oxygen vacancies on optical properties of anatase tio$_2$ thin films. *Journal of Luminescence*, 128(5–6):911 – 913, 2008. ISSN 0022-2313. doi: https://doi.org/10.1016/j.jlumin.2007.11.071. URL http://www.sciencedirect.com/science/article/pii/S0022231307003882. Proceedings of the 16th International Conference on Dynamical Processes in Excited States of Solids16th International Conference on Dynamical Processes in Excited States of Solids.

[1097] Taketoshi Minato, Yasuyuki Sainoo, Yousoo Kim, Hiroyuki S. Kato, Ken ichi Aika, Maki Kawai, Jin Zhao, Hrvoje Petek, Tian Huang, Wei He, Bing Wang, Zhuo Wang, Yan Zhao, Jinlong Yang, and J. G. Hou. The electronic structure of oxygen atom vacancy and hydroxyl impurity defects on titanium dioxide (110) surface. *The Journal of Chemical Physics*, 130(12):124502, 2009. doi: 10.1063/1.3082408. URL http://dx.doi.org/10.1063/1.3082408.

[1098] Sajith Kurian, Hyungtak Seo, and Hyeongtag Jeon. Significant enhancement in visible light absorption of tio$_2$ nanotube arrays by surface band gap tuning. *The Journal of Physical Chemistry C*, 117(33):16811–16819, 2013. doi: 10.1021/jp405207e. URL http://dx.doi.org/10.1021/jp405207e.

[1099] Vaidyanathan Subramanian, Eduardo E. Wolf, and Prashant V. Kamat. Catalysis with tio$_2$/gold nanocomposites. effect of metal particle size on the fermi level equilibration. *Journal of the American Chemical Society*, 126(15):4943–4950, 2004. doi: 10.1021/ja0315199. URL http://dx.doi.org/10.1021/ja0315199. PMID: 15080700.

[1100] Sergio Pinilla, Abniel Machin, Sang Hoon Park, Juan Camilo Arango, Valeria Nicolosi, Francisco M. Marquez-Linares, and Carmen Morant. Tio$_2$-based nanomaterials for the production of hydrogen and the development of lithium-ion batteries. *The Journal of Physical Chemistry B*, 0(ja):null, 2017. doi: 10.1021/acs.jpcb.7b07130. URL http://dx.doi.org/10.1021/acs.jpcb.7b07130. PMID: 29058914.

[1101] P. Pandi and C. Gopinathan. Synthesis and characterization of tio$_2$–nio and tio$_2$–wo$_3$ nanocomposites. *Journal of Materials Science: Materials in Electronics*, 28(7):5222–5234, Apr 2017. ISSN 1573-482X. doi: 10.1007/s10854-016-6179-5. URL https://doi.org/10.1007/s10854-016-6179-5.

[1102] Elham Rahmanian, Rasoul Malekfar, and Martin Pumera. Nanohybrids of two-dimensional transition metal dichalcogenides and titanium dioxide for photocatalytic applications. *Chemistry - A European Journal*, 24(1):18–31, 2018. doi: 10.1002/chem.201703434. URL http://dx.doi.org/10.1002/chem.201703434.

[1103] Arash Mohammadpour, Samira Farsinezhad, Benjamin D. Wiltshire, and Karthik Shankar. Majority carrier transport in single crystal rutile nanowire arrays. *physica status solidi (RRL) – Rapid Research Letters*, 8(6):512–516, 2014. ISSN 1862-6270. doi: 10.1002/pssr.201308296. URL http://dx.doi.org/10.1002/pssr.201308296.

[1104] Xia Sheng, Liping Chen, Tao Xu, Kai Zhu, and Xinjian Feng. Understanding and removing surface states limiting charge transport in tio$_2$ nanowire arrays for enhanced optoelectronic device performance. *Chem. Sci.*, 7:1910–1913, 2016. doi: 10.1039/C5SC04076K. URL http://dx.doi.org/10.1039/C5SC04076K.

[1105] A Mohammadpour, B D Wiltshire, Y Zhang, S Farsinezhad, A M Askar, R Kisslinger, Y Ren, P Kar, and K Shankar. 100-fold improvement in carrier drift mobilities in alkanephosphonate-passivated monocrystalline tio$_2$ nanowire arrays. *Nanotechnology*, 28(14):144001, 2017. URL http://stacks.iop.org/0957-4484/28/i=14/a=144001.

[1106] C. Divya, B. Janarthanan, and J. Chandrasekaran. Review on recent advances in titanium dye oxide nanoparticles for dye sensitized solar cell applications. *International Journal of Advances in Applied Sciences (IJAAS)*, 6(2):135–144, 2017.

[1107] K.L. Kliewer and J.S. Koehler. Space charge in ionic crystals. i. general approach with application to nacl. *Phys. Rev.*, 140:A1226–A1240, Nov 1965. doi: 10.1103/PhysRev.140.A1226. URL https://link.aps.org/doi/10.1103/PhysRev.140.A1226.

[1108] Roger A. De Souza and Manfred Martin. Using ^{18}o/^{16}o exchange to probe an equilibrium space-charge layer at the surface of a crystalline oxide: method and application. *Physical Chemistry Chemical Physics*, 10:2356–2367, 2008. doi: 10.1039/B719618K. URL http://dx.doi.org/10.1039/B719618K.

[1109] Joachim Maier. Ionic conduction in space charge regions. *Progress in solid state chemistry*, 23(3):171–263, 1995.

[1110] J. Jamnik, J. Maier, and S. Pejovnik. Interfaces in solid ionic conductors: Equilibrium and small signal picture. *Solid State Ionics*, 75:51 – 58, 1995. ISSN 0167-2738. doi: http://dx.doi.org/10.1016/0167-2738(94)00184-T. URL http://www.sciencedirect.com/science/article/pii/016727389400184T.

References

[1111] J. D. Eshelby, C. W. A. Newey, P. L. Pratt, and A. B. Lidiard. Charged dislocations and the strength of ionic crystals. *Philosophical Magazine*, 3(25):75–89, 1958. doi: 10.1080/14786435808243228. URL http://dx.doi.org/10.1080/14786435808243228.

[1112] R.W. Whitworth. Charged dislocations in ionic crystals. *Advances in Physics*, 24(2):203–304, 1975. doi: 10.1080/00018737500101401. URL http://dx.doi.org/10.1080/00018737500101401.

[1113] Kiran Kumar Adepalli, Marion Kelsch, Rotraut Merkle, and Joachim Maier. Influence of line defects on the electrical properties of single crystal tio₂. *Advanced Functional Materials*, 23(14):1798–1806, 2013. ISSN 1616-3028. doi: 10.1002/adfm.201202256. URL http://dx.doi.org/10.1002/adfm.201202256.

[1114] B.O. Aduda, P. Ravirajan, K.L. Choy, and J. Nelson. Effect of morphology on electron drift mobility in porous Tio₂. *International Journal of Photoenergy*, 6(3):141–147, 2004. doi: 10.1155/S1110662X04000170. URL http://dx.doi.org/10.1155/S1110662X04000170.

[1115] Meagen A. Gillispie, Maikel F.A.M. van Hest, Matthew S. Dabney, John D. Perkins, and David S. Ginley. Sputtered nb- and ta-doped tio₂ transparent conducting oxide films on glass. *Journal of Materials Research*, 22(10):2832–2837, 2007. doi: 10.1557/JMR.2007.0353.

[1116] A. A. Bolzan, C. Fong, B. J. Kennedy, and C. J. Howard. Structural studies of rutile-type metal dioxides. *Acta Crystallographica Section B*, 53(3):373–380, Jun 1997. doi: 10.1107/S0108768197001468. URL https://doi.org/10.1107/S0108768197001468.

[1117] E. Hendry, M. Koeberg, B. O'Regan, and M. Bonn. Local field effects on electron transport in nanostructured tio₂ revealed by terahertz spectroscopy. *Nano Lett.*, 6(4):755–759, 2006. doi: 10.1021/nl0600225.

[1118] A. Kirchner, K. Busch, and C. M. Soukoulis. Transport properties of random arrays of dielectric cylinders. *Phys. Rev. B*, 57:277–288, Jan 1998. doi: 10.1103/PhysRevB.57.277. URL https://link.aps.org/doi/10.1103/PhysRevB.57.277.

[1119] Jonathan E. Spanier and Irving P. Herman. Use of hybrid phenomenological and statistical effective-medium theories of dielectric functions to model the infrared reflectance of porous sic films. *Phys. Rev. B*, 61:10437–10450, Apr 2000. doi: 10.1103/PhysRevB.61. 10437. URL https://link.aps.org/doi/10.1103/PhysRevB.61.10437.

[1120] C. G. Granqvist and O. Hunderi. Optical properties of ag-sio₂ cermet films: A comparison of effective-medium theories. *Phys. Rev. B*, 18:2897–2906, Sep 1978. doi: 10.1103/PhysRevB.18.2897. URL https://link.aps.org/doi/10.1103/PhysRevB.18.2897.

[1121] J. I. Gittleman and B. Abeles. Comparison of the effective medium and the maxwell-garnett predictions for the dielectric constants of granular metals. *Phys. Rev. B*, 15:3273–3275, Mar 1977. doi: 10.1103/PhysRevB.15.3273. URL https://link.aps.org/doi/10.1103/PhysRevB.15.3273.

[1122] Th. Dittrich, E. A. Lebedev, and J. Weidmann. Electron drift mobility in porous tio₂ (anatase). *physica status solidi (a)*, 165(2):R5–R6, 1998. ISSN 1521-396X. doi: 10.1002/(SICI)1521-396X(199802)165:23.0.CO;2-9. URL http://dx.doi.org/10.1002/(SICI)1521-396X(199802)165:23.0.CO;2-9.

[1123] N. Kopidakis, E. A. Schiff, N.-G. Park, J. van de Lagemaat, and A. J. Frank. Ambipolar diffusion of photocarriers in electrolyte-filled, nanoporous tio₂. *The Journal of Physical Chemistry B*, 104(16):3930–3936, 2000. doi: 10.1021/jp9936603. URL http://dx.doi.org/10. 1021/jp9936603.

[1124] Juan A. Anta, Jenny Nelson, and N. Quirke. Charge transport model for disordered materials: Application to sensitized tio₂. *Phys. Rev. B*, 65:125324, Mar 2002. doi: 10.1103/PhysRevB.65.125324. URL https://link.aps.org/doi/10.1103/PhysRevB.65.125324.

[1125] Gordon M. Turner, Matthew C. Beard, and Charles A. Schmuttenmaer. Carrier localization and cooling in dye-sensitized nanocrystalline titanium dioxide. *The Journal of Physical Chemistry B*, 106(45):11716–11719, 2002. doi: 10.1021/jp025844e. URL http://dx.doi.org/10.1021/jp025844e.

[1126] K. Shimakawa, T. Itoh, H. Naito, and S. O. Kasap. The origin of non-drude terahertz conductivity in nanomaterials. *Applied Physics Letters*, 100(13):132102, 2012. doi: 10.1063/1.3697404. URL http://dx.doi.org/10.1063/1.3697404.

[1127] Hynek Němec, Petr Kužel, and Villy Sundström. Far-infrared response of free charge carriers localized in semiconductor nanoparticles. *Phys. Rev. B*, 79:115309, Mar 2009. doi: 10.1103/PhysRevB.79.115309. URL https://link.aps.org/doi/10.1103/PhysRevB.79.115309.

[1128] Kostiantyn Shportko, Stephan Kremers, Michael Woda, Dominic Lencer, John Robertson, and Matthias Wuttig. Resonant bonding in crystalline phase-change materials. *Nature Publishing Group*, 7(8):653 – 658, 2008. doi: 10.1038/nmat2226. URL http://dx.doi.org/10. 1038/nmat2226.

[1129] N. V. Smith. Classical generalization of the drude formula for the optical conductivity. *Phys. Rev. B*, 64:155106, Sep 2001. doi: 10.1103/PhysRevB.64.155106. URL https://link.aps.org/doi/10.1103/PhysRevB.64.155106.

[1130] S. Kasap, H. Ruda, and Y. Boucher. *Cambridge Illustrated Handbook of Optoelectronics and Photonics*. Cambridge University Press, 2009. ISBN 9781139643726. URL https://books.google.de/books?id=18P3CwAAQBAJ.

[1131] Frederick Wooten. *Optical Properties of Solids*. Academic Press, 1972.

[1132] Jeppe C. Dyre and Thomas B. Schrøder. Universality of ac conduction in disordered solids. *Rev. Mod. Phys.*, 72:873–892, Jul 2000. doi: 10.1103/RevModPhys.72.873. URL https://link.aps.org/doi/10.1103/RevModPhys.72.873.

[1133] V.N. Prigodin and A.J. Epstein. Quantum hopping in metallic polymers. *Physica B: Condensed Matter*, 338(1–4):310 – 317, 2003. ISSN 0921-4526. doi: http://dx.doi.org/10.1016/j.physb.2003.08.011. URL http://www.sciencedirect.com/science/article/pii/S0921452603005167. Proceedings of the Sixth International Conference on Electrical Transport and Optical Properties of Inhomogeneous Media.

[1134] P. W. Anderson. Absence of diffusion in certain random lattices. *Phys. Rev.*, 109:1492–1505, Mar 1958. doi: 10.1103/PhysRev.109.1492. URL https://link.aps.org/doi/10.1103/PhysRev.109.1492.

[1135] N. F. Mott. Conduction in non-crystalline materials. *The Philosophical Magazine: A Journal of Theoretical Experimental and Applied Physics*, 19(160):835–852, 1969. doi: 10.1080/14786436908216338. URL http://dx.doi.org/10.1080/14786436908216338.

[1136] Harvey Scher and Elliott W. Montroll. Anomalous transit-time dispersion in amorphous solids. *Phys. Rev. B*, 12:2455–2477, Sep 1975. doi: 10.1103/PhysRevB.12.2455. URL https://link.aps.org/doi/10.1103/PhysRevB.12.2455.

[1137] L. E. Brus. Electron-electron and electron-hole interactions in small semiconductor crystallites: The size dependence of the lowest excited electronic state. *The Journal of Chemical Physics*, 80(9):4403–4409, 1984. doi: 10.1063/1.447218. URL http://dx.doi.org/10. 1063/1.447218.

[1138] Yosuke Kayanuma. Quantum-size effects of interacting electrons and holes in semiconductor microcrystals with spherical shape. *Phys. Rev. B*, 38:9797–9805, Nov 1988. doi: 10.1103/PhysRevB.38.9797. URL https://link.aps.org/doi/10.1103/PhysRevB.38.9797.

[1139] Aline L. Schoenhalz and Gustavo M. Dalpian. Cobalt-doped zno nanocrystals: quantum confinement and surface effects from ab initio methods. *Phys. Chem. Chem. Phys.*, 15:15863–15868, 2013. doi: 10.1039/C3CP51395E. URL http://dx.doi.org/10.1039/C3CP51395E.

[1140] Haowei Peng and Jingbo Li. Quantum confinement and electronic properties of rutile tio2 nanowires. *The Journal of Physical Chemistry C*, 112(51):20241–20245, 2008. doi: 10.1021/jp807439q. URL http://dx.doi.org/10.1021/jp807439q.

[1141] H. Lin, C.P. Huang, W. Li, C. Ni, S. Ismat Shah, and Yao-Hsuan Tseng. Size dependency of nanocrystalline tio2 on its optical property and photocatalytic reactivity exemplified by 2-chlorophenol. *Applied Catalysis B: Environmental*, 68(1):1 – 11, 2006. ISSN 0926-3373. doi: https://doi.org/10.1016/j.apcatb.2006.07.018. URL http://www.sciencedirect.com/science/article/pii/S0926337306003328.

[1142] L. E. Brus. A simple model for the ionization potential, electron affinity, and aqueous redox potentials of small semiconductor crystallites. *The Journal of Chemical Physics*, 79(11):5566–5571, 1983. doi: 10.1063/1.445676. URL http://dx.doi.org/10.1063/1.445676.

[1143] A. Fojtik, H. Weller, U. Koch, and A. Henglein. Photo-chemistry of colloidal metal sulfides 8. photo-physics of extremely small cds particles: Q-state cds and magic agglomeration numbers. *Berichte der Bunsengesellschaft für physikalische Chemie*, 88(10):969–977, 1984. ISSN 0005-9021. doi: 10.1002/bbpc.19840881010. URL http://dx.doi.org/10.1002/bbpc.19840881010.

[1144] H. Weller, H.M. Schmidt, U. Koch, A. Fojtik, S. Baral, A. Henglein, W. Kunath, K. Weiss, and E. Dieman. Photochemistry of colloidal semiconductors. onset of light absorption as a function of size of extremely small cds particles. *Chemical Physics Letters*, 124(6):557 – 560, 1986. ISSN 0009-2614. doi: https://doi.org/10.1016/0009-2614(86)85075-8. URL http://www.sciencedirect.com/science/article/pii/0009261486850758.

[1145] H.M. Schmidt and H. Weller. Quantum size effects in semiconductor crystallites: Calculation of the energy spectrum for the confined exciton. *Chemical Physics Letters*, 129(6):615 – 618, 1986. ISSN 0009-2614. doi: https://doi.org/10.1016/0009-2614(86)80410-9. URL http://www.sciencedirect.com/science/article/pii/0009261486804109.

[1146] Rustam Singh, Rudheer Bapat, Lijun Qin, Hao Feng, and Vivek Polshettiwar. Atomic layer deposited (ald) tio2 on fibrous nano-silica (kcc-1) for photocatalysis: Nanoparticle formation and size quantization effect. *ACS Catalysis*, 6(5):2770–2784, 2016. doi: 10.1021/acscatal.6b00418. URL http://dx.doi.org/10.1021/acscatal.6b00418.

[1147] Jana Drbohlavova, Marina Vorozhtsova, Radim Hrdy, Rene Kizek, Ota Salyk, and Jaromir Hubalek. Self-ordered tio2 quantum dot array prepared via anodic oxidation. *Nanoscale Research Letters*, 7(1):123, Feb 2012. ISSN 1556-276X. doi: 10.1186/1556-276X-7-123. URL https://doi.org/10.1186/1556-276X-7-123.

[1148] Carena P. Church, Elayaraja Muthuswamy, Guangmei Zhai, Susan M. Kauzlarich, and Sue A. Carter. Quantum dot ge/tio2 hetero-junction photoconductor fabrication and performance. *Applied Physics Letters*, 103(22):223506, 2013. doi: 10.1063/1.4826916. URL http://dx.doi.org/10.1063/1.4826916.

[1149] Abdul Faheem Khan, Mazhar Mehmood, Muhammad Aslam, and Syed Ismat Shah. Nanostructured multilayer tio2-ge films with quantum confinement effects for photovoltaic applications. *Journal of Colloid and Interface Science*, 343(1):271 – 280, 2010. ISSN 0021-9797. doi: https://doi.org/10.1016/j.jcis.2009.11.045. URL http://www.sciencedirect.com/science/article/pii/S0021979709014908.

[1150] Yuan Zhi-Hao, Tang Cheng-Chun, and Fan Shou-Shan. Optical absorption of sol-gel derived zno/tio2 nanocomposite films. *Chinese Physics Letters*, 18(11):1520, 2001. URL http://stacks.iop.org/0256-307X/18/i=11/a=330.

[1151] A.N. Banerjee, C.K. Ghosh, and K.K. Chattopadhyay. Effect of excess oxygen on the electrical properties of transparent p-type conducting cualo2+x thin films. *Solar Energy Materials and Solar Cells*, 89(1):75 – 83, 2005. ISSN 0927-0248. doi: https://doi.org/10.1016/j.solmat.2005.01.003. URL http://www.sciencedirect.com/science/article/pii/S0927024805000164.

[1152] J. Bisquert, G. Garcia-Belmonte, and F. Fabregat-Santiago. Modelling the electric potential distribution in the dark in nanoporous semiconductor electrodes. *J. Solid State Electrochem.*, 3:337, 1999.

[1153] Gordon M. Turner, Matthew C. Beard, and Charles A. Schmuttenmaer. Carrier localization and cooling in dye-sensitized nanocrys-talline titanium dioxide. *The Journal of Physical Chemistry B*, 106(45):11716–11719, 2002. doi: 10.1021/jp025844e. URL http://dx.doi.org/10.1021/jp025844e.

[1154] Raffaella Calarco, Michel Marso, Thomas Richter, Ali I. Aykanat, Ralph Meijers, André v.d. Hart, Toma Stoica, and Hans Lüth. Size-dependent photoconductivity in nide-grown gan-nanowires. *Nano Letters*, 5(5):981–984, 2005. doi: 10.1021/nl0500306. URL http://dx.doi.org/10.1021/nl0500306. PMID: 15884906.

[1155] Cheng-Ying Chen, Jose Ramon Duran Retamal, I-Wen Wu, Der-Hsien Lien, Ming-Wei Chen, Yong Ding, Yu-Lun Chueh, Chih-I Wu, and Jr-Hau He. Probing surface band bending of surface-engineered metal oxide nanowires. *ACS Nano*, 6(11):9366–9372, 2012. doi: 10.1021/nn205097e. URL http://dx.doi.org/10.1021/nn205097e. PMID: 23092152.

[1156] Federico M. Pesci, Gongming Wang, David R. Klug, Yat Li, and Alexander J. Cowan. Efficient suppression of electron-hole recombi-nation in oxygen-deficient hydrogen-treated tio2 nanowires for photoelectrochemical water splitting. *The Journal of Physical Chemistry C*, 117(48):25837–25844, 2013. doi: 10.1021/jp4099914. URL http://dx.doi.org/10.1021/jp4099914. PMID: 24376902.

[1157] Kai Zhu, Nikos Kopidakis, Nathan R. Neale, Jao van de Lagemaat, and Arthur J. Frank. Influence of surface area on charge transport and recombination in dye-sensitized tio2 solar cells. *The Journal of Physical Chemistry B*, 110(50):25174–25180, 2006. doi: 10.1021/jp065284+. URL http://dx.doi.org/10.1021/jp065284+. PMID: 17165961.

[1158] S. Krischok, J. Günster, D.W. Goodman, O. Höfft, and V. Kempter. Mies and ups(hei) studies on reduced tio2(110). *Surface and Interface Analysis*, 37(1):77–82, 2005. ISSN 1096-9918. doi: 10.1002/sia.2013. URL http://dx.doi.org/10.1002/sia.2013.

[1159] N. Shibata, A. Goto, S.-Y. Choi, T. Mizoguchi, S. D. Findlay, T. Yamamoto, and Y. Ikuhara. Direct imaging of reconstructed atoms on tio2(110) surfaces. *Science*, 322(5901):570–573, 2008. ISSN 0036-8075. doi: 10.1126/science.1165044. URL http://science.sciencemag.org/content/322/5901/570.

[1160] Zhen Zhang, Junseok Lee, John T. Yates, Ralf Bechstein, Estephania Lira, Jonas Hansen, Stefan Wendt, and Flemming Besenbacher. Unraveling the diffusion of bulk ti interstitials in rutile tio2(110) by monitoring their reaction with o adatoms. *The Journal of Physical Chemistry C*, 114(7):3059–3062, 2010. doi: 10.1021/jp910358w. URL http://dx.doi.org/10.1021/jp910358w.

[1161] K. T. Park, M. H. Pan, V. Meunier, and E. W. Plummer. Surface reconstructions of tio2(110) driven by suboxides. *Phys. Rev. Lett.*, 96:226105, Jun 2006. doi: 10.1103/PhysRevLett.96.226105. URL https://link.aps.org/doi/10.1103/PhysRevLett.96.226105.

[1162] Albert Goossens, Barbara van der Zanden, and Joop Schoonman. Single-electron migration in nanostructured tio$_2$. *Chemical Physics Letters*, 331(1):1 – 6, 2000. ISSN 0009-2614. doi: https://doi.org/10.1016/S0009-2614(00)01159-3. URL http://www.sciencedirect.com/science/article/pii/S0009261400011593.

[1163] George Eisenberg. Colorimetric determination of hydrogen peroxide. *Industrial & Engineering Chemistry Analytical Edition*, 15(5): 327–328, 1943.

[1164] Julio Villanueva-Cab, Song-Rim Jang, Adam F. Halverson, Kai Zhu, and Arthur J. Frank. Trap-free transport in ordered and disordered tio$_2$ nanostructures. *Nano Letters*, 14(5):2305–2309, 2014. doi: 10.1021/nl4046087. URL http://dx.doi.org/10.1021/nl4046087. PMID: 24758307.

[1165] Jeong Min Baik, Myung Hwa Kim, Christopher Larson, Xihong Chen, Shujing Guo, Alec M. Wodtke, and Martin Moskovits. High-yield tio$_2$ nanowire synthesis and single nanowire field-effect transistor fabrication. *Applied Physics Letters*, 92(24):242111, 2008. doi: 10.1063/1.2949086. URL http://dx.doi.org/10.1063/1.2949086.

[1166] Ramazan Asmatulu, Annamalai Karthikeyan, David C. Bell, Shriram Ramanathan, and Michael J. Aziz. Synthesis and variable temperature electrical conductivity studies of highly ordered tio$_2$(110) nanotubes. *Journal of Materials Science*, 44(17):4613–4616, Sep 2009. ISSN 1573-4803. doi: 10.1007/s10853-009-3703-5. URL https://doi.org/10.1007/s10853-009-3703-5.

[1167] R. S. Chen, C. A. Chen, W. C. Wang, H. Y. Tsai, and Y. S. Huang. Transport properties in single-crystalline rutile tio$_2$ nanorods. *Applied Physics Letters*, 99(22):222107, 2011. doi: 10.1063/1.3665635. URL http://dx.doi.org/10.1063/1.3665635.

[1168] Naveen Kumar Elumalai, Rajan Jose, Panikar Sathyaseelan Archana, Vijila Chellappan, and Seeram Ramakrishna. Charge transport through electrospun sno$_2$ nanoflowers and nanofibers: Role of surface trap density on electron transport dynamics. *The Journal of Physical Chemistry C*, 116(42):22112–22120, 2012. doi: 10.1021/jp304876j. URL http://dx.doi.org/10.1021/jp304876j.

[1169] Michael Grätzel. Solar energy conversion by dye-sensitized photovoltaic cells. *Inorganic Chemistry*, 44(20):6841–6851, 2005. doi: 10.1021/ic0508371. URL http://dx.doi.org/10.1021/ic0508371. PMID: 16180840.

[1170] S. Schmitt-Rink, D. A. B. Miller, and D. S. Chemla. Theory of the linear and nonlinear optical properties of semiconductor microcrystallites. *Phys. Rev. B*, 35:8113–8125, May 1987. doi: 10.1103/PhysRevB.35.8113. URL https://link.aps.org/doi/10.1103/PhysRevB.35.8113.

[1171] Masami Kumagai and Toshihide Takagahara. Excitonic and nonlinear-optical properties of dielectric quantum-well structures. *Phys. Rev. B*, 40:12359–12381, Dec 1989. doi: 10.1103/PhysRevB.40.12359. URL https://link.aps.org/doi/10.1103/PhysRevB.40.12359.

[1172] L. Brus. Quantum crystallites and nonlinear optics. *Applied Physics A*, 53(6):465–474, 1991. ISSN 1432-0630. doi: 10.1007/BF00331535. URL http://dx.doi.org/10.1007/BF00331535.

[1173] Bingsuo Zou, Yan Zhang, Liangzhi Xiao, and Tiejin Li. Self-trapped state and phonon localization in tio$_2$ quantum dot with a dipole layer. *Journal of Applied Physics*, 73(9):4689–4690, 1993. doi: 10.1063/1.352766. URL http://dx.doi.org/10.1063/1.352766.

[1174] LGJ De Haart and G Blasse. The observation of exciton emission from rutile single crystals. *Journal of Solid State Chemistry*, 61(1): 135–136, 1986.

[1175] Bingsuo Zhou, Liangzhi Xiao, Tie Jin Li, Jialong Zhao, Zhuyou Lai, and Shiwei Gu. Absorption redshift in tio$_2$ ultrafine particles with surfacial dipole layer. *Applied Physics Letters*, 59(15):1826–1828, 1991. doi: 10.1063/1.106211. URL http://dx.doi.org/10.1063/1.106211.

[1176] W. Schottky. Halbleitertheorie der sperrschicht. *Naturwissenschaften*, 26(52):843–843, 1938. ISSN 1432-1904. doi: 10.1007/BF01774216. URL http://dx.doi.org/10.1007/BF01774216.

[1177] W. Schottky. Vereinfachte und erweiterte theorie der randschicht-gleichrichter. *Zeitschrift für Physik*, 118(9):539–592, Feb 1942. ISSN 0044-3328. doi: 10.1007/BF01329843. URL https://doi.org/10.1007/BF01329843.

[1178] Roland Dietmüller. *Hybrid organic-inorganic heterojunctions for photovoltaic applications*. PhD thesis, Universität München, 2012.

[1179] I. n. Mora-Seró, F. Fabregat-Santiago, B. Denier, J. Bisquert, R. n. Tena-Zaera, J. Elias, and C. Lévy-Clément. Determination of carrier density of zno nanowires by electrochemical techniques. *Appl. Phys. Lett.*, 89:203117, 2006.

[1180] J. Tornow and K. Schwarzburg. Transient electrical response of dye-sensitized zno nanorod solar cells. *J. Phys. Chem. C*, 111:8692, 2007.

[1181] D.K. Schroder. *Semiconductor Material and Device Characterization*. Wiley - IEEE. Wiley, 2006. ISBN 9780471749080. URL https://books.google.de/books?id=0X2cHKJWCKgC.

[1182] J. Chan, N. Y. Martinez, J. J. D. Fitzgerald, A. V. Walker, R. A. Chapman, D. Riley, A. Jain, C. L. Hinkle, and E. M. Vogel. Extraction of correct schottky barrier height of sulfur implanted nisi/n-si junctions: Junction doping rather than barrier height lowering. *Applied Physics Letters*, 99(1):012114, 2011. doi: 10.1063/1.3609874. URL http://dx.doi.org/10.1063/1.3609874.

[1183] Hans Albrecht Bethe. *Theory of the boundary layer of crystal rectifiers*. Radiation Laboratory, Massachusetts Institute of Technology, 1942.

[1184] Heinz K Henisch. *Rectifying semi-conductor contacts*. Clarendon Press, 1957.

[1185] S. M. Sze, C. R. Crowell, and D. Kahng. Photoelectric determination of the image force dielectric constant for hot electrons in schottky barriers. *Journal of Applied Physics*, 35(8):2534–2536, 1964. doi: 10.1063/1.1702894. URL http://dx.doi.org/10.1063/1.1702894.

[1186] Fu-Chien Chiu and Chih-Ming Lai. Optical and electrical characterizations of cerium oxide thin films. *Journal of Physics D: Applied Physics*, 43(7):075104, 2010. URL http://stacks.iop.org/0022-3727/43/i=7/a=075104.

[1187] JYM Lee, FC Chiu, and PC Juan. The application of high-dielectric-constant and ferroelectric thin films in integrated circuit technology. *Handbook of Nanoceramics and Their Based Nanodevices*, 4, 2009.

[1188] A. M. Cowley and S. M. Sze. Surface states and barrier height of metal-semiconductor systems. *Journal of Applied Physics*, 36(10): 3212–3220, 1965. doi: 10.1063/1.1702952. URL http://dx.doi.org/10.1063/1.1702952.

[1189] D. K. Schroder and D. L. Meier. Solar cell contact resistance: A review. *IEEE Transactions on Electron Devices*, 31(5):637–647, May 1984. ISSN 0018-9383. doi: 10.1109/T-ED.1984.21583.

[1190]   H. Fox, K. E. Newman, W. F. Schneider, and S. A. Corcelli. Bulk and surface properties of rutile tio$_2$ from self-consistent-charge density functional tight binding. *Journal of Chemical Theory and Computation*, 6(2):499–507, 2010. doi: 10.1021/ct900665a. URL http://dx.doi.org/10.1021/ct900665a. PMID: 26617305.

[1191]   S.M. Sze and K.K. Ng. *Physics of Semiconductor Devices*. Wiley, 2006. ISBN 9780470068304. URL https://books.google.de/books?id=o4unkmHBHb8C.

[1192]   Volker Heine. Theory of surface states. *Phys. Rev.*, 138:A1689–A1696, Jun 1965. doi: 10.1103/PhysRev.138.A1689. URL http://link.aps.org/doi/10.1103/PhysRev.138.A1689.

[1193]   J. Tersoff. Schottky barriers and semiconductor band structures. *Phys. Rev. B*, 32:6968–6971, Nov 1985. doi: 10.1103/PhysRevB.32.6968. URL http://link.aps.org/doi/10.1103/PhysRevB.32.6968.

[1194]   R Schlaf, H Murata, and Z.H Kafafi. Work function measurements on indium tin oxide films. *Journal of Electron Spectroscopy and Related Phenomena*, 120(1):149 – 154, 2001. ISSN 0368-2048. doi: https://doi.org/10.1016/S0368-2048(01)00310-3. URL http://www.sciencedirect.com/science/article/pii/S0368204801003103.

[1195]   S. Gutmann, M. A. Wolak, M. Conrad, M. M. Beerbom, and R. Schlaf. Effect of ultraviolet and x-ray radiation on the work function of tio$_2$ surfaces. *Journal of Applied Physics*, 107(10):103705, 2010. doi: 10.1063/1.3410677. URL http://dx.doi.org/10.1063/1.3410677.

[1196]   Lukas Schmidt-Mende and Jonas Weickert. *Organic and Hybrid Solar Cells. An Introduction*. Berlin, Boston: De Gruyter, 2016. URL http://www.degruyter.com/view/product/182200.

[1197]   Gill Sang Han, Sangwook Lee, Jun Hong Noh, Hyun Suk Chung, Jong Hoon Park, Bhabani Sankar Swain, Jeong-Hyeok Im, Nam-Gyu Park, and Hyun Suk Jung. 3-d tio$_2$ nanoparticle/ito nanowire nanocomposite antenna for efficient charge collection in solid state dye-sensitized solar cells. *Nanoscale*, 6:6127–6132, 2014. doi: 10.1039/C4NR00621F. URL http://dx.doi.org/10.1039/C4NR00621F.

[1198]   Yung-Huang Chang, Chien-Min Liu, Yuan-Chieh Tseng, Chih Chen, Chia-Chuan Chen, and Hsyi-En Cheng. Direct probe of heterojunction effects upon photoconductive properties of tio$_2$ nanotubes fabricated by atomic layer deposition. *Nanotechnology*, 21(22):225602, 2010. URL http://stacks.iop.org/0957-4484/21/i=22/a=225602.

[1199]   Meng Yang, Zhang Pei-Jian, Liu Zi-Yu, Liao Zhao-Liang, Pan Xin-Yu, Liang Xue-Jin, Zhao Hong-Wu, and Chen Dong-Min. Enhanced resistance switching stability of transparent ito/tio$_2$/ito sandwiches. *Chinese Physics B*, 19(3), 2010. doi: 10.1088/1674-1056/19/3/037304.

[1200]   R. Könenkamp and R. Henninger. Recombination in nanophase tio$_2$ films. *Applied Physics A*, 58(1):87–90, Jan 1994. ISSN 1432-0630. doi: 10.1007/BF00331523. URL https://doi.org/10.1007/BF00331523.

[1201]   Boris Levy, Wang Liu, and Scott E. Gilbert. Directed photocurrents in nanostructured tio$_2$/sno$_2$ heterojunction diodes. *The Journal of Physical Chemistry B*, 101(10):1810–1816, 1997. doi: 10.1021/jp962105n. URL http://dx.doi.org/10.1021/jp962105n.

[1202]   Dewei Chu, Adnan Younis, and Sean Li. Direct growth of tio$_2$ nanotubes on transparent substrates and their resistive switching characteristics. *Journal of Physics D: Applied Physics*, 45(35):355306, 2012. URL http://stacks.iop.org/0022-3727/45/i=35/a=355306.

[1203]   Ming Xiao, Kevin P. Musselman, Walter W. Duley, and Y. Norman Zhou. Reliable and low-power multilevel resistive switching in tio$_2$ nanorod arrays structured with a tio$_2$ seed layer. *ACS Applied Materials & Interfaces*, 9(5):4808–4817, 2017. doi: 10.1021/acsami.6b14206. URL http://dx.doi.org/10.1021/acsami.6b14206. PMID: 28098978.

[1204]   N. Szydlo and R. Poirier. I-v and c-v characteristics of au/tio$_2$ schottky diodes. *Journal of Applied Physics*, 51(6):3310–3312, 1980. doi: 10.1063/1.328037. URL http://dx.doi.org/10.1063/1.328037.

[1205]   Y. L. Zhao, W. M. Lv, Z. Q. Liu, S. W. Zeng, M. Motapothula, S. Dhar, Ariando, Q. Wang, and T. Venkatesan. Variable range hopping in tio$_2$ insulating layers for oxide electronic devices. *AIP Advances*, 2(1):012129, 2012. doi: 10.1063/1.3682346. URL http://dx.doi.org/10.1063/1.3682346.

[1206]   Doo Seok Jeong, Herbert Schroeder, Uwe Breuer, and Rainer Waser. Characteristic electroforming behavior in pt/tio$_2$/pt resistive switching cells depending on atmosphere. *Journal of Applied Physics*, 104(12):123716, 2008. doi: 10.1063/1.3043879. URL http://dx.doi.org/10.1063/1.3043879.

[1207]   Min Hwan Lee and Cheol Seong Hwang. Resistive switching memory: observations with scanning probe microscopy. *Nanoscale*, 3:490–502, 2011. doi: 10.1039/C0NR00580K. URL http://dx.doi.org/10.1039/C0NR00580K.

[1208]   Hisashi Shima, Fumiyoshi Takano, Hidenobu Muramatsu, Hiro Akinaga, Isao H. Inoue, and Hidenori Takagi. Control of resistance switching voltages in rectifying pt/tiox/pt trilayer. *Applied Physics Letters*, 92(4):043510, 2008. doi: 10.1063/1.2838350. URL http://dx.doi.org/10.1063/1.2838350.

[1209]   Yuanmin Du, Amit Kumar, Hui Pan, Kaiyang Zeng, Shijie Wang, Ping Yang, and Andrew Thye Shen Wee. The resistive switching in tio$_2$ films studied by conductive atomic force microscopy and kelvin probe force microscopy. *AIP Advances*, 3(8):082107, 2013. doi: http://dx.doi.org/10.1063/1.4818119. URL http://scitation.aip.org/content/aip/journal/adva/3/8/10.1063/1.4818119.

[1210]   Jung Ho Yoon, Kyung Min Kim, Min Hwan Lee, Seong Keun Kim, Gun Hwan Kim, Seul Ji Song, Jun Yeong Seok, and Cheol Seong Hwang. Role of ru nano-dots embedded in tio$_2$ thin films for improving the resistive switching behavior. *Applied Physics Letters*, 97(23):232904, 2010. doi: 10.1063/1.3525801. URL http://dx.doi.org/10.1063/1.3525801.

[1211]   J. Joshua Yang, Matthew D. Pickett, Xuema Li, Douglas A. A. Ohlberg, Duncan R. Stewart, and R. Stanley Williams. Memristive switching mechanism for metal/oxide/metal nanodevices. *Nature Nanotechnology*, 3:429–433, 2008. doi: 10.1038/nnano.2008.160.

[1212]   Ni Zhong, Hisashi Shima, and Hiro Akinaga. Transient current study on pt/tio$_{2-x}$/pt capacitor. *Japanese Journal of Applied Physics*, 49(4S):04DJ15, 2010. URL http://stacks.iop.org/1347-4065/49/i=4S/a=04DJ15.

[1213]   W.B. Luo, P. Zhang, Y. Shuai, X.Q. Pan, Q.Q. Wu, C. Yang, and W.L. Zhang. Forming free resistive switching in au/tio$_2$/pt stack structure. *Thin Solid Films*, 617(Part B):63 – 66, 2016. ISSN 0040-6090. doi: https://doi.org/10.1016/j.tsf.2016.01.019. URL http://www.sciencedirect.com/science/article/pii/S0040609016000328. E-MRS Spring 2015 Multifunctional binary and complex oxide films and nanostructures for nanoelectronics and energy applications – II Symposia.

[1214]   Th. Dittrich, V. Zinchuk, V. Skryshevskyy, I. Urban, and O. Hilt. Electrical transport in passivated pt tio$_2$ ti schottky diodes. *Journal of Applied Physics*, 98(10):104501, 2005. doi: http://dx.doi.org/10.1063/1.2135890. URL http://scitation.aip.org/content/aip/journal/jap/98/10/10.1063/1.2135890.

[1215] Jiun-Jia Huang, Chih-Wei Kuo, Wei-Chen Chang, and Tuo-Hung Hou. Transition of stable rectification to resistive-switching in ti/tio$_2$/pt oxide diode. *Applied Physics Letters*, 96(26):262901, 2010. doi: 10.1063/1.3457866. URL http://dx.doi.org/10.1063/1.3457866.

[1216] C. Nauenheim, C. Kuegeler, A. Ruediger, and R. Waser. Investigation of the electroforming process in resistively switching tio$_2$ nanocrosspoint junctions. *Applied Physics Letters*, 96(12):122902, 2010. doi: http://dx.doi.org/10.1063/1.3367752. URL http://scitation.aip.org/content/aip/journal/apl/96/12/10.1063/1.3367752.

[1217] Jiun-Jia Huang, Chih-Wei Kuo, Wei-Chen Chang, and Tuo-Hung Hou. Transition of stable rectification to resistive-switching in ti/tio$_2$/pt oxide diode. *Applied Physics Letters*, 96(26):262901, 2010. doi: 10.1063/1.3457866. URL http://scitation.aip.org/content/aip/journal/apl/96/26/10.1063/1.3457866.

[1218] Yahya Alivov, Vivek Singh, Yuchen Ding, Logan Jerome Cerkovnik, and Prashant Nagpal. Doping of wide-bandgap titanium-dioxide nanotubes: optical, electronic and magnetic properties. *Nanoscale*, 6:10839–10849, 2014. doi: 10.1039/C4NR02417F. URL http://dx.doi.org/10.1039/C4NR02417F.

[1219] Byoung Hun Lee, Yongjoo Jeon, Keith Zawadzki, Wen-Jie Qi, and Jack Lee. Effects of interfacial layer growth on the electrical characteristics of thin titanium oxide films on silicon. *Applied Physics Letters*, 74(21):3143–3145, 1999. doi: http://dx.doi.org/10.1063/1.124089. URL http://scitation.aip.org/content/aip/journal/apl/74/21/10.1063/1.124089.

[1220] A. Barman, C. P. Saini, P. Sarkar, B. Satpati, S. R. Bhattacharyya, D. Kabiraj, D. Kanjilal, S. Dhar, and A. Kanjilal. Self-organized titanium oxide nano-channels for resistive memory application. *Journal of Applied Physics*, 118(22):224903, 2015. doi: 10.1063/1.4936961. URL http://dx.doi.org/10.1063/1.4936961.

[1221] Jonghan Kwon, Abhishek A. Sharma, James A. Bain, Yoosuf N. Picard, and Marek Skowronski. Oxygen vacancy creation, drift, and aggregation in tio$_2$-based resistive switches at low temperature and voltage. *Advanced Functional Materials*, 25(19):2876–2883, 2015. ISSN 1616-3028. doi: 10.1002/adfm.201500444. URL http://dx.doi.org/10.1002/adfm.201500444.

[1222] Ming Xiao, Kevin P. Musselman, Walter W. Duley, and Norman Y. Zhou. Resistive switching memory of tio$_2$ nanowire networks grown on ti foil by a single hydrothermal method. *Nano-Micro Letters*, 9(2):15, Nov 2016. ISSN 2150-5551. doi: 10.1007/s40820-016-0116-2. URL https://doi.org/10.1007/s40820-016-0116-2.

[1223] V. Mikhelashvili and G. Eisenstein. Effects of annealing conditions on optical and electrical characteristics of titanium dioxide films deposited by electron beam evaporation. *Journal of Applied Physics*, 89(6):3256–3269, 2001. doi: 10.1063/1.1349860. URL http://scitation.aip.org/content/aip/journal/jap/89/6/10.1063/1.1349860.

[1224] Rifat Capan and Asim K. Ray. Dielectric measurements on sol–gel derived titania films. *Journal of Electronic Materials*, 46(11):6646–6652, Nov 2017. ISSN 1543-186X. doi: 10.1007/s11664-017-5670-y. URL https://doi.org/10.1007/s11664-017-5670-y.

[1225] N. Rausch and E.P. Burte. Thin high-dielectric tio$_2$ films prepared by low pressure mocvd. *Microelectronic Engineering*, 19(1):725 – 728, 1992. ISSN 0167-9317. doi: http://dx.doi.org/10.1016/0167-9317(92)90531-U. URL http://www.sciencedirect.com/science/article/pii/016793179290531U.

[1226] Kanchana Vydianathan, Guillermo Nuesca, Gregory Peterson, Eric T. Eisenbraun, Alain E. Kaloyeros, John J. Sullivan, and Bin Han. Metalorganic chemical vapor deposition of titanium oxide for microelectronics applications. *Journal of Materials Research*, 16(6):1838–1849, 2001. doi: 10.1557/JMR.2001.0253.

[1227] Lit Ho Chong, Kanad Mallik, C H de Groot, and Reinhard Kersting. The structural and electrical properties of thermally grown tio$_2$ thin films. *Journal of Physics: Condensed Matter*, 18(2), 2006. doi: 10.1088/0953-8984/18/2/020.

[1228] Wenli Yang, Joseph Marino, Alexander Monson, and Colin A Wolden. An investigation of annealing on the dielectric performance of tio$_2$ thin films. *Semiconductor Science and Technology*, 21(12):1573, 2006. URL http://stacks.iop.org/0268-1242/21/i=12/a=012.

[1229] S Dueñas, H Castán, H García, E San Andrés, M Toledano-Luque, I Mártil, G González-Díaz, K Kukli, T Uustare, and J Aarik. A comparative study of the electrical properties of tio$_2$ films grown by high-pressure reactive sputtering and atomic layer deposition. *Semiconductor Science and Technology*, 20(10):1044, 2005. URL http://stacks.iop.org/0268-1242/20/i=10/a=011.

[1230] Aniruddha Mondal, Jay Chandra Dhar, P. Chinnamuthu, Naorem Khelchand Singh, Kalyan Kumar Chattopadhyay, Sanat Kumar Das, Santosh Ch Das, and Anirban Bhattacharyya. Electrical properties of vertically oriented tio$_2$ nanowire arrays synthesized by glancing angle deposition technique. *Electronic Materials Letters*, 9(2):213–217, 2013. ISSN 2093-6788. doi: 10.1007/s13391-012-2136-5. URL http://dx.doi.org/10.1007/s13391-012-2136-5.

[1231] Sunil Kumar, Kamal Sindhu, Ajay Shankar, and Nawal Kishore. A study on laser induced damage threshold of sio$_2$ and tio$_2$ thin films. *Journal of Integrated Science and Technology*, 5(1):5–8, 2016. ISSN 2321 – 4635. URL http://pubs.iscience.in/journal/index.php/jist/article/view/441.

[1232] M. Houssa, A. Stesmans, M. Naili, and M. M. Heyns. Charge trapping in very thin high-permittivity gate dielectric layers. *Applied Physics Letters*, 77(9):1381–1383, 2000. doi: 10.1063/1.1290138. URL http://scitation.aip.org/content/aip/journal/apl/77/9/10.1063/1.1290138.

[1233] Kumar Shubham and RU Khan. Electrical characterization of tio$_2$ insulator based pd/tio$_2$/si mis structure deposited by sol-gel process. *Journal of Nano-and Electronic Physics*, 5(1):1021–1, 2013.

[1234] Yun Jeong Hwang, Akram Boukai, and Peidong Yang. High density n-si/n-tio$_2$ core/shell nanowire arrays with enhanced photoactivity. *Lawrence Berkeley National Laboratory*, July 2009. URL http://escholarship.org/uc/item/8jp9f8zb.

[1235] Hyeon-Seag Kim, D. C. Gilmer, S. A. Campbell, and D. L. Polla. Leakage current and electrical breakdown in metal-organic chemical vapor deposition of tio$_2$ dielectrics on silicon substrates. *Applied Physics Letters*, 69(25):3860–3862, 1996. doi: 10.1063/1.117129. URL http://dx.doi.org/10.1063/1.117129.

[1236] Lori E. Greene, Matt Law, Benjamin D. Yuhas, and Peidong Yang. Zno-tio$_2$ core-shell nanorod/p3ht solar cells. *J. Phys. Chem. C*, 111(50):18451–18456, 2007. doi: 10.1021/jp0775931.

[1237] Guanglu Shang, Jihuai Wu, Shen Tang, Lu Liu, and Xiaopei Zhang. Enhancement of photovoltaic performance of dye-sensitized solar cells by modifying tin oxide nanorods with titanium oxide layer. *The Journal of Physical Chemistry C*, 117(9):4345–4350, 2013. doi: 10.1021/jp309193n. URL http://dx.doi.org/10.1021/jp309193n.

[1238] Nobuhiro Fuke, Atsushi Fukui, Ashraful Islam, Ryoichi Komiya, Ryohsuke Yamanaka, Hiroshi Harima, and Liyuan Han. Influence of tio$_2$/electrode interface on electron transport properties in back contact dye-sensitized solar cells. *Solar Energy Materials and Solar Cells*, 93(6–7):720 – 724, 2009. ISSN 0927-0248. doi: http://dx.doi.org/10.1016/j.solmat.2008.09.037. URL http://www.sciencedirect.com/science/article/pii/S0927024808002894. 17th International Photovoltaic Science and Engineering Conference.

[1239] David R. Lide. *CRC Handbook of Chemistry and Physics 83rd edn.* Boca Raton, FL: CRC Press, 2003.

[1240] R. Könenkamp and R. Henninger. Recombination in nanophase tio$_2$ films. *Applied Physics A*, 58(1):87–90, 1994. ISSN 1432-0630. doi: 10.1007/BF00331523. URL http://dx.doi.org/10.1007/BF00331523.

[1241] R. Könenkamp, R. Henninger, and P. Hoyer. Photocarrier transport in colloidal titanium dioxide films. *The Journal of Physical Chemistry*, 97(28):7328–7330, 1993. doi: 10.1021/j100130a034. URL http://dx.doi.org/10.1021/j100130a034.

[1242] Gonghu Li and Kimberly A. Gray. The solid-solid interface: Explaining the high and unique photocatalytic reactivity of tio$_2$-based nanocomposite materials. *Chemical Physics*, 339(1-3):173 – 187, 2007. ISSN 0301-0104. doi: http://dx.doi.org/10.1016/j.chemphys. 2007.05.023. URL http://www.sciencedirect.com/science/article/pii/S0301010407001954. Doping and Functionalization of Photoactive Semiconducting Metal Oxides.

[1243] Toshiaki Ozawa, Mitsunobu Iwasaki, Hiroaki Tada, Tomoki Akita, Koji Tanaka, and Seishiro Ito. Low-temperature synthesis of anatase-brookite composite nanocrystals: the junction effect on photocatalytic activity. *Journal of Colloid and Interface Science*, 281 (2):510 – 513, 2005. ISSN 0021-9797. doi: http://dx.doi.org/10.1016/j.jcis.2004.08.137. URL http://www.sciencedirect.com/science/article/pii/S0021979704008835.

[1244] Zhaoyue Liu, Xintong Zhang, Shunsuke Nishimoto, Ming Jin, Donald A. Tryk, Taketoshi Murakami, and Akira Fujishima. Anatase tio$_2$ nanoparticles on rutile tio$_2$ nanorods: A heterogeneous nanostructure via layer-by-layer assembly. *Langmuir*, 23(22):10916–10919, 2007. doi: 10.1021/la7018023. URL http://dx.doi.org/10.1021/la7018023. PMID: 17892314.

[1245] F.A. Padovani and R. Stratton. Field and thermionic-field emission in schottky barriers. *Solid-State Electronics*, 9(7):695 – 707, 1966. ISSN 0038-1101. doi: http://dx.doi.org/10.1016/0038-1101(66)90097-9. URL http://www.sciencedirect.com/science/article/pii/ 0038110166900979.

[1246] C.Y. Chang, Y.K. Fang, and S.M. Sze. Specific contact resistance of metal-semiconductor barriers. *Solid-State Electronics*, 14(7):541 – 550, 1971. ISSN 0038-1101. doi: http://dx.doi.org/10.1016/0038-1101(71)90129-8. URL http://www.sciencedirect.com/science/article/ pii/0038110171901298.

[1247] Tuncay Tunç, Şemsettin Altindal, İbrahim Uslu, İlbilge Dökme, and Habibe Uslu. Temperature dependent current-voltage (i-v) characteristics of au/n-si (111) schottky barrier diodes with pva(ni,zn-doped) interfacial layer. *Materials Science in Semiconductor Processing*, 14(2):139 – 145, 2011. ISSN 1369-8001. doi: https://doi.org/10.1016/j.mssp.2011.01.018. URL http://www.sciencedirect. com/science/article/pii/S1369800111000412.

[1248] Fu-Chien Chiu. A review on conduction mechanisms in dielectric films. *Advances in Materials Science and Engineering*, 2014(578168): 18, 2014. doi: 10.1155/2014/578168.

[1249] O. Pakma, N. Serin, T. Serin, and Ş. Altindal. The double gaussian distribution of barrier heights in al/tio$_2$/p-si (metal-insulator-semiconductor) structures at low temperatures. *Journal of Applied Physics*, 104(1):014501, 2008. doi: 10.1063/1.2952028. URL http://dx.doi.org/10.1063/1.2952028.

[1250] Z. Y. Zhang, C. H. Jin, X. L. Liang, Q. Chen, and L.-M. Peng. Current-voltage characteristics and parameter retrieval of semiconducting nanowires. *Applied Physics Letters*, 88(7):073102, 2006. doi: 10.1063/1.2177362. URL http://dx.doi.org/10.1063/1.2177362.

[1251] Xuedong Bai, Dmitri Golberg, Yoshio Bando, Chunyi Zhi, Chengchun Tang, Masanori Mitome, and Keiji Kurashima. Deformation-driven electrical transport of individual boron nitride nanotubes. *Nano Letters*, 7(3):632–637, 2007. doi: 10.1021/nl0625401. URL http://dx.doi.org/10.1021/nl0625401. PMID: 17288485.

[1252] Fu-Chien Chiu, Zhi-Hong Lin, Che-Wei Chang, Chen-Chih Wang, Kun-Fu Chuang, Chih-Yao Huang, Joseph Ya min Lee, and Huey-Liang Hwang. Electron conduction mechanism and band diagram of sputter-deposited al/zro$_2$/si structure. *Journal of Applied Physics*, 97(3):034506, 2005. doi: 10.1063/1.1846131. URL http://dx.doi.org/10.1063/1.1846131.

[1253] J. G. Simmons. Richardson-schottky effect in solids. *Phys. Rev. Lett.*, 15:967–968, Dec 1965. doi: 10.1103/PhysRevLett.15.967. URL https://link.aps.org/doi/10.1103/PhysRevLett.15.967.

[1254] Sufi Zafar, Robert E. Jones, Bo Jiang, Bruce White, V. Kaushik, and S. Gillespie. The electronic conduction mechanism in barium strontium titanate thin films. *Applied Physics Letters*, 73(24):3533–3535, 1998. doi: 10.1063/1.122827. URL http://dx.doi.org/10.1063/ 1.122827.

[1255] C.R. Crowell and S.M. Sze. Electron-optical-phonon scattering in the emitter and collector barriers of semiconductor-metal-semiconductor structures. *Solid-State Electronics*, 8(12):979 – 990, 1965. ISSN 0038-1101. doi: http://dx.doi.org/10.1016/0038-1101(65) 90164-4. URL http://www.sciencedirect.com/science/article/pii/0038110165901644.

[1256] Chung-Whei Kao, C.Lawrence Anderson, and C.R. Crowell. Photoelectron injection at metal-semiconductor interfaces. *Surface Science*, 95(1):321 – 339, 1980. ISSN 0039-6028. doi: http://dx.doi.org/10.1016/0039-6028(80)90145-4. URL http://www.sciencedirect.com/ science/article/pii/0039602880901454.

[1257] C. R. Crowell and S. M. Sze. Quantum-mechanical reflection of electrons at metal-semiconductor barriers: Electron transport in semiconductor-metal-semiconductor structures. *Journal of Applied Physics*, 37(7):2683–2689, 1966. doi: 10.1063/1.1782103. URL http://dx.doi.org/10.1063/1.1782103.

[1258] C.Y. Chang and S.M. Sze. Carrier transport across metal-semiconductor barriers. *Solid-State Electronics*, 13(6):727 – 740, 1970. ISSN 0038-1101. doi: http://dx.doi.org/10.1016/0038-1101(70)90060-2. URL http://www.sciencedirect.com/science/article/pii/ 0038110170900602.

[1259] Fu-Chien Chiu. Interface characterization and carrier transportation in metal/hfo$_2$/silicon structure. *Journal of Applied Physics*, 100 (11):114102, 2006. doi: 10.1063/1.2401657. URL http://aip.scitation.org/doi/abs/10.1063/1.2401657.

[1260] Claus Hamann, Hubert Burghardt, and Thomas Frauenheim. *Electrical conduction mechanisms in solids.* VEB Deutscher Verlag der Wissenschaften, Berlin, Germany, 1988.

[1261] Andrea Ghetti. *Gate oxide reliability: Physical and computational models*, volume 72 of *0933-033X*, chapter Gate Oxide Reliability: Physical and Computational Models, pages 201–258. Springer Berlin Heidelberg, 2004. doi: 10.1007/978-3-662-09432-7_6.

[1262] Shimeng Yu, Ximeng Guan, and H.-S. Philip Wong. Conduction mechanism of tin/hfox/pt resistive switching memory: A trap-assisted-tunneling model. *Applied Physics Letters*, 99(6):063507, 2011. doi: 10.1063/1.3624472. URL http://dx.doi.org/10.1063/1.3624472.

[1263] J. Y. M. Lee, F. C. Chiu, and P. C. Juan. *The application of high-dielectric-constant and ferroelectric thin films in integrated circuit technology*, volume 4. American Scientific Publishers, Stevenson Ranch, Calif, USA, 2009.

References

[1264] M. Nakano, A. Tsukazaki, R. Y. Gunji, K. Ueno, A. Ohtomo, T. Fukumura, and M. Kawasaki. Schottky contact on a zno (0001) single crystal with conducting polymer. *Applied Physics Letters*, 91(14):142113, 2007. doi: 10.1063/1.2789697. URL http://dx.doi.org/10.1063/1.2789697.

[1265] Fu-Chien Chiu, Hong-Wen Chou, and Joseph Ya min Lee. Electrical conduction mechanisms of metal/la$_2$o$_3$/si structure. *Journal of Applied Physics*, 97(10):103503, 2005. doi: 10.1063/1.1896435. URL http://dx.doi.org/10.1063/1.1896435.

[1266] S. Denda and M-A. Nicolet. Pure space-charge-limited electron current in silicon. *Journal of Applied Physics*, 37(6):2412–2424, 1966. doi: 10.1063/1.1708829. URL http://dx.doi.org/10.1063/1.1708829.

[1267] J.L. Chu, G. Persky, and S.M. Sze. Thermionic injection and space-charge-limited current in reach-through p+np+ structures. *Journal of Applied Physics*, 43(8):3510–3515, 1972. doi: 10.1063/1.1661745. URL http://dx.doi.org/10.1063/1.1661745.

[1268] J. Bisquert and V. S. Vikhrenko. Interpretation of the time constants measured by kinetic techniques in nanostructured semiconductor electrodes and dye-sensitized solar cells. *J. Phys. Chem. B*, 108:2313, 2004.

[1269] H. Wittmer, St. Holten, H. Kliem, and H.D. Breuer. Detection of space charge limited currents in nanoscaled titania. *physica status solidi (a)*, 181(2):461–469, 2000. ISSN 1521-396X. doi: 10.1002/1521-396X(200010)181:2<461::AID-PSSA461>3.0.CO;2-X. URL http://dx.doi.org/10.1002/1521-396X(200010)181:2<461::AID-PSSA461>3.0.CO;2-X.

[1270] D. Vanmaekelbergh and P. E. de Jongh. Electron transport in disordered semiconductors studied by a small harmonic modulation of the steady state. *Phys. Rev. B*, 61:4699, 2000.

[1271] K. P. Wang and H. Teng. Zinc-doping in tio$_2$ films to enhance electron transport in dye-sensitized solar cells under low-intensity illumination. *Phys. Chem. Chem. Phys.*, 11:9489, 2009.

[1272] P.-T. Hsiao and H. Teng. Coordination of tio^{4+} sites in nanocrystalline tio$_2$ films used for photoinduced electron conduction: Influence of nanoparticle synthesis and thermal necking. *J. Am. Ceram. Soc.*, 92:888, 2009.

[1273] L. Peter and K. Wijayantha. Electron transport and back reaction in dye sensitised nanocrystalline photovoltaic cells. *Electrochim. Acta*, 45:4543, 2000.

[1274] N. Kopidakis, K. D. Benkstein, J. van de Lagemaat, and A. J. Frank. Transport-limited recombination of photocarriers in dye-sensitized nanocrystalline tio$_2$ solar cells. *J. Phys. Chem. B*, 107:11307, 2003.

[1275] Kwan C Kao and Wei Hwang. Electrical transport in solids, international series in the science of the solid state. *Electrical Transport in solids, International series in the science of the solid state*, 14, 1981.

[1276] Jidong Jin. *Metal-oxide-based electronic devices*. PhD thesis, The University of Manchester, 2013.

[1277] N.F. Mott and E.A. Davis. *Electronic Processes in Non-Crystalline Materials*. Oxford Classic Texts in the Physical Sciences. OUP Oxford, 1979. ISBN 9780191023286. URL https://books.google.de/books?id=YCpoAgAAQBAJ.

[1278] P. Hesto. *The nature of electronic conduction in thin insulating films*, volume 1. Elsevier Science, North Holland, Amsterdam, The Netherlands, 1986.

[1279] Murray A Lampert and Peter Mark. *Current injection in solids: The regional approximation method*. Electrical science series. Academic Press, New York, NY, 1970. URL http://cds.cern.ch/record/112986.

[1280] A. Rose. Space-charge-limited currents in solids. *Phys. Rev.*, 97:1538–1544, Mar 1955. doi: 10.1103/PhysRev.97.1538. URL https://link.aps.org/doi/10.1103/PhysRev.97.1538.

[1281] Murray A. Lampert. Simplified theory of space-charge-limited currents in an insulator with traps. *Phys. Rev.*, 103:1648–1656, Sep 1956. doi: 10.1103/PhysRev.103.1648. URL https://link.aps.org/doi/10.1103/PhysRev.103.1648.

[1282] S. Bubel, N. Mechau, H. Hahn, and R. Schmechel. Trap states and space charge limited current in dispersion processed zinc oxide thin films. *Journal of Applied Physics*, 108(12):124502, 2010. doi: 10.1063/1.3524184. URL http://dx.doi.org/10.1063/1.3524184.

[1283] G. S. Oehrlein. Oxidation temperature dependence of the dc electrical conduction characteristics and dielectric strength of thin ta$_2$o$_5$ films on silicon. *Journal of Applied Physics*, 59(5):1587–1595, 1986. doi: 10.1063/1.336468. URL http://dx.doi.org/10.1063/1.336468.

[1284] Fu-Chien Chiu, Chun-Yen Lee, and Tung-Ming Pan. Current conduction mechanisms in pr$_2$o$_3$/oxynitride laminated gate dielectrics. *Journal of Applied Physics*, 105(7):074103, 2009. doi: 10.1063/1.3103282. URL http://dx.doi.org/10.1063/1.3103282.

[1285] R. L. Angle and H. E. Talley. Electrical and charge storage characteristics of the tantalum oxide-silicon dioxide device. *IEEE Transactions on Electron Devices*, 25(11):1277–1283, Nov 1978. ISSN 0018-9383. doi: 10.1109/T-ED.1978.19266.

[1286] R. Hill. The essential structure of constitutive laws for metal composites and polycrystals. *Journal of the Mechanics and Physics of Solids*, 15(2):79 – 95, 1967. ISSN 0022-5096. doi: https://doi.org/10.1016/0022-5096(67)90018-X. URL http://www.sciencedirect.com/science/article/pii/002250966790018X.

[1287] Fu-Chien Chiu, Wen-Chieh Shih, and Jun-Jea Feng. Conduction mechanism of resistive switching films in mgo memory devices. *Journal of Applied Physics*, 111(9):094104, 2012. doi: 10.1063/1.4712628. URL http://dx.doi.org/10.1063/1.4712628.

[1288] Yong-Jun Ma, Feng Zhou, Li Lu, and Ze Zhang. Low-temperature transport properties of individual sno$_2$ nanowires. *Solid State Communications*, 130(5):313 – 316, 2004. ISSN 0038-1098. doi: https://doi.org/10.1016/j.ssc.2004.02.013. URL http://www.sciencedirect.com/science/article/pii/S0038109804001097.

[1289] Yen-Fu Lin, Tzu-Han Chen, Chia-Hung Chang, Yu-Wei Chang, Yi-Cheng Chiu, Hsiang-Chih Hung, Ji-Jung Kai, Zhaoping Liu, Jiye Fang, and Wen-Bin Jian. Electron transport in high-resistance semiconductor nanowires through two-probe measurements. *Phys. Chem. Chem. Phys.*, 12:10928–10932, 2010. doi: 10.1039/C0CP00038H. URL http://dx.doi.org/10.1039/C0CP00038H.

[1290] T. K. Kundu and Joseph Ya-min Lee. Thickness-dependent electrical properties of pb(zr, ti)o$_3$ thin film capacitors for memory device applications. *Journal of The Electrochemical Society*, 147(1):326–329, 2000. doi: 10.1149/1.1393194. URL http://jes.ecsdl.org/content/147/1/326.abstract.

[1291] I. Shalish, L. Kronik, G. Segal, Yoram Shapira, S. Zamir, B. Meyler, and J. Salzman. Grain-boundary-controlled transport in gan layers. *Phys. Rev. B*, 61:15573–15576, Jun 2000. doi: 10.1103/PhysRevB.61.15573. URL https://link.aps.org/doi/10.1103/PhysRevB.61.15573.

[1292] M. J. Cohen, J. S. Harris, and J. R. Waldrop. *Grain Boundary Resistance Measurements in Polycrystalline GaAs*. 1979.

[1293] Harry C. Gatos and Jacek Lagowski. Surface photovoltage spectroscopy - a new approach to the study of high-gap semiconductor surfaces. *Journal of Vacuum Science and Technology*, 10(1):130–135, 1973. doi: 10.1116/1.1317922. URL http://dx.doi.org/10.1116/1.1317922.

[1294] Leeor Kronik and Yoram Shapira. Surface photovoltage phenomena: theory, experiment, and applications. *Surface Science Reports*, 37(1):1 – 206, 1999. ISSN 0167-5729. doi: https://doi.org/10.1016/S0167-5729(99)00002-3. URL http://www.sciencedirect.com/science/article/pii/S0167572999000023.

[1295] R. Tsu and L. Esaki. Tunneling in a finite superlattice. *Applied Physics Letters*, 22(11):562–564, 1973. doi: 10.1063/1.1654509. URL http://dx.doi.org/10.1063/1.1654509.

[1296] M. Houssa, M. Tuominen, M. Naili, V. Afanas'ev, A. Stesmans, S. Haukka, and M. M. Heyns. Trap-assisted tunneling in high permittivity gate dielectric stacks. *Journal of Applied Physics*, 87(12):8615–8620, 2000. doi: 10.1063/1.373587. URL http://dx.doi.org/10.1063/1.373587.

[1297] Mau Phon Houng, Yeong Her Wang, and Wai Jyh Chang. Current transport mechanism in trapped oxides: A generalized trap-assisted tunneling model. *Journal of Applied Physics*, 86(3):1488–1491, 1999. doi: 10.1063/1.370918. URL http://dx.doi.org/10.1063/1.370918.

[1298] Rohana Perera, Akihiro Ikeda, Reiji Hattori, and Yukinori Kuroki. Trap assisted leakage current conduction in thin silicon oxynitride films grown by rapid thermal oxidation combined microwave excited plasma nitridation. *Microelectronic Engineering*, 65(4):357 – 370, 2003. ISSN 0167-9317. doi: http://dx.doi.org/10.1016/S0167-9317(02)01025-0. URL http://www.sciencedirect.com/science/article/pii/S0167931702010250.

[1299] R. Moazzami and Chenming Hu. Stress-induced current in thin silicon dioxide films. In *1992 International Technical Digest on Electron Devices Meeting*, pages 139–142, Dec 1992. doi: 10.1109/IEDM.1992.307327.

[1300] B. Riccó, G. Gozzi, and M. Lanzoni. Modeling and simulation of stress-induced leakage current in ultrathin sio2 films. *IEEE Transactions on Electron Devices*, 45(7):1554–1560, Jul 1998. ISSN 0018-9383. doi: 10.1109/16.701488.

[1301] A. Gehring. *Simulation of Tunneling in Semiconductor Devices*. PhD thesis, Technische Universität Wien, 2003. URL http://www.iue.tuwien.ac.at/phd/gehring.

[1302] Robert Entner. *Modeling and simulation of negative bias temperature instability*. na, 2007.

[1303] M. Herrmann and A. Schenk. Field and high-temperature dependence of the long term charge loss in erasable programmable read only memories: Measurements and modeling. *Journal of Applied Physics*, 77(9):4522–4540, 1995. doi: 10.1063/1.359414. URL http://dx.doi.org/10.1063/1.359414.

[1304] Wai Jyh Chang, Mau Phon Houng, and Yeong Her Wang. Simulation of stress-induced leakage current in silicon dioxides: A modified trap-assisted tunneling model considering gaussian-distributed traps and electron energy loss. *Journal of Applied Physics*, 89(11):6285–6293, 2001. doi: 10.1063/1.1367399. URL http://dx.doi.org/10.1063/1.1367399.

[1305] F. Jiménez-Molinos, A. Palma, F. Gámiz, J. Banqueri, and J. A. López-Villanueva. Physical model for trap-assisted inelastic tunneling in metal-oxide-semiconductor structures. *Journal of Applied Physics*, 90(7):3396–3404, 2001. doi: 10.1063/1.1398603. URL http://dx.doi.org/10.1063/1.1398603.

[1306] F. Schuler, R. Degraeve, P. Hendrickx, and D. Wellekens. Physical description of anomalous charge loss in floating gate based nvm's and identification of its dominant parameter. In *2002 IEEE International Reliability Physics Symposium. Proceedings. 40th Annual (Cat. No.02CH37320)*, pages 26–33, 2002. doi: 10.1109/RELPHY.2002.996606.

[1307] Daniele Ielmini, Alessandro S Spinelli, Andrea L Lacaita, and Alberto Modelli. Modeling of anomalous {SILC} in flash memories based on tunneling via multiple defects. *Solid-State Electronics*, 46(11):1749 – 1756, 2002. ISSN 0038-1101. doi: https://doi.org/10.1016/S0038-1101(02)00144-2. URL http://www.sciencedirect.com/science/article/pii/S0038110102000144-2.

[1308] D. Ielmini, A. S. Spinelli, A. L. Lacaita, and M. J. van Duuren. Defect generation statistics in thin gate oxides. *IEEE Transactions on Electron Devices*, 51(8):1288–1295, Aug 2004. ISSN 0018-9383. doi: 10.1109/TED.2004.832104.

[1309] L. Larcher. Statistical simulation of leakage currents in mos and flash memory devices with a new multiphonon trap-assisted tunneling model. *IEEE Transactions on Electron Devices*, 50(5):1246–1253, May 2003. ISSN 0018-9383. doi: 10.1109/TED.2003.813236.

[1310] Juraj Racko, Juraj Pecháček, Miroslav Mikolášek, Peter Benko, Alena Grmanova, Ladislav Harmatha, and Juraj Breza. Trap-assisted tunneling in the schottky barrier. *Radioengineering*, 22(1):241, 2013.

[1311] Robert Entner, Andreas Gehring, Hans Kosina, Tibor Grasser, and Siegfried Selberherr. Impact of multi-trap assisted tunneling on gate leakage of cmos memory devices. *NSTI-Nanotech 2005*, 2005.

[1312] R. Entner, T. Grasser, S. Selberherr, A. Gehring, and H. Kosina. Modeling of tunneling currents for highly degraded cmos devices. In *2005 International Conference On Simulation of Semiconductor Processes and Devices*, pages 219–222, Sept 2005. doi: 10.1109/SISPAD.2005.201512.

[1313] Gunther Jegert, Dan Popescu, Paolo Lugli, Martin Johannes Häufel, Wenke Weinreich, and Alfred Kersch. Role of defect relaxation for trap-assisted tunneling in high-κ thin films: A first-principles kinetic monte carlo study. *Phys. Rev. B*, 85:045303, Jan 2012. doi: 10.1103/PhysRevB.85.045303. URL https://link.aps.org/doi/10.1103/PhysRevB.85.045303.

[1314] X. R. Cheng, Y. C. Cheng, and B. Y. Liu. Nitridation-enhanced conductivity behavior and current transport mechanism in thin thermally nitrided sio2. *Journal of Applied Physics*, 63(3):797–802, 1988. doi: 10.1063/1.340072. URL http://dx.doi.org/10.1063/1.340072.

[1315] M.K. Bera and C.K. Maiti. Electrical properties of sio2/tio2 high-k gate dielectric stack. *Materials Science in Semiconductor Processing*, 9(6):909 – 917, 2006. ISSN 1369-8001. doi: https://doi.org/10.1016/j.mssp.2006.10.008. URL http://www.sciencedirect.com/science/article/pii/S1369800106002253. E-MRS 2006 Spring Meeting - Symposium I: Characterization of high-k dielectric materials.

[1316] G D Pitt. The conduction band structures of gaas and inp. *Journal of Physics C: Solid State Physics*, 6(9):1586, 1973. URL http://stacks.iop.org/0022-3719/6/i=9/a=016.

[1317] A.K. Singh. *Electronic Devices and Integrated Circuits*. PHI Learning, 2011. ISBN 9788120344716. URL https://books.google.de/books?id=ZBW5PeHCITcC.

References

[1318] Arno Förster, Jürgen Stock, Simone Montanari, Mihail Ion Lepsa, and Hans Lüth. Fabrication and characterisation of gaas gunn diode chips for applications at 77 ghz in automotive industry. *Sensors (Basel)*, 6(4):350–360, 2006.

[1319] Sungho Kim, Hu Young Jeong, Sung-Yool Choi, and Yang-Kyu Choi. Comprehensive modeling of resistive switching in the al/tio$_x$/tio$_2$/al heterostructure based on space-charge-limited conduction. *Applied Physics Letters*, 2010. doi: 10.1063/1.3467461. URL http://scholarworks.unist.ac.kr/handle/201301/9002.

[1320] Kuyyadi P Biju, XinJun Liu, El Mostafa Bourim, Insung Kim, Seungjae Jung, Manzar Siddik, Joonmyoung Lee, and Hyunsang Hwang. Asymmetric bipolar resistive switching in solution-processed pt/tio$_2$/w devices. *Journal of Physics D: Applied Physics*, 43(49):495104, 2010. URL http://stacks.iop.org/0022-3727/43/i=49/a=495104.

[1321] Yuanmin Du, Hui Pan, Shijie Wang, Tom Wu, Yuan Ping Feng, Jisheng Pan, and Andrew Thye Shen Wee. Symmetrical negative differential resistance behavior of a resistive switching device. *ACS Nano*, 6(3):2517–2523, 2012. doi: 10.1021/nn204907t. URL http://dx.doi.org/10.1021/nn204907t. PMID: 22309136.

[1322] Kyung Min Kim, Seungwu Han, and Cheol Seong Hwang. Electronic bipolar resistance switching in an anti-serially connected pt/tio$_2$/pt structure for improved reliability. *Nanotechnology*, 23(3):035201, 2012. URL http://stacks.iop.org/0957-4484/23/i=3/a=035201.

[1323] Kyung Jean Yoon, Min Hwan Lee, Gun Hwan Kim, Seul Ji Song, Jun Yeong Seok, Sora Han, Jung Ho Yoon, Kyung Min Kim, and Cheol Seong Hwang. Memristive tri-stable resistive switching at ruptured conducting filaments of a pt/tio$_2$/pt cell. *Nanotechnology*, 23(18):185202, 2012. URL http://stacks.iop.org/0957-4484/23/i=18/a=185202.

[1324] Kyung Min Kim, Byung Joon Choi, Min Hwan Lee, Gun Hwan Kim, Seul Ji Song, Jun Yeong Seok, Jeong Ho Yoon, Seungwu Han, and Cheol Seong Hwang. A detailed understanding of the electronic bipolar resistance switching behavior in pt/tio$_2$/pt structure. *Nanotechnology*, 22(25), 2011. doi: 10.1088/0957-4484/22/25/254010.

[1325] Sujaya Kumar Vishwanath and Jihoon Kim. Resistive switching characteristics of all-solution-based ag/tio$_2$/mo-doped in$_2$o$_3$devices for non-volatile memory applications. *J. Mater. Chem. C*, 4:10967–10972, 2016. doi: 10.1039/C6TC03607D. URL http://dx.doi.org/10.1039/C6TC03607D.

[1326] F. Cardon. On the electrical properties of nearly stoichiometric rutile single crystals. *physica status solidi (b)*, 8:1415, 1963.

[1327] Jun-Seok Hwang, Fabrice Donatini, Julien Pernot, Robin Thierry, Pierre Ferret, and Le Si Dang. Carrier depletion and exciton diffusion in a single zno nanowire. *Nanotechnology*, 22(47):475704, 2011. URL http://stacks.iop.org/0957-4484/22/i=47/a=475704.

[1328] Oliver Hayden, Gengfeng Zheng, Prabhat Agarwal, and Charles M. Lieber. Visualization of carrier depletion in semiconducting nanowires. *Small*, 3(12):2048–2052, 2007. ISSN 1613-6829. doi: 10.1002/smll.200700600. URL http://dx.doi.org/10.1002/smll.200700600.

[1329] Cheng-Han Hsu, Qiaoming Wang, Xin Tao, and Yi Gu. Electrostatics and electrical transport in semiconductor nanowire schottky diodes. *Applied Physics Letters*, 101(18):183103, 2012. doi: 10.1063/1.4765653. URL http://dx.doi.org/10.1063/1.4765653.

[1330] J.C. Tinoco, M. Estrada, and G. Romero. Room temperature plasma oxidation mechanism to obtain ultrathin silicon oxide and titanium oxide layers. *Microelectronics Reliability*, 43(6):895 – 903, 2003. ISSN 0026-2714. doi: http://dx.doi.org/10.1016/S0026-2714(03)00098-2. URL http://www.sciencedirect.com/science/article/pii/S0026271403000982.

[1331] S. Chakraborty, M.K. Bera, S. Bhattacharya, and C.K. Maiti. Current conduction mechanism in tio$_2$ gate dielectrics. *Microelectronic Engineering*, 81(2–4):188 – 193, 2005. doi: http://dx.doi.org/10.1016/j.mee.2005.03.005. URL http://www.sciencedirect.com/science/article/pii/S0167931705000081X. The Proceedings of the 2nd International Symposium on Nano- and Giga-Challenges in MicroelectronicsThe Proceedings of the 2nd International Symposium on Nano- and Giga-Challenges in Microelectronics.

[1332] J. Yan, D. C. Gilmer, S. A. Campbell, W. L. Gladfelter, and P. G. Schmid. Structural and electrical characterization of tio$_2$ grown from titanium tetrakis-isopropoxide (ttip) and ttip/h$_2$o ambients. *Journal of Vacuum Science & Technology B: Microelectronics and Nanometer Structures Processing, Measurement, and Phenomena*, 14(3):1706–1711, 1996. doi: 10.1116/1.589214. URL http://avs.scitation.org/doi/abs/10.1116/1.589214.

[1333] Chich-Shang Chang, Tzu-Ping Liu, and Tai-Bor Wu. Effects of postannealing on the electrical properties of ta$_2$o$_5$ thin films deposited on tin/t. *Journal of Applied Physics*, 88(12):7242–7248, 2000. doi: 10.1063/1.1326464. URL http://dx.doi.org/10.1063/1.1326464.

[1334] J.S Lee, S.J Chang, J.F Chen, S.C Sun, C.H Liu, and U.H Liaw. Effects of o$_2$ thermal annealing on the properties of {CVD} ta$_2$o$_5$ thin films. *Materials Chemistry and Physics*, 77(1):242 – 247, 2003. ISSN 0254-0584. doi: https://doi.org/10.1016/S0254-0584(01)00559-4. URL http://www.sciencedirect.com/science/article/pii/S0254058401005594.

[1335] Dong Heon Lee, Yong Soo Cho, Woul In Yi, Tae Song Kim, Jeon Kook Lee, and Hyung Jin Jung. Metalorganic chemical vapor deposition of tio$_2$:n anatase thin film on si substrate. *Applied Physics Letters*, 66(7):815–816, 1995. doi: http://dx.doi.org/10.1063/1.113430. URL http://scitation.aip.org/content/aip/journal/apl/66/7/10.1063/1.113430.

[1336] S. Due nas, R. Pelaez, E. Castan, R. Pinacho, L. Quintanilla, J. Barbolla, I. Martil, and G. Gonzalez-Diaz. Experimental observation of conductance transients in al/sinx:h/si metal-insulator-semiconductor structures. *Applied Physics Letters*, 71(6):826–828, 1997. doi: 10.1063/1.119658. URL http://dx.doi.org/10.1063/1.119658.

[1337] W. S. Lau, S. J. Fonash, and J. Kanicki. Stability of electrical properties of nitrogen-rich, silicon-rich, and stoichiometric silicon nitride films. *Journal of Applied Physics*, 66(6):2765–2767, 1989. doi: 10.1063/1.344202. URL http://dx.doi.org/10.1063/1.344202.

[1338] Li He, Hideki Hasegawa, Takayuki Sawada, and Hideo Ohno. A self-consistent computer simulation of compound semiconductor metal-insulator-semiconductor c-v curves based on the disorder-induced gap-state model. *Journal of Applied Physics*, 63(6):2120–2130, 1988. doi: 10.1063/1.341067. URL http://dx.doi.org/10.1063/1.341067.

[1339] Li He, Hideki Hasegawa, Takayuki Sawada, and Hideo Ohno. A computer analysis of effects of annealing on inp insulator-semiconductor interface properties using mis c-v curves. *Japanese Journal of Applied Physics*, 27(4R):512, 1988. URL http://stacks.iop.org/1347-4065/27/i=4R/a=512.

[1340] K. L. Siefering and G. L. Griffin. Kinetics of low-pressure chemical vapor deposition of tio$_2$ from titanium tetraisopropoxide. *Journal of The Electrochemical Society*, 137(3):814–818, 1990. doi: 10.1149/1.2086561. URL http://jes.ecsdl.org/content/137/3/814.abstract.

[1341] K F Schuegraf and Chenming Hu. Reliability of thin sio$_2$. *Semiconductor Science and Technology*, 9(5):989, 1994. URL http://stacks.iop.org/0268-1242/9/i=5/a=002.

[1342] B E Weir, M A Alam, J D Bude, P J Silverman, A Ghetti, F Baumann, P Diodato, D Monroe, T Sorsch, G L Timp, Y Ma, M M Brown, A Hamad, D Hwang, and P Mason. Gate oxide reliability projection to the sub-2 nm regime. *Semiconductor Science and Technology*, 15(5):455–461, 2000. doi: 10.1088/0268-1242/15/5/304. URL http://stacks.iop.org/0268-1242/15/i=5/a=304.

[1343] A. Baikalov, Y. Q. Wang, B. Shen, B. Lorenz, S. Tsui, Y. Y. Sun, Y. Y. Xue, and C. W. Chu. Field-driven hysteretic and reversible resistive switch at theag-pr$_{0.7}$ca$_{0.3}$mno$_3$ interface. *Applied Physics Letters*, 83(5):957–959, 2003. doi: 10.1063/1.1590741. URL http://dx.doi.org/10.1063/1.1590741.

[1344] S. Tsui, A. Baikalov, J. Cmaidalka, Y. Y. Sun, Y. Q. Wang, Y. Y. Xue, C. W. Chu, L. Chen, and A. J. Jacobson. Field-induced resistive switching in metal-oxide interfaces. *Applied Physics Letters*, 85(2):317–319, 2004. doi: 10.1063/1.1768305. URL http://dx.doi.org/10.1063/1.1768305.

[1345] A. Sawa, T. Fujii, M. Kawasaki, and Y. Tokura. Hysteretic current-voltage characteristics and resistance switching at a rectifying ti/pr$_{0.7}$ca$_{0.3}$mno$_3$ interface. *Applied Physics Letters*, 85(18):4073–4075, 2004. doi: 10.1063/1.1812580. URL http://dx.doi.org/10.1063/1.1812580.

[1346] T. Fujii, M. Kawasaki, A. Sawa, H. Akoh, Y. Kawazoe, and Y. Tokura. Hysteretic current–voltage characteristics and resistance switching at an epitaxial oxide schottky junction srruo$_3$/srtio$_{0.99}$nb$_{0.01}$o$_3$. *Applied Physics Letters*, 86(1):012107, 2005. doi: 10.1063/1.1845598. URL http://dx.doi.org/10.1063/1.1845598.

[1347] A. Ghetti. *Gate Oxide Reliability: Physical and Computational Models*, pages 201–258. Springer Berlin Heidelberg, Berlin, Heidelberg, 2004. ISBN 978-3-662-09432-7. doi: 10.1007/978-3-662-09432-7_6. URL http://dx.doi.org/10.1007/978-3-662-09432-7_6.

[1348] Navid Azizi and Peter Yiannacouras. Gate oxide breakdown. *ECE1768–Reliability of Integrated Circuits*, 2003.

[1349] Andrew K Jonscher. Dielectric relaxation in solids. *Journal of Physics D: Applied Physics*, 32(14):R57, 1999. URL http://stacks.iop.org/0022-3727/32/i=14/a=201.

[1350] D S Campbell and A R Morley. Electrical conduction in thin metallic, dielectric and metallic-dielectric films. *Reports on Progress in Physics*, 34(1):283, 1971. URL http://stacks.iop.org/0034-4885/34/i=1/a=305.

[1351] Kohei Uosaki and Hideaki Kita. Effects of the helmholtz layer capacitance on the potential distribution at semiconductor/electrolyte interface and the linearity of the mott-schottky plot. *Journal of The Electrochemical Society*, 130(4):895–897, 1983.

[1352] Oleg V. Prezhdo, Walter R. Duncan, and Victor V. Prezhdo. Photoinduced electron dynamics at the chromophore–semiconductor interface: A time-domain ab initio perspective. *Progress in Surface Science*, 84(1–2):30 – 68, 2009. ISSN 0079-6816. doi: http://dx.doi.org/10.1016/j.progsurf.2008.10.005. URL http://www.sciencedirect.com/science/article/pii/S0079681608000671.

[1353] Eugen Merzbacher. Quantum mechanics john wiley & sons. *Inc., New York*, 1998.

[1354] K. Schwarzburg and F. Willig. Influence of trap filling on photocurrent transients in polycrystalline tio$_2$. *Applied Physics Letters*, 58 (22):2520–2522, 1991. doi: 10.1063/1.104839. URL http://dx.doi.org/10.1063/1.104839.

[1355] S. Nakade, Y. Saito, W. Kubo, T. Kanzaki, T. Kitamura, Y. Wada, and S. Yanagida. Laser-induced photovoltage transient studies on nanoporous tio$_2$ electrodes. *The Journal of Physical Chemistry B*, 108(5):1628–1633, 2004. doi: 10.1021/jp036786f. URL http://dx.doi.org/10.1021/jp036786f.

[1356] E. Rosenbaum, J.C. King, and Chenming Hu. Accelerated testing of sio$_2$ reliability. *IEEE Transactions on Electron Devices*, 43(1): 70–80, January 1996. ISSN 0018-9383. doi: 10.1109/16.477595.

[1357] T. H. Ning. High-field capture of electrons by coulomb-attractive centers in silicon dioxide. *Journal of Applied Physics*, 47(7):3203–3208, 1976. doi: 10.1063/1.323116. URL http://dx.doi.org/10.1063/1.323116.

[1358] Eli Harari. Dielectric breakdown in electrically stressed thin films of thermal sio$_2$. *Journal of Applied Physics*, 49(4):2478–2489, 1978. doi: 10.1063/1.325096. URL http://aip.scitation.org/doi/abs/10.1063/1.325096.

[1359] J. C. Lee, Chen Ih-Chin, and Hu Chenming. Modeling and characterization of gate oxide reliability. *IEEE Transactions on Electron Devices*, 35(12):2268–2278, Dec 1988. ISSN 0018-9383. doi: 10.1109/16.8802.

[1360] D. J. DiMaria, E. Cartier, and D. Arnold. Impact ionization, trap creation, degradation, and breakdown in silicon dioxide films on silicon. *Journal of Applied Physics*, 73(7):3367–3384, 1993. doi: 10.1063/1.352936. URL http://dx.doi.org/10.1063/1.352936.

[1361] M. Depas, B. Vermeire, and M. M. Heyns. Breakdown and defect generation in ultrathin gate oxide. *Journal of Applied Physics*, 80(1): 382–387, 1996. doi: 10.1063/1.362794. URL http://dx.doi.org/10.1063/1.362794.

[1362] T. Nigam, R. Degraeve, G. Groeseneken, M. M. Heyns, and H. E. Maes. A fast and simple methodology for lifetime prediction of ultra-thin oxides. In *1999 IEEE International Reliability Physics Symposium Proceedings. 37th Annual (Cat. No.99CH36296)*, pages 381–388, 1999. doi: 10.1109/RELPHY.1999.761643.

[1363] Edward H Nicollian, John R Brews, and Edward H Nicollian. *MOS (metal oxide semiconductor) physics and technology*, volume 1987. Wiley New York et al., 1982.

[1364] K. Okada, H. Kubo, A. Ishinaga, and K. Yoneda. A new prediction method for oxide lifetime and its application to study dielectric breakdown mechanism. In *1998 Symposium on VLSI Technology Digest of Technical Papers (Cat. No.98CH36216)*, pages 158–159, June 1998. doi: 10.1109/VLSIT.1998.689239.

[1365] M Houssa, P W Mertens, and M M Heyns. Relation between stress-induced leakage current and time-dependent dielectric breakdown in ultra-thin gate oxides. *Semiconductor Science and Technology*, 14(10):892, 1999. URL http://stacks.iop.org/0268-1242/14/i=10/a=302.

[1366] J. Rudolph. über den leitungsmechanismus oxydischer halbleiter bei hohen temperaturen. *Zeitschrift für Naturforschung A*, 1959.

[1367] Fei Cao, Gerko Oskam, Peter C. Searson, Jeremy M. Stipkala, Todd A. Heimer, Fereshteh Farzad, and Gerald J. Meyer. Electrical and optical properties of porous nanocrystalline tio$_2$ films. *The Journal of Physical Chemistry*, 99(31):11974–11980, 1995. doi: 10.1021/j100031a027. URL http://dx.doi.org/10.1021/j100031a027.

[1368] Christophe J. Barbé, Francine Arendse, Pascal Comte, Marie Jirousek, Frank Lenzmann, Valery Shklover, and Michael Grätzel. Nanocrystalline titanium oxide electrodes for photovoltaic applications. *Journal of the American Ceramic Society*, 80(12):3157–3171, 1997. ISSN 1551-2916. doi: 10.1111/j.1151-2916.1997.tb03245.x. URL http://dx.doi.org/10.1111/j.1151-2916.1997.tb03245.x.

[1369] R. Degraeve, G. Groeseneken, R. Bellens, J. L. Ogier, M. Depas, P. J. Roussel, and H. E. Maes. New insights in the relation between electron trap generation and the statistical properties of oxide breakdown. *IEEE Transactions on Electron Devices*, 45(4):904–911, Apr 1998. ISSN 0018-9383. doi: 10.1109/16.662800.

[1370] E. F. Runnion, S. M. Gladstone, R. S. Scott, D. J. Dumin, L. Lie, and J. C. Mitros. Thickness dependence of stress-induced leakage currents in silicon oxide. *IEEE Transactions on Electron Devices*, 44(6):993–1001, Jun 1997. ISSN 0018-9383. doi: 10.1109/16.585556.

[1371] Juan A. Anta. Random walk numerical simulation for solar cell applications. *Energy Environ. Sci.*, 2:387–392, 2009. doi: 10.1039/B819979E. URL http://dx.doi.org/10.1039/B819979E.

[1372] J. P. Gonzalez-Vazquez, Juan A. Anta, and Juan Bisquert. Random walk numerical simulation for hopping transport at finite carrier concentrations: diffusion coefficient and transport energy concept. *Phys. Chem. Chem. Phys.*, 11:10359–10367, 2009. doi: 10.1039/B912935A. URL http://dx.doi.org/10.1039/B912935A.

[1373] Jenny Nelson. Continuous-time random-walk model of electron transport in nanocrystalline tio2 electrodes. *Phys. Rev. B*, 59:15374–15380, Jun 1999. doi: 10.1103/PhysRevB.59.15374. URL https://link.aps.org/doi/10.1103/PhysRevB.59.15374.

[1374] Ryuzi Katoh, Akihiro Furube, Ken-ichi Yamanaka, and Takeshi Morikawa. Charge separation and trapping in n-doped tio2 photocatalysts: A time-resolved microwave conductivity study. *The Journal of Physical Chemistry Letters*, 1(22):3261–3265, 2010. doi: 10.1021/jz1011548. URL http://dx.doi.org/10.1021/jz1011548.

[1375] Luchao Du, Akihiro Furube, Kazuhiro Yamamoto, Kohjiro Hara, Ryuzi Katoh, and M. Tachiya. Plasmon-induced charge separation and recombination dynamics in gold-tio2 nanoparticle systems: Dependence on tio2 particle size. *The Journal of Physical Chemistry C*, 113(16):6454–6462, 2009. doi: 10.1021/jp810576s. URL http://dx.doi.org/10.1021/jp810576s.

[1376] Xiuli Wang, Zhaochi Feng, Jianying Shi, Guoqing Jia, Shuai Shen, Jun Zhou, and Can Li. Trap states and carrier dynamics of tio2 studied by photoluminescence spectroscopy under weak excitation condition. *Phys. Chem. Chem. Phys.*, 12:7083–7090, 2010. doi: 10.1039/B925277K. URL http://dx.doi.org/10.1039/B925277K.

[1377] T. Tiedje and A. Rose. A physical interpretation of dispersive transport in disordered semiconductors. *Solid State Communications*, 37(1):49 – 52, 1981. ISSN 0038-1098. doi: https://doi.org/10.1016/0038-1098(81)90886-3. URL http://www.sciencedirect.com/science/article/pii/0038109881908863.

[1378] I. C. Chen, S. Holland, K. K. Young, C. Chang, and C. Hu. Substrate hole current and oxide breakdown. *Applied Physics Letters*, 49 (11):669–671, 1986. doi: 10.1063/1.97563. URL http://scitation.aip.org/content/aip/journal/apl/49/11/10.1063/1.97563.

[1379] H. Satake and A. Toriumi. Substrate hole current generation and oxide breakdown in si mosfets under fowler-nordheim electron tunneling injection. In *Proceedings of IEEE International Electron Devices Meeting*, pages 337–340, Dec 1993. doi: 10.1109/IEDM.1993. 347339.

[1380] R. Degraeve, G. Groeseneken, R. Bellens, J.L. Ogier, M. Depas, P.J. Roussel, and H.E. Maes. New insights in the relation between electron trap generation and the statistical properties of oxide breakdown. *IEEE Transactions on Electron Devices*, 45(4):904–911, April 1998. ISSN 0018-9383. doi: 10.1109/16.662800.

[1381] Takayuki Tomita, Hiroto Utsunomiya, Yoshinari Kamakura, and Kenji Taniguchi. Hot hole induced breakdown of thin silicon dioxide films. *Applied Physics Letters*, 71(25):3664–3666, 1997. doi: 10.1063/1.120474. URL http://dx.doi.org/10.1063/1.120474.

[1382] D. Ielmini, A.S. Spinelli, M.A. Rigamonti, and A.L. Lacaita. Modeling of silc based on electron and hole tunneling. i. transient effects. *IEEE Transactions on Electron Devices*, 47(6):1258–1265, June 2000. ISSN 0018-9383. doi: 10.1109/16.842971.

[1383] Akinobu Teramoto, Kiyoteru Kobayashi, Yasuji Matsui, and Makoto Hirayama. Dielectric breakdown caused by hole-induced-defect in thin siog films. *Applied Surface Science*, 117:245 – 248, 1997. ISSN 0169-4332. doi: http://dx.doi.org/10.1016/S0169-4332(97)80088-8. URL http://www.sciencedirect.com/science/article/pii/S0169433297800888.

[1384] A. V. Schwerin, M. M. Heyns, and W. Weber. Investigation on the oxide field dependence of hole trapping and interface state generation in sio2 layers using homogeneous nonavalanche injection of holes. *Journal of Applied Physics*, 67(12):7595–7601, 1990. doi: 10.1063/1.345827. URL http://dx.doi.org/10.1063/1.345827.

[1385] Quazi Deen Mohd Khosru, Naoki Yasuda, Kenji Taniguchi, and Chihiro Hamaguchi. Generation and relaxation phenomena of positive charge and interface trap in a metal-oxide-semiconductor structure. *Journal of Applied Physics*, 77(9):4494–4503, 1995. doi: 10.1063/1.359445. URL http://dx.doi.org/10.1063/1.359445.

[1386] S.A. Lyon. Interface states generated by the injection of electrons and holes into siog. *Applied Surface Science*, 39(1):552 – 564, 1989. ISSN 0169-4332. doi: https://doi.org/10.1016/0169-4332(89)90471-6. URL http://www.sciencedirect.com/science/article/pii/0169433289904716.

[1387] Ih-Chin Chen, S.E. Holland, and Chenming Hu. Electrical breakdown in thin gate and tunneling oxides. *IEEE Journal of Solid-State Circuits*, 20(1):333–342, February 1985. ISSN 0018-9200. doi: 10.1109/JSSC.1985.1052311.

[1388] Mohammed T Quddus, Thomas A DeMassa, Dieter K Schroder, and Julian J Sanchez. Modeling of time dependence of hole current and prediction of qbd and tbd for thin gate mos devices based upon anode hole injection. *Solid-State Electronics*, 45(10):1773 – 1785, 2001. ISSN 0038-1101. doi: https://doi.org/10.1016/S0038-1101(01)00199-X. URL http://www.sciencedirect.com/science/article/pii/S003811010100199X.

[1389] J. R. Brews. Limitations upon photoinjection studies of charge distributions close to interfaces in mos capacitors. *Journal of Applied Physics*, 44(1):379–384, 1973. doi: 10.1063/1.1661891. URL http://dx.doi.org/10.1063/1.1661891.

[1390] A. v. Schwerin and M. M. Heyns. Homogeneous hole injection into gate layers of mosfet: injection efficiency, hole trapping and si/sio2 interface state generation. In *INFOS 91, Liverpool, April 1991, Contributed Papers, Section 7*, 1991.

[1391] Yee-Chia Yeo. Mosfet gate oxide reliability: Anode hole injection model and its applications. *International Journal of High Speed Electronics and Systems*, 11(03):849–886, 2001. doi: 10.1142/S0129156401001015. URL http://www.worldscientific.com/doi/abs/10.1142/S0129156401001015.

[1392] M.M. Heyns, D. Krishna Rao, and R.F. De Keersmaecker. Oxide field dependence of si-sio2 interface state generation and charge trapping during electron injection. *Applied Surface Science*, 39(1):327 – 338, 1989. ISSN 0169-4332. doi: https://doi.org/10.1016/0169-4332(89)90447-9. URL http://www.sciencedirect.com/science/article/pii/0169433289904479.

[1393] Pierre Bonhôte, Jacques-E. Moser, Robin Humphry-Baker, Nicolas Vlachopoulos, Shaik M. Zakeeruddin, Lorenz Walder, and Michael Grätzel. Long-lived photoinduced charge separation and redox-type photochromism on mesoporous oxide films sensitized by molecular dyads. *Journal of the American Chemical Society*, 121(6):1324–1336, 1999. doi: 10.1021/ja981742j. URL http://dx.doi.org/10.1021/ja981742j.

[1394] V.Yu Timoshenko, P.K Kashkarov, A.B Matveeva, E.A Konstantinova, H Flietner, and Th Dittrich. Influence of photoluminescence and trapping on the photovoltage at the por-si/p-si structure. *Thin Solid Films*, 276(1):216 – 218, 1996. ISSN 0040-6090. doi: https://doi.org/10.1016/0040-6090(95)08056-2. URL http://www.sciencedirect.com/science/article/pii/0040609095080562. Papers presented at the European Materials Research Society 1995 Spring Conference, Symposium I: Porous Silicon: Material, Technology and Devices.

[1395] Rainer Waser and Masakazu Aono. Nanoionics-based resistive switching memories. *Nat Mater*, 6(11):833–840, 2007. doi: 10.1038/nmat2023. URL http://dx.doi.org/10.1038/nmat2023.

[1396] M. A. Alam, J. Bude, and A. Ghetti. Field acceleration for oxide breakdown-can an accurate anode hole injection model resolve the e vs. 1/e controversy? In *2000 IEEE International Reliability Physics Symposium Proceedings. 38th Annual (Cat. No.00CH37059)*, pages 21–26, 2000. doi: 10.1109/RELPHY.2000.843886.

[1397] Ching-Wu Wang, Shih-Fang Chen, and Guan-Ting Chen. Gamma-ray-irradiation effects on the leakage current and reliability of sputtered Tio₂ gate oxide in metal-oxide-semiconductor capacitors. *Journal of Applied Physics*, 91(11):9198–9203, 2002. doi: 10.1063/1.1473668. URL http://scitation.aip.org/content/aip/journal/jap/91/11/10.1063/1.1473668.

[1398] A. Cester. Wear-out and breakdown of ultra-thin gate oxides after irradiation. *Electronics Letters*, 38:1137–1139(2), September 2002. ISSN 0013-5194. URL http://digital-library.theiet.org/content/journals/10.1049/el_20020757.

[1399] D. J. DiMaria and J. H. Stathis. Non-arrhenius temperature dependence of reliability in utlrathin silicon dioxide films. *Appl. Phys. Lett.*, 74(12):1752, 1999. doi: 10.1063/1.123677.

[1400] R. Degraeve, N. Pangon, B. Kaczer, T. Nigam, G. Groeseneken, and A. Naem. Temperature acceleration of oxide breakdown and its impact on ultra-thin gate oxide reliability. *Symposium on VLSI Technology*, pages 59–60, 1999. doi: 10.1109/VLSIT.1999.799339.

[1401] J Wu, L Register, and E Rosenbaum. Trap-assisted tunneling current through ultra-thin oxide. *Reliability Physics Symposium Proceedings*, pages 389–395, 1999. doi: 10.1109/RELPHY.1999.761644. URL http://ieeexplore.ieee.org/stamp/stamp.jsp?tp=&arnumber=761644&isnumber=16472.

[1402] S Takagi, N Yasuda, and A Toriumi. A new i-v model for stress-induced leakage current including inelastic tunneling. *IEEE Transactions on Electron Devices*, 46(2):348–354, February 1999. doi: 10.1109/16.740901.

[1403] B E Weir, P J Silverman, D Monroe, K S Krisch, M A Alam, G B Alers, T W Sorsch, G L Timp, F Baumann, C T Liu, Y Ma, and D Huang. Ultra-thin gate dielectrics: they break down, but do they fail? In *International Electron Devices Meeting. Technical Digest.*, pages 73–76, Dec 1997. doi: 10.1109/IEDM.1997.649463.

[1404] A. Cester, L. Bandiera, G. Ghidini, I. Bloom, and A. Paccagnella. Soft breakdown current noise in ultra-thin gate oxides. *Solid-State Electronics*, 46(7):1019 – 1025, 2002. ISSN 0038-1101. doi: http://dx.doi.org/10.1016/S0038-1101(02)00036-9. URL http://www.sciencedirect.com/science/article/pii/S0038110102000369.

[1405] S. Lombardo, F. Crupi, A. La Magna, C. Spinella, A. Terrasi, A. La Mantia, and B. Neri. Electrical and thermal transient during dielectric breakdown of thin oxides in metal-sio₂-silicon capacitors. *Journal of Applied Physics*, 84(1):472–479, 1998. doi: 10.1063/1.368050. URL http://dx.doi.org/10.1063/1.368050.

[1406] M Depas, T Nigam, and M.M Heyns. Definition of dielectric breakdown for ultra thin (<2 nm) gate oxides. *Solid-State Electronics*, 41(5):725 – 728, 1997. ISSN 0038-1101. doi: http://dx.doi.org/10.1016/S0038-1101(96)00111-6. URL http://www.sciencedirect.com/science/article/pii/S0038110196001116.

[1407] B. Neri, P. Olivo, and B. Riccó. Low-frequency noise in silicon-gate metal-oxide-silicon capacitors before oxide breakdown. *Applied Physics Letters*, 51(25):2167–2169, 1987. doi: 10.1063/1.98930. URL http://dx.doi.org/10.1063/1.98930.

[1408] Arnost Neugroschel, Lingquan Wang, and Gennadi Bersuker. Trapped charge induced gate oxide breakdown. *Journal of Applied Physics*, 96(6):3388–3398, 2004. doi: 10.1063/1.1781766. URL http://dx.doi.org/10.1063/1.1781766.

[1409] Seok-Hee Lee, Byung-Jin Cho, Jong-Choul Kim, and Soo-Han Choi. Quasi-breakdown of ultrathin gate oxide under high field stress. In *International Electron Devices Meeting. Technical Digest.*, pages 605–608, Dec 1994. doi: 10.1109/IEDM.1994.383337.

[1410] Kenji Okada. The gate oxide lifetime limited by 'b-mode' stress induced leakage current and the scaling limit of silicon dioxides in the direct tunnelling regime. *Semiconductor Science and Technology*, 15(5):478, 2000. URL http://stacks.iop.org/0268-1242/15/i=5/a=307.

[1411] M Depas, Tanya Nigam, and Marc M Heyns. Soft breakdown of ultra-thin gate oxide layers. *IEEE Transactions on Electron Devices*, 43(9):1499–1504, September 1996. ISSN 0018-9383. doi: 10.1109/16.535341.

[1412] Enrique Miranda, J. Sune, R. Rodriguez, M. Nafria, and X. Aymerich. A function-fit model for the soft breakdown failure mode. *Electron Device Letters, IEEE*, 20(6):265–267, June 1999. ISSN 0741-3106. doi: 10.1109/55.767093.

[1413] M. Houssa, T. Nigam, P. W. Mertens, and M. M. Heyns. Model for the current-voltage characteristics of ultrathin gate oxides after soft breakdown. *Journal of Applied Physics*, 84(8):4351–4355, 1998. doi: 10.1063/1.368654. URL http://scitation.aip.org/content/aip/journal/jap/84/8/10.1063/1.368654.

[1414] M.A. Alam, Bonnie E. Weir, and Paul J. Silverman. A study of soft and hard breakdown - part i: Analysis of statistical percolation conductance. *IEEE Transactions on Electron Devices*, 49(2):232–238, February 2002. ISSN 0018-9383. doi: 10.1109/16.981212.

[1415] Andrea Ghetti. Characterization and modeling of the tunneling current in si-Sio₂-si structures with ultra-thin oxide layer. *Microelectronic Engineering*, 59(1-4):127 – 136, 2001. ISSN 0167-9317. doi: 10.1016/S0167-9317(01)00656-6. URL http://www.sciencedirect.com/science/article/pii/S0167931701006566. 12th Biannual Conference on Insulating Films on Semiconductors.

[1416] A. Yassine, H. E. Nariman, and K. Olasupo. Field and temperature dependence of tddb of ultrathin gate oxide. *IEEE Electron Device Letters*, 20(8):390–392, Aug 1999. ISSN 0741-3106. doi: 10.1109/55.778152.

[1417] E. Vincent, N. Revil, C. Papadas, and G. Ghibaudo. Electric field dependence of tddb activation energy in ultrathin oxides. *Microelectronics Reliability*, 36(11):1643 – 1646, 1996. ISSN 0026-2714. doi: http://dx.doi.org/10.1016/0026-2714(96)00164-3. URL http://www.sciencedirect.com/science/article/pii/0026271496001643.

[1418] J. McPherson, J-Y. Kim, A. Shanware, and H. Mogul. Thermochemical description of dielectric breakdown in high dielectric constant materials. *Applied Physics Letters*, 82(13):2121–2123, 2003. doi: 10.1063/1.1565180. URL http://dx.doi.org/10.1063/1.1565180.

[1419] Tao Fenfen, Yang Hong, Tang Bo, Tang Zhaoyun, Xu Yefeng, Xu Jing, Wang Qingpu, and Yan Jiang. Tddb characteristic and breakdown mechanism of ultra-thin sio_2/hfo_2 bilayer gate dielectrics. *Journal of Semiconductors*, 35(6):064003, 2014. URL http://stacks.iop.org/1674-4926/35/i=6/a=064003.

[1420] J. McPherson, J. Kim, A. Shanware, H. Mogul, and J. Rodriguez. Proposed universal relationship between dielectric breakdown and dielectric constant. In *Digest. International Electron Devices Meeting,*, pages 633–636, Dec 2002. doi: 10.1109/IEDM.2002.1175919.

[1421] E. Rymaszewski, T. Lu, M. Nielsen, and J. Kim. Integrated discrete components. In *Proc. Electrochem. Soc., 1st Symposium on Dielectric Materials for Advanced Electronic Packaging, San Diego, CA, May*, pages 212–266, 1998.

[1422] S E Savel'ev, A S Alexandrov, A M Bratkovsky, and R Stanley Williams. Molecular dynamics simulations of oxide memory resistors (memristors). *Nanotechnology*, 22(25):254011, 2011. URL http://stacks.iop.org/0957-4484/22/i=25/a=254011.

[1423] Dmitri B. Strukov, Julien L. Borghetti, and R. Stanley Williams. Coupled ionic and electronic transport model of thin-film semiconductor memristive behavior. *Small*, 5(9):1058–1063, 2009. ISSN 1613-6829. doi: 10.1002/smll.200801323. URL http://dx.doi.org/10.1002/smll.200801323.

[1424] Dmitri B. Strukov, Gregory S. Snider, Duncan R. Stewart, and R. Stanley Williams. The missing memristor found. *Nature*, 453:80–83, 2008. doi: 10.1038/nature06932. URL http://dx.doi.org/10.1038/nature06932.

[1425] Zvi Rosenstock, Irena Feldman, Yoram Gil, and Ilan Riess. Semi-conductors with mobile ions show a new type of i-v relations. *Journal of Electroceramics*, 14(3):205–212, Jul 2005. ISSN 1573-8663. doi: 10.1007/s10832-005-0959-2. URL https://doi.org/10.1007/s10832-005-0959-2.

[1426] Y. Gil, O.M. Umurhan, Y. Tsur, and I. Riess. I–v relations in nano thin semi-conductors with mobile acceptors or donors. *Solid State Ionics*, 179(1):24 – 32, 2008. ISSN 0167-2738. doi: https://doi.org/10.1016/j.ssi.2007.12.039. URL http://www.sciencedirect.com/science/article/pii/S0167273807004572. 8th International Symposium on Systems with Fast Ionic Transport.

[1427] P. Bordewijk. Defect-diffusion models of dielectric relaxation. *Chemical Physics Letters*, 32(3):592 – 596, 1975. ISSN 0009-2614. doi: http://dx.doi.org/10.1016/0009-2614(75)85248-1. URL http://www.sciencedirect.com/science/article/pii/0009261475852481.

[1428] Michael F. Shlesinger and Elliott W. Montroll. On the williams-watts function of dielectric relaxation. *Proceedings of the National Academy of Sciences*, 81(4):1280–1283, 1984. URL http://www.pnas.org/content/81/4/1280.abstract.

[1429] J. Kakalios, R. A. Street, and W. B. Jackson. Stretched-exponential relaxation arising from dispersive diffusion of hydrogen in amorphous silicon. *Phys. Rev. Lett.*, 59:1037–1040, Aug 1987. doi: 10.1103/PhysRevLett.59.1037. URL https://link.aps.org/doi/10.1103/PhysRevLett.59.1037.

[1430] K. L. Ngai, G. N. Greaves, and C. T. Moynihan. Correlation between the activation energies for ionic conductivity for short and long time scales and the kohlrausch stretching parameter β for ionically conducting solids and melts. *Phys. Rev. Lett.*, 80:1018–1021, Feb 1998. doi: 10.1103/PhysRevLett.80.1018. URL https://link.aps.org/doi/10.1103/PhysRevLett.80.1018.

[1431] K. L. Ngai and S. W. Martin. Correlation between the activation enthalpy and kohlrausch exponent for ionic conductivity in oxide glasses. *Phys. Rev. B*, 40:10550–10556, Nov 1989. doi: 10.1103/PhysRevB.40.10550. URL https://link.aps.org/doi/10.1103/PhysRevB.40.10550.

[1432] G. Balzer-Jöllenbeck, O. Kanert, H. Jain, and K. L. Ngai. New interpretation of activation enthalpies for electrical conductivity and nuclear spin relaxation in glassy ionic conductors. *Phys. Rev. B*, 39:6071–6075, Mar 1989. doi: 10.1103/PhysRevB.39.6071. URL https://link.aps.org/doi/10.1103/PhysRevB.39.6071.

[1433] Dmitri B. Strukov and R. Stanley Williams. Exponential ionic drift: fast switching and low volatility of thin-film memristors. *Applied Physics A*, 94(3):515–519, Mar 2009. ISSN 1432-0630. doi: 10.1007/s00339-008-4975-3. URL https://doi.org/10.1007/s00339-008-4975-3.

[1434] Feng Miao, J Joshua Yang, Julien Borghetti, Gilberto Medeiros-Ribeiro, and R Stanley Williams. Observation of two resistance switching modes in tio₂ memristive devices electroformed at low current. *Nanotechnology*, 22(25):254007, 2011. URL http://stacks.iop.org/0957-4484/22/i=25/a=254007.

[1435] R. Metzler, G. Oshanin, and S. Redner. *First-Passage Phenomena and Their Applications*. World Scientific Studies in International Economics. World Scientific Publishing Company, 2014. ISBN 9789814590303. URL https://books.google.de/books?id=dAS3CgAAQBAJ.

[1436] Yoshitaka Aoki, Carsten Wiemann, Vitaliy Feyer, Hong-Seok Kim, Claus Michael Schneider, Han Ill-Yoo, and Manfred Martin. Bulk mixed ion electron conduction in amorphous gallium oxide causes memristive behaviour. *Nature communications*, 5(3473), 2014. doi: 10.1038/ncomms4473. URL http://dx.doi.org/10.1038/ncomms4473.

[1437] Mikko Ritala, Markku Leskelä, Erja Nykfinen, Pekka Soininen, and Lauri Niinistö. Growth of titanium dioxide thin films by atomic layer epitaxy. *Thin Solid Films*, 225(1-2):288 – 295, 1993. doi: 10.1016/0040-6090(93)90172-L.

[1438] Akira Yamada, Baosheng Sang, and Makoto Konagai. Atomic layer deposition of zno transparent conducting oxides. *Applied Surface Science*, 112:216–222, 1997. doi: 10.1016/S0169-4332(96)01002-7.

[1439] Victor V. Zhirnov and Ralph K. Cavin. Nanodevices: Charge of the heavy brigade. *Nature Nanotechnology*, 3:377 – 378, 2008.

[1440] Lilyana Kolaklieva and Roumen Kakanakov. *Ohmic Contacts for High Power and High Temperature Microelectronics*, chapter Ohmic Contacts for High Power and High Temperature Microelectronics, page 386. InTech, 2009.

[1441] S. Yu and H. S. P. Wong. A phenomenological model for the reset mechanism of metal oxide rram. *IEEE Electron Device Letters*, 31 (12):1455–1457, Dec 2010. ISSN 0741-3106. doi: 10.1109/LED.2010.2078794.

[1442] C. Cagli, F. Nardi, and D. Ielmini. Modeling of set/reset operations in nio-based resistive-switching memory devices. *IEEE Transactions on Electron Devices*, 56(8):1712–1720, Aug 2009. ISSN 0018-9383. doi: 10.1109/TED.2009.2024046.

[1443] R. T. P. Lee, A. E. J. Lim, K. M. Tan, T. Y. Liow, D. Z. Chi, and Y. C. Yeo. Sulfur-induced ptsi:c/si:c schottky barrier height lowering for realizing n-channel finfets with reduced external resistance. *IEEE Electron Device Letters*, 30(5):472–474, May 2009. ISSN 0741-3106. doi: 10.1109/LED.2009.2017213.

[1444] Ch. Walczyk, Ch. Wenger, R. Sohal, M. Lukosius, A. Fox, J. Dąbrowski, D. Wolansky, B. Tillack, H.-J. Müssig, and T. Schroeder. Pulse-induced low-power resistive switching in hfo₂ metal-insulator-metal diodes for nonvolatile memory applications. *Journal of Applied Physics*, 105(11):114103, 2009. doi: 10.1063/1.3139282. URL http://dx.doi.org/10.1063/1.3139282.

[1445] Yoshihiro Sato, Kentaro Kinoshita, Masaki Aoki, and Yoshihiro Sugiyama. Consideration of switching mechanism of binary metal oxide resistive junctions using a thermal reaction model. *Applied Physics Letters*, 90(3):033503, 2007. doi: 10.1063/1.2431792. URL http://dx.doi.org/10.1063/1.2431792.

[1446] U. Russo, D. Ielmini, C. Cagli, and A. L. Lacaita. Self-accelerated thermal dissolution model for reset programming in unipolar resistive-switching memory (rram) devices. *IEEE Transactions on Electron Devices*, 56(2):193–200, Feb 2009. ISSN 0018-9383. doi: 10.1109/TED.2008.2010584.

[1447] P. Bousoulas, I. Michelakaki, and D. Tsoukalas. Influence of oxygen content of room temperature tio₂₋ₓ deposited films for enhanced resistive switching memory performance. *Journal of Applied Physics*, 115(3):034516, 2014. doi: 10.1063/1.4862797. URL http://dx.doi.org/10.1063/1.4862797.

[1448] Qibiao Lv, Shuxiang Wu, Jingquan Lu, Mei Yang, Ping Hu, and Shuwei Li. Conducting nanofilaments formed by oxygen vacancy migration in ti/tio₂/tin/mgo memristive device. *Journal of Applied Physics*, 110(10):104511, 2011. doi: 10.1063/1.3662922. URL http://dx.doi.org/10.1063/1.3662922.

[1449] D.H. Kwon, J.M. Jeon, J.H. Jang, K.M. Kim, C.S. Hwang, and M. Kim. Direct observation of conducting paths in tio₂ thin film by transmission electron microscopy. *Microscopy and Microanalysis*, 15:996–997, 2009.

[1450] S E Savel'ev, A S Alexandrov, A M Bratkovsky, and R Stanley Williams. Molecular dynamics simulations of oxide memory resistors (memristors). *Nanotechnology*, 22(25):254011, 2011. URL http://stacks.iop.org/0957-4484/22/i=25/a=254011.

[1451] J. H. Stathis. Percolation models for gate oxide breakdown. *Journal of Applied Physics*, 86(10):5757–5766, 1999. doi: 10.1063/1.371590. URL http://dx.doi.org/10.1063/1.371590.

[1452] Young-Hee Kim and Jack C Lee. Reliability characteristics of high-k dielectrics. *Microelectronics Reliability*, 44(2):183 – 193, 2004. ISSN 0026-2714. doi: http://dx.doi.org/10.1016/j.microrel.2003.10.008. URL http://www.sciencedirect.com/science/article/pii/S0026271403004311.

[1453] R. Degraeve, G. Groeseneken, R. Bellens, M. Depas, and H. E. Maes. A consistent model for the thickness dependence of intrinsic breakdown in ultra-thin oxides. In *Proceedings of International Electron Devices Meeting*, pages 863–866, Dec 1995. doi: 10.1109/IEDM.1995.499353.

[1454] J. H. Stathis and D. J. DiMaria. Reliability projection for ultra-thin oxides at low voltage. In *International Electron Devices Meeting 1998. Technical Digest (Cat. No.98CH36217)*, pages 167–170, Dec 1998. doi: 10.1109/IEDM.1998.746309.

[1455] K. R. Farmer, R. Saletti, and R. A. Buhrman. Current fluctuations and silicon oxide wear-out in metal-oxide-semiconductor tunnel diodes. *Applied Physics Letters*, 52(20):1749–1751, 1988. doi: 10.1063/1.99029. URL http://dx.doi.org/10.1063/1.99029.

[1456] D. M. Fleetwood, P. S. Winokur, R. A. Reber Jr., T. L. Meisenheimer, J. R. Schwank, M. R. Shaneyfelt, and L. C. Riewe. Effects of oxide traps, interface traps, and "border traps" on metal-oxide-semiconductor devices. *Journal of Applied Physics*, 73(10):5058–5074, 1993. doi: 10.1063/1.353777. URL http://dx.doi.org/10.1063/1.353777.

[1457] Stephan Menzel, Matthias Waters, Astrid Marchewka, Ulrich Böttger, Regina Dittmann, and Rainer Waser. Origin of the ultra-nonlinear switching kinetics in oxide-based resistive switches. *Advanced Functional Materials*, 21(23):4487–4492, 2011. ISSN 1616-3028. doi: 10.1002/adfm.201101117. URL http://dx.doi.org/10.1002/adfm.201101117.

[1458] Fabien Alibart, Ligang Gao, Brian D Hoskins, and Dmitri B Strukov. High precision tuning of state for memristive devices by adaptable variation-tolerant algorithm. *Nanotechnology*, 23(7):075201, 2012. URL http://stacks.iop.org/0957-4484/23/i=7/a=075201.

[1459] Y. L. Sandler, *Progress Rep.ort No. XVII, Laboratory of Insulation Research, Massachusetts Institute .of Technology, p.16.*, 1955.

[1460] Michael L. Hair. Hydroxyl groups on silica surface. *Journal of Non-Crystalline Solids*, 19:299 – 309, 1975. ISSN 0022-3093. doi: http://dx.doi.org/10.1016/0022-3093(75)90095-2. URL http://www.sciencedirect.com/science/article/pii/0022309375900952. Glass Surfaces.

[1461] J Bo Peri and R B_ Hannan. Surface hydroxyl groups on γ-alumina1. *The Journal of Physical Chemistry*, 64(11):1526–1530, 1960.

[1462] E. McCafferty and J. P. Wightman. Determination of the concentration of surface hydroxyl groups on metal oxide films by a quantitative xps method. *Surface and Interface Analysis*, 26(8):549–564, 1998. ISSN 1096-9918. doi: 10.1002/(SICI)1096-9918(199807)26:8<549::AID-SIA396>3.0.CO;2-Q. URL http://dx.doi.org/10.1002/(SICI)1096-9918(199807)26:8<549::AID-SIA396>3.0.CO;2-Q.

[1463] F. B. McLean. A framework for understanding radiation-induced interface states in sio₂ mos structures. *IEEE Transactions on Nuclear Science*, 27(6):1651–1657, Dec 1980. ISSN 0018-9499. doi: 10.1109/TNS.1980.4331084.

[1464] M. Houssa, M. Naili, M. M. Heyns, and A. Stesmans. Model for the charge trapping in high permittivity gate dielectric stacks. *Journal of Applied Physics*, 89(1):792–794, 2001. doi: 10.1063/1.1330757. URL http://dx.doi.org/10.1063/1.1330757.

[1465] D. J. DiMaria and J. W. Stasiak. Trap creation in silicon dioxide produced by hot electrons. *Journal of Applied Physics*, 65(6):2342–2356, 1989. doi: 10.1063/1.342824. URL http://scitation.aip.org/content/aip/journal/jap/65/6/10.1063/1.342824.

[1466] R.E. Stahlbush and E. Cartier. Interface defect formation in mosfets by atomic hydrogen exposure. *IEEE Transactions on Nuclear Science*, 41(6):1844–1853, December 1994. ISSN 0018-9499. doi: 10.1109/23.340516.

[1467] D. J. DiMaria. Electron energy dependence of metal-oxide-semiconductor degradation. *Applied Physics Letters*, 75(16):2427–2428, 1999. doi: 10.1063/1.125036. URL http://scitation.aip.org/content/aip/journal/apl/75/16/10.1063/1.125036.

[1468] Dao-Bao Chu, Xiao-Hua Li, Xin-Yuan Liu, and Wen-Li Yao. Electrolytic fixation of co₂ by electrocarboxylation of rx on nanocrystalline tio₂-pt cathode. *Chinese Journal of Chemistry*, 22(11):1231–1234, 2004. ISSN 1614-7065. doi: 10.1002/cjoc.20040221104. URL http://dx.doi.org/10.1002/cjoc.20040221104.

[1469] Robert E. Rettew, Nageh K. Allam, and Faisal M. Alamgir. Interface architecture determined electrocatalytic activity of pt on vertically oriented tio₂ nanotubes. *ACS Applied Materials & Interfaces*, 3(2):147–151, 2011. doi: 10.1021/am1012563. URL http://dx.doi.org/10.1021/am1012563. PMID: 21268611.

References

[1470] Adam Lewera, Laure Timperman, Agata Roguska, and Nicolas Alonso-Vante. Metal-support interactions between nanosized pt and metal oxides (wo₃ and tio₂) studied using x-ray photoelectron spectroscopy. *The Journal of Physical Chemistry C*, 115(41):20153–20159, 2011. doi: 10.1021/jp2068446. URL http://dx.doi.org/10.1021/jp2068446.

[1471] Sajeev John. Localization of light. *Phys. Today*, 44(5):32–40, 1991.

[1472] P. Yu and M. Cardona. *Fundamentals of Semiconductors: Physics and Materials Properties*. Graduate Texts in Physics. Springer Berlin Heidelberg, 1996. ISBN 9783642007101. URL https://books.google.de/books?id=5aBuKYBT_hsC.

[1473] Paweł Zawadzki. Absorption spectra of trapped holes in anatase tio₂. *The Journal of Physical Chemistry C*, 117(17):8647–8651, 2013. doi: 10.1021/jp400082u. URL http://dx.doi.org/10.1021/jp400082u.

[1474] Jianjun Dong and D. A. Drabold. Band-tail states and the localized-to-extended transition in amorphous diamond. *Phys. Rev. B*, 54: 10284–10287, Oct 1996. doi: 10.1103/PhysRevB.54.10284. URL https://link.aps.org/doi/10.1103/PhysRevB.54.10284.

[1475] Y. Pan, F. Inam, M. Zhang, and D. A. Drabold. Atomistic origin of urbach tails in amorphous silicon. *Phys. Rev. Lett.*, 100:206403, May 2008. doi: 10.1103/PhysRevLett.100.206403. URL https://link.aps.org/doi/10.1103/PhysRevLett.100.206403.

[1476] Shadia J Ikhmayies and Riyad N Ahmad-Bitar. An investigation of the bandgap and urbach tail of spray-deposited sno₂:f thin films. *Physica Scripta*, 84(5):055801, 2011. URL http://stacks.iop.org/1402-4896/84/i=5/a=055801.

[1477] Tracy L. Thompson and John T. Yates. Tio₂-based photocatalysis: Surface defects, oxygen and charge transfer. *Topics in Catalysis*, 35(3):197–210, Jul 2005. ISSN 1572-9028. doi: 10.1007/s11244-005-3825-1. URL https://doi.org/10.1007/s11244-005-3825-1.

[1478] Ying-Chih Pu, Yichuan Ling, Kao-Der Chang, Chia-Ming Liu, Jin Z. Zhang, Yung-Jung Hsu, and Yat Li. Surface passivation of tio₂ nanowires using a facile precursor-treatment approach for photoelectrochemical water oxidation. *The Journal of Physical Chemistry C*, 118(27):15086–15094, 2014. doi: 10.1021/jp5041019. URL http://dx.doi.org/10.1021/jp5041019.

[1479] T Sekiya, S Kamei, and S Kurita. Luminescence of anatase tio₂ single crystals annealed in oxygen atmosphere. *Journal of Luminescence*, 87-89(Supplement C):1140 – 1142, 2000. ISSN 0022-2313. doi: https://doi.org/10.1016/S0022-2313(99)00570-0. URL http://www.sciencedirect.com/science/article/pii/S0022231399005700.

[1480] Takashi Tachikawa and Tetsuro Majima. Exploring the spatial distribution and transport behavior of charge carriers in a single titania nanowire. *Journal of the American Chemical Society*, 131(24):8485–8495, 2009. doi: 10.1021/ja900194m. URL http://dx.doi.org/10.1021/ja900194m. PMID: 19480455.

[1481] Candy C. Mercado, Fritz J. Knorr, Jeanne L. McHale, Shirin M. Usmani, Andrew S. Ichimura, and Laxmikant V. Saraf. Location of hole and electron traps on nanocrystalline anatase tio₂. *The Journal of Physical Chemistry C*, 116(19):10796–10804, 2012. doi: 10.1021/jp301680d. URL http://dx.doi.org/10.1021/jp301680d.

[1482] Fan Zuo, Le Wang, Tao Wu, Zhenyu Zhang, Dan Borchardt, and Pingyun Feng. Self-doped ti³⁺ enhanced photocatalyst for hydrogen production under visible light. *Journal of the American Chemical Society*, 132(34):11856–11857, 2010. doi: 10.1021/ja103843d. URL http://dx.doi.org/10.1021/ja103843d. PMID: 20687606.

[1483] Candy C. Mercado, Fritz J. Knorr, and Jeanne L. McHale. Observation of charge transport in single titanium dioxide nanotubes by micro-photoluminescence imaging and spectroscopy. *ACS Nano*, 6(8):7270€"7280, 2012. doi: 10.1021/nn302392p.

[1484] Lu-Lu Long, Ai-Yong Zhang, Jun Yang, Xing Zhang, and Han-Qing Yu. A green approach for preparing doped tio₂ single crystals. *ACS Applied Materials & Interfaces*, 6(19):16712–16720, 2014. doi: 10.1021/am503661w. URL http://dx.doi.org/10.1021/am503661w. PMID: 25188022.

[1485] Ruei-San Chen, Yi-Ling Liu, Ching-Hsiang Chan, and Ying-Sheng Huang. Photoconductivities in anatase tio₂ nanorods. *Applied Physics Letters*, 105(15):153107, 2014. doi: 10.1063/1.4898004. URL https://doi.org/10.1063/1.4898004.

[1486] Nobuyuki Sakai, Yasuo Ebina, Kazunori Takada, and Takayoshi Sasaki. Electronic band structure of titania semiconductor nanosheets revealed by electrochemical and photoelectrochemical studies. *Journal of the American Chemical Society*, 126(18):5851–5858, 2004. doi: 10.1021/ja0394582. URL http://dx.doi.org/10.1021/ja0394582. PMID: 15125677.

[1487] Yoshiaki Tamaki, Kohjiro Hara, Ryuzi Katoh, M. Tachiya, and Akihiro Furube. Femtosecond visible-to-ir spectroscopy of tio₂ nanocrystalline films: Elucidation of the electron mobility before deep trapping. *The Journal of Physical Chemistry C*, 113(27):11741–11746, 2009. doi: 10.1021/jp901833j. URL http://dx.doi.org/10.1021/jp901833j.

[1488] Juan Bisquert and Iván Mora-Seró. Simulation of steady-state characteristics of dye-sensitized solar cells and the interpretation of the diffusion length. *The Journal of Physical Chemistry Letters*, 1(1):450–456, 2010. doi: 10.1021/jz900297b. URL http://dx.doi.org/10.1021/jz900297b.

[1489] Yasuhiro Yamada and Yoshihiko Kanemitsu. Determination of electron and hole lifetimes of rutile and anatase tio₂ single crystals. *Applied Physics Letters*, 101(13):133907, 2012. doi: http://dx.doi.org/10.1063/1.4754831. URL http://scitation.aip.org/content/aip/journal/apl/101/13/10.1063/1.4754831.

[1490] Mingchun Xu, Youkun Gao, Elias Martinez Moreno, Marinus Kunst, Martin Muhler, Yuemin Wang, Hicham Idriss, and Christof Wöll. Photocatalytic activity of bulk tio₂ anatase and rutile single crystals using infrared absorption spectroscopy. *Phys. Rev. Lett.*, 106: 138302, Mar 2011. doi: 10.1103/PhysRevLett.106.138302. URL https://link.aps.org/doi/10.1103/PhysRevLett.106.138302.

[1491] Maria C. Fravventura, Laurens D. A. Siebbeles, and Tom J. Savenije. Mechanisms of photogeneration and relaxation of excitons and mobile carriers in anatase tio₂. *The Journal of Physical Chemistry C*, 118(14):7337–7343, 2014. doi: 10.1021/jp500132w. URL http://dx.doi.org/10.1021/jp500132w.

[1492] Alex Henning, Gino Günzburger, Res Jöhr, Yossi Rosenwaks, Biljana Bozic-Weber, Catherine E Housecroft, Edwin C Constable, Ernst Meyer, and Thilo Glatzel. Kelvin probe force microscopy of nanocrystalline tio₂ photoelectrodes. *Beilstein Journal of Nanotechnology*, 4:418–428, 2013. doi: 10.3762/bjnano.4.49.

[1493] R. S. Chen, C. A. Chen, H. Y. Tsai, W. C. Wang, and Y. S. Huang. Ultrahigh efficient single-crystalline tio₂ nanorod photoconductors. *Applied Physics Letters*, 100(12):123108, 2012. doi: 10.1063/1.3694926. URL https://doi.org/10.1063/1.3694926.

[1494] Wardatun Nadrah, MOHD AMIN, ZAINAL Zulkarnain, TALIB Zainal Abidin, LIM Hong Ngee, and Chang Sook Keng. Quantum dot-sensitized solar cell based on nano-tio₂ electrodes. *Pertanika Journal of Scholarly Research Reviews*, 2015.

394

[1495] Zhaoke Zheng, Baibiao Huang, Xiaoyan Qin, Xiaoyang Zhang, Ying Dai, and Myung-Hwan Whangbo. Facile in situ synthesis of visible-light plasmonic photocatalysts m@tio$_2$ (m = au, pt, ag) and evaluation of their photocatalytic oxidation of benzene to phenol. *J. Mater. Chem.*, 21:9079–9087, 2011. doi: 10.1039/C1JM10983A. URL http://dx.doi.org/10.1039/C1JM10983A.

[1496] Scott C. Warren, David A. Walker, and Bartosz A. Grzybowski. Plasmoelectronics: Coupling plasmonic excitation with electron flow. *Langmuir*, 28(24):9093–9102, 2012. doi: 10.1021/la300377j. URL http://dx.doi.org/10.1021/la300377j. PMID: 22385329.

[1497] Hyunsu Kim, Chulmin Choi, Jirapon Khamwannah, Sun Young Noh, Yanyan Zhang, Tae-Yeon Seong, and Sungho Jin. Plasmonic au nanoparticles on 8nm tio$_2$ nanotubes for enhanced photocatalytic water splitting. *Journal of Renewable and Sustainable Energy*, 5(5): 053104, 2013. doi: 10.1063/1.4821177. URL https://doi.org/10.1063/1.4821177.

[1498] Di Zhang, Wallace C. H. Choy, Fengxian Xie, Wei E. I. Sha, Xinchen Li, Baofu Ding, Kai Zhang, Fei Huang, and Yong Cao. Plasmonic electrically functionalized tio$_2$ for high-performance organic solar cells. *Advanced Functional Materials*, 23(34):4255–4261, 2013. ISSN 1616-3028. doi: 10.1002/adfm.201203776. URL http://dx.doi.org/10.1002/adfm.201203776.

[1499] Xu Shi, Kosei Ueno, Naoki Takabayashi, and Hiroaki Misawa. Plasmon-enhanced photocurrent generation and water oxidation with a gold nanoisland-loaded titanium dioxide photoelectrode. *The Journal of Physical Chemistry C*, 117(6):2494–2499, 2013. doi: 10.1021/ jp3064086. URL http://dx.doi.org/10.1021/jp3064086.

[1500] Syed Mubeen, Joun Lee, Nirala Singh, Stephan Krämer, Galen D. Stucky, and Martin Moskovits. An autonomous photosynthetic device in which all charge carriers derive from surface plasmons. *Nature Nanotechnology*, 8:247–251, 2013.

[1501] Anitha Devadoss, Asako Kuragano, Chiaki Terashima, P. Sudhagar, Kazuya Nakata, Takeshi Kondo, Makoto Yuasa, and Akira Fujishima. Single-step electrospun tio$_2$-au hybrid electrodes for high selectivity photoelectrocatalytic glutathione bioanalysis. *J. Mater. Chem. B*, 4:220–228, 2016. doi: 10.1039/C5TB01740H. URL http://dx.doi.org/10.1039/C5TB01740H.

[1502] Yang Tian and Tetsu Tatsuma. Mechanisms and applications of plasmon-induced charge separation at tio$_2$ films loaded with gold nanoparticles. *Journal of the American Chemical Society*, 127(20):7632–7637, 2005. doi: 10.1021/ja042192u. URL http://dx.doi.org/10. 1021/ja042192u. PMID: 15898815.

[1503] Meidan Ye, Jiaojiao Gong, Yuekun Lai, Changjian Lin, and Zhiqun Lin. High-efficiency photoelectrocatalytic hydrogen generation enabled by palladium quantum dots-sensitized tio$_2$ nanotube arrays. *Journal of the American Chemical Society*, 134(38):15720–15723, 2012. doi: 10.1021/ja307449z. URL http://dx.doi.org/10.1021/ja307449z. PMID: 22963520.

[1504] JJ Gong, YK Lai, and CJ Lin. Palladium quantum dots sensitized tio$_2$ nanotube arrays for highly efficient photoelectrocatalytic hydrogen generation. *NSTI-Nanotech*, 2011.

[1505] Ronen Gottesman, Shay Tirosh, Hannah-Noa Barad, and Arie Zaban. Direct imaging of the recombination/reduction sites in porous tio$_2$ electrodes. *The Journal of Physical Chemistry Letters*, 4(17):2822–2828, 2013. doi: 10.1021/jz401549e. URL http://dx.doi.org/10. 1021/jz401549e. PMID: 26706647.

[1506] M Li, FY Shao, QQ Ban, and JW Yang. Cds quantum dots sensitized single-crystalline tio$_2$ nanowire array films: Photovoltaic performance and photoelectron transport properties. *Journal of Ovonic Research*, 9(6):157–165, 2013.

[1507] Juncao Bian, Chao Huang, Lingyun Wang, TakFu Hung, Walid A. Daoud, and Ruiqin Zhang. Carbon dot loading and tio$_2$ nanorod length dependence of photoelectrochemical properties in carbon dot/tio$_2$ nanorod array nanocomposites. *ACS Applied Materials & Interfaces*, 6(7):4883–4890, 2014. doi: 10.1021/am4059183. URL http://dx.doi.org/10.1021/am4059183. PMID: 24601482.

[1508] Shelia N. Baker and Gary A. Baker. Luminescent carbon nanodots: Emergent nanolights. *Angewandte Chemie International Edition*, 49(38):6726–6744, 2010. ISSN 1521-3773. doi: 10.1002/anie.200906623. URL http://dx.doi.org/10.1002/anie.200906623.

[1509] Ya-Ping Sun, Bing Zhou, Yi Lin, Wei Wang, K. A. Shiral Fernando, Pankaj Pathak, Mohammed Jaouad Meziani, Barbara A. Harruff, Xin Wang, Haifang Wang, Pengju G. Luo, Hua Yang, Muhammet Erkan Kose, Bailin Chen, L. Monica Veca, and Su-Yuan Xie. Quantum-sized carbon dots for bright and colorful photoluminescence. *Journal of the American Chemical Society*, 128(24):7756–7757, 2006. doi: 10.1021/ja062677d. URL http://dx.doi.org/10.1021/ja062677d. PMID: 16771487.

[1510] Vanthan Nguyen, Jinhai Si, Lihe Yan, and Xun Hou. Electron-hole recombination dynamics in carbon nanodots. *Carbon*, 95(Supplement C):659 – 663, 2015. ISSN 0008-6223. doi: https://doi.org/10.1016/j.carbon.2015.08.066. URL http://www.sciencedirect.com/science/ article/pii/S0008622315301834.

[1511] Vanthan Nguyen, Lihe Yan, Jinhai Si, and Xun Hou. Femtosecond laser-induced size reduction of carbon nanodots in solution: Effect of laser fluence, spot size, and irradiation time. *Journal of Applied Physics*, 117(8):084304, 2015. doi: 10.1063/1.4909506. URL https://doi.org/10.1063/1.4909506.

[1512] Hui Liu, Nan Gao, Meiyong Liao, and Xiaosheng Fang. Hexagonal-like nb$_2$o$_5$ nanoplates-based photodetectors and photocatalyst with high performances. *Scientific Reports*, 5(7716), 2014. doi: doi:10.1038/srep07716.

[1513] J.D. Perkins, C.W. Teplin, M.F.A.M van Hest, J.L. Alleman, X. Li, M.S. Dabney, B.M. Keyes, L.M. Gedvilas, D.S. Ginley, Y. Lin, and Y. Lu. Optical analysis of thin film combinatorial libraries. *Applied Surface Science*, 223(1):124 – 132, 2004. ISSN 0169-4332. doi: https://doi.org/10.1016/S0169-4332(03)00917-6. URL http://www.sciencedirect.com/science/article/pii/S0169433203009176. Proceedings of the Second Japan-US Workshop on Combinatorial Materials Science and Technology.

[1514] Masahide Takahashi, Kaori Tsukigi, Takashi Uchino, and Toshinobu Yoko. Enhanced photocurrent in thin film tio$_2$ electrodes prepared by sol-gel method. *Thin Solid Films*, 388(1–2):231 – 236, 2001. ISSN 0040-6090. doi: https://doi.org/10.1016/S0040-6090(01)00811-2. URL http://www.sciencedirect.com/science/article/pii/S0040609001008112.

[1515] Jean M. Bennett, Emile Pelletier, G. Albrand, J. P. Borgogno, B. Lazarides, Charles K. Carniglia, R. A. Schmell, Thomas H. Allen, Trudy Tuttle-Hart, Karl H. Guenther, and Andreas Saxer. Comparison of the properties of titanium dioxide films prepared by various techniques. *Appl. Opt.*, 28(16):3303–3317, Aug 1989. doi: 10.1364/AO.28.003303. URL http://ao.osa.org/abstract.cfm?URI= ao-28-16-3303.

[1516] Rand Dannenberg and Phil Greene. Reactive sputter deposition of titanium dioxide. *Thin Solid Films*, 360(1–2):122 – 127, 2000. ISSN 0040-6090. doi: http://dx.doi.org/10.1016/S0040-6090(99)00938-4. URL http://www.sciencedirect.com/science/article/ pii/S0040609099009384.

[1517] B. Karunagaran, R.T. Rajendra Kumar, D. Mangalaraj, Sa. K. Narayandass, and G. Mohan Rao. Influence of thermal annealing on the composition and structural parameters of dc magnetron sputtered titanium dioxide thin films. *Crystal Research and Technology*, 37 (12):1285–1292, 2002. ISSN 1521-4079. doi: 10.1002/crat.200290004. URL http://dx.doi.org/10.1002/crat.200290004.

References

[1518] B. Karunagaran, R. T. Rajendra Kumar, C. Viswanathan, D. Mangalaraj, Sa. K. Narayandass, and G. Mohan Rao. Optical constants of dc magnetron sputtered titanium dioxide thin films measured by spectroscopic ellipsometry. *Crystal Research and Technology*, 38(9): 773–778, 2003. ISSN 1521-4079. doi: 10.1002/crat.200310094. URL http://dx.doi.org/10.1002/crat.200310094.

[1519] B. Karunagaran, S.J. Chung, E.-K. Suh, and D. Mangalaraj. Dielectric and transport properties of magnetron sputtered titanium dioxide thin films. *Physica B: Condensed Matter*, 369(1–4):129 – 134, 2005. ISSN 0921-4526. doi: http://dx.doi.org/10.1016/j.physb.2005.08.006. URL http://www.sciencedirect.com/science/article/pii/S0921452605009336.

[1520] B. Karunagaran, Kyunghae Kim, D. Mangalaraj, Junsin Yi, and S. Velumani. Structural, optical and raman scattering studies on dc magnetron sputtered titanium dioxide thin films. *Solar Energy Materials and Solar Cells*, 88(2):199 – 208, 2005. ISSN 0927-0248. doi: http://dx.doi.org/10.1016/j.solmat.2004.03.008. URL http://www.sciencedirect.com/science/article/pii/S0927024804004763. International Symposium on Solar-Hydrogen-Fuel CellInternational Symposium on Solar-Hydrogen-Fuel Cell.

[1521] A.A. Akl, H. Kamal, and K. Abdel-Hady. Fabrication and characterization of sputtered titanium dioxide films. *Applied Surface Science*, 252(24):8651 – 8656, 2006. ISSN 0169-4332. doi: http://dx.doi.org/10.1016/j.apsusc.2005.12.001. URL http://www.sciencedirect.com/science/article/pii/S0169433205016491.

[1522] H. Sankur and J. T. Cheung. Formation of dielectric and semiconductor thin films by laser-assisted evaporation. *Applied Physics A*, 47(3):271–284, 1988. ISSN 1432-0630. doi: 10.1007/BF00615933. URL http://dx.doi.org/10.1007/BF00615933.

[1523] C. M. Dai, C. S. Su, and D. S. Chuu. Composition and chemical reactions of titanium oxide films deposited by laser evaporation. *Journal of Applied Physics*, 69(6):3766–3768, 1991. doi: http://dx.doi.org/10.1063/1.348473. URL http://scitation.aip.org/content/aip/journal/jap/69/6/10.1063/1.348473.

[1524] N. Lobstein, E. Millon, A. Hachimi, J.F. Muller, M. Alnot, and J.J. Ehrhardt. Deposition by laser ablation and characterization of titanium dioxide films on polyethylene-terephthalate. *Applied Surface Science*, 89(3):307 – 321, 1995. ISSN 0169-4332. doi: http://dx.doi.org/10.1016/0169-4332(95)00046-1. URL http://www.sciencedirect.com/science/article/pii/0169433295000461.

[1525] Takahiro Nakamura, Tetsu Ichitsubo, Eiichiro Matsubara, Atsushi Muramatsu, Nobuaki Sato, and Hideyuki Takahashi. Preferential formation of anatase in laser-ablated titanium dioxide films. *Acta Materialia*, 53(2):323 – 329, 2005. ISSN 1359-6454. doi: http://dx.doi.org/10.1016/j.actamat.2004.09.026. URL http://www.sciencedirect.com/science/article/pii/S1359645404005695.

[1526] Meng-Hsiu Tsai, Shuei-Yuan Chen, and Pouyan Shen. Laser ablation condensation of particles: Effects of laser energy, oxygen flow rate and phase transformation. *Journal of Aerosol Science*, 36(1):13 – 25, 2005. ISSN 0021-8502. doi: http://dx.doi.org/10.1016/j.jaerosci.2004.08.007. URL http://www.sciencedirect.com/science/article/pii/S0021850204003271.

[1527] P. Ayyub, R. Chandra, P. Taneja, A.K. Sharma, and R. Pinto. Synthesis of nanocrystalline material by sputtering and laser ablation at low temperatures. *Applied Physics A*, 73(1):67–73, 2001. ISSN 1432-0630. doi: 10.1007/s003390100833. URL http://dx.doi.org/10.1007/s003390100833.

[1528] C.H. Liang, Y. Shimizu, T. Sasaki, and N. Koshizaki. Preparation of ultrafine tio2 nanocrystals via pulsed-laser ablation of titanium metal in surfactant solution. *Applied Physics A*, 80(4):819–822, 2005. ISSN 1432-0630. doi: 10.1007/s00339-003-2489-6. URL http://dx.doi.org/10.1007/s00339-003-2489-6.

[1529] S C Singh, R K Swarnkar, and R Gopal. Synthesis of titanium dioxide nanomaterial by pulsed laser ablation in water. *Journal of Nanoscience and Nanotechnology*, 9(9):5367–5371, 2009.

[1530] Edward B. Graper. Evaporation characteristics of materials from an electron-beam gun. *Journal of Vacuum Science & Technology*, 8(1):333–337, 1971. doi: http://dx.doi.org/10.1116/1.1316331. URL http://scitation.aip.org/content/avs/journal/jvst/8/1/10.1116/1.1316331.

[1531] K. Balasubramanian, X. F. Han, and K. H. Guenther. Comparative study of titanium dioxide thin films produced by electron-beam evaporation and by reactive low-voltage ion plating. *Appl. Opt.*, 32(28):5594–5600, Oct 1993. doi: 10.1364/AO.32.005594. URL http://ao.osa.org/abstract.cfm?URI=ao-32-28-5594.

[1532] R. J. Nemanich, C. C. Tsai, and G. A. N. Connell. Interference-enhanced raman scattering of very thin titanium and titanium oxide films. *Phys. Rev. Lett.*, 44:273–276, Jan 1980. doi: 10.1103/PhysRevLett.44.273. URL http://link.aps.org/doi/10.1103/PhysRevLett.44.273.

[1533] S. Y. Kim. Simultaneous determination of refractive index, extinction coefficient, and void distribution of titanium dioxide thin film by optical methods. *Appl. Opt.*, 35(34):6703–6707, Dec 1996. doi: 10.1364/AO.35.006703. URL http://ao.osa.org/abstract.cfm?URI=ao-35-34-6703.

[1534] Yiming Zhang, Julian R. G. Evans, and Shoufeng Yang. Corrected values for boiling points and enthalpies of vaporization of elements in handbooks. *Journal of Chemical & Engineering Data*, 56(2):328–337, 2011. doi: 10.1021/je1011086. URL http://dx.doi.org/10.1021/je1011086.

[1535] M.C. Bartelt and J.W. Evans. Nucleation and growth of square islands during deposition: Sizes, coalescence, separations and correlations. *Surface Science*, 298(2):421 – 431, 1993. ISSN 0039-6028. doi: http://dx.doi.org/10.1016/0039-6028(93)90057-Q. URL http://www.sciencedirect.com/science/article/pii/003960289390057Q.

[1536] H. K. Jang, S. W. Whangbo, H. B. Kim, K. Y. Im, Y. S. Lee, I. W. Lyo, C. N. Whang, G. Kim, H.-S. Lee, and J. M. Lee. Titanium oxide films on si(100) deposited by electron-beam evaporation at 250°c. *Journal of Vacuum Science & Technology A*, 18(3):917–921, 2000. doi: http://dx.doi.org/10.1116/1.582275. URL http://scitation.aip.org/content/avs/journal/jvsta/18/3/10.1116/1.582275.

[1537] S. Zhang, Y.F. Zhu, and D.E. Brodie. Photoconducting tio2 prepared by spray pyrolysis using ticl4. *Thin Solid Films*, 213(2):265 – 270, 1992. ISSN 0040-6090. doi: http://dx.doi.org/10.1016/0040-6090(92)90292-J. URL http://www.sciencedirect.com/science/article/pii/004060909290292J.

[1538] P. Murugavel, M. Kalaiselvam, A. R. Raju, and C. N. R. Rao. Sub-micrometre spherical particles of tio2, zro2 and pzt by nebulized spray pyrolysis of metal-organic precursors. *Journal of Materials Chemistry*, 7:1433–1438, 1997. doi: 10.1039/A700301C. URL http://dx.doi.org/10.1039/A700301C.

[1539] M.A Rashti and D.E Brodie. The photoresponse of high resistance anatase tio2 films prepared by the decomposition of titanium isopropoxide. *Thin Solid Films*, 240(1):163 – 167, 1994. ISSN 0040-6090. doi: http://dx.doi.org/10.1016/0040-6090(94)90715-3. URL http://www.sciencedirect.com/science/article/pii/0040609094907153.

[1540] Makoto Yoshida and Paras N. Prasad. Sol-gel-processed sio2/tio2/poly(vinylpyrrolidone) composite materials for optical waveguides. *Chemistry of Materials*, 8(1):235–241, 1996. doi: 10.1021/cm950331o. URL http://dx.doi.org/10.1021/cm950331o.

[1541] Yongxiang Li, Jürgen Hagen, Winfried Schaffrath, Peter Otschik, and Dietrich Haarer. Titanium dioxide films for photovoltaic cells derived from a sol–gel process. *Solar Energy Materials and Solar Cells*, 56(2):167 – 174, 1999. ISSN 0927-0248. doi: http://dx.doi.org/10.1016/S0927-0248(98)00157-3. URL http://www.sciencedirect.com/science/article/pii/S0927024898001573.

[1542] Yasutaka Takahashi, Ayako Ohsugi, Takeshi Arafuka, Tomokazu Ohya, Takayuki Ban, and Yutaka Ohya. Development of new modifiers for titanium alkoxide-based sol-gel process. *Journal of Sol-Gel Science and Technology*, 17(3):227–238, 2000. ISSN 1573-4846. doi: 10.1023/A:1008716122654. URL http://dx.doi.org/10.1023/A:1008716122654.

[1543] Mikko Ritala, Markku Leskela, Lauri Niinisto, and Pekka Haussalo. Titanium isopropoxide as a precursor in atomic layer epitaxy of titanium dioxide thin films. *Chemistry of Materials*, 5(8):1174–1181, 1993. doi: 10.1021/cm00032a023. URL http://dx.doi.org/10.1021/cm00032a023.

[1544] Mikko Ritala, Markku Leskelä, Erja Nykänen, Pekka Soininen, and Lauri Niinistö. Growth of titanium dioxide thin films by atomic layer epitaxy. *Thin Solid Films*, 225(1):288 – 295, 1993. ISSN 0040-6090. doi: http://dx.doi.org/10.1016/0040-6090(93)90172-L. URL http://www.sciencedirect.com/science/article/pii/004060909390172L.

[1545] Michael C. Langston, Neil P. Dasgupta, Hee Joon Jung, Manca Logar, Yu Huang, Robert Sinclair, and Fritz B. Prinz. In situ cycle-by-cycle flash annealing of atomic layer deposited materials. *The Journal of Physical Chemistry C*, 116(45):24177–24183, 2012. doi: 10.1021/jp308895e. URL http://dx.doi.org/10.1021/jp308895e.

[1546] Markku Leskelä and Mikko Ritala. Atomic layer deposition (ald): from precursors to thin film structures. *Thin Solid Films*, 409(1): 138 – 146, 2002. ISSN 0040-6090. doi: http://dx.doi.org/10.1016/S0040-6090(02)00117-7. URL http://www.sciencedirect.com/science/article/pii/S0040609002001177. Proceedings of the 2nd Asian Conference on Chemical Vapour Deposition.

[1547] Steven M. George. Atomic layer deposition: An overview. *Chem. Rev.*, 110(1):111–131, 2010. doi: 10.1021/cr900056b.

[1548] Kaupo Kukli, Aleks Aidla, Jaan Aarik, Mikael Schuisky, Anders Hårsta, Mikko Ritala, and Markku Leskelä. Real-time monitoring in atomic layer deposition of Tio2 from Tii4 and H2O-H2O2. *Langmuir*, 16(21):8122–8128, 2000. doi: 10.1021/la0004451. URL http://dx.doi.org/10.1021/la0004451.

[1549] Samuel Chen, M. G. Mason, H. J. Gysling, G. R. Paz-Pujalt, T. N. Blanton, T. Castro, K. M. Chen, C. P. Fictorie, W. L. Gladfelter, A. Franciosi, P. I. Cohen, and J. F. Evans. Ultrahigh vacuum metalorganic chemical vapor deposition growth and insitu characterization of epitaxial Tio2 films. *Journal of Vacuum Science & Technology A*, 11(5):2419–2429, 1993. doi: http://dx.doi.org/10.1116/1.578587. URL http://scitation.aip.org/content/avs/journal/jvsta/11/5/10.1116/1.578587.

[1550] Mikael Schuisky, Anders Hårsta, Aleks Aidla, Kaupo Kukli, Alma-Asta Kiisler, and Jaan Aarik. Atomic layer chemical vapor deposition of tio2 low temperature epitaxy of rutile and anatase. *Journal of The Electrochemical Society*, 147(9):3319–3325, 2000. doi: 10.1149/1.1393901. URL http://jes.ecsdl.org/content/147/9/3319.abstract.

[1551] Jaan Aarik, Aleks Aidla, Teet Uustare, Mikko Ritala, and Markku Leskelä. Titanium isopropoxide as a precursor for atomic layer deposition: characterization of titanium dioxide growth process. *Applied Surface Science*, 161(3-4):385–395, 2000. ISSN 0169-4332. doi: http://dx.doi.org/10.1016/S0169-4332(00)00274-9. URL http://www.sciencedirect.com/science/article/pii/S0169433200002749.

[1552] E.-L. Lakomaa, S. Haukka, and T. Suntola. Atomic layer growth of tio2 on silica. *Applied Surface Science*, 60:742 – 748, 1992. ISSN 0169-4332. doi: http://dx.doi.org/10.1016/0169-4332(92)90506-S. URL http://www.sciencedirect.com/science/article/pii/0169433292905068.

[1553] Olivera B. Milošević, Mirjana K. Mirković, and Dragan P. Uskoković. Characteristics and formation mechanism of batio2 powders prepared by twin-fluid and ultrasonic spray-pyrolysis methods. *Journal of the American Ceramic Society*, 79(6):1720–1722, 1996. ISSN 1551-2916. doi: 10.1111/j.1151-2916.1996.tb08794.x. URL http://dx.doi.org/10.1111/j.1151-2916.1996.tb08794.x.

[1554] Najme Sarvari and Mohammad Reza Mohammadi. Enhanced electron collection efficiency of nanostructured dye-sensitized solar cells by incorporating tio2 cubes. *Journal of the American Ceramic Society*, pages n/a–n/a, 2017. ISSN 1551-2916. doi: 10.1111/jace.15184. URL http://dx.doi.org/10.1111/jace.15184.

[1555] Ju-Hwan Shin, Ji-Hwan Kang, Woo-Min Jin, Jong Hyeok Park, Young-Sang Cho, and Jun Hyuk Moon. Facile synthesis of tio2 inverse opal electrodes for dye-sensitized solar cells. *Langmuir*, 27(2):856–860, 2011. doi: 10.1021/la104512c. URL http://dx.doi.org/10.1021/la104512c. PMID: 21155579.

[1556] Changzheng Wu, Lanyu Lei, Xi Zhu, Jinlong Yang, and Yi Xie. Large-scale synthesis of titanate and anatase tubular hierarchities. *Small*, 3(9):1518–1522, 2007. ISSN 1613-6829. doi: 10.1002/smll.200700179. URL http://dx.doi.org/10.1002/smll.200700179.

[1557] T. Sugiura, T. Yoshida, and H. Minoura. Designing a tio2 nano-honeycomb structure using photoelectrochemical etching. *Electrochemical and Solid-State Letters*, 1(4):175–177, 1998. doi: 10.1149/1.1390676. URL http://esl.ecsdl.org/content/1/4/175.abstract.

[1558] Kevin P. Musselman, Gregory J. Mulholland, Adam P. Robinson, Lukas Schmidt-Mende, and Judith L. MacManus-Driscoll. Low-temperature synthesis of large-area, free-standing nanorod arrays on ito/glass and other conducting substrates. *Advanced Materials*, 20(23):4470–4475, 2008. ISSN 1521-4095. doi: 10.1002/adma.200801253. URL http://dx.doi.org/10.1002/adma.200801253.

[1559] Han Gil Na, Dong Sub Kwak, Yong Jung Kwon, Hong Yeon Cho, Chongmu Lee, and Hyoun Woo Kim. Tio2/sio2 core-shell nanowires generated by heating the multilayered substrates. *Metals and Materials International*, 19(4):861–867, 2013. ISSN 2005-4149. doi: 10.1007/s12540-013-4030-6. URL http://dx.doi.org/10.1007/s12540-013-4030-6.

[1560] S. M. Liu, L. M. Gan, L. H. Liu, W. D. Zhang, and H. C. Zeng. Synthesis of single-crystalline tio2 nanotubes. *Chemistry of Materials*, 14(3):1391–1397, 2002. doi: 10.1021/cm0115057. URL http://dx.doi.org/10.1021/cm0115057.

[1561] Jonas Weickert, Claudia Palumbiny, Mihaela Nedelcu, Thomas Bein, and Lukas Schmidt-Mende. Controlled growth of tio2 nanotubes on conducting glass. *Chemistry of Materials*, 23(2):155–162, 2011. doi: 10.1021/cm102389m. URL http://dx.doi.org/10.1021/cm102389m.

[1562] Gopal K. Mor, Oomman K. Varghese, Maggie Paulose, Karthik Shankar, and Craig A. Grimes. A review on highly ordered, vertically oriented tio2 nanotube arrays: Fabrication, material properties, and solar energy applications. *Solar Energy Materials and Solar Cells*, 90(14):2011 – 2075, 2006. ISSN 0927-0248. doi: https://doi.org/10.1016/j.solmat.2006.04.007. URL http://www.sciencedirect.com/science/article/pii/S0927024806001693.

[1563] Il-Doo Kim, Avner Rothschild, Byong Hong Lee, Dong Young Kim, Seong Mu Jo, , and Harry L. Tuller. Ultrasensitive chemiresistors based on electrospun tio2 nanofibers. *Nano Letters*, 6(9):2009–2013, 2006. doi: 10.1021/nl061197h. URL http://dx.doi.org/10.1021/nl061197h. PMID: 16968017.

[1564] Surawut Chuangchote, Takashi Sagawa, and Susumu Yoshikawa. Efficient dye-sensitized solar cells using electrospun tio2 nanofibers as a light harvesting layer. *Applied Physics Letters*, 93(3):033310, 2008. doi: http://dx.doi.org/10.1063/1.2958347. URL http://scitation.aip.org/content/aip/journal/apl/93/3/10.1063/1.2958347.

References

[1565] Haidar Abdul Razaq Abdul Hussian, Marwa Abdul Muhsien Hassan, and Ibrahim R. Agool. Synthesis of titanium dioxide (tio_2) nanofiber and nanotube using different chemical method. *Optik - International Journal for Light and Electron Optics*, 127(5):2996 – 2999, 2016. ISSN 0030-4026. doi: http://dx.doi.org/10.1016/j.ijleo.2015.12.012. URL http://www.sciencedirect.com/science/article/pii/S003040261501921X.

[1566] D Reyes-Coronado, G Rodríguez-Gattorno, M E Espinosa-Pesqueira, C Cab, R de Coss, and G Oskam. Phase-pure tio_2 nanoparticles: anatase, brookite and rutile. *Nanotechnology*, 19(14):145605, 2008. URL http://stacks.iop.org/0957-4484/19/i=14/a=145605.

[1567] Claudius Kormann, Detlef W. Bahnemann, and Michael R. Hoffmann. Preparation and characterization of quantum-size titanium dioxide. *The Journal of Physical Chemistry*, 92(18):5196–5201, 1988. doi: 10.1021/j100329a027. URL http://dx.doi.org/10.1021/j100329a027.

[1568] PL Chen, CT Kuo, FM Pan, and TG Tsai. Preparation and phase transformation of highly ordered tio_2 nanodot arrays on sapphire substrates. *WOS:000222121010056*, 2004. URL http://hdl.handle.net/11536/26780?locale=en.

[1569] Jiwon Lee, Dai Hong Kim, Seong-Hyeon Hong, and Jae Young Jho. A hydrogen gas sensor employing vertically aligned tio_2 nanotube arrays prepared by template-assisted method. *Sensors and Actuators B: Chemical*, 160(1):1494 – 1498, 2011. ISSN 0925-4005. doi: http://dx.doi.org/10.1016/j.snb.2011.08.001. URL http://www.sciencedirect.com/science/article/pii/S0925400511007234.

[1570] Jian Shi and Xudong Wang. Growth of rutile titanium dioxide nanowires by pulsed chemical vapor deposition. *Crystal Growth & Design*, 11(4):949–954, 2011. doi: 10.1021/cg200140k. URL http://dx.doi.org/10.1021/cg200140k.

[1571] C A Chen, Y M Chen, A Korotcov, Y S Huang, D S Tsai, and K K Tiong. Growth and characterization of well-aligned densely-packed rutile tio_2 nanocrystals on sapphire substrates via metal–organic chemical vapor deposition. *Nanotechnology*, 19(7):075611, 2008. URL http://stacks.iop.org/0957-4484/19/i=7/a=075611.

[1572] Maggie Paulose, Karthik Shankar, Sorachon Yoriya, Haripriya E. Prakasam, Oomman K. Varghese, Gopal K. Mor, Thomas A. Latempa, Adriana Fitzgerald, , and Craig A. Grimes. Anodic growth of highly ordered tio_2 nanotube arrays to 134 μm in length. *The Journal of Physical Chemistry B*, 110(33):16179–16184, 2006. doi: 10.1021/jp064020k. URL http://dx.doi.org/10.1021/jp064020k. PMID: 16913737.

[1573] Mi Yeon Song, Do Kyun Kim, Kyo Jin Ihn, Seong Mu Jo, and Dong Young Kim. Electrospun tio_2 electrodes for dye-sensitized solar cells. *Nanotechnology*, 15(12):1861, 2004. URL http://stacks.iop.org/0957-4484/15/i=12/a=030.

[1574] Masoud Iraj, Fatemeh Dehghan Nayeri, Ebrahim Asl-Soleimani, and Keyvan Narimani. Controlled growth of vertically aligned tio_2 nanorod arrays using the improved hydrothermal method and their application to dye-sensitized solar cells. *Journal of Alloys and Compounds*, 659:44 – 50, 2016. ISSN 0925-8388. doi: http://dx.doi.org/10.1016/j.jallcom.2015.11.004. URL http://www.sciencedirect.com/science/article/pii/S0925838815315607.

[1575] Pavel A. Sedach, Terry J. Gordon, Sayed Y. Sayed, Tobias Furstenhaupt, Ruohong Sui, Thomas Baumgartner, and Curtis P. Berlinguette. Solution growth of anatase tio_2 nanowires from transparent conducting glass substrates. *J. Mater. Chem.*, 20:5063–5069, 2010. doi: 10.1039/C0JM00266F. URL http://dx.doi.org/10.1039/C0JM00266F.

[1576] Xiao Yu, Hai Wang, Yong Liu, Xiang Zhou, Baojun Li, Ling Xin, Yu Zhou, and Hui Shen. One-step ammonia hydrothermal synthesis of single crystal anatase tio_2 nanowires for highly efficient dye-sensitized solar cells. *J. Mater. Chem. A*, 1:2110–2117, 2013. doi: 10.1039/C2TA00494A. URL http://dx.doi.org/10.1039/C2TA00494A.

[1577] Zheng ji Zhou, Jun qi Fan, Xia Wang, Wen hui Zhou, Zu liang Du, and Si xin Wu. Effect of highly ordered single-crystalline tio_2 nanowire length on the photovoltaic performance of dye-sensitized solar cells. *ACS Applied Materials & Interfaces*, 3(11):4349–4353, 2011. doi: 10.1021/am201001t. URL http://dx.doi.org/10.1021/am201001t. PMID: 21966998.

[1578] G. Wang and G. Li. Titania from nanoclusters to nanowires and nanoforks. *The European Physical Journal D - Atomic, Molecular, Optical and Plasma Physics*, 24(1):355–360, 2003. ISSN 1434-6060. doi: 10.1140/epjd/e2003-00172-y. URL http://dx.doi.org/10.1140/epjd/e2003-00172-y.

[1579] Haimei Zheng, Rachel K. Smith, Young-wook Jun, Christian Kisielowski, Ulrich Dahmen, and A. Paul Alivisatos. Observation of single colloidal platinum nanocrystal growth trajectories. *Science*, 324(5932):1309–1312, 2009. doi: 10.1126/science.1172104. URL http://www.sciencemag.org/content/324/5932/1309.abstract.

[1580] Yuanyuan Li, Jinping Liu, and Zhijie Jia. Morphological control and photodegradation behavior of rutile tio_2 prepared by a low-temperature process. *Materials Letters*, 60(13–14):1753 – 1757, 2006. ISSN 0167-577X. doi: http://dx.doi.org/10.1016/j.matlet.2005.12.012. URL http://www.sciencedirect.com/science/article/pii/S0167577X05012565.

[1581] Rachel V. Zucker, Dominique Chatain, Ulrich Dahmen, Serge Hagège, and W. Craig Carter. New software tools for the calculation and display of isolated and attached interfacial-energy minimizing particle shapes. *Journal of Materials Science*, 47(24):8290–8302, 2012. ISSN 0022-2461. doi: 10.1007/s10853-012-6739-x. URL http://dx.doi.org/10.1007/s10853-012-6739-x.

[1582] Victor Goldschmidt. *Atlas der Krystallformen*. Carl Winters Universitätsbuchhandlung: Heidelberg, 1913.

[1583] Keita Kakiuchi, Eiji Hosono, Hiroaki Imai, Toshio Kimura, and Shinobu Fujihara. 111-faceting of low-temperature processed rutile tio_2 rods. *Journal of Crystal Growth*, 293(2):541 – 545, 2006. ISSN 0022-0248. doi: 10.1016/j.jcrysgro.2006.06.004. URL http://www.sciencedirect.com/science/article/pii/S0022024806005719.

[1584] Xiangping Huang and Chunxu Pan. Large-scale synthesis of single-crystalline rutile tio_2 nanorods via a one-step solution route. *Journal of Crystal Growth*, 306(1):117 – 122, 2007. ISSN 0022-0248. doi: 10.1016/j.jcrysgro.2007.04.018. URL http://www.sciencedirect.com/science/article/pii/S0022024807004113.

[1585] Naoya Murakami, Asami Ono, Misa Nakamura, Toshiki Tsubota, and Teruhisa Ohno. Development of a visible-light-responsive rutile rod by site-selective modification of iron(iii) ion on 111 exposed crystal faces. *Applied Catalysis B: Environmental*, 97(1–2):115 – 119, 2010. ISSN 0926-3373. doi: 10.1016/j.apcatb.2010.03.030. URL http://www.sciencedirect.com/science/article/pii/S0926337310001463.

[1586] Peter M. Oliver, Graeme W. Watson, E. Toby Kelsey, and Stephen C. Parker. Atomistic simulation of the surface structure of the Tio_2 polymorphs rutileand anatase. *Journal of Materials Chemistry*, 7:563–568, 1997. doi: 10.1039/A606353E. URL http://dx.doi.org/10.1039/A606353E.

[1587] Frédéric Labat, Philippe Baranek, Christophe Domain, Christian Minot, and Carlo Adamo. Density functional theory analysis of the structural and electronic properties of tio_2 rutile and anatase polytypes: Performances of different exchange-correlation functionals. *The Journal of Chemical Physics*, 126(15):154703, 2007. doi: 10.1063/1.2717168. URL http://scitation.aip.org/content/aip/journal/jcp/126/15/10.1063/1.2717168.

[1588] Trevor P. Hardcastle, Che R. Seabourne, Rik M. D. Brydson, Ken J. T. Livi, and Andrew J. Scott. Energy of step defects on the tio_2 rutile (110) surface: An ab initio dft methodology. *The Journal of Physical Chemistry C*, 117(45):23766–23780, 2013. doi: 10.1021/jp4078135. URL http://dx.doi.org/10.1021/jp4078135.

[1589] H. Perron, C. Domain, J. Roques, R. Drot, E. Simoni, and H. Catalette. Optimisation of accurate rutile tio_2 (110), (100), (101) and (001) surface models from periodic dft calculations. *Theoretical Chemistry Accounts*, 117(4):565–574, 2007. ISSN 1432-881X. doi: 10.1007/s00214-006-0189-y. URL http://dx.doi.org/10.1007/s00214-006-0189-y.

[1590] Benjamin J. Morgan and Graeme W. Watson. A density functional theory + u study of oxygen vacancy formation at the (110), (100), (101), and (001) surfaces of rutile tio_2. *The Journal of Physical Chemistry C*, 113(17):7322–7328, 2009. doi: 10.1021/jp811288n. URL http://dx.doi.org/10.1021/jp811288n.

[1591] A Kiejna, T Pabisiak, and S W Gao. The energetics and structure of rutile tio_2(110). *Journal of Physics: Condensed Matter*, 18(17): 4207, 2006. URL http://stacks.iop.org/0953-8984/18/i=17/a=009.

[1592] Hua Gui Yang, Cheng Hua Sun, Shi Zhang Qiao, Jin Zou, Gang Liu, Sean Campbell Smith, Hui Ming Cheng, and Gao Qing Lu. Anatase tio_2 single crystals with a large percentage of reactive facets. *Nature*, 453(7195):638–641, 2008. doi: 10.1038/nature06964. URL http://www.nature.com/nature/journal/v453/n7195/suppinfo/nature06964_S1.html.

[1593] Ju Seong Kim, Seong Sik Shin, Hyun Soo Han, Sun Shin, Jae Ho Suk, Kisuk Kang, Kug Sun Hong, and In Sun Cho. Facile preparation of tio_2 nanobranch/nanoparticle hybrid architecture with enhanced light harvesting properties for dye-sensitized solar cells. *Journal of Nanomaterials*, 2015(139715), 2015.

[1594] Hai Wang, Yong Liu, Zhong Liu, Hongmei Xu, Youjun Deng, and Hui Shen. Hierarchical rutile tio_2 mesocrystals assembled by nanocrystals-oriented attachment mechanism. *CrystEngComm*, 14:2278–2282, 2012. doi: 10.1039/C2CE06314J. URL http://dx.doi.org/10.1039/C2CE06314J.

[1595] Xia Sheng, Dongqing He, Jie Yang, Kai Zhu, and Xinjian Feng. Oriented assembled tio_2 hierarchical nanowire arrays with fast electron transport properties. *Nano Letters*, 14(4):1848–1852, 2014. doi: 10.1021/nl4046262. URL http://dx.doi.org/10.1021/nl4046262. PMID: 24628675.

[1596] Vanja Jordan, Uroš Javornik, Janez Plavec, Aleš Podgornik, and Aleksander Rečnik. Self-assembly of multilevel branched rutile-type tio_2 structures via oriented lateral and twin attachment. *Nature Scientific Reports*, 6:24216, 2016. doi: 10.1038/srep24216. URL http://dx.doi.org/10.1038/srep24216.

[1597] Debabrata Sarkar, Chandan. K. Ghosh, and Kalyan K. Chattopadhyay. Morphology control of rutile tio_2 hierarchical architectures and their excellent field emission properties. *CrystEngComm*, 14:2683–2690, 2012. doi: 10.1039/C2CE06392A. URL http://dx.doi.org/10.1039/C2CE06392A.

[1598] Dongsheng Li, Frank Soberanis, Jia Fu, Wenting Hou, Jianzhong Wu, and David Kisailus. Growth mechanism of highly branched titanium dioxide nanowires via oriented attachment. *Crystal Growth & Design*, 13(2):422–428, 2013.

[1599] Yuan-Sheng Huang and Hong-Wei Liu. Growth morphologies of nanostructured rutile tio_2. *Journal of Materials Engineering and Performance*, 23(4):1240–1246, Apr 2014. ISSN 1544-1024. doi: 10.1007/s11665-014-0895-x. URL https://doi.org/10.1007/s11665-014-0895-x.

[1600] Ke Yang, Jianmin Zhu, Junjie Zhu, Shisong Huang, Xinhua Zhu, and Guobin Ma. Sonochemical synthesis and microstructure investigation of rod-like nanocrystalline rutile titania. *Materials Letters*, 57(30):4639 – 4642, 2003. ISSN 0167-577X. doi: https://doi.org/10.1016/S0167-577X(03)00376-8. URL http://www.sciencedirect.com/science/article/pii/S0167577X03003768.

[1601] Li Guang-Lai, Wang Guang-Hou, and Hong Jian-Ming. Morphologies of rutile form tio_2 twins crystals. *Journal of materials science letters*, 18(15):1243–1246, 1999.

[1602] Weigang Lu, Britain Bruner, Gilberto Casillas, Jibao He, Miguel Jose-Yacaman, and Patrick J. Farmer. Large scale synthesis of v-shaped rutile twinned nanorods. *CrystEngComm*, 14:3120–3124, 2012. doi: 10.1039/C2CE06564A. URL http://dx.doi.org/10.1039/C2CE06564A.

[1603] Weijia Zhou, Xiaoyan Liu, Jingjie Cui, Duo Liu, Jing Li, Huaidong Jiang, Jiyang Wang, and Hong Liu. Control synthesis of rutile tio_2 microspheres, nanoflowers, nanotrees and nanobelts via acid-hydrothermal method and their optical properties. *CrystEngComm*, 13: 4557–4563, 2011. doi: 10.1039/C1CE05186E. URL http://dx.doi.org/10.1039/C1CE05186E.

[1604] Makoto Kobayashi, Valery Petrykin, Masato Kakihana, and Koji Tomita. Hydrothermal synthesis and photocatalytic activity of whisker-like rutile-type titanium dioxide. *Journal of the American Ceramic Society*, 92:S21–S26, 2009. ISSN 1551-2916. doi: 10.1111/j. 1551-2916.2008.02641.x. URL http://dx.doi.org/10.1111/j.1551-2916.2008.02641.x.

[1605] Dong Kyu Roh, Won Seok Chi, Harim Jeon, Sang Jin Kim, and Jong Hak Kim. High efficiency solid-state dye-sensitized solar cells assembled with hierarchical anatase pine tree-like tio_2 nanotubes. *Advanced Functional Materials*, 24(3):379–386, 2014. ISSN 1616-3028. doi: 10.1002/adfm.201301562. URL http://dx.doi.org/10.1002/adfm.201301562.

[1606] Seare A. Berhe, Soumya Nag, Zachary Molinets, and W. Justin Youngblood. Influence of seeding and bath conditions in hydrothermal growth of very thin (20 nm) single-crystalline rutile tio_2 nanorod films. *ACS Applied Materials & Interfaces*, 5(4):1181–1185, 2013. doi: 10.1021/am302315q. URL http://dx.doi.org/10.1021/am302315q. PMID: 23387875.

[1607] Baoyuan Wang, Jingshu Wan, Qingyun Liu, Jun Zhang, and Hao Wang. Optimizing the prepared condition of tio_2 1d/3d network structure films to enhance the efficiency of dye-sensitized solar cells. *RSC Advances*, 5:82968–82976, 2015. doi: 10.1039/C5RA16458C. URL http://dx.doi.org/10.1039/C5RA16458C.

[1608] A. Wisnet, K. Bader, S. B. Betzler, M. Handloser, P. Ehrenreich, T. Pfadler, J. Weickert, A. Hartschuh, L. Schmidt-Mende, C. Scheu, and J. A. Dorman. Defeating loss mechanisms in 1d tio_2-based hybrid solar cells. *Advanced Functional Materials*, 25(17):2601–2608, 2015. ISSN 1616-3028. doi: 10.1002/adfm.201404010. URL http://dx.doi.org/10.1002/adfm.201404010.

[1609] Hailiang Li, Qingjiang Yu, Yuewu Huang, Cuiling Yu, Renzhi Li, Jinzhong Wang, Fengyun Guo, Shujie Jiao, Shiyong Gao, Yong Zhang, Xitian Zhang, Peng Wang, and Liancheng Zhao. Ultralong rutile tio_2 nanowire arrays for highly efficient dye-sensitized solar cells. *ACS Applied Materials & Interfaces*, 8(21):13384–13391, 2016. doi: 10.1021/acsami.6b01508. URL http://dx.doi.org/10.1021/acsami.6b01508. PMID: 27097727.

[1610] Serena A. Corr, Madeleine Grossman, Yifeng Shi, Kevin R. Heier, Galen D. Stucky, and Ram Seshadri. vo_2(b) nanorods: solvothermal preparation, electrical properties and conversion to rutilevo_2 and v_2o_3. *J. Mater. Chem.*, 19:4362–4367, 2009. doi: 10.1039/B900982E. URL http://dx.doi.org/10.1039/B900982E.

[1611] Latha Kumari, Wenzhi Li, and Dezhi Wang. Monoclinic zirconium oxide nanostructures synthesized by a hydrothermal route. *Nanotechnology*, 19(19):195602, 2008.

[1612] Li-Xia Yang, Ying-Jie Zhu, Hua Tong, Liang Li, and Ling Zhang. Multistep synthesis of cuo nanorod bundles and interconnected nanosheets using cu$_2$(oh)$_3$cl plates as precursor. *Materials Chemistry and Physics*, 112(2):442 – 447, 2008. ISSN 0254-0584. doi: http://dx.doi.org/10.1016/j.matchemphys.2008.05.071. URL http://www.sciencedirect.com/science/article/pii/S0254058408003611.

[1613] Xiaofei Yang, Chunyu Lu, Jieling Qin, Rongxian Zhang, Hua Tang, and Haojie Song. A facile one-step hydrothermal method to produce graphene-moo$_3$ nanorod bundle composites. *Materials Letters*, 65(15–16):2341 – 2344, 2011. ISSN 0167-577X. doi: http://dx.doi.org/10.1016/j.matlet.2011.05.019. URL http://www.sciencedirect.com/science/article/pii/S0167577X11005234.

[1614] Jun Yang, Cuikun Lin, and Zhenling Wangand Jun Lin. In(oh)$_3$ and in$_2$o$_3$ nanorod bundles and spheres: Microemulsion-mediated hydrothermal synthesis and luminescence properties. *Inorganic Chemistry*, 45(22):8973–8979, 2006. doi: 10.1021/ic060934+. URL http://dx.doi.org/10.1021/ic060934+.

[1615] Qifei Lu, Haibo Zeng, Zhenyang Wang, Xueli Cao, and Lide Zhang. Design of sb$_2$s$_3$ nanorod-bundles: imperfect oriented attachment. *Nanotechnology*, 17(9):2098, 2006. URL http://stacks.iop.org/0957-4484/17/i=9/a=004.

[1616] Jin Du, Liqiang Xu, Guifu Zou, Lanlan Chai, and Yitai Qian. Solvothermal synthesis of single crystalline znte nanorod bundles in a mixed solvent of ethylenediamine and hydrazine hydrate. *Journal of Crystal Growth*, 291(1):183 – 186, 2006. ISSN 0022-0248. doi: http://dx.doi.org/10.1016/j.jcrysgro.2006.02.040. URL http://www.sciencedirect.com/science/article/pii/S0022024806001886.

[1617] Lianshan Li, Nijuan Sun, Youyuan Huang, Yao Qin, Nana Zhao, Jining Gao, Meixian Li, Henghui Zhou, and Limin Qi. Topotactic transformation of single-crystalline precursor discs into disc-like bi$_2$s$_3$ nanorod networks. *Advanced Functional Materials*, 18(8):1194–1201, 2008. ISSN 1616-3028. doi: 10.1002/adfm.200701467. URL http://dx.doi.org/10.1002/adfm.200701467.

[1618] Arup Purkayastha, Qingyu Yan, Makala S. Raghuveer, Darshan D. Gandhi, Huafang Li, Zhong W. Liu, Raju V. Ramanujan, Theodorian Borca-Tasciuc, and Ganapathiraman Ramanath. Surfactant-directed synthesis of branched bismuth telluride/sulfide core/shell nanorods. *Advanced Materials*, 20(14):2679–2683, 2008. ISSN 1521-4095. doi: 10.1002/adma.200702572. URL http://dx.doi.org/10.1002/adma.200702572.

[1619] Lionel Vayssieres, Niclas Beermann, Sten-Eric Lindquist, and Anders Hagfeldt. Controlled aqueous chemical growth of oriented three-dimensional crystalline nanorod arrays: Application to iron(iii) oxides. *Chemistry of Materials*, 13(2):233–235, 2001. doi: 10.1021/cm001202x. URL http://dx.doi.org/10.1021/cm001202x.

[1620] L. Vayssieres, C. Sathe, S.M. Butorin, D.K. Shuh, J. Nordgren, and J. Guo. One-dimensional quantum-confinement effect in α-fe$_2$o$_3$ ultrafine nanorod arrays. *Advanced Materials*, 17(19):2320–2323, 2005. ISSN 1521-4095. doi: 10.1002/adma.200500992. URL http://dx.doi.org/10.1002/adma.200500992.

[1621] Peng Sun, Lu You, Dawei Wang, Yanfeng Sun, Jian Ma, and Geyu Lu. Synthesis and gas sensing properties of bundle-like α-fe$_2$o$_3$ nanorods. *Sensors and Actuators B: Chemical*, 156(1):368 – 374, 2011. ISSN 0925-4005. doi: http://dx.doi.org/10.1016/j.snb.2011.04.050. URL http://www.sciencedirect.com/science/article/pii/S0925400511003479.

[1622] Rüdiger Kniep, Paul Simon, and Elena Rosseeva. Structural complexity of hexagonal prismatic crystal specimens of fluorapatite-gelatine nanocomposites: A case study in biomimetic crystal research. *Crystal Research and Technology*, 49(1):4–13, 2014. ISSN 1521-4079. doi: 10.1002/crat.201300207. URL http://dx.doi.org/10.1002/crat.201300207.

[1623] Yang Zhang, Wei Liu, Lang Jiang, Louzhen Fan, Chunru Wang, Wenping Hu, Haizheng Zhong, Yongfang Li, and Shihe Yang. Template-free solution growth of highly regular, crystal orientation-ordered c$_{60}$ nanorod bundles. *J. Mater. Chem.*, 20:953–956, 2010. doi: 10.1039/B913897H. URL http://dx.doi.org/10.1039/B913897H.

[1624] Cherng-Yuh Su and Hsuan-Ching Lin. Direct route to tungsten oxide nanorod bundles: Microstructures and electro-optical properties. *The Journal of Physical Chemistry C*, 113(10):4042–4046, 2009. doi: 10.1021/jp809458j. URL http://dx.doi.org/10.1021/jp809458j.

[1625] Xiaoping Shen, Guoxiu Wang, and David Wexler. Large-scale synthesis and gas sensing application of vertically aligned and double-sided tungsten oxide nanorod arrays. *Sensors and Actuators B: Chemical*, 143(1):325 – 332, 2009. ISSN 0925-4005. doi: http://dx.doi.org/10.1016/j.snb.2009.09.015. URL http://www.sciencedirect.com/science/article/pii/S0925400509007023.

[1626] Lingfei Chi, Ningsheng Xu, Shaozhi Deng, Jun Chen, and Juncong She. An approach for synthesizing various types of tungsten oxide nanostructure. *Nanotechnology*, 17(22):5590, 2006. URL http://stacks.iop.org/0957-4484/17/i=22/a=011.

[1627] Hong-En Wang, Zhenhua Chen, Yu Hang Leung, Chunyan Luan, Chaoping Liu, Yongbing Tang, Ce Yan, Wenjun Zhang, Juan Antonio Zapien, Igor Bello, and Shuit-Tong Lee. Hydrothermal synthesis of ordered single-crystalline rutile tio$_2$ nanorod arrays on different substrates. *Applied Physics Letters*, 96(26):263104, 2010. doi: http://dx.doi.org/10.1063/1.3442913. URL http://scitation.aip.org/content/aip/journal/apl/96/26/10.1063/1.3442913.

[1628] Jung-Chul Lee, Kyung-Soo Park, Tae-Geun Kim, Heon-Jin Choi, and Yun-Mo Sung. Controlled growth of high-quality tio$_2$ nanowires on sapphire and silica. *Nanotechnology*, 17(17):4317, 2006. URL http://stacks.iop.org/0957-4484/17/i=17/a=006.

[1629] Sharipah Nadzirah, Uda Hashim, and Tijjani Adam. Hydrothermal growth of titanium dioxide nanowires on different seed. In *Fifth International Conference on Intelligent Systems, Modelling and Simulation*, 2014.

[1630] Seong Keun Kim, Wan-Don Kim, Kyung-Min Kim, Cheol Seong Hwang, and Jaehack Jeong. High dielectric constant tio$_2$ thin films on a ru electrode grown at 250°C by atomic-layer deposition. *Applied Physics Letters*, 85(18):4112–4114, 2004. doi: http://dx.doi.org/10.1063/1.1812832. URL http://scitation.aip.org/content/aip/journal/apl/85/18/10.1063/1.1812832.

[1631] Mitarbeiter. Gestis-stoffdatenbank - titan(iv)-oxid. Technical report, Institut für Arbeitssicherheit der Deutschen Gesetzlichen Unfallversicherung, June 2016.

[1632] Chan Jin, Ying Tang, F. Guang Yang, X. Lin Li, Shan Xu, X. Yan Fan, Y. Ying Huang, and Y. Ji Yang. Cellular toxicity of tio$_2$ nanoparticles in anatase and rutile crystal phase. *Biological Trace Element Research*, 141(1):3–15, 2011. ISSN 1559-0720. doi: 10.1007/s12011-010-8707-0. URL http://dx.doi.org/10.1007/s12011-010-8707-0.

[1633] Brooke T. Mossman and Andrew Churg. Mechanisms in the pathogenesis of asbestosis and silicosis. *American Journal of Respiratory and Critical Care Medicine*, 157:1666–1680, 1998. doi:10.1164/ajrccm.157.5.9707141. URL http://dx.doi.org/10.1164/ajrccm.157.5.9707141.

[1634] Abderrahim Nemmar, Khaled Melghit, and Badreldin H. Ali. The acute proinflammatory and prothrombotic effects of pulmonary exposure to rutile tio$_2$ nanorods in rats. *Experimental Biology and Medicine*, 233(5):610–619, 2008. doi: 10.3181/0706-RM-165. URL https://doi.org/10.3181/0706-RM-165.

[1635] L. Avril, J.M. Decams, and L. Imhoff. Pulsed direct liquid injection ald of tio₂ films using titanium tetraisopropoxide precursor. *Physics Procedia*, 46:33 – 39, 2013. ISSN 1875-3892. doi: http://dx.doi.org/10.1016/j.phpro.2013.07.063. URL http://www.sciencedirect.com/science/article/pii/S1875389213005211.

[1636] A. Rahtu and M. Ritala. Reaction mechanism studies on titanium isopropoxide-water atomic layer deposition process. *Chemical Vapor Deposition*, 8(1):21–28, 2002. ISSN 1521-3862. doi: 10.1002/1521-3862(20020116)8:1<21::AID-CVDE21>3.0.CO;2-0. URL http://dx.doi.org/10.1002/1521-3862(20020116)8:1<21::AID-CVDE21>3.0.CO;2-0.

[1637] Paul Poodt, David C. Cameron, Eric Dickey, Steven M. George, Vladimir Kuznetsov, Gregory N. Parsons, Fred Roozeboom, Ganesh Sundaram, and Ad Vermeer. Spatial atomic layer deposition: A route towards further industrialization of atomic layer deposition. *Journal of Vacuum Science & Technology A: Vacuum, Surfaces, and Films*, 30(1):010802, 2012. doi: 10.1116/1.3670745. URL http://dx.doi.org/10.1116/1.3670745.

[1638] David Muñoz-Rojas and Judith MacManus-Driscoll. Spatial atmospheric atomic layer deposition: a new laboratory and industrial tool for low-cost photovoltaics. *Mater. Horiz.*, 1:314–320, 2014. doi: 10.1039/C3MH00136A. URL http://dx.doi.org/10.1039/C3MH00136A.

[1639] David Muñoz-Rojas, Haiyan Sun, Diana C. Iza, Jonas Weickert, Li Chen, Haiyan Wang, Lukas Schmidt-Mende, and Judith L. MacManus-Driscoll. High-speed atmospheric atomic layer deposition of ultra thin amorphous tio₂ blocking layers at 100°c for inverted bulk heterojunction solar cells. *Progress in Photovoltaics: Research and Applications*, 21(4):393–400, 2013. ISSN 1099-159X. doi: 10.1002/pip.2380. URL http://dx.doi.org/10.1002/pip.2380.

[1640] Eva L. Unger, Francesca Spadavecchia, Kazuteru Nonomura, Pal Palmgren, Giuseppe Cappelletti, Anders Hagfeldt, Erik M. J. Johansson, and Gerrit Boschloo. Effect of the preparation procedure on the morphology of thin tio₂ films and their device performance in small-molecule bilayer hybrid solar cells. *ACS Applied Materials & Interfaces*, 4(11):5997–6004, 2012. doi: 10.1021/am301604x. URL http://dx.doi.org/10.1021/am301604x. PMID: 23066994.

[1641] G. P. Burns. Titanium dioxide dielectric films formed by rapid thermal oxidation. *Journal of Applied Physics*, 65(5):2095–2097, 1989. doi: http://dx.doi.org/10.1063/1.342856. URL http://scitation.aip.org/content/aip/journal/jap/65/5/10.1063/1.342856.

[1642] Gang Lu, Steven L. Bernasek, and Jeffrey Schwartz. Oxidation of a polycrystalline titanium surface by oxygen and water. *Surface Science*, 458(1–3):80 – 90, 2000. ISSN 0039-6028. doi: http://dx.doi.org/10.1016/S0039-6028(00)00420-9. URL http://www.sciencedirect.com/science/article/pii/S0039602800004209.

[1643] Nicolas Vogel, Sebastian Goerres, Katharina Landfester, and Clemens K. Weiss. A convenient method to produce close- and non-close-packed monolayers using direct assembly at the air-water interface and subsequent plasma-induced size reduction. *Macromolecular Chemistry and Physics*, 212(16):1719–1734, 2011. ISSN 1521-3935. doi: 10.1002/macp.201100187. URL http://dx.doi.org/10.1002/macp.201100187.

[1644] Martin Stärk, Frank Schlickeiser, Dennis Nissen, Birgit Hebler, Philipp Graus, Denise Hinzke, Elke Scheer, Paul Leiderer, Mikhail Fonin, Manfred Albrecht, Ulrich Nowak, and Johannes Boneberg. Controlling the magnetic structure of co/pd thin films by direct laser interference patterning. *Nanotechnology*, 26(20):205302, 2015. URL http://stacks.iop.org/0957-4484/26/i=20/a=205302.

[1645] Stephen Riedel, Markus Schmotz, Paul Leiderer, and Johannes Boneberg. Nanostructuring of thin films by ns pulsed laser interference. *Applied Physics A*, 101(2):309–312, 2010. ISSN 1432-0630. doi: 10.1007/s00339-010-5822-x. URL http://dx.doi.org/10.1007/s00339-010-5822-x.

[1646] T. Geldhauser, P. Leiderer, J. Boneberg, S. Walheim, and Th. Schimmel. Generation of surface energy patterns by single pulse laser interference on self-assembled monolayers. *Langmuir*, 24(22):13155–13160, 2008. doi: 10.1021/la801812j. URL http://dx.doi.org/10.1021/la801812j. PMID: 18950211.

[1647] A. Lasagni and F. Mücklich. Study of the multilayer metallic films topography modified by laser interference irradiation. *Applied Surface Science*, 240(1–4):214 – 221, 2005. ISSN 0169-4332. doi: http://dx.doi.org/10.1016/j.apsusc.2004.06.143. URL http://www.sciencedirect.com/science/article/pii/S0169433204010992.

[1648] Stephen Riedel. *Oberflächenstrukturierung mittelsgepulster Laserinterferenz: Grundlagen und Anwendungen*. PhD thesis, University of Konstanz, 2012.

[1649] M. A. Lantz, S. J. O'Shea, and M. E. Welland. Characterization of tips for conducting atomic force microscopy in ultrahigh vacuum. *Review of Scientific Instruments*, 69(4):1757–1764, 1998. doi: http://dx.doi.org/10.1063/1.1148838. URL http://scitation.aip.org/content/aip/journal/rsi/69/4/10.1063/1.1148838.

[1650] Ranjit A. Patil, Rupesh S. Devan, Jin-Han Lin, Yung Liou, and Yuan-Ron Ma. An efficient methodology for measurement of the average electrical properties of single one-dimensional nio nanorods. *Scientific Reports*, 3, 2013. doi: 10.1038/srep03070.

[1651] Eugen Zimmermann. Nanostructured extremely thin absorber solar cells. Master's thesis, University of Konstanz, 2013.

[1652] Michael Puls. Impedance spectroscopy for nanostructured solar cells. Master's thesis, Ludwig-Maximilians-Universität München, 2011.

[1653] Kazuhiko Shimizu, Hiroaki Imai, Hiroshi Hirashima, and Koji Tsukuma. Low-temperature synthesis of anatase thin films on glass and organic substrates by direct deposition from aqueous solutions. *Thin Solid Films*, 351(1–2):220 – 224, 1999. ISSN 0040-6090. doi: http://dx.doi.org/10.1016/S0040-6090(99)00084-X. URL http://www.sciencedirect.com/science/article/pii/S004060909900084X.

[1654] H Lin, H Kozuka, and T Yoko. Preparation of tio₂ films on self-assembled monolayers by sol–gel method. *Thin Solid Films*, 315(1–2):111 – 117, 1998. ISSN 0040-6090. doi: http://dx.doi.org/10.1016/S0040-6090(97)00759-1. URL http://www.sciencedirect.com/science/article/pii/S0040609097007591.

[1655] M. D. Banus, T. B. Reed, and A. J. Strauss. Electrical and magnetic properties of tio and vo. *Phys. Rev. B*, 5:2775–2784, Apr 1972. doi: 10.1103/PhysRevB.5.2775. URL http://link.aps.org/doi/10.1103/PhysRevB.5.2775.

[1656] Sergey V Ovsyannikov, Xiang Wu, Vladimir V Shchennikov, Alexander E Karkin, Natalia Dubrovinskaia, Gaston Garbarino, and Leonid Dubrovinsky. Structural stability of a golden semiconducting orthorhombic polymorph of tio₃ under high pressures and high temperatures. *Journal of Physics: Condensed Matter*, 22(37):375402, 2010. URL http://stacks.iop.org/0953-8984/22/i=37/a=375402.

[1657] A. W. Czanderna, C. N. Ramachandra Rao, and J. M. Honig. The anatase-rutile transition. part 1.-kinetics of the transformation of pure anatase. *Transactions of the Faraday Society*, 54:1069–1073, 1958. doi: 10.1039/TF9585401069. URL http://dx.doi.org/10.1039/TF9585401069.

[1658] J. K. Yao, H. L. Huang, J. Y. Ma, Y. X. Jin, Y. A. Zhao, J. D. Shao, H. B. He, K. Yi, Z. X. Fan, F. Zhang, and Z. Y. Wu. High refractive index tio₂ film deposited by electron beam evaporation. *Surface Engineering*, 25(3):257–260, 2009. doi: 10.1179/026708408X329498. URL http://dx.doi.org/10.1179/026708408X329498.

References

[1659] L. Calvert. Private communication. Lakes Entrance, Victoria, Australia., 1993.

[1660] J. Häglund, A. Fernández Guillermet, G. Grimvall, and M. Körling. Theory of bonding in transition-metal carbides and nitrides. *Phys. Rev. B*, 48:11685–11691, Oct 1993. doi: 10.1103/PhysRevB.48.11685. URL http://link.aps.org/doi/10.1103/PhysRevB.48.11685.

[1661] Daniel S Eppelsheimer and Robert R Penman. Accurate determination of the lattice of beta-titanium at 900°C. *Nature*, 166(4231):960, 1950. doi: 10.1038/166960a0N1. URL http://dx.doi.org/10.1038/166960a0N3.

[1662] Perry G. Cotter, J. A. Kohn, and R. A. Potter. Physical and x-ray study of the disilicides of titanium, zirconium, and hafnium. *Journal of the American Ceramic Society*, 39(1):11–12, 1956. ISSN 1551-2916. doi: 10.1111/j.1151-2916.1956.tb15590.x. URL http://dx.doi.org/10.1111/j.1151-2916.1956.tb15590.x.

[1663] Jong Hyun Shim, In-Tae Bae, and Junghyun Cho. Microstructure development of hydrothermally grown tio2 thin films with vertically aligned nanorods. *Journal of the American Ceramic Society*, 101(1):50–60, 2018. doi: 10.1111/jace.15110. URL http://dx.doi.org/10.1111/jace.15110.

[1664] Julian Kalb, James A. Dorman, Alena Folger, Melanie Gerigk, Vanessa Knittel, Claudia S. Plüisch, Bastian Trepka, Daniela Lehr, Emily Chua, Berit H. Goodge, Alexander Wittemann, Christina Scheu, Sebastian Polarz, and Lukas Schmidt-Mende. Influence of substrates and rutile seed layers on the assembly of hydrothermally grown rutile tio2 nanorod arrays. *Journal of Crystal Growth*, 494: 26–35, 2018. doi: 10.1016/j.jcrysgro.2018.05.004.

[1665] Julian Kalb, James A. Dorman, Stephan Siroky, and Lukas Schmidt-Mende. Controlling the spatial direction of hydrothermally grown rutile tio2 nanocrystals by the orientation of seed crystals. *Crystals*, 9(2):64, 2019. doi: 10.3390/cryst9020064.

[1666] Julian Kalb, Alena Folger, Eugen Zimmermann, Melanie Gerigk, Bastian Trepka, Christina Scheu, Sebastian Polarz, and Lukas Schmidt-Mende. Controlling the density of hydrothermally grown rutile tio2 nanorods on anatase tio2 films. *Surfaces and Interfaces*, 15:141–147, 2019. doi: 10.1016/j.surfin.2019.02.010.

[1667] H. W. P. Koops, R. Weiel, D. P. Kern, and T. H. Baum. High resolution electron beam induced deposition. *Journal of Vacuum Science & Technology B*, 6(1):477–481, 1988. doi: 10.1116/1.584045. URL http://scitation.aip.org/content/avs/journal/jvstb/6/1/10.1116/1.584045.

[1668] U. Bach, D. Lupo, P. Comte, J.E. Moser, F. Weissörtel, J. Salbeck, H. Spreitzer, and M. Grätzel. Solid-state dye-sensitized mesoporous Tio2 solar cells with high photon-to-electron conversion efficiencies. *Nature*, 395:583–585, 1998. doi: 10.1038/26936M3.

[1669] Seigo Ito, Paul Liska, Pascal Comte, Raphael Charvet, Peter Pechy, Udo Bach, Lukas Schmidt-Mende, Shaik Mohammed Zakeeruddin, Andreas Kay, Mohammad K. Nazeeruddin, and Michael Grätzel. Control of dark current in photoelectrochemical (tio2/i⁻ -i3⁻) and dye-sensitized solar cells. *Chem. Commun.*, pages 4351–4353, 2005. doi: 10.1039/B505718C. URL http://dx.doi.org/10.1039/B505718C.

[1670] K.S. K.S Sree Harsha. *Principles of Vapor Deposition of Thin Films*. Elsevier Science, 2005. URL https://books.google.de/books?id=k8fI2BH1KVEC.

[1671] Victor K LaMer and Robert H Dinegar. Theory, production and mechanism of formation of monodispersed hydrosols. *Journal of the American Chemical Society*, 72(11):4847–4854, 1950.

[1672] Victor K La Mer. Nucleation in phase transitions. *Industrial & Engineering Chemistry*, 44(6):1270–1277, 1952.

[1673] Fenghua Zhao, Xiuyan Li, Jian-Guo Zheng, Xianfeng Yang, Fuli Zhao, Kam Sing Wong, Jing Wang, Wenjiao Lin, Mingmei Wu, and Qiang Su. Zno pine-nanotree arrays grown from facile metal chemical corrosion and oxidation. *Chemistry of Materials*, 20(4):1197–1199, 2008. doi: 10.1021/cm702598r. URL http://dx.doi.org/10.1021/cm702598r.

[1674] Pham Van Tong, Nguyen Duc Hoa, Vu Van Quang, Nguyen Van Duy, and Nguyen Van Hieu. Diameter controlled synthesis of tungsten oxide nanorod bundles for highly sensitive no2 gas sensors. *Sensors and Actuators B: Chemical*, 183:372 – 380, 2013. ISSN 0925-4005. doi: http://dx.doi.org/10.1016/j.snb.2013.03.086. URL http://www.sciencedirect.com/science/article/pii/S0925400513003675.

[1675] Fan Zuo, Le Wang, Tao Wu, Zhenyu Zhang, Dan Borchardt, and Pingyun Feng. Self-doped ti³⁺ enhanced photocatalyst for hydrogen production under visible light. *Journal of the American Chemical Society*, 132(34):11856–11857, 2010. doi: 10.1021/ja103843d. URL http://dx.doi.org/10.1021/ja103843d. PMID: 20687606.

[1676] Gongming Wang, Hanyu Wang, Yichuan Ling, Yuechao Tang, Xunyu Yang, Robert C. Fitzmorris, Changchun Wang, Jin Z. Zhang, and Yat Li. Hydrogen-treated tio2 nanowire arrays for photoelectrochemical water splitting. *Nano Letters*, 11(7):3026–3033, 2011. doi: 10.1021/nl201766h. URL http://dx.doi.org/10.1021/nl201766h. PMID: 21710974.

[1677] Julian Kalb, Alena Folger, Christina Scheu, and Lukas Schmidt-Mende. Non-equilibrium growth model of fibrous mesocrystalline rutile tio2 nanorods. *Journal of Crystal Growth*, 511:8–14, 2019. doi: 10.1016/j.jcrysgro.2019.01.024.

[1678] George V. Wulff. Zur frage der geschwindigkeit des wachstums und der auflösung der krystallflächen. *Zeitschrift für Krystallographie und Mineralogie*, 34:449–530, 1901.

[1679] Hua Gui Yang, Cheng Hua Sun, Shi Zhang Qiao, Jin Zou, Gang Liu, Sean Campbell Smith, Hui Ming Cheng, and Gao Qing Lu. Anatase tio2 single crystals with a large percentage of reactive facets. *Nature*, 453:638–641, 2008. doi: 10.1038/nature06964.

[1680] James J. De Yoreo and Peter G. Vekilov. Principles of crystal nucleation and growth. *Reviews in Mineralogy and Geochemistry*, 54(1): 57–93, 2003. ISSN 1529-6466. doi: 10.2113/0540057. URL http://rimg.geoscienceworld.org/content/54/1/57.

[1681] Ying Pang and Paul Wynblatt. Effects of nb doping and segregation on the grain boundary plane distribution in tio2. *Journal of the American Ceramic Society*, 89(2):666–671, 2006. ISSN 1551-2916. doi: 10.1111/j.1551-2916.2005.00759.x. URL http://dx.doi.org/10.1111/j.1551-2916.2005.00759.x.

[1682] Changhui Ye, Xiaosheng Fang, Yufeng Hao, Xuemei Teng, , and Lide Zhang. Zinc oxide nanostructures: Morphology derivation and evolution. *The Journal of Physical Chemistry B*, 109(42):19758–19765, 2005. doi: 10.1021/jp0509358. URL http://dx.doi.org/10.1021/jp0509358. PMID: 16853555.

[1683] M.K. Singh, Arup Banerjee, and P.K. Gupta. Simulating vapour growth morphology of crystalline urea using modified attachment energy model. *Journal of Crystal Growth*, 343(1):77 – 85, 2012. ISSN 0022-0248. doi: http://dx.doi.org/10.1016/j.jcrysgro.2012.01.032. URL http://www.sciencedirect.com/science/article/pii/S0022024812000814.

[1684] Xianmiao Sun, Qiong Sun, Qian Zhang, Qianqian Zhu, Hongzhou Dong, and Lifeng Dong. Significant effects of reaction temperature on morphology, crystallinity, and photoelectrical properties of rutile tio2 nanorod array films. *Journal of Physics D: Applied Physics*, 46(9):095102, 2013. URL http://stacks.iop.org/0022-3727/46/i=9/a=095102.

[1685] Reui-San Chen, Alexandru Korotcov, Ying-Sheng Huang, and Dah-Shyang Tsai. One-dimensional conductive iro$_2$ nanocrystals. *Nanotechnology*, 17(9):R67, 2006. URL http://stacks.iop.org/0957-4484/17/i=9/a=R01.

[1686] Chih-Chieh Wang and Po-Hsun Chou. Effects of various hydrogenated treatments on formation and photocatalytic activity of black tio$_2$ nanowire arrays. *Nanotechnology*, 27(32):325401, 2016. URL http://stacks.iop.org/0957-4484/27/i=32/a=325401.

[1687] Zhifeng Jiang, Jianjun Zhu, Dong Liu, Wei Wei, Jimin Xie, and Min Chen. In situ synthesis of bimetallic ag/pt loaded single-crystalline anatase tio$_2$ hollow nano-hemispheres and their improved photocatalytic properties. *CrystEngComm*, 16:2384–2394, 2014. doi: 10.1039/C3CE41949E. URL http://dx.doi.org/10.1039/C3CE41949E.

[1688] Huilin Hou, Minghui Shang, Fengmei Gao, Lin Wang, Qiao Liu, Jinju Zheng, Zuobao Yang, and Weiyou Yang. Highly efficient photocatalytic hydrogen evolution in ternary hybrid tio$_2$/cuo/cu thoroughly mesoporous nanofibers. *ACS Applied Materials & Interfaces*, 8(31):20128–20137, 2016. doi: 10.1021/acsami.6b06644. URL http://dx.doi.org/10.1021/acsami.6b06644. PMID: 27430307.

[1689] Wei Wen, Jin-Ming Wu, Yin-Zhu Jiang, Jun-Qiang Bai, and Lu-Lu Lai. Titanium dioxide nanotrees for high-capacity lithium-ion microbatteries. *J. Mater. Chem. A*, 4:10593–10600, 2016. doi: 10.1039/C6TA03331H. URL http://dx.doi.org/10.1039/C6TA03331H.

[1690] Wu-Qiang Wu, Hao-Lin Feng, Hong-Yan Chen, Dai-Bin Kuang, and Cheng-Yong Su. Recent advances in hierarchical three-dimensional titanium dioxide nanotree arrays for high-performance solar cells. *J. Mater. Chem. A*, 5:12699–12717, 2017. doi: 10.1039/C7TA03521G. URL http://dx.doi.org/10.1039/C7TA03521G.

[1691] Hui Liu, Mengyan Li, Ting Lv, and Chunkui Zhu. Synthesis and improved dye-sensitized solar cells performance of tio$_2$ nanowires/nanospheres composites. *Journal of Materials Science: Materials in Electronics*, 27(12):12591–12598, Dec 2016. ISSN 1573-482X. doi: 10.1007/s10854-016-5390-8. URL https://doi.org/10.1007/s10854-016-5390-8.

[1692] S. Norasetthekul, P.Y. Park, K.H. Baik, K.P. Lee, J.H. Shin, B.S. Jeong, V. Shishodia, E.S. Lambers, D.P. Norton, and S.J. Pearton. Dry etch chemistries for tio$_2$ thin films. *Applied Surface Science*, 185(1–2):27 – 33, 2001. ISSN 0169-4332. doi: http://dx.doi.org/10.1016/S0169-4332(01)00562-1. URL http://www.sciencedirect.com/science/article/pii/S0169433201005621.

[1693] Ashwini Sinha, Dennis W. Hess, and Clifford L. Henderson. Area-selective ald of titanium dioxide using lithographically defined poly(methyl methacrylate) films. *Journal of The Electrochemical Society*, 153(5):G465–G469, 2006. doi: 10.1149/1.2184068. URL http://jes.ecsdl.org/content/153/5/G465.abstract.

[1694] Kyung S. Park, Eun K. Seo, Young R. Do, Kwan Kim, and Myung M. Sung. Light stamping lithography: Microcontact printing without inks. *Journal of the American Chemical Society*, 128(3):858–865, 2006. doi: 10.1021/ja055377p. URL http://pubs.acs.org/doi/abs/10.1021/ja055377p.

[1695] Xudong Wang and Jian Shi. Evolution of titanium dioxide one-dimensional nanostructures from surface-reaction-limited pulsed chemical vapor deposition. *Journal of Materials Research*, 28(3):270–279, 002 2013. doi: 10.1557/jmr.2012.356. URL https://www.cambridge.org/core/article/evolution-of-titanium-dioxide-one-dimensional-nanostructures-from-surface-reaction-limited-pulsed-chemical-vapor-deposition/9DEF7FBD0569C67BD60B95C5C598B50B.

[1696] Steven J. Barcelo, Ansoon Kim, Wei Wu, and Zhiyong Li. Fabrication of deterministic nanostructure assemblies with sub-nanometer spacing using a nanoimprinting transfer technique. *ACS Nano*, 6(7):6446–6452, 2012. doi: 10.1021/nn3020807. URL http://dx.doi.org/10.1021/nn3020807. PMID: 22735072.

[1697] Feng Shan, Xuemin Lu, Qian Zhang, Jun Wu, Yuzhu Wang, Fenggang Bian, Qinghua Lu, Zhaofu Fei, and Paul J. Dyson. A facile approach for controlling the orientation of one-dimensional mesochannels in mesoporous titania films. *Journal of the American Chemical Society*, 134(50):20238–20241, 2012. doi: 10.1021/ja309168f. URL http://dx.doi.org/10.1021/ja309168f. PMID: 23214929.

[1698] Babak Nikoobakht, Xudong Wang, Andrew Herzinga, and Jian Shib. Scalable synthesis and device integration of self-registered one-dimensional zinc oxide nanostructures and related materials. *Chemical Society Reviews*, 2012. doi: 10.1039/c2cs35164a. URL http://dx.doi.org/10.1039/C2CS35164A.

[1699] J. C. Fisher. Calculation of diffusion penetration curves for surface and grain boundary diffusion. *Journal of Applied Physics*, 22(1):74–77, 1951. doi: 10.1063/1.1699825. URL http://dx.doi.org/10.1063/1.1699825.

[1700] Zhong-Ze Gu, Akira Fujishima, and Osamu Sato. Biomimetic titanium dioxide film with structural color and extremely stable hydrophilicity. *Applied Physics Letters*, 85(21):5067–5069, 2004. doi: http://dx.doi.org/10.1063/1.1825052. URL http://scitation.aip.org/content/aip/journal/apl/85/21/10.1063/1.1825052.

[1701] Xu Dong Wang, Elton Graugnard, Jeffrey S. King, Zhong Lin Wang, and Christopher J. Summers. Large-scale fabrication of ordered nanobowl arrays. *Nano Letters*, 4(11):2223–2226, 2004. doi: 10.1021/nl048589d. URL http://dx.doi.org/10.1021/nl048589d.

[1702] Yun Gun, Gwang Yeom Song, Vu Hong Vinh Quy, Jaeyeong Heo, Hyunjung Lee, Kwang-Soon Ahn, and Soon Hyung Kang. Joint effects of photoactive tio$_2$ and fluoride-doping on sno$_2$ inverse opal nanoarchitecture for solar water splitting. *ACS Applied Materials & Interfaces*, 7(36):20292–20303, 2015. doi: 10.1021/acsami.5b05914. URL http://dx.doi.org/10.1021/acsami.5b05914. PMID: 26322646.

[1703] Kanti Jain. *Excimer Laser Lithography*, volume PM03. SPIE PRESS BOOK, 1990.

[1704] Julian Kalb, Vanessa Knittel, and Lukas Schmidt-Mende. Advanced scanning probe lithography using anatase-to-rutile transition to create localized tio$_2$ nanorods. *Beilstein Journal of Nanotechnology*, 10(1):412–418, 2019. doi: 10.3762/bjnano.10.40.

[1705] Eui-Hyun Kong, Yong-June Chang, and Hyun Myung Jang. A tri-functional tio$_2$ photoelectrode: single crystalline nanowires directly grown on nanoparticles for dye-sensitized solar cells. *RSC Adv.*, 4:943–947, 2014. doi: 10.1039/C3RA44394A. URL http://dx.doi.org/10.1039/C3RA44394A.

[1706] Chenyang Zha, Liming Shen, Xiaoyan Zhang, Yifeng Wang, Brian A. Korgel, Arunava Gupta, and Ningzhong Bao. Double-sided brush-shaped tio$_2$ nanostructure assemblies with highly ordered nanowires for dye-sensitized solar cells. *ACS Applied Materials & Interfaces*, 6(1):122–129, 2014. doi: 10.1021/am404942n. URL http://dx.doi.org/10.1021/am404942n.

[1707] A. Sclafani, L. Palmisano, and M. Schiavello. Influence of the preparation methods of titanium dioxide on the photocatalytic degradation of phenol in aqueous dispersion. *The Journal of Physical Chemistry*, 94(2):829–832, 1990. doi: 10.1021/j100365a058. URL http://dx.doi.org/10.1021/j100365a058.

[1708] D Strauss, G Müller, G Schumacher, V Engelko, W Stamm, D Clemens, and W.J Quaddakers. Oxide scale growth on mcraly bond coatings after pulsed electron beam treatment and deposition of ebpvd-tbc. *Surface and Coatings Technology*, 135(2–3):196 – 201, 2001. ISSN 0257-8972. doi: http://dx.doi.org/10.1016/S0257-8972(00)00916-6. URL http://www.sciencedirect.com/science/article/pii/S0257897200009166.

References

[1709] Yoshiyuki Uno, Akira Okada, Kensuke Uemura, Purwadi Raharjo, Toshihiko Furukawa, and Kosaku Karato. High-efficiency finishing process for metal mold by large-area electron beam irradiation. *Precision Engineering*, 29(4):449 – 455, 2005. ISSN 0141-6359. doi: http://dx.doi.org/10.1016/j.precisioneng.2004.12.005. URL http://www.sciencedirect.com/science/article/pii/S0141635905000127.

[1710] Michael A. Henderson, William S. Epling, Charles H. F. Peden, and Craig L. Perkins. Insights into photoexcited electron scavenging processes on Tio$_2$ obtained from studies of the reaction of o$_2$ with oh groups adsorbed at electronic defects on Tio$_2$(110). *The Journal of Physical Chemistry B*, 107(2):534–545, 2003. doi: 10.1021/jp0262113. URL http://dx.doi.org/10.1021/jp0262113.

[1711] Dong Ming Guo, Wei Si Li, Zhu Ji Jin, Zhe Wang, and Ze Wei Yuan. Effects of high dielectric constant abrasives on ecmp. In *Frontiers of Manufacturing and Design Science II*, volume 121 of *Applied Mechanics and Materials*, pages 3263–3267. Trans Tech Publications, 1 2012. doi: 10.4028/www.scientific.net/AMM.121-126.3263.

[1712] Julian Kalb, Fabian Weller, Lukas Irmler, Vanessa Knittel, Philipp Graus, Johannes Boneberg, and Lukas Schmidt-Mende. Position-controlled laser-induced creation of rutile tio$_2$ nanostructures. *Nanotechnology*, 30(33):335302, 2019. doi: 10.1088/1361-6528/ab1964.

[1713] A. D. Kudryavtseva, N. V. Tcherniega, M. I. Samoylovich, and A. S. Shevchuk. Photon–phonon interactions in nanostructured systems. *International Journal of Thermophysics*, 33(10):2194–2202, 2012. ISSN 1572-9567. doi: 10.1007/s10765-012-1259-0. URL http://dx.doi.org/10.1007/s10765-012-1259-0.

[1714] Yihua Ren, Shuiqing Li, Yiyang Zhang, Stephen D. Tse, and Marshall B. Long. Absorption-ablation-excitation mechanism of laser-cluster interactions in a nanoaerosol system. *Phys. Rev. Lett.*, 114:093401, Mar 2015. doi: 10.1103/PhysRevLett.114.093401. URL http://link.aps.org/doi/10.1103/PhysRevLett.114.093401.

[1715] Christian Lejon and Lars Österlund. Influence of phonon confinement, surface stress, and zirconium doping on the raman vibrational properties of anatase tio$_2$ nanoparticles. *Journal of Raman Spectroscopy*, 42(11):2026–2035, 2011. ISSN 1097-4555. doi: 10.1002/jrs.2956. URL http://dx.doi.org/10.1002/jrs.2956.

[1716] Martin A. Green and Mark J. Keevers. Optical properties of intrinsic silicon at 300 k. *Progress in Photovoltaics: Research and Applications*, 3(3):189–192, 1995. ISSN 1099-159X. doi: 10.1002/pip.4670030303. URL http://dx.doi.org/10.1002/pip.4670030303.

[1717] G.E. Jellison. Optical functions of silicon determined by two-channel polarization modulation ellipsometry. *Optical Materials*, 1(1):41 – 47, 1992. ISSN 0925-3467. doi: http://dx.doi.org/10.1016/0925-3467(92)90015-F. URL http://www.sciencedirect.com/science/article/pii/092534679290015F.

[1718] A.D. Barros, K.F. Albertin, J. Miyoshi, I. Doi, and J.A. Diniz. Thin titanium oxide films deposited by e-beam evaporation with additional rapid thermal oxidation and annealing for isfet applications. *Microelectronic Engineering*, 87(3):443 – 446, 2010. ISSN 0167-9317. doi: http://dx.doi.org/10.1016/j.mee.2009.06.020. URL http://www.sciencedirect.com/science/article/pii/S0167931709004742. Materials for Advanced Metallization 2009Proceedings of the eighteenth European Workshop on Materials for Advanced Metallization 2009.

[1719] R. J. Nemanich, R. W. Fiordalice, and H. Jeon. Raman scattering characterization of titanium silicide formation. *IEEE Journal of Quantum Electronics*, 25(5):997–1002, May 1989. ISSN 0018-9197. doi: 10.1109/3.27991.

[1720] H. Bracht and N.A. Stolwijk. *Diffusion in Si, Ge, and their alloys*, volume 33, chapter 2. Springer, Berlin, Heidelberg, 1998. doi: 10.1007/b53031.

[1721] A. E. Morgan, E. K. Broadbent, K. N. Ritz, D. K. Sadana, and B. J. Burrow. Interactions of thin ti films with si, sio$_2$, si$_3$n$_4$, and sio$_x$n$_y$ under rapid thermal annealing. *Journal of Applied Physics*, 64(1):344–353, 1988. doi: http://dx.doi.org/10.1063/1.341434. URL http://scitation.aip.org/content/aip/journal/jap/64/1/10.1063/1.341434.

[1722] Yunji L. Corcoran, Alexander H. King, Nimal de Lanerolle, and Bonggi Kim. Grain boundary diffusion and growth of titanium silicide layers on silicon. *Journal of Electronic Materials*, 19(11):1177–1183, 1990. ISSN 1543-186X. doi: 10.1007/BF02673330. URL http://dx.doi.org/10.1007/BF02673330.

[1723] Robert Beyers and Robert Sinclair. Metastable phase formation in titanium-silicon thin films. *Journal of Applied Physics*, 57(12):5240–5245, 1985. doi: http://dx.doi.org/10.1063/1.335263. URL http://scitation.aip.org/content/aip/journal/jap/57/12/10.1063/1.335263.

[1724] G. G. Bentini, R. Nipoti, A. Armigliato, M. Berti, A. V. Drigo, and C. Cohen. Growth and structure of titanium silicide phases formed by thin ti films on si crystals. *Journal of Applied Physics*, 57(2):270–275, 1985. doi: http://dx.doi.org/10.1063/1.334799. URL http://scitation.aip.org/content/aip/journal/jap/57/2/10.1063/1.334799.

[1725] M. H. Wang and L. J. Chen. Simultaneous occurrence of multiphases in interfacial reactions of ultrahigh vacuum deposited ti thin films on (111)si. *Applied Physics Letters*, 59(19):2460–2462, 1991. doi: http://dx.doi.org/10.1063/1.105995. URL http://scitation.aip.org/content/aip/journal/apl/59/19/10.1063/1.105995.

[1726] D. G. Howitt and A. B. Harker. The oriented growth of anatase in thin films of amorphous titania. *Journal of Materials Research*, 2: 201–210, 4 1987. ISSN 2044-5326. doi: 10.1557/JMR.1987.0201. URL http://journals.cambridge.org/article_S0884291400001382.

[1727] N. S. Gluck, H. Sankur, J. Heuer, J. DeNatale, and W. J. Gunning. Microstructure and composition of composite sio$_2$/tio$_2$ thin films. *Journal of Applied Physics*, 69(5):3037–3045, 1991. doi: http://dx.doi.org/10.1063/1.348591. URL http://scitation.aip.org/content/aip/journal/jap/69/5/10.1063/1.348591.

[1728] P Alexandrov, J Koprinarova, and D Todorov. Dielectric properties of tio$_2$-films reactively sputtered from ti in an rf magnetron. *Vacuum*, 47(11):1333 – 1336, 1996. ISSN 0042-207X. doi: http://dx.doi.org/10.1016/0042-207X(96)00196-0. URL http://www.sciencedirect.com/science/article/pii/S0042207X96001960.

[1729] Md. Mosaddeq-ur Rahman, Guolin Yu, Tetsuo Soga, Takashi Jimbo, Hiroshi Ebisu, and Masayoshi Umeno. Refractive index and degree of inhomogeneity of nanocrystalline tio$_2$ thin films: Effects of substrate and annealing temperature. *Journal of Applied Physics*, 88(8):4634–4641, 2000. doi: http://dx.doi.org/10.1063/1.1290456. URL http://scitation.aip.org/content/aip/journal/jap/88/8/10.1063/1.1290456.

[1730] C. Hauf, R. Kniep, and G. Pfaff. Preparation of various titanium suboxide powders by reduction of tio$_2$ with silicon. *Journal of Materials Science*, 34(6):1287–1292, 1999. ISSN 1573-4803. doi: 10.1023/A:1004589813050. URL http://dx.doi.org/10.1023/A:1004589813050.

[1731] KG Grigorov, GI Grigorov, L. Drajeva, D Bouchier, R Sporken, and R Caudano. Synthesis and characterization of conductive titanium monoxide films. diffusion of silicon in titanium monoxide films. *Vacuum*, 51(2):153 – 155, 1998. ISSN 0042-207X. doi: http://dx.doi.org/10.1016/S0042-207X(98)00149-3. URL http://www.sciencedirect.com/science/article/pii/S0042207X98001493.

[1732] S. Hocine and D. Mathiot. Titanium diffusion in silicon. *Applied Physics Letters*, 53(14):1269–1271, 1988. doi: http://dx.doi.org/10.1063/1.100446. URL http://scitation.aip.org/content/aip/journal/apl/53/14/10.1063/1.100446.

[1733] Krishan L. Luthra. Stability of protective oxide films on ti-base alloys. *Oxidation of Metals*, 36(5):475–490, 1991. ISSN 1573-4889. doi: 10.1007/BF01151593. URL http://dx.doi.org/10.1007/BF01151593.

[1734] J. L. Murray and H. A. Wriedt. The o-ti (oxygen-titanium) system. *Journal of Phase Equilibria*, 8(2):148–165, 1987. ISSN 1054-9714. doi: 10.1007/BF02873201. URL http://dx.doi.org/10.1007/BF02873201.

[1735] Philipp G. Wahlbeck and Paul W. Gilles. Reinvestigation of the phase diagram for the system titanium–oxygen. *Journal of the American Ceramic Society*, 49(4):180–183, 1966. ISSN 1551-2916. doi: 10.1111/j.1151-2916.1966.tb13229.x. URL http://dx.doi.org/10.1111/j.1151-2916.1966.tb13229.x.

[1736] C.L. Yeh, H.J. Wang, and W.H. Chen. A comparative study on combustion synthesis of ti-si compounds. *Journal of Alloys and Compounds*, 450(1–2):200 – 207, 2008. ISSN 0925-8388. doi: http://dx.doi.org/10.1016/j.jallcom.2006.10.074. URL http://www.sciencedirect.com/science/article/pii/S0925838806016628.

[1737] G. Guisbiers, O. Van Overschelde, and M. Wautelet. Theoretical investigation of size and shape effects on the melting temperature and energy bandgap of tio2 nanostructures. *Applied Physics Letters*, 92(10):103121, 2008. doi: http://dx.doi.org/10.1063/1.2897297. URL http://scitation.aip.org/content/aip/journal/apl/92/10/10.1063/1.2897297.

[1738] M Haynes William, DR Lide, and TJ Bruno. *CRC Handbook of Chemistry and Physics*. CRC Press: Boca Raton, FL, 92 edition, 2011.

[1739] Spectra Software. Sopra sa material library. Technical report, Software Spectra, Inc., 2016.

[1740] Kuen Lee and Juh Tzeng Lue. Formation of titanium silicides and their refractive index measurements. *Physics Letters A*, 125(5):271 – 275, 1987. ISSN 0375-9601. doi: http://dx.doi.org/10.1016/0375-9601(87)90208-8. URL http://www.sciencedirect.com/science/article/pii/0375960187902088.

[1741] G. E. Jellison, L. A. Boatner, J. D. Budai, B.-S. Jeong, and D. P. Norton. Spectroscopic ellipsometry of thin film and bulk anatase (tio2). *Journal of Applied Physics*, 93(12):9537–9541, 2003. doi: http://dx.doi.org/10.1063/1.1573737. URL http://scitation.aip.org/content/aip/journal/jap/93/12/10.1063/1.1573737.

[1742] L Miao, P Jin, K Kaneko, A Terai, N Nabatova-Gabain, and S Tanemura. Preparation and characterization of polycrystalline anatase and rutile tio2 thin films by rf magnetron sputtering. *Applied Surface Science*, 212-213:255 – 263, 2003. ISSN 0169-4332. doi: http://dx.doi.org/10.1016/S0169-4332(03)00106-5. URL http://www.sciencedirect.com/science/article/pii/S0169433203001055. 11th International Conference on Solid Films and Surfaces.

[1743] George F. Burkhard, Eric T. Hoke, and Michael D. McGehee. Accounting for interference, scattering, and electrode absorption to make accurate internal quantum efficiency measurements in organic and other thin solid cells. *Advanced Materials*, 22(30):3293–3297, 2010. ISSN 1521-4095. doi: 10.1002/adma.201000883. URL http://dx.doi.org/10.1002/adma.201000883.

[1744] Leif A. A. Pettersson, Lucimara S. Roman, and Olle Inganäs. Modeling photocurrent action spectra of photovoltaic devices based on organic thin films. *Journal of Applied Physics*, 86(1):487–496, 1999. doi: 10.1063/1.370757. URL http://aip.scitation.org/doi/abs/10.1063/1.370757.

[1745] Peter Peumans, Aharon Yakimov, and Stephen R. Forrest. Small molecular weight organic thin-film photodetectors and solar cells. *Journal of Applied Physics*, 93(7):3693–3723, 2003. doi: 10.1063/1.1534621. URL http://aip.scitation.org/doi/abs/10.1063/1.1534621.

[1746] Gao Lihong, F. Lemarchand, and M. Lequime. Refractive index determination of sio2 layer in the uv/vis/nir range: spectrophotometric reverse engineering on single and bi-layer designs. *Journal of the European Optical Society-Rapid publications*, 8, 2013.

[1747] Aleksandar D. Rakić, Aleksandra B. Djurišić, Jovan M. Elazar, and Marian L. Majewski. Optical properties of metallic films for vertical-cavity optoelectronic devices. *Appl. Opt.*, 37(22):5271–5283, Aug 1998. doi: 10.1364/AO.37.005271. URL http://ao.osa.org/abstract.cfm?URI=ao-37-22-5271.

[1748] W. Beyer, J. Hüpkes, and H. Stiebig. Transparent conducting oxide films for thin film silicon photovoltaics. *Thin Solid Films*, 516(2-4):147 – 154, 2007. ISSN 0040-6090. doi: http://dx.doi.org/10.1016/j.tsf.2007.08.110. URL http://www.sciencedirect.com/science/article/pii/S004060900701512X.

[1749] R.A. Arndt, J.F. Allison, A. Meulenberg, Jr., and J.G. Haynos. Optical properties of the comsat non-reflective cell. In *11th Photovoltaic Specialists Conference*, pages 40–43, May 1975.

[1750] H. K. Pulker, G. Paesold, and E. Ritter. Refractive indices of tio2 films produced by reactive evaporation of various titanium-oxygen phases. *Appl. Opt.*, 15(12):2986–2991, Dec 1976. doi: 10.1364/AO.15.002986. URL http://ao.osa.org/abstract.cfm?URI=ao-15-12-2986.

[1751] Peter Ritterskamp, Andriy Kuklya, Marc-André Wüstkamp, Klaus Kerpen, Claudia Weidenthaler, and Martin Demuth. A titanium disilicide derived semiconducting catalyst for water splitting under solar radiation-reversible storage of oxygen and hydrogen. *Angewandte Chemie International Edition*, 46(41):7770–7774, 2007. ISSN 1521-3773. doi: 10.1002/anie.200701626. URL http://dx.doi.org/10.1002/anie.200701626.

[1752] Zhigang Mou, Shunli Yin, Mingshan Zhu, Yukou Du, Xiaomei Wang, Ping Yang, Junwei Zheng, and Cheng Lu. Ruo2/tisi2/graphene composite for enhanced photocatalytic hydrogen generation under visible light irradiation. *Phys. Chem. Phys.*, 15:2793–2799, 2013. doi: 10.1039/C2CP44270A. URL http://dx.doi.org/10.1039/C2CP44270A.

[1753] Yongjing Lin, Sa Zhou, Xiaohua Liu, Stafford Sheehan, and Dunwei Wang. Tio2/tisi2 heterostructures for high-efficiency photoelectrochemical h2o splitting. *Journal of the American Chemical Society*, 131(8):2772–2773, 2009. doi: 10.1021/ja808426h. URL http://dx.doi.org/10.1021/ja808426h. PMID: 19209858.

[1754] Che-Ming Chang, Yu-Cheng Chang, Chung-Yang Lee, Ping-Hung Yeh, Wei-Fan Lee, and Lih-Juann Chen. ti5si4 nanobats with excellent field emission properties. *The Journal of Physical Chemistry C*, 113(21):9153–9156, 2009. doi: 10.1021/jp902082x. URL http://dx.doi.org/10.1021/jp902082x.

[1755] J.-W. Chen, A.G. Milnes, and A. Rohatgi. Titanium in silicon as a deep level impurity. *Solid-State Electronics*, 22(9):801 – 808, 1979. ISSN 0038-1101. doi: http://dx.doi.org/10.1016/0038-1101(79)90130-8. URL http://www.sciencedirect.com/science/article/pii/0038110179901308.

[1756] D. Mathiot and S. Hocine. Titanium-related deep levels in silicon: A reexamination. *Journal of Applied Physics*, 66(12):5862–5867, 1989. doi: http://dx.doi.org/10.1063/1.343608. URL http://scitation.aip.org/content/aip/journal/jap/66/12/10.1063/1.343608.

[1757] J. Olea, M. Toledano-Luque, D. Pastor, G. González-Díaz, and I. Mártil. Titanium doped silicon layers with very high concentration. *Journal of Applied Physics*, 104(1):016105, 2008. doi: http://dx.doi.org/10.1063/1.2949258. URL http://scitation.aip.org/content/aip/journal/jap/104/1/10.1063/1.2949258;jsessionid=8N3kEsXQn8OvzacJkmQS5Nro.x-aip-live-01.

[1758] T. E. Seidel, D. J. Lischner, C. S. Pai, and S. S. Lau. Temperature transients in heavily doped and undoped silicon using rapid thermal annealing. *Journal of Applied Physics*, 57(4):1317–1321, 1985. doi: http://dx.doi.org/10.1063/1.334532. URL http://scitation.aip.org/content/aip/journal/jap/57/4/10.1063/1.334532.

[1759] Y.S.P. Touloukian, W. R., C.Y. Ho, and P.G. Klemens. Thermal conductivity of non-metallic solids. In *Thermophyiscal Properties of Matter*, volume 2. New York: IFI/Plenum, 1970.

[1760] Xuhui Feng, Xiaopeng Huang, and Xinwei Wang. Thermal conductivity and secondary porosity of single anatase tio$_2$ nanowire. *Nanotechnology*, 23(18):185701, 2012. URL http://stacks.iop.org/0957-4484/23/i=18/a=185701.

[1761] C. J. Glassbrenner and Glen A. Slack. Thermal conductivity of silicon and germanium from 3k to the melting point. *Phys. Rev.*, 134: A1058–A1069, May 1964. doi: 10.1103/PhysRev.134.A1058. URL http://link.aps.org/doi/10.1103/PhysRev.134.A1058.

[1762] David G. Cahill and Thomas H. Allen. Thermal conductivity of sputtered and evaporated sio$_2$ and tio$_2$ optical coatings. *Applied Physics Letters*, 65(3):309–311, 1994. doi: http://dx.doi.org/10.1063/1.112355. URL http://scitation.aip.org/content/aip/journal/apl/65/3/10.1063/1.112355.

[1763] R. McPherson. Formation of metastable phases in flame- and plasma-prepared alumina. *Journal of Materials Science*, 8(6):851–858, 1973. ISSN 1573-4803. doi: 10.1007/BF00553735. URL http://dx.doi.org/10.1007/BF00553735.

[1764] Xiaobao Fan and Takamasa Ishigaki. Critical free energy for nucleation from the congruent melt of mosi$_2$. *Journal of Crystal Growth*, 171(1):166 – 173, 1997. ISSN 0022-0248. doi: http://dx.doi.org/10.1016/S0022-0248(96)00428-9. URL http://www.sciencedirect.com/science/article/pii/S0022024896004289.

[1765] Yali Li and Takamasa Ishigaki. Thermodynamic analysis of nucleation of anatase and rutile from tio$_2$ melt. *Journal of Crystal Growth*, 242(3–4):511 – 516, 2002. ISSN 0022-0248. doi: http://dx.doi.org/10.1016/S0022-0248(02)01438-0. URL http://www.sciencedirect.com/science/article/pii/S0022024802014380.

[1766] R.A. Laudise, J.B. Mullin, B. Mutaftschiev, Rustum Roy, and William B. White. Third international conference on crystal growth growth of titanium oxide crystals of controlled stoichiometry and order. *Journal of Crystal Growth*, 13:78 – 83, 1972. ISSN 0022-0248. doi: http://dx.doi.org/10.1016/0022-0248(72)90066-8. URL http://www.sciencedirect.com/science/article/pii/0022024872900668.

[1767] Sotiris E. Pratsinis and Srinivas Vemury. First international particle technology forum particle formation in gases: A review. *Powder Technology*, 88(3):267 – 273, 1996. ISSN 0032-5910. doi: http://dx.doi.org/10.1016/S0032-5910(96)03130-0. URL http://www.sciencedirect.com/science/article/pii/S0032591096031300.

[1768] Karl A. Kusters and Sotiris E. Pratsinis. Strategies for control of ceramic powder synthesis by gas-to-particle conversion. *Powder Technology*, 82(1):79 – 91, 1995. ISSN 0032-5910. doi: http://dx.doi.org/10.1016/0032-5910(94)02892-R. URL http://www.sciencedirect.com/science/article/pii/003259109402892R.

[1769] Ya-Li Li and Takamasa Ishigaki. Synthesis of crystalline micron spheres of titanium dioxide by thermal plasma oxidation of titanium carbide. *Chemistry of Materials*, 13(5):1577–1584, 2001. doi: 10.1021/cm000893u. URL http://dx.doi.org/10.1021/cm000893u.

[1770] Takahiro Nakamura, Tetsu Ichitsubo, Eiichiro Matsubara, Atsushi Muramatsu, Nobuaki Sato, and Hideyuki Takahashi. Preferential formation of anatase in laser-ablated titanium dioxide films. *Acta Materialia*, 53(2):323 – 329, 2005. ISSN 1359-6454. doi: http://dx.doi.org/10.1016/j.actamat.2004.09.026. URL http://www.sciencedirect.com/science/article/pii/S1359645404005695.

[1771] SR Yoganarasimhan and CN Ramachandra Rao. Mechanism of crystal structure transformations. part 3. - factors affecting the anatase-rutile transformation. *Transactions of the Faraday Society*, 58:1579–1589, 1962.

[1772] Limin Guo, Xiaohui Wang, Hui Zhang, and Longtu Li. Photoelectrochemical properties of tio$_2$/srtio$_3$ combined nanotube arrays. *Ceramics International*, 2012. doi: 10.1016/j.ceramint.2012.10.151. URL http://dx.doi.org/10.1016/j.ceramint.2012.10.151.

[1773] P. Avouris, R. Martel, T. Hertel, and R. Sandstrom. Afm-tip-induced and current-induced local oxidation of silicon and metals. *Applied Physics A*, 66(1):S659–S667, 1998. ISSN 1432-0630. doi: 10.1007/s003390051218. URL http://dx.doi.org/10.1007/s003390051218.

[1774] Christian Sämann, Jürgen R. Köhler, Morris Dahlinger, Markus B. Schubert, and Jürgen H. Werner. Pulsed laser porosification of silicon thin films. *Materials*, (7), 2016. URL http://www.mdpi.com/1996-1944/9/7/509.

[1775] A.J. Glass, A.H. Guenther, and United States. National Bureau of Standards. *Laser induced damage in optical materials, 1976: proceeding of a symposium*. ASTM special technical publication. U.S. Dept. of Commerce, National Bureau of Standards : for sale by the Supt. of Docs., U.S. Govt. Print. Off., 1976. URL https://books.google.de/books?id=AAlk39hmfOoC.

[1776] P. Baumgart, D. J. Krajnovich, T. A. Nguyen, and A. G. Tam. A new laser texturing technique for high performance magnetic disk drives. *IEEE Transactions on Magnetics*, 31(6):2946–2951, Nov 1995. ISSN 0018-9464. doi: 10.1109/20.490199.

[1777] G. Wysocki, R. Denk, K. Piglmayer, N. Arnold, and D. Bäuerle. Single-step fabrication of silicon-cone arrays. *Applied Physics Letters*, 82(5):692–693, 2003. doi: 10.1063/1.1538347. URL https://doi.org/10.1063/1.1538347.

[1778] Won-Kyu Rhim and Kenichi Ohsaka. Thermophysical properties measurement of molten silicon by high-temperature electrostatic levitator: density, volume expansion, specific heat capacity, emissivity, surface tension and viscosity. *Journal of Crystal Growth*, 208 (1–4):313 – 321, 2000. ISSN 0022-0248. doi: http://dx.doi.org/10.1016/S0022-0248(99)00437-6. URL http://www.sciencedirect.com/science/article/pii/S0022024899004376.

[1779] J Bonse, K.-W Brzezinka, and A.J Meixner. Modifying single-crystalline silicon by femtosecond laser pulses: an analysis by micro raman spectroscopy, scanning laser microscopy and atomic force microscopy. *Applied Surface Science*, 221(1–4):215 – 230, 2004. ISSN 0169-4332. doi: http://dx.doi.org/10.1016/S0169-4332(03)00881-X. URL http://www.sciencedirect.com/science/article/pii/S016943320300881X.

[1780] R. O. Bell, M. Toulemonde, and P. Siffert. Calculated temperature distribution during laser annealing in silicon and cadmium telluride. *Applied physics*, 19(3):313–319, Jul 1979. ISSN 1432-0630. doi: 10.1007/BF00900475. URL https://doi.org/10.1007/BF00900475.

[1781] M. Müllenborn, H. Dirac, and J. W. Petersen. Silicon nanostructures produced by laser direct etching. *Applied Physics Letters*, 66(22): 3001–3003, 1995. doi: 10.1063/1.114257. URL http://dx.doi.org/10.1063/1.114257.

[1782] G. E. Jellison Jr. and D. H. Lowndes. Measurements of the optical properties of liquid silicon and germanium using nanosecond time-resolved ellipsometry. *Applied Physics Letters*, 51(5):352–354, 1987. doi: 10.1063/1.98438. URL http://aip.scitation.org/doi/abs/10.1063/1.98438.

[1783] R. Paniagua-Dominguez, G. Grzela, J. Gomez Rivas, and J. A. Sanchez-Gil. Enhanced and directional emission of semiconductor nanowires tailored through leaky/guided modes. *Nanoscale*, 5:10582–10590, 2013. doi: 10.1039/C3NR03001F. URL http://dx.doi.org/10.1039/C3NR03001F.

[1784] Zhao-Hui Liu, Xun-Jia Su, Gen-Liang Hou, Song Bi, Zhou Xiao, and Hai-Peng Jia. Enhanced performance for dye-sensitized solar cells based on spherical tio$_2$ nanorod-aggregate light-scattering layer. *Journal of Power Sources*, 218:280 – 285, 2012. ISSN 0378-7753. doi: http://dx.doi.org/10.1016/j.jpowsour.2012.06.104. URL http://www.sciencedirect.com/science/article/pii/S0378775312011159.

[1785] Min Hwan Lee, Seul Ji Song, Kyung Min Kim, Gun Hwan Kim, Jun Yeong Seok, Jung Ho Yoon, and Cheol Seong Hwang. Scanning probe based observation of bipolar resistive switching nio films. *Applied Physics Letters*, 97(6):062909, 2010. doi: 10.1063/1.3479526. URL https://doi.org/10.1063/1.3479526.

[1786] Chawloon Thu, Philipp Ehrenreich, Ka Kan Wong, Eugen Zimmermann, James Dorman, Wei Wang, Azhar Fakharuddin, Martin Putnik, Charalampos Drivas, Aimilios Koutsoubelitisand Maria Vasilopoulou, Leonidas C. Palilis, Stella Kennou, Julian Kalb, Thomas Pfadler, and Lukas Schmidt-Mende. Role of the metal-oxide work function on photocurrent generation in hybrid solar cells. 2017.

[1787] John R. Jameson, Yoshiaki Fukuzumi, Zheng Wang, Peter Griffin, Koji Tsunoda, G. Ingmar Meijer, and Yoshio Nishi. Field-programmable rectification in rutile tio$_2$ crystals. *Applied Physics Letters*, 91(11):112101, 2007. doi: 10.1063/1.2769961. URL http://doi.org/10.1063/1.2769961.

[1788] Min Hwan Lee, Kyung Min Kim, Seul Ji Song, Sang Ho Rha, Jun Yeong Seok, Ji Sim Jung, Gun Hwan Kim, Jung Ho Yoon, and Cheol Seong Hwang. Surface redox induced bipolar switching of transition metal oxide films examined by scanning probe microscopy. *Applied Physics A*, 102(4):827–834, Mar 2011. ISSN 1432-0630. doi: 10.1007/s00339-011-6266-7. URL https://doi.org/10.1007/s00339-011-6266-7.

[1789] Yingtao Li, Peng Yuan, Liping Fu, Rongrong Li, Xiaoping Gao, and Chunlan Tao. Coexistence of diode-like volatile and multilevel non-volatile resistive switching in a zro$_2$/tio$_2$ stack structure. *Nanotechnology*, 26(39):391001, 2015. URL http://stacks.iop.org/0957-4484/26/i=39/a=391001.

[1790] Zhensen Tang, Liang Fang, Nuo Xu, and Rulin Liu. Forming compliance dominated memristive switching through interfacial reaction in ti/tio$_2$/au structure. *Journal of Applied Physics*, 118(18):185309, 2015. doi: 10.1063/1.4935622. URL https://doi.org/10.1063/1.4935622.

[1791] S. Stille, Ch. Lenser, R. Dittmann, A. Koehl, I. Krug, R. Muenstermann, J. Perlich, C. M. Schneider, U. Klemradt, and R. Waser. Detection of filament formation in forming-free resistive switching srtio$_3$ devices with ti top electrodes. *Applied Physics Letters*, 100 (22):223503, 2012. doi: 10.1063/1.4724108. URL https://doi.org/10.1063/1.4724108.

[1792] J. Zhu, L.-H. Hu, and S.-Y. Dai. Theoretical modeling of the impact of electrode work function on the performance of tio$_2$/pbs planar heterojunction excitonic solar cells. *Chemical Journal of Chinese Universities*, 33:55–559, 2012.

[1793] Yao Shuai, Shengqiang Zhou, Danilo Bürger, Manfred Helm, and Heidemarie Schmidt. Nonvolatile bipolar resistive switching in au/bifeo$_3$/pt. *Journal of Applied Physics*, 109(12):124117, 2011. doi: 10.1063/1.3601113. URL https://doi.org/10.1063/1.3601113.

[1794] Shang Da-Shan, Sun Ji-Rong, Shen Bao-Gen, and Wuttig Matthias. Resistance switching in oxides with inhomogeneous conductivity. *Chinese Physics B*, 22(6):067202, 2013. URL http://stacks.iop.org/1674-1056/22/i=6/a=067202.

[1795] Sefa B.K. Aydin, Dilber E. Yildiz, Hatice Kanbur Çavuş, and Recep Şahingöz. Ald tio$_2$ thin film as dielectric for al/p-si schottky diode. *Bulletin of Materials Science*, 37(7):1563–1568, 2014. ISSN 0973-7669. doi: 10.1007/s12034-014-0726-6. URL http://dx.doi.org/10.1007/s12034-014-0726-6.

[1796] R. Rodriguez, J. H. Stathis, and B. P. Linder. A model for gate-oxide breakdown in cmos inverters. *IEEE Electron Device Letters*, 24 (2):114–116, Feb 2003. ISSN 0741-3106. doi: 10.1109/LED.2002.808155.

[1797] Ricardo Garcia, Armin W. Knoll, and Elisa Riedo. Advanced scanning probe lithography. *Nature Nanotechnology*, 9(8):577 – 587, 2014. doi: 10.1038/nnano.2014.157. URL http://dx.doi.org/10.1038/nnano.2014.157.

[1798] Valerio Zardetto, Thomas M. Brown, Andrea Reale, and Aldo Di Carlo. Substrates for flexible electronics: A practical investigation on the electrical, film flexibility, optical, temperature, and solvent resistance properties. *Journal of Polymer Science Part B: Polymer Physics*, 49(9):638–648, 2011. ISSN 1099-0488. doi: 10.1002/polb.22227. URL http://dx.doi.org/10.1002/polb.22227.

[1799] F. F. Xu, Y. Ebina, Y. Bando, and T. Sasaki. Structural characterization of (tba, h)ca2nb3o10 nanosheets formed by delamination of a precursor-layered perovskite. *The Journal of Physical Chemistry B*, 107(36):9638–9645, 2003. doi: 10.1021/jp030136u. URL http://dx.doi.org/10.1021/jp030136u.

[1800] Chenmin Liu and Shihe Yang. Synthesis of angstrom-scale anatase titania atomic wires. *ACS Nano*, 3(4):1025–1031, 2009. doi: 10.1021/nn900157r. URL http://dx.doi.org/10.1021/nn900157r.

[1801] Alena Folger, Julian Kalb, Lukas Schmidt-Mende, and Christina Scheu. Fabrication and characterization of abrupt tio$_2$-sio$_x$ core-shell nanowires by a simple heat treatment. *APL Materials*, 5(8):086101, 2017. doi: 10.1063/1.4996211. URL http://dx.doi.org/10.1063/1.4996211.

[1802] M. H. White, D. A. Adams, and J. Bu. On the go with sonos. *IEEE Circuits and Devices Magazine*, 16(4):22–31, Jul 2000. ISSN 8755-3996. doi: 10.1109/101.857747.

[1803] T. Sugizaki, M. Kobayashi, M. Ishidao, H. Minakata, M. Yamaguchi, Y. Tamura, Y. Sugiyama, T. Nakanishi, and H. Tanaka. Novel multi-bit sonos type flash memory using a high-k charge trapping layer. In *2003 Symposium on VLSI Technology. Digest of Technical Papers (IEEE Cat. No.03CH37407)*, pages 27–28, June 2003. doi: 10.1109/VLSIT.2003.1221069.

[1804] International Center for Diffraction Data. Diffraction data v2.1, 2000.

[1805] Kuan Ting Lee and Shih Yuan Lu. Porous fto thin layers created with a facile one step n^{4+} based anodic deposition process and their potential applications in ion sensing. *J. Mater. Chem.*, 22:16259–16268, 2012. doi: 10.1039/C2JM33060A. URL http://dx.doi.org/10.1039/C2JM33060A.

[1806] Seon Mi Kong, Yubin Xiao, Kyung Ha Kim, Wan In Lee, and Chee Won Chung. Performance improvement of dye-sensitized solar cells by surface patterning of fluorine-doped tin oxide transparent electrodes. *Thin Solid Films*, 519(10):3173 – 3176, 2011. ISSN 0040-6090. doi: http://dx.doi.org/10.1016/j.tsf.2011.01.251. URL http://www.sciencedirect.com/science/article/pii/S0040609011003142.

[1807] Zhenzhen Yang, Tao Xu, Shanmin Gao, Ulrich Welp, and Wai-Kwong Kwok. Enhanced electron collection in tio$_2$ nanoparticle-based dye-sensitized solar cells by an array of metal micropillars on a planar fluorinated tin oxide anode. *The Journal of Physical Chemistry C*, 114(44):19151–19156, 2010. doi: 10.1021/jp108761k. URL http://dx.doi.org/10.1021/jp108761k.

[1808] Danni Lei, Ming Zhang, Quanyi Hao, Libao Chen, Qiuhong Li, Endi Zhang, and Taihong Wang. Morphology effect on the performances of sno2 nanorod arrays as anodes for li-ion batteries. *Materials Letters*, 65(8):1154 – 1156, 2011. ISSN 0167-577X. doi: http: //dx.doi.org/10.1016/j.matlet.2011.01.012. URL http://www.sciencedirect.com/science/article/pii/S0167577X11000255.

[1809] X. Wang, W. Liu, H. Yang, X. Li, N. Li, R. Shi, H. Zhao, and J. Yu. Low-temperature vapor-solid growth and excellent field emission performance of highly oriented sno2 nanorod arrays. *Acta Materialia*, 59(3):1291 – 1299, 2011. ISSN 1359-6454. doi: http://dx.doi.org/10.1016/j.actamat.2010.10.061. URL http://www.sciencedirect.com/science/article/pii/S1359645410007354.

[1810] Le Viet Thong, Le Thi Ngoc Loan, and Nguyen Van Hieu. Comparative study of gas sensor performance of sno2 nanowires and their hierarchical nanostructures. *Sensors and Actuators B: Chemical*, 150(1):112 – 119, 2010. ISSN 0925-4005. doi: http://dx.doi.org/10. 1016/j.snb.2010.07.033. URL http://www.sciencedirect.com/science/article/pii/S0925400510006234.

[1811] R.S. Roth and J.L. Waring. Phase equilibria as related to crystal structure in the system niobium pentoxide-tungsten trioxide. *Journal of Research of the National Bureau of Standards*, 70A(4):281, 1966.

[1812] I.J. McColm, R. Steadman, and S.J. Wilson. Iron-promoted phases in the tungsten-oxygen system. *Journal of Solid State Chemistry*, 23 (1):33 – 42, 1978. ISSN 0022-4596. doi: http://dx.doi.org/10.1016/0022-4596(78)90051-8. URL http://www.sciencedirect.com/science/article/pii/0022459678900518.

[1813] O. Glemser, J. Weidelt, and F. Freund. Genotypische oxidhydrate des wolframs. zur frage der wolframblauverbindungen. *Zeitschrift für anorganische und allgemeine Chemie*, 332(5-6):299–313, 1964. ISSN 1521-3749. doi: 10.1002/zaac.19643320511. URL http://dx.doi. org/10.1002/zaac.19643320511.

[1814] Franklin J. Wong and Shriram Ramanathan. Heteroepitaxy of distorted rutile-structure wo2 and nbo2 thin films. *Journal of Materials Research*, 28:2555–2563, 9 2013. ISSN 2044-5326. doi: 10.1557/jmr.2013.247. URL http://journals.cambridge.org/article_ S0884291413002471.

[1815] D. J. Palmer and P. G. Dickens. Tungsten dioxide: structure refinement by powder neutron diffraction. *Acta Crystallographica Section B*, 35(9):2199–2201, Sep 1979. doi: 10.1107/S0567740879008785. URL http://dx.doi.org/10.1107/S0567740879008785.

[1816] Hideaki Yoshitake, Tae Sugihara, and Takashi Tatsumi. Preparation of wormhole-like mesoporous tio2 with an extremely large surface area and stabilization of its surface by chemical vapor deposition. *Chemistry of Materials*, 14(3):1023–1029, 2002. doi: 10.1021/ cm010539b. URL http://dx.doi.org/10.1021/cm010539b.

[1817] Verena Pfeifer, Paul Erhart, Shunyi Li, Karsten Rachut, Jan Morasch, Joachim Brötz, Philip Reckers, Thomas Mayer, Sven Rühle, Arie Zaban, Iván Mora Seró, Juan Bisquert, Wolfram Jaegermann, and Andreas Klein. Energy band alignment between anatase and rutile tio2. *The Journal of Physical Chemistry Letters*, 4(23):4182–4187, 2013. doi: 10.1021/jz402165b. URL http://dx.doi.org/10.1021/ jz402165b.

[1818] Akrajas Ali Umar, Siti Khatijah Md Saad, Marjoni Imamora Ali Umar, Mohd Yusri Abd Rahman, and Munetaka Oyama. Advances in porous and high-energy (001)-faceted anatase tio2 nanostructures. *Optical Materials*, 75(Supplement C):390 – 430, 2018. ISSN 0925-3467. doi: https://doi.org/10.1016/j.optmat.2017.10.002. URL http://www.sciencedirect.com/science/article/pii/S0925346717306158.

[1819] Wei Wen, Jin-Cheng Yao, Yi-Jie Gu, Tu-Lai Sun, He Tian, Qi-Lai Zhou, and Jin-Ming Wu. Balsam-pear-like rutile/anatase core/shell titania nanorod arrays for photoelectrochemical water splitting. *Nanotechnology*, 28(46):465602, 2017. URL http://stacks.iop.org/ 0957-4484/28/i=46/a=465602.

[1820] Physik Instrumente (PI) GmbH & Co KG. *Corvus eco - high resolution positioning controller*, 2015.

[1821] J. E. Moody and R. H. Hendel. Temperature profiles induced by a scanning cw laser beam. *Journal of Applied Physics*, 53(6):4364–4371, 1982. doi: 10.1063/1.331217. URL http://dx.doi.org/10.1063/1.331217.

[1822] R.F. Egerton, P. Li, and M. Malac. Radiation damage in the {TEM} and {SEM}. *Micron*, 35(6):399 – 409, 2004. ISSN 0968-4328. doi: http://dx.doi.org/10.1016/j.micron.2004.02.003. URL http://www.sciencedirect.com/science/article/pii/S0968432804000381. International Wuhan Symposium on Advanced Electron Microscopy.

[1823] Nan Jiang. Electron beam damage in oxides: a review. *Reports on Progress in Physics*, 79(1):016501, 2016. URL http://stacks.iop.org/ 0034-4885/79/i=1/a=016501.

[1824] Omkaram Nalamasu, Frank A. Baiocchi, and Gary N. Taylor. *Photooxidation of Polymers*, volume 412, chapter 12, pages 189–209. 1989. doi: 10.1021/bk-1989-0412.ch012. URL http://pubs.acs.org/doi/abs/10.1021/bk-1989-0412.ch012.

[1825] J. M. Aitken, D. R. Young, and K. Pan. Electron trapping in electron-beam irradiated sio2. *Journal of Applied Physics*, 49(6):3386–3391, 1978. doi: http://dx.doi.org/10.1063/1.325241. URL http://scitation.aip.org/content/aip/journal/jap/49/6/10.1063/1.325241.

[1826] Sujin Baek, Su-Jin Ha, Heechul Lee, Kiwon Kim, Dongchoul Kim, and Jun Hyuk Moon. Monolithic two-dimensional photonic crystal reflectors for the fabrication of highly efficient and highly transparent dye-sensitized solar cells. *ACS Applied Materials & Interfaces*, 0 (0):null, 2017. doi: 10.1021/acsami.7b09885. URL http://dx.doi.org/10.1021/acsami.7b09885. PMID: 29022691.

[1827] Jeethendra Kumar. Coherence length of a light emitting diode. Technical report, Lab Experiments, KamalJeeth Instrumentation & Service Unit, JRD Tata Nagar, Bangalore-560 092, India, 2010.

[1828] A A Talin, F Léonard, A M Katzenmeyer, B S Swartzentruber, S T Picraux, M E Toimil-Molares, J G Cederberg, X Wang, S D Hersee, and A Rishinaramangalum. Transport characterization in nanowires using an electrical nanoprobe. *Semiconductor Science and Technology*, 25(2):024015, 2010. URL http://stacks.iop.org/0268-1242/25/i=2/a=024015.

[1829] S. J. O'Shea, R. M. Atta, and M. E. Welland. Characterization of tips for conducting atomic force microscopy. *Review of Scientific Instruments*, 66(3):2508–2512, 1995. doi: 10.1063/1.1145649. URL https://doi.org/10.1063/1.1145649.

[1830] Nancy A. Burnham, Richard J. Colton, and Hubert M. Pollock. Interpretation issues in force microscopy. *Journal of Vacuum Science & Technology A: Vacuum, Surfaces, and Films*, 9(4):2548–2556, 1991. doi: 10.1116/1.577271. URL https://doi.org/10.1116/1.577271.

[1831] Peng Zhong, Xinpeng Chen, Qiaoying Jia, Gangqiang Zhu, Yimin Lei, He Xi, Yong Xie, Xuejiao Zhou, and Xiaohua Ma. Annealing temperature-dependent electron transfer properties in hydrothermal tio2 nanorod arrays. *Journal of Solid State Electrochemistry*, Oct 2017. ISSN 1433-0768. doi: 10.1007/s10008-017-3786-x. URL https://doi.org/10.1007/s10008-017-3786-x.

[1832] Hamidreza Arab Bafrani, Mahdi Ebrahimi, Saeed Bagheri Shouraki, and Alireza Z. Zaker Moshfegh. A facile approach for reducing the working voltage of au/tio2/au nanostructured memristor by enhancing the local electric field. *Nanotechnology*, 2017. URL http: //iopscience.iop.org/10.1088/1361-6528/aa99b7.

[1833] Ryan O'Hayre, Minhwan Lee, and Fritz B. Prinz. Ionic and electronic impedance imaging using atomic force microscopy. *Journal of Applied Physics*, 95(12):8382–8392, 2004. doi: 10.1063/1.1737047. URL https://doi.org/10.1063/1.1737047.

[1834] M. A. Lantz, S. J. O'Shea, and M. E. Welland. Simultaneous force and conduction measurements in atomic force microscopy. *Phys. Rev. B*, 56:15345–15352, Dec 1997. doi: 10.1103/PhysRevB.56.15345. URL https://link.aps.org/doi/10.1103/PhysRevB.56.15345.

[1835] Raymond A. Serway. *Principles of Physics 2nd.* 1997. ISBN ISBN-13: 978-0030204579, ISBN-10: 0030204577.

Subject Index

List of scientific Output and Projects

List of Publications

(15) **Kalb, Julian**; Stärk, Martin; Boneberg, Johannes & Schmidt-Mende, Lukas, TiO_2-*assisted Silicon Patterning with Pulsed-Laser Lithography*, **2020** (in preparation)

(14) **Kalb, Julian**; Weller, Fabian; Irmler, Lukas; Knittel, Vanessa; Graus, Philipp; Boneberg, Johannes & Schmidt-Mende, Lukas, *Position-controlled laser-induced creation of rutile TiO_2 nanostructures*, Nanotechnology, **2019**, 30(33), 335302, doi: 10.1088/1361-6528/ab1964

(13) **Kalb, Julian**; Folger, Alena; Zimmermann, Eugen; Gerigk, Melanie; Trepka, Bastian; Scheu, Christina; Polarz, Sebastian & Schmidt-Mende, Lukas, *Controlling the density of hydrothermally grown rutile TiO_2 nanorods on anatase TiO_2 films*, Surfaces and Interfaces, **2019**, 15, 141–147, doi: 10.1016/j.surfin.2019.02.010

(12) **Kalb, Julian**; Knittel, Vanessa & Schmidt-Mende, Lukas, *Advanced scanning probe lithography using anatase-to-rutile transition to create localized TiO_2 nanorods*, Beilstein journal of nanotechnology, **2019**, 10(1), 412–418, doi: 10.3762/bjnano.10.40

(11) **Kalb, Julian**; Dorman, James A.; Siroky, Stephan & Schmidt-Mende, Lukas, *Controlling the spatial direction of hydrothermally grown rutile TiO_2 nanocrystals by the orientation of seed crystals*, Crystals, **2019**, 9(2), 64, doi: 10.3390/cryst9020064

(10) **Kalb, Julian**; Folger, Alena; Scheu, Christina & Schmidt-Mende, Lukas, *Non-equilibrium growth model of fibrous mesocrystalline rutile TiO_2 nanorods*, Journal of Crystal Growth, **2019**, 511, 8–14, doi: 10.1016/j.jcrysgro.2019.01.024

(9) **Kalb, Julian**; Dorman, James A.; Folger, Alena; Gerigk, Melanie; Knittel, Vanessa; Plüisch, Claudia S.; Trepka, Bastian; Lehr, Daniela; Chua, Emily; Goodge, Berit H.; Wittemann, Alexander; Scheu, Christina; Polarz, Sebastian & Schmidt-Mende, Lukas, *Influence of substrates and rutile seed layers on the assembly of hydrothermally grown rutile TiO_2 nanorod arrays*, Journal of Crystal Growth, **2018**, 494, 26–35, doi: 10.1016/j.jcrysgro.2018.05.004

(8) Nawaz, Asmat; Wong, Ka Kan; Ebenhoch, Carola; Zimmermann, Eugen; Zheng, Zhaoke; Akram, Muhammad Nadeem; **Kalb, Julian**; Wang, Kaiying; Fakharuddin, Azhar & Schmidt-Mende, Lukas *Improving pore-filling in TiO_2 nanorods and nanotubes scaffolds for perovskite solar cells via methylamine gas healing*, Solar Energy, **2018**, 170, 541–548, doi: 10.1016/j.solener.2018.05.092

(7) Thu, Chawloon; Ehrenreich, Philipp; Wong, Ka Kan; Zimmermann, Eugen; Dorman, James; Wang, Wei; Fakharuddin, Azhar; Putnik, Martin; Drivas, Charalampos; Koutsoubelitis, Aimilios; Vasilopoulou, Maria; Palilis, Leonidas C.; Kennou, Stella; **Kalb, Julian**; Pfadler, Thomas & Schmidt-Mende, Lukas *Role of the Metal-Oxide Work Function on Photocurrent Generation in Hybrid Solar Cells*, Scientific reports, **2018**, 8(1), 3559, doi: 10.1038/s41598-018-21721-2

(6) Liang, Qijun; **Kalb, Julian**; Schmidt-Mende, Lukas & Dekorsy, Thomas *Conductivity Measurements of TiO_2 Nanowires via Terahertz Time-Domain Spectroscopy*, Conference Contribution, **2017**

(5) Grupp, Alexander; Ehrenreich, Philipp; **Kalb, Julian**; Budweg, Arne; Schmidt-Mende, Lukas & Brida, Daniele *Incoherent Pathways of Charge Separation in Organic and Hybrid Solar Cells*, The Journal of Physical Chemistry Letters, **2017**, 8(19), 4848–4864, doi: 10.1021/acs.jpclett.7b01873

(4) Folger, Alena; **Kalb, Julian**; Scheu, Christina & Schmidt-Mende, Lukas, *Tuning the Electronic Conductivity in Hydrothermally Grown Rutile TiO_2 Nanowires: Effect of Heat Treatment in different Environments*, Nanomaterials, **2017**, 7(10), 289, doi:10.3390/nano7100289

(3) Folger, Alena; **Kalb, Julian**; Scheu, Christina & Schmidt-Mende, Lukas, *Fabrication and characterization of abrupt TiO_2–SiO_x core-shell nanowires by a simple heat treatment*, APL Materials, **2017**, 5(8), 086101, doi: 10.1063/1.4996211.

(2) Kollek, Tom; Wurmbrand, Daniel; Birkhold, Susanne T.; Zimmermann, Eugen; **Kalb, Julian**; Schmidt-Mende, Lukas & Polarz, Sebastian, *Thiophene-Functionalized Hybrid Perovskite Microrods and their Application in Photodetector Devices for Investigating Charge Transport Through Interfaces in Particle-Based Materials*, ACS Applied Materials & Interfaces, **2017**, 9(1), 1077-1085.

(1) Zimmermann, Eugen; Pfadler, Thomas; **Kalb, Julian**; Dorman, James A.; Sommer, Daniel; Hahn, Giso; Weickert, Jonas & Schmidt-Mende, L., *Toward High-Efficiency Solution-Processed Planar Heterojunction Sb_2S_3 Solar Cells*, Advanced Science, **2015**, 2(5), 1500059.

Awards and Grants

(3) **Nano-Preis der Universitätsgesellschaft Konstanz e.V.** for outstanding achievements in research in the areas of nano-technology and analytics at the University of Konstanz (**2016**)

(2) **2x Travel grant**; Phantoms Foundation, Madrid, Spain (**2016, 2017**)

(1) **3x DAAD RISE internship grant** for incoming students; Deutscher Akademischer Austauschdienst (DFG), Bonn, Germany (**2013, 2014, 2015**)

Conferences, Workshops, and Talks

2017 Conference: **Trends in Nanoscience** (SFB767), *Nanomechanics, Nanooptics, Nanoelectronics*; Bad Irrsee, Germany; **Posters**: *Optoelectronic Properties on Interfaces of Metal Oxide Nanostructures, Designing and Characterizing Optoelectronic Properties of Metal Oxide Nanostructures*

2017 Conference: **NanoPT**, *Characterization, Fabrication, & Applications of Nanomaterials, Nanooptics, Nanoelectronics*; Porto, Portugal; **Talk**: *Positions-controlled growth of rutile TiO_2 nanorods and their optical and electronic properties*

2017 Conference: **Future Technologies** (Tagesspiegel/Sachsen), *Applied Nanomaterials*; Dresden, Germany

2016 Workshop: **Kompetenznetz Funktionelle Nanostrukturen** (Baden-Württemberg Stiftung), *Characterization, Fabrication, & Applications of Nanomaterials*; Bad Herrenalb, Germany; Posters: *Images of the nanoworld*, **Talk**: *Position-controlled hydrothermal growth of rutile TiO_2 nanorods for Lab-on-a-Chip devices*

2016 Conference: **Trends in Nanotechnology (TNT)**, *Characterization, Fabrication, & Applications of Nanomaterials, Nanooptics, Nanoelectronics*; Fribourg, Switzerland; **Talk**: *Positions-controlled growth of rutile TiO_2 nanorods and their optical and electronic properties*

2016 Workshop: **Refine Summer School** (Carl Zeiss Stiftung), *Hybrid Solar Cells*; Oberstdorf, Germany; **Main organizer, Talk**: *Position controlled Hydrothermal Growth of Rutile TiO_2 Nanorods on Silicon Substrates*

2015 Invited talk: Colloid Seminar (Department of Chemistry); Konstanz, Germany; *Rutile TiO_2 nanostructures: Controlled growth and electronic characterization*

2014 Talk: **Lange Nacht der Wissenschaft** (University of Konstanz); Konstanz, Germany; *TiO_2 – Vom Kaugummi zum Datenspeicher*

2013 Conference: **Nanowires**, *Fabrication of Nanowires, Nanooptics, Nanoelectronics*; Tel Aviv, Israel; **Poster**: *Growth and charge transport in single titania nanowires*

2013 Conference: **Trends in Nanoscience** (SFB767), *Nanomechanics, Nanooptics, Nanoelectronics*; Bad Irrsee, Germany; **Poster**: *Charge transport in TiO_2 films and nanowires used for hybrid solar cells*

Co-operation Partners

Dr. Alena Folger, Prof. Christina Scheu; MPI für Eisenforschung, Düsseldorf, Germany; *HR TEM characterization of TiO_2 nanostructures*

Dr. Alexexander Littig, Prof. Alf Mews; Institute for Physical Chemistry, University of Hamburg, Germany; *CdSe nanowire sensitized TiO_2 nanorods for hybrid solar cell applications*

Prof. Andre ten Elshof; *$Ca_2Nb_3O_{10}$-TiO_2 nanosheets as seed for conductive, flexible, and transparent seed layers for TiO_2 nanorods*

Dr. Daniela Lehr, Dr. Melanie Gerigk, Prof. Sebastian Polarz; Department of Chemistry, University of Konstanz; *XRD measurements on polycrystalline TiO_2 films*

Eun Ji Park, Prof. Gerd Ganteför; Department of Physics, University of Konstanz; *Photocatalysis of NRAs on PSML*

Lukas Irmler, Fabian Weller, Dr. Philipp Graus, Dr. Martin Stärk, Prof. Johannes Boneberg; Department of Physics, University of Konstanz; *Laser-induced melting of TiO_2 films*

Dr. Maximilian Seitner, Dr. Katrin Gajo, Prof. Eva Weig; Department of Physics, University of Konstanz; *Electron-beam evaporated platinum and titanium films; optical lithography*

Dr. Philipp Leicht, Prof. Mikhail Fonin, Prof. Ulrich Rüdiger; Department of Physics, University of Konstanz; *Low-Energy Electron Diffraction (LEED) on polycrystalline TiO_2 films*

Dr. Qijun Liang, Prof. Thomas Dekorsy; Department of Physics, University of Konstanz; *Thz spectroscopy on TiO_2 nanorods*

Simon Bretschneider, Dr. Frédéric Laquai; MPI for Polymer Research, Mainz, Germany; *TiO_2 nanorod arrays for hybrid solar cells*

Dr. Simone Plüisch, Maxim Schlegel, Prof. Alexander Wittemann; Department of Chemistry, University of Konstanz; *Fabrication of polystyrene sphere monolayers (PSMLs)*

Dr. Vanessa Knittel, Porf. Alfred Leitenstorfer; Department of Physics, University of Konstanz; *Electron-beam lithography*

Supervision

Doctoral thesis: Carola Ebenhoch, University of Konstanz; *Understanding and controlling charge carrier dynamics in TiO_2 and Nb_2O_5 nanocrystals and crystal ensembles with confined geometries* (2016-2018)

Doctoral thesis: Sohaila Zaghloul Nabi Mohammed, University of Konstanz; *Position-controlled growth and optoelectronic properties of MoO_3 and HfO_2 nanocrystals* (2016-2018)

Doctoral thesis: Shaista Andleeb, University of Konstanz; *Optical, electronic, and optoelectronic properties of metal oxide mesocrystals*, SFB1214 (2016-2018)

Bachelor thesis: Elise Sirotti, University of Konstanz; *Electronic characterization of rutile TiO_2 nanorods annealed in different atmosphere* (2017, 4 months)

Bachelor thesis: Lukas Krumbein, University of Konstanz; *Structure dependent investigation of electronic properties of TiO_2 nanowires* (2015, 4 months)

Bachelor thesis: Philipp Beck, University of Konstanz; *Investigation of charge transport in single TiO_2 nanowires* (2014, 4 months)

DAAD awardee: Sara Sand; Ohio University, Athens (Ohio), United States; *Exploring fundamental charge transport effects in nanostructured titania for extensive electronic device applications* (2015, 3 months)

DAAD awardee: Berit Goodge; Carleton College, Northfield (Minnesota), United States; *Manipulation of electron mobility in titania nanostructures for hybrid solar cell applications* (2014, 3 months)

DAAD awardee: Emily Chua; Dalhousie University, Halifax (Nova Scotia), Canada; *Characterization of charge transport in titania nanowires* (2013, 3 months)

State examination thesis: Sabrina Rueß, University of Konstanz; *Fabrication and characterization of plasmonic TiO_2/Au core-shell structures (working title)* (2017, 2 months)

Teaching

SS2017 Lecture: *Integrierter Kurs II (physics lecture for 2nd term, hydro- and electrodynamics)*, Prof. Schmidt-Mende/ Prof. Ulrich Nowak, University of Konstanz (tutor)

WS2016/17 Lecture: *Integrierter Kurs I (physics lecture for 1st term, classical mechanics)*, Prof. Schmidt-Mende/ Prof. Guido Burkard, University of Konstanz (tutor)

SS2016 Junior Lab: *Franck-Hertz experiment and Vis Spectroscopy (lab for 4th term)*, Dr. Bernd-Uwe Runge, University of Konstanz (tutor)

WS2015/16 Lecture: *Solid State Physics (lecture for 5th term)*, Prof. Elke Scheer, University of Konstanz (tutor)

SS2015 Lecture: *Integrierter Kurs II (physics lecture for 2nd term, hydro- and electrodynamics)*, Prof. Schmidt-Mende/ Prof. Ulrich Nowak, University of Konstanz (tutor)

WS2014/15 Lecture: *Integrierter Kurs I (physics lecture for 1st term, classical mechanics)*, Prof. Schmidt-Mende/ Prof. Ulrich Nowak, University of Konstanz (tutor)

SS2014 Special Junior Lab: *Detection of Myons and Laser Guitar (lab for 4th term)*, Dr. Bernd-Uwe Runge, University of Konstanz (tutor)

WS2013/14 Lecture: *Physics I for Chemists (lecture for 1st term)*, Prof. Johannes Boneberg, University of Konstanz (tutor)

SS2013 Lecture: *Integrierter Kurs II (physics lecture for 2nd term, hydro- and electrodynamics)*, Prof. Schmidt-Mende/ Prof. Matthias Fuchs, University of Konstanz (tutor)

WS2012/13 Lecture: *Integrierter Kurs I (physics lecture for 1st term, classical mechanics)*, Prof. Schmidt-Mende/ Prof. Matthias Fuchs, University of Konstanz (tutor)

2008-2011 Junior Lab: *Moment of inertia, Freezing point depression, Critical point, Fresnel's formulas, Diffractive optics, Franck-Hertz experiment, vis spectroscopy (lab for 4th term)*, Dr. Bernd-Uwe Runge, University of Konstanz (tutor)

Acknowledgment

I would like to thank all of the people, who supported me during my time at the University of Konstanz and especially in the **hybrid nanostructure group** between 2012 and 2016. For me, it was a pleasure to work with this group and I enjoyed the ties of friendship being the basis for all the professional and private conversations. In particular, I would like to thank **Prof. Lukas Schmidt-Mende** for giving me the chance to perform my Ph.D. thesis in his group. Although the financial situation was difficult in the beginning and it took a while for the first utilizable results, he was always convinced by my projects and ideas. I was always able to work freely and to manage my research independently. Doing so, I came up with unexpected scientific results, which was a great benefit for both of us and the scientific community. I wish him and his family the very best for the future, health, satisfaction, and further scientific success. Additionally, I thank **Prof. Paul Leiderer** for reviewing my thesis, discussing any issues that arose during my thesis, and supporting me to get this Ph.D. position. Furthermore, I enjoyed the discussions and experiments concerning low-dimensional electron systems on liquid helium in confined geometries, although my time for that was quite limited during the work on my Ph.D. thesis. I thank also **Prof. Ulrich Nowak** for the chairmanship of the board of examiners during my defense. Besides that, I appreciated his advice for my theory talk and preparation of my defense.

In the following, I would like to thank some current and former group members in particular. My work was based on the hydrothermal growth of rutile TiO_2 nanostructures. Thus, I am indebted particularly to **Dr. James Dorman**, who taught me the basic techniques needed for my experiments. During frequent discussions, I could benefit from his great experience as a chemical engineer and scientist. I wish the very best for his future career and life in general. Besides James, **Dr. Jonas Weickert** was a great supporter of my work, since he had always an open ear for my questions and contributed immeasurable advice to my research. Furthermore, he taught me to love our sputter deposition device. I will miss his way of keeping things in perspective. Beyond that, I thank **Dr. Thomas Pfadler** for his advice concerning optoelectronic questions, thesis writing, publishing, job-seeking advice, and introductions to some experimental setups. I appreciated his very obliging, kind, and honest character. Thanks a lot to **Dr. Chaw Loon Thu** and **Dr. Kwang-Dea Kim** for having shared some projects with me and introducing me to the exotic meals of south east Asia.

It was one of the happiest moments during my work, when I heard that my research will be continued by **Carola Ebenhoch** as a Ph.D. student. During the last months of my time in this group, we discussed a lot of interesting questions and I am pretty sure that Caro will deliver some fascinating research results in a few years. Besides her systematic approach to experimental problems, I got to know her as a very kind person. Caro, I would have appreciated so much if we could have worked longer on these projects together. For me, the most constructive preparation for conferences, publications, and my Ph.D. exam was the critical view on experimental results from **Dr. Philipp Ehrenreich**. Independent where we met – in the seminar room or in the beer garden – I was always a little bit wiser afterward. In particular, discussions about the optoelectronic measurements were very fruitful and it was a pleasure for me to work on some publications with him. I received always a valuable support concerning device software from **Dr. Eugen Zimmermann**. Additionally, we will never forget his frequent preparation of

427

waffle coated Nutella (or was it the other way round?). Good luck with your Ph.D. thesis! Although there was hardly an overlap of our research topics, I appreciated the positive attitude to work and life of **Dr. Susanne Birkhold** a lot. In addition, we had also some interesting discussions about our research and she was always a pleasant and kind interlocutor. Thanks and good luck for your Ph.D. thesis, Susi! I also thank my further colleagues **Dr. Yuyi Feng, Shaista Andleeb** and **Sohaila Zaghloul Nabi Mohammed** for some basic discussions about the electronic characterization of related nanostructures.

During my work, I supervised three Bachelor students and three scholarship students. **Philipp Beck** performed his Bachelor thesis in 2014 about the charge transport across rutile TiO_2 nanorod/PtIr interfaces. By implementing a new and simple measurements techniques we were able to determine the charge characteristics on different samples quickly. Thanks a lot for this great job. **Lukas Krumbein** investigated during his Bachelor thesis in 2015 the influence of crystal defects, temperature, and light on the charge transport of rutile TiO_2 nanorods. I thank Lukas for his comprehensive work and remarkable results such as the negative differential resistance in certain rutile TiO_2 systems. **Elise Sirotti** performed during her Bachelor thesis in 2017 an extensive study on the transient current through nanorod arrays that were fabricated in different growth conditions and exposed to various chemical and thermal post-treatments. Doing so, she managed to handle and analyze big data volumes and brought a great success to my research. Besides that, I always appreciated her curiosity and kindness. A great thank-you to all my former Bachelor students and I wish the very best for their further life. **Emily Chua** was a Canadian scholarship student founded by the DAAD (Deutscher Akademischer Austauschdienst) in 2013. She worked hard on the understanding of hydrothermal growth processes on rutile TiO_2 nanorods on different substrates. I thank Emily a lot for her contribution, which was the groundwork of my following research and I was happy to return the favor by writing some reference letters for her further scientific career. I appreciated her calm and kind character very much. **Berit Goodge** was a US scholarship holder founded by the DAAD in 2014 and she investigated basic questions concerning the hydrothermal growth process as well. I thank Berit for this work and for determining the potential landscape and structure of incorporated high-temperature treated gold and platinum electrodes using electrostatic force microscopy (EFM). Her independence and lab skills were outstanding for such a young researcher. Finally, **Sara Sand** supported me as a US scholarship student in 2015 as well. She investigated the I-V characteristics of thin TiO_2 films fabricated with optical contact lithography. I thank her for contributing to my final experiments on optoelectronic properties of TiO_2. I valued highly her kind and honest character. Hence, I was happy to hear that she is coming back back to our group for a one-year research stay.

Besides the supervised students, I was supported by four scientific assistants. **Felix Rochau** helped to optimize the recipe for the optical lithography and analyzed film properties using an ellipsometer. I thank him for having a share in my research projects with his above-average lab skills. **Dr. Bastian Trepka** promoted the investigations of the hydrothermal growth process with his outstanding motivation and exploratory spirit. I thank Bastian for his numerous findings. **Sabrina Rueß** helped me for several months with preparing samples for various projects so that I could focus on writing and analyzing data. Additionally, she joined in a small research project with the aim to determine plasmonic effects on nanostructured TiO_2 nanowalls. It was a pleasure to receive help from such a kind and conscientious person. Good luck on your way becoming a teacher! **David Nabben** joined me in the late stage of my work and performed measurements on levitating NRA membranes. I thank David for this new, flexible measurement technique that avoids the diffusion of substrate atoms into TiO_2 during post-annealing treatments. This technique will certainly help Carola in her work as well.

A special thank to all my collaboration partners helping me to establish and perform versatile scientific projects. A deep insight into the structure of rutile TiO_2 nanorods and their seed layers using HR TEM was realized by **Dr. Alena Folger** and **Prof. Dr. Christina Scheu** (Max-Planck-Institut für Eisenforschung GmbH, Düsseldorf). Furthermore, I appreciated the discussions about the fine structure of the investigated nanocrystals a lot, since it helped me to establish a new growth model presented in this thesis. The number of common publications is certainly a measure of this excellent collaboration. During the numberless emails, telephone calls, and visits I cherished Alena's helpful and kind character. The first application of nanorod arrays on a superlattice was realized by a monolayer of polystyrene spheres (PSML) covered with the seed layer for the nanorods. Without the concentrated efforts of **Dr. Simone Plüisch** and detailed recommendations of **Prof. Dr. Alexander Wittemann** (Department of Chemistry, University of Konstanz), this project would not have been possible. In addition, **Maxim Schlegel** from the AG Wittemann never got tired of performing reactions to gain the optimal sphere diameters. Great job! I received valuable help for performing photocatalysis on the PSML based NRAs from **Eun Ji Park** from the AG Ganteför (Department of Physics, University of Konstanz). One main challenge of this theses was to find routes to achieve well defined and locally confined NRAs. One route to achieve this goal was optical lithography. Here, **Dr. Maximilian Seitner** (AG Weig, Department of Physics, University of Konstanz) helped me to optimize the exposure and development time. Another established but technically ambitious lithography technique was electron-beam lithography, which was supervised by **Dr. Vanessa Knittel** (LS Leitenstorfer, Department of Physics, University of Konstanz). But beyond that, I thank Vanessa for a great friendship during my time in Konstanz, which began with a few icy challenges during the junior lab in our second term and last but not least the evil thing in the nutshell. I thank Dr. Maximilian Seitner, **Dr. Katrin Gajo** (AG Weig, Department of Physics, University of Konstanz), **Julian Braun**, and **Chris Espy** (both AG Scheer, Department of Physics, University of Konstanz) for depositing platinum and titanium using electron-beam evaporators. A novel structuring technique for rutile seed layers was introduced by using continuous and pulsed lasers. I thank **Prof. Dr. Johannes Boneberg** (Department of Physics, University of Konstanz) for numerous, intensive, and fruitful discussions resulting in outstanding facilities to create locally confined nanorods on our samples. It was due to the outstanding technical ability of **Dr. Martin Stärk, Dr. Philipp Graus** (both pulsed laser), **Lukas Irmler**, and **Fabian Weller** (both cw-laser) (all AG Scheer and Boneberg group, Department of Physics, University of Konstanz). I thank **Dr. Daniela Lehr** and **Dr. Melanie Gerigk** (both AG Polarz, Department of Chemistry, University of Konstanz) for performing XRD measurements on my samples. With their work, it was possible to minimize the samples for the time-consuming HR TEM studies significantly. A special thank to **Prof. Dr. Thomas Dekorsy** and **Dr. Liang Qijun** (Department of Physics, University of Konstanz) for determining the THz-conductivity of my rutile TiO_2 NRAs. Their complementary method of measurement was a good way to construe the results from the contact measurements such as I-V, I-t, or IS. Interesting projects ideas were born, when I met **Dr. Alexander Littig** and **Dr. Dino Behn** on my first international conference in Israel (Nanowires 2013). I like to remember this time. Later, I learned the basics for contacting and characterizing single nanorods from Dino and Alexander motivated me to gain control over the density of NRAs in order to introduce CdSe nanowires into TiO_2 NRAs for hybrid solar cell applications. I am very thankful for that. On the search for flexible, conductive substrates that are able to resist the seed fabrication and the subsequent hydrothermal growth of TiO_2 nanorods, I met **Prof. André ten Elshof** on the REFINE summer school I organized in 2016. Thanks to him I was able to test the adhesion of my nanostructure on highly innovative nanosheets that were satisfying all requirements. From the STM and LEED investigations performed by **Dr. Philipp Leicht**,

Acknowledgment

I learned that as-prepared titania is covered with dirt and amorphous material. This was an important reference, which I remembered in my later work. Thanks a lot for your patience!

Besides partners in different research groups, I was supported by the nanostructure laboratory team as well. A very special thank to **Matthias Hagner**, who taught me a lot of techniques such as SEM and FIB. Furthermore, he spent plenty of hours cutting TEM lamella and establishing electronic contacts to my nanostructures by ion beam induced platinum deposition. And last but not least, a lot of technical support was needed for the used setups during my work. The first insights into the nanorods using HR TEM were performed by **Marina Krumova**. I thank Marina for having so much patience and endurance with my samples – and me.

Behind good scientific research, there are always professional craftsmen, technicians, and administration secretaries. Whenever a complex construction was to be made, **Louis Kukk** was the man who realized it. Without his ability and intelligence as well as his spontaneous operational readiness, many experiments could not have been performed. Additionally, I thank him for the detailed introduction into his workshop and I wish him always a satisfying and good time while constructing and fishing. Beyond that, I thank our **workshop team** in general. But I would like to bring out the very kind and helpful advice and setups I received from the electronic workshop under the direction of **Gerd Sulger**. For taking care about our laboratories, some advice, and administration work, I thank **Hamidreza Riazi-Nejad** very much. For managing and helping with administrative issues, I thank our secretaries **Nicole Frederick, Friederike Stuckenbrock**, and **Stefanie Fischer**. I always appreciated their obliging and cooperative kind a lot. Although working with our administration is complex and a lot of administration secretaries are involved, I want to point out one colleague in particular. Since I was responsible for ordering, I had to work closely with the supply department of our university. In this context, I thank **Gabriele Sims** (supply department, University of Konstanz) for an outstanding and efficient teamwork especially concerning all urgent, time-critical, and difficult order issues.

Many thanks to the **Deutsche Forschungsgesellschaft** (DFG) for supporting me within the framework of the SFB767, SFB1214, DAAD-programs, and further research projects. I am also grateful to the **Carl Zeiss Stiftung**, that founded the great and fruitful REFINE summer school amongst other research. A very special thanks to the **Universitätsgesellschaft Konstanz e.V.** and its association chairman **Björn Graf Bernadotte**, who gave me the Nano-award in 2016.

Last but not least, I thank all my friends and, in particular, **Catharina** for supporting me so much during the work on my Ph.D. thesis. Finally, I would like to thank my parents, **Ulrike** and **Roland Kalb**, who enabled my studies by providing me financial support. I wish them health and satisfaction in their new home.